THE MARINE ENCYCLOPAEDIC DICTIONARY

FIFTH EDITION

THE MARINE ENCYCLOPAEDIC DICTIONARY

FIFTH EDITION

By the late
ERIC SULLIVAN, F.I.C.S.
Sometime Chairman and Managing Director
Mediterranean Trading Shipping Co. Ltd.

|L|L|P|

LONDON   NEW YORK   HONG KONG
1996

LLP Ltd
Legal and Business Publishing Division
27 Swinton Street
London WC1X 9NW

USA AND CANADA
LLP Inc
Suite 308, 611 Broadway
New York, NY 10012 USA

SOUTH EAST ASIA
LLP Asia Ltd
Room 1101, Hollywood Centre
233 Hollywood Road
Hong Kong

First published in Malta, 1980
Second edition first published
in Great Britain, 1988
Third edition, 1992
Fourth edition, 1995
Fifth Edition, 1996

British Library Cataloguing in Publication Data
A catalogue record for
this book is available from the
British Library

ISBN 1–85978–043–1

Typeset in 9pt on 10pt Bembo by
Interactive Sciences, Gloucester
Printed in Great Britain by
WBC Ltd.
Bridgend, Mid-Glamorgan

DEDICATION

The Marine Encyclopædic Dictionary needs little introduction around the shipping and business world. This fifth edition is another fine achievement by Eric Sullivan who unhappily did not live long enough to see it published. Mr Sullivan underwent serious operations late last year from which he was recuperating but complications set in and he succumbed rather suddenly last November.

The fifth edition, is dedicated to Eric Sullivan by his widow and children, Charlotte, Alexander, Ursula and Peter, who were determined to undertake the remaining stages of the book.

The late Mr Sullivan enjoyed a very busy and hectic life. Up at the crack of dawn he began work every day while most people were still asleep. Chairman and Managing Director of three shipping companies he did not spare himself from the many problems arising from ship management. But he was also a close family man and keen gardener. This can be attested to by his widow, May, and children and their undaunted efforts to complete this book to his memory. This fifth edition, like all his other editions will surely find a place on the desks of the world's shipping and business executives.

March 1996

FOREWORD

In consequence of the undoubted popularity of previous editions of *The Marine Encyclopædic Dictionary* the publishers are now proud to present the Fifth Edition of this unique compendium of knowledge.

This volume contains the explanation of many thousands of terms, expressions, acronyms and abbreviations of daily currency in the worlds of shipping, transport, freighting, insurance, finance and, now, offshore oil and gas. For the general reader's edification added spice is provided by the author's gleanings in wider fields of academic and technological interest.

In our present age no day passes without the creation of fresh organisations, procedures and techniques, which bring with them new definitions and a plethora of the initials which are now so much in vogue. Throughout all the editions the text has been extensively revised and updated, and many new entries have been incorporated. The author's work in compilation has been edited by Commander Brian Wainwright, O.B.E., M.N.I., R.N., and the book has benefited from the input of a number of other highly experienced shipping professionals, including Robert H. Brown, F. Inst. A.M., A.C.I.I., author of several marine insurance publications.

We are confident that the reader will find this a most valuable and practical book of reference.

March 1996   THE PUBLISHERS

A

✠ Symbol used by Lloyd's Register of Shipping (Classification Society) to denote that a vessel was built under the supervision of Lloyd's Register surveyors in accordance with the Rules and Regulations of the society.

+ Addition; plus.

& *Sign for* and.

& Co. *Sign for* and Company.

& Co. Ltd. *Sign for* and Company Limited.

100 Al Character used in Lloyd's Register denoting that a sea-going ship is constructed of iron or steel and is classed with Lloyd's Register of Shipping.

al Second Polaris Correction *nautical*.

Al First Class—a symbol used in Lloyd's Register. It is assigned to ships intended to trade within sheltered waters such as harbours, rivers, estuaries.

Al Conference Clauses US shipment conditions of frozen or chilled cargoes.

a2 Third Polaris Correction *nautical*.

a About; Acre; Advance; Advanced; Aft *Lloyd's Register notation*; Anno *Italian/Latin*—year; Ante *Latin*—before; First Polaris Correction *nautical*.

A Aberdeen; Absolute; Ampere; Atomic Weight; Australian; Austria.

A° Angstrom unit (after Swedish physicist 1814–74).

aa Absolute Alcohol; Acting Appointment; Author's Alteration; Always Afloat *or* Always Safety Afloat, *q.v.*

a.a. Acting Appointment; After Arrival, *q.v.* Always Afloat *or* Always Safety Afloat, *q.v.*; Author's Alteration.

AA Aerolineas Argentinas *Argentine Airlines*; Alcoholics Anonymous; American Airlines; Average Adjusters; Average Agents.

AA and **AAA** American financial credit ratings for corporations.

AA *or* **A.A.** Associate of Arts.

A.A. Advertising Association; Anti-Aircraft; Associate in Accounting; Architectural Association; Apparent Altitude *nautical*. Automobile Association *UK*.

A&A Additions and Amendments.

AAA *or* **A.A.A.** Agriculture Adjustment Act *USA*; American Automobile Association; Association of Average Adjusters.

AAAA Always Accessible and Always Afloat *charter-party term*.

AAACC Association of Asian-American Chambers of Commerce.

AAAE American Association of Airport Executives.

AAAS American Association for the Advancement of Science.

A.A.B. Air Authority Board.

AAC Aeronautical Advisory Council; Anno Ante Christum *Latin*—in the year before Christ.

AACC Airport Association Co-ordinating Council.

A.A.C.C.A. Associate of the Association of Corporate and Certified Accountants, now A.C.C.A, *q.v.*

AACJC American Association of Community and Junior Colleges.

AACLA Association of American Chambers of Commerce in Latin America.

AACO Arab Air Carriers Organisation.

AACRAO American Association of Collegiate Registrars and Admissions Officers.

AAD Annual Aggregate Deductible *insurance*; Arab Accounting Dinar.

AAE Association of Aircraft Enthusiasts.

A.A.I.A. Associate of the Association of International Accountants.

AAIB *or* **A.A.I.B.** Air Accident Investigation Board; Associate of the Association of Insurance Brokers.

AAIE American Association of Industrial Engineers.

A.A.I.E.C. Afro-Asian Organisation of Economical Co-operation.

A.A.I.I. Associate of the Australian Insurance Institute.

AALC Amphibious Assault Landing Craft *USA*.

AAM Association of American Management.

AAMS *or* **A.A.M.S.** All American Marine Slip, *q.v.*

a.a.O. am angeführten Orte *German*—At the place quoted.

AAPA American Association of Port Authorities.

AAPG American Association of Petroleum Geologists.

AAPMA Association of Australian Port and Marine Authorities.

a.a.r. After Action Report; Against All Risks *or* All and Any Risks, *q.v.*; Aircraft Accident Report/Record; Average Annual Rainfall.

AAR Aircraft Accident Report.

AARA *or* **A.A.R.A.** Always Accessible or Reachable on Arrival.

AAS Academiae Americanae Socius *Latin*—Fellow of the American Academy; American Arbitration Society; Annual Automated Survey *of a Vessel*.

A.A.S.A. Associate of the Australian Society of Accountants.

AASC Aerospace Applications Study Committee *NATO.*

AASMM Associated African States, Madagascar and Mauritius (*now absorbed into Association of African, Caribbean and Pacific States*).

AASO Assocation of American Shipowners, *New York.*

AASR Airport and Airways Surveillance Radar.

AAT Airport Authority of Thailand.

AATCR Air and Air Traffic Control Regulations 1981.

A.A.T.T.A. Air Association of Tourism and Travel Agents.

AAUP American Association of University Professors.

AAW Anti-Air Warfare.

a.b. *or* **AB** *or* **A.B.** Average Bond, *q.v.*

AB *or* **A.B.** Able-Bodied Seaman, *q.v., also abbrev.* ABS; Advisory Board; Air Board; American Bureau Classification Society, which is similar to American Record AR; Artium Baccalaureus *Latin*—Bachelor of Arts; Asthmatic Bronchitis.

A/B *or* **Ab** Aktiebolag, *q.v.*

a/b *or* **abn** Airborne.

ABA *or* **A.B.A.** American Bankers Association, founded in 1875; American Bar Association; Associate of British Archeologists; American Booksellers' Association; Associate in Business Administration.

Aback Describes square sails when pressed against the mast by a wind ahead. Said also of a ship in such a situation.

Abaft Stern or towards the stern of the ship; Behind.

Abandonment The right a marine assured has to abandon property in order to establish a Constructive Total Loss, *q.v.* An underwriter is not obliged to accept abandonment, but if he does he accepts responsibility for the property and liabilities attaching thereto, in addition to being liable for the full sum insured. *See* **Notice of Abandonment**.

Abandon Ship An order, usually only given verbally by the master, for all personnel on a vessel to leave it during an emergency.

Abandon Ship Ladder A rope ladder used either with the lifeboats or the sides of the ship in the event of abandoning a ship while in distress. *Abbrev.* A.S.L.

à bas *French*—Down with.

ABBA American Board of Bio-Analysis.

Abbr *or* **Abbrev.** Abbreviated; Abbreviation.

Abrasion of Coins The ordinary wear and tear occurring in the daily use of coinage resulting in the loss of the original mint weight.

ABC Atomic Biological Chemical.

ABC *or* **A.B.C.** Advance Booking Charter; Advanced Biomedical Capsule; Air Bridge to Canada; Alcoholic Beverage Control; Alphabet; Alphabetical Time Table for Air and Train Schedules; Already Been Converted; American Broadcasting Company; American Bureau of Circulation; Audit Bureau of Circulation; the rudiments of any subject; America-Britain-Canada; Book-prices Current; Animal Birth Control; Argentina-Brazil-Chile; Automated Binary Computer; Automatic Brake Control.

ABCA Army Bureau of Current Affairs, *obsolete.*

A.B.C.B. Association of Birmingham Clearing Banks.

A.B.C.C. Arab/British Chamber of Commerce; Association of The British Chambers of Commerce.

ABCD America, Britain, China, Dutch East Indies. World War II powers in Pacific.

ABC Islands Aruba, Bonaire, Curacao, *Netherlands Antilles.*

ABCM Association of British Chemical Manufacturers.

ABCU Automatic Bridge Control Unattended. System where an unmanned Engine Room is controlled from the Bridge.

Abdnt Abandonment, *q.v.*

ABDP Association of British Directory Publishers.

Abeam A direction at right angles to a ship's fore-and-aft line.

ABECOR Associated Banks of Europe. An association of seven member banks and three associates established in 1974 cooperating in banking and financial services to customers.

Aber *or* **Abr.** Aberdeen, *UK.*

ab extra *Latin*—From without.

Abeyance Uncertain right of ownership or title of dignity or estate.

A.B.F. Associated British Foods.

Abf Abfahrt *German*—Departure.

ABFM American Board of Foreign Missions.

Abft. Abaft, *q.v.*

abgk Abgekürzt *German*—Abbreviated.

ABH American Bureau of Shipping Hellas, *in Piraeus.*

ABI Association of British Insurers.

A.B.I. Associate of the Institute of Book-keepers; Associazione Bibliotecari Italiani *Italian* Association of Book-Keepers.

A.B.I.A. Associate of the Bankers' Institute of Australasia.

Ability to Pay Condition which may in some circumstances be taken into account when demanding taxes, rents, damages or fines.

A.B.I.M. American Board of International Missions; Association of British Insecticide Manufacturers.

ab initio *Latin*—From the beginning.

ABIRS *or* **ABSIRS** American Bureau's Information Retrieval System.

abk Abkkurzung *German*—Abbreviation.

ABLA American Business Law Association.

Able-Bodied Seaman *or* **Able Seaman** A seaman, certificated by examinations, who must have had at least three years' service at sea. *Abbrev.* A.B. An unqualified seaman who has nine months' sea service or more is called an Ordinary Seaman, *q.v.*, and *abbrev.* O.S. *See also* **Before the Mast**.

ABLE Activity Balance Line Evaluation.

ABLS Bachelor of Arts in Library Science.

A.B.M. Anti-Ballistic Missile; Associate in Business Management.

ABMEC Association of British Mining Equipment Companies.

ABMEX Association of British Mining Equipment Exporters.

ABMS American Bureau of Metal Statistics.

ABMTM Associated British Marine Tool Makers.

abn *or* **a/b** Airborne.

Abn *or* **Aber.** Aberdeen, *UK*.

Aboard On board a ship.

A.B.O.F. Association of British Organic Fertilizers.

A.B.O.G. American Board of Obstetrics and Gynaecology.

A.B.O.I. Association of British Offshore Industries.

à bon marché *French*—Cheap; a good bargain.

Abort To abandon or cancel a take-off already started by an aircraft.

A.B.O.S. American Board of Oral and/or Orthopaedic Surgery.

About To change tack; To go about *nautical*.

Above Par When the market value of the shares are higher than that of the normal price they are called above par; At a premium.

Above the Norm Above normal; Above the average.

ABP American Business Press; Associated British Ports.

ABR American Board of Radiology.

abr. Abridge; Abridged.

Abreast Side by side; Alongside.

abs. Absolute.

ABS *or* **A.B.S.** American Bureau of Shipping *or* American Bureau of Shipping Standards; Auxiliary Boiler Survey *Lloyd's Register notation*; Aeronautical Broadcast Service *aviation*.

ABSCOMP American Bureau of Shipping Computers Inc.

Absence Indicator A Pendant, *q.v.*, hoisted on the mast to indicate that the Commanding Officer or Staff Officer is not on board.

Absence Flag A square blue flag of small dimensions hoisted on a yacht when the owner is not on board. This is substituted by a blue light signal during darkness.

Abse.re. Absente reo *Latin*—The defendant being absent.

Absolute Humidity Mass of water vapour per unit volume of air. Expressed in grams per cubic metre.

Absolute Insurance *See* **Assurance**.

Absorption An International Air Transport Association term practised by associated air traffic companies when airfreight cargoes are accepted in accordance with air traffic conditions; The result of a small company merging with a larger one. It becomes absorbed.

Absqua ulla nota *Abbrev.* a.u.n., *q.v.*

ABSTECH ABS Worldwide Technical Services Inc., *q.v.*

Abstract of Title Summary of facts concerning ownership. Required by the buyer before he acquires property from the seller.

A.B.S.W. Association of British Science Writers.

ABS Worldwide Technical Services Inc. A wholly-owned subsidiary of the American Bureau of Shipping providing inspection and technical services throughout the world for land and offshore structures, equipment, machinery, etc. *Abbrev.* ABSTECH.

A.B.T. Association of Building Technicians.

ABTA *or* **A.B.T.A.** Allied Brewery Traders' Association; Association of British Travel Agents; Australian British Trade Association.

ABTAC Australian Book Trade Advisory Committee.

ABU *or* **A.B.U.** American Board of Urology.

Abudul-Bagi al Bakri *Arabic*—Civil Law of Iraq.

Ab uno disce omnes *Latin*—From one specimen judge the rest.

Ab Urbe Condita *Latin*—From the founding of the City *referring to Rome. Abbrev.* AUC *or* A.U.C.

A-burton Manner of stowing casks in a hold when laid Athwartships.

abv *or* **ab.** Above.

ABWMAC Australian Ballast Water Management Advisory Council.

ABYA *or* **A.B.Y.A.** Association of British Yacht Agents.

Abyssalbenthic Zone Part of the sea which is deeper than a thousand metres. *See* **Arechibenthic Zone**.

ac Acre, *q.v.*

Ac Admiralty Coefficient; Altocumulus Clouds, *q.v.*

AC *or* **A.C.** Aero Club; Air Command; Air Commodore; Air Control; Air Council; Alternating Current *electricity*; Altocumulus Clouds, *q.v.*; Analogue Computer; Analytical Chemist; Annual Conference; Ante Cibum *Latin*—Before Meals; Appeal Case; Appeal Court; Army Corps; Artillery College; Arts Council; Assistant Commissioner.

a.c. *or* **A.C.** *or* **A/C** A capo *Italian*—Fresh line; Account; A compte *French*—On account; Account Current; Aircraft; Aircraft Only; All Conveniences; Altered Course *nautical*; Alternating Current *electricity*; Author's Corrections.

ACA *or* **A.C.A.** Accession or Accessory Compulsory Account *Customs Procedures*; Adjacent Channel Attenuation; Advanced Combat Aircraft; Agricultural Credit Administration; American Composers Alliance; Arts Council of America; Associate of the Institution of Chartered Accountants.

A.C.A.A. Associate of the Australasian Institution of Chartered Accountants.

ACAAI Air Cargo Agents Association of India.

A. Cant. After Cant Frames *ship construction*.

Acap. Acapulco, *Mexico*.

a capite ad calcem *Latin*—From head to heel.

ACAS *or* **A.C.A.S.** Advisory Conciliation and Arbitration Service; Established under the Employment Protection Act 1975.

ACB Amphibious Construction Battalion *USA*.

ACBI *or* **A.C.B.I.** Associate of the Institute of Book-Keepers.

ACC *or* **A.C.C.** Acceptable Container Condition; Acceptance; Accepted; American Chamber of Commerce; Area Control Centre *aviation*; Army Catering Corps; Automatic Control Certified.

acc *or* **acce** Account, *q.v.* Acceptance, *q.v.*, Accepted.

ACCA *or* **A.C.C.A.** Aeronautical Chamber of Commerce of America; Agricultural Central Co-operative Association; Air Charter Carriers Association; Association of Certified and Corporate Accountants.

Acc Claims Accident Claims Department *insurance*.

Acc. Dmge. Accidental Damage.

Ace Said of an air pilot who has shot down many enemy aircraft during combat.

Accelerated Depreciation Writing off the cost of capital assets in the early year(s) of their life, usually to take advantage of capital allowances against tax.

Acceleration Increase of speed, rate of change of velocity (e.g. of an aircraft).

Acceleration Acceptance During the initial stages of a business deal offers or counter offers are exchanged between the prospective buyers, the sellers and, on some occasions, by the brokers. If agreement is reached that would lead to the introduction for the approval of the contract. *Abbrev.* ACP *or* A.C.P. *or* ACC *or* A.C.C. *or* acpt.

Accept/Except A terminology in the negotiations between a broker and an owner or principal accepting certain points of the contract and at the same time deleting and/or altering others.

Acceptance *or* **Accepting Houses** Merchant bankers specialising and dealing in foreign trade transactions and possessing a high degree of experience in financial business affairs. One of their functions is the acceptance of Bill of Exchange, *q.v.*, on behalf of customers for a nominal fee.

Acceptance Supra-Protest An acceptance for the sake of saving honour due to a protest such as: A protested Bill of Exchange, *q.v.*, is accepted by another party so as to avoid dishonour to the drawee. *See* **Accommodation Bill**.

Acceptilation A legal Scottish term meaning the release from a debt.

Accepting Authority Senior naval officer or senior civilian contractor who is designated to deal on behalf of the government in accepting a naval ship *USA*.

Acceptor That person who has accepted liability of a Bill of Exchange, *q.v. See* **Acceptance Supra-Protest, Accommodation Bill** and **Bill of Exchange**.

ACCESS Air Canada Cargo Enquiry Service System; Name of a credit card used in the UK.

Accessories Clause A motor vehicle insurance clause. Accessories of a motor vehicle are not covered by the insurance unless the whole vehicle is stolen or lost.

Accident Insurance An insurance covering accidents while travelling by land, sea or air.

Accident Prevention Board *or* **Bureau** An American organisation established in San Francisco in 1927 devoted to the prevention of accidents involved in cargo handling operations. All accidents are studied and recommendations made to the Pacific Maritime Association. *Abbrev.* APB *or* A.P.B. *See* **P.A.B.** *or* **PAB** *or* **PMA**.

Accidental Vanishing Point Any vanishing point in the perspective which is not in the Horizontal Line, i.e. the line drawn on the picture plane at eye level. *Abbrev.* AVP *or* A.V.P.

accom. Accommodation.

Accommodation Bill This is an endorsed Bill of Exchange *q.v.*, to safeguard another party. The person signing it is holding himself responsible as guarantor if the other party does not pay at the proper time. Sometimes, though not very often, it is termed Kite. *See* **Acceptance Supra-Protest**.

Accommodation Ladder A Gangway, *q.v.*, ladder with flat steps and handrails on either side enabling passengers and members of the ship's crew to embark or disembark. *Abbrev.* A.L.

Accommodation Party The person or company signing and accepting the Accommodation Bill, *q.v.*

Accomplished Bill of Lading The consignee presents the original Bill of Lading, *q.v.*, to the ship's agents duly endorsed which is then retained for record purposes. This endorsed bill of lading is called 'accomplished'.

According to Customs of Port *See* **Customary Despatch**.

Account Arithmetical calculations; Statement of business transactions.

Account Current *see* **Current Account**.

Account Interest Accumulated Interest.

Account Payee Only *or* **Acc. Payee Only** A phrase on a crossed cheque meaning that it cannot be transferred by endorsement to an individual other than as named on the cheque.

Account Sales When merchandise is sent on consignment, credit or stock, the receiver or agent periodically (normally every month) gives a detailed account of the sales effected, including his commission and the remaining stock.

Account of Wages A full account of the ship's crew's wages including pension contribution, national insurance, holidays, allotments, withdrawals etc. The account of wages is signed by the crew and countersigned by the master.

Accountant A person engaged in book-keeping.

ACCP *or* **A.C.C.P.** American College of Chest Physicians.

Acc (Prop) Accident (Property) Department *insurance*.

accred Accredited.

Accrued Interests Accumulated interest on securities or interest due but still unpaid.

A.C.C.S. Associate of the Corporation and Corporate Secretaries. Now A.C.I.S., *q.v.*

Acct. Account; Accountant.

ACCU Automatic Control Certified Unattended. It refers to Bridge control of an unmanned Engine Room.

ACCU-O Automatic Control Certified Unattended—Open Sea. Same as ACCU, *q.v.*, but only applicable in the open sea.

Accumulative Hours A coal Charterparty, *q.v.*, stipulation that allows extra time to the charterers for loading in the event that the vessel arrives after her expected date of readiness. *Also called* Hours Accumulative.

ACD Adaptive Compass Drive *computer*; Automatic Call Distributor *telecommunications*.

ACDA Arms Control and Disarmament Agency *USA*.

ACE *or* **A.C.E.** Advisory Centre of Education *UK*; Alcohol-Chloroform-Ether; American Council of Education; Association of Circulation Executives; Association of Consulting Engineers; Association of Cultural Exchange; Australian College of Education; Automatic Computing Engine.

ACEA Action Committee of European Aerospace.

Acey-deucy Nautical slang for backgammon.

A.C.F.A. Air Charter Forwarders Association.

ACFN American Committee for Flags of Necessity. *Now known as* Federation of American Controlled Shipping. *Abbrev.* FACS.

ac. ft. Acre foot.

A.C.G.B. Arts Council of Great Britain.

A.C.G.I. Associate of City & Guilds of London Institute.

ACGIH *or* **A.C.G.I.H.** American Conference of Government Industrial Hygiene.

ACHR American Society of Human Rights.

A.Ch.S. Associate of the Society of Chiropodists.

ACI Airport Council International.

ACI *or* **A.C.I.** Associate of Clothing Industry; Associate of the Institute of Commerce; Assuré Contre l'Incendie *French*—Insured against fire.

A.C.I.A. Associate of the Corporation of Insurance Agents.

A.C.I.A.A. Australian Commercial and Industrial Artists Association.

A.C.I.Arb. Associate of the Chartered Institute of Arbitrators.

A.C.I.B. Associate of the Corporation of Insurance Brokers.

ACIC Aeronautical Chartering and Information Center *USA*.

ACICAFE Association du Commerce et/de l'Industrie du Café dans la C.E.E. *French*—Association for the Coffee Trade and Industry in the European Economic Community.

Acid A chemical solution with a low pH value, *q.v.*, pH 7 is neutral and 0 is extreme acidity.

Acid Test Ratio An indication of a company's ability to meet its immediate obligations. It is obtained by comparing Liquid Assets, *q.v.*, against Current Liabilities. Also known as the Quick Ratio. *See also* **Current Ratio**.

A.C.I.I. Associate of the Chartered Insurance Institute.

ACILA *or* **A.C.I.L.A.** Associate of the Chartered Institute of Loss Adjusters.

A.C.I.S. Associate of the Chartered Institute of Secretaries.

A.C.I.T. Associate of the Chartered Institute of Transport.

A.C.I.V. Associate of the Commonwealth Institute of Valuers.

ack. Acknowledge; Acknowledgement.

ACL Acknowledgement; Associated Container Lines—group of internationally known shipowners who formed the first company to use container ships.

ACLC Air Cadet League of Canada.

ACLS American Council of Learned Societies; Automatic Carrier Landing System.

ACM Association for Computing Machinery *USA*; Asian Currency Market; Authorised Control Material.

ACMA Associate of the Institute of Cost and Management Accountants. Formerly **ACWA** *or* **A.C.W.A.**, *q.v.*

ACMC Association of Canadian Medical Colleges.

ACME Accreditation Council for Graduate Medical Education *USA*; Advanced Computer for Medical Research; Advisory Council on Medical Education *USA*; Association of Canadian Medical Education; Association of Consulting Management Engineers *USA*.

ACMET *or* **A.C.M.E.T.** Advisory Council on Middle East Trade.

ACMF *or* **A.C.M.F.** Air Corps Medical Forces *USA*; Australian Commonwealth Military Forces.

A.C.M.P. Assistant Commissioner of the Metropolitan Police.

A.C.M.R. The Advisory Committee on Marine Resources Research of the Food and Agriculture Organisation (FAO).

ACMT American College of Medical Technologists.

A.C.N. Ante Christum Natum *Latin*—Before the Birth of Christ.

acn *or* **a.c.n.** All concerned notified.

ACNA *or* **A.C.N.A.** Arctic Institute of North America.

ACO *or* **A.C.O.** Admiralty Compass Observatory; Administrative Contracting Officer *USA*.

acog *or* **ACOG** Aircraft on Ground.

A.Com. *or* **A.Comm.** Associate in Commerce.

A.Comm.A. Associate of the Society of Commercial Accountants.

à compte *French*—On account; Part payment; Payment by instalments.

ACOP Approved Code of Practice.

ACORP Alaska Co-operative Oil Spill Responsive Planning Committee. Established to draft and design a joint oil spill contingency plan for the state of Alaska.

Acoustic Science of sound concerned with the production, reception, propagation, properties and uses of sound.

Acoustic Sounding *Similar to* **Echo Sounding**, *q.v.* The use of a special sounding device emitting sound waves through sea water which averages 4900 ft. per second. The depth is automatically recorded on a luminous scale or printed out on a strip of paper. This useful marine instrument enables the master to record the depth of the sea and thus avoid any accident involving his ship touching ground.

ACP *or* **A.C.P.** Acceptance, *q.v.* African Caribbean–Pacific Countries; American College of Physicians.

A&CP *or* **A. & C.P.** *Lloyd's Register notation* for Anchors and Chains Proved.

acpt. Acceptance, *q.v.*

Acquittance Written discharge of liability, contract, debt, etc.

ACR Airfield Control Radar/s.

Acre An area of 4,840 square yards.

Acre ft. Acre foot.

ACRI Air-Conditioning and Refrigeration Institute *USA*.

ACRL Association of College and Research Libraries *USA*.

ACS *or* **A.C.S.** Admiralty Computing Service; Airport Catering Services; American Cancer Society; American College of Surgeons; Assembly Control System; Associate in Commericial Science; Australian Computer System; Automatic Control System in Air Cushion Vessels; Auxiliary Crane Ship *USA*.

ACSC Australian Coastal Surveillance Centre, Canberra, Australia.

a/cs Accounts.

A/cs Accident Claims Department *insurance*.

A/cs Pay Accounts Payable.

A/cs Rec Accounts Receivable.

Act Acceptance; Accepted; Decree; Law; American College Test.

ACT *or* **A.C.T.** Advanced Corporation Tax; Advisory Council on Technology; Associated Container Transportation; Australian Capital Territory; Australian College of Technology.

a. cta. a cuenta *Spanish*—By instalment; On account; Part payment.

Acta imperii *Latin*—Acts of the Sovereign authority. Exercise of the royal prerogative.

Acta gestionis *Latin*—Acts of management.

Actg. Acting.

Adrenocorticotropic Hormone Gland stimulator.

Actio debit *or* **debitoris mei** *Latin*—My debtor's debtor.

Actio in personam *Latin*—An action against a person. *See* **In rem**.

Actio in rem *Latin*—*See* **In rem**.

Actio personalis moritur cum persona *Latin*—A personal action dies with the person.

Actio redhibitoria *Latin*—An action to avoid a sale within 6 months from conclusion of a contract where the thing sold is so defective that the buyer would not have purchased it had he known.

Active Balance of Payments International financial and trade transactions between a country and others. It is more or less similar to a statement of accounts of income and expenditure between one country and all the others. It is also termed as Favourable Balance of Payments or Trade.

Active Bonds Bonds which are paid from the date of issue at a fixed rate of interest.

Active Debt A debt upon which interest is actually paid. *See* **Passive Debt**.

Active Market Certain types of popular Stocks and Shares, *q.v.*, which are continually on the market.

Active Partner A shareholder in a company or a firm who participates in the running of the business for which service he is paid. If a partner or a shareholder does not take an active part he is called Dormant or Sleeping Partner, *q.v.*

Act of Bankruptcy When a firm or company becomes insolvent and is officially declared as such it publishes or declares an act of bankruptcy. By so doing its normal activities virtually come to a halt and the creditors start claiming moneys due through official or legal quarters. The procedure takes some time as the auditors in charge of the dissolution of the company, due to its insolvency, will have to assess the relative and proportional payment for each and every claim. Sometimes it is called Keeping House.

Act of God An inevitable event occurring without the intervention of man—such as flood, tempest, or death—operating in case of certain contracts, such as those of Insurers, *q.v.* or Carriers.

Act of Omission of Shipper or Owner of the Goods or Agents An excepted peril of the sea in the Carriage of Goods by Sea Act 1924, subsequently amended. The Carrier is exonerated from any consequences due to omission or carelessness on the part of the shippers or his/their agents.

Act of Public Enemies An excepted peril of the sea in the Carriage of Goods by Sea Act 1924, subsequently amended. This refers to an act of piracy.

Act of War An excepted peril of the sea in the Carriage of Goods by Sea Act 1924, subsequently amended, which includes wars of all nations. The nationality of the ship is immaterial.

Actual Total Loss *Abbrev.* ATL or A.T.L. This relates to an insurance policy and can occur in any of four ways: (1) The property is completely destroyed; (2) The owner is irretrievably deprived of the property; (3) Goods change their character to such a degree that they can be said to be no longer the thing insured by the policy; (4) The subject matter of the insurance, be it ship or goods on board the ship, is recorded as 'missing' at Lloyd's.

Actual Payload Net Weight. The difference between the gross weight and the Tare, *q.v.*

Actuary An expert in the insurance business who estimates or assesses premiums or various types of insurance risks.

Actus reus *Latin*—The act of a guilty man.

ACUE American Committee of United Europe.

A cuenta *Abbrev.* a.cta., *q.v.*

ACURIL Association of Caribbean University Research and Institute Libraries.

ACUS Administrative Customs Union; Atlantic Council of the United States.

acv *or* **a.c.v.** *or* **A.C.V.** Actual Cash Value; Air Cushion Vehicle *Hovercraft*.

A.C.V. Form A Lloyd's policy form of insurance on an Air Cushion Vessel Hovercraft, consisting of three forms. *A.C.V. Form 1* covers damage incurred to the hull, but excluding fire, theft, burglary, wear and tear as well as machinery and mechanical parts. *A.C.V. Form 2* refers to the Returns Clause or to the allowance of return of premium to the assured should the policy be mutually withdrawn or if the hovercraft is laid up for at least 30 days *A.C.V. Form 3* refers to the recovery of legal expenses; third party liability; personal injuries to the passengers and baggage damage and loss or damage to cargo.

A.C.V. 1 *See* **A.C.V. Form.**

A.C.V. 2 *See* **A.C.V. Form.**

A.C.V. 3 *See* **A.C.V. Form.**

acw *or* **ACW** *or* **A.C.W.** Air Contact and Warning; Alternating Continuous Waves; Automatic Car Wash.

ACWA *or* **A.C.W.A.** Amalgamated Clothing Works of America; Associate of the Institute of Cost and Works Accountants, now ACMA, *q.v.*

ACWRRE *or* **A.C.W.R.R.E.** American Cargo War Risk Reinsurance Exchange.

ACWS Aircraft Control and Warning System.

acy *or* **ACY** Accuracy; Average Crop Yield.

ad Addatur *Latin*—Let there be added; Average Deviation *nautical*; Advertisement.

a.d. *or* **a/d** After date, *q.v.* Ante diem *Latin*—Before the day.

AD *or* **A.D.** Accidental Damage *insurance*; Above Deck; After Draught (Draft); Anno Domini *Latin*—In the Year of Christ or After Christ; Average Deviation; Assigned Draught (Draft) *nautical*; Average Demurrage, *q.v.* Norwegian Limited Company; Advise Duration.

A/D Anno Domini *Latin*—In the Year of Christ or After Christ; After Date, *q.v.*; Alongside Date, *q.v.*

AD&C Advise duration and charge.

Ada Adelaide, *Australia*.

ADA *or* **A.D.A.** Action Data Automation; Agriculture Development Association; Air Defence Agency *USA*; Aluminium Development Association; American/Australian Dental Association; Americans for Democratic Action; Atomic Development Authority.

ADAA American Dental Assistants Association.

ad absurdum *Latin*—To the point of absurdity.

adaml *or* **ADAML** Advice by Airmail.

ADAPSO Association of Data Processing Organizations *USA*

ADAPTS Air Deliverable Anti-Pollution Transfer System, *US Coast Guard System*.

Ad arbitrium *Latin*—At pleasure.

ADAS Action Data Automation System *US Navy electronics*

ad astra per aspera *Latin*—To the stars through difficulties. Motto of Kansas, USA.

a dato *Latin*—From the date.

A. Day Assault or Attack Day.

ADB *or* **A.D.B.** Accidental Death Benefit, *insurance*; Asian Development Bank.

ADC Association of Diving Contractors; Aide-de-Camp; Automatic Density Control.

Adcom *or* **Ad. Com** *or* **ADCOM** Address Commission, *q.v.* Administrative Command *USA*.

Ad commodatum populi *Latin*—In the public service.

ADD *or* **A.D.D.** Average Due Date, *q.v.*

add. To be added; Addenda; Addendum; Addition; Address.

ad die *Latin*—From that day.

Additional Expense Clause Strike Risks.

Additional Premium Occasionally no sufficiently detailed information can be furnished by the insurance broker to the Underwriters, *q.v.*, about the insured client. In this case and always subject to good faith on the part of the insured, pending the particulars required, an additional premium is accordingly assessed and submitted. *Abbrev.* A/P *or* A.P.

ADDIZ Air Defence Identification Zone.

Add'l Additional.

Addns. Additions.

Address Commission A remuneration allowed to the charterers in some charterparties. This virtually reduces the rate of freight. If no similar remuneration or address commission is given to the Charterers, *q.v.*, it is referred to as Free of Address, *q.v. Abbrev.* Adcom or Ad. Com or ADCOM.

Addressograph Brand of printing machine which produces addressed labels.

ADE Automatic Drafting Equipment *computer*.

à demi *French*—Half; by halves; in part.

Ademption (1) Implied revocation of a Legacy *q.v.* where a Testator's subsequent action makes a willed bequest irrelevant. (2) Reduction of a Total Marine Loss to a partial loss for legal proceedings due to a change of circumstances since the notice of abandonment.

A des. A destra *Italian*—To the right.

Ad eund. Ad eundem gradum *Latin*—To the same degree.

Ad extremum *Latin*—To the extreme; at last.

Adenosine Triphosphate Energy carrying chemical. *Abbrev.* ATP.

ADF Automatic Document Feeder.

ADF *or* **A.D.F.** Automatic Direction Finder.

adf *or* **ADF** After deducting Freight; Air/Automatic Direction Finder.

A.D.F.A. Australian Dried Fruit Association.

ad finem *Latin*—Near or towards the end.

ADFOA Australian Duty Free Operators Association.

ad gustum *Latin*—To one's taste.

ad.h.l. Ad hunc locum *Latin*—To or at this place; On this passage.

ad hoc *Latin*—For a particular occasion; For a special purpose.

ad hominem *Latin*—Personal; To the man; used of an argument in which a person is confronted with his own words.

ad hunc locum *Latin. Abbrev.* ad.h.l., *q.v.*

ad hunc sub judice lis est *Latin*—It is still under judicial determination. *See* **Sub Judice**.

ad idem *Latin*—At the same point. Essentially in agreement.

ad inf. Ad infinitum *Latin*—Without end; To infinity.

ad infinitum *Latin. Abbrev.* ad inf., *q.v.*

ad init. Ad initium *Latin*—To or at the beginning.

ad initium *Latin. Abbrev.* ad init., *q.v.*

ad int. Ad interim *Latin*—In the meantime; Meanwhile.

ad interim *Latin. Abbrev.* ad int., *q.v.*

ad internecionem *Latin*—To extermination.

ADIS *or* **A.D.I.S.** Automated Computer Based Data Interchange System.

ADIZ Air Defence Identification Zone.

Adj. Adjourned; Adjustable; Adjuster/s, *q.v.*; Adjustment; Adjutant.

Adjacent Zone Alternatively known as 'Contiguous Zone'. It is the area beyond the International Territorial Zone. Some nations still impose their rights to a further 150/200 miles over the three or 12 miles limit to safeguard their national security so as to protect fisheries, offshore oil exploration etc.

ADJAG *or* **A.D.J.A.G.** Assistant Deputy Judge Advocate General.

Adjble. Adjustable.

Adjr. Adjuster, *q.v.*

Adjt. Adjuster; Adjustment; Adjutant.

Adjudication Order A demand or order by law declaring one's property to be held under power of a trustee when one has filed for bankruptcy.

Adjuster/s In insurance claims these are appointed persons or accountants nominated by the owners or their representatives/brokers whose duties are to compile all the relative expenses in relation to repairs or damage sustained by the ship. These adjusters are generally known as 'Average Adjusters'. *Abbrev.* Adj.

a.dk. *or* **A.dk.** Awning Deck *Lloyd's Register notation.*

A.dk. After Deck.

ADL *or* **A.D.L.** Automatic Data Logging.

a.d.l. Activities of daily living.

ad lib *or* **ad libit** Ad libitum *Latin*—At pleasure; To any extent; To the extent required.

ad libitum *Latin. Abbrev.* ad lib *or* ad libit, *q.v.*

ad loc Ad locum *Latin*—At the place.

Adm *or* **ADM** Admiral; Admiralty.

ADM *or* **A.D.M.** Annual Delegated Meeting.

ADMA American Drug Manufacturers Association; Aviation Distributors and Manufacturers Association *USA.*

ADMATT Advanced Mobile Torpedo Target *USA.*

Admeasurement The confirmed or official dimensions of a ship.

admin. Administer; Administration; Administrator, *q.v.*

Administrative Lead Time The time or interval taken in the formalities before an actual contract or order is concluded. This term is generally used in America.

Administrator A person appointed by law to act and administer estates when no testators are mentioned in the will.

Admiral of the Ocean Sea Award An American award assigned annually to the person who has rendered outstanding service to shipping. The award consists of a silver statuette of Christopher Columbus. The recipient is also entitled to the honorary distinction of Admiral of the Ocean Sea. The committee responsible for the selection comprises US government officials, presidents, managers of private shipping companies, officers and many others engaged in the marine field. The first award was granted by Queen Isabella of Spain in 1493 to Christopher Columbus after he had discovered America. *Abbrev.* AOTOS.

Admiralty Droit Money obtained by selling wrecks or derelicts or from ships seized by the Admiralty, especially during time of war.

Admiralty Law Maritime Law.

Admiralty Measured Mile 6,080 feet or 1,853.184 metres as distinguished from the nautical mile of 6,045.95 ft. or 1,842.8055 m. The mean nautical mile is 6,076.91 ft. or 1,852. 2421 m. *Abbrev.* AMM *or* A.M.M.

A.D.M.O. Assistant Director of Meteorological Office *Civil Aviation Section.*

ad modum *Latin*—In the manner of.

admov. Admoveatur *Latin*—Let it apply.

A.D.M.T. Association of Dental Manufacturers and Traders.

ad nauseam *Latin*—To the point of disgust; Enough to make a person sick.

ADO *or* **A.D.O.** Advanced Development Objective; Association of Dispensing Opticians.

ADP Automatic Data Processing *computer.*

ADPLAN Advanced Planning.

ADR *or* **A.D.R.** Agreement of Carriage of Dangerous Goods by Road; Advisory Routes, *q.v.*; Advise by Return; Alternative Dispute Resolution, *q.v.*; American Depository Receipt.

ad rem *Latin*—To the purpose; To the point.

ad referendum *Latin*—For further consideration. Although a contract may have been signed, there will be a period of time for final consideration by the party or parties concerned when this phrase is applied.

Adrift An object floating at sea and moving with the current, tide or wind. If a ship is at anchor and the anchors are loosened from the seabed she starts drifting unless she is prevented from so doing by her own engines or outside help, such as salvage tugs.

adr *or* **ADRN** Advise by Return.

ad.s Ad sectum *Latin*—At the suit of.

a.d.s. Additional Duty on Sugar; Autograph Document Signed.

ADS Actual Defined System *computer*; Atmospheric Diving Suits *used in Oil Rigging Operations*; Advanced Dressing Station; Additional Duty on Sugar.

ad saec. Ad saeculum *Latin*—To the century.

ad saeculum *Latin. Abbrev.* ad saec., *q.v.*

ad sectum *Latin. Abbrev.* ad s., *q.v.*

ADSL Assymetric Digital Subscriber Loop. Technology for video transmission over telephone lines.

ad sum *Latin*—Here; I am present.

ad summam *Latin*—In short; in a word.

ad summum *Latin*—To the utmost.

ad unguem *Latin*—On the nail; to the point.

ad unum omnes *Latin*—All to a man.

ad us. Ad usum *Latin*—As customary.

ad us. ext. Ad usum externum *Latin*—For external use.

ad usum *Latin. Abbrev.* ad us., *q.v.*

ad usum externum *Latin. Abbrev.* ad us. ext., *q.v.*

ad utrumque paratus *Latin*—Prepared for either event.

adv *or* **Adv.** Adversus *Latin*—Against; Advice. Advancement of Special Survey *Lloyd's Register notation.*

Ad Valorem *Latin*—According to the value. An ad valorem stamp on deeds or documents is one fixed in proportion to the amount of rent reserved or other element of value expressed in the deed.

Advance Freight *or* **Freight Paid in Advance** *or* **Freight Prepaid** Freight which is paid against the delivery or handing over of the original Bills of Lading, *q.v.*, to the shipper or consignor by the carrier or the agent. In most cases freight is due for payment on presentation of the Shipping Order, *q.v.*, signed by the chief officer or the master of the vessel. This is a formal receipt that the goods have been received on board by a responsible officer representing the Carrier. In return for this receipt or for this signed shipping order the original bill of lading, together with other 'non-negotiable' copies of bills of lading, are given to the shippers. The carriers or shipowners have an advantageous position on such freight as in almost all cases the freight thus paid is not refundable whether 'ship or cargo is lost or not lost' provided, of course, the loss is not due to any fault of the carriers, such as unseaworthiness etc. *Abbrev.* AF *or* A.F. or Adv. Frt. *See* **Freight Paid in Advance.**

Advancement Money advanced by an administrator to a beneficiary. *See* **Administrator.**

Advance Note *or* **Allotment Note** A draft on a shipowner for wages, given to a seaman on signing Articles of Agreement and redeemable after the ship has sailed with the seaman on board.

Advance Notice A Charterparty, *q.v.* clause whereby owners undertake to keep the Charterers, *q.v.*, closely informed in relation to the expected time of loading or alteration thereto.

Advance Purchase Excursion An international pre-air passage booking for 14 days up to 60 days, with a minimum and maximum day allowable. *Abrev.* Apex *or* APEX.

Advancing Tide *Similar to* **Flood tide,** *q.v.*

Advection Fog Fog which occurs when warm, damp air moves over a surface which is cooler than the air dewpoint. The frequent fogs on the Newfoundland-Grand Banks occur when warm moist air overlying the Gulf Stream is blown northwards to the Labrador region. *See* **Radiation Fog.**

ad verbum *Latin*—Word for word; exactly.

Adverse Balance Unfavourable or passive balance of a payment in the international market between one country *vis-à-vis* others.

Adverse Possession Holding of land by a person who is not the true owner.

Adversus *Latin. Abbrev.* adv, *q.v.*

Advise Fate A banking term commonly used when enquiring about whether there is enough money to meet the demand on the cheque about to be drawn. If the receiving bank entertains any doubt before honouring the cheque it communicates with the issuing source (by telex or cable) to be advised of the said cheque's fate.

Advisory Message A message by the master of a ship to a local government agency giving advance notice of a situation that may arise following a distress signal for assistance if and when the ship's condition continues to deteriorate.

Advisory Routes The air passage or air corridor at a specific level wherefrom aircraft in flight can obtain the necessary service from the control towers. More or less similar to **Advisory Service Area**, *q.v. Abbrev.* ADR.

Advisory Service Area Specific area where advice to the aircraft is available, *Abbrev.* ASA. *See also* **Advisory Routes.**

ad vitam aut culpam *Latin*—For life or until there is misconduct.

Advocate Barrister; Solicitor A person who holds a law degree and who pleads the cause of others in law courts.

Advocatus diabli *Latin*—Devil's advocate.

Adv. pmt. Advance payment.

Advt. Advertisement.

ae *or* **aet** Aetatis *Latin*—Of age; At the age of.

AE *or* **A.E.** Adult Education; Aeronautical Engineering; Air Escape; Atomic Energy; Automatic Exposure *photography.*

A&E Accident and Emergency, *medical organisation.*

AEA *or* **A.E.A.** Agricultural Education Association; Agricultural Engineers' Association; Airline Entertainment Association; American Economic Association; Association of European Airlines; Atomic Energy Authority *established in 1945 to control research and development.*

A.E.B. Associated Examining Board.

AEC *or* **A.E.C.** Agricultural Executive Company; American Engineering Council; American Express Company; Association of Education Committees; Atomic Energy Commission *USA.*

A.E.D. Association of Engineering Distributors; Automated Engineering Design.

A.Ed. Associate of Education.

AEE Airborne Evaluation Equipment; Atomic Energy Establishment *USA.*

AEF *or* **A.E.F.** Afrique Equatoriale Française, French Equatorial Africa; Aviation Environment Federation, formerly Airfields Environment Federation, founded in 1975. A UK non-governmental organisation.

AEG Active Element Group.

AEGIS Arab-Euro-Greek Investment Symposium.

AEI Automotive Engineering Inc. *A US society of automobile engineers*; Amalgamated Union of Engineering and Foundry Workers.

A.E.L.E. Association Europeéne de Libre-échange *French*—European Free Trade Association.

A.E.&M.P. Ambassador Extraordinary and Minister Plenipotentiary: *similar to* **A.E.&P.,** *q.v.*

AEM Acoustic Emission Monitor.

AEMEURO African, Eastern Mediterranean and European.

A.E.M.T. Association of Electrical Machinery Trades.

A. En. Associate in English.

A.Eng. Associate in Engineering.

A.E.O.S. Astronomical, Earth and Ocean Sciences *USA.*

A.E.&P. Ambassador Extraordinary and Plenipotentiary: *similar to* **A.E.&M.P.,** *q.v.*

aeq. Aequales *Latin*—Equal; Equals.

aequabiliter et diligenter *Latin*—Equably and carefully.

aequales *French. Abbrev.* aeq., *q.v.*

aequam servare mentem *Latin*—To keep an open mind.

aer. Aeronautic/s; Aeroplane.

A.E.R.E. Atomic Energy Research Establishment.

Aer. E *or* **Aero. E.** Aeronautical Engineer.

Aer Lingus Irish Republic International Airlines.

Aerial Lights These come under the category of lighthouses. They are generally posted on high poles to provide navigational directions to aircraft during the course of the flight, landing and take-off. They are also commonly seen as beacons at airports.

Aerodynamics The scientific effects of a moving body while passing through the air and gases.

Aerometer An instrument used to measure the specific gravity of fluids.

Aeron Aeronautics.

AEROSAT Aeronautical Communications Satellite.

AES *or* **A.E.S.** Aerospace Electrical Society *USA*; Airways Engineering Society *USA*; Association of Engineering Schools, *in the marine field.*

AEST Allgemeiner Europaische Stückguttarif *German—similar to* **EPLT,** *q.v.*

aet *or* **ae** Aetatis *Latin*—Of age; At the age of.

aetatis *Latin. Abbrev.* aet *or* ae, *q.v.*

aetatis suis *Latin*—At his(her) age.

A.E.T. Associate in Electrical/Electronic Technology.

AEW *or* **A.E.W.** Airborne Early Warning. *See* **Airborne Early Warning and Control System.**

AEW&CO *or* **A.E.W.&Co.** Airborne Early Warning and Control System.

AEWL Association of Employers of Waterside Labour.

AEWS *or* **A.E.W.S.** Aircraft Early Warning System.

af *or* **alf.** A favour *Spanish*—In Favour; *Italian*—Next Year; All faults.

a.f. Audio frequency; anno futuro.

a/f Also for *Ports of Call.*

Af. Africa; African; Afrikaans; Afghani *Afghanistan unit of currency.*

AF *or* **A.F.** *or* **A/F** Advanced Freight. *See* **Freight Prepaid; All Faults; Anti-flooding; Anti-fouling.**

AF *or* **A.F.** *or* **af** Admiral of the Fleet; Anglo French; Audio Frequency *radio*; Air France *French National Airlines*; All Freighter *aircraft.* Also for *Ports of Call.*

AFA Air Force Association *USA*; American Foundrymen's Association; Associate of the Faculty of Actuaries.

A.F.A.A. *or* **a.f.a.a.** As far as applicable.

A.F.A.I.M. Associate Fellow of the Australian Institute of Management.

AFAR Azores Fixed Acoustic Range.

A.F.A.S. Associate of the Faculty of Architects and Surveyors.

AFB *or* **A.F.B.** Air Fare Bureau; Air Force Base; Air Freight Bill: *similar to* **Airway Bill** *or* **Air Waybill.**

a.f.c. Automatic Frequency Control.

AFC Automatic Flight Control; Automatic Frequency Control; Air Force Cross.

A.F.C. Australian Flying Corps, *obsolete.*

AFCAC African Civil Aviation Commission.

A.F.C.A.I. Associate Fellow of the Canadian Aeronautical Institute.

AFCAN French Shipmasters' Association.

AFCE *or* **A.F.C.E.** Automatic Flight Control Equipment; Associate in Fuel Technology and Chemical Engineering.

AFCENT Allied Forces in Central Europe *NATO.*

AFCS *or* **A.F.C.S.** *or* **a.f.c.s.** Adaptive/Automatic Flight Control System.

AFD Auxiliary Floating Drydock; Actual Forward Distance.

AFDB African Development Bank.

AFDC Aid to Families with Dependent Children.

AFDM Auxiliary Floating Drydock Medium.

AFEXMAR Association Française des Experts Maritimes—Association of French Maritime Experts.

Aff.AIB Affiliate of Associated Insurance Brokers.

AFFF Aqueous Film-forming Foam.

Affidavit A written statement on oath witnessed in the presence of an authorised or legal personality, i.e. a magistrate, senior clerk of the law courts, Notary Public, Public Major, Consul, Commissioner of Oaths etc. *Abbrev.* Afft.

Affiliate An associated or subsidiary company; A firm joined with other firms.

Affirmant A person making a solemn declaration in giving *viva voce* evidence instead of an oath or affidavit.

Affirmative Standard word in radio telephone communication signifying 'Yes'. The opposite is Negative meaning 'No', *q.v.*

Affirmative Flag An international letter Code Flag 'C' signifying 'Yes'. *A small flag having blue, white and red striped horizontal colours.*

Affluent Quality of society with a high standard of living; a tributary stream or river.

Affreightment A contract to carry goods by sea or air. Charterparties and bills of lading are contracts of affreightment. *See* **Charterparty.**

Afft. Affidavit, *q.v.*

Afg *or* **Afgh** Afghanistan.

Afghani Unit of currency of Afghanistan. *Abbrev.* Af.

AFI *or* **A.F.I.** American Film Institute; Air Freight Institute; Associate of the Faculty of Insurance.

AFIA *or* **A.F.I.A.** American Foreign Insurance Association; Associate of the Federal Institute of Accountants *Australia.*

AFIAS Associate Fellow of the Institute of Aerospace Science *USA.*

AFII American Federation of International Institutes.

A.F.I.I.M. Associate Fellow of the Institute of Industrial Managers.

AFISCO Aerodrome Flight Information Service Officer *aviation UK.*

AFL Aeroflot *Soviet Airlines*; American Federation of Labor.

AFL CIO American Federation of Labor-Congress of Industrial Organizations.

Afloat The act of remaining on the surface of a liquid and not touching the bottom or ground. Opposite of Aground, *q.v.* In most of the charterparties one of the conditions enumerated is that the vessel is to load always afloat. In some other charterparties the phrase 'touching on soft ground' is tolerated. *Abbrev.* Aflt. *See* **Always Afloat.**

Aflt. Afloat, *q.v.*

AFM Air Force Medal; American Federation of Musicians; Armed Forces Malta.

AFMENCO Africa and Middle East Management Consultancy.

AFN Association Française de Normalisation *French standards institute.*

AFNOR Association Française de Normalisation. French Standards Association.

à fond *French*—In depth; thoroughly.

a fortiori *Latin*—With stronger reason.

AFR *or* **A.F.R.** Air Fuel Ratio.

A.Fr. Algerian franc. *An Algerian unit of currency*; Anglo French.

Afr. Africa.

AFRA *or* **A.F.R.A.** Average Freight Rate Assessment, *q.v.*

AFRAMAX Describes a tanker of 79,999 dwt, i.e. at the top of the range 45,000 to 79,999 dwt which is one of the ranges used in Average Freight Rate Assessment, *q.v.* A ship of this measurement is of advantage economically to the oil shipper.

AFRAS *or* **A.F.R.A.S.** Associate for Rescue at Sea.

A.F.R.Ae.S. Associate Fellow of the Royal Aeronautical Society.

AFRASEC. Afro-Asian Organisation for Economic Co-operation.

AFS *or* **A.F.S.** Air Force Station; American Field Service; Arab Federation of Shipping; Army/Auxiliary Fire Service; Associate Surveyor, Member of the Faculty of Architects and Surveyors; Aeronautical Fixed Services *aviation.*

AFSBO American Federation of Small Business Organizations.

AFT *or* **A.F.T.** American Federation of Teachers.

Aft. B. After Body, *q.v.*

AFTE American Federation of Technical Engineers.

After Arrival In marine insurance this refers to the day or date of the insured merchandise or ship after arrival at destination. *Abbrev.* a.a. and is normally inserted on the slip by the broker extending the policy against an additional and proportional Premium, *q.v.*, covering the time extended, i.e. the additional period covered after the arrival of the merchandise or the ship.

After Back-Spring A mooring line running from the after end of a vessel to a point on the quay some distance forward of this, to prevent her from moving astern. Also important as a ship-handling aid during berthing and unberthing operations.

After Body The after part of a ship, abaft the midship point. *Abbrev.* Aft. B.

After Date A Bill of Exchange term referring to the payment to be settled at a certain time after the date shown on the bill of exchange or Promissory Note. *Abbrev.* A/D *or* a.d. *or* a/d.

After End The poop or stern section of a ship.

Aftermost Nearest to the stern.

Afternoon Watch The seaman's watch starting from midday (the time when a nautical day starts) up to 4 p.m.

After Peak *or* **Aft Peak Tank** A tank positioned at the after end of a ship often filled either with fresh water for the daily use of the officers and crew, or with sea water ballast. This tank, together with the deep or bottom tanks as well as the fore peak tanks, are considered essential when taking into account the stability and seaworthiness of a ship. *Abbrev.* APT *or* A.P.T.

After Sight A Bill of Exchange phrase in connection with the acceptance of the draft by the drawee which is to be paid by a fixed date. *Abbrev.* A.S. *or* a/s *or* A/S.

AFTI Automated Freight Tariff Information *USA.*

Aftn. Afternoon.

AFTN Aeronautical Fixed Telecommunication Network *aviation.*

AFTRA American Federation of Television and Radio Artists.

AFVOA *or* **A.F.V.O.A.** Aberdeen Fishing Vessel Owners Association.

AG *or* **A.G.** Adjutant General; Aktiengesellschaft *German* Limited Company; Armour Grating; Symbol for silver.

AG Argentum. *Latin*—Silver.

A.G. Arabian Gulf.

aga or **a.g.a.** Air-to-ground-to-air.

AGA or **A.G.A.** American Gas Association

AGACS Automatic Ground-to-Air Communications.

Against All Risks or **All and Any Risks** An insurance phrase. The insured is covered comprehensively within the terms of the contract. The insurance is subject to the final clause appearing at the end of the policy known as the Memorandum, *q.v.* *Abbrev.* a.a.r.

Against Documents *See* **Documents Against Payment.**

AGARD Advisory Group for Airspace Research and Development *NATO.*

agb or **AGB** or **a.g.b.** A Good Brand; Any Good Brand.

AGC African Groundnut Council; Automatic Grain Control.

AGCA Automatic Ground Control Approach *aviation.*

Agd. Agreed.

Age Admitted In life insurance (assurance) this clause refers to the actual age acceptance or admittance by the insurance company which requests a birth certificate or similar document to ascertain the age declared by the proposed assured.

Agency The right given by an individual or a body to another person or company to act on his or its behalf or to market or sell his or its products. An agency may be classified in three categories; (1) Specific or Special Agency where the duties are restricted to a nominated transaction. (2) General Agency where actions can be undertaken on behalf of the principals with a certain amount of responsibility. (3) Universal Agency which involves the full power of attorney provided all actions will be within the law. Agency appointments are normally given on the basis of a certain period of time or indefinitely by means of a written contract or a letter of intent. There is a wide scope in agency work such as (a) Representing a shipping company (b) Contracting agreements for the booking (or stemming) of ships for drydocking and/or repairing in the marine engineering, drydocking field (c) Personal representation on official matters in the absence of the principals (d) Importation of materials and/or products. *Abbrev.* Agy. *See* **Agent.**

Agency Commission Clause An insurance clause whereby the insurer will not reimburse the assured for expenses and time incurred in obtaining and supplying information and documents in support of a claim, nor any fees or commissions involved in doing so.

Agency Fee A remuneration allowed to an agent in a port by a shipowner for work performed during loading and/or discharging operations or any other agency work performed during the time that the vessel has entered and left the port. The agency fees vary in every port and official tariffs are invariably laid down by either the government or other authorities. There is also a very similar agency fee where the importer is allowed a certain percentage against the amount of the imported merchandise by the exporter or the manufacturer or his principal. This is termed as 'Commission Agency'. *See also* **Commission Agent.**

Agenda A pre-arranged list of business topic items, circulated to members for discussion when meetings are convened. The circulated papers are called the Agenda Papers, *q.v.*

Agenda Papers *See* **Agenda**.

Agent A person or a company authorised to act on behalf of his or its principal. The appointed agent is very important as he will be the man on the spot who has to act promptly and cautiously to deal with any problems on behalf of his principal. The agent is duty bound to follow the instructions of his principal very carefully and, in many instances, he indicates on whose authority he is acting. It is also common to find under the signature 'As Agent(s) Only' to signify in what capacity he is signing. There are many classes of representation involving agents, some of which are: Chartering Agent/Broker, Commission Agent, Distributing Agent, Forwarding Agent, Loading Agent, Operating Agent, Owner's Agent, Managing Owner, Master for the Owners, Ship Agents and others. *Abbrev.* Agt. *See* **Agency.**

Agent de Change. The French meaning for stockbroker or Mercantile Broker.

Agents of Input Similar to Factors of Production, *q.v.*

Age quod agis *Latin*—Do what you are doing.

ager publicus *Latin*—Public property.

agfo or **A.G.F.O.** Arabian Gulf for Orders.

Agg Aggregate.

AGHS Active Gas Handling System *oil and gas industry.*

AGI or **A.G.I.** American Geographical Institute.

Agio The difference in value between the paper and the metallic moneys of a country.

Agiotage Commercial securities in Foreign or Stock Exchange.

AGIP or **A.G.I.P.** Azienda Generale Italiana Petroli. Italian National Oil Company.

agl Above Ground-Level.

AGM or **A.G.M.** Air to Ground Missile; Annual General Meeting.

AGN Articles for the Government of the Navy *USA.*

AGNIS Approach Guidance Nose-In to Stand. *Aviation term used while taxiing.*

Agnus Dei *Latin*—The Lamb of God.

ago. agosto *Italian*—August.

Agony of the moment A 'state of emergency' or 'in view of the state of emergency'. During a sea peril the master of a ship is expected to act reasonably, although spontaneously, to protect life as well as collective property in the General Average act. He should be guided by (1) The voluntary sacrifice of the property. (2) Any extraordinary expenditure incurred during the time of peril. (3) The preservation of the entire common maritime adventure. *See* **General Average**, **General Average Act**, **General Average Fact**.

AGP *or* **A.G.P.** Academy of General Practice; Aviation General Policy.

AGPA American Group of Psychotherapy Association.

AGR Advanced Gas-Cooled Reactor.

à grand frais *French*—At great expense.

Agreed Returns The return of premiums by the underwriters as per policy conditions to the insured. *Ex.* Laying Up Returns, *q.v.*

Agreed Value Clause A bill of lading clause limiting the carriers' liability to a certain sum of money or a limited sum per package or per shipping unit, in case of subsequent claims by the shippers or their agents.

Agronomy The cultivation of land, soil management and crop production.

Aground A ship is termed 'aground' when she touches hard ground. *See* **Afloat**.

Agrt Agreement.

Agst Against.

Agt/s Agent/s, *q.v.*

AGU American Geophysical Union.

AGV Automated Guided Vehicle.

AGW Atlantic Gulf, West Indies *Insurance limits*; All Goes Well; All Going Well.

AGW *or* **a.g.w.** Actual Gross Weight.

AGWI *or* **A.G.W.I.** Atlantic Gulf West Indies Warranty. *Warranty limit areas in marine insurance.*

Agwalla Asiatic seaman under Lascar's Agreement doing the job of a fireman.

Agy Agency, *q.v.*

A.H. After hatch; Anno Hegirae *Latin*—Mohammedan Calendar; Antwerp/Hamburg.

Ah After Hatch; Ampere-hour.

AHA American Historical Association; American Hospital Association.

AHASC Air Handling Agreement Sub-Committee *IATA*.

à haute voix *French*—Out loud.

A.H.C. Accepting Houses Committee; Air Handling Committee.

ahd Ahead, *q.v.*

AHD *or* **A.H.D.** American Heritage Dictionary.

A.H.E. Associate of Home Economics.

Ahead A forward direction in the line of a vessel's fore-and aft line. When a vessel moves forward it is said to move ahead, while if it moves backwards it is said to move astern. *Abbrev.* ahd.

AHESC Air Handling Equipment Sub-Committee *IATA*.

AHF *or* **A.H.F.** American Hull Form, *q.v.*

AHFDA *or* **A.H.F.D.A.** American Hull Form Deductible Average *marine insurance policy on hull.*

AHI American Health Insurance; American Hospital Institute.

AHIS *or* **A.H.I.S.** American Hull Insurance Syndicate, *q.v.*

AHM Air Handling Manual *IATA*.

Ahoy A nautical exclamation used to hail a person or persons from a distance.

AHP *or* **A.H.P.** Air Horse Power. *Also abbrev.* air h.p.

AHPSC Air Handling Procedures Sub-Committee *IATA*.

ahr *or* **AHR** Acceptable Hazard Rate *insurance*. Antwerp/Hamburg Range.

AHS *or* **A.H.S.** American Helicopter Society; Annual Hull Survey.

AHST Anchor Handling Salvage Tug *oil rig tender.*

AHT Anchor Handling Tug *oil rig tender.*

AHTS Anchor Handling Tug Supply *oil rig tender.*

a.h.v. Ad hanc vocem *Latin*—At this word.

AI Airborne Interceptor System *aviation*; Air India *India State Airlines*; American Institute; First Class.

A.I. Air India *India State Airlines*; Anno Inventionis *Latin*—In the year of discovery; Auctioneers Institute.

a.i. Ad interim *Latin*—For the meantime. *Also abbrev.* Ad int.

AIA *or* **A.I.A.** Abrasive Industries Association; Aerospace Industries Association; American Institute of Architects/Aeronautics; American Investors Association; Associate of the Institute of Actuaries; Associate of International Accountants; Aviation Industry Association *New Zealand*.

AIAA American Institute of Aeronautics and Astronautics.

A.I.A.A. Associate Architect, Member of the Incorporated Association of Architects and Surveyors; Association of International Advertising Agencies.

AIAC *or* **A.I.A.C.** Air Industries Association of Canada; Associate of the Institute of Company Accountants.

A.I.A.E. Associate of the Institute of Automobile Engineers.

A.I.A.N.Z. Associate of the Incorporated Institute of Accountants of New Zealand.

A.I.A. Quan. Quantity Surveyor, Member of the Incorporated Association of Architects and Surveyors.

A.I. Arb. Associate of the Institute of Arbitrators.

A.I.A.S. Associate Surveyor, Member of the Incorporated Association of Architects and Surveyors; Australian Institute of Agriculture and Science.

A.I.B. Accidents Investigation Branch; American Institute of Banking; Associate of the Institute of

Bankers; Associate of the Institute of Building; Associazione Italiana Biblioteche *Italian Library Association.*

A.I.B.C. Architectural Institute of British Columbia.

AIBL Atlantic International Bank Ltd., *q.v.*

AIBS American Institute of Biological Sciences.

AIC *or* **A.I.C.** American Institute of Chemists; Associate Institute of Chemistry.

A.I.C.A. Associate Member of the Commonwealth Institute of Accountants.

AICE *or* **A. Inst. C.E.** American Institute of Chemical Engineers; American Institute of Consulting Engineers; Associate of the Institute of Civil Engineers.

A.I.C.E.A. Associate Member of the Association of Industrial and Commercial Executive Accountants.

AICH *or* **A.I.C.H.** American Institute Hull Clauses *marine insurance.*

A.I. Chem. E. Associate Member of the Institute of Chemical Engineers.

A.I. Conference Clause American Shipment Conditions of Frozen or Chilled Cargoes.

AICPA American Institute of Certified Public Accountants.

A.I.C.S. Associate of the Institute of Chartered Shipbrokers. Now changed to Member of the Institute of Chartered Shipbrokers. *Abbrev.* M.I.C.S.

AID *or* **A.I.D.** Agency for International Development; American Institute of Decorators; Army Intelligence Department; Artificial Insemination By Donor.

AIDA Attention, Interest, Desire and Action, *q.v.*

AIDC Australian Industry Development Corporation.

Aide-de-Camp An officer who assists superior officers. *Abbrev.* AD *or* ADC *or* A.D.C.

AIDE Association Internationale des Dispacheurs Européens *French*—International Association of European General Average Adjusters.

AIDEC Association for European Industrial Development and Economic Co-operation, The Hague.

Aide-toi, et le ciel t'aidera *French*—Help yourself, and Heaven will help you.

AIDS Acquired Immune Deficiency Syndrome.

Aids to Navigation Instruments, seamarks and landmarks which may be used in the process of navigating a vessel from one point to another. These may include electronic devices such as radar, Loran and Omega, optical instruments such as the sextant, and objects outside a vessel such as lighthouses, beacons and buoys.

A.I.E.E. Associate of the Institution of Electrical Engineers.

A.I.F.T.A. Anglo-Irish Free Trade Area Agreement; Associate of the Institute of Freight Trades Associations.

A.I.G.M. Associate of the Institute of General Managers.

AIGSS Annual Inert Gas System Survey.

AIHA American Industrial Hygiene Association.

AIHC American Institute of Hull Clauses *marine insurance.*

A.I.I. Air India International *India State Airlines.*

AIIE American Institute of Industrial Engineers.

A.I.L. Associate of the Institute of Linguists.

AILAS Automatic Instrument Landing Approach System.

Ailerons Flaps attached to the trailing edges of the wings of an aircraft which are adjustable upwards or downwards for balance control.

AIM Airman's Information Manual.

AIMA *or* **A.I.M.A.** *or* **a.i.m.a.** As Interest May Appear.

A.I.M.E. Associate of the Institute of Mining Engineers.

A.I.Mar.E. Associate of the Institute of Marine Engineers.

A.I.Min.E. Associate of the Institute of Mineral Engineers.

AIM&ME *or* **A.I.M.& M.E.** American Institute of Mining and Metallurgical Engineers.

A.I.M.O. Associate of Industrial Medical Officers.

AIMS Airborne Income Management System; American Institute of Merchant Shipping.

A.I.M.T.A. Associate of the Institute of Municipal Treasurers and Accountants.

AIMU *or* **A.I.M.U.** American Institute of Marine Underwriters. *See* **US Marine Underwriters.**

AIN *or* **A.I.N.** Australian Institute of Navigation.

AINA *or* **A.I.N.A.** American Institute of Nautical Archaeology; Associate of the Institute of Naval Architects.

AINE *or* **A.I.N.E.** Associate of the Institute of Naval Engineers.

A.Inst.C.E. Associate of the Institute of Civil Engineers.

A Inst M *or* **A.Inst.M.** Associate of the Institute of Marketing.

AIOA Aviation Insurance Officers' Association.

A.I.O.B. Associate of the Institute of Buildings.

AIP Aeronautical Information Publication.

AIPE American Institute of Plant Engineers.

A.I.Pet. Associate of the Institute of Petroleum.

A.I.Q.S. Associate of the Institute of Quantity Surveyors.

Air Africa *or* **Air Afrique** Africa International Airlines.

Air Algérie Algeria International Airlines.

Air Bahama Bahama International Airlines.

Airbase *or* **Airdrome** *or* **Aerodrome** A base where aircraft may take off or land, and are stationed or maintained for all operations. *See* **Landing Field.**

Airborne Anything carried on an aircraft or anything supported in the air.

Airborne Early Warning and Control System. A long-range control surveillance system effectively used for both low and high altitudes by aircraft during all inclement weather. It provides an extensive amount of information concerning air control, close direct communications, search and rescue and many other reconnaissance operations. This radar system has proved to be useful in the field of air safety and communications. *Abbrev.* AEWS *or* A.E.W.S. Also termed as Aircraft Early Warning System.

Airborne Radiation Thermometer An instrument working by radiation which measures ocean surface temperatures.

Air Botswana National airline of Botswana.

Air brake The aircraft speed brake.

Air Canada Canada International Airlines.

Air Carrier An aircraft carrying passengers, mails and cargo; Company involved in the air transportation of passengers and cargo.

AIRCAT Automated Integrated Radar Control for Air Traffic.

Aircoach A bus engaged on a regular schedule carrying air passengers to and from an airport.

Air Cock A safety valve normally placed on top of boilers or air cylinders to release extra air pressure.

AIRCOM Airways Communications System.

Air Control A section at or near to an airport which controls the whole operations of the flights at that airport and within its international air control area. It is also termed Air Control Centre or Air Control Center.

Air Control Centre *or* **Air Control Center** *Similar to* **Air Control**, *q.v.*

Air Controller A person engaged on land at the Air Control, *q.v.*, who controls aircraft within his national and international area by means of radio, radar and other sophisticated electronic devices.

Air Corridors *or* **Air Routes** *or* **Airways** International air routes or channels specifically allotted to aircraft for their flights.

Aircraft Any kind of flying machine.

Aircraft Early Warning System *See* **Airborne Early Warning and Control System**.

Aircraft Report A routine weather report transmitted every hour to some stations by aircraft flying within the same area. *Abbrev.* AIREP.

Aircraft Surface Movement Indicator A radar screen in an airport control tower displaying all the movements of the aircraft within its area. *Abbrev.* ASMI.

Aircrew The personnel engaged to fly and operate the service from the time the passengers enter the aircraft up to the time they leave.

Air Cushion Craft Hovercraft, *q.v.*

Air Draft (Draught) (1) Air Ventilation (2) The height of a ship taken from the waterline to the top of the mast. This calculation is most important if the ship is meant to trade or navigate under bridges, especially along rivers. There are in fact certain river boats which have collapsible masts and also hydraulic bridges to allow them to be raised or lowered accordingly while navigating under bridges. *See* **Low Profile vessels**. (3) The distance between the waterline and the top of hatchcoamings on vessels which need to manoeuvre beneath shore loading or discharging apparatus—e.g. Panamax bulk carriers.

Airdrome *or* **Aerodrome** *Similar to* **Airbase**, *q.v.*

Airdrying Paints Paints which dry when exposed to air. These are conventional paints composed of various oils and chemicals.

Air Duct An air tube providing circulated fresh air in a mine by expelling the bad air.

Air Engine An engine actuated by heated air.

AIREP Aircraft Report, *q.v.*

Air Express Method of sending goods by fast air transportation; Express Airmail.

Air foil Rudder, wind or other device counteracting the moving airstream of an aircraft.

Air Frame The whole structural part of an aircraft less the engines.

Air freight Cargo sent by air.

Air Freighter An aircraft specially meant to carry air cargo.

Air Gauge Apparatus to measure air pressure.

Air Glow Radiant emission seen at high atmosphere during darkness and at lower areas at dawn.

Air Guinea National airline of Guinea.

Air Hostess Stewardess on an aircraft, serving the passengers during flight.

air hp. Air Horse Power. *Also abbrev.* AHP *or* A.H.P.

Air jacket Inflatable lifejacket of a buoyant type used in case of emergency in an aircraft. Each passenger and crew member on board is supplied with one.

Air Jamaica Jamaica International Airlines.

Air Lanka Sri Lanka International Airlines.

Air Launch To launch from an aircraft, such as firing a missile.

Air Letter A light paper form supplied by Post Offices with an impressed stamp and easily folded into the form of a closed envelope. With its uniform stamp it can be posted to various countries abroad and is normally treated as an airmail letter.

Air lift Large scale air transport operation.

Airline Prefix Unique number allocated by IATA to each airline, which is applied to the beginning of an Air Waybill, *q.v.* Numbers issued by that airline.

Airliner A fairly large passenger aircraft flying to and fro on a scheduled service.

Airload All merchandise carried on an aircraft.

Airman's Information Manual A handbook or manual published by the Federal Agency of America providing general information on civil flights in the National Airspace System. *Abbrev.* AIM.

Airmic *or* **AIRMIC** Association of Insurance and Risk Managers in Industry and Commerce.

Air Pipe A pipe allowing air to escape from a ship's tank containing fuel, water or other liquid. It is connected from the top of the tank up to the deck or upper deck. *Also* **Vent Pipe.**

Airport Terminal base for passengers and aircraft. *Similar to* **Landing Field,** *q.v.*

Airport Handling Charges Charges levied by airlines for the handling of cargo carried by their aircraft.

Airport Insurance This is generally called for short Ariel, *q.v.*

Airport Tax A levy on boarding air passengers in some countries.

Air Post Airmail Letter.

Air Route/s *Similar to* **Air Corridors,** *q.v.*

Air Route Surveillance Radar *or* **Airport Surveillance Radar** A long-range air traffic control radar positioned at airports which records the location and bearing of aircraft as far as approximately 30 nautical miles. *Abbrev.* ARSV *or* ASR.

Air Scoop Device in front of an aircraft to suck in air during the flight. An air scoop is also used in some engines for engine cooling purposes.

Air/Sea Interchange Refers to merchandise which is partly despatched by air and partly by sea. Normally two documents are needed, an air waybill for the air despatch and a transhipment bill of lading for the sea freight.

Airscrew Aircraft propeller.

Air Sendings Air cargo in transit.

Airspeed Speed of aircraft considering air pressure which is different to the ground speed. There are three types of air speeds, i.e., Indicated Airspeed, *q.v.*; Calibrated Airspeed, *q.v.*, and True Airspeed, *q.v.*

Airspeed Indicator An instrument marking the aircraft's speed in knots and Mach Number, *q.v. Abbrev.* ASI.

Airstop Helicopter passenger station.

Airstrip A runway on an airfield usually constructed of concrete.

Air Ticket Flight Coupon or Ticket.

Air Transport Method of transportation whereby goods and/or passengers are carried by air.

Air Transport Undertaking Legally defined as an undertaking whose business includes the carriage by air of passengers and cargo for valuable consideration.

Airwaybill or Air Waybill *or* **Air Freight Bill** A document similar to a bill of lading, *q.v.*, but utilised for the transportation of merchandise by air freight. It shows the title of ownership and the details of goods being airfreighted, *Abbrev.* AWB *or* A.W.B. for Airwaybill *or* Air Waybill and AFB *or* A.F.B. for Air Freight Bill.

Airworthiness The general operational safety of an aircraft.

Airworthiness Certificate An official certificate regulated on an international level issued by the competent authorities certifying that the aircraft conforms with all the air safety requirements before it is permitted to operate.

AIS Aeronautical Information Services.

A.I.S.A. Associate of the Incorporated Secretaries Association.

AISI *or* **A.I.S.I.** American Iron and Steel Institute.

Aisle Passageway the length of an aircraft between the seats.

AISM *or* **A.I.S.M.** Association Internationale de Signalisation Maritime *French*—International Association of Maritime Signalling.

A.I.S.T. Associate of the Institute of Science and Technology.

A.I.Struct.E. Associate of the Institution of Structural Engineers.

AIT *or* **A.I.T.** Alliance Internationale de Turisme *French*—International Tourist Alliance; American Institute of Technology; Association of H.M. Inspectors of Taxes; Association of Investment Trusts.

AITH Form *or* **A.I.T.H. Form** American Institute Time Hull Form *marine insurance policy form.*

AIU American International Underwriters.

A.I.W.E. Associate of the Institution of Water Engineers.

AIWM *or* **A.I.W.M.** American Institute of Weights and Measures.

A.I.W.M.A. Associate Member of the Institute of Weights and Measures Administration.

AJ Air Belgium International *airline designator*; American Jurisprudence; Associate in Journalism; Associate Justice.

A.J.A. Australian Journalists Association.

AJAG *or* **A.J.A.G.** Assistant Judge Advocate General.

AJB Associated Japanese Bank (International) Ltd.

A.J.M.S.A. All Japan Marine Suppliers Association. A member group of the International Ship Suppliers Association.

AJSU Association of Japanese Seamen's Union.

AK Alaska, *USA.*

aka *or* **a.k.a.** *or* **AKA** Also known as.

A.K.C. American Kennel Club; Associate of King's College; Association of King's College.

Aksjerederi *Norwegian*—Shipping Company Ltd. *Abbrev.* A/R.

Aksjeselskap *Norwegian and Finnish*—Limited Company. *Abbrev.* A/S.

Akt *or* **Aktb** *or* **Aktieb** Aktiebolaget *Swedish Limited Company.*

Akt. Ges. Aktiengesellschaft *German Corporation of Joint Stock Company.*

Aktiebolag *Swedish*—Limited Company, *Abbrev.* A/B *or* Ab.

Aktiebolaget *Abbrev.* Akt. *or* Aktb, *q.v.*

Aktieselskab *Danish*—Joint Stock Company Ltd., *Abbrev.* A/S.

Aktiengessellschaft *German*—Joint Stock Company, *Abbrev.* AG.

AL Alabama, *USA*.

al Alias *Latin*—Under another name, *see* **Alias**; Alcohol.

Al Aluminium.

A.L. Accommodation Ladder; *q.v.* American Legion; Anno Lucis *Latin*—In the year of the light.

a.l. Allotment Letter; Après Livraison *French*——After Delivery of Goods; Assumed Latitude *nautical*; Autograph Letter.

ALA *or* **A.L.A.** Air Licensing Authority *UK*; American Library Association; Associate of the Library Association; Association of Lecturers in Accountancy.

à l'abandon *French*—In disorder; left uncared for; at random.

à la belle étoile *French*—Under the stars; at night in the open.

à l'abri *French*—In shelter; under cover.

à la bonne heure *French*—As you please; In good time; All right.

à la mode *French*—In the fashion.

à la morte *French*—To the death.

Ala *or* **Alb.** Alabama, *USA*.

A.L.A.A. *or* **A.L.C.A.** Associate of the London Association of Certified Accountants *or* Associate of London Certified Accountants.

ALAM *or* **A.L.A.M.** Associate of the London Academy of Music.

A.L.A.MAR. Association of Latin American Shipowners.

ALARA *or* **Alara** As low as reasonably achievable *level of human exposure to hazardous substances*.

ALAS *or* **A.L.A.S.** Associate of the Land Agents' Society.

Alas Alaska *USA*.

ALB *or* **A.L.B.** Air Licensing Board.

Alb. Albania.

Alba Alberta, *Canada*.

ALBM. Air-Launched Ballistic Missile.

a.l.c. A la carte *French*—Menu *or* According to the list/menu.

ALCA American Leather Chemists Association.

ALCD *or* **A.L.C.D.** Associate of the London College of Divinity.

ALCM *or* **A.L.C.M.** Air-launched Cruise Missile; Associate of the London College of Music.

ALCS Authors' Licensing and Collecting Society.

Ald Alderman.

Aldis Lamp A handy portable lamp with heavy duty battery which is currently used on board ships and aircraft for Morse Code signalling.

ALE Association of Liberal Education.

A-lee *Similar to* **Leeward**, *q.v.*

ALESCO *or* **A.L.E.S.C.O.** Arab League Educational, Cultural and Scientific Organisation.

A.Level Advanced Level. Examination in the General Certificate of Education. *Abbrev.* G.C.E. in the advanced standard.

Alex. Alexandria, *Egypt*.

ALFA Adriatic Levant Freight Agreement.

Alfa International Radio Telephone phonetic alphabet for 'A'.

al fresco *Italian*—In the open air.

Alg. Algeria; Algiers.

Alg. *or* **a.l.g.** Advanced Landing Ground *aviation*.

Algae Aquatic group of plants living on rocks, trees and underwater parts of a ship. Their colours vary from green, yellow/green, blue/green, brown and sometimes red.

ALGES *or* **A.L.G.E.S.** Association of Local Government Engineers and Surveyors.

ALGFO Association of Local Government Financial Officers.

Algorithmic Language used in computers for mathematical calculations.

ALI *or* **A.L.I.** American Library Institute.

Alias *Latin*—Otherwise known as; When a person assumes another name. *Ex.* Joseph Smith alias Peter Smith.

Alicon Air Lift Containers system. Platforms with air cushions used in fast handling of unit loads on and off Ro-Ro, *q.v.*, vessels without using trucks or trailers. One tractor vehicle can move a string of air-cushioned units.

Alidade Revolving index ring carrying the sight or telescope of an astrolabe or quadrant which was used in olden times in navigation for observing the altitude of celestial bodies. The word is from the Arabic.

Alien A foreigner.

Alieni juris *Latin*—A legal expression meaning the power of a father, husband or master, as opposed to sui juris. It concerns the inability of certain types of people who are unable to cope by themselves in any situation. Consequently they have to be represented by a responsible protector in their daily way of life. A child or a sick person is looked upon as an alieni juris.

Alimeter An aneroid barometer, *q.v.*

AILSA American League for International Security Assistance. *US Labor Management Organization*.

Alist A listing ship; List over.

ALITALIA Italian International State Airlines.

Aliter *Abbrev.* alr, *q.v.*

Alkali Any chemical compound having a pH value from 7 (as neutral) upwards. The numeral 14 represents extreme alkalinity.

All American Marine Slip An American marine insurance policy covering expensive risks on sea/air pollution, oil rigs, etc. *Abbrev.* AAMS *or* A.A.M.S.

All and Any Risks *Similar to* **Against All Risks**, *q.v.*

All E.R. All England Reports *Law Reports*.

Alleyway A ship's internal passageway or corridor.

All Found Generally refers to remuneration where food, accommodation and other amenities are included as well as wages.

All Hands All personnel on board a vessel.

Allision The act of striking or collision of a moving vessel against a stationary object.

All Japan Marine Suppliers Association Members of the International Ship Suppliers Association of Japan *Abbrev.* AJMSA *or* A.J.M.S.A.

Allocatur *Latin*—It is allowed. The certificate issued by a court's taxing master approving the amount of a solicitor's bill after an action, and when they have been 'taxed'.

Allonge An additional slip attached to Bills of Exchange, *q.v.*, and used for further endorsement in case no more space is available after the previous signatures.

Allotment Note *See* **Allotment** and **Advance Note**.

All Outside Cabin Concept A concept introduced by Finnish shipbuilders where all passenger cabins in a cruise liner have to obtain natural light. *Abbrev.* AOC.

All Purposes Voyage C/P Term This works in conjunction with Laytime, *q.v.*, or Laydays, *q.v.*, where loading and discharging are to be accounted for en block. *Similar to* **All Purposes**.

All Rights Reserved Book Publishers Copyright. Also expressed as Copyright Reserved.

All Risks. An insurance term which means that the policy covers the insured property for loss caused by any fortuity. The policy does not cover inevitable loss.

All-Round-Light Nautical term for light showing unobscured through a horizontal of 360°.

All Round Price The price which covers the cost of the material and all the other extras over the basic price.

All-Terrain Vehicle Car designed to go on any rough road including bumpy and difficult fields. *Abbrev.* ATV.

All Time Saved A saving on the total time allowed for loading and/or discharging. The charterers are entitled to despatch money on all time saved, making due allowance for weekends, holidays and other charterparty exceptions. In a charter the expression is used in connection with Despatch Money or Dispatch Money, *q.v. Abbrev.* ATS *or* A.T.S.

All Told The total deadweight of the ship. This includes the full weight capacity of the holds, bunkers, spares, provisions, water, etc., on board.

Alluvium Very fine sand deposited in the bottom of running water.

All-Weather Air Stations Specific airports or air stations which have the means of Instrument Flight Rules (IFR) to clear private or single pilot aircraft on arrival. *Abbrev.* AWAS.

All Working Time Saved *Similar to* **All time Saved** but Laytime is calculated during working time only and not during Sundays/Saturdays and holidays.

ALM Antilliaanse Luchvaart Maatschappij. *Dutch*—Netherland Antilles Airline; Association of Lloyd's Members.

Almanac A nautical book, published annually for seafarers and others giving useful and general information on many aspects including abbreviations, tides, weather, navigation, astronomy, etc.

Alnico Trade name for an alloy of alluminium, nickel, cobalt and iron. It is mostly used for magnets.

Aloft Above decks; On a mast or rigging. *Nautical*.

Alongside A common word used in contracts of affreightments regarding the delivery of the cargo both in loading and discharging. It protects the owner of the ship in case of any claims by the consignors and consignees of the cargo. Such cargo is at the charterer's risk and expense until lying within reach of the ship's tackle. Also, the position of a ship when she is securely moored on her berth in a port. *Abbrev.* A.S. *or* a/s *or* A/S. *See* **Within Reach of Tackle**.

Alongside Date The date by which the ship accepts the merchandise for loading.

ALP American Labour Party; Articulated Loading Platform.

ALPA *or* **A.L.P.A.** Airline Pilots' Association.

Alpha Time Greenwich Mean Time or Universal Time plus one hour. *Abbrev.* A.Time.

Alphanumeric Display The writing shown in letters and figures on an electronic meter or monitor.

ALPURCOMS All Purpose Communications System.

ALR *or* **A.L.R.** American Law Reports.

Alr Aliter *Latin*—Otherwise.

ALS *or* **A.L.S.** Automatic Landing System *aviation*.

a.l.s. Autograph Letter Signed.

ALT Aer Lingus *Irish State International Airlines*; Articulated Landing Platform *oil rig*.

alt Alteration; Altered *Lloyd's Register notation*; Alternating *electricity*; Altitude, *q.v.*

Alta Alberta, *Canada*.

Altars Dry docks generally have steps constructed all around the sides from top to bottom for the use of those persons attending to the ships in the dock. Moreover, these altars or steps are used to support the Shores, *q.v.*

Alter ego *Latin*—One's second self; A close friend.

Alternative Director A person appointed at a general meeting of a company to act as a director in lieu of another.

Alternative Dispute Resolution The settlement of contested claims by any means short of going to law i.e., arbitration, agreement or compromise.

alterum tantum *Latin*—As much again; twice as much.

Altitude The angular distance of an object above the horizon, measured in navigation to assist in fixing a vessel's position. The instrument most commonly used for this purpose is the Sextant, *q.v. Abbrev.* alt or A/T.

Altitude Circle Parallel of altitude *nautical.*

Altimeter An aneroid barometer in an aircraft showing the altitude and the surface pressure.

Altocumulus Clouds White and/or grey patch, sheet or layer of clouds, composed of laminae or rather flattened globular masses, the smallest elements of the regularly arranged layer being fairly small and thin, with or without shading. They appear between 6,500 and 18,000 ft. approx. *Abbrev.* Ac.

Altostratus Clouds Striated or fibrous veil, more or less grey or bluish in colour, appearing between 6,500 and 18,000 ft. approx. The sun or moon may show vaguely through them. *Abbrev.* As.

Aluminium Bronze An alloy of copper, iron and nickel usually used for normal sized propellers of ships.

À l'usine *French*—Ex-Works, *q.v.*

Always Accessible or Reachable on Arrival A Charter Party, *q.v.*, term in respect of the loading and/or discharging position when berthed which must be accessible without difficulty. *Abbrev.* AARA or A.A.R.A.

Always Afloat *or* **Always Safely Afloat** A charter-party clause which stipulates that the ship is to berth for loading or discharging without touching the bottom of the sea/river/lake, etc. In many cases owners may agree that the vessel can touch on harmless ground whenever low tides occur. *Abbrev.* a.a. *See* **Floating Clause** which is similar, *also* **Sitting Aground**.

Always Safely Afloat *See* **Always Afloat**.

ALWT Advanced Lightweight Torpedo *USA.*

a.m. Ammeter; Ante Meridiem *Latin*—Before Noon.

AM Automatic Magnification.

AM *or* **A.M.** Academy of Management; Aero-mexico *airline designator*; Amplitude Modulation *nautical*; Anno Mundi *Latin*—In the year of the World; Ante Meridiem *Latin*—Before Noon; Area Manager; Artium Magister *Latin*—Master of Arts; Assistant Manager; Assurance Mutuelle *French*—Mutual Insurance. *See* **Assurance Mutuelle.**

A.M. Air-Mail.

Am *or* **Amer.** America; American.

AMA *or* **A.M.A.** American Maritime Association *New York*; American Medical Association.

A.M.A.I.M.M. *or* **A.M.Aus.I.M.M.** Associate Member of the Australian Institute of Mining and Metallurgy.

à main armée *French* carrying arms.

Amalgam An alloy of mercury with other metals.

AM.Am.I.E.E. *or* **A.M.Am.I.E.E.** Associate Member of the American Institute of Electrical Engineers.

A.M.A.S.C.E. Associate Member of the American Society of Civil Engineers.

Amat. Amateur.

Amatal A high explosive in munitions.

a maximis ad minima *Latin*—from the greatest to the least.

AMB *or* **A.M.B.** Australian Meat Board.

Amb. Ambassador.

A.M.B.A.C. Associate Member of the British Association of Chemists.

Ambient Surrounding—as in 'ambient temperature'.

A.M.B.I.M. Associate Member of the British Institute of Management, *obsolete.*

AMC *or* **A.M.C.** Agricultural Mortgage Corporation; American Maritime Cases; Armed Merchant Cruiser; Association of Municipal Corporations; Australian Maritime College.

AMCBH *or* **A.M.C.B.H.** Auxiliary Marine Casing Bulkhead.

A.M.C.I.A. Associate Member of the Association of Cost and Industrial Accountants.

A.M.C.I.B. Associate Member of the Corporation of Insurance Brokers.

AMCS *or* **A.M.C.S.** Airborne Missile Control System; Association of Marine Catering Superintendents.

am.cur. Amicus Curiae *Latin*—A friend of the court; a third party with a brief on the defendant's behalf.

A.M.D.G. Ad Majorem Dei Gloriam *Latin*—To the Greater Glory of God.

AMDP Aircraft Maintenance Delayed for Parts.

A.M.E.I.C. Associate Member of the Engineering Institute of Canada.

Amende honorable *French*—Public apology; reparation.

Amended Jason Clause *See* **Jason Clause.**

Amer *or* **Am.** America; American.

AMERC Australian Maritime Engineering Research Centre, established 1992.

American Bureau's Information Retrieval System A computer system practised in New York

City headquarters of the American Bureau of Shipping which covers extensive details of all vessels classed with ABS worldwide. These include characteristics, structural equipment, technical notes, damage incurred, machinery failures, answers to any technical or engineering problems, etc. *Abbrev.* ABIRS or ABSIRS.

American Bureau of Shipping Computers Inc. This is the System Service Department of the American Bureau of Shipping. It operates as a separate entity but plays a tangible part in technical and informative work. The department issues certificates for offshore and land based structures, machinery and processing plant equipment. *Abbrev.* ABSCOMP.

American Clause A marine insurance clause referring to double insurance which is usually inserted in American insurance policies. In the event of such a claim, the underwriter who effected the insurance is responsible for meeting the claim. If unable to do so, the second underwriter will have to cover this liability instead, in part or in full, whichever the case may be.

American Cloth Textile with a glazed coating.

American Club *Similar to* **American Steam-Ship Owners Mutual Protection and Indemnity Association**, *q.v.*

American Hull Form Uniform clauses sanctioned by American insurance underwriters which although basically in line with the Institute Time Clauses (Hull) have some different provisions; such as the 4/4th collision instead of 3/4th Collision Clause. *Abbrev.* AHF or A.H.F.

American Hull Insurance Syndicate An association of American insurance corporations who act as combined insurance underwriters and also issue Marine Hull insurance policies.

American Offshore Insurance Syndicate American representative of over 60 insurance companies and management offices providing worldwide insurance coverage against losses, damage liabilities and expenses in relation to the construction, movement and operation of facilities encountered within related offshore exploration and other similar work. *Abbrev.* AOIS.

American Organ Harmonium.

American Petroleum Institute To obtain the standard American Petroleum Institute Gravity (API) it is necessary to establish the specific gravity of crude oil by means of a Hydrometer, *q.v.*, and then apply the following formula:—

141.5 to be divided by specific gravity and the resulting figure to be subtracted by 131.5

i.e.

$$\frac{141.5}{\text{Specific Gravity}} - 131.5 = \text{API Gravity}$$

Abbrev. API.

American Selling Price A United States method of assessing import duty. Duty is paid on a value arrived at by comparing prices at which the article(s) are already offered in the domestic market. *Abbrev.* ASP.

American Short Ton Equivalent to a weight of 2000 lbs.

American Steamship Owners Mutual Protection & Indemnity Association Inc. Popularly known as The American Club. It services its client shipowners both in America and abroad. Originally established in 1917 as a non-profit Mutual Club or P. & I. Club in the United States. Its functions were then restricted to US clients but by 1973 it had developed into an international club with members all over the world. All administration and daily claims are handled by personnel of the Shipowners Claims Bureau Inc., considered to be the oldest Mutual Club and Insurance Underwriters in the United States. *Abbrev.* ASOMPIA.

American Stock Exchange A voluntary incorporated association originally called the New York Curb Exchange until 5 January 1953. Its members are regular stockbrokers engaged in the free sale and purchase of all types of securities, varying from industrial and commercial to railway shares. *See* **The New York Stock Exchange**; **Curb Market**; **Street Market.**

American Tanker Rate *or* **American Tanker Rate Schedule** The rates of freight applicable to oil tankers on the main world routes. Started in 1956 and revised periodically, these rates are usually quoted in US currency. In 1963 the International Tanker Normal Freight Scale Association Ltd., *abbrev. Intascale*, was formed and it collaborated on the 1969 World Scale. Complicated calculations concerning demurrage, tonnage, port running expenses, laytime etc. are required to arrive at these scales. This organisation, with offices in London and New York, works on a non-profitable basis and its overall expenses are covered by the subscribers involved in the trade.

American Waterways Association *or* **American Waterways Operators** The national trade association for the barge and towing industry and the shipyards employed in the repair and construction of these craft in the USA. *Abbrev.* AWA for American Waterways Association and AWO for American Waterways Operators.

Amer.Std. American Standard.

Amer T.L. & Ex. American Total Loss and Excess *marine insurance.*

à merveille *French*—Wonderfully.

Amex *Abbrev.* for American Stock Exchange; American Export; London based banking subsidiary, American Express.

AMF Arab Monetary Fund, in Abu Dhabi; ACE Mobile Force *NATO.*

AMG *or* **A.M.G.** Amount Made Good, *q.v.*

AMHCI *or* **A.M.H.C.I.** Associate Member of the Hotel and Catering Institute.

ami *or* **AMI** Air Mileage Indicator; Advanced Manned Interceptor.

A.M.I.Ae.E. Associate Member of the Institute of Aeronautical Engineers.

A.M.I.B.E. Associate Member of the Institute of British Foundrymen.

A.M.I.C.E. Associate Member of the Institute of Civil Engineers.

A.M.I.Chem.E. Associate Member of the Institute of Chemical Engineers.

A.M.I.C.I. Associate Member of the Hotel and Catering Institute.

Amici probantur rebus adversis *Latin*—Friends are tested by adversity.

Amicus Curiae *Abbrev.* am. cur., *q.v.*

amicus usque ad aras *Latin*—A friend as far as the altars i.e., in everything not contrary to one's religion.

AMIDEAST American—Mideast Educational and Training Services.

Amidships Central position of helm or rudder; an order to the helmsman to centre the helm. *See also* **Midship/s**.

A.M.I.E.E. Associate Member of the Institute of Electrical Engineers.

A.M.I.Ex. Associate Member of the Institute of Exporters.

AMIM American Insurance Market.

A.M.I.M.E. Associate Member of the Institute of Marine Engineers.

A.M.I.Mech.E. Associate Member of the Institute of Mechanical Engineers.

A.M.Inst.M. Associate Member of the Institute of Marketing.

A.M.Inst.M.S.M. Associate Member of the Institute of Marketing and Sales Management.

A.M.I.S.A.S.I. Associate Member of the International Society of Air Safety Investigation.

A.M.Inst.T. Associate Member of the Institute of Transport.

A.M.I.W.S.P. Associate Member of the Institute of Works Study Practitioners.

AML Absolute Maximum Loss *reinsurance term*; Absolute Maximum Load *aviation*.

A.M.L.B.O. Associate of Master Lightermen and Barge Owners (Port of London).

AMLS Association of Maritime Learned Societies *USA*.

AMM *or* **A.M.M.** Admiralty Measured Mile, *q.v.*; American Merchant Marine; Ammunition.

Ammeter An instrument to measure the intensity of current through an electrical circuit. This is used to measure the amperes of the electrical current.

AMMI American Merchant Marine Institute.

AMMLA American Merchant Marine Library Association.

A.M.M.R.S. Associate Member of the Market Research Society.

Amm Syst Ammonia System *refrigeration*.

A.M.N.I. Associate Member of the Nautical Institute.

A.M.O.B. Automatic Meteorological Oceanography Buoy.

à moitié *French*—By halves; Half.

amor patriae *Latin*—Love of one's country.

Amor vincit omnia *Latin*—Love conquers all.

Amortization The act of reducing a sum of money loaned by gradual payment over a period of time.

AMOSUP Associated Marine Officers' and Seamen's Union of the Philippines.

Amount Made Good In General Average, *q.v.*, the property sacrificed is eligible for claims in relation to the conditions made. It can also be termed 'Made Good'. *See also* **General Average Contribution**. *Abbrev.* AMG *or* A.M.G.

Amounts Differ A banking term. When a cheque is refused because the figures and wording do not tally.

amour propre *French*—Self respect; Keen interest.

Amp. Ampere, *q.v.*

AMPAS Academy of Motion Picture Arts and Sciences *USA*.

Ampere A uniform unit of current of electrical energy that passes through the conductor or the current that passes through a conductor with one ohm of resistance when at its extreme a tension of one volt is applied. *Abbrev.* Amp.

Amphibian An object which operates on land and sea.

Amp. hr. Ampere-hour.

Amplitude The difference between the observed bearing of the rising or setting sun and due east or west.

AMPM Amsterdam Municipal Port Management.

AMPS Association of Management and Professional Staffs.

AMPTC Arab Maritime Petroleum Transport Company.

AMQ *or* **A.M.Q.** American Medical Qualifications.

AMRIE Alliance of the Maritime Interests in Europe. Formed by MEP's, *q.v.* in 1993 to promote EU, *q.v.* co-operation in and co-ordination of maritime policy.

AMS *or* **A.M.S.** Administration Management Society *USA*; Aeronautical Mobile Service *aviation UK*; American Mathematical Society; Australian Medical Science.

amsl *or* **AMSL.** Above mean sea level.

Amstel Club A Swiss finance company registered as Amstel Finance International AG represented by

Austria, Belgium, Denmark, Finland, France, Germany, Italy, the Netherlands, Norway, Portugal, Spain, Sweden, Switzerland, UK and USA.

A.M.R.I.N.A. Associate Member of the Royal Institution of Naval Architects.

AMS Armed Merchant Ship.

AMSA Australian Government Marine Safety Authority.

AMT *or* **A.M.T.** Air Mail Transfer *banking*; Amsterdam Municipal Transport; Association of Marine Traders.

amt *or* **Amt.** Amount.

AMTC American Marine Tourist Coalition, formed in 1995 to promote US-flagged passenger vessel activities.

A.M.T.E. Admiralty Marine Technology Establishment *UK*.

Am.T.L.O.& Exs *or* **Am.TLD & Exs.** American Total Loss Only Disbursement Clauses and Including Excess Liabilities *US marine insurance*.

A.M.T.P.I. Associate Member of Town Planning Institute.

amtrac *or* **AMTRAC** Amphibious Tractor.

AMTS Automatic Telex Unit. *Also* ATS.

amu Atomic Mass Unit.

AMU Arab Magreb Union.

AMV *or* **A.M.V.** Armed Merchant Vessel; Association Mondiale Vétérinaire *French*—World Veterinary Association.

AMVER The US Coast Guard's Automated Mutual-assistance Vessel Rescue System, which began in 1958 as a computerized search and rescue aid for merchant ships in the North Atlantic Ocean. Today, AMVER is the only worldwide safety network, safeguarding ships from more than 134 nations, in every ocean of the world. AMVER participation is free, voluntary and available to ships of all nations, making it a cost-effective and efficient search and rescue tool. AMVER is fully endorsed by the International Maritime Organization (IMO) and is available to the world's cruise and shipping communities. All ships of all flags are welcome to participate. Every day more than 2,600 vessels worldwide participate in the AMVER Safety Network.

The heart of the AMVER system is a computer-generated data base that in 1993 plotted the voyages of more than 14,416 ships. Voyage information such as time of departure, destination, turn points, radio call sign, medical personnel and other relevant data are transmitted to the computer facility in Martinsburg, West Virginia. The computer can dead reckon the locations of AMVER vessels in relation to a distress signal, as well as indicate their direction of sailing. This gives rescue planners a 'real time' window on the high seas around the world, providing information which enables rescue authorities to make decisions rapidly and with confidence. In 1993 the AMVER network was directly responsible for saving more than 126 lives.

Maximum participation is vital to AMVER's support of search and rescue around the world. The more ships enrolled in the system, the better the chances of emergency assistance being quickly located and dispatched to a vessel in distress. A network of 130 coastal radio stations worldwide relay AMVER position reports from participants at no cost. AMVER information is available only to recognized search and rescue agencies around the world.

AMVETS American Veterans of World War II and Korea.

AMWELSH 93 Americanised Welsh Coal Charter, BIMCO.

AMWSU Amalgamated Metal Workers and Shipwrights Union *Australia*.

AN *or* **A.N.** Also notify; Associate in Nursing.

an Anno *Latin/Italian*—In the year; Anonymous; Ante *Latin*—Before.

a.n. Above named.

ANA *or* **A.N.A.** Air Navigation Act 1963; All Nippon Airways *Japanese airline*; Article Numbering Association.

ANAFC Association of North Atlantic Freight Conferences.

Analyst In the financial world, one who studies performance and trends with a view to forecasting investment prospects.

ANAOR Air Navigation and Airworthiness Orders and Requirements.

ANC African National Congress; Army Nurse Corps.

ANC *or* **A.N.C.** Air Navigation Commission; Air Navigation Conference *ICAO*.

Anch. Anchor, *q.v.*; Anchored.

Anchor A heavy hooked device, usually made of steel, used in conjunction with the Anchor Cable, *q.v.*, to hold the vessel to the seabed, *q.v. Abbrev.* Anch.

Anchor's Aweigh Seaman's term to denote that an anchor which is being Weighed has left the Seabed, *q.v.*, and is hanging vertically from the vessel.

Anchor Ball *or* **Black Ball** A black ball or circular shape, not less than 0.6 metre in diameter on a vessel 20 metres or more in length, which is hoisted in the fore part to show that she is at anchor.

Anchoring Berth An area in a roadstead, river or port where the depth of the water enables vessels to lie safely At Anchor, *q.v.*

Anchor Buoy A buoy connected by a line or pennant to an anchor to mark the latter's position, and used sometimes to assist retrieval of the anchor.

Anchor Cable Chain, wire or rope connecting a vessel to her anchor or anchors. *See* **Anchor**. *Also termed* **Anchor Chain**, *q.v.*

Anchor and Chain Clause A marine insurance clause not commonly used, freeing the underwriters from expenses to recover lost anchors and chains.

Anchor Chain *Similar to* **Anchor Cable,** *q.v.*

Anchor Currency A dominating currency within the European Monetary System, e.g. the German Mark.

Anchorage A place where a ship may be anchored or berthed.

Anchor Detail A vessel's crew members whose responsibility is to operate her anchor-handling equipment.

Anchor Lights A light shown all around the horizon between sunset and sunrise and during other times of poor visibility to indicate that a vessel is At Anchor. During daylight an Anchor Ball or Black Ball, *q.v.*, is shown.

Anchor Policy Lloyd's Policy Form which has an imprinted anchor sign on the top.

Anchor Position Specific place where a vessel is anchored or about to anchor.

And. Andorra *Spain.*

And Arrival A term relating to return of premiums on a hull policy. The ship must be safe at expiry of the policy; otherwise no return of premium will be paid.

ANDB Air Navigation Development Board *USA.*

And/or Used in conjunction with two combined operations, either in addition or separately. *Ex.* The damage occurred during loading and/or discharging. *Abbrev.* A/or.

Anemograph Instrument used to record the direction and force of the wind.

Anemometer Meteorological instrument for measuring the velocity of the wind.

Anemoscope Instrument for showing the direction of the wind.

ANERA *or* **Anera** Asia North America Eastbound Rate Agreement.

Aneroid Barometer Instrument for measuring atmospheric pressure without the use of mercury or other fluids.

ANF Allied Nuclear Force *USA*; Arrival Notification Form.

ANFE Aircraft Not Fully Equipped.

Ang. Angola *Africa.*

ANG Air National Guard.

Angary Right of a belligerent to use or destroy material property subject to the obligation to compensate the owners.

Angbats Aktiebolag *Swedish*—Steamship Company Ltd.

Angfartygs *Swedish*—Steamship Company.

Angle of Attack The angle at which the air is met by the wing of an aircraft while in flight.

Angle of Repose The angle between the horizontal plane and the cone shape obtained when bulk cargo is emptied into this plane. A low angle of repose characterises a bulk cargo which is particularly liable to dry surface movement aboard ship. In general a bulk cargo whose angle of repose is less than 35 degrees

needs levelling out or trimming to reduce the risk of the cargo shifting and thus ensuring that the ship remains seaworthy during the voyage. The smaller the angle of repose the more the cargo is liable to shift.

Angslups *Swedish*—Steamship Company.

ANGTS Alaska Natural Gas Transportation System.

ANGUS Air National Guard of the United States; Australia, New Guinea and the United States.

Anhyd. Anhydrous.

ANI Automatic Number Identification.

animo et fide *Latin*—By courage and faith.

Animus opibusque parati *Latin*—Prepared in mind and resources. Motto of South Carolina, USA.

Ank. Ankunft. *German*—Arrival.

Anker A Dutch and German liquid measure equivalent to approximately 10 gallons or 46 litres.

Ankunft *Abbrev.* ank, *q.v.*

Anna Pakistani or Indian coin amounting to 1/16th of a rupee.

Anneal The process of strengthening metals by gradually diminishing heat.

anno *Abbrev.* an, *q.v.*

anno inventionis *Abbrev.* AI, *q.v.*

anno lucis *Abbrev.* AL, *q.v.*

anno mundi *Abbrev.* A.M., *q.v.*

anno ordinis *Abbrev.* A.O., *q.v.*

anno regni *Abbrev.* A.R., *q.v.*

anno regni *or* **anno reginae** *Abbrev.* A.R.R., *q.v.*

Annual Load Line Certificate An annual certificate issued by shipping authorities or the Classification Society of the registered ship giving the various load line markings on either side amidships, beyond which she cannot be submerged when loaded. The mark is commonly known as the Plimsoll Line after Samuel Plimsoll who worked incessantly until he convinced the authorities of the day that they should observe these rules for the protection and safety of the crew members and the ship. Today the load line is internationally recognised by all ships carrying passengers and cargoes.

Annuitant An individual who receives an annual lump sum of money for a fixed period of years or for a lifetime. *See* **Annuity.**

Annuity A sum of money advanced every year or paid by instalments for a period of time. Life assurance is a case in point. A person can buy a life policy by paying a premium per year. In return he is assured of a substantial sum per year or a lump sum after a fixed number of years. In the event of his death before fulfilling all his premium payments his relatives or testators are assured of the payment of annuity agreed upon before his death. *See* **Assurance.**

Annus mirabilis *Latin*—Year of Wonders —applied to 1666 (year of plague, fire of London) etc.

ANO Air or Aircraft Navigation Order 1980.

Anode The electrode at which oxydisation occurs which is the positive pole of an electronic cell but this will be the negative part of a battery. In the case of preventing wear and deterioration of metal zinc is oxidised and this protects the steel with which it is in contact. Zinc plates or anodes are invariably attached to various parts of the underwater sections of a ship to protect against wear, including the propellers and propeller shafts. *See also* **Sacrificial anode.**

anon. Anonymous.

Anonima Responsibilita e Limitada *Spanish*——Limited Company *Abbrev.* A.R.L.

Anonim Sirketi *Turkish*—Limited Company *Abbrev.* A.S.

Anonyme French Limited Company.

ANPA American Newspaper Publishers Association.

A.N.P.A.N. Associazione Nazionale Proveditori Appaltatori Navali *similar to* I.S.A. the Italian Suppliers Association and members of ISSA.

Anpartssellskap *Danish*—Limited Company *Abbrev.* ApS.

ANR Air Navigation Regulations.

anrac *or* **ANRAC** Aid Navigation Radio Control.

a.n.s. Autograph Note Signed.

Ans. Answer.

ANS Alaska North Slope, *crude oil.*

A.N.S. Association of Navigation Schools.

ANSI American National Standards Institute.

ANSS American National Standards Specification.

Ant. Antarctica.

ANT *or* **A.N.T.** Aids to Navigation Teams *USA*; Average Neap Tide/s.

ANTA American National Theatre and Academy.

Ante *Abbrev* an, *q.v.*

Ante–date To advance the date of a document.

Anthracite A brilliant lustre black coal containing over 90° of carbon deposits. When anthracite is burnt no smoke or flame is emitted.

Anticipated Prices A businessman who is in the trade of buying and selling merchandise has to be able to forecast or anticipate the fluctuation of commodity prices or any other business enterprise he intends to negotiate at a later date.

Anti-Corrosive Paint A paint composition applied to bare metal as a first protective coat against corrosion. It is also referred to as a primer. There are various kinds of anti-corrosive paints or primers. Other coatings are then applied over the primer to give the metal the required finishing touch, colour and protection.

ANTM *or* **A.N.T.M.** Admiralty Notices to Mariners.

Anti-Cyclone A meteorological term referring to a high pressure system wherein the pressure in the central region is higher than in the surrounding parts. It is thus the opposite of cyclone.

Antifouling *or* **Antifouling Composition** *or* **Antifouling Paint** A marine paint composition containing poisonous ingredients which prevent or retard fouling or marine underwater growth on ships' bottoms etc. Mercuric oxides are the chemicals generally blended to act as repellants to the undesirable growth. Paint manufacturers are continuously studying the possibility of introducing improved and sophisticated antifouling material and the formulas used are mostly kept secret by the blenders or manufacturers. *See* **Foulings** and **Bottom Composition.**

Antilog Antilogarithm.

Antipodean Day The crossing of the meridian line from East to West and gaining one day in the process. *See* **Date Line** and **International Date Line.**

Anti-Trades Alternative name for the Westerlies, *q.v.*, which blow on the polar sides of oceanic anti-cyclones.

Anti-Trust Laws Laws enacted by the Federal Government of the USA to prevent the formation of monopolies in the business world. The merging of large companies was considered detrimental both to the consumer and to the state as a whole.

Any One Bottom Refers to one vessel or ship. *Abbrev.* a.o.b. *or* A.O.B.

Antwerp/Hamburg Range *Abbrev.* ahr *or* AHR *or* A.H.R. and is used in charterparties with reference to the sequence or range of loading/discharging operations including the named ports.

ANU Australian National University.

Anw *or* **a.n.w.** Actual Net Weight.

ANWR Arctic National Wildlife Refuge.

ANWTA Association of the National Waterway and Transport Association *USA.*

Any Charter *or* **Wet Charter** *or* **Wet Hire** An aircraft charter or lease or hire where the full complement of the crew, including stewards, are provided by the charterer.

ANZ Australia and New Zealand Bank.

Anzac *or* **ANZAC** Australia & New Zealand Corps.

A.N.Z.A.M. Australia, New Zealand and Malaya.

ANZMSA Australian and New Zealand Merchants' and Shippers' Association.

ANZTAC Australia and New Zealand Trade Advisory Committee.

Anzus *or* **ANZUS** Australia, New Zealand and United States.

A.O. Anno ordinis *Latin*—In the year of the order; After Office.

a.o.a. *or* **A.O.A.** *or* **A.O.Acc.** Any One Accident *insurance.*

a/o *or* **A/O** Account of.

a.o.b. *or* **A.O.B.** Any One Bottom, *q.v.* At or Below.

AOB Any Other Business *meeting agenda*.

AOC Airport Operators Council, Washington. All Outside Cabin Concept, *q.v.*; All other Contents *insurance*.

AOCI Airport Operators Council International.

AOCP Airport Out of Commission for Parts.

AOCT/S *or* **A.O.C.T/S.** Associated Overseas Countries and Territories.

AODC Association of Offshore Diving Contractors; merged within IMCA, *q.v.*, in 1995.

AODS Association of Diesel Specialists.

AOE Airport of Entry *for Customs*; Any One Event.

A of F Admiral of the Fleet.

A of S Academy of Science; Agreement of Sale.

AOG *or* **A.O.G.** Aircraft on Ground.

a.o.h. *or* **A.O.H.** After Office Hours.

AOIS American Offshore Insurance Syndicate, *q.v.*

aol *or* **AOL** *or* **A.O.L.** Absent Over Leave; Any One Loss *insurance*.

A.O.Loc. Any One Location.

a.o.o. *or* **A.O.O.** Any One Occurrence *insurance*.

A.O.P. Associate of Optical Practitioners.

a/or And/or, *q.v.*

a.o.r. *or* **A.O.R.** Any One Risk *insurance*.

AOS *or* **A.O.S.** *or* **a.o.s.** Any One Sending; Any One Steamer; Agreement of Sale.

A.O.S. Apostleship of the Sea, *q.v.*

AOSS Airborne Oil Surveillance System *Branch of US Coast Guard*.

AOT All Ocean Tug.

AOTOS Admiral of the Ocean Sea Award, *q.v.*

AOTS American Overseas Tourist Service.

à outrance *French*—To the extreme; without limit.

aov *or* **A.O.V.** Any One Vessel.

AO Voy *or* **A.O.Voy** Any One Voyage.

ap apud *Latin*—In the writings of.

a.p. Above Proof; Author's Proof.

AP Accommodation Platform, *oil drilling*; Aft Peak; Aft Part; After Perpendicular; Arbitrazhnaia Praktika *Russian*—Arbitration Practice.

Ap *or* **Apl** April.

A/P *or* **A.P.** Additional Premium, *q.v.*; A protestor *Latin*—Bills to be protested; After Particulars; Air Port; Armour Piercing; Associated Press; Assumed Position *nautical*; Authority to Pay; Authority to Purchase; American Pharmacopoeia; Aircraft Procurement *USA*; Average Product *economics*.

APA *or* **A.P.A.** Additional Personal Allowances *income tax*; Amsterdam Port Authority; As per Advice, *q.v.*

Apatite A mineral form of calcium phosphate of which there are large deposits in the Kola Peninsula.

Used in fertilisers and as a source of phosphorus. Also the chief component of human tooth enamel.

APB *or* **A.P.B.** Admiralty Pilot Book; Accident Prevention Board or Bureau, *q.v.*; All Points Bulletin.

APBH *or* **A.P.B.H.** Aft Peak Bulkhead.

APC Armoured Personnel Carrier.

APCA American Pollution Control Association.

APE Annual Pack Efficiency *oil drilling*.

APEC *or* **Apec** Asia-Pacific Economic Co-operation.

Aperture Envelope *Similar to* **Window Envelope.**

Apex *or* **APEX** Advance Purchase Excursion, *q.v.*; Analysis of Petroleum Exports *Lloyd's Maritime Consultancy Division, SEAGROUP.*

APFA Association of Professional Flight Attendants.

API *or* **A.P.I.** Air Pollution Indicator; Air Position Indicator; American Paper Industry; American Petroleum Institute.

APICORP Arab Petroleum Investment Corporation.

Apita *or* **APITA** *or* **A.P.I.T.A.** Australian Pulp and Paper Industries Technical Association.

A. Pk Aft or After Peak.

apl *or* **APL** *or* **A.P.L.** Applied Physics Laboratory *USA*; As Per List.

APM Automatic Positioning Mooring *oil drilling*.

A.P.O. Army Post Office.

Apogee A heavenly body's point of greatest distance from the earth. Opposite to Perigee, *q.v.*, *nautical*.

à point *French*—To a nicety; just right.

a posse ad esse *Latin*—From possibility to reality.

Apostleship of the Sea A Christian organisation, founded in Glasgow in 1920, which cares for seafarers and their families of all colours and creeds worldwide. It provides hospitality and chaplaincy services for seamen in a great many ports and has some chaplains afloat. It assists the sick and injured and those suffering injustice in their employment. *Abbrev.* A.O.S.

Apota *or* **A.P.O.T.A.** Automatic Positioning Telemotoring Antenna.

Apoth Apothecary.

Apothecary/ies Weight The chemist's and physician's metric weight measure used for their prescriptions. The English and American systems are *Avoirdupois*.

app Approved.

App. Apparel, *q.v.*; Apparatus; Appeal; Appendix; Applied; Approval; Approve; Approximate.

APP *or* **A.P.P.** Air Parcel Post.

App. alt Apparent Altitude, *q.v.*

Apparel The equipment of a ship such as anchors, chain cables; derricks; lifeboats; ropes and other ship's gear.

Apparel and Tackle A charterparty term referring to the equipment necessary to make the ship entirely seaworthy for the trade in which she is going to be employed.

Apparent Altitude The observed, uncorrected, altitude of a heavenly body above the horizon.

Apparent Good Order and Condition Paragraph 3, Article 3, of the Carriage of Goods by Sea Act 1924 stipulates that as soon as the cargo is loaded the shipper is entitled to a Bill of Lading, *q.v.*, from the Carrier, *q.v.*, or his agents at the loading port, showing the marks and particulars to be in Apparent Good Order and Condition. The carrier is bound to deliver the goods in the same order and also subject to the other exceptions and conditions contained in the Act, as amended.

Apparent Sun The sun as viewed in a motionless position by an observer on earth.

Apparent Time Time calculated by the observed position of the sun. Noon is when the sun crosses the observers' Meridian, *q.v.*

Apparent Wind Not the actual wind, but the wind as felt by an observer on a moving object, such as a ship.

APPB Advanced Payment Purchase Bond.

Appd. Approved.

Appendages Objects protruding from the underwater part of a vessel, such as bilge keels, stabilising fins, etc.

Apple Isle Another name for Tasmania. This is because of the enormous amount of apples that are grown on the island.

Application for Cargo Insurance An established form of application in American cargo insurance similar to Lloyd's Non-Marine Insurance Form.

Application for Hull Insurance A standard form of application in American ship insurance similar to Lloyd's Non-Marine Insurance Form.

Appraise To value goods or property.

Appraisement A written report produced by appraisers in connection with their estimated value of goods.

Appraiser An expert valuer of property authorised to act as an auctioneer or estate agent.

Appreciation An increase in the value of an object including Stocks and Shares, *q.v.* This is the opposite to Depreciation, *q.v.*

Apprentice A beginner in a trade or art. Similar to a cadet. As a beginner he or she cannot be expected to earn the full wages paid to a skilled worker.

Appro On Approval; Appropriate; Approval.

Approach Guidance Nose-in to Stand System for guiding air pilots when approaching a stand. *Abbrev.* AGNIS *or* Agnis.

Approportion Substituted merchandise offered to the consignee by the agents or carriers to make up for what might have been shortlanded.

Approportion Account The net profit of a business concern which is calculated on dividends, reserves, pension funds, annual bonuses and so on.

Approportion of Payment A creditor owed two or more debts, who is paid for one unspecified account, is legally authorised to appropriate or choose which account he wants settled first. However, this does not apply if the bill being paid refers to any significant account.

Approx Approximate; Approximately.

Appt Appoint; Appointment.

Appurtenances An indispensable word inserted in a ship sale contract referring to all the tackle and furniture needed to make the ship useful for her purpose.

APR *or* **A.P.R.** Annual Progress Report.

Apr. April.

APRC American Port Risk Clauses.

a prima vista *Italian*—At first sight.

a priori *Latin*—From what was before. From cause to effect. By deduction. Refers to an argument where the premises logically account for the conclusion.

Apron An American term referring to that area of the loading/unloading quay between the cargo sheds and the ship which is used in most cases for stacking outward and inward bound cargoes.

à propos de bottes *French*—Beside the point; Irrelevant.

à propos de rien *French*—Of no consequence.

A protestor *Abbrev.* A.P., *q.v.*

APRT Advance Petroleum Revenue Tax.

APS *or* **A.P.S.** American Philatelic Society; American Philosophical Society; American Physical Society; As per sample; Arrival Pilot Station, *q.v.*, Average Propensity to Save, *q.v.*

APSA *or* **A.P.S.A.** Aerolineas Peruanas SA. *Peru International State Airlines*; American Political Science Association; Australian Political Association.

APT Annulus Pressure Transducer *oil drilling*.

APT *or* **A.P.T.** Advanced Passenger Train; Aft Peak Tank, *q.v.*

Aptdo Apartado *Spanish*—Post Office Box.

Aptitude Tests Tests undertaken by various organisations or companies to determine the fitness, natural propensity or ability of a person to undertake a particular type of work.

APU *or* **A.P.U.** Arab Postal Union; Auxiliary Power Unit *q.v.*, *aviation*.

APV Anti Pollution Valve.

aq. Aqua *Latin*—Water.

aq. dest. aqua destillata *Latin*—Distilled water.

à quai *French*—term for Ex-Quay

A.Q.C. Associate of Queen's College, London.

AQIS Australian Quarantine and Inspection Service.

AQL *or* **A.Q.L.** Acceptable Quality Level; Aeroglisseur à Quille Laterale *French*—Lateral Keel Hovercraft.

AR Aspect Ratio, *q.v.*

Ar. Argentum *Latin*—Silver.

AR *or* **A.R.** *or* **A/R** Account Receivable; Acid Resistance; Advice Receipt; Advice of Receipt; All Risks *similar to* a.a.r. *or* Against all Risks *or* All and Any Risks, *q.v.*; American Record *similar to* AB, meaning American Bureau of Shipping, *q.v.*; Anno Regni *Latin*—In the year of reign; Annual Return; Australian Registered *shares.* Aerolineas Argentinas *airline designator.*

A.R. *or* **A/R** Air Remittance; Airmail Remittance.

ARA *or* **A.R.A.** Agriculture Research Association; Air Reserve Association; American Railway Association; Antwerp/Rotterdam/Amsterdam range of loading/discharging ports in a charterparty; American Radio Association; Association of River Authorities; Associate of the Royal Academy.

Aracon Code name for the Baltic and International Maritime Council charterparty approved in 1937 for the carriage of groundnuts.

ARAM *or* **A.R.A.M.** Associate of the Royal Academy of Music.

ARB Agriculture Research Board.

Arbitrage Buying securities in one country or currency or market and selling in another to take advantage of price differentials.

Arbitrageur A trader engaged in Arbitrage, *q.v.*

Arbitration In a dispute or argument judges or impartial persons are appointed as arbitrators to settle the differences existing between parties rather than undertaking the long and complicated methods of legal action through the law courts. There are clauses in certain contracts which refer to arbitration as a way of settling disputes. However, in certain instances it may be preferable for the case to be heard in a High Court because of the considerable amount of money involved.

Arbitration Award. The decision or award given in an arbitration. *See* **Arbitration**.

Arbitration Clause A clause commonly found in charterparties, ship sale contracts and others, stating where arbitration is to take place in the event of disputes between the parties. *See* **Arbitration**.

Arbitrator *See* **Arbitration**.

ARC Aircraft Radio and Control; American Red Cross; Automobile Racing Club.

Archibenthic Zone That part of the area of sea water from 200 to 1,000 metres in depth. *See* **Abyssalbenthic Zone.**

Architect A qualified person who prepares plans and superintends building work; designer of complex structures etc., including ships.

Archipelago A group of islands.

ARCAP Adaptive Radar Controlled Autopilot.

ARCM *or* **A.R.C.M.** Associate of the Royal College of Music.

Arc of Visibility The arc of a navigation light seen from the side of a ship on the horizon and which is represented by degrees.

ARCS Admiralty Raster Chart Service. Digital charts and weekly up-dates, introduced in 1994 by UK Hydrographic Office, the first to do so.

ARCS *or* **A.R.C.S.** Associate of the Royal College of Science; Autonomous Remotely Controlled Submersible.

A.R.C.S. Associate of the Royal College of Surgeons.

A.R.E. Arab Republic Emirates. *Also* U.A.R., *q.v.*

Are A surface unit measurement of 100 square metres, corresponding to 119.6033 square yards.

Arg. Argentina.

argumentum ad hominem *Latin*—An argument that attacks or prejudices the man himself rather than his reasoning.

argumentum ad ignorantiam *Latin*—An argument based on the ignorance of the opponent.

argumentum ad judicium *Latin*—An argument of common sense and judgement.

ARI *or* **A.R.I.** Aerospace Research Institute Munich, *Germany. See* **Satellite SOS System Aids.**

A.R.I.B.A. Associate of the Royal Institute of British Architects.

A.R.I.C. Associate of the Royal Institute of Chemistry.

A.R.I.C.S. Professional Associate of the Royal Institution of Chartered Surveyors.

Ariel Short term for Airport Insurance. It covers (a) premises and machinery, food and drink, (b) servicing and repairs undertaken, and (c) workshop hangar fees for the duration of repairs.

Arig Arab Insurance Group.

Arigato *Japanese*—Thank you.

A.R.I.N.A. Associate of the Royal Institution of Naval Architects.

ARINC Aeronautic Radio Inc. *USA.*

ARIS Advanced Range Instrumentation Ship.

Ariz. Arizona *USA.*

Ark. Arkansas *USA.*

ARM Anti-Radar Missile.

Arm Armenia; Armenian.

Ar.M. Architecturae Magister *Latin*—Master of Architecture.

Armco A trade name for pure low carbon iron used for bars, quadrantals, etc.

Armé en flûte Term formerly used to describe a warship with armament reduced to accommodate troops or cargo.

A.R.M.I.T. Associate of the Royal Melbourne Institute of Technology.

ARNO or **A.R.N.O.** Association of Royal Navy Officers.

ARNS Aeronautical Radio Navigation Service *aviation UK.*

ARPA Advance Research Projects Agency *US Government;* Automatic Radar Plotting Aids *electronics.*

ARPASIM Automatic Radar Plotting Aids Simulator *electronics.*

ARPS or **A.R.P.S.** Associate of the Royal Photographic Society; Automatic Radar Planning System.

A.R.R. Anno Regni Regis or Reginae *Latin*—In the years of the King's or Queen's Reign.

Arranged Total Loss or **Compromised Total Loss** Whenever the insured article has suffered a substantial loss, but not to the extent of an apparent actual or constructive total loss, there may be a possibility that the underwriters and the assured mutually agree to settle the claim on a Total Loss basis. *Abbrev.* ATL or A.T.L. or Arr. T.L. and CTL or C.T.L.

Arrd. Arrived.

Arrears Bills or moneys which are due but have not as yet been paid.

Arrest or **Restraint of Princes, Rulers** or **Seizure under Legal Process.** An excepted peril of the sea in the Carriage of Goods by Sea Act 1924, amended in 1968 and 1971, in relation to government restrictions, Royal Proclamations and the like.

Arrest *and/or* **Seizure** An official order of arrest or seizure of a ship, empowered by the state authorities, for contravening the laws of a country or to exercise a Maritime Lien, *q.v.*, on a ship. The word seizure has a stronger meaning than arrest as it may be interpreted to be a forcible possession by an overpowering force or by the authority of a state. *See* **Seizure.**

Arrestment A legal Scottish term in connection with the stoppage of any money transaction until the money owed has been settled by the debtor.

Arrgt/s Arrangement/s.

arrière pensée *French*—Ulterior motive.

Arrival Notification Form Notification in writing to the consignee/s of the arrival of merchandise. *Abbrev.* ANF.

Arrival Pilot Station. A Time Charterparty term. A vessel is considered to be delivered to the charterers as soon as she has arrived at the pilot station, irrespective of bad weather. A drawback for the charterers. *Abbrev.* APS.

Arrived Ship A shipping phrase in conjunction with the agreed terms of the Charterparty, *q.v.* A ship has 'arrived' when she is within the precincts of the port. This phrase is regularly inserted in charterparties.

Arr/s Arrival/s; Arrive/s.

Arr. T.L. Arranged Total Loss, *q.v.*

Arroba *Spanish* or *Portuguese*—Unit of weight equivalent to 25.35 lbs.

Ars longa, vita brevis *Latin*—Art is long, life is short.

ARS or **A.R.S.** Agriculture Research Service *USA.*

A.R.S.A. Associate of the Royal Society of Arts.

A.R.San.I. Associate of the Royal Sanitary Institution.

ARSR Air Route Surveillance Radar, *q.v.*

Art. Article.

ART Airborne Radar Technician.

ARTA Automatic Radar Target Aids.

A.R.T.C. Air Route Traffic Control; Associate of the Royal Technical College.

Articled Clerk A clerk under articles (i.e., heads and particulars) of an agreement to serve a solicitor or acountant (the principal) in consideration of being initiated into the routine and mystery of the profession; a person may not be admitted as a solicitor unless he or she has served as an articled clerk. Where the clerk is at the date of the articles under age, his parent or guardian is usually made a party to the articles as well as himself.

Articles of Association Any seven or more persons (or, in the case of a private company, any two or more persons) may, by subscribing their names to a memorandum and otherwise complying with the statutory requirements as to registration, form an incorporated company with or without limited liability. Whatever be the kind of company which is formed, the memorandum must state (a) its name, (b) whether the registered office is to be in England, or Scotland, and (c) its objects; and every subscriber must subscribe it, if it has a share capital, with the number of shares he takes. If liability is limited, by shares or guarantee, that fact must by stated. Articles of association are regulations for the management of a company. The articles are freely alterable by special resolution. *See* **Memorandum of Association.**

Articles of Partnership A detailed documented contract setting out the individual responsibility of the active partners of the company. The articles can be altered by agreement among the partners.

Articulated Tug Barge A tug-barge combination system capable of operating on the high seas, coastwise and further inland. It has been so well designed that it has proved to be seaworthy in waves of 30 ft. and over. It combines a normal barge, with a bow resembling that of a ship, but having a deep indent at the stern to accommodate the bow of a pusher tug. The fit is such that the resulting combination behaves almost like a single vessel at sea as well as while manoeuvring. This system can be compared to the Ocean Going Pusher Barge or Tug Combination or Sea Link System. The barge can be disconnected and left behind in a port for loading and/or discharging. In the meantime the tug can undertake another voyage with another barge sailing either empty or full, thus saving considerable expense and time on such a dual purpose voyage. *Abbrev.* Artubar.

Artificial Horizon An aircraft gyro-stabilised instrument displaying pitch and roll movements.

Artubar Short for Articulated Tug Barge, *q.v.*

ARVA *or* **A.R.V.A.** Associate of the Rating and Valuation Association.

As Asia; Altostratus Clouds, *q.v.*

AS All Section *insurance*; Anonim Siriket *Turkish*——Joint Stock Company.

AS *or* **A.S.** Associate of Arts *USA.*

A.S. *or* **a/s** *or* **A/S** Account Sales; After Sight, *q.v.*; Air Scoop; Alongside, *q.v.* Angle Stiffener; Anno salutis *Latin*—In the year of Salvation; Annual Survey *Lloyd's Register notation.*

A/S *Abbrev.* for Aksjeselskap *(Norwegian)* or Aktieselskab *(Danish)* Ltd. Co.

ASA *or* **A.S.A.** Air Lanka/Singapore Airlines. A merger signed on 10 October 1979 between the governments of the Republic of Singapore and the Democratic Socialist Republic of Sri Lanka; Always Safely Afloat, *q.v.*; American Shipbuilding Association, Washington USA, formed in 1994; American Shipmasters Association; American Standards Association; Association of Shipbrokers and Agents Inc. *USA*; Advisory Service Area, *q.v.*; American Soyabean Association.

A.S.A.A. Associate of the Society of Incorporated Accountants and Auditors.

ASAAS Association of Suppliers to Airlines, Airports and Shipping.

ASABOSA *or* **A.S.A.B.O.S.A.** Association of Ships' Agents and Brokers of South Africa.

As Agreed by Leading Underwriters The insurers agree to the leading underwriters' patterns and conditions of the policy which has still to be drafted. *Abbrev.* As Agreed L/U.

ASAEN Association of South-East Asian Nations.

As Agreed L/U As Agreed by Leading Underwriters, *q.v.*

As and Where Lies A phrase often used in the sale of second-hand ships and other objects. The sale is effected as the ship or other objects stand, irrespective of the condition of the vessel or object. The seller is therefore exonerated from any liability arising from latent defects, but not from hidden defects known by the seller which would have to be reckoned with provided that proof is submitted.

asap *or* **ASAP** *or* **A.S.A.P.** As Soon As Possible.

ASAPHA *or* **A.S.A.P.H.A.** Association of Sea and Air Port Health Authorities.

asb Aircraft Safety Beacon; Asbestos.

ASB Alternate Side Band.

ASBA Association of Shipbrokers and Agents (New York).

ASC Austrian Shippers Council.

ASCAP American Society of Composers, Authors and Publishers.

ASCE *or* **A.S.C.E.** American Society of Civil Engineers.

ASCEA *or* **A.S.C.E.A.** American Society of Civil Engineers and Architects.

ASCII American Standard Code for Information Interchange *computers.*

ASCOPE Asian Council on Petroleum.

ASCU Association of State Colleges and Universities *USA.*

As Customary A condition under the Laytime, *q.v.*, or Laydays, *q.v.*, in a charterparty. The ship is to load or unload according to the usual practice and custom of the port. Thus the charterer or the consignee is not obliged to hasten despatch or advance the completion of cargo so long as the rate of loading/unloading is being followed in accordance with that prevalent at the port. This is of course a disadvantage to the carrier as his ship is at the mercy of the normal time taken. *Abbrev.* COP *or* C.O.P. meaning Custom of the Port. *See also* **With Customary Despatch** *or* **Customary Despatch**.

As Customary Always Afloat *See* **At All Times of Tide** and **Always Afloat.**

A.Sc.W. Association of Scientific Workers.

A.S.D. Admiralty Salvage Department.

ASDE Airborne Surface Detection Equipment.

ASDIC *or* **A.S.D.I.C.** Allied Submarine Detection Investigation Committee. Hence the Royal Navy's Submarine detection equipment was originally called Asdic. Now known as Sonar, *q.v.*

a.s.e. *or* **A.S.E.** Air Standard Efficiency; Amalgamated Society of Engineers; American Stock Exchange; Associate of the Society of Engineers; Association of Science Education.

ASEAN *or* **A.S.E.A.N.** Association of South East Asian Nations, comprising Indonesia, Malaysia, Philippines, Singapore and Thailand.

As Expiring An insurance expression giving advance information to the insured regarding his intention of renewing his existing insurance on the same terms and conditions so as to avoid undue delays.

ASF *or* **A.S.F.** Associate of the Institute of Shipping and Forwarding Agents; Automatic Sheet Feeder *computer.*

As Fast As Can An expression in a charterparty making it an obligation on the part of the charterer, towards the owner, to load the cargo as fast as the ship's lifting gear can function. This is also subject to the custom of the port where the cargo handling is taking place. *Abbrev.* afac or AFAC or A.F.A.C. *See also* **Fast as Can.**

ASFINAVAL Association Française des Industries Navales *French*—French Shipyards Association.

A.S.G.B. Aeronautic Society of Great Britain.

ASH Action on Smoking and Health *USA.*

'A' Shares Some merged and public companies make distinctions among their types of shares. 'A'

Shares belong to the small shareholders who are not entitled to vote.

Ash Breeze Old sailors' expression for oar propulsion of a becalmed sailing ship when rowed or towed by rowing boats, wooden oars being commonly made of ash.

Ashore On the land.

ASHRAE American Society of Heating, Refrigerating and Air Conditioning Engineers.

Ashram A hermitage; A house for destitute people; A place of religious retreat.

ASI or **A.S.I.** Actuaries Share Indices *finance*; Air Safety Investigators; Air Speed Indicator.

ASIA or **A.S.I.A.** Airlines Staff International Association.

ASIAS Airline Schedules and Interline Availability Study *IATA*.

a/side Alongside.

asinus ad lyram *Latin*—An ass at the lyre. Unsusceptible to music. Awkward, clumsy, inept.

ASIS Association for the Structural Improvement of Shipbuilding *Japan*; As Is Where Is, *q.v.*

ASIS or **A.S.I.S.** Amphibious Support Information System *USA*.

As Is Where Is A common phrase used in the purchase and sale contracts of ships, the buyers accepting that the purchase of the vessel is unconditional. However, a Latent Defect, *q.v.*, could render the seller responsible for damages if it could be proved that he actually knew of the hidden defect. *Abbrev.* ASIS. *See also* **As and Where Lies**.

ASL or **A.S.L.** Abandon Ship Ladder, *q.v.*

asl Above Sea Level.

ASLE American Society of Lubrication Engineers.

ASLEF or **A.S.L.E.F.** Associated Society of Locomotive Engineers and Firemen.

ASLIB or **A.S.L.I.B.** Association of Special Libraries and Information Bureaux.

A.S.L.O. Associate of Scottish Life Offices *Scottish life assurance*.

ASM Air Surface Missile; Anti-Submarine *USA*.

ASME or **A.S.M.E.** American Society of Mechanical Engineers. *See* **PVQA**.

ASMI Aircraft Surface Movement Indicator, *q.v.*; Association of Singapore Marine Industries.

ASMS Advanced Surface Missile System; American Society of Mechnical Engineers.

ASN or **A.S.N.** American Standard Nuts *screw thread system*.

ASNA American Society of Naval Architects.

ASNE American Society of Naval Engineers.

As Near As She Can Safely Get A charterparty term naming the port of discharge. In the event that the vessel is prevented from entering the port then the master has the option to proceed to the nearest safe port of discharge. Also termed as: As Near to As She Can Safely Get, Always Afloat. *See* **Always Afloat**.

ASO Area Survey Organisation *insurance*.

a.s.o And so on.

ASOF Australian Steamship Owners' Federation.

ASOMPIA The American Steamship Owners Mutual Protection & Indemnity Association. *See* **American Steamship Owners Mutual Protection & Indemnity Assn. Inc**.

As Ordered A charterparty term relating to Laydays, *q.v.* The laydays start to count when the ship is actually berthed at the loading/discharging berth which has been allocated and not when she simply arrives in the port.

As Orig. As Original, *q.v.*

As Original An insurance expression commonly found in reinsurance policies where the reinsurers agree to the original conditions of the policy. *Abbrev.* As Orig.

ASOS Automatic Storm Observation Service.

ASP or **A.S.P.** Accelerated Surface Post; Aerospace Plane *USAF*; American Selling Price, *q.v.*

ASPAC Asian and Pacific Council.

ASPCA American Society for the Prevention of Cruelty to Animals.

ASPE American Society of Petroleum Engineers.

Aspect Ratio The square of the wing span of an aircraft divided by the area. *Abbrev.* AR.

As Per Advice Words inserted on bills of exchange meaning that the drawee has been advised that the bills were drawn on him. *Abbrev.* APA or A.P.A.

As Per Foreign Statement A marine insurance clause binding the underwriters in respect of any General Average which might occur and where the adjusters are in a foreign country. The underwriters agree to accept the foreign adjustments arrived at.

ASPR Armed Services Procurement Regulations *USA*.

ASR Airport Surveillance Radar, *see* **Air Route Surveillance Radar**; Air Rescue Service; Altimeter Setting Region *aviation*.

ASROC Anti-Submarine Rocket *USA*.

ASRV or **A.S.R.V.** Angle Stop Radiator Valve.

ASS or **A.S.S.** Admiralty Standard Stockless *ship's anchor*.

Ass *See* **Assisted Into**.

Assay Checking or testing the alloy or contents of coins and bullion.

Assay Master A person authorised to carry out an Assay, *q.v.*

Assce. Assurance, *q.v.*

Assd or **assd.** Assured.

ASSD or **A.S.S.D.** Association of Stocks and Shares Dealers, *q.v.*

Assembly line System of production where the items to be manufactured move continously through the workshop from one process to the next, each worker having a separate operation to perform. It enables division of labour while setting a uniform pace of output.

Assent The Stock Exchange, *q.v.*, term for the agreement of the holder of a bond, loan stock or debenture to vary his rights with regard to the payment of interest or its repayment. An Assented Bond can normally still be dealt in on the Stock Market but its price is likely to differ from that of its non-assented counterpart.

Assessment The valuation of property, income and outgoings for tax or insurance purposes.

Assessor One who sits as assistant, adviser, to a judge or magistrate; one who assesses taxes or estimates, value of property for taxation. *See also* **Valuer**.

Assets Credits belonging to any individual, firm or association, etc. The actual goods and properties which are in possession can be converted or exchanged into money to make good for any debts incurred by the company or individual. Opposite to Liabilities, *q.v.*

Assigned Freeboards Minimum freeboards assigned to a ship by an 'assigning authority' (usually the state shipping authority) after satisfactory drawings and plans have been produced, a load line survey passed and stability tests completed to the authority's requirements. *See* **Freeboard**.

Assignee A person appointed by another to do an act in his own right, or to take over rights or title by assignment.

Assignor One who transfers or assigns property to another.

Assignment The passing of beneficial rights from one party to another. A policy or certificate of insurance cannot be assigned after interest has passed, unless an agreement to assign was made, or implied, prior to the passing of interest. An assignee acquires no greater rights than were held by the assignor, and a breach of good faith by the assignor is deemed to be a breach on the part of the assignee.

Assignment Clause A marine insurance clause authorising the transfer of the insurance interests to another by endorsement.

Assisted Into An insurance expression used in towage and referring to towing into the port of refuge. *Abbrev.* Ass.

Assn. Association.

Assoc. Associate; Associated; Association.

Association A society the members of which are united by mutual interests or for a common purpose. *Abbrev.* Assn *or* Assoc.

Association de Services Transports Informatiques *French*—Association of Transport Information Services, concerned with worldwide freight, transport, insurance, financing and management of computer services for its members, with headquarters in Switzerland. *Abbrev.* ASTI.

Association of Stock and Share Dealers Stock Exchange members who are authorised to deal in the business of stocks and shares outside the Stock Exchange. They are often called Outside Brokers. *Abbrev.* ASSD *or* A.S.S.D.

Associazione Armatori Giuliani Italian Shipowners' Association Trieste.

ASSR *or* **A.S.S.R.** Autonomous Soviet Socialist Republic.

Asst. Assistant.

Assurance *or* **Life Insurance** Personal insurance or an assurance of a person during his or her lifetime or in the case of death. There are two types of assurances: (1) Absolute Assurance where the underwriters pay a sum of money in the event of death before the expiration of the policy (2) Contingent Assurance covering the assured's next of kin or those he leaves in his will in case of death before the policy expires or during his lifetime when the policy time expires, in which case the underwriters pay a lump sum of money, with or without bonuses. This is called 'Endowment'. *See* **Life Insurance**.

Assurance Mutuelle A French expression meaning Mutual Insurance or Protection & Indemnity, *q.v.* It has become the practice of shipowners to form a sort of mutual insurance club, often called Protection & Indemnity Club, in which each and every member has to contribute to the upkeep in originating and paying out claims arising and not normally covered by the insurance policy, such as a quarter of collision liability; payments for legal proceedings in claims for or against; customs fines; cargo shortages; cost of personal accidents etc. Other firms in business such as Shipbrokers and Agents are also forming their own mutual clubs today, covering their members against a wide range of expenses and discrepancies. *Abbrev.* AM *or* A.M.

Assured *See* **Insured**.

Ast *or* **AST** Astronomical Time.

ASTA *or* **A.S.T.A.** American Sail Training Association, which is located in Newport, Rhode Island, USA; American Society of Travel Agents.

ASTEC Advanced Solar Turbo-Electric Conversion.

Astern A backward direction in the line of a vessel's fore-and aft line; behind. If a vessel moves backwards it is said to move astern. Opposite to ahead, *q.v.*

Asteroids Minor planets. *See* **Planets**.

ASTM *or* **A.S.T.M.** American Society of Testing Materials; American Society of Tropical Medicine.

ASTMS *or* **A.S.T.M.S.** Association of Scientific, Technical and Managerial Staffs.

Astn Astern.

ASTRO Association of State Trading Organizations.

Astro-hatch The overhead cockpit window of an aircraft.

Astrolabe An astronomical instrument used to obtain accurate positions in survey work ashore. Formerly it was also used by seamen to take the altitude of the sun or stars but it was replaced by the quadrant and, more recently, the Sextant.

Astrology The art of judging the reputed occult influence of stars and planets upon human affairs.

Astrometer An instrument to measure the stars and the light of the stars.

Astro-Navigation Navigation using heavenly bodies, ie. sun, moon, stars and planets.

Astronomy The science dealing with celestial bodies, their magnitudes, nature, motions, distances and behaviour.

ASU *or* **A.S.U.** Arab Socialist Union; Australian Seamen's Union.

ASV *or* **A.S.V.** Angle Stop Valve; Air-to-Surface Vessel *radar*; Auto-start Value.

ASVAB Armed Services Vocational Aptitude Battery *USA*.

ASW *or* **A.S.W.** Anti-Submarine Warfare *USA*.

ASWAG Anti-Submarine Warfare Systems Analysis Group *USA*.

at Airtight.

At Alteration.

A.t. Atlantic Time.

AT Achievement Test/s *USA*.

AT *or* **A.T.** Airfreight; American Terms in the *grain trade*; Apparent Time. *q.v.*; Atlantic Time; Air Tight.

A/T Altitude, *q.v.*; American Terms *grain trade*.

ATA *or* **A.T.A.** Actual Time of Arrival; Air Transport Association; American Translators' Association.

ATAC *or* **A.T.A.C.** Air Transport Advisory Council; Air Transport Association of Canada.

At All Times of the Tide and Always Afloat This charterparty clause relieves the ship from waiting in a tidal river or harbour until the tide allows her to proceed to the dock or wharf where she is to load or discharge. The charterer will be required to name a loading or discharging berth where she can lie Always Afloat, *q.v.*, At All Times of the Tide. The clause Always Afloat alone will not justify a vessel in declining to go to a berth when she cannot lie continuously afloat as she can do so partly before and partly after neap tides. So where the ship is chartered to load or discharge As Customary Always Afloat at such a wharf or anchorage, the charterers may direct her to a berth where she can load or discharge part of her cargo afloat, but will then load/discharge the rest of the cargo by means of lighters at an anchorage. That being a customary method of loading in that port, the charterer will commit no breach of contract by using the different methods mentioned.

At Anchor A vessel is said to be at anchor when her anchor is down and holding. When weighing anchor a vessel continues to be at anchor until the anchor is out of the ground. A vessel moored to buoys may be considered to be at anchor, but a vessel whose anchor is dragged is considered to be 'Under way', *q.v.*

At and From In marine insurance when concerning cargo this covers the shipped merchandise immediately it is shipped; i.e., at the port and from the port. When concerning the ship it means that the vessel is immediately insured from the port of the commencement of the voyage. If not yet in port, the insurance still operates until she arrives safely irrespective of whether another policy exists. In the case of the insurance being on chartered freight the risk attaches on arrival at the named port.

ATAS *or* **A.T.A.S.** Air Transport Auxiliary Service.

ATB Advanced Turboprop *aviation*; Automated Ticket Boarding *boarding card*.

ATBA Area to be avoided, in safe routeing of tankers.

ATC *or* **A.T.C.** Air Traffic Conference *USA*; Air Traffic Control; Air Transport Command; Association of Translations Companies.

ATCA Air Traffic Control Association *USA*.

At Call Money to be paid when requested. This often happens when money is borrowed for a very short period, at practically short notice and at very low interest. *Also known as* Near Money; Money at Call and At Short Notice.

ATCC *or* **A.T.C.C.** Air Traffic Control Centre.

A.T.C.L. Associate of Trinity College of Music, London.

ATCO *or* **A.T.C.O.** Air Traffic Control Officer *aviation*.

ATCU *or* **A.T.C.U.** Air Traffic Control Unit *aviation*.

ATD *or* **A.T.D.** Actual Time of Departure; Air Tight Door; Australian Transport Department; Automatic Target Detection.

ATDN *or* **atdn** *or* **A.t.d.n.** Any Time Day or Night.

ATEC *or* **A.T.E.C.** Air Transport Electronics Council *aviation*.

ATF *or* **A.T.F.** American Trust Fund.

ATFP Arab Trade Financing Programme.

Athwart Across.

ATIC Association Internationale des Interprêtes de Conférence *French*—International Association of Conference Interpreters.

A.T.I.I. Associate of the Taxation Institute Incorporated.

A.Time Alpha Time, *q.v.*

ATIS Automatic Terminal Information Service *aviation*.

ATIT *or* **A.T.I.T.** Associate of the Institute of Taxation.

ATK Available Tonne Kilometres.

Atl. Atlantic.

ATL *or* **A.T.L.** Arranged Total Loss. *See* **Compromised Total Loss.**

ATLA *or* **A.T.L.A.** Air Transport Licensing Authority.

Atlantic International Bank Ltd. A consortium of the following banks; Banco di Napoli, Italy; Charter House Japhet and Thomasson; De Neuflize Schumberger Mallet et Cie, France; Fran Lanchot; First Pennsylvania Bank and Trust Co; Manufacturers National Bank of Detroit; National Shawmut Bank of Boston and United California Bank. *Abbrev.* AIBL.

Atlantic Merchant Vessel Report A system originated by the United States Coast Guard (USCG) called the Maritime Mutual Assistance Programme which arranged Search and Rescue (SAR) along most of the American coastline, including the Equator and West of Prime Meridian, all messages and information being passed through electronic computers to trace swiftly and efficiently the exact position of vessels and persons in distress. Nowadays *see* **AMVER**.

ATLB Air Transport Licensing Board *UK.*

atm Atmosphere.

ATM Asynchronous Transfer Mode, *data handling technology.*

ATM *or* **A.T.M.** Air Traffic Management *aviation;* Automatic Teller Machine or Cash Dispenser *banking.*

At Merchant's Risk *or* **Merchant's Risk** Occasionally found in contracts for the Carriage of Goods by Sea, *q.v.*, and is similar to the term Advanced Freight, *q.v.*; this method of payment is advantageous to the owners or carriers because should the vessel be lost due to some unforeseen circumstances such as an Act of God, *q.v.*, the freight thus paid is not recoverable. When the carriage is effected by sea the term used is At Owner's Risk. The word owner refers to the merchandise. Similar expressions are Advance Freight, *q.v.*, and Freight Paid in Advance, *q.v.*

ATMC *or* **A.T.M.C.** Admiralty Transmitting Magnetic Compass.

Atmospherics Radio reception interference noise.

ATMR Advanced Technology Medium Range *aviation.*

At No Atomic Number.

ATO *or* **A.T.O.** Art Teachers Organisation; Assisted take-off *aviation.*

A to a *or* **A to A** Air to Air.

Atoll A coral reef in the shape of a ring which grows around a submerged island. Some large atolls are inhabited.

A to N Aid(s) to Navigation.

à tort et à travers *French*—at random; without distinction.

à toute force *French*—At all costs; absolutely.

à tout prix *French*—At any cost or price.

At par When stocks and shares are paid at the normal value or at cost price.

atrima *or* **a.t.r.i.m.a.** As their respective interests may appear *legal term.*

ATRS *or* **A.T.R.S.** American Tanker Rate Schedule, *q.v.*

ATS Air Traffic Service.

ATS *or* **AMTS** Automatic Telex System.

ATSBE *or* **A.T.S.B.E.** All Time Saved Both Ends, *q.v.*

ATSDO *or* **A.T.S.D.O.** All Time Saved Discharging Only, *q.v.*

At Sea In marine insurance this phrase applies to a ship which is free from its moorings and ready to sail.

At Short Notice *See* **At Call**.

At Sight This refers to the immediate payment against the presentation of the documents; even with a Bill of Exchange, *q.v.*

ATSU Air Traffic Service Unit.

Att. Attaché; Attention; Attorney.

ATT *or* **A.T.T.** Admiralty Tide Tables.

AT&T American Telephone and Telegraph.

Attachment The enforcement of a debtor to ensure that he will not dispose of the money or articles in his possession until full settlement is effected. *See* **Arrestment**.

attd Attached.

Attendance Money Payment to dockers for attending, or standing-by, for work which in the event may not be, or was not, required.

Attention, Interest, Desire and Action Motto for a salesman in his relations with a buyer. *Abbrev* AIDA.

Attestation The signing by a witness to the signature of another of a statement that a document (notably wills) was signed in the presence of the witness. Regarding wills, the witness must have no interest in the matter.

Attestation Clause A marine insurance clause confirming that the underwriters have officially signed the policy of marine insurance.

Attested Copy *Similar to* **Certified Copy**, *q.v.*

Attorney Solicitor; Advocate; Barrister or a Trustee empowered by an official declaration to act on behalf of other(s) for a specific purpose.

Atty Attorney.

ATU Air Transport Users Committee *USA.*

At.Wt. Atomic Weight.

Au. Aurum *Latin*—Gold element.

AU *or* **A.U.** Angstrom Unit *The International Wave Length Unit;* Astronomical Unit.

AUA Austria International State Airlines.

AUC Air Transport Users' Committee *UK. under the auspices of the Civil Aviation Authority.*

AUC *or* **A.U.C.** Ab Urbe Condita, *q.v.*

AUCCTU All Union Central Council Trade Unions *South Africa.*

au contraire *French*—On the contrary.

au courant *French*—Acquainted; Aware of; Well informed.

Auction The sale in most cases of second-hand objects to the public by an auctioneer, *q.v.* Whoever offers the highest bid will get the object.

Auctioneer A licensed person offering goods, in most cases second-hand, for sale to the public. *See* **Auction**, also **Appraiser**.

aud. Audit, *q.v.*; Auditor, *q.v.*

aud. gen. Auditor General.

Audi alteram partem *Latin*—Hear both sides; Hear the other side.

Audio-frequency Oscillation of sound waves audible in the range 20 to 20,000 cycles per second.

Audiometer An instrument for measuring the power of hearing.

Audit The official checking of accounts of a company by authorised accountants. *Abbrev.* aud.

Auditor An authorised person who audits accounts. *Abbrev.* aud.

AUEW Amalgamated Union of Engineering Workers.

au fait *French*—Well acquainted with.

Aufl Auflage *German*—Edition.

au fond *French*—Basically.

Aug August.

A.U.I.B. Association of Underwriters and Insurance Brokers. Established in Glasgow in April 1918.

AULD Air Unit Load Devices.

au naturel *French*—In the natural or original state.

a.u.n. Absqua ulla nota *Latin*—With no identification mark.

au pied de la lettre *French*—Quite literally; Close to the letter.

AUP *or* **A.U.P.** Accepted Under Protest.

Aurum *Latin*—Gold. *Abbrev.* Au.

Aus. Austria; Austrian; Australia.

Ausg. Ausgabe *German*—Edition.

AUSREP Australian Ship Reporting System.

Aust Australia.

Austral *or* **AUSTRAL** Code name form for Chamber of Shipping. Australian Grain Charterparty as amended on 20 July 1928; Adjective relating to Australasia or lands southeast of Asia; Currency unit of Argentina.

Austwheat 1990 *or* **AUSTWHEAT 1990** Code name for the Baltic and International Council form for the carriage of wheat to Australia, as amended in 1991.

aut *or* **AUT** Automatic Unmanned Machinery Space.

A.U.T. Association of University Teachers.

autant d'hommes autant d'avis *French*—as many men as many ideas.

aut mors aut victoria *Latin*—Death or victory.

AUTEC Atlantic and Undersea Test and Evaluation Center *USA*.

Auth. Author; Authorised.

Authorised Capital Alternatively known as nominal or registered capital. It is the capital of a company or association when authorised by the Memorandum of Agreement giving details of the number of shares and the types or classes of the shares as to how the capital is divided.

Authorised Clerk A stockbroker's clerk who has access to the Stock Exchange, *q.v.*, and at the same time acts on behalf of his principals in buying and selling.

Author's Proof A proof of a printed matter referred to the author of a book or other publication for correction and acknowledgement.

Auth. R/I Authorised Re-Insurance, *q.v.*

Authorised Re-Insurance A reputable firm authorised to deal in re-insurance, *abbrev.* Auth. R/I.

AUTIF Association of Unit Trusts and Investment Funds.

Auto Automatic Control.

Auto Ferry A ferry boat for carrying vehicles.

Autograph A person's own work, signature or writing.

Autogyro Helicopter or similar aircraft having a horizontal screw to ascend or descend in a vertical position.

Autoignition Point The lowest temperature at which a substance will ignite without the introduction of a flame or other ignition source.

Autoland An electronic device fitted to an aircraft which enables it to land safely in dense fog.

Automatic Pilot An instrument designed to control automatically a vessel's steering gear so that she follows a pre-determined track through the water; also applicable to aircraft.

Automatic Selling The selling of goods by means of a coin-operated machine.

Automatic Teller Machine A machine provided at Banks to facilitate withdrawal of money, and other services by using a card. *Abbrev.* ATM. Also called Cash Dispenser.

Automatic Watchkeeper Automatic service on a ship monitoring and supervising the routine working system. This new system eliminates a large amount of normal manual watchkeeping throughout the navigation and also in port.

Autrefois acquit *or* **Autrefois convict** Norman French—Previously acquitted *or* Previously convicted. Plea of a person accused of the same offence for a second time.

AUV Autonomous Underwater Vehicle.

AUW *or* **A.U.W.** *or* **a.u.w.** All-up-Weight *aviation.*

Aux. Auxiliary.

Aux.B. Auxiliary Boiler/s *Lloyd's Register notation.*

Auxiliary A vessel's machinery, apart from her main engine(s). This may include generators, windlasses, winches, evaporators etc.

Auxiliary Power Unit Aircraft installation to provide heat and ventilation as well as for starting engines, when main engines have been shut down.

a.v. annos vixit *Latin*—He or she lived (so many) years.

Av. *or* **AV** Average. In insurance this can refer to partial loss.

A/V *or* **a.v.** Ad Valorem, *q.v.*; At Valuation; Asset Value.

AVAD Automatic Voice Alerting Device.

Aval Guarantee An obligation of a guarantee by a third party *Also called* Suritry par aval.

Avast An order to stop moving or heaving *nautical.* A term for 'hold fast'.

AVC *or* **A.V.C.** American Veterans Committee.

AVCAT Aviation Category *high flashpoint aviation fuel.*

Av. Cert. Aviator's Certificate.

Av. Disbs. Average Disbursements, *q.v.*

avdp Avoirdupois, *q.v.*

Ave. Avenue.

AVENAVE Asociacion de Navieros Españos, *Spanish Shipowners' Association.*

AVENSA Aerovias Venezolanas *Venezuela International State Airlines.*

Average Partial Loss *in insurance. Abbrev.* Av. or Avg.

Average Adjuster A person appointed by a shipowner to collect data, guarantees, etc. in relation to general average, and to calculate contributions due from the parties concerned to make good general average losses. The adjuster may also adjust claims on hull insurance policies on behalf of underwriters. *See* **Adjusters**.

Average Bond Any extraordinary expenditure or sacrifice of ship, freight or cargo, intentionally and reasonably incurred or made in time of peril, in order to secure the common or general safety of the property involved in a maritime adventure may lead to a claim for General Average. Those who incur such expenditure as well as the owners of any property sacrificed are entitled, subject to the conditions imposed by maritime law and contract, to a rateable contribution from the other parties interested. The apportionment is made by a Statement or Adjustment of General Average which is usually prepared by an Average Adjuster whose appointment rests primarily with the shipowners. The shipowners are entitled to refuse delivery of cargo until adequate security has been provided by the receivers or cargo owners for the payment which shall eventually be found due from

each. This security often takes the form of an Average Bond and, in addition, shipowners may demand a cash payment as a deposit in advance, or a Guarantee of the Corporation of Lloyd's, or other approved security.

It is usual for underwriters (although not legally liable to do so) to refund to their assured any amounts thus deposited if Lloyd's Form of General Average Deposit Receipt is used and the funds have been deposited in a special account in a Bank. On completion of the General Average Statement the underwriters then claim any sum by which the amount deposited may be found to exceed the actual liability. The Deposit Receipt is produced in proof of the claimant's right to a refund. It is essential that Deposit Receipts are carefully safeguarded as the issue of duplicates is undesirable. Lloyd's Form of General Average Deposit Receipt is a bearer document and does not require endorsement.

In cases where, owing to the nature of the average expenditure or sacrifice as, for example, where a ship is towed or assisted and a salvage award or settlement has yet to be arrived at, or where damage to a probably serious extent has been caused by flooding the hold to extinguish a fire, but the extent of the damage cannot yet be ascertained—it is sometimes necessary for the shipowners to collect as a deposit from the cargo interests who have benefited from the expenditure or sacrifice a sum which may eventually prove to have been over-estimated. In order to avoid such collections, and as a convenience generally, Lloyd's has a system, applicable to insured cargo, under which the Guarantee of the Corporation of Lloyd's may be given in a form acceptable to shipowners in lieu of a cash deposit.

The Guarantees of the Corporation of Lloyd's are issued in London by the General Average, Salvage and Collision Section at Lloyd's, which, together with the Recoveries against Carriers Section, also acts for insurers in the exercise of their subrogation rights, and in General Average and Salvage cases investigates their rights and liabilities, on receipt of Deposit Receipts, documents of title, or details of any undertaking or security which has been provided to shipowners or to salvors by cargo interests.

Lloyd's Form of General Average Deposit Receipt and Lloyd's Average Bond Forms (LAB 77) with a detachable Valuation Form can be obtained from the General Average Section of Lloyd's Underwriters' Claims and Recoveries Office.

Lloyd's Average Bond was completely revised in 1977 and the code title LAB 77 is incorporated to facilitate reference in telegraphic messages and computerised documents. The form produced at Lloyd's consists of two pages. The top page is the Average Bond part and the bottom page is for recording the value of the goods received. The particulars common to both pages coincide, so that by the use of a sheet of carbon only one typing is necessary. The completion and return of the Average Bond part of the form may result in the early release of the goods and when their condition has been ascertained the information asked for in the bottom page (the Valuation Form) should be provided together with a copy of the commercial invoice to assist the Average Adjuster to draw up his Statement or Adjustment promptly. *See also* **Average Statement** and **General Average Deposit**. *Abbrev.* a.b. *or* AB *or* A.B.

Average Clause A marine insurance clause whereby the insurers or underwriters are liable to certain perishable goods in Particular Average either in part (percentage) or in all. The Memorandum of Marine Insurance, *see* **Memorandum**, refers to the Average Clause where certain commodities are insured over a percentage of say 5% franchise unless of course the vessel is under a General Average act where the full amount is accepted for payment by the underwriters or insurers. *Ex.* of these commodities are: Flax, fish, hemp, hides, skins, sugar, tobacco and other perishables.

Average Cost The total expenses of production in a period of time divided by the unit number.

Average Damage Affecting Class This is generally a condition applied on the delivery of a vessel on charter whereunder the owner is committed to keeping his vessel within her society class, fully maintained, free of all recommendations, and free of average damage which could affect her class.

Average Demurrage The agreement in a charterparty for despatch money due to the charterers. *See* **Despatch Money** and **Despatch Days,** *q.v.*

Average Deposit *See* **General Deposit** *or* **Average Bond**.

Average Disbursements Expenditure incurred by the shipowner in connection with a general average act or an act of salvage. Such expenditure, when properly incurred, is recoverable from the G.A. or salvage fund created by the average adjuster, not from hull underwriters.

Average Due Date The average date arrived at from outstanding money due from various dates. *See* **Equated Time.**

Average Freight Rate Assessment First created by the London Tanker Brokers' Panel in 1954 as an independent assessment of international tanker shipping rates. Since 1959, it has been a monthly calculation of average rates for all chartered tankers in six size ranges. Product carriers and protected cabotage tonnage are excluded. The AFRA, taken as shipping cost, when added to the FAB cost of oil, can be used in negotiating import price.

Average Irrespective of Percentage Partial loss is covered irrespective of any franchise in the marine policy under this clause.

Average Propensity to Save That amount of income or consumption of food saved by an individual or by a whole community. *Abbrev.* APS *or* A.P.S.

Average Starter *Similar to* **Average Adjuster**, *q.v.*

Average Statement A general statement of accounts prepared by average adjusters showing the detailed contribution in relation to the values arrived at in the General Average Contribution, *q.v. See* **Average Bond**.

Average Unless General *Similar to* **Particular Average**, *q.v.*

Average Warranty An express warranty in the marine insurance policy in which the franchise is accounted for in the General Average, *q.v.*

Avg. Average.

AVGAS Aviation Gasoline, *q.v.*

A.V.I. Association of Veterinary Inspectors.

Aviation The act of air navigation; mechanical flying.

Aviation Gasoline *or* **Spirit** High octane spirit or gasoline suitable for piston-type aircraft engines. *Abbrev.* AVGAS for Aviation Gasoline.

AVIACO Aviacion y Comercio SA. *Spanish* commercial charter operator.

Aviator Air pilot; flying man.

Avionics Electronics in aviation such as computers, radar, navigation aids, automatic pilots etc.

AVIT *or* **A.V.I.T.** Audio Visual Instruction Technique.

AVL Automatic Vehicle Loading.

Av. Lub. Aviation Lubrication.

AVM Automatic Vehicle Monitoring.

A/Vm Unit of Electrical Conductivity.

AVMA American Veterinary Medical Association.

AVO Amplitude-Versus Profiling *geophysics*.

Avoidance The right of an underwriter to avoid a contract of marine insurance. This can occur in the event of a breach of good faith by the assured or by his broker or, in the case of a voyage policy, where the voyage does not commence within a reasonable time after acceptance of the risk by the Underwriter, *q.v.*

Avoir Avoirdupois *or* avoirdupois, *q.v. See* **Apothecaries Weight.**

Avoirdupois *or* **avoirdupois** *Middle English*—The English and American prescription weight system used in commerce. *Abbrev.* Av *or* avdp. *See* **Apothecaries Weight.**

à volonté *French*—At will. At pleasure.

AVP *or* **A.V.P.** Accidental Vanishing Point, *q.v.*

AVPA Automatic Visual Plotting Aid *computer.*

AVR Automatic Voltage Regulator.

AVZ *or* **A.V.Z.** Algemene Vereniging Van Zeevarenden *Dutch*—General Association of Seafarers.

AW *or* **A.W.** Actual Weight; Atomic Weight.

A/W Actual Weight; Airworthiness; Airworthy.

AWA American Waterways Association, *q.v.*; American World Airlines.

AWACS *or* **A.W.A.C.S.** Airborne Warning and Control System, *q.v. Similar to* **Airborne Early Warning Control System**, *q.v.*

Award The decision given by an arbitrator, to whom a matter in dispute has been referred. An arbitrator states only the effect of his decision, without reasons—thus differing from a judge, who usually states the grounds of his judgment. *See* **Arbitration.**

Awash Washed by the waves; touched by the water surface *nautical.*

AWB *or* **A.W.B.** Airwaybill *or* Air Waybill, *q.v.*; Australian Wheat Board.

AWEA American Wind Energy Association.

Aweigh When a vessel's anchor is loosened off from the seabed prior to her proceeding to sail.

AWES Association of West European Shipbuilders.

AWG *or* **A.W.G.** American Wire Gauge.

awl *or* **A.W.L.** Absent Without Leave; *Abbrev. also* AWOL *or* A.W.O.L.

A Weather A sail sheeted to windward or fastened by a rope in a position to catch the wind.

Awning Canvas or other material spread over deck areas on ships to provide shade from the sun.

AWO All Weather Operators *aviation;* American Waterways Operators. *See* **American Waterways Association.**

AWOL *or* **A.W.O.L.** Absent Without Leave. *Also abbrev.* AWL *or* A.W.L.

Awp *or* **AWP** *or* **A.W.P.** Area of Waterplane; Amusements with prizes; machines installed in casino vessels; *gaming law.*

AWPPA Arctic Waters Pollution Prevention Conference.

AWRIS Arab War Risk Insurance Syndicate.

AWS *or* **A.W.S.** Automatic Warning System.

AWSC American Waterways Shipyard Conference.

AWTS *or* **A.W.T.S.** All Working Time Saved.

AWTSBE *or* **A.W.T.S.B.E.** All Working Time Saved Discharging Only, *q.v.*

AWTSLO *or* **A.W.T.S.L.O.** All Working Time Saved Loading Only, *q.v.*

A.W.U. Atomic Weight Unit; Australian Workers' Union.

A/WUK Anywhere in the United Kingdom.

AWWF Australian Waterside Workers' Federation.

ax. Axion.

Axis Imaginary line passing through the centre of the earth from the North Pole to the South Pole.

AY Code flight for Finnair OY *Finland International State Airlines.*

AYH American Youth Hostels.

AYOP Amsterdam Ymuiden Offshore Port.

AZ Code flight for Alitalia, *Italian International State Airlines;* Arizona, *USA.*

az Azimuth, *q.v.*

Az. Arizona, *USA*

A.Z.C. Australian Zone Charges.

Azcon. Code name for the Chamber of Shipping Grain Charterparty 1910 for Azoff Berth Contracts as amended in 1931.

Azimuth The azimuth of a heavenly body is the angle at the zenith between the observer's meridian and the vertical circle through the heavenly body, measured east or west from 0° to 180°, and named 'N' or 'S' from elevated pole. *Abbrev.* az.

Azimuth Mirror An optical instrument used by navigators for taking azimuths, or bearings of heavenly bodies, or bearings of objects on land or sea, by day or night.

Azo. Azores.

B

b Bag; Bale; Bay; Berth; Blue Skin; Blue Sky *nautical*; Breadth; Bridge; Lightning *weather observation*.

B Atmospheric Pressure Condition *altitude*; Bolivar *Venezuelan currency*; Baccalaureus *Latin*—Bachelor; Bag; Bale Capacity in cubic metres *Lloyd's Register notation*; Bearing Angle *nautical*; Breadth; Bridge; British Thermal Unit; Soft Lead Pencil; Blood group.

B 26 Stands for a ship of about 27,000 deadweight tons capacity which can competitively trade on tramp bulk cargoes at an economic speed of about 15 knots.

B 30 Similar to B 26, *q.v.*, with the difference that the deadweight carrying capacity would refer to 30/35,000 tons.

B/— Bag; Bale; Balboa *Panamanian currency*.

Ba Barium *metallic element*.

B.A. Baccalaureus Artium *Latin*—Bachelor of Arts; Breathing Apparatus; British Airways; Buenos Aires *Argentina*; British Association (*standard thread formulated by the British Standards Institution*).

BAA or B.A.A. Bachelor of Applied Arts.

B.A.A. Bachelor of Arts and Architecture; Booking Agents Association; British Airports Authority; British Army Agents; British Archaeological Association; British Astronomical Association.

B.A.A. & A. British Association of Accountants and Auditors.

B.A.A.S. British Association for Advancement of Science.

B.A.B. British Airways Board.

Babbit-metal A soft alloy of tin, antimony and copper.

BABS Blind Approach Beam System *electronics*; British Airways Booking System.

Bac Baccalaureus *Latin*—Bachelor.

BAC Blood Alcohol Content.

B.A.C. British Aircraft Corporation; British Atlantic Committee; British Automatic Company.

BACA or B.A.C.A. Baltic Air Chartering Association.

B.A.C.A.H. British Association of Consultants in Agriculture and Horticulture.

Bacat or BACAT Barge Aboard Catamaran *or* Barge Aboard Carrying Catamaran, *q.v.*

B. Acc. Bachelor of Accountancy.

BACC British-American Chamber of Commerce.

BACCHUS British Aircraft Corporation Commercial Habitat Under the Sea.

B.A.C.I.E. British Association for Commercial and Industrial Education.

Back (1) Backwardation, *q.v.* (2) To put the engines astern. (3) To reverse the rowing action of an oar. (4) To haul the clew of a sail to windward *nautical*. (5) To lay an additional anchor to the first for extra security. (6) Wind changing direction anti-clockwise.

Back A Bill Guaranteeing the payment of a Bill of Exchange, *q.v.*, by endorsing it. *See* **Acceptance Supra-Protest** and **Accommodation Bill**.

Back Around The wind is said to 'back around' when it changes direction anti-clockwise, e.g., from north-east to north-west via north.

Back Door The purchase of Treasury Bills by the Bank of England at a special discount through the bank's brokers. Method by which the Bank of England injects cash into the money market.

Backed Note A note issued by a shipping company authorising the ship's tally clerk at the warehouse to deliver the specified goods therein indicated. *Similar to* **Delivery Order**, *q.v.*

Backfreight *or* **Back Freight** The owners of a ship are entitled to payment as freight for merchandise returned through the fault of either the Consignees, *q.v.*, or the Consignors, *q.v.*, such payment, which is over and above the normal freight, is called backfreight. If this is due to the fault of the carrier such as bad stowage, no Back Freight is paid. If freight is not recoverable within a reasonable time on merchandise payable at destination, the master or agents are entitled to dispose of the goods at the expense of the consignors and/or consignees by either selling the goods at the port of destination or forwarding them to an alternative port. If any funds remain they are to be returned to the owners of the goods.

Back-Haul Sea or air cargo returned to destination or any other place without affecting the original rate of freight.

Backhoe Dredger A Dredger, *q.v.*, fitted with a hydraulically operated arm and a single bucket.

Backing (1) Propaganda given for a country's gold and securities. That part of note issue is termed Fiduciary Issue, *q.v.* (2) An anti-clockwise direction of wind *similar to* **Back**, *q.v.*

Backing Wind Wind whose direction is changing anti-clockwise. *See* **Back Around**.

Backlog Pile of work assembled but not finished to time. Total of undelivered or unfilled orders. Goods remaining unsold beyond the expected date. It may be due to holidays, sickness, industrial unrest, or overloading of resources.

Back Pay Arrears of salary or wages.

Back Post *Similar to* **Rudder Post**, *q.v.*

Backshore That part of the lower shore covered by water during heavy storms.

Back Spring A mooring line running from the forepart or afterpart of a vessel to a point on the shore adjacent to the vessel's midship body, to prevent her moving ahead or astern.

Back-to-Back Credit The seller of a beneficiary in the transaction requesting the confirmation of the paying bank to open a credit in favour of a third party being provided with the merchandise. On presentation of the proper documents to the confirming or paying bank the credit amount is automatically released to the third party after bank expenses and commission have been charged.

Back Up Force A force of workers detailed to work on an extra shift outside normal hours (generally between 4 p.m. and 9 p.m.) on critical and essential work. The workmen engaged are paid an extra full normal pay for this period or they may be offered other conditions. *Abbrev.* BUF or B.U.F.

Backwardation The way a stock and share speculator acts in the buying of securities so as to say 'on account procedure' with a view to selling them at a good profit against cash. In this case his first buying acquisition does not involve any actual payment until he succeeds in selling at a good profit, this time against payment. This speculator is transacting the deal with other people's money on a profit basis. *Abbrev.* bk.

Backward Letter A form of supplementary contract where a special arrangement or an extra clause to the original contract is agreed upon by the parties concerned. A backward letter is identical to a Letter of Indemnity against withdrawal of the clean Bill of Lading, *q.v.*

Backwash Turbulent water thrown astern by oars, paddle-wheels or ships' propellers.

Back Water Reversing oar strokes to give astern movement to a boat.

Backwater Water retained by a dam; minor arm of a river.

BACO Barge Container, *q.v.*

BACS or B.A.C.S. Bankers Automated Clearing System; Bankers Automated Services Ltd.

B.A.C.U.K. British Airways Cargo United Kingdom.

B & D Bad and Doubtful. Referring to a debt unlikely to be recovered.

B.A.D. Base Air Depot; British Association of Dermatology.

B.A.D.A. British Antique Dealer's Association.

Bad Debts Debts or money due which are not recovered *accountancy.*

B.A.(Econ.) Bachelor of Arts *economics.*

B. Admin. Bachelor of Administration.

Bad Weather Occasionally used in a Charterparty, *q.v.*, in respect of loading and unloading operations. In this manner the charterer or consignee is freed from counting laytime for stoppage of work due to either heavy rains or rough seas rendering the operation impossible.

BAE *or* **B.A.E.** Bachelor of Aeronautical Engineering; Bachelor of Agricultural Engineering; Bachelor of Art Education; Bachelor of Arts in Education.

B.A.(Ed.) Bachelor of Arts (Education).

BAF *or* **B.A.F.** British Air Ferries; Bunker Adjustment Factor, *q.v.*

BAFF Bankers Association for Foreign Trade. A United States group of bankers engaged in international banking activities.

Baffle An expression generally used in aviation, when any object or structure stops the fuel.

B.A.F.M.A. British and Foreign Maritime Agencies.

Bag *or* **Bagged Cargo** Various kinds of commodities usually packed in sacks or in bags, such as sugar, cement, milk powder, onion, potatoes, grains, flour, coffee, etc. The bags or sacks are made of either gunny cloth, hard brown paper or are of a synthetic composition. Such cargo should first be tallied on board while loading and any defective bags should be rejected or clearly noted on the Bills of Lading, *q.v.* By so doing the carrier is cleared from eventual deficiencies/damage and subsequent claims.

Baggage/s Suitable container(s) in a form of a handy and manageable case wherein the personal effects of travellers are packed and carried mostly by hand and on small carriers.

Baggage Handling The carrying of passengers' baggage by porters at air, rail, and sea terminals.

Baggage Master A person who may be a Purser, *q.v.*, of a ship, generally of a passenger ship, or one who works in the section or department which cares for the passengers' baggage stowage and security.

Baggage Reclaim Designated area where passengers' baggage can be collected and withdrawn.

Baggage Room A room, normally in the customs area, where the contents of passengers' baggage is inspected and cleared by the customs officers. The baggage room can also serve as a place for storing baggage on a passenger ship.

Baggage Tag Identification label attached to baggage to ensure correct direction and assist in tracing it if lost.

Baggage Trolley Wheeled receptacle in which passengers' baggage is carried for ease of transport and comfort.

Baggyrinkle Rope sennit (braided cordage) sheathing a stay to protect it from chafing another rope or sail.

B.Agr. Bachelor of Agriculture.

BAH British Airways Helicopters.

Bah. Bahamas.

Bah. Is. Bahamas Island.

BAHREP British Association of Hotel Reservation Representatives.

Baht Unit of currency used in Thailand.

B.A.I. Baccalaureus Artis Ingeniariae *Latin*—Bachelor of Engineering; Book Association of Ireland.

B.A.I.E. British Association of Industrial Editors.

Bail To remove unwanted water from a boat.

Bailee The holder of good trust who must observe the instructions for which the delivery to him is made. An agent, trustee, warehousekeeper, etc.

Bailee Clause Institute Cargo Clause (W.A.). The insurer's responsibility ends up to the extent of the carrier's obligation. The former is not liable to loss or damages if the subject-matter is in the care of another person or Bailee, *q.v.*

Bailer A scoop or other utensil used to Bail, *q.v.*, out water from a boat. This is one of the items included in a ship's lifeboat kit. Also spelt 'baler'.

Bail Out To eject from the seat of an aircraft.

B.A.(J) Bachelor of Arts in Journalism.

Baker's Dozen Thirteen.

Bal. Balance; Baluchistan.

Balanced Budget Financial forecast or plan which on inspection will show that projected disbursements will not exceed receipts.

Balance of Payment *or* **Balance of Trade** Statements of a country showing estimates of the following: (1) The balance difference in the value of imports and exports. (2) The cost of the services that other countries render to the country concerned. (3) Drawings and payments of loans made by a country regarding overseas governments etc.

Balance Sheet A book-keeping term referring to the prepared statement of accounts from where one can deduce the financial condition of a company resulting: (1) In the first year's trading. (2) The comparison of the past and present. (3) The financial standing. The balance sheet is usually presented to the directors for their attention by the auditors or accountants of the company for the necessary confirmation of the contents.

Balboa Unit of currency used in Panama.

BALDRIA Baltic and Alexandria, *referring to the timber trade.*

Bale Soft wrapped package of a round, oval, square, rectangular or irregular shape.

Bale cubic capacity An indication of space available in a ship's cargo compartments for other than bulk cargoes—e.g. for bags or scrap metal. Measured to the inside of cargo battens, to the top of the ceiling, (*q.v.*), or tank top, (*q.v.*), and to the underside of the underdeck beams, including spaces in the hatchways.

Baling Form of packaging of merchandise for transport and shipment. *See* **Bale**.

Balk A piece of timber from 4 to 10 inches square section. *Also* Baulk.

BALKAN Bulgaria International State Airlines.

Ballast Heavy substances loaded by a vessel to improve Stability, *q.v.*, trimming, sea-keeping and to increase the immersion at the propeller. Sea water ballast is commonly loaded in most vessels in the Ballast Tanks, *q.v.*, positioned in compartments right at the bottom and in some cases on the sides, called wing tanks. Solid ballast may be used as an alternative in suitable vessels, but much less commonly than in former days. Some vessels are specially adapted to carry liquids like edible or other oils in place of water. This sort of ballast no doubt is useful to the owners because they can earn additional Freight, *q.v.*

Ballast Bonus A lump sum payment to cover hire and bunkers and voyage costs (*e.g.* canal tolls) for positioning a vessel at a delivery place for a time charter.

Ballast Keel A heavy keel fitted to sailing vessels to lower the centre of gravity and improve stability, *q.v.*

Ballast Tanks Compartments at the bottom of a ship or, in unusual cases (on bulk ore/oil carriers for example) on the sides which are filled with liquids for stability and to make the ship seaworthy. *See* **Ballast** and also **Fore** *or* **Aft Peak Tank**.

Ballot A small bale weighing from 80 up to 125 lbs or up to 58 kilos.

Ballot Paper Slip used for casting a vote or ballot in an election.

Balpa *or* **BALPA** *or* **B.A.L.P.A.** British Airline Pilots' Association.

Baltic and International Maritime Council, The Usually referred to as BIMCO. Founded in Copenhagen in 1905 with 102 members owning 1,056 vessels as The Baltic and White Sea Conference. In 1927 it became The Baltic and International Conference and in 1985 it took its present name. Private but non-profit making, it is the largest international forum in the shipping industry with a membership of 2,600 companies in 108 countries (1994). There are 950 owner-members, 1,575 broker-members, 60 club members including P&I clubs, defence associations and others, and associate members from the banks, insurance companies, classification societies and maritime lawyers. BIMCO writes and speaks up for the interests of shipowners and others in the industry in national and international circles and has consultative status with the International Maritime Organisation (IMO). It seeks to ensure that regulations are sensible and cost effective. It is active in providing members with advice on contractual commitments, enabling them to reduce exposure to costly omissions or mistakes, e.g. by warning against risky charterparty clauses. It assists in the resolution of disbursement problems. BIMCO has developed hundreds of documents—charterparties, bills of lading, memoranda of agreement, etc. which are universally adopted as standard, and which constitute a first defence against disputes arising from fixtures. It maintains a register of unreliable companies. It gives advice on the menaces of drug trafficking, piracy and stowaways, and on security matters. BIMCO issues hundreds of publications to assist owners with their operations and to improve their financial results. Its subsidiary BIMCO Informatique Association (BIAS) provides an on-line 24-hour information service on port conditions, ice conditions, port working hours, taxes, bunker prices and much other data. It sells software support for such things as voyage estimates and laytime calculations as well as databases. In 1989 it established a global satellite communications network, BIMCOM, for international transport.

Baltic Exchange, The Known also as The Baltic for short, it is the only international shipping exchange in the world and a major earner of foreign currency for Britain. Its origins can be traced to the 18th century, when shipowners and merchants met in London coffee houses. Foremost among these were the Jerusalem Coffee House and the Virginia and Maryland Coffee House (known from 1744 as the Virginia and Baltic as the cargoes dealt with came from the American colonies or from the countries on the Baltic seaboard). Today membership of the Baltic includes nearly 600 companies on whose behalf about 1,500 individual men and women are entitled to trade on the 'Floor' of the Exchange. Matching cargoes to available ships worldwide at the best price is the main activity of the brokers at the Baltic Exchange. It has been estimated that approximately half of all the world's open market bulk cargo movement is at some stage handled by members of the Baltic. Other important activities include sale and purchase of ships and aircraft, trading in grain, vegetable oils and oil seeds and the chartering of aircraft. An index of freight rates (Baltic Freight Index) is published daily by the Baltic Exchange. This index is traded on a dry bulk cargo freight futures market called BIFFEX, offering 'hedging' against freight rate movements. The Corn Exchange holds its weekly market session at the Baltic every Monday.

Baltic Freight Index An index of international dry bulk 'spot market' voyages, produced by the Baltic International Freight Futures Market (BIFFEX), *q.v.*, to protect shipowners, operators and charterers against the risks of volatile freight rates by means of standardised contracts. The unit of trading is an index point valued at US$10. *Abbrev.* BFI.

Baltic International Freight Futures Market, The *See* **BIFFEX**.

BALTIME 1939 Code name for Baltic & International Maritime Council Charterparty. It serves as a Time charterparty. Amended 1 March 1939 and adopted by the Documentary Committee.

BALTWOOD Chamber of Shipping Baltic Charterparty. Carriage of wood from Scandinavian ports to UK and Continent.

BAM *or* **B.A.M.** Bulk Air Mail, *q.v.*

BAMBI Ballistic Missile Boost Intercept.

Bank An establishment from which transactions of various types of finance take place, from money deposits to overseas transfer of moneys on behalf of clients. *See* **Banking**.

Bank Acceptance An accepted Bill of Exchange, *q.v.*, by the bank.

Bank Accounts Fixed or current deposit, loan, and other accounts.

Bank Advance A banking system in relation to the advancing or financing or lending to clients against some sort of security. When such money is advanced the terms of interest to be charged are agreed upon. Normally bank interest is paid in full against the money loaned.

Bank Americard Credit card issued by the Bank of America.

Bankasi *Turkish* Bank.

Bank Bills Bills of Exchange, *q.v.*, are produced and/or accepted by a bank. The discount rates of such bills are in most cases quoted less than those of other trade bills.

Bank Cash Card Card issued to a bank customer which he can use to obtain cash by inserting it in an Automatic Teller Machine, *q.v.*, or Cash Dispenser.

Bank Charges Charges submitted by a bank for work done.

Bank Credit Information A banking system servicing its clients by providing full financial details in confidence about other prospective foreign or local persons or companies as a guideline to creditworthiness and reputation.

Bank Deposit Account A banking system whereby the customer deposits money and is given a percentage interest if it is deposited for a minimum of one year. If money is withdrawn before the period the interest is paid pro rata. There are two types of deposit accounts: Fixed Deposit where money deposited cannot be withdrawn before a period of one year and an encouraging interest rate is given, and Savings Deposit Account where money can be withdrawn before one year and the interest per year is less than that of the Fixed Deposit Account.

Bank Disclosure A judicial order to a bank to supply information regarding a customer's accounts or interests. This is generally used in taxation cases.

Bank Draft A chit in the form of a cheque made out and signed by a bank on behalf of its clients to pay a certain amount of money to a named person or company living locally or overseas on presentation to the representative bank at the receiving place. It has become a practice to send money abroad by means of

Bank Drafts instead of by credit transfers, possibly because less bank charges are involved and the certainty that the actual date the draft has been posted to the receiver will be clearly shown on the draft itself. *Similar to* **Banker's Draft**, *q.v.*

Bankers Proprietors or partners of a private bank; governors, directors etc. of a joint-stock bank. Those personnel working in a bank as cashiers, tellers, etc., are usually referred to as bank clerks.

Banker's Card Alternatively known as a Cheque Card. This is given by the banks to their customers. *See* **BarclayCard**, **Credit Card**.

Banker's Draft A Bill of Exchange, *q.v.*, corresponding to a cheque drawn out by a bank (on its name) at the request of the client in lieu of a personal cheque generally to be presented to overseas creditors who will in turn be able to cash it at all local issuing banks. Very similar to Bank Draft, *q.v. Abbrev.* BD *or* B.D. *or* b/d *or* B/D.

Bank for International Settlements An international bank organisation established after the First World War mainly to act as an agent for the payment of reparations by the Germans. The bank really started operating fully in May 1930, backed by France, Germany, Great Britain, Italy, Japan and the United States. Its head office is in Basle, Switzerland. The objects of the bank are to: (1) Promote and facilitate operations between all the central banks. (2) Facilitate financial matters whenever necessary. (3) Act as trustee or agent in international settlements within their sphere.

Banking The act of commercial work routeing, facilitating in the distribution and exchange of commodities by collecting deposits and guaranteeing payments when requested, acting as the source of loans to clients and for payments and accounts. *Abbrev.* Bnkg. *See* **Bank**.

Bank Money Money deposited in a bank to be used in a current account and to be withdrawn by means of cheques. *See* **Cheque** and **Cheque Book**, *also* **Current Account**.

Bank Note A bank promissory note expressed to be payable on demand. It may be issued again after it has been paid, as opposed to an ordinary promissory note.

Bank Overdraft The money due to the bank by its client when the amount of cheque(s) given out by the client fails to balance the actual credit deposited in the Current Account, *q.v.*, or Bank Money.

Bank Rate The minimum rate is the percentage for money lent by the bank to its clients. This evaluation varies according to the standing of the bank and the currency of the state. Commercial banks often give encouragement and favourable rates to attract more clients. It is generally the case that bankers have mutual agreements in regard to the commercial bank rate quotations.

Bankrupt Officially insolvent. *See* **Act of Bankruptcy**.

Bank Settlement Plan A combined system and passenger procedure in the issuing of air tickets, control, booking remittances, selling, reportings, etc. This of course includes the unified system of Air Contracting Agents.

Banks, Chief of The Originally National Westminster, Lloyds. Barclays International, National Provincial, and Midland. They are now merged and named Barclays Bank Group, National Westminster Bank, Lloyds Bank and Midland Bank.

b.a.p. Billets à payer *French*—Bills Payable, *q.v.*

BAPA *or* **B.A.P.A.** British Airline Pilots' Association.

BAPCO Bahrain Petroleum Company.

B.A. (P.E.) Bachelor of Arts in Physical Education.

Bar (1) An obstruction under the sea made up of rocks, mud or sand which would prevent certain deep draught (draft) vessels from entering ports. *See* **Bar Draft** *or* **Bar Draught**. (2) A unit or atmospheric pressure calculated at one million dynes per square centimetre.

BAR *or* **B.A.R.** Board of Airline Representatives; British Association of Removers; Blade Area Ratio.

bar *or* **Bar** Barometer; Barrel; Barque.

b.a.r. Billets à Recevoir *French*—Bills Receivable, *q.v.*

Barb Barbados.

Bar Bound A ship is said to be bar bound when she cannot cross the bar due to shallow water caused by the bar of obstructing material while approaching the loading/discharging port. *See* **Bar Draft** *or* **Bar Draught**.

Bar Buoy A buoy located at a point near the mouth of a river or a port showing the bar area.

BARCAP Barrier Combat Air Patrol *USA*.

B. Arch. Bachelor of Architecture.

B. Arch. & T. Bachelor of Architecture and Town Planning.

Barclaycard Small plastic card issued by Barclays Bank International to its clients. On presenting it one can obtain money, services or supplies without paying on the spot. It is a form of credit card. Any person rendering the service will register the number, name of the holder and other details shown thereon. The credit has a limit of the amount of money allowed to be honoured. Other similar credit cards include Diners Club, Europa Card, American Express etc. *See* **Credit Card** *and* **Banker's Card**.

Bar Codes A combination of short black stripes or bars printed in a block on packaging, labels and documents which represent identifying characters and cyphers. These codes are machine readable to permit automatic sorting, pricing, inventory and other processes. Originating in the USA in 1974, they are used in many countries, notably in retailing.

Bar Draft *or* **Bar Draught** The draught (or draft) to which a vessel is restricted by the presence of a Bar, *q.v.*, at a port entrance. There are some ports, rivers, lakes, etc. where bars do not allow ships to load a full cargo. The vessels are then compelled to finish the full load either at a nearby port or from lighters outside the bar draft (draught) port, weather permitting. *Abbrev.* b/d *or* B/D. *See* **Bar Bound** *also* **Bar**.

BARDS Baseland Radar Docking System, *q.v.*

Bareboat Charter *or* **Demise Charter** The charterers hire or charter the ship for a long period, appoint the master and crew, and pay all running expenses. The hire is calculated at so much per ton deadweight per calendar month. The charter hire may also be on a Lump Sum, *q.v.*, daily basis. During this hire period the charterers are to maintain the ship in drydocking, painting, repairing and all expenses incidental thereto. In certain instances the insurance of the ship is for the account of the owners though not necessarily so. During the Second World War all ships of a country were requisitioned on a bareboat charter basis.

BARECON 89 The BIMCO, *q.v.* Standard Bareboat Charter.

Bare Ice Ice without snow cover.

Bargain This word can have various meanings, i.e. (1) A purchase which, when compared with the normal price, is found to be relatively cheap. (2) Any commercial transaction in stocks and shares. (3) An arrangement between two parties to negotiate on the sale or purchase of an article or other business. (4) Any object in the act of purchase or sale.

Barge A flat-bottomed boat, with or without sails, used on rivers or canals or close in-shore.

Barge Aboard Catamaran *or* **Barge Aboard Carrying Catamaran** A way of loading cargo into large barges and then in turn loading the barges into a ship. The barge is loaded and unloaded through the prongs on the stern of the ship's own giant travelling gantry crane. This ship can carry about 85 barges of 400 tons cargo capacity. The Bacat or BACAT, as it is abbreviated, does not need special terminals and she normally anchors at the Fairway, *q.v.*, where the barges are unloaded and towed away by waiting local tugs to be shifted on to the unloading quays of the port. The empty barges can then be reloaded on to the Bacat for her return journey. Appreciable time and port expenses are saved through this operation. A similar system is used in Artubar, B.C.S., Capricorn, Catug, EBCS, Lash, (all *q.v.*), Velment etc. *See* **Barge Carrying Vessel**.

Barge Carrier Similar to Barge Carrying Vessel, *q.v.*

Barge Carrying Vessel A general term for vessels which carry barges together with cargo. The barges are unloaded inside or outside the harbour, taken over by local tugs and towed into the port alongside the unloading quays. *Abbrev.* BCV *or* B.C.V. These barge carrying vessels are also known as Artubar, Bacat, B.C.S., Capricorn, Catug, EBCS, Lash, Velment etc. *See* **Barge Aboard Catamaran**.

Barge Container A barge for carrying containers and designed for short sea crossings and navigable rivers. *Abbrev.* BACO.

Bargee or Barge Man A crew member of a barge or a seaman who is engaged to sail on a barge.

Barge Forwarding Unloading containers on to barges for onward despatch to the landing quay.

Bar Hours Time allotted for bars on board vessels and aircraft to be open to serve officers, crew and passengers.

Bar of Cyclone Sometimes referred to as a 'hurricane cloud' which is a heavy rain cloud in a storm and which appears to be touching the horizon.

Barograph A self-recording barometer with a rotating cylinder covered by a chart with a pen attached to a pointer to register the atmospheric pressure.

Barometer An instrument for measuring the pressure of the atmosphere. Aneroid and mercury types are in use at sea. The barometer is an important aid to weather forecasting. An old term is Weather Glass.

Barometric Altimeter An instrument employed to determine height. It is mostly used by aircraft.

Bar Port A port which has bar draughts or drafts. *See* **Bar Bound**, **Bar Draft** *or* **Bar Draught**.

Barr Barrister, *q.v.*

Barratry Fraudulent practices on behalf of the master or crew of a vessel against the interests of the shipowners.

Barrel (1) Cask or receptacle measuring 34.97261 Imperial or 42 US gallons. (2) Barrel of butter containing 224 lbs or 101.6 kgs. (3) The generally accepted measurement in the oil business, one barrel of oil being 35 Imperial gallons or 42 US gallons while one ton of oil is approximately 7.3 barrels. *Abbrev.* bbl.

Barrel Hooks *or* **Can Hooks** Two hooks attached to each end of a moderate length of chain. A ring is attached to the centre of the chain to which the hook of the crane lift is secured. The two hooks are fitted to both ends of the Chines of the barrels or casks which tighten when lifting takes place. This method provides a quick and safe system of handling. Also called **Dog Hooks**, *q.v.*

Barrier Reef A reef separated from the coastline by a marsh or shallow pond or lagoon.

Barrels per Day The amount of oil obtained or extracted from an oil well in a day. *Abbrev.* b.p.d. *or* b/d.

Barrister A counsellor or advocate learned in the law, admitted to plead at the bar, and there to take upon himself (or herself) the protection and defence of clients. He is termed *jurisconsultus* and *licentiatus in jure*. A barrister is a member of one of the four Inns of Court who has been called to the Bar by his Inn. That makes him a barrister, and gives him, along with other barristers, the exclusive right of audience in the House of Lords sitting as a tribunal of appeal, the Privy Council and the Supreme Court (except at sittings of the High Court in bankruptcy and at matters heard in chambers).

Bart. Baronet.

Bart Bay Area Rapid Transit. The underground railway system beneath San Francisco Bay.

BART Bacterial Activity and Reaction Test.

Barter The exchange of one commodity for another instead of a money transaction between one country and another, or one party and another.

Barouk or **BAROUK** Board of Airline Representatives United Kingdom.

BAS or **B.A.S.** Bachelor of Agricultural Science; Bachelor of Applied Science; British Antarctic Survey; British Air Services.

BASATA British and South Asian Trade Association.

B.A. Sc. Bachelor of Agricultural Science; Bachelor of Applied Science.

BASC or **B.A.S.C.** Business Applications and Standards Committee.

Base Container home port.

BASE Barge Accommodation Services.

Base Land Radar Docking System A computerised instrument used within about half-a-mile radius of a mooring point which activates equipment designed to aid berthing operations. *Abbrev.* BARDS.

BASEEFA British Approved Standard for Electrical Equipment in Flammable Atmospheres.

Base Line A line drawn on a chart between two points of land from which a nation's territorial limits are measured seawards. The outer parallel limit becomes the Territorial Waters, *q.v.*, and a further extension of about 12 miles or more (since a large number of states vary their claim) is considered to be International Waters or High Seas, *q.v. Abbrev.* Bas.L. *or* B.L.

Base Metal Metal other than a precious metal. It can be altered on exposure to air, moisture or heat.

Base Plate *Similar to* **Bed Plate**, *q.v.*

Base-Point System The American system of quoting and exercising uniform freight charges to customers irrespective of the different ports of destination and the distances covered for each and every port.

Base Rate The rate of interest on loans and deposits of a Commercial Bank which is the basis from which it establishes the actual rates offered to companies or individuals.

Base Stock The relationship between the value of falling money and the valuation of current assets which may present commercial problems. Base stock is roughly calculated from the selling price of an item less the anticipated profit on it and the selling charges.

BASIC Beginners' All Purpose Symbolic Instruction Code *electronics*.

Basic Rate Minimum wage for each hour of a standard working day or week, ignoring overtime and allowances.

Basic Rate of Income Tax The rate applied to all taxpayers. Higher rates may be applied according to actual income.

Basis Price Net Price.

Basket Trade The practice of trading houses and stockbrokers, in currency and futures markets, of selling one item while negotiating the purchase of another in the same sector.

Bas L. Base Line, *q.v.*

BASS or **B.A.S.S.** British Association of Ship Suppliers who form part of the International Association of Ships Suppliers.

B.A.(S.S.) Bachelor of Arts in Social Science.

Bat Batavia; Batten(s), *q.v.*

Bate In the frozen meat trade a reduction of about 1 lb for every quarter is allowed to compensate for the extra weight of the packages. This is probably a contraction from Rebate.

BATF Bureau of Alcohol, Tobacco and Firearms *USA*.

Bathythermograph An instrument to obtain the temperature at different depths of sea water when a ship is either at anchor or navigating.

B.A.(Theol) Bachelor of Arts in Theology.

BATO or **B.A.T.O.** Balloon Assisted Take Off *aviation*.

bat/s Batten(s). *See* **Battens**.

Batten Down A nautical term meaning to close and secure all hatches and other openings before departure of a ship to prevent any possible water leakage into the vessel.

Batten Ends *See* **Battens**.

Battens (1) Long, narrow strips of wood or metal employed on ships for various purposes, one of which is to secure hatch tarpaulins against the hatch coamings. (2) Lengths of sawn timber from 5 to 7 inches wide and from 2 to 4 inches thick.

B.A.U. British Association Unit.

B.A.U.A. Business Aircraft Users' Association.

Bauxite Ore from which aluminium is obtained by electrolysis.

Baulk A rather heavy beam of either hewn or sawn timber measuring 4 inches $\times$ $4\frac{1}{2}$ inches or 10.16 centimetres $\times$ 11.43 centimetres in cross section or over. *Also* Balk.

Bav. Bavaria, Bavarian.

b.a.v. Bon à vue *French*—Good at sight.

Baxter Bolt A hook which is attached to the deck to enable an aircraft to be fastened or secured.

B.A.Y.S. British Association of Young Scientists.

bb Ballast Bonus *chartering*.

BB Bulbous Bow.

BB *or* **b.b.** Below Bridges, *q.v.* Both to Blame. *See* **Both to Blame Clause**; Branch Bill; Bulbous Bow *Lloyd's Register notation*; United Kingdom Customs certificate for inward clearance of vessels coming from foreign ports; Ballast Bonus: Ball Bearing; Bearer Bonds; Beer Barrel; Bill Book.

b & b Bed and Breakfast. Hotel board rates inclusive of breakfast.

B&B Brandy and Benedictine; Bed and Breakfast.

B.B.A. Bachelor of Business Administration; British Bankers Association.

BBB *or* **B.B.B.** Bed, Breakfast and Bath (Hotel board rates inclusive of Bed, Breakfast and Bath); Better Business Bureau (A USA organisation investigating the integrity of any business concern); Before Breaking Bulk, *q.v.*

BBC British Broadcasting Corporation.

BB Clause *or* **B to B** *or* **B to Blame Clause** Both to Blame Clause. *See* **Both to Blame Collision Clause**.

B.Bd. Bulletin Board.

bbl Barrel, *q.v.*

B. Bisc. Bay of Biscay.

BBP Built-up Non-ferrous Propeller.

B.B.S. Bachelor of Business Science; Belgische Beroepsvereniging der Scheepsbevooraders, Belgian Ship Suppliers Association. *See* **B.S.A.** and **U.P.B.A.N.**

Bbt *or* **BBT** Basal Body Temperature.

B. Build Bachelor of Building.

BBW Birmingham Wire Gauge.

bc A meteorological sign for a partly cloudy sky.

BC British Channel; British Council.

B.C. Bachelor of Surgery; Bachelor of Commerce; Before Christ; Bills of Collection; Bristol Channel; British Corporation Classification Society; British Columbia; Board of Control; Bankruptcy Court.

B.C.A. Bachelor of Commerce and Administration; British Cement Association; British Columbia Aviation; Bureau of Current Affairs.

B.C.A.C. British Conference on Automation and Computation.

B. Cal. British Caledonian.

B.C.A.R. British Civil Airworthiness Requirements; British Council for Aid to Refugees.

BCBH *or* **B.C.B.H.** Boiler Casing Bulkhead.

BCC Brazilian Chamber of Commerce; Bulk Carrier Crane.

BCC Bulk Cargo Carrier.

BCCB British Chamber of Commerce Bangkok.

BCCI Bank of Credit and Commerce International *Moslem*.

BCCS British Columbia Chamber of Shipping.

B.C.C.U.S. British Commonwealth Chamber of Commerce in United States.

BCD Bad Conduct Discharge *USA*; Binary Coded Decimal *computer*.

B.C.E. Bachelor of Chemical Engineering; Bachelor of Civil Engineering.

B.C.E.C.C. British and Central European Chamber of Commerce.

BCF Bureau of Commercial Fisheries *USA*.

Bch. Branch; Bunch.

B. Ch. Baccalaureus Chirurgiae *Latin*—Bachelor of Surgery.

B. Ch. *or* **B/Ch.** British Channel.

BCHC British Crane Hire Corporation.

BCH Code Code for the construction and equipment of ships carrying dangerous chemicals in bulk *IMO.*

B. Ch. D. Baccalaureus Chirurgiae Dentalis *Latin*—Bachelor of Dental Surgery.

B. Ch. E. Bachelor of Chemical Engineering.

B.C.I. Banca Commerciale Italiana; *Italian Commercial Bank.*

BCIC Birmingham Chamber of Industry and Commerce.

BCL *or* **B.C.L.** Bachelor of Canon Law; Bachelor of Civil Law; Business Corporation Law *USA*.

Bcn *or* **BCN** Beacon.

B.C.N. British Commonwealth of Nations.

B. Com. Bachelor of Commerce.

B. Com. Sc. Bachelor of Commercial Science.

BCMEA British Columbia Maritime Employers Association.

BCP *or* **B.C.P.** Bachelor of City Planning; Book of Common Prayer; Built-up Cast-Iron Propeller.

BCPIT *or* **B.C.P.I.T.** British Council for the Promotion of International Trade.

BCR Bow Crossing Range *navigation.*

BCS Barge Carrier System or Barge Carrying Ship. *See* **Barge Carrying Vessel**; British Calibration Service.

B.C.S. Bachelor of Chemical Science.

BCSP Built-up Cast Steel Propeller.

B.C.T. Belfast Chamber of Trade.

B.C.T.A. British Canadian Trade Association.

B.C.U. British Commonwealth Union.

BCV Barge Carrying Vessel, *q.v.*

BCWOA British Columbia Wharf Operators' Association.

bd *or* **Bd** Board; Bond; Bound; Bundle.

b/d *or* **B/D** Bank Draft, *q.v.*; Bar Draft (Draught), *q.v.*; Barrels per Day *oil*; Brought Down *book-keeping.*

BD *or* **B.D.** Bachelor of Divinity; Bahrain Dinar Unit of Currency; Bank Draft, *q.v.*; Brought Down *book-keeping.*

BDA Battle Damage Assessment.

B & D 1 Births and deaths. Form 1.

b.d.c. Bottom Dead Centre.

B.D.E. British Department of Energy.

B. Dft. Bank Draft, *q.v.*

bd. ft. Boardfoot. *See* **Board Measure**.

BDI *or* **b.d.i.** *or* **B.D.I.** Bearing Deviation Indicator *nautical*; Both Days Inclusive, *q.v.*

B. dk. British Deck.

B'down Breakdown (referring to machinery).

Bdry Boundary.

BDR *or* **B.D.R.** Bearer Depository Account.

B.D.S. Bachelor of Dental Surgery; Brokers Daily Statement.

Bds Boards.

BDSA *or* **B.D.S.A.** Business and Defence Service Administration.

BDST *or* **B.D.S.T.** British Double Summer Time.

BE Bevelled Edge; Bill of Exchange, *q.v.*

Bé Baumé *French*—Hydrometer named after its inventor.

B.E. Bachelor of Education; Bachelor of Engineering; Bill of Exchange.

B/E Bill of Exchange, *q.v.*; Bill of Entry, *q.v.*

b & e Beginning and Ending.

BEA *or* **B.E.A.** British East Africa; British European Airways now British Airways. *Abbrev.* BA or B.A.

Beach (1) A marine expression meaning to strand. In order to avoid the ship foundering, the master attempts to steer her towards a beach or shallow water. By so doing he will enable her to be floated at a later date. (2) To beach a vessel for the purpose of underwater cleaning and for painting during low and high tides. (3) The land nearest to the sea or lake, etc.

Beachcomber (1) A long rolling wave. (2) An idle person who is seen scrounging on the wharves of sea ports. (2) Slang expression for a hoisted flag in port to show that crew members are at their mess rooms and off duty. This is generally an American term.

Beach Gear Heavy anchor equipment used to secure a stranded ship and stop her from moving under the influence of currents and waves. This gear is also commonly in use in conjunction with the salvage operation of a beached or stranded ship. *See* **Beach**.

Beacon A seamark, aid to navigation.

Beak Metal point or ram, often of brass, projecting from the bows of ancient war galleys.

Beam (1) The registered breadth of a ship, measured from the outside of the hull amidships where the breadth of the tonnage deck is widest. (2) Greatest width of a vessel. (3) Stout length of wood, iron or steel. Structural member extending across a vessel to strengthen the hull and support the deck. (4) Continuous radio signal serving as a directional guide to navigators. (5) Fixed or revolving path of light emitted from a lighthouse.

BEAMA British Electrical and Allied Manufacturers' Association.

Beam Bearing The bearing of an object taken at the moment when the observer's vessel has it abeam, i.e., at right angles to the fore-and-aft line of the ship.

Beam Ends A vessel is said to be 'on her beam ends' when she is lying over so much that her deck beams are almost vertical; transverse girders running between opposite frames at the sides of a ship, supporting the sides from collapse, and bearing the weight of the deck above.

Bean Rag Slang expression for a hoisted flag in port to show crew members are at their mess rooms and off duty. This is generally an American expression.

Bear An investor who has sold a security in the hope of buying it back at a lower price.

Bear Away Nautical term to turn away from the wind by putting up the helm.

Bearer This refers to the presenter of a cheque. When a cheque is paid to bearer and not crossed the banker may pay it to anyone who presents it, unless its negotiability is restrained.

Bearer Bonds UK Government securities which are handled by the banks. Since the banks are the holders of these Bearer Bonds they automatically become legal holders or owners. As such they can be transferred from one person to another without registration.

Bearer Securities A title in stocks and shares. Debentures and the like which are made out to bearer have no names on them. Therefore they can be transferred from hand to hand without any formalities.

Bearing The direction in which an object lies from an observer, usually defined by angular measurement between the observer's meridian and a line from the observer to the object. May be 'true', 'relative' or 'compass' bearing. The angle is measured in degrees from 0 up to 360 in a clockwise direction. The bearings are stated in terms of North, South, East and West. *Abbrev.* Bg or Brng.

Bear Market A market in which Bears, *q.v.* may prosper; a falling market.

Bear up To put the helm to windward, and so turn to leeward.

Beating Nautical term for sailing close-hauled to windward.

Beaudaux System The method of encouraging workers by giving extra bonuses in relation to the output of work undertaken in favour of the employer.

Becket A rope handle of a container carrying water; a small rope ring; a small eye.

B.Ec. *or* **B. Econ.** Bachelor of Economics.

B. Ed. Bachelor of Education.

Bed and Breakfast A Stock Exchange term referring to selling shares and repurchasing them the next day to realise a capital gain or to establish a new market value for tax purposes.

Bed Plate *or* **Base Plate** *or* **Sale Plate** The foundation platform to which the engines of vehicles, factories, ships, etc., are fastened in a permanent position and to prevent them from shifting while they are working.

Beaufort Scale A standard scale invented by Sir Francis Beaufort during the 18th century giving the condition and force of the wind in numerical scale ranging from 0 indicating calm weather up to force 12 being a hurricane.

Bed of the Sea Land submerged under the water. The land lying at the bottom of the sea which is called 'res nullius' namely 'a thing which has no owner', as opposed to 'res communes' that is 'something which is owned in common by all.' *Also called* Sea Bed, *q.v.*

Beds Bedfordshire, *UK.*

B.E.E. Bachelor of Electrical Engineering.

Beef Boat American slang word for cargo or supply boat.

Beer Muster American slang for Beer Party Ashore.

bef. Before.

Before Fore or towards the bow of a ship.

Before Breaking Bulk Before a ship starts discharging or before opening the cargo hatches prior to discharging. This term generally refers to the payment of freight or charter freight which has to be paid to owners before discharging takes place. *Abbrev.* BBB *or* B.B.B.

Before the Mast An old expression signifying the accommodation of seamen in the forecastle of a ship. Nowadays seamen are accommodated amidships or aft so this expression does not really apply as far as accommodation is concerned. 'Three Years Before The Mast' indicated the period of three years continuous service which a seaman served before assuming the rating of an AB or A.B., namely Able Bodied Seaman, *q.v.*

Befrachtung *German*—Chartering.

Beggar my Neighbour Policy A reciprocal policy. Generally referred to reciprocal concessions enjoyed in mutual trade and general agreements among nations or companies. Colloquially it could be assimilated to the phrase 'you scratch my back and I will scratch yours'. *See* **Fair Trade Policy.**

BEHA *or* **B.E.H.A.** British Export Houses Association.

BEI Biological Exposure Index.

beil Beiliegend *German*—Enclosed.

Being Full Signed *or* **Being Full Written** *or* **Full Written Line** The signature of an underwriter confirming he is accepting a line on a Slip, *q.v.* If the underwritten amount is exceeded the signed lines are reduced in proportion to the required or requested insurance cover. *Abbrev.* b.f.s.l. *or* B.F.S.L. It is also termed Being Full Written Line *abbrev.* b.f.w.l. *or* B.F.W.L. *or* Full Written Line *abbrev.* f.w.l. *or* F.W.L.

Being Full Written Similar to Being Full Signed Line, *q.v.*

bel Below.

Bel Belgium.

Belay To make fast or to secure; To fasten a rope to a Cleat, *q.v.*, or Belaying Pin, *q.v.*; Knot done after first turn.

Belaying Pin A movable wooden or metal pin in a rail of a ship to Belay, *q.v.*, ropes.

Belg. Belgium.

Belgische Beroepsvereniging der Scheepsbevooraders Belgian Ship Suppliers Association, being members of the International Ship Suppliers Association.

Bell A compulsory fitting on a sea-going ship. It is rung when the ship is at anchor in restricted visibility. Time on board ship was formerly indicated by the half-hourly ringing of the bell. The bell has other uses on ships. It is rung when the anchor is being weighed or dropped to indicate the number of shackles of chain let out. One stroke is rung for every shackle. On some ships the chimes of bells indicate the time for breakfast, lunch and dinner. Passenger ships, too, make good use of them to denote that the passengers are allowed to go ashore.

Bell Boy *or* **Bellhop** An employee of an hotel who attends to the general needs of the guests.

Bell Buoy A Buoy with a bell mounted on it. The bell rings when the buoy is rocked by the waves or the wash of passing ships.

Bellum internecionem *Latin*—War of extermination.

Belly (1) The swelling of a sail caused by the wind. (2) The bottom part of an aircraft's fuselage having a protruding resemblance to the human body. *See* **Belly Hold**.

Belly Hold Space for cargo, baggage, and mail in the Belly, *q.v.*, of an aircraft.

Belly Landing Crash landing of an aircraft on its Belly, *q.v.*, without the Undercarriage, *q.v.*

Belly Robber An American slang word for Chief Steward.

Below Nautical term for downstairs.

Below Bridges The expression refers to the movements of ships which have to pass under bridges so as to load or unload their cargo. *Abbrev.* b.b. *or* B.B.

Below Par When stocks and shares or currencies are paid below the market or nominal value. Opposite to Above Par.

BEM *or* **B.E.M.** Bachelor of Engineering of Mines; British Empire Medal; Bachelor of Ecclesiastical Music.

BEMAS Buoy Monitor and Alarm System.

Benacon A code name for the Chamber of Shipping British North American Charterparty 1914. Carriage of wood from the Atlantic coast of North America to the United Kingdom and Continent. (Amended on 28 October 1973.)

Benchmark A standard against which something can be measured. A datum for surveying. In the financial market it usually means a share price index.

B. Ency. Business Encylopedia.

Bend A nautical name for the intertwining of two ropes to secure one to the other.

Bends *See* **Staggers**.

Beneficial Interest A person who does not have any legal title to a property yet he still acts as a trustee. *See* **Beneficial Owner**.

Beneficial Owner A trustee. A person who has no legal title to a property but has tangible interest in it, insofar as he is acting as a trustee. *See* **Beneficial Interest**.

Beneficiary A person who is entitled to receive any money or other advantages or a holder of certain rights or a person gaining under a trustee or a will. *See* **Beneficial Interest** and **Beneficial Owner**.

Benefit in Kind Reward in the form of a favour, advantage, utility or service in place of money. It could, for example, consist of payment of life assurance, home telephone bills, cost of a company car, or house rent.

Benefit of Face In advanced cargo bookings freight is paid as at the time of the booking irrespective of any increase or decrease in Freight, *q.v.* If the exporter is given the Benefit of Face he is charged at the reduced freight but not before the time the Freight Rates, *q.v.*, are reduced.

Benefit of Fall In forward cargo bookings at current rates of freight, shipments already booked will not alter the rates irrespective of the fact that there may be an increase or decrease in the rates of freights. This system is often practised in the liner trade. Some special arrangements are made either for the advantage of the carriers or the shippers in case freights are altered accordingly in the meantime.

Benefit of Insurance Clause A clause in a contract of carriage by which the bailee of goods claims the benefit of any insurance policy effected by the cargo owner on the goods in care of the bailee. Such a clause in a contract of carriage, issued in accordance with the Carriage of Goods by Sea Act, is void at law.

BENELUX A contraction for Belgium/Netherlands/Luxembourg.

Bene merenti *Latin*—to the well deserving.

B.Eng. Bachelor of Engineering.

Beng. Bengal.

Bent Slang word for a person who had the Bends or Staggers, *q.v.*

Bentonite A mineral fluid used for rock drilling as a ground supporting agent instead of air.

Benzine *Similar to* **Gasoline**, *q.v.*

BEP *or* **B.E.P.** Bachelor of Engineering Physics.

BEPA British Edible Pulse Association.

Bequest A personal allowance or gift left by a person to another or others through a will.

Bergings Aktiebolag *Swedish*—Salvage Company Limited.

Bergings Maatschappij *Dutch*—Salvage Company.

Berks Berkshire, *UK*.

BERLIN Berlin Agreement Carriers.

Berm. Bermuda.

Bermuda Rig A boat rigged with a high tapering sail.

Bermuda Triangle An area in the Atlantic Ocean and in the vicinity of the Bermuda Islands in which numerous unhappy accidents to ships and aircraft have occurred. Some have been rumoured to have 'disappeared' and others sank.

Berne List A nautical book listing various information relating to radio stations and call signs by the International Union of Telecommunications in Geneva, Switzerland.

Berth (1) A place in which a vessel is moored or secured. (2) Allotted accommodation in a ship. (3) The space around a vessel at anchor in which she will swing.

Berthage Charges for the use of a Berth, *q.v.*

Berth Cargo When a Liner cargo vessel accepts extra cargo to fill up the empty space remaining. In some cases special reduced Freights, *q.v.*, are allowed as an incentive.

Berth Charterparty An agreement whereby the ship is chartered to load at a berth but the exact nature of the cargo is not known before the charterer has booked it.

Berth Clause A Charterparty, *q.v.*, clause where a ship is to await her turn to berth before Laydays, *q.v.*, are to commence to count.

Berth & Dock (1) Charterparty, *q.v.*, where a ship must be at her discharging/loading berth before she is an 'Arrived Ship', *q.v.*, for Laydays, *q.v.*, to count. (2) Charterparty, *q.v.*, phrase known as On the Berth, *q.v.* The ship is hired at the sole risk of the charterers insofar as the whole payment of freight is concerned. Therefore if no full cargo is available, the Deadfreight, *q.v.*, due is attributed to the sole responsibility of the charterers.

Berthing (1) The external planking positioned above the sheerstrake of a ship. (2) The operation of putting a vessel on her berth. *See* **Berth**.

Berthing Master A person responsible for berthing and mooring ships in a harbour, including the supervision of port operations.

Berth no Berth Similar to whether in Berth or Not.

Berth Note *or* **Booking Note** In booking individual part cargo the master signs a form of contract giving details of tonnage, the kind of cargo, loading/unloading ports, laydays, etc. The Note is endorsed by owners, agents or charterers.

Berth Owner *or* **Shipowner** A common carrier accepting the transportation of merchandise in a specialised trade.

Berth Rates Alternatively known as Liner Rates. Regular liners apply standard rates of freight on each trade route.

Berth Shipowner *See* **Berth Owner**.

Berth Terms Also called Liner Terms. This applies to the freight charges, including the payment of stevedores and winchmen by the carriers or owners in the loading and discharging operations. The other handling charges are borne by the consignors and the consignees respectively. *Abbrev.* B.T. or b.t.

Berth Traffic Tramp shipping and ships on Berth Charter, which carry cargoes available on the Berth, *q.v. See* **Berth Charterparty**.

BES Bureau of Employments Securities; Business Expansion Scheme.

BESA *or* **B.E.S.A.** British Engineering Standards Association, now **BSI** *or* **B.S.I.**

Beset In reference to a ship trapped in the ice in such a way that she is unable to manoeuvre and has to drift with the ice. This may take her into shallow waters or into contact with submerged objects.

Besloten Vennootschap Dutch Private Liability Company. *Abbrev.* BV *or* B.V.

BET British Electric Traction.

Bet. Between.

BETA Business Equipment Trade Association.

B.E.T.R.O. British Export Trade Research Organisation.

Between Decks Between the lower hold and the main deck of a cargo vessel; the space between the lower and upper decks.

Between Wind and Water During loading and discharging and also when pitching and rolling some underwater parts of the ship's hull become exposed. These exposures are called 'between wind and water'.

B.E.U. Benelux Economic Union; British Empire Union, now changed to B.C.U., *q.v.*

Bev. Billion Electron Volts.

Bev. Bd. Bevel Board.

Bevrachting *Dutch*—Chartering.

B.E.W. Board of Economic Warfare.

B.F. Bachelor of Finance; Bachelor of Forestry.

b/f *or* **B/F** Base Frequency; Beer Firkin; Brought Forward; Bankruptcy Fee.

B.F.A. Bachelor of Fine Arts.

BFC *or* **B.F.C.** Baltimore Berth Grain Charterparty 1913; Bank Fire Credit.

BFDC *or* **B.F.D.C.** Bureau of Foreign and Domestic Commerce.

BFEC British Food Export Council.

BFF British Fishing Federation Ltd.

BFI Baltic Freight Index, *q.v.*

BFO Beat-frequency oscillator; Blended Fuel Oil System; British Fuel Oil; Bunker Fuel Oil.

BFPO British Forces Post Office.

BFS *or* **B.F.S.** Bachelor of Foreign Service.

B.F.S.C. Bachelor of Science in the Foreign Service.

b.f.s.l. *or* **B.F.S.L.** Being Full Signed Line, *q.v. Alternatively* Full Signed Line, *q.v.*

BFSSM *or* **B.F.S.S.M.** British Federation of Ship Stores Merchants.

BFT *or* **B.F.T.** Bachelor of Foreign Trade.

b.f.w.l. *or* **B.F.W.L.** Being Full Written Line. *Alternatively* Full Signed Line, *q.v.*, and Being Full Signed Line, *q.v.*

BFSVEA British Fishing and Small Vessel Equipment Association.

B.G. *or* **b/g** *or* **B/G** Birmingham Gauge; Bondage or Bonded Goods. *See* **Bonded Stores.**

Bg. Bag; bearing; brigantine.

BG Biman Bangladesh Airlines *airline designator.*

BGC *or* **B.G.C.** Bank Giro Credit; British Gas Corporation.

BGE *or* **B.G.E.** Bachelor of Geological Engineering.

B. Gen. Ed. Bachelor of General Education.

bgl Below Ground Level.

B.G.L. Bachelor of General Laws.

BGM *or* **B.G.M.** Biennial General Meeting.

bgt. Bought.

bh Barrels per hour; Bougie-heure *French*—Candle hour.

Bh *or* **Bhd** Bulkhead.

B.H. Bulkhead(s) *Lloyd's Register notation*; Bill of Health. British Honduras.

B/H Bill of Health, *q.v.*; Bordeaux/Hamburg Range.

BHA Bottomhole Assembly *oil drilling*; Bulk Handling Authority Australia.

B'ham Birmingham, *UK.*

Bhandari An Indian cook.

BHC British High Commission.

BHCEC British Health Care Export Council.

BHEC British Hospitals Export Council.

Bhf Bahnhof *German*—Railway Station.

BHFP Bottomhole Flow Pressure *oil drilling*.

Bhn Brinell Hardness Number.

bhp-hr Brake Horse Power-hour.

BHT Bottomhole Temperature *oil drilling*.

BI *or* **B.I.** Background Information; Bahama Isles; Baltic Isles; Bermuda Isles; Board of Investigation; Bulk Issue; Royal Brunei Airlines *airline designator.*

B.I.A. British Insurance Association. *Obsolete on merging with* **ABI**, *q.v.*, in 1985.

B.I.B.A. British Insurance Brokers Association.

BIBO Bulk In-Bulk Out, *q.v.*

BIC *or* **B.I.C.** British Importers' Corporation; Bureau International des Containers *French*—International Container Office or Bureau.

B.I.C.C. Berne International Copyright Convention; British Insulated Callender's Cables Ltd.

B.I.C.E.M.A. British International Combustion Engine Manufacturers Association.

B.I.C.E.R.A. British International Combustion Engine Research Association.

b.i.d. Bis in die *Latin*—Twice a day.

Bid (1) An offer in an Auction, *q.v.*, sale. (2) A price offer to buy an article. (3) To tender for a contract, to supply or obtain a contract.

Bid Price The selling price of stocks and shares at a particular time as opposed to Offer Price which is the buying price. *See* **Offer Price.**

BIE *or* **B.I.E.** Bachelor of Industrial Engineering; Bureau International d'Education *French*—International Bureau of Education; Bureau International des Expositions *French*—Organisation for International Exhibitions, Paris.

bien *French*—Well or good.

bien entendu *French*—All right; Of course; Surely.

B.I.E.T. British Institute of Educational Technology; British Institute of Engineering Technology.

BIF *or* **B.I.F.** Banded Iron Formation *geophysics*; British Industries Federation.

BIFA *or* **B.I.F.A.** British International Freight Association.

BIFFEX The usual term for the Baltic International Freight Futures Market, part of the London Commodity Exchange, which was founded in 1985. BIFFEX offers shipowners, operators and charterers a method of protection against the risks of volatile freight rates, by means of a standardised contract that is settled against an index of international dry bulk 'spot market' voyages, the Baltic Freight Index (BFI). A Charterer who has sold a commodity forward is concerned that freight rates will *rise*, eroding or eliminating his trading profit. To protect himself he *buys* BIFFEX futures contracts. If freight rates do *rise*, the loss incurred in the 'cash' shipping market, when he actually secures a ship, will be offset by a profit on his futures position. A shipowner's risk is the exact reverse of the charterer's risk, hence he *sells* futures contracts to protect himself against *falling* freight rates.

By using BIFFEX, it is possible to hedge forward freight rates at the level of the futures price traded for up to 18 months forward. Futures prices are broadcast through data vending companies worldwide and are obtainable from futures brokers. With an index-based futures contract it is not possible to deliver a cargo or ship in settlement of a futures position. All contracts traded on BIFFEX that remain open on the Last Trading Day are settled in cash, based on the average of the BFI over the last five business days of the Settlement Month. Cash settlement is a most convenient method of settling futures contracts.

BIFN 'Bye for Now' Telex *abbrev.*

BIFU Banking, Insurance and Finance Union, a UK trade union.

B.I.G. British Insurance Group, *q.v.*

Big Apple American expression for New York.

Big Five *See* **Banks, Chief of The**.

Big D Dallas, Texas, *USA*.

Bight (1) A loop in a rope when bent. (2) Length between two ends of a rope. (3) Long indented coastline. (4) An open bay.

B.I.I.B.A. British Insurance and Investment Brokers Association.

BILA *or* **B.I.L.A.** British Insurance Law Association.

Bilateral Trade Trade both ways between two countries.

Bilge The part of a ship's hull where the side curves around towards the bottom.

Bilge Blocks Roughly squared or rectangular pieces of wood used as supports under ships when drydocked.

Bilge Free Stowage A method of stowing barrels or casks so that their 'Bilges', i.e., the rounded belly part, are kept clear of the deck by Chocks, *q.v.*.

Bilge Keel A fin protruding from the side of the bilge Strake, *q.v.*, to minimise rolling as much as possible. A bilge keel is positioned on each side of the ship. *Abbrev.* Bil.K.

Bilge Pump Pump which removes water and sediment from the Bilge, *q.v.*, of a ship.

Bilge Water Accumulated water in Bilges which has no use.

Bilge Water Damage Attributed to the overflow from the Bilge, *q.v.*, or waste water pipe. It covers the marine insurance clause Fresh Water Damage, *q.v.*

Bilgeway *Similar to* **Bilge**, *q.v.*

Bil.K. Bilge Keel, *q.v.*

Bill *or* **B.E.** *or* **B/E.** Bill of Exchange, *q.v.*

Bill Broker A person engaged in the purchase and sale of Bills of Exchange, *q.v.* He sells bills for those drawing on foreign countries, and buys bills for those remitting to them.

Bill Discounted A bill of exchange upon which a banker or other person has advanced money for a percentage called discount.

Billing Preparing and sending invoices to customers for settlement.

Billingsgate One of the oldest wholesale fish markets, re-located from the City of London.

Bill in Parliament (1) Public, affecting England, Wales, Scotland or Northern Ireland generally, or a very important part of them; (2) Local and personal, affecting particular areas only, as railway construction bills, water or gas supply bills, etc.; or (3) Private, as bills settling estates, divorce or naturalisation bills.

Bill of Costs An account of fees, charges and disbursements by a solicitor in a legal business.

Bill of Cravings A list of expenses incurred by a sheriff for judges' lodgings, executions, etc., during his year of office: he had to furnish it when passing his accounts.

Bill of Credit A licence or authority given in writing from one person to another, very common among merchants, bankers and those who travel, empowering a person to receive or take up money of their correspondents abroad; more usually termed a letter of credit.

Bill of Entry A certificate given by importers of goods to the Customs authorities setting out particulars of the goods and stating the port, etc., from which they were imported, and the names of the parties to the transaction; also the certificate given to the Customs authority by an exporter of goods from a British port.

Bill of Exchange An unconditional order in writing, addressed by one person (A) to another (B), signed by the person giving it, requiring the person to whom it is addressed to pay, on demand, or at a fixed or determinable future time, a sum certain in money to, or to the order of, a specified person (C), or to bearer. A is called the drawer, B the drawee, and C the payee. Sometimes A, the drawer, is himself the payee. The holder of a bill may treat it as a promissory note if the drawer and drawee are the same person. When B, the drawee, has, by accepting the bill, undertaken to pay it, he is called the acceptor.

If a bill is made payable to C without more, that makes it payable to 'C or order', unless it contains words indicating that it is not transferable. Usually, however, a bill is made payable either to 'C or order', or to 'C or bearer.' Whether the bill is made payable to 'C' or to 'C or order', C can transfer the bill by a written order to pay to someone else (D), who, after the bill has been delivered to him, becomes the holder of it, and can transfer it to E, and so on. This order is generally written on the back of the bill, and is called an indorsement, C being then the indorser, and D the indorsee. If the bill is payable to 'C or bearer' C can transfer it to D by merely delivering it to him. A bill payable to order may be converted into a bill to bearer if the payee or indorsee indorses it in blank, that is to say, without specifying an indorsee; but after a bill has been indorsed in blank any holder may convert the blank indorsement into a special indorsement by writing above the indorser's signature a direction to pay the bill to or to the order of any person. 'Holder' is a general word applied to anyone in actual or constructive possession of a bill, and entitled to recover and receive its contents from the parties to it, viz., the acceptor, drawer and indorsers. A payee is not a holder in due course.

The legal effect of drawing a bill is a conditional contract by the drawer to pay the bill if the drawee dishonours it, either by failing to accept it or, having accepted it, by failing to pay it at maturity. The effect of accepting a bill is an absolute contract by the acceptor to pay the bill. The effect of indorsing a bill is a contract by the indorser to pay his immediate or any succeeding indorsee, or the bearer, in case of the acceptor's default. But if the bill is not presented for payment at the proper time, all the antecedent parties, except the acceptor, are discharged from liability, and the liability of an acceptor may be qualified by the terms of his acceptance, though the payer of a bill may treat a conditional indorsement as if it were unconditional.

When the drawee of a bill fails to accept it on its being presented to him for that purpose, it is said to be dishonoured by non-acceptance; when the acceptor of a bill fails to pay it on presentment at the proper time, it is said to be dishonoured by non-payment. As a general rule, it is incumbent on the holder of a bill which has been dishonoured to give prompt notice of the fact to the antecedent parties, otherwise they will be discharged from all liability.

When a bill is made payable at a certain time (e.g. 30 days after date or sight), on that time arriving the bill is said to be at maturity or due. After that time it is said to be overdue or afterdue. The negotiability of an overdue bill is qualified to the extent that it can only be negotiated subject to any defect of title affecting it at maturity.

Bills of exchange, like promissory notes and a few other instruments, differ from other simple contracts, first, in their negotiability, and secondly, in being presumed to have been given or transferred for valuable consideration until the contrary is proved. A bill which originally was negotiable remains negotiable unless it is restrictively indorsed.

Bills of exchange are either inland or foreign. Inland bills are those which are both drawn and payable within the limits of the British Islands; all others are foreign bills. Any holder of a bill may treat it as an inland bill unless the contrary appears on its face. Foreign bills differ from inland bills principally in being frequently drawn in sets or parts, and in requiring to be protested on dishonour in order to charge the drawer.

Bills of exchange are choses in action, but for the encouragement of commerce they were assignable at common law by mere indorsement. Also termed Bank Bill, Demand Draft, Long Draft or Documentary Bill.

Bill of Gross Adventure An instrument in writing which contains a contract of bottomry, respondentia and every species of maritime loan (French Law). *See* **Bottomry Bond**.

Bill of Health A document given to the master of a ship by the consul or other proper authority of the port from which he comes, describing the sanitary state of the place at the time when the ship sailed. It may be a clean, suspected or foul bill. The first is given where no disease of an infectious or contagious kind is known to exist; the second (also called a touched patent or bill) is given where, though no such disease has appeared, there is reason to fear it; and the last is given when such a disease actually exists at the time of the ship's departure. The latter subjects the ship to the full period of quarantine.

Bill of Lading Where a ship is not chartered wholly to one person, but the owners offer her generally to carry the goods of any merchants who may choose to employ her, or where, if chartered to one merchant, he offers her to several sub-freighters for the conveyance of their goods, she is called a general ship. In these cases the contract entered into by and with the owners, or the master on their behalf, is evidenced by the bill of lading. This is a document which is signed and delivered by the master to the shippers on the goods being shipped. In practice, when goods are shipped an acknowledgment is given by the mate, known as the 'mate's receipt'. This is afterwards exchanged by the captain or the broker of the ship for the bill of lading. Several parts, that is to say copies, of the bill of lading are commonly made out; one or more of these is sent by the shipper of the goods to the consignee, one is retained by the shipper himself, and another is kept by the master for his own guidance.

A bill of lading specifies the name of the master, the port and destination of the ship, the goods, the consignee, and the rate of freight.

A bill of lading is not a negotiable instrument, and a bona fide transferee for value obtains no better title than the transferor, though he may defeat the right of stoppage in transit of an unpaid seller. It resembles a negotiable instrument in that if it is drawn 'to the order' or 'to the assigns' of a person it may be indorsed and transferred by delivery. There are various kinds of bills of lading; Shipped Bill of Lading; Received Bill of Lading; Clean Bill of Lading; Direct Bill of Lading; Through Bill of Lading; Open Bill of Lading; Order Bill of Lading; Straight Bill of Lading; Bearer Bill of Lading; Outward Bill of Lading; Homeward Bill of Lading; Liner Bill of Lading; Foul or Dirty Bill of Lading; Ocean Bill of Lading; Port Bill of Lading.

Bill of Parcels An account given by a seller to a buyer containing particulars of the goods bought and their price.

Bill of Sale A deed assigning personal property either absolutely or by way of security.

A bill of sale of ordinary chattels by way of absolute assignment is not very common, except in cases where the thing sold is of some importance, as where a sheriff sells goods under an execution.

A bill of sale is the usual, and in the case of registered ships the only, mode of transferring ships; a bill of sale of a registered ship, under the Merchant Shipping Act, 1894, is to be in the form given in the Act, or as near thereto as possible, specifying the number and date of registry, the description of the ship, etc.; it is under seal and must be registered. Formerly bills of sale of ships were divided into two kinds, the grand bill of sale, which conveyed the ship from the builder to the purchaser or first owner, and the ordinary bill of sale by which any subsequent transfer was made; but these terms are not now used.

The most usual kind of bill of sale, and that to which the term is in practice applied, is a bill of sale of chattels (e.g., furniture, horses, stock-in-trade, etc.) by way of mortgage to secure a debt, being an assignment with a covenant for reconveyance on payment of the debt. Unlike in the case of a pledge or pawn, possession of the chattels is not given to the person who lends money on the chattels. It is with reference to bills of sale by way of mortgage that the Bills of Sale Acts are of importance; they are passed to prevent frauds 'committed upon creditors by secret bills of sale of personal chattels, whereby persons are enabled to keep up the appearance or being in good circumstances and possessed of property' which really belongs to someone else. *Abbrev.* BS *or* B.S. *or* B/S.

Bill of Sale Absolute A contract of sale which is registered and witnessed by a notary public or a solicitor. The title of ownership is permanently transferred and cannot be retained by the seller.

Bill of Sale Conditional A sale system which is not necessarily contracted by a notary public or a solicitor though it must be witnessed. The title of the ownership is conditional upon a return of the sold object to the original proprietor or seller if the conditions laid down are not adhered to.

Bill of Sight When a merchant is ignorant of the real quantities or qualities of any goods assigned to him, so that he is unable to make a perfect entry of them, he must acquaint the Customs authorities of the circumstance; and he is authorised, on making a declaration that he cannot, for want of full information, make a perfect entry, to receive an entry by bill of sight for the goods by the best description which can be given, and to grant warrant that the same may be landed and examined by the importer in the presence of the Customs officers. The importer must later make a perfect entry and either pay the duties or warehouse the goods. In default of perfect entry within the time allowed the goods are taken to the Queen's warehouse and if they are not cleared within a month they are sold to pay the duties. *Abbrev.* B/S *or* B/St.

Bill of Stores A document for the entry duty free into the United Kingdom of dutiable goods which have been exported therefrom prior to their reimportation.

Bill of Sufferance A licence issued by the customs to a merchant under the repealed statute 1662, 14 Car. 2, c. 11, authorising him to trade between English ports without paying dues.

Bill Payable A bill drawn on a drawee and accepted by him; used in connection with the accounts or affairs of a debtor to describe a bill accepted by him.

Bill Receivable A bill remitted to a trader in payment for value received, or drawn by a trader on his customer; used in connection with the accounts or affairs of a creditor to describe a bill of which he is the holder.

Bills Retired American term for withdrawn Bills of Exchange, q.v., either disposed before Due Date, q.v., or withdrawn from circulation.

Bill Victualling or **Victualling Bill** A customs house form authorising suppliers or ship chandlers to supply items duty free on board ships. See **Victualling Bill**.

B.I.M. Bachelor of Indian Medicine.

Biman Bangladesh Bangladesh National Airline.

BIMCHEMTIME or **Bimchemtime** Code name for Baltic and International Maritime Council Charterparty for vessels carrying chemicals in bulk.

BIMCO The Baltic and International Maritime Council, q.v.

Binary Stars Pairs of stars which are held together by their correlative or mutual attraction and revolve about their common centre of mass.

Bin Boards Moving or dividing boards used in separating one type of bulk grain cargo from others in granaries.

Bin Card A card left in a storehouse to record the actual quantities of articles used and remaining.

Binder (1) Components consisting of drying oils, etc., used in the manufacture of ordinary or conventional paints. (2) The American expression for a slip in the insurance business.

Binnacle A casing on the bridge of a ship which accommodates the standard Compass, q.v. It is fixed and positioned in front of the wheel and contains magnetic corrector rods for making compass adjustments.

B.I.P. Baggage Improvement Programme; British Institute in Paris.

BIPAR European Union Insurance Brokers' Association, in Brussels.

BIRD Banque Internationale pour la Reconstruction et le Développement French—International Bank for Reconstruction and Development.

Birdfarm American slang for aircraft carrier.

Bird Strike A collision between a flying bird or flock of birds and an aircraft in the air.

Birr Currency unit of Ethiopia.

Birth Marks Builder's irregularities in the construction of a ship.

BIS or **B.I.S.** Bank for International Settlements; British Information Services.

Bis Latin—Again; Encore, Twice.

BISCOILVOYBILL or **Biscoilvoybill** Bill of Lading, q.v., code name used in conjunction with Biscoilvoy Charterparty.

Bise A French word for a cold dry wind from the North West or North East followed by heavy clouds, generally met with in Switzerland and France.

bis in die Abbrev. b.i.d., q.v.

BISRA or **B.I.S.R.A.** British Iron and Steel Research Association.

54

bit Binary Unit.

BIT or **B.I.T.** Built-in Test electronics; Bureau International du Travail French—International Labour Office.

BITE Built-in Test Equipment.

Bitt Stout post to which ropes are made fast. Bitts are two posts near to each other around which mooring ropes are fastened. Similar to Bollard.

Bitumen Carrier A tanker which is specially constructed to carry bitumen in bulk. Specialised heating coils are installed around the holds to melt the bitumen and enable it to flow freely in loading and unloading operations. Wing tanks are in most cases used for ballast purposes.

Bituminous Paint A paint composition having a large percentage of bitumen. Mostly used for rust protection on internal steel in ships, as in the chain lockers. Also called Black Varnish.

B.J. Bachelor of Journalism.

bk Backwardation, q.v.

Bk. Back; Backwardation, q.v.; Book.

B.K. Bank Book; Bar Keel.

Bkg Banking.

Bkge Breakage; Brokerage, q.v.

BKI Biro Klassifikasi Indonesia—Indonesia Classification Society.

BKK Bangkok Thailand.

Bklyn Brooklyn USA.

BKN Broken Clouds weather observation.

Bkrpt Bankrupt, q.v.

Bkt Bracket; Bucket.

Bkt. Dgr. Bucket Dredger.

B/l or **B/L** or **Bl** Bale(s); Bill of Lading, q.v.

B.L. Bachelor of Law; Bachelor of Literature; Barrister-at-Law, Base Line, q.v.; Bill Lodge; Bow Line; Breaking Load; Below Line.

BLA or **B.L.A.** Bachelor of Landscape Architecture.

BLACC or **B.L.A.C.C.** British and Latin American Chamber of Commerce.

Black Ball See **Anchor Ball**.

Black Balled See **Black Balling**.

Black Balling Voting against a candidate who applies for membership in a society, club, etc. Usually the negative vote is presented in a form of black colour. The expression Black Balled means that one of the candidates has been turned down.

Black Bourse Foreign exchange term for black market; a manner of illegal trading by obtaining an exaggerated amount of profit for any item offered for sale.

Black Box or **Flight Data Recorder** or **Flight Recorder** Nicknamed 'Box of Tricks' it is usually painted bright red to be traced or seen from afar. This gives or records the flight details of an aircraft such as: altitude; airspeed; pitch; acceleration behaviour, etc., and is of great use in case of an accident. It is built to resist an impact of over 4000 lbs and temperatures of over 2000°F.

Black Bread Bread made of rye.

Black Cargo Cargo banned by general cargo workers for some reason. This ban could be because the cargo is dangerous or hazardous to health.

Black Economy Commercial activity which is unreported to evade taxation.

Black Exchange Rates This is similar to Black Bourse, *q.v.* When foreign exchange rates are controlled by the state and traders deal on the black market without the knowledge of officials.

Black Gold Nickname for Oil.

Blackleg (1) Defiance in a strike when one or more members ignore the directives of their unions. (2) May also apply to a person who refuses to join a union.

Blacklist *or* **Black List** (1) A list composed of persons, companies, countries or others which is held on record for the purpose of avoiding or banning any dealings whatsoever with all those named on the blacklist. Also termed Black Star. (2) To ban or to avoid doing any business or dealings whatsoever.

Black Maria Prison van mostly used in transporting criminals and prisoners.

Black Plate Uncoated mild steel plate.

Black Sea Berth Terms A contract form of charterparty for a ship to load grains 'On the Berth', *q.v.*, with an option of general cargo with conditions attached.

BLACKSEAWOOD Code name for Baltic and International Maritime Council with Soviet Organisation 1973 for the carriage of wood from the USSR and Romanian ports of the Black Sea/Danube to all parts of the world.

BLACKSEAWOODBILL Bill of Lading code name used in conjunction with Blackseawood Charterparty.

Black Star On the Black List, *q.v.*

Black Varnish Identical to Bituminous Paint, *q.v.*

Blade Leaf of a rotating fan, turbine or propeller converting energy to forcing the movement of air or driving a ship or aircraft.

Blank Bill A Bill of Exchange, *q.v.*, without the name of the buyer.

Blank Bill of Lading A Bill of Lading, *q.v.*, where the name of the receiver or consignee is not inserted and is replaced by the word Bearer, *q.v.*

Blank Cheque A Letter of Credit, *q.v.*, which has no amount of money inserted.

Blank Credit A Letter of Credit, *q.v.*, with the amount of money being left unspecified.

Blanket Policy An insurance policy indemnifying the insured against damage or loss regarding a group of household contents, including fire and theft. It also covers fines from the authorities through any unforeseen or premeditated acts by people or pets in the house, including any malignant action committed by a third party.

Blast Signal given on a ship's whistle. *See* **Short Blast** and **Long Blast**.

BLCC *or* **B.L.C.C.** Belgium—Luxumbourg Chamber of Commerce.

Bldg Building.

Bleed Air Pressurised air coming from the main engine of an aircraft to the cabins and other areas. It is termed 'Bleed' for short.

Bleeders Screw plugs inserted at the bottom of a ship which can be opened when the ship is in dry dock to drain out water from the bottom tanks.

Bleeding The act of opening up bags of grain and allowing the contents to drop into the broken stowage of the other bags in the hold. By so doing the cargo space capacity of the holds is entirely utilised.

Bleeding Wing-Tanks Upper wing tanks in bulk carriers for water ballast are sometimes dried out and used for carrying cargo. They are loaded through openings in the main decks and discharged by being bled through slots into the hold below. *Abbrev.* BWT.

BLEU *or* **B.L.E.U.** Belgo–Luxembourg Economic Union; Blind Landing Experimental Unit.

Bleve Boiling Liquid Expanding Vapour Expansion.

Blind Alley A situation in which there is little or no opportunity of advancement, a dead end. Sometimes applied to occupations, i.e., a blind alley job.

Blind Hatch An extra hatch situated away from a 'tween deck hatch on the same deck, and which has no direct access to or from the main hatch. May serve as an additional hatch to allow general cargo from the 'tween decks to be stowed in the innermost part of the bottom hold when such stevedoring cannot be carried out directly through the main hatch.

Blind Holes Round holes on metal sheets which do not coincide and will therefore obstruct the rivets from passing through. Opposite to Fair Holes.

Blind Pool Often referred to in the USA as persons, firms or corporations who entrust to the care of an individual the funds and interest they hold without interference. Such an individual has the advantage of negotiating with the utmost secrecy and freedom on his part.

Blind Test Trying the market by offering a product to the general public without revealing the manufacturer's name or brand.

Blind Underwriter An insurer who undertakes widespread coverage irrespective of the individual risks, however hazardous, that might be involved.

Blinker/s A Morse Code light signal which is either transmitted by hand or by a telegraph device. *See* **Morse Lamp**.

B. Lit. Baccalaureus Literarum *Latin*—Bachelor of Literature.

Blizzard Strong cold wind with falling snow.

Blk *or* **BLK** Bulk.

B.L.L. Baccalaureus Legum *Latin*—Bachelor of Law.

Block A ship's pulley; a grooved wheel (sheave) working in a frame or shell on which the rope passes through. A hook may be attached to the top of the block as a support.

Blockade The prevention by a naval force of entry to, or exit from, a port. A belligerent power has the right to blockade an enemy's ports. A neutral power must respect this or risk confiscation of a ship or her cargo. Such a cargo would be deemed Contraband. Blockade has been exercised by the United Nations in support of sanctions.

Block Booking The booking of a quantity of known cargo to be carried at a stipulated time in various shipments on named vessels owned or managed by the same particular company.

Block Coefficient The ratio of the immersed volume of a vessel to a rectangular prism similar in length, breadth and draught. A fast ship with fine lines has a small coefficient of say 0.50°, while a slow bulky ship has a bigger Block Coefficient of 0.70° and upwards. To find the block coefficient the following formula may be used:

$$\frac{\text{Displacement in Tons} \times 35}{\text{Length BP} \times \text{Breadth} \times \text{Draught}}$$

Blocked Currency Money that can only be used in its own country.

Blocking and Trapping Clause In marine insurance an extension of War and Strike clauses to cover the perils of a ship becoming blocked and detained in a port or waterway or prevented from sailing by hostile actions for a long period.

Blocking Off The act of fastening the cargo in the holds when not completely full, to prevent it shifting during the voyage. *See* **Lashings**.

Blocks, Chains and Shackles Part of the equipment or apparel of the ship. *See* **Apparel**.

Block Ship A sunken or wrecked ship obstructing vessels while manoeuvring.

Block Speed Result obtained by dividing the total distance covered by an aircraft from the time taken. An air transport term.

Block Transfer *or* **General Transfer** Transfer of Shut Out cargo from a vessel into another vessel because the holds are full.

Blood Money A law practised generally in Middle East countries such as Saudi Arabia whereby blood money is forced to be paid if a person is found guilty of a fatal accident by a vehicle. The driver is liable to pay a sum of money assessed by the authorities to the victim's relatives.

Blown-Up American expression said of a ship having an excessively high percentage of broken stowage.

Blow Oil Oil leaks which go out of control in oil producing areas.

Blow Out Preventer A shut-off device used by oil rigs capable of closing wells when there is danger of oil or gas leaks. This normally has a hydraulic power system. *Abbrev.* BOP *or* B.O.P.

BL/S *or* **Bl/s** Bale(s); Barrel(s).

BLS *or* **B.L.S.** Bachelor of Library Science.

B.L.S. Bureau of Labour Statistics.

B/L Ton Bill of Lading, *q.v.*, Ton.

Blue Book Blue-bound book first published in 1957 by H.M.S.O under the title Carriage of Dangerous Goods in Ships. In 1967 and thereafter the rules were revised jointly with the Standing Advisory Committee of the Inter-Governmental Maritime Consultative Organization, which is now called the International Maritime Organization. A very useful book for shipping executives and merchant navy officers dealing in the transportation of dangerous goods.

Blue Button An authorised clerk of the Stock Exchange, *q.v.*

Blue Chip Investment An investment considered to be quite safe for the investor since the company issuing the shares has a good standing and sound financial position.

Blue Chips High Class Industrial Shares. *See* **Blue Chip Investment**.

Blue Collar Worker A manual or factory worker as opposed to a White Collar Worker, *q.v.*

Bluenose Colloquial for a Nova Scotian; a person who has sailed beyond the Arctic Sea Circle.

BLT Battalion Landing Team *USA*.

Blue Peter A rectangular flag, blue with a white square in the centre and the 'P' flag in the International Code of Signals. It is displayed to indicate that the vessel is about to sail and all persons should therefore be on board.

Blue Pigeon A nautical slang term for the hand lead-line, Tape Line or Ullage Tape.

Blue Sky Less than a quarter clouded according to the Beaufort notation b.

Blue Sky Laws American law originally enacted in Kansas City in 1911 and thereafter adopted in all the states. Principally aimed at combating fraud committed by some stockbrokers in the course of their transactions with gullible clients. Also to obtain bank commissioners' endorsements prior to the purchase and sale of stocks. Failure to do so resulted in imprisonment. As time passes this law is gradually being disregarded.

Blue Water Expression used in matters pertaining to deep sea areas.

BLV Biologic Limit Values (the biological measurements in exposing human health hazards).

Blvd. Boulevard.

b.m. Bene merenti—*Latin*.

B.m. Bench Mark; Board Measure, *q.v.*

BM Height of Metacentre above the centre of buoyancy.

BM *or* **B.M.** Beatae Memoriae *Latin*—Blessed Memory; Bachelor of Medicine; Bachelor of Music; Barge Master; Bench Master; Berthing Master; Bermuda; Bending Moment *Technical*; Boatswain's Mate; British Museum; Bureau of Mines.

B.M. *or* **BMA** Bermuda.

BMA *or* **B.M.A.** British Marine Association; British Medical Association.

B. Mar. E Bachelor of Marine Engineering.

BMB *or* **B.M.B.** Baltic Marine Biologist(s).

BMCF British Maritime Charitable Foundation, *q.v.*

BMCN *or* **B.M.C.N.** BIMCO Maritime Community Network.

B.M.E. Bachelor of Mechanical Engineering; Bachelor of Mining Engineering; Bachelor of Music Education.

B.M.E.C. British Marine Equipment Council.

B. Mech. E. Bachelor of Mechanical Engineering.

BMEG Building Materials Export Group, *UK.*

BMEP *or* **b.m.e.p.** BrakeMean Effective Pressure, *q.v.*

B. Met. Bachelor of Metallurgy.

B. Mgt. E. Bachelor of Management in Engineering.

BMEWS Ballistic Missile Early Warning System.

B. Min. E. Bachelor of Mining Engineering.

B.M.L.A. British Maritime Law Association, *q.v.*

B.M.L.A.A. British Maritime Law Association's Agreement (referring to Gold Clause Agreement, *q.v.*).

BMMEA British Merchant Marine Equipment Association.

BMP *or* **b.m.p.** Brake Mean Power.

BMR *or* **B.M.R.** Basal Metabolic Rate; Bureau of Mineral Resources, *USA.*

B.M.S.E. Baltic Mercantile and Shipping Exchange.

BMSOA British Motorship Owners' Association Ltd., London.

B.M.S.S.O.A. British Motor and Sailing Ship Owners' Association.

BMT British Maritime Technology, *q.v.*

BMT *or* **B.M.T.** British Mean Time *or* Greenwich Mean Time *or* Universal Time. *See* **Greenwich Mean Time**.

BMTA *or* **B.M.T.A.** British Motor Trade Association.

BMTEC British Management Training Export Council.

B.M.U. *or* **B.M.** Bermuda.

B. Mus. Bachelor of Music.

B.M.V. Beata Maria Virgo *Latin*—Blessed Virgin Mary.

Bn. Beacon, *q.v.*

BN Brussels Nomenclature. Also known as Customs Co-operative Council Nomenclature.

B.N. Bachelor of Nursing; Bank Note, *q.v.*

B.N.A. British North America; British North Atlantic (Marine Insurance Limited Area Liability).

BNC/ICC *or* **BNCICC** British National Committee of the International Chamber of Commerce.

BNEA *or* **B.N.E.A.** British Naval Equipment Association.

BNEC *or* **B.N.E.C.** British National Export Council.

BNEP Basic Naval Establishment Plan *USA.*

BNF British National Formulary (Drug Index Book); British Nuclear Fuels.

B.N.I.T.A. *or* **BNITA** British Nautical Instrument Trade Association, Glasgow.

Bnkg Banking, *q.v.*

BNL Banca Nazionale de Lavoro. *An Italian Bank.*

BNOC *or* **B.N.O.C.** British National Oil Corporation.

BNS *or* **B.N.S.** Bachelor of Natural Science.

b/o *or* **b.o.** *or* **B/O** *or* **B.O.** Bank Order; Bad Order; Black Out; Board of Ordinance; Box Office; Branch Office; Broker's Order; Brought Over; Buyer's Option.

B.O.A. Baccalaureus Obstetrice *Latin*—Bachelor of Obstetrics; Bankruptcy Annulment Order, British American Oil.

Board The collective word for directors, members, presidents, vice-presidents of companies, corporations, institutes and other groups who meet periodically with a view to discuss and decide certain pertinent and objective details concerning their work. *See* **Board Meeting**.

Boarding Card *or* **Pass** A card given to a passenger about to depart by sea, land, or air authorising him to embark.

Boarding Vessel *or* **Boarding a Vessel** A shipping term referring to going on board a ship. Another meaning for customs officers to go on board is 'clearing vessel' or 'Rummaging', *q.v.*

Board Measure Measurements used in the American Timber trade *abbrev.* B.M. A Board Foot is the unit of measurement equivalent to 1 foot × 1 foot × 1 inch or 144 cubic inches. Twelve Board Feet are equivalent to 1728 cubic inches. A Mille is equivalent to 1000 board feet and equals 83.333 cubic feet.

Board Meeting A meeting held by directors, managers, presidents, vice-presidents and other high executives of companies, institutions, corporations and other concerns. *See* **Board**.

Board of Directors A number of people or a committee elected by the shareholders of a company to direct the business.

Board of Trade The name is now changed to Department of Trade and Industry in the United Kingdom.

Boards Pieces of sawn timber less than two inches thickness and of any length.

Board Wages The additional payment made to industrial and hotel workers to provide for their own food.

Boat Box Another word for 'First Aid Kit' which is part of the equipment of a lifeboat.

Boat Drill *See* **Life Boat Drill**.

Boat Note This is usually presented for a mate's signature when cargo is being delivered from alongside a quay or from a barge or other means into a bulk cargo ship.

Boat Station The specific duty allotted to every officer and crew member during a lifeboat drill and when lowering lifeboats in an emergency. *See* **Life Boat Drill**.

Boatswain A petty officer who is in charge of the deck crew members of a ship and who in turn comes under the direct orders of the master or chief officer or mate. The word is shortened to Bosun and *abbrev.* Bosn or Bo'sun.

BOB *or* **B.O.B.** Barge on Board. *See* **L.A.S.H.**.

B.O.B.A. British Overseas Banks Association.

BOC British Oxygen Co Ltd.

B.O.C.E. Board of Customs and Excise.

BOD Biochemical Oxygen Demand (referring to pollution treatment in the removal of solid materials by dissolution, especially in sewage).

Body Corporate An entity, in the form of a company, fellowship, society or other association, which has a legal personality, enjoying, and subject to, legal rights and duties.

Body Matter The main part of a document ignoring introduction, recitals and signature.

BOE *or* **B.O.E.** Barrel of Oil Equivalent—conversion of gas at 5.8 billion cubic feet equals 1 million barrels; Board of Education.

BOEA British Offshore Equipment Association.

Boeki *Japanese*—Foreign Trade or Commerce.

B of E Bank of England.

Boffer Best Offer *in quotations.*

B of H. Board of Health.

Bogie Set of wheels or trolley placed under containers, vehicles and other heavy weights for manoeuvrability in movement and stowage; set of wheels upon which a waggon runs.

Bogsering *Swedish*—Towing.

B.O.H. Board of Health.

Boh. Bohemia, *Czechoslovakia.*

Boiler Mountings The essential parts accessory to the steam boiler consisting of water and pressure gauges; feed check, test, stop, auxiliary blowdown and warning or whistling, valves; salinometer cock; and others.

Boiler Pick *See* **Chipping Hammer**.

Boiling Liquid Expanding Vapour When liquefied gas is carried under pressure and is exposed to heat from fire. The contents may rise to boiling point and will result in a spontaneous explosion. On LNG ships this should not happen as liquefied gas is contained in tanks at atmospheric pressure. *Abbrev.* Bleve.

Boil Off Liquid Natural *or* **Liquid Pressure Gas** Heat evaporation which occurs while in transit. This cannot happen if adequate insulation is provided.

Bol. Bolivar, *South America*; Bolivia.

BOLAM Bank of London and Montreal.

Bolivar Unit of currency used in Venezuela. One hundred centimos equals one bolivar.

Bollard Pull The universal calculation of the pull or the strain a tug is capable of producing when towing. This is generally tested by having moorings around a bollard and is measured from zero to actual towing speed. Thus the derivation of 'Bollard' pull.

B.O.L.S.A. Bank of London and South America.

Bolt Standard piece or roll of canvas as supplied, 39 yards long and 22–30 inches wide.

Boltrope Rope sewn along the edges of sails for strength.

BOM Bureau of Mines, *USA.*

BOMAX Barbados Oceanographic and Atmospheric Experiment.

Bona vacantia *Latin*—Goods without apparant owner. Unclaimed property. This can refer to a property in the hands of a trustee or a liquidator in a dissolved company. After a certain time of bona vacantia the authorities generally claim the automatic transfer of the property to the government.

Bond (1) Obligation; Guarantee (an obligation against free bond re-exportation or an obligation to pay General Average, *q.v.*). (2) Warehouse. Chiefly referring to a government or private warehouse in which merchandise is stored for ships' use or re-exportation. Naturally goods taken out of these warehouses are free of tax and/or customs duty. *See* **Bonded Stores**.

Bond Premium The action taken to release an arrested ship by the Bonding Company or Security Company, *q.v.*

Bonded Carman A person who with the bonded lighterman is in charge of transferring goods from one bonded store to another.

Bonded Stores *or* **Bonded Goods** Goods deposited in government or private warehouses which are to be exempt from the payment of customs duty and/or other taxes pending the presentation of documentation to the customs authorities. *See* **Bonded Warehouse**.

Bonded Values *or* **Values in Bond** (1) General Average adjustments in claims against bonded merchandise. (2) The sound and damaged values before the payment of duty. This relates to the customary claims on merchandise sold from the Bonded Stores.

Bonded Vaults A bonded warehouse, *q.v.*, specifically used for wines and spirits.

Bonded Warehouse *or* **Bonded Stores** Imported merchandise is stored directly as soon as it is landed into the Government Bonded Warehouse pending withdrawal to be used for home consumption by its respective importers after custom's formalities have taken place. *See* **Bonded Stores** *or* **Bonded Goods** and **Bond**.

Bonding Company *or* **Surety Company** A group of financiers or associations who bind themselves as fully responsible, on behalf of others with whom they had previously entered into agreement with, against loss, default of payment, damage, bail, etc. Such a bond could release a vessel under Arrest, *q.v. See* **Bond Premium**.

Bonds Certificates issued by a government or corporation confirming the amount of money invested and in some cases the interest payable at stated intervals.

Bon mot *French*—good saying.

Bonne Arrivée *French*—Safe arrival.

bonos mores *Latin*—Good manners.

Bon Ton *French*—Height of fashion.

Bonus A special allowance premium or present over normal wages or payments.

Bon Vivant *French*—Person who lives well.

Booby Hatch A round or square-shaped raised cabin hatch on a small vessel, having a sliding door.

Boodle Counterfeit money or other credits to obtain undeserved credentials or business. A corruption or fraud concerning money.

Book Accounts Debit and credit accounts in book-keeping. *Book-keeping term.*

Book Debts Money which is forthcoming to the trader. *Book-keeping term.*

Booking Note *Similar to* Berth Note, *q.v.*

Book of Rules The Protection and Indemnity Clubs, *q.v.*, *or* Mutual Clubs periodically issue 'Books of Rules' to their members outlining general rules and conditions.

Boom A Derrick, *q.v.*, for handling cargo; that period of time when business is at its best; a long spar for extending foot of sail; a floating barrier of timber across the mouth of a river or harbour.

Boom Carpet The ground strip area from where a Supersonic, *q.v.*, bang is heard.

Boot American slang for Marine Recruit.

Bootlegging Importing or selling alcoholic liquor illicitly.

Boot Top *or* **Boot-top** The waterline sides of the ship which are immersed when she is loaded with cargo. The side areas between the Light Loadline, *q.v.*, and Load Waterline, *q.v.*, are called Boot-top.

BOP *or* **B.O.P.** Blow Out Preventor, *q.v.*

Borax Antiseptic compound used in surgery and is also a preservative of food as well as being useful as a flux in welding to obtain clean metal surfaces. It is a form of white powder.

Bord Bordeaux, *France.*

Bordereau *French*—Memorandum; Invoice; Bill of Lading; Bill of Exchange and similar documents.

Bore A relatively high tidal wave that develops in certain rivers and passes up ahead of the normal tidal flood.

Boreas *Greek*—The North wind.

Borehole A hole drilled into land or into the seabed for the purpose of extracting oil or gas.

BORO Bulk Oil/Roll On/Roll Off. Relating to a ship specially designed for multi-purpose cargo operations.

Bort A useful abrasive agent. An ordinary diamond, alternatively pure carbon which is not suitable for cutting purposes.

BOSCA British Oil Spill Control Association.

Bosn *or* **B'son** *or* **'Bosun** Boatswain, *q.v.*

BOSS *or* **Boss** Bureau of State Security, South Africa; the head of the propeller to which the blades are connected.

Boss Fr. Aft Bossed Frames Aft.

Boss Plt. Boss Plate.

Bosun's Chair A kind of wooden chair or seat suspended by ropes in such a way as to allow a seaman to be lifted up or down any high part of a ship.

BOSVA British Offshore Support Vessels Association.

B.O.T. Board of Trade (now Department of Trade and Industry).

B.O.T.A.C. Board of Trade Advisory Committee.

B.O.T.B. British Overseas Trade Board.

Both Days Inclusive A Charterparty, *q.v.*, phrase where both the first and the last day in the Laydays, *q.v.*, are taken into account in the calculation of expenses to be charged to the Charterer, *q.v.*, by the owner. *Abbrev.* BDI *or* B.D.I. *or* b.d.i.

Both Ends A Charterparty, *q.v.*, clause in relation to the overall counting system of loading and discharging in the Laydays, *q.v.* Thus any time lost either way (while loading or discharging) can be made up or balanced in the next operation. *Abbrev.* BENDS *or* Bends. It also expressed as Reversible Laytime.

Both-to-Blame Collision Clause *or* **Both-to-Blame Clause** A shipping expression quite often inserted in the contracts of affreightment, most especially those relating to trading in the US areas. *Abbrev.* B.B. Clause *or* B. Blame Clause. Under the Collision Convention 1910, if two ships collide they are both to blame for the cargo each are carrying in proportion to their respective blame in the accident. This is likewise accepted where there is a possibility of American law governing the liability for such collision since the USA did not ratify this convention. In US jurisdiction where both vessels are found to be equally to blame for the collision the cargo owners are entitled to recover in full from the other vessel, irrespective of the ratio of the blame involved. The inclusion of the Both-to-Blame Clause in respect of cargo means that the owners are legally bound that claims submitted under US law will later be fully recovered by the ship in accordance with the Collision Convention 1910.

BOTI Board of Trade Inflatable Boat(s).

Bottlescrew *or* **Bottle-screw** A hollowed pipe having two eyes (screwed to either side of the pipe) which when screwed inwards towards each other enables the ropes attached to them to become fastened to the adjustment required. It is used for Rigging, *q.v.*, purposes or for cargo Lashing, *q.v.* Also called Turnbuckle, *q.v.*

Bottom Another maritime word for a 'ship'; the side of a ship below the Boot-top area.

Bottom Composition This is a general term concerning marine paints which are used for the underwater parts of a ship. These paint compositions are today invariably applied to metal ships in order to prevent corrosion and as Antifouling, *q.v.* Thus the ship is painted with an anticorrosive as a first coating and then antifouling paints are applied. The latter paints have certain ingredients which kill the organisms which cling to the bottom of the ship. *See* **Antifouling**.

Bottom Hole Pressure The pressure obtained at the bottom of holes drilled for oil or gas exploration.

Bottom Plating Shell plating positioned below the waterline of a ship.

Bottomry Bond *or* **Bottomry Bill** Also known as Respondentia Bond. A guarantee for the ship or cargo lodged by the master to obtain the necessary funds and continue with the scheduled voyage. This sort of action hardly exists nowadays. Money settlements are easily remitted quickly via airmail, cables or telexes.

Bottom Scraping *or* **Scraping the Bottom** Removing Algae, weeds, barnacles, etc., from under the waterline of a ship by means of scrapers and hard wire brushes and also by high pressure steam cleaning. This is done while the ship is in dry dock. Bottom scraping is also effected afloat by divers.

Bottom Treatment Clause A hull insurance clause which specifies liability for underwater repairs including painting.

B.O.T.U. Board of Trade Unit (now Department of Trade and Industry).

Bought Day Book A book, in book-keeping, kept to record a list of items bought daily.

Bought-out Parts Finished parts not made in the factory but purchased from other sources. Also called Bought-in Parts.

Bounce/Bounced Cheque Terms applying to a dishonoured cheque which cannot be met from the drawer's available funds.

Bound Destined; Anticipated destination.

BOUNDS Buy-Write Option Unitary Market Association *USA*.

Bourse Meeting place where commercial business is transacted and commonly called 'Chamber of Commerce'. The word 'Bourse' is derived from the French.

Bow The fore part of a ship; opposite to stern.

Bow Door Hinged door in the fore end of a ship which is opened to allow vehicles and cargoes to be loaded or discharged. The bow door is sometimes called a Ramp. Mostly used on Ro/ro ships, *q.v.*

Bower A main Anchor, *q.v.*, positioned in the fore part of a ship.

Bowsprit A projecting spar from the bow of a ship.

Bow Stopper A device used to prevent chain cable from moving, fitted to the deck ahead of the windlass.

Bow Thrusters Small propellers fitted to the port and starboard side of a ship's bow area adjacent to the collision bulkhead; used to propel the vessel's bows to port or to starboard, providing greater vessel handling precision. Some vessels also have stern thrusters, *q.v.*

Box The position of the Underwriters, *q.v.*, and their personnel in the underwriting room at Lloyd's *insurance*.

Box Hold A term used to describe the shape of a cargo hold without upper wing tanks or lower hoppers. Useful for unitised cargo such as packaged lumber or containers. Usually found in modern coasters and with open hatches, *q.v.*

Box K Box Keel *Lloyd's Register notation.*

Box Layout Written agreement laid out in printed clauses and others enclosed in a box system. The latter method is mostly used in Charterparties, *q.v.*, where the entire specifications of the contracted vessel and voyage routes are 'enclosed' or inserted.

Box of Tricks Nickname for Black Box, *q.v.*, or Flight Data Recorder on an aircraft.

Box Rate Lump sum freight for FCL, *q.v.*, traffic. So much per container 'Box'.

Box Ship *or* **Box Vessel** Container Vessel, *q.v.*

Box the Compass Nautical term for naming the compass points in regular order.

Boxtime Charter The BIMCO Uniform Charter Party for container vessels.

b/p Boiler Pressure; Boiling Point.

b.p. *or* **B/P** Below Proof; Bill of Parcels; Bills Payable, *q.v.*

BP *or* **B.P.** Bachelor of Pharmacy; Bachelor of Philosophy; Between perpendiculars *Lloyd's Register notation*; Bills Payable, *q.v.*; Blood Pressure; Boiling Point.

BPA *or* **B.P.A.** Bachelor of Professional Arts; Bristol Ports Authority; British Ports Association, London; Business Publications Audit of Circulation Inc., *USA*; Blanket Purchase Agreement.

BPB *or* **B.P.B.** Bank Post Bill.

b.p.c.d. Barrels per Calendar Day *oil well output.*

bpd or **b.p.d.** Barrels per Day, *q.v.*, *oil output production*.

B.P.E. Bachelor of Physical Education.

B. Pet. E. Bachelor of Petroleum Engineering.

b.p.f. Bon pour francs *French*—Value in francs.

B.Pharm. Bachelor of Pharmacy.

Bpgc Bearing per Gyro Compass *nautical*.

bph or **b.p.h.** Barrels per Hour *oil output production*.

B.P.H. Bachelor of Public Health.

B. Phil. Bachelor of Philosophy.

b.p.i. Bytes per Inch *computer*.

B.P.I. Booksellers Provident Institution; British Pacific Islands; Bureau of Public Enquiries.

bpl Birthplace.

bpm or **b.p.m.** Barrels per Minute.

B.Ps. Bachelor of Philosophy.

BPS Base Point System, *q.v.*

bpsd or **B.P.S.D.** Barrels per Steam Day.

BPSS Barge-mounted Production and Storage System *oil drilling*.

bque or **Bqe.** Barque.

b.r. Bank Rate; Bills Receivable, *q.v.*, Block Release.

Br. Brazil; Britain; British; Difference between heading and wind direction.

BR British Rail (*formerly British Railways*); Bulgarski Koraben Registar—Bulgarian Register of Shipping.

B/R or **B.R.** Bill of Rights; Bills Receivable, *q.v.*, Boiler Room *Lloyd's Register notation*; Bordeaux/Rouen Range *grain trade*.

BRA By what right and authority.

Brackish A mixture of salt and fresh water, as may be experienced in a river estuary.

Brake Horse Power *Similar to* **Shaft Horse Power**, *q.v.*

Brake Mean Effective Pressure Average pressure on piston working stroke in an engine. *Abbrev.* BMEP or b.m.e.p.

Branch Pilot American expression for a local pilot who familiarises himself with all the characteristics of the port area.

Branded Goods Goods that bear the trade mark, name or brand of the manufacturers or suppliers to identify them from similar goods produced by others.

BRAS Ballistic Rocket Air Suppression; British Research Advisory Service.

Brass An alloy of 50/70 per cent of copper and zinc with a small percentage of lead.

Brassage Charges made for the minting of coins. Also known as 'Mintage'.

Brass Plate Merchant An importer who uses the space of the dock company or the port authorities until such time as he sells the merchandise to his clients. This saves the cost of using his own storage and transportation.

Bravo International Radio Telephone phonetic alphabet for 'B'.

Braz. Brazil.

B.R.B. British Railways Board.

Br. C. British Columbia, *Canada*.

BRC or **B.R.C.** Broker Regulatory Committee *Lloyd's*.

B.R.C.S. British Red Cross Society.

BRD or **B.R.D.** Building Research Division.

Br. Dk. Bridge Deck.

BRE British Research Establishment.

Breach of Condition A condition which cannot be fulfilled due to a discrepancy in the concluded contract. *Ex.* a vessel in a Charterparty, *q.v.*, is delivered unseaworthy. The charterer has the option of cancelling the contract because of a 'Breach of Condition'.

Breach of Contract Infringement of contract when concluded. A company committing a breach of contract or term of contract is liable for damages. *See* **Strike**.

Breach of Warranty Principally refers to a frustration or non-fulfilment of an implied warranty in a contract such as unreasonable deviation from the normal course of a contracted voyage charter or unreasonable despatch of a ship. *See* **Implied Warranties**.

Breakage Proportion or percentage allowance for breakages in merchandise. *Abbrev.* Brkg.

Breakage of Shafts *See* **Inchmaree Clause**.

Break Bulk Cargo General cargoes carried in a ship which is shown on separate Bills of Lading, *q.v.*, to distinguish the merchandise from the whole shipment. Opposite to Bulk Cargo.

Break Bulker A cargo ship which calls at various ports to pick up different kinds of cargoes.

Breakdown (1) An expression used in the passenger department of a shipping line referring to the detailed number of passengers embarked on a passenger or emigrant ship with corresponding information as to the individual number of males, females, adults and infants. (2) Engine damage or breakdown.

Breakdown Clause A Charterparty, *q.v.*, clause which is in short an 'Off Hire Clause', *q.v.* If time is lost due to the ship becoming inoperative and unseaworthy the vessel automatically goes off hire. The vessel is off from the charter until she is again seaworthy. There is an allowance of a limited time for this breakdown prearranged by the owners and the charterers, say up to 24/48 hours, after which time no charter fees (charter rate fees) are paid until the vessel is in commission again. Also expressed as Loss of Hire.

Break Even When a business is not flourishing or trending downward the managers will endeavour to work on a no profit/no loss basis for the sake of carrying on without interruption.

Break Even Analysis *or* **Break Even Point** The exact equalising part of income derived from the sale with that of the total cost, in which case no profit or loss is involved.

Break Ground Setting the anchors free from the seabed. Also called Break Out, *q.v.*

Breaking Bulk The opening of a Container, *q.v.*, or a consignment to sell part of the contents or to obtain samples. A common word in shipping terminology signifying the opening of the cargo holds of a ship. In some Charterparties, *q.v.*, a clause to this effect is inserted in reference to the commencement of payment of the charter freight on Breaking Bulk. The usual words inserted which have the same meaning are 'Freight to be paid before breaking bulk'. Cargo will not be discharged on arrival of the vessel before the freight agreed is paid to the owners.

Breaking Out (1) The act of removing the legs from the jack-up oil drilling rig from the seabed. The opposite word is 'Spudding', *q.v.* (2) Starting to unload merchandise from the hold of a ship. (3) Removing part of a stock or section of merchandise from a transit warehouse.

Breaking Strength The weight of load or pull on a rope or wire which will destroy it.

Break Out Similar to Break Ground, *q.v.*

Break-up Voyage A Marine Insurance, *q.v.* Policy, *q.v.* on a vessel scheduled to sail from one port to another (normally direct) for breaking up or scrapping. In some instances, the vessel is allowed to load scrap metal cargo. The insurance thus effected will generally be on a flat total loss basis. This could lead to the underwriter's assessment of the actual cost of the scrap value to be insured for the voyage.

Breakwater Strong structural material intended to break the force of the sea waves. A breakwater is constructed outside an open harbour to protect it from large waves and make it possible for ships to berth safely. This word can also refer to protection on the weather deck of a ship to stop waves from overriding the deck.

Breeches Buoy A canvas appliance slung along a rope which is used to convey men or stores from ship to ship and vice versa whenever sea conditions do not permit an easy and safe approach.

Breasthook (1) A bracket in a form of a hook on both sides at the far end or stern of a ship. (2) Flat plates in a triangular form attached to the bow plating of a ship.

Breast Line Mooring Line, *q.v.*, or Hawser, *q.v.*, to hold a ship alongside the quay. Leads roughly at right angles to fore-and-aft line of the ship.

BREEMA British Radio and Electronic Equipment Manufacturers' Association.

BRF British Road Federation.

Brg. Bearing, *q.v.*

BRH *or* **B.R.H.** British Race Hovercraft.

Bribe Money, etc., offered to procure (often illegal or dishonest) action in favour of the giver.

Bridge (1) Erected structure amidships or aft or very rarely fore over the main deck of a ship to accommodate the wheelhouse. The crew accommodation is generally situated below that of the officers under the bridge or sometimes aft, depending on the structure and type of ship. (2) A structure carrying a road or path or rail tracks across a stream, ravine, road, river, valley or narrow seas.

Bridge Deck Ship A ship with a deck above the shelter deck.

Bridge Policy An American marine policy covering accidents and damage to bridges.

Bridging Loan Borrowed money sufficient to cover essential requirements in keeping a business going; or to assist in buying article while waiting to dispose of another.

Bridles Ropes, chains or wire ropes used for towing purposes. These are tied around the two Bollards of a tug and the other ends are fixed to an Eye, *q.v.*, or Shackle, *q.v.*, where the towed Ship's Tackles, *q.v.*, are attached.

Brief A document containing the materials or instructions furnished by a solicitor to a barrister to enable him or her to represent the client in the trial of a criminal case or of an action, etc.

Briefing An outline or summary of a situation and intended actions; Formally passing on essential information.

brigandage sur mer *French*—Maritime Crime; Piracy, *q.v.*

Brig. Gen. Brigadier General.

Brightwork A general and collective word for polished brass on a ship or yacht.

Britannia Definition of silver of 95.8% fineness.

Brit. British; Britain.

Brit. Ass. British Association for the Advancement of Science.

Britcont *or* **BRITCONT** Code name for General Home Trade Charterparty 1928 for use in trades for which there is no Chamber of Shipping C/P.

Britconbill *or* **BRITCONBILL** Bill of Lading, *q.v.*, code name used in conjunction with the Chamber of Shipping General Home Trade Charter 1928.

BRITE Basic Research in Industrial Technology for Europe; Bright Radar Indicator Tower Equipment.

British Funds British Government Stocks, or Securities, *q.v.*

British Insurance Association A British insurance association whose aims are to publicise the insurance industry and its activities. *Abbrev.* BIA *or* B.I.A.

British Insurance Group A British group of companies who are engaged in insurance in Japan. *Abbrev.* BIG *or* B.I.G.

British Management Training Export Council An organisation formed by a group of professionals during 1977 to provide a unified front in international marketing and expertise in the training of British management. *Abbrev.* BMTEC.

British Marine Equipment Council *Abbrev.* BMEC. A federation of the British Naval Equipment Association (BNEA), the British Marine Equipment Association (BMEA), the Association of British Off-shore Industries (ABOI), and the British Oil Spill Control Association (BOSCA), which acts as the collective voice of the marine equipment industry in representation to government and in active co-operation with other maritime organisations in the UK and Europe. Beside promoting the interests of its 150 members, it is involved in the harmonising of European equipment standards and procurement procedures.

British Maritime Charitable Foundation, The Established in 1982, its aim is to promote Britain's maritime industry, commerce and defence for the benefit of the nation through education, training and research. Its charitable objectives were eventually enlarged to combine with those of the UK Centre for Maritime Policy Studies, originally known as the British Maritime League.

British Maritime Law Association An organisation that keeps all aspects of maritime law under review, assists in formulating UK policy and makes a contribution to the Comité Maritime International in Antwerp. It was established by several organisations such as Lloyd's, the Institute of London Underwriters and the Chamber of Shipping assisted by some eminent judges, and its membership extends to maritime law firms, P&I Clubs, average adjusters, university faculties, cargo underwriters and others, including shipowners and salvage interests. *Abbrev.* BMLA.

British Maritime Technology (BMT) was formed in 1985, with UK Government support, by the joint privatisation of two of the UK's leading maritime organisations; the National Maritime Institute (NMI) and the British Ship Research Association (BSRA). NMI had strong historical links with the National Physical Laboratory and represented the national centre for large scale experimental facilities for vessel hydrodynamics and wind aerodynamics. NMI gained a worldwide reputation in the fields of vessel movement and simulation techniques and was closely associated with the development of the North Sea offshore structures throughout the 1970s and early 1980s. NMI produced the first authoritative atlas of global wave statistics which has been substantially refined in recent years by BMT. BSRA had a long and complementary history in fields aligned to the design, build and performance of vessels, and provided the underpinning research technology for the British ship-building industry. In particular, BSRA was responsible for the development of major vesel geometry design and production control software which is presently used world wide. As an independent consultant and international centre of technical excellence for the maritime industry, BMT operates in four main markets: commercial shipping, maritime defence, ports and harbours and offshore oil and gas. Specialist services are provided in such related fields as coastal and estuarial hydraulics, applied oceanography, marine science, navigation studies, ship manoeuvring simulation, and maritime risk assessment. BMT draws on a pool of over 500 highly qualified scientists, technicians, engineers and project managers.

British Pharmacopoeia A comprehensive medical list of measures, mixtures, preparations and weights.

British Sailors' Society The British Sailors' Society was established in 1818 by a group of London Christian businessmen who were concerned with the spiritual state of sailors entering the Port of London. From this beginning grew the worldwide Society as it is today, with seafarers' clubs in many countries and dedicated chaplaincy and other staff supported by voluntary workers caring for seafarers of all nationalities. The Personal Affairs Department assists many families of seafarers in the United Kingdom and overseas each year where the need arises. A children's home and two seafarers' homes are maintained. The Society is dependent on voluntary contributions.

British Thermal Unit The amount of heat required to raise the temperature of one pound weight of pure water to one degree Fahrenheit (American Standard *abbrev* Btu; Conventional British system B. Th. U. or BTU or B.T.U.) at a pressure of 30 inches of mercury. One Thermal Unit equals the French equivalent of 0.252016 Kilogram Calorie. One Kilogram Calorie equals 3.968 British Thermal Units or 10,000 gram calories.

Brkg. Breakage, *q.v.*; Brokerage, *q.v.*

brkt Bracket.

Brl. Barrel.

Broach to Said of a vessel under sail when she is running free (before the wind) if, through bad steering or a large sea striking the Quarter, *q.v.*, she turns suddenly into the wind.

BROB *or* **B.R.O.B.** Bunkers Remaining on Board. *Also abbrev.* BOB *or* B.O.B. Bunkers on Board.

Broderna *Swedish*—Brothers.

Brodogradiliste *Yugoslavian*—Shipyard.

Brodren *Danish & Norwegian*—Brothers.

Brng. Bearing, *q.v.*

Brng rel. Relative Bearing *nautical.*

Broken Deck A ship which has the main deck not flush, such as a ship with a raised quarter deck; An amount of dunnage used to steady the cargo. *See* **Grain Capacity** *or* **Grain Space** and **Bale Capacity**.

Broken Stowage The spaces lost and unoccupied between each side of an individual parcel of cargo while being stowed in the hold of a ship. Loss of space in this broken stowage accounts for the uneven shape of the parcels but can also be due to the shape of the holds. Dunnage, *q.v.*, is used between the spaces to keep the parcels steady and stop them from shifting. If the packages are regular in size they are easily stacked or stowed with practically little if any broken stowage, as opposed to the uneven ones which need a substantial amount of dunnage. *Abbrev.* BS or B.S.

Broker Agent employed (at a customary or an agreed rate of commission or remuneration) to buy or sell goods, merchandise or marketable securities, or to negotiate insurances, freight rates or other matters, for a principal; the sales or transactions being negotiated not in his own name but in that of the principal.

Broker/Dealer A London Stock Exchange member firm which provides advice and dealing services to the public and which can deal on its own account. *See* **Broker**.

Brokerage *See* **Commission Brokerage**.

Brokerage Clause A charterparty clause stipulating the brokerage rate allowed to the broker at the successful conclusion of business. *See* **Broker** *and also* **Brokerage**.

Broker of Record An American expression for a licensed insurance broker nominated by the assured.

Broker's Cancellation A hull insurance clause protecting the broker in case the assured is slow in paying the Premium, *q.v.*, on a time policy. In this case the broker has the right to cancel the policy.

Broker's Cover Sometimes called Master Cover. An open cover form in the name of the agent or broker binding a broad degree of insurance from Underwriters, *q.v.*

Bronze An alloy of copper and tin.

Bros. Brothers.

Brot *or* **Brt** Brought.

Brought to and Taken from Alongside A shipping term signifying that the ship only accepts cargo for loading if brought to the ship's side. It also applies to the discharge of cargo and therefore the responsibility for bringing or discharging cargo alongside is that of the consignors and consignees respectively.

Brought–Up When a ship is anchored securely *nautical*.

Brow The passageway or Gangway, *q.v.*, from ship to quay.

Brown Water Expression used in matters pertaining to coastal and estuarial sea areas.

BRPC Basic Radiation Protection Criteria.

brs Brass.

BRS *or* **B.R.S.** British Road Services.

brz Bronze.

brzg Brazing.

BS *or* **B.S.** Bachelor of Science; Bachelor of Surgery; Bill of Sale; Boiler Survey *Lloyd's Register notation*; British Standard.

B.S. *or* **B/S** Balance Sheet; Broken Stowage, *q.v.*

B/S Balance Sheet; Biennial Survey *Lloyd's Register notation*; Bill of Sale, *q.v.*; Bill of Sight, *q.v.*; Bill of Store; Butt Strap.

BSA *or* **B.S.A.** Bachelor of Science in Agriculture; Belgium Ship Suppliers, an Association *similar to* **B.B.S.** and **U.P.A.N.**; Boy Scouts of America; British Shipowners' Association.

BSAA *or* **B.S.A.A.** Bachelor of Science in Applied Arts.

BS Adv *or* **B.S. Adv.** Bachelor of Science in Advertising.

BSAE *or* **B.S.A.E.** Bachelor of Science in Aeronautical Engineering; Bachelor of Science in Agricultural Engineering; Bachelor of Science in Architectural Engineering.

BS. Arch. *or* **B.S. Arch.** Bachelor of Architecture.

BS Art. Ed. *or* **B.S. Art. Ed.** Bachelor of Science in Art Education.

BSBA *or* **B.S.B.A.** Bachelor of Science in Business Administration.

BS. Bus *or* **B.S. Bus** Bachelor of Science in Business.

B. Sc. Bachelor of Science.

B. Sc. (Econ). Bachelor of Science (Economics).

BSC Breakbuild Service Charges.

B.S.C. British Shippers' Council; British Sugar Corporation; British Steel Corporation.

BSC *or* **B.S.C.** Bachelor of Science in Commerce; Basic Sea Charge.

B.S.C.C. British-Swiss Chamber of Commerce; British-Swedish Chamber of Commerce in Sweden; British Soviet Chamber of Commerce.

BSCE *or* **B.S.C.E.** Bachelor of Science in Civil Engineering.

BS Ch *or* **B.S.Ch.** Bachelor of Science in Chemistry.

BS Ch E *or* **B.S.Ch. E.** Bachelor of Science in Chemical Engineering.

B. Scheme Balance Settlement Scheme.

BSCP *or* **B.S.C.P.** British Standard Code of Practice (issued by the B.S.I., i.e. British Standards Institution).

BSCS *or* **B.S.C.S.** British Shipping Careers Service.

BSc (Soc) *or* **B.Sc. (Soc).** Bachelor of Science in Social Science.

BSD *or* **B.S.D.** Bachelor of Science in Design.
Bsd. Barrels per Steam Day.
BSDA *or* **B.S.D.A.** Bonded Store Dealers of America; British Shipowners' Demurrage Association.
BSE *or* **B.S.E.** Bachelor of Science in Education; Bachelor of Science in Engineering.
BS Ec *or* **B.S.Ec.** Bachelor of Science in Economics.
B.S. Ed. Bachelor of Science (Education).
BSEM *or* **B.S.E.M.** Bachelor of Science in Engineering of Mines.
BSEP *or* **B.S.E.P.** Bachelor of Science in Engineering Physics.
BSF *or* **B.S.F.** Bachelor of Science in Forestry; British Shipping Federation; British Standard Fine.
B.S.F. British Shipping Federation Limited.
b.s.f.c. Brake-Specific Fuel Consumption.
BSFSSF Business Stabilization Foundation for Specific Shipbuilding Firms, *q.v.*
BSG *or* **B.S.G.** British Standard Gauge.
b.s.g.d.g. Breveté sans garantie du Gouvernment *French*—Patented Without Government Guarantee.
bsh Bushel.
BSHA *or* **B.S.H.A.** Bachelor of Science in Hospital Administration.
BSHS *or* **B.S.H.S.** Bachelor of Science in Home Studies.
B.S.I. British Standards Institution, previously B.E.S.A., *q.v.*
BSIE *or* **B.S.I.E.** Bachelor of Science in Industrial Education.
BSIR *or* **B.S.I.R.** Bachelor of Science in Industrial Relations.
BSJ *or* **B.S.J.** Bachelor of Science in Journalism.
BSJC *or* **B.S.J.C.** British Seafarers' Joint Council.
BSL *or* **B.S.L.** Bachelor of Science in Law.
BSLS *or* **B.S.L.S.** Bachelor of Science in Library Science.
BSM *or* **B.S.M.** Bachelor of Science in Music; Bachelor of Science in Sacred Music; Battery Sergeant Major.
BS Met E *or* **B.S. Met. E.** Bachelor of Science in Metallurgical Engineering.
BS Min *or* **B.S. Min.** Bachelor of Science in Mineralogy.
BSMT *or* **B.S.M.T.** Bachelor of Science in Medical Technology.
BSN *or* **B.S.N.** Bachelor of Science in Nursing.
BSO Business Statistics Office.
BSOA British Steamship Owners' Association.
BSOT *or* **B.S.O.T.** Bachelor of Science in Occupational Therapy.
BSP *or* **B.S.P.** Bachelor of Science in Pharmacy; Black Short Period *insurance*.
BSP Bank Settlement Plan (air passenger ticket document form), *q.v.*; Bronze Spare Propeller.
B.S.P. British Standard Pipe-thread.

BSPA *or* **B.S.P.A.** Bachelor of Science in Public Administration.
BSPE *or* **B.S.P.E.** Bachelor of Science in Physical Education.
BSPH *or* **B.S.P.H.** Bachelor of Science in Public Health.
BSPHN *or* **B.S.P.H.N.** Bachelor of Science in Public Health Nursing.
BSPT *or* **B.S.P.T.** Bachelor of Science in Physio-Therapy.
BSR *or* **B.S.R.** Basic Sea Rate.
BSRT *or* **B.S.R.T.** Bachelor of Science in Radiological Technology.
B.S.S. Bachelor in Social Science; British Standards Specification (relating to aluminium as a metal, aluminium doors, windows and similar metal structures).
B.S.S.A. Bachelor of Science in Secretarial Administration.
BSSS *or* **B.S.S.S.** Bachelor of Science in Social Science.
bsst *or* **BSST** Brick or Stone, Slated or Tiled. *Insurance.*
BST *or* **B.S.T.** British Standard Time; British Summer Time.
B/st *or* **B/St.** Bill of Sight, *q.v.*
B. Surv. Bachelor of Surveying.
BSW *or* **B.S.W.** British Standard Whitworth (screw threads system).
BS & W *or* **B.S.& W.** Basic Sediment and Water; Bottom Settling and Water.
b.s.w. Barrels of Salt Water.
B.S.W.I.A. British Steel Wire Industries Association.
B.T. Bachelor of Teaching; Bachelor of Theology.
Bt. Baronet; Bought.
BT *or* **B.T.** Bow Thruster, *q.v.*
BTA *or* **B.T.A.** Blood Transfusion Association *USA*; Bulgarian Telegraph Agency; British Tourist Authority; British Travel Association; British Trade Association; British Tug Owners' Association.
BTC Blackpool Corporation Transport.
BTC *or* **B.T.C.** Bankers' Trust Company; British Textile Confederation; British Transport Commission.
B.T.C.C. Board of Transportation Commission of Canada.
B & T Cl. Blocking and Trapping Clause *insurance*.
b.t.d. Bomb Testing Device.
BTD *or* **B.T.D.** British Transport Dock.
B.T.D.B. Bermuda Trade Development Board; British Transport Docks Board.
b.t.d.c. Before top dead centre.
Bt.Dk Boat Deck.
bté Breveté *French*—Patented.

BTEC Business and Technician Education Council.

B. Tech. Bachelor in Technology.

B.T.F. British Trawlers Federation.

BTF *or* **b.t.f.** Barrels of total fuel.

BTG *or* **B.T.G.** British Technology Group.

B. Th. Bachelor of Theology.

bth/s Berths *Lloyd's Register notation.*

B.T.H.A. British Travel and Holiday Association.

B. Th.U. British Thermal Unit, *q.v.*

Btk.L. Buttock Line.

btm *or* **B.T.M.** Bottom, *q.v.*

BTM Buoyant Turret Mooring *oil drilling.*

BTMA British Textile Machinery Association.

BTN *or* **B.T.N.** Brussels Tariff Nomenclature.

Btn Between.

BTO *or* **B.T.O.** British Tug Owners Association.

B. to Blame *or* **B to B** *or* **B. Blame Clause** Both-to-Blame or Both-to-Blame Collision Clause, *q.v.*

B.T.P. Bachelor of Town Planning.

B.T.R.A. Bombay Textile Industry's Research Association.

BTS *or* **B.T.S.** Brokers Telex Service.

BTTMC British Truck Trailer Manufacturers Council.

BTU *or* **B.T.U.** Board of Trade Unit; Bow Thrust Unit(s). British Thermal Unit, *q.v.*

B.T.U.C. Bahamas Trade Union Congress.

Btwn Between.

bu. Base Unit; Bureau; Bushel, *q.v.*; Bushel Volume.

Bu *or* **Bur.** Bureau.

Bu. Aer. Bureau of Aeronautics *USA.*

BUAF *or* **B.U.A.F.** British United Air Ferries.

Bubble Used in connection with excessive speculation in the prices of Stocks and Shares, *q.v.*, by a Bubble Company, *q.v.*, and whose real value could be much less, to the detriment of clients.

Bubble Company An ordinary company which has never succeeded in establishing itself due to its dishonest policy in trading. *See* **Bubble.**

Bucket Elevator A fast loading/discharging method used in the bulk cargo trades. A mechanical rotation belt is attached to its various adjacent buckets which fill in from one end (the shore or silo) and discharge at the other end (the holds of the ship) or vice versa. Bulk grains are today discharged by means of suction hoses which give an easier and quicker despatch.

Bucket Shop (1) Slang for an unlicensed trader in commodities or stocks. (2) A dealer in discounted air tickets.

Bucks Buckinghamshire, *UK.*

Budget Account An account whereby a customer pays a certain amount of money per month after which he is permitted to buy articles which exceed the deposited amount. In return the customer pays an agreed amount of interest on the balance outstanding per month.

Budget Controller A person responsible for the budget in a company.

BuDocks Bureau of Yards and Docks *USA.*

BUE British Underwater Engineering.

BUF *or* **B.U.F.** Back-Up Force, *q.v.*

Bugserselskab *Danish*—Towing Company.

Builders Certificate Certificate produced by the builders of a ship as soon as she is launched and delivered to her new owners. It contains a detailed description of the hull, engines, holds, stability, speed, deck machinery etc.

Builders' Policy *Similar to* **Construction Policy,** *q.v.*

Builders' Risk Standard clause inserted in the Institute Clauses for shipbuilders' risk covering any damage occurring during construction up to the launch of a new ship.

Bulbous Bow A design of bow favoured for modern vessels intended to reduce the drag effect of water along a vessel's hull.

Bulcon *or* **BULCON** Code name for Bulgarian Berth Contract 1911. Baltic and White Sea Conference Coal charterparty. Carriage of coal from the Baltic to the White Sea and Scandinavian Ports. Amended 1931.

Bulg. Bulgaria.

Bulgarski Koaben Registar Bulgarian Register of Shipping. *Abbrev.* BR.

Bulge *Similar to* **Bilge,** *q.v.*

Bulk Air Mail Palletised mail service provided by some private couriers.

Bulk Buying It is always the case when buying goods in large quantities in one bulk the price charged will be less than the unit price. Some governments or organisations undertake to buy for the whole population or for a group of businessmen to achieve this economical price. When the goods arrive at their destination they are in turn handed or allotted to the various distributors or to the general public.

Bulk Carrier A ship designed to carry dry bulk cargoes. Larger sized bulk carriers from Panamax, *q.v.*, size upwards tend to be gearless, relying on shore equipment to load and discharge. Smaller bulk carriers are usually geared—most modern ones with cranes rather than derricks.

Bulk Container Ship or Vessel A ship which can carry containers in the holds or on deck as well as bulk cargoes. *See* **Conbulker.**

Bulkhead A ship's partition dividing the interior into various compartments for reinforcement, strength and safety. *Abbrev.* Bhd.

Bulkhead Draft *or* **Bulkhead Draught** The maximum permissible mean Draught (Draft), *q.v.*, of the seasonal loading of a passenger vessel.

Bulk In–Bulk Out New concept of Bulkers which significantly reduces the period of loading and discharging especially meant for sugar handling. The vessel is loaded with bulk sugar with a small percentage being bagged. The sugar in bags is directly unloaded into trucks alongside the quay. It is presumed that 12,000 tons of bulk sugar can be loaded and discharged in about five days. *Abbrev.* BIBO.

Bulk Liquid Bag Collapsible Container. *See* **Collapsible** under **Container**.

Bulk Liquid Container A **Container**, *q.v.*, adaptable to contain bulk shipments of chemicals, wine, spirits and oils.

Bull An investor who buys shares for a short time with the intention of disposing of them as soon as their value increases, aiming for big profits.

Bull. Bulletin.

bull Bulla *Latin*—Papal Seal.

Bulling Pouring water into a cask that has held spirits to obtain a weak grog.

Bullion Uncoined gold and silver shipped in ingots or bars and packed in very strong cases for security. The vessel receiving bullion must be fitted with a proper strongroom. In general gold bars are not to exceed 400 ozs each in weight.

Bullish Describes the stock market when share values are expected to rise.

Bullrope A hawser passed from the stern or bow of a ship to protect her from hitting the mooring buoy.

Bulwark Plating around the outboard edge of the upper deck of a ship to prevent entry of the sea. *Abbrev.* bwk.

Bumbershoot or **Umbrella Liabilities** Underwriters extend cover to the entire liability of the assured, including mental disability, personal injury, etc.

Bumboat A boat engaged in selling and supplying stores alongside ships in a harbour. In some countries this system is being done away with. *See* **Bumboatman**.

Bumboatman A person selling or supplying articles from a Bumboat, *q.v.*

Bumes or **BUMES** Bureau of Medicine and Surgery *USA*.

BUNAV Bureau of Navigation *USA*.

Bundesbank The Central Bank of Germany, established 1957. It aims to protect the German Mark and control inflation. It is shielded from government interference and so can resist political pressure. It plays an important role in establishing monetary policy for the European Union.

Bunk A sleeping berth in a ship which is a built-in bed to save space. Also rectangular box style furniture used as a seat during the day and converted into a bed during sleeping or rest time.

Bunker or **Bunkering** To replenish or replenishing the ship with fuel. Bunker also means a tank or compartment for the stowage of fuel.

Bunker Adjustment Factor Adjustment to freight rates to compensate for fluctuations in the market price of fuel; Bunker Surcharge. *Abbrev.* BAF.

Bunker Clause Charterparty, *q.v.*, (1) clause obliging the owners and the charterers to pay for the vessel's remaining bunkers at the port of delivery (by the charterers) and re-delivery (by the owners). (2) Clause describing the quality of bunkers to be supplied to a vessel.

Bunkers Fuel consumed by the engines of a ship; compartments or tanks in a ship for fuel storage.

Bunker Tanks Tanks which are positioned at the bottom compartment under or near to the engine room for the purpose of holding fuel Bunkers, *q.v.*, for the ship's use. *See* **Ballast**.

Bunt Middle part, the belly, of a square sail; Centre section of a yard where it crosses the mast.

Bunting Thin woollen stuff of which flags, pennants and ensigns are made. The term may refer to a ship's stock of flags.

BUNY Board of Underwriters of New York.

Bu. Ord. Bureau of Ordinance *USA*.

Buoy Charges Charges imposed on ships for the use of buoys while entering harbours in the UK.

Buoy Dues *Similar to* **Buoy Charges**, *q.v.*

Buoyancy The difference between the weight of an immersed, or partly immersed, object and the upward pressure of the liquid in which it lies. *Abbrev.* buoy.

Buoyancy Tank An empty and hermetically closed tank, usually of metal, either rectangular, square, or oval which is positioned around the sides or fore and aft of lifeboats to prevent them foundering during rough weather.

Buoyant The property of floating on the surface.

Buoy Jumper Seaman who goes onto a mooring buoy to secure a ship's picking-up rope and the cable.

Buoy Rope A wire or rope rove through a rung on a buoy.

Bupers or **BUPERS** or **Bu. Pers.** Bureau of Naval Personnel *USA*.

Bu. Pub. Aff. Bureau of Public Affairs *USA*.

Burden or **Burthen** The weight or carrying capacity of a ship in tons deadweight. *See* **Deadweight Tonnage**.

Burden of Proof In law it is the obligation on the part of the plaintiff or the prosecutor to prove what he is alleging.

Bureau A French word for Office.

Burdened Vessel or **Burdened Ship** The navigating vessel responsible for altering course or 'giving way' as per Rule of the Road at Sea. Alternatively, the other vessel which has the right of maintaining her original course is called 'Privileged Vessel'.

Bureaucracy Government by officials responsible only to their departmental chiefs. Thought of as a system where petty regulations assume too much importance and decisions are unnecessarily delayed.

Bureau-de-Change An office which deals in the exchange of foreign currencies.

Bu.Rec. Bureau of Reclamation *USA*.

Burgee A tapered flag in a triangular shape hoisted as a house flag and more usually used by a yacht club; a swallow-tailed flag.

Burning Oil Kerosene or paraffin.

Burning Ratio A reinsurance phrase being the ratio arrived at when collating the sum of money insured against the corresponding losses or claims.

Burn-Off Weight The fuel consumption of an aircraft during the flight. *See* **Take-Off Weight**.

Bursting of Boilers In marine insurance this refers to the latent or hidden defects in the boilers which would result in bursting when in use. *See* **Inchmaree Clause**.

Burthen *or* **Burthen Cargo Tonnage** The carrying capacity of a ship. *See* **Burden** *or* **Burthen**.

Burtoning The use of two adjacent derricks for a quicker loading and unloading operation. One of the derricks is placed at an angle on top of the hold and the other horizontally (or at an angle) pointing to the side of the ship. The wires from the two derricks are shackled together in such a way that the lifted cargo, whether alongside the ship or in the holds, can be shifted and lowered in a quick and easy way. This is a common American term. Double Whip *or* Split is another expression.

Busand A. Bureau of Supply and Accounts *USA*.

B.U.S.F. British Universities Sports Federation.

Bushel A standard grain measure also used for dried fruits. The American bushel measures 2149.21 cubic inches, equivalent to 77.5 lbs or 9.7 gallons and the English bushel measures 2218.2 cubic inches, equivalent to 80 lbs or 10 gallons of water. *Abbrev.* bu.

Bu. Ships Bureau of Ships *USA*.

Business General commercial activity performed by a person, firm, corporation or a company, etc., on an intended profit basis. This can involve extensive transactions or a single trade.

Business Enterprise Company; Corporation; Firm. Foresight in a business project with anticipation of a successful product and/or result.

Bus. Mgr. Business Manager.

Business Reply Service A Post Office reply system where a person wishes to obtain a reply from another person without putting that person to the inconvenience of paying postage for the reply. This entails paying the postage for the reply. No postage stamp is necessary. The person provided with the service must obtain a licence to do it from the Post Office and has to pay the relevant cost.

Business Stabilization Foundation for Specific Shipbuilding Firms A body set up by the Japanese Government in order to buy redundant shipbuilding yards and real estate during depression periods. *Abbrev.* BSFSSF.

Butane A colourless inflammable hydrocarbon gas (C_4H_{10}) obtained naturally from wells or extracted from petroleum. It is widely used as a refrigerant, a solvent and a fuel.

Butt *or* **Pipe** A wine or ale cask having a capacity of three barrels or 108–140 gallons.

Buyer's Market When goods are plentiful and low prices favour consumers.

Buyers' Over A market term meaning when more buyers are available in the market than sellers. This is the opposite to Sellers' Over.

Buyers' Representative Clause This commonly refers to the sale agreement of a vessel. The sellers allow the buyers' representative to be present on board the vessel prior to the delivery date to familiarise himself with the manoeuvrability and general operation of the ship. The representative is put on board at the sole risk and expense of the buyers.

Buying In On a stock exchange when the promised shares are not presented on time the buyer has the option to buy from elsewhere and charges the extra expense to the seller who was unable to meet his commitments.

Buying Rate of Exchange The price at which the bank or other company will purchase local or foreign currency.

BV *or* **B.V.** Belloten Vennootschap, Dutch Private Liability Company; Bureau Veritas.

BVI *or* **B.V.I.** British Virgin Islands.

BVM Business Visitors Memorandum; Beata Virgo Maria *Latin*—Blessed Virgin Mary.

B.V.M.A. British Value Manufacturers' Association.

B.V.M.S. *or* **B.V.M. & S.** Bachelor of Veterinary Medicine & Surgery.

B.V.S. Bachelor of Veterinary Surgeons.

bw *or* **b.w.** Barrels of Water.

b.w. Bitte wenden *German*—Please Turn Over; Bridgeways.

BW *or* **B.W.** Board of Works; Bonded Warehouse, *q.v.*; British Waterways; Butt Weld.

BWA *or* **B.W.A.** Backward Wave Amplifier.

BWAD. Brackish Water Arrival Draft (Draught).

B-Way Broadway *USA*.

BWB *or* **B.W.B.** British Waterways Board.

B.W.G. *or* **B.w.g.** Birmingham Wire Gauge.

B.W.I. British West Indies, now West Indies.

bwk. Brickwork; Bulwark, *q.v.*

B.W.M.A. British Woodwork Manufacterers' Association.

BWP Bronze Working Propeller.

bwpd *or* **b.w.p.d.** Barrels of Water Per Day.

bwph *or* **b.w.p.h.** Barrels of Water Per Hour.

bwr *or* **B.W.R.** Boiler Water Reactor.

BWT Bleeding Wing Tank(s), *q.v.*

bwv Back Water Valve.

Bx. K. Box Keel.

bx/s Box(es).

By *or* **BY** Blowing Spray *weather observations.*

By-laws Conditions or regulations set out by a public body, society, company, corporation or any other private company. Also spelt Bylaws, Byelaws or Bye-laws.

BYO Bring your own.

By-Pass Engine Jet Engine *aviation.*

By-Product The waste of a product forming another useful product in manufacturing.

Byr/s Buyers.

By the Head When the Draught (Draft), *q.v,* forward of a ship is greater than that aft. *Opposite to* **By the Stern**, *q.v.*

By the Lee The situation of a vessel going before the wind when she has fallen off so much as to let the wind go past the stern and her sails may be taken aback.

By the Stern When the draught (draft) aft of a ship is greater than that forward. *Opposite to* By the Head, *q.v.*

By the Wind A vessal sailing as close as possible to the wind. *Similar to* Full and Bye, *q.v.*

Bz Benzine.

bzw. Beziehungsweise *German*—Respectively.

C

c *or* **C** Capacity; Carat; Carton; Case; Cent; Centigram; Centimetre; Centre; Chapter; Curie, *q.v.*; Currency; Current.

C Candle; Capacity; Capsule; Carbon; Cathode; Celsius; Centigrade; Centre *Lloyd's Register notation*; Chancery; Chapter; Chief; Chronometer time *nautical*; Church; Circa; Circus; City; Cloudy *nautical*; Collected; Compass Direction; Congress; Constant; Consul; Convertible; Correction *altitude*; Coulomb *electric charge*; Course; Course Angle *nautical*; Electrical capacity measured in Farad; Detached Clouds *meteorology*; Total capacity of containers; *Lloyd's Register notation* signifying that the ship's engines are compound; Roman numeral for the figure 100.

C-1 Vessel A United States Maritime Commission term in regard to standardisation of a cargo vessel having a single screw with diesel or turbine engines.

C-2 Vessel A United States Maritime Commission term in regard to standardisation of a cargo shelterdeck vessel with diesel or turbine engines.

C-3 Vessel A United States Maritime Commission term in regard to standardisation of a cargo shelterdeck vessel or a cargo passenger vessel having a single screw with diesel or turbine engines.

C.II Form Lloyd's Register of Shipping Form related to a vessel's Load Line.

C/- Coupon; Currency.

C/- *or* **Cs.** Case.

ca Circa *Latin/Italian*—About.

Ca Companhia *Portuguese*, Compania *Spanish*, Compagnia *Italian*—Company.

CA California, *USA*; Calcium; Controlled Atmosphere.

C.A. Central America; Controlled Atmosphere.

C/A Charterers' Agent/s.

C/A *or* **C.A.** Capital Allowances Tax; Capital; Current Account; Caterers' Association; Central Accounting; Central America/Africa; Certificate of Airworthiness; Certifying Authority; Chartered Accountant; Chemical Abstracts; Chief Accountant; Child Allowances Tax; Chronological Age; Church Assembly; City Attorney; Civil Affairs; Civil Aviation; Claim Agents *insurance*; Claim Adjustment; Classical Association; College of Arms; Command Accountant; Commercial Agent; Community Association; Companies Act; Comptroller of Accounts; Constituent Assembly; Consular Agents; Consumer's Association;

Current Alternating; Current Assets; Court of Appeal; Cruising Association.

CAA *or* **C.A.A.** Capital Allowance Act; Central African Airways Corporation; Civil Aviation Aeronautics *USA*; Civil Aviation Authority; Collision Advance Aids; Community Action Agency *USA*; Cost Accountants' Association.

CAAA Commuter Airline Association of America.

CAACE Comité des Associations d'Armateurs des Comunautes Européenes *French*—Committee of the Common Market Owners' Association.

CAADRP Civil Aircraft Airworthiness Data Recording Programme.

CAAIS Canadian Assisted Action Information Systems.

CAAS Civil Aviation Authority Singapore.

CAB *or* **C.A.B.** Captured Air Bubble referring to Air Cushion Vessel; Civil Aeronautics Board *USA*; Co-operative Analysis Broadcasting; Consumers' Advisory Board.

Cab (1) The visual control room *aviation*. (2) Sheltered corner position of a bridge. (3) Cabriolet. All types of wheeled carriage with a number plate for hire to the general public for transportation purposes—a taxi. The legal expression for cab is 'hackney carriage'.

CABA *or* **C.A.B.A.** Compressed Air Breathing Apparatus.

Cabby Cab Driver; a taxi driver.

CABE *or* **C.A.B.E.** *or* **c.a.b.e.** Charterers' Agents Both Ends. Charterparty expression.

Cabin A small room or compartment in a ship or aircraft to accommodate officers, crew or passengers.

Cabin Boy A youth employed on a passenger ship whose duty is to wait on officers and passengers.

Cabin Class An air or sea passenger seat or accommodation costing less than 1st or special class.

Cabin Passengers Passengers on ships who are accommodated in cabins as opposed to Steerage Passengers, *q.v.*

Cabin Staff Stewards, *q.v.*, and stewardesses of the airline crew attending to the needs of passengers and other crew members.

Cable Telegram; A nautical length of 600 feet or 100 fathoms.

Cable *or* **Telex Transfer** A money remittance from one bank to another overseas bank by means of cable or telex. The telex system accelerates the payment procedure.

Cable Car *or* **Cableway** A form of car or carriage which moves along a cable carrying passengers or materials.

Cablegram A telegram via underwater cable. Commercially this is being replaced by the telex system. *See* **Cable** *or* **Telex Transfer**.

Cable Laid *or* **Hawser Laid** A rope pattern consisting of three strands of left hand twisted rope, each strand being an individual rope.

Cabo da Buona Speranza *See* **Cabo Tormentoso**.

Cabo Tormentoso Cape of Good Hope or Stormy Cape or Cabo da Buona Speranza.

Cableway *See* **Cable Car**.

Caboose (1) Last freight train car which is used by the crew. (2) Deck house on a ship particularly used for cooking. (3) Ship's galley.

Cabotage *or* **Cabotage Fare** Certain countries give priority to commerce and tourism. On this understanding mutual agreements come into operation to develop freight and passenger services on a relatively cheap basis by way of encouragement. Nowadays cabotage air fares are applicable to groups of air travellers between countries with mutual agreements and where state licences are issued to operate these special flights.

Caboteur *French*—Coaster; Coasting Vessel.

Cab't Cabinet.

CAC Central Arbitration Committee; Customs Additional Code.

CACA or C.A.C.A. Canadian Agricultural Chemicals Association; Cement and Concrete Association.

CACAS Civil Aviation Council of Arab States.

C.A.C.C. Civil Aviation Communications Centre.

CACD Computer-Aided Circuit Design.

Cacex Carteira do Comercio Exterior, Banco de Brazil *Portuguese*. Overseas Commercial Bank of Brazil.

Cache *French*—A hiding-place for provisions, ammunition etc.

CACM Central American Common Market consisting of Guatemala, El Salvador, Honduras, Nicaragua and Costa Rica. A treaty signed in 1960 and put into operation in 1961.

cacoethes loquindi *Latin*—Mania for talking.

cacoethes scribendi *Latin*—Mania for writing.

CA Conf Cargo Agency Conference.

CACTLVO *or* **C.A.C.T.L.V.O.** Comprised and/or Arranged and/or Constructive Total Loss of Vessel Only *marine insurance*.

CACUL Canadian Association of College and University Libraries.

CAC & W Continental Aircraft Control and Warning.

CAD *or* **C.A.D.** *or* **c.a.d.** Cash Against Documents or Disbursements; Contract Award Date; Cash After Delivery; Computer Aided Design.

c-à-d C'est-à-dire *French*—That is to say; Namely.

CADC Central Air Data Computer; Colour Analysis Display Computer; Commuted Antenna Direction Communicator.

CAD/CAM Computer-Aided Design/Computer-Aided Manufacturing.

C & D Canal Chesapeake and Delaware Canal.

CADE Computer-Assisted Data Evaluation.

Cadet *Similar to* **Apprentice**, *q.v.*

C & E Customs and Excise.

CADV Cash advance/ed.

CAE *or* **C.A.E.** Canadian Aviation Electronics; Chartered Automobile Engineer; Cobrese al Entregar *Spanish*—Cash on Delivery.

CAEA *or* **C.A.E.A.** Central American Economics Association.

C.A.E.A.I. Chartered Auctioneers' and Estate Agents' Institute.

CAF Coût, Assurance, Fret *French*—Cost, Insurance and Freight; Currency Adjustment Factor.

CAF *or* **C.A.F.** *or* **c.a.f.** Ceylon (Sri Lanka) Air Force; Central African Federation;

CAFFA-ICC Commission on Asian and Far Eastern Affairs of the International Chamber of Commerce.

C.A.F.M. *or* **c.a.f.m.** Commercial Air Freight Movement.

CAFOD Catholic Fund for Overseas Development.

CAFSO Council of American Flag Ship Operators, *q.v.*

CAFU *or* **C.A.F.U.** Civil Aviation Flying Unit.

Cage-tainer Container, *q.v.*, made with steel gratings on its four sides and roof.

CAI *or* **C.A.I.** Canadian Aeronautical Institute; Club Alpino Italiano *Italian*—Italian Alpine Club.

C.A.I.B. Certified Associate of the Institute of Bankers.

Caique Sailing or rowing boat used in Turkey and in Malta.

Caisson Watertight chambers mainly used to raise wreckage from the bottom of the sea and also in the construction of bridges to keep them resting on the floating dock until both ends are fastened. Caissons are also used at canal entrances and repair docks.

CAL China Air Lines of Taiwan; Continental Air Lines.

cal Calendar; Calends; Calibre; Calorie; Colonial.

C.A.L.A. Civil Aviation Licensing Act.

CALC Cargo Acceptance and Load Control.

Calcined Reduced to powder by heat, during which process water and volatile substances have been expelled, e.g. calcined petcoke.

Calculable Laytime Calculation of Laytime. So much per tonnage, i.e. rate of tons per hour or per day or so many tons per hatch per hour or per day.

Calendar Day *Similar to* **Civil Day**, *q.v.*

Calendar Month In a Charterparty, *q.v.*, the calendar month refers to the date specified during delivery resulting in the same day being used as the date for the redelivery. *Ex.* If a time charter is concluded for three calendar months beginning on 1 May, the redelivery date would fall on 2 August.

Calf-Dozer Adaptable-sized bulldozer kept in the hold of a ship to trim bulk cargo.

Calibrated Airspeed The determined and corrected speed of an aircraft in conjunction with the instruments on board. *See* **Airspeed**.

Calibration. The act of checking and ascertaining graduation of scales with measuring instruments or by physical trials; checking the performance of equipment against known or designed values.

Calibration of Tanks Oil term meaning the measuring of tanks to check or establish their oil capacity.

Calibration Tables Kept on board all tankers showing cargo volume at each depth of ullage, *q.v.*, often expressed in gross barrels and in cubic metres.

Call *or* **Calls** A request for payment due for shares when requested by a company if and when additional capital is required. Also a word used on the American and English stock exchanges meaning an option of a contract signed by a person to another named person to 'buy' or to 'call' a certain amount of money for a fixed number of stocks at an appointed day at a price against a percentage. The double option to 'sell' or to 'buy' is alternatively termed 'Put and Call'. Another word used is 'Straddle' and the American corresponding word is 'Spreadeagle'.

Callable Stocks An American term for fixed interest stocks and Debentures, *q.v.*

Call Account A savings account in a bank bearing interest and from which money may be withdrawn at any time on request.

Call-Back Pay, *or* **Call-In Pay** *or* **Call-Out Pay** Remuneration of an employee who is sent for to carry out, or stand by for, unscheduled work.

Call Frequency Similar to Calling Cycle, *q.v.*

Calling Card American term for Business Card.

Calling Cycle The intervals at which a salesman visits a particular customer.

Call Letters *Similar to* **Call Sign**, *q.v.*

Call Money Money which has to be paid on demand. *Similar to* **At Call**, *q.v.*

Call of More A stock exchange term. The right of purchasing the same amount of stock. *See also* **Call** *or* **Calls**. *Abbrev.* c/m.

Call Report Information about the customer given in writing by a Salesman to his superior.

Call Sign A group of figures and letters to identify a station transmitting or receiving radio messages.

Ships' use their 4-letter Signal Letters, *q.v.*, as call signs.

Call-up Capital *or* **Called-up Capital** The actual total shares of a formed company paid in or in possession of the holder. The part of the shares that are left in abeyance for any eventual payments are named 'Uncalled Capital'.

CALM Catenary Anchor Leg Mooring, *q.v.* Type of mooring for an offshore buoy used by tankers. Campaign Against the Lorry Menace, *q.v.*

Calorie A unit of heat. The amount of heat required to raise one gram of water through one degree Celsius. This is now being replaced by the Joule, *q.v.* A calorie equals 4.1868 joules. A kilo calorie equals 1000 calories.

Calorimeter. Apparatus for measuring the Specific Heat, *q.v.*, of a substance.

CALPA Canadian Air Line Pilots' Association.

Calpers California Public Employees Retirement System.

Calz. Calzada *Spanish*—Boulevard.

Cam *or* **Camb** Cambridge, *UK.*

CAM *or* **C.A.M.** Commercial Air Movement; Communication Advertising and Marketing; Computer-Aided Manufacturing.

CAMAL *or* **C.A.M.A.L.** Continuous Airborne Missile-Launched with Low Level System.

Camber (1) The convex part of curvature of the deck of a ship or bridge or road to allow water to run off. (2) *Similar to* **Round of Beam**, *q.v.*

CAME *or* **C.A.M.E.** Conference of Allied Ministers of Education.

CAMEL Cunard Arabian Middle East Line.

CAMM *or* **C.A.M.M.** Council of American Master Mariners.

Campaign Against the Lorry Menace A UK association having a connection with the movements of goods by road. *Abbrev.* CALM.

CAMS Central Atmosphere Monitor System, *q.v.*

CAN *or* **C.A.N.** Customs Assignment Number *or* Customs Assigned Number.

Can Canada.

Canadian Trust Fund This refers to the Canadian dollar transfer of payments against premiums, etc., to Lloyd's *Abbrev* C.T.F.

Canadian Water Carriage of Goods Act 1936 The carriage of merchandise by sea which was enacted in 1936 relating to the transportation from any port in Canada to other Canadian or overseas ports.

Canals *See* **Ship Canals**.

Canalock Double-ended shipbuilding dock where simultaneous construction of two very large crude oil tankers can be undertaken.

Can Buoy A buoy with a flat top in the form of a can showing above the water.

Canc Cancelled; Cancelling.

Cancellation Clause A clause in marine insurance protecting both parties in case of unforeseen prejudice in the contract. The cancelling party is to advise the other party of his grievances and intention to cancel the contract. Cancellation may take effect within 30 days of his notice.

Cancelling Clause A clause which allows the Charterer to cancel a Charterparty if the vessel is late in being ready at the loading port or delivery place. The date agreed is the Cancelling Date.

Cancelling Returns If a ship is sold or transferred to a new owner underwriters have the option to cancel the insurance. The underwriters will return the unused days by giving a pro rata return of premium or cancelling returns. The underwriters also return pro rata premium or cancellation returns of a laid-up ship's duration. Also called **Laid-up Returns**.

Cancelling War Risk *Similar to* **War Cancellation Clause**, *q.v.*

CANCIRCO Cancer International Research Co-Operative.

Cancl Cancellation *Charterparty term*; Cancelled.

Candela Unit of measurement of luminous intensity and corresponding to the 20th part of the intensity emitted by 1 sq.cm. of platinum at its melting point of 1775°C. It is used in the scale of artificial lights obtained by luminous producers from candles, paraffin lamps or electricity. The latter is measured in Watts, *q.v. See* **SI** *or* **S.I.** *Abbrev.* cd *or* Cd.

CANDF Cost and Freight, *q.v.*

CANDF SEA/AIR Cost and Freight by Sea and/or Air.

C and/or J China and/or Japan.

CANEL Connecticut Advanced Nuclear Engineering Laboratory *USA*.

Can. Fr. Canadian French.

C.A.N.G.O. Committee for Air Navigation and Ground Organisation.

Can Hooks *or* **Barrel Hooks** Two hooks attached to both ends of the chain of the lifting gears to lift or hook barrels or casks while loading or discharging. *See* **Chain Hooks** *and also* **Barrel Hooks**.

Can. I. Canary Islands.

Canned Goods Another phrase for Tinned Goods.

Cannibalize The removing of component parts (used as spare parts) from one unit to another or similar repair unit, namely taking out a part or parts from one engine to put into another in order to make the latter work.

Canon Law The law appertaining to the Christian Churches particularly to the Roman Catholic, Protestant and Orthodox Churches.

C.A.N.S.G. Civil Aviation Navigational Aviation Group.

Cant. Canterbury.

Cantiere *Italian*—Shipyard.

Cant Hook A wooden lever having an adjustable iron hook at one end for the purpose of handling wood logs.

CANTRAN Cancel in Transmission; Cancel in Transit.

CANUKUS Canada, United Kingdom and United States.

Canvas Nets Canvas cargo nets bound at the edges by ropes for strengthening. Used to lift cargo or provisions to or from the ship. Rope nets made of manilla, synthetic, and wire ropes are commonly in use.

Canvas Sling *See* **Cargo Sling**.

Canyon Deep narrow gorge with steep slopes separated by a river.

C.A.O. Chief Accountant Officer; Crimean Astrophysical Observatory *in USSR*.

C.A.O.R.B. Civil Aviation Operational Research Branch.

C.A.O.R.F. Computer-Aided Operations Research Facility—Referring to a *Maritime Administered Department USA*.

cap Capacity.

c.a.p. Codice di arriviamento postale *Italian*—Mail Code Number.

Cap Capitolo *Italian*—Chapter; Capitano *Italian*—Captain; Capitulum *Latin*—Chapter; section.

CAP *or* **C.A.P.** Canadian Association of Physicists; Code of Advertising Practice; College of American Physicists; College of American Pathologists; Common Agricultural Policy *Common Market*; Community Action Program *USA*; Cash Against Policy *insurance*; Civil Air Patrol *USA*; Combat Air Patrol *USA*.

CAPA *or* **C.A.P.A.** Canadian Airline Pilots' Association; Canadian Association of Purchasing Agents; Computer-Aided Performance Analysis.

Capacity Plan Plan of a ship showing her loading capacity. This is one of the essential items currently in use when the ship is loading cargo for good Stowage and easy reference regarding the delivery of various cargoes in every port of discharge. *See* **Cargo Plan**.

Capacity Ton Mile *or* **Capacity Tonne Mile** *See* **Capacity Tonne Kilometre**.

Capacity Tonne Kilometre *or* **Capacity Ton Mile** The actual freight payload availability in an aircraft.

CAPAL *Lloyd's Register notation* for Cable & Protection Analysis by Lloyd's.

Cape Land protruding into the sea.

Cape of Good Hope *See* **Cabo Tormentoso**.

Cape Size A description given to large bulk carriers too wide to transit the Panama Canal and obliged to move between the Atlantic Basin and the Indian Ocean via either the Suez Canal or the Cape of Good Hope.

Capital Clause In the Memorandum of Association of a company there is a clause outlining the company's nominal capital as well as the amount of shares allowed.

Capital Punishment *or* **Capital Sentence** Death sentence.

Capital Constructive Fund A United States Department of the Maritime Administrative and Internal Revenue Service which is responsible for a method of cost recovery while investigating whether ships built against these funds are eligible for investment tax credit. *Abbrev.* CCF *or* C.C.F.

Capital Employed Business capital which is actually in use.

Capitalist Economy *Similar to* **Free Enterprise**, *q.v.*

Capital Structure Composition of a company's finance; equity, debentures, borrowings, grants etc.

Capital Tax Income Tax imposed on the Capital Outlay.

Capricorn A cargo carrier used for a dual purpose as Roll On/Roll Off, *q.v.*, and B.C.V., Barge Carrying Vessel, *q.v.* The barges are carried and manoeuvred into the ship's water-flooded hold by the vessel's own small tug when the hinged bow doors are opened. When the barges are set into the right place in the hold the bow doors are locked into the vertical position and the hold is pumped dry, the barges resting on the tank top. For unloading, the opposite system is used whereby the barges are made to float in the hold and pulled out by the tug. This is a similar method to the LSD, Landing Ship Dock, *q.v.*, and the Trimariner.

Capstan A solid steel drum positioned on mooring quays and also mostly at the aft end of a ship. It enables the mooring lines to pull the ship to her berth. The capstan can be operated either manually or mechanically.

Capt. Captain.

Captain Finds the Ship When the victualling is looked after by the ship's master it is said to be 'Master Finds the Ship'. Some shipowners have a contract with the master and/or the Chief Steward, *q.v.*, and they are allowed a fixed amount of money per man per day to cater for the officers and crew members. In turn either the master or the chief steward will do his own buying of the victuals at his own risk. If the contract is done with the chief steward it is termed as 'Steward Finds the Ship'.

Captain of Port *or* **Captain of the Port** A government official. A person well acquainted with shipping navigation who would generally be a Certified Master Mariner who would be responsible for the inward and outward movements of vessels in port. He is also responsible for port regulations.

Captain's Entry Filling up and signing of a customs' form by the master of a ship on arrival, giving details of cargo to be unloaded.

Captain's Protest *or* **Notary Protest** *or* **Sea Protest** *or* **Writ of Protest** *or* **Maritime Declaration** An official protest produced or written by the master of a ship whereby he gives a detailed account of any accidents or suspected damage sustained by his ship or cargo, with copies to the shipowners and

agents. This legal protest, which is registered by a Notary Public, serves as advance protection against any eventual cargo claims and will cover the owners in the case of a claim being submitted to Underwriters, *q.v.*

Caption Arrest; authority for indictment; warrant for arrest in Scotland until 1837; heading or title for text or illustrations in formulated matter and correspondence.

Captive Insurance Company A self-insurance company. A very large company that sets up its own insurance business on a separate footing. Its main object is to have all the consequent benefits remaining within their insurance perimeter and at the same time provides self-insurance for re-insurance within its own organisation. Some countries such as Bermuda exempt such companies from tax.

Capture Products Products for which the buyer has no option but to obtain spares from the original manufacturer or his agents.

Capture One of the exceptions in the Carriage of Goods by Sea Act. It covers the taking of a ship by force during war as a prize or a reprisal. *Similar to* **Capture and Seizure**, *q.v.*

Capture and Seizure A war peril. One of the exceptions in the Carriage of Goods by Sea Act and in Marine Insurance, *q.v. Similar to* **Capture**, *q.v.*

car. Carat, *q.v.*; Carpenter.

CAR *or* **C.A.R.** Canadian Association of Radiologists; Chart of Airspace Restrictions; Civil Air Regulations *UK*; Contractors All Risks, *q.v.*

CARA Combat Air Rescue Aircraft *USA.*

CARAC *or* **C.A.R.A.C.** Civil Aviation Radio Advisory Committee.

Carat or Karat A word used to assess the value of precious stones and gold. Standard unit of 200 milligrams to weigh precious stones. As for pure gold it consists of 24 carats/karats, so when gold is declared as nine and 18 carats it is meant that the remaining 15 and six units respectively represent alloy. *Abbrev.* ct. or kt.

CARAT Close Approach Radar and Thermal Imaging System.

Caravan (1) Enclosed van serving as mobile living quarters with space for sleeping, eating and cooking. It may be driven or towed. (2) Convoy of pack animals carrying merchandise across country.

Carbon Dioxide One of the Inert Gases, *q.v.*, used in Crude Oil Washing, *q.v.*.

Carboxyhemoglobin A by-product of carbon monoxide which reduces the oxygen carrying capacity of the blood.

Carboy A large glass bottle generally containing corrosive acids. When carboys are shipped as cargo they are always stowed on deck at 'Shippers Risk'. The glass of the carboy is protected by steel or a thick plastic grill and cushioned by hay or straw. *Abbrev.* cb.

CARD Campaign Against Racial Discrimination.

Cardinal Point One of the four main points of the compass, namely any one of North, South, East, West. *See* **Rhumb**.

Card Index Box(es) or Tray(s) of cards titled sequentially providing simple office references.

CARDIS Cargo Data Interchange System. Established by the United States National Committee on International Trade Documentation.

CARE Co-operative for American Relief Everywhere.

Careen The act of listing a ship for the purpose of cleaning and painting the exposed side. This is also done to repair the sides of the hull, thus avoiding drydocking.

Car Ferries Fairly fast passenger and car vessels that ply from one place to another crossing channels, rivers, lakes, canals, etc. The car ferries have mostly a roll-on/roll-off (ro/ro) system which eases and hastens the loading of vehicles and cargo.

Cargadoorskantoor *Dutch*—Shipbrokers.

CARGO Consolidated Afloat Requisitioning Guide Overseas.

Cargo The general meaning of merchandise transported on a ship or aircraft or by land vehicles. Cargo is accepted on production of a Bill of Lading, *q.v.*, or Air Waybill, *q.v.*, as a form of Contract of Affreightment, *q.v.*, showing the full details of the cargo carried, including the Freight, *q.v.*, charges involved. *Abbrev.* Cgo *or* cgo.

Cargo All Risks Clauses Marine insurance policy which covers the insured comprehensively.

Cargo Assembly Point/Area Place where individual cargoes and parcels are collected for later transport or shipment.

Cargo Book Documentary book held by shipbrokers or ships' agents showing details of the merchandise including marks, numbers, types of packages, weights and/or measurements, shippers, consignees, carried by either land, sea or air.

Cargo Carrying Capacity This is a rather vague word in a way. The capacity could either mean the deadweight capacity or measurement in tons, freight tons, standards, quarters etc.

Cargo Door *Similar* to Cargo Port, *q.v.*

Cargo Gear *See* **Gear**.

Cargo General Average Cargo's proportion in General Average Contribution, *q.v.*, by the cargo owners who have to pay their part in proportion to the extraordinary expenses incurred in General Average, *q.v.*

Cargo-Handling The act of loading and discharging a cargo ship.

Cargo-Handling Machinery The machines or winches that drive the cranes and derricks used in loading, discharging or handling of the cargo of ships. Part of the quick turn-round of a ship depends on these factors.

Cargo Hatch *See* **Hatch**.

Cargo Information Card When a US merchant ship carries hazardous cargoes (*see* **Dangerous Cargo**) she is bound by law to carry this card in a very conspicuous place. It provides a clear description of the cargo and gives instructions to follow in case of fire, spillage or leakage, explosions and other hazards. It is generally termed as CIC-86 May 1971 Form or MCA Cargo Information Card. *Abbrev.* MCA Cargo.

Cargo Liner A cargo vessel which is employed in regular scheduled runs from one or more ports to others. Also known as Liner Vessel or Liner Ship.

Cargo Net Square or rectangular net of wire or rope with a loop in each corner to which the hook of a crane can be attached for hoisting and lowering cargo. *See* **Cargo Sling**.

Cargo Papers Documents needed for a ship to load and/or unload cargoes. These consist of ship's stowage plan, manifest, bills of lading, tally clerk sheets and Shipping Order, *q.v.*

Cargo Plan A sketch or a plan of the ship showing the holds with detailed distribution of all cargoes on board for use by the Chief Officer, *q.v.*, customs and port workers. It enables anyone to know the cargo distribution in the holds. *See* **Capacity Plan** and **Stowage Plan**.

Cargo Plane *or* **Cargo Aircraft** Aircraft designed to carry merchandise rather than passengers.

Cargo Port A watertight opening door on the side of the ship from which cargo can be carried. This opening is very useful for cattle transportation purposes as well as mobile articles and vehicles.

Cargo Ship A ship having Holds, *q.v.*, with Derricks, *q.v.*, above to carry cargoes. There are cargo ships which are purposely used for Bulk Cargo and others that are Tween Deckers for general cargo and for better distribution of various types of cargo. In some instances Single Deckers are also used for general cargoes. Also termed Cargo Vessel.

Cargo Ship Construction Certificate A certificate issued by the Classification Societies which has then to be renewed every year by the authorities of the state where the ship is registered.

Cargo Sling Canvas strap, rope, wire or chain with a loop or ring at each end to engage in the hook of a crane for lifting or lowering merchandise. *See* **Cargo Net** *and* **Sling**.

Cargo Space The cargo space available on a ship.

Cargo Tray A wooden platform with four round hooks at every corner from where the lifting tackle can be hooked and hoisted for loading/discharging ship's provisions, stores and/or cargo.

Cargo Underwriter/s Insurer(s) or underwriter(s) who issue(s) policy of insurance covering cargo.

Cargo Vessel *See* **Cargo Ship**.

Cargoworthiness *or* **Cargoworthy** The ship is to be fully equipped in relation to the type of cargo she is carrying apart from the fact she has to be always fully classed. If an accident occurs due to deficiency of the ship's equipment or gear this is said to be the cause of unseaworthiness.

Cargoworthy *Similar to* **Cargoworthiness**, *q.v.*

Carib Caribbean.

Caribbean Area Includes Colombia, Venezuela, Trinidad and the Dutch islands of Aruba and Curaçao.

CARICOM Caribbean Community and Common Market.

CARIFTA Caribbean Free Trade Area.

Carnet French word for log book. It is an international customs document very commonly used for the clearance of cars while crossing frontiers or from a ship or ferry when loaded and unloaded. The carnet exempts customs import duty when shown to the customs authorities. It is also used for items temporarily imported for exhibition. In most cases a carnet is issued as a green form by the Chambers of Commerce.

Car Pallet A flat tray to accommodate a vehicle for loading on to a cargo ship.

Carpe diem *Latin*—Enjoy the present.

CARQUAL Carrier Qualification *USA*.

Carriage The freight charges in reference to the carriage of goods by rail.

Carriage and Insurance Paid An INCOTERM 1990, *q.v.* expression. The seller is responsible for the shipment and insurance of the merchandise up to the named destined port. *Abbrev.* CIP *or* C.I.P.

Carriage Forward (1) Cost of the goods excludes the expenses for forwarding and transportation. *See also* **Carriage Paid**. (2) Carriage or freight payable at destination. *Abbrev.* Cge. Fwd.

Carriage Free Delivery of the goods to the Consignee, *q.v.*, is effected free of all charges.

Carriage of Goods Some nation's legislatures refer to the Carriage of Goods Acts and are more or less based on the same lines. (1) The Harter Act (1893) approved on 13 February 1893 by the Senate and House of Representatives of the USA. This refers to merchandise or property transported from or between ports of the USA and foreign ports. (2) Canadian Water Carriage of Goods Act 1936. This is in relation to and in connection with the carriage of goods by water in ships from any port in Canada to any other port whether in or outside Canada. (3) Sea Carriage of Goods Act 1924 in relation to and in connection with the carriage of goods by sea in ships from any port in the Commonwealth of Australia to any other port whether in or outside the Commonwealth. (4) New Zealand Sea Carriage of Goods Act 1940. This Act solely refers to the carriage from any port in New Zealand. *Abbrev.* COG *or* C.O.G. *or* C of G.

Carriage of Goods by Sea Act 1924 This Act was introduced after many shipping conferences were held in Brussels among various European nations interested in shipping transportation and it was finally formulated and termed as the Carriage of Goods by Sea Act 1924. The American Carriage of Goods by Sea Act (COGSA) was passed in 1936 and is the counterpart of the 1924 Act. It contains six sections and is subdivided into nine Articles. The object of this Act is that carriers and shippers are bound by the Bill of Lading, *q.v.*, covering both the shippers and the carriers with responsibilities and exceptions in the course of the carriage of cargo from one port to another.

Carriage of Goods by Sea Act 1936 of the USA *or* **US Carriage of Goods by Sea Act 1936 (COGSA)** A law enacted on 16 April 1936 covering the transportation of merchandise by sea to or from ports of the United States and in foreign trades. *See also* the **Harter Act**.

Carriage Paid Similar to Carriage Free, *q.v.*, but delivery is to an agreed point at the place of destination, or if one is not agreed, to a point suitable to the seller, which could be different from the location of the Consignee's, *q.v.*, stores. *Abbrev.* Carr pd. *Opposite* to Carriage Forward, *q.v.*

Carriage Paid To An INCOTERM 1990, *q.v.*, expression. The seller is to pay for the freight up to the end of the voyage trailer and container transport method. Insurance is for account of the consignees. *Abbrev.* CPT *or* C.P.T.

Carrier's Lien The shipowner's right to withhold cargo from being delivered to the Consignee, *q.v.*, as a guarantee against the collection of Freight, *q.v.*, and other charges for the shipment.

Carr. pd. Carriage Paid, *q.v.*

Carry Away To break *nautical*.

Carrying Forward Dispatch Time This may be construed as Reversible Laydays namely any hours saved in loading to be added to the hours allowed for discharging. Alternatively days or parts of days not used in loading may be calculated as a deduction from the time of discharging. *Abbrev.* cfdt. *or* CFDT *or* C.F.D.T.

Carrying Trade Trade consisting of commercial articles carried on land, sea and air from one country to another.

Carrying Vessel The opposite of Non Carrying Vessel, *q.v.*

Carry Over (1) The retaining of merchandise beyond the time limit or season which will be utilised for the next selling period or season. (2) Term for the postponement of settlement to the next account.

Carry-Over-Day A Stock Exchange, *q.v.*, term meaning the first day of each new account, *obsolete*.

Cars Knocked Down Crated but otherwise complete motor cars for assembly elsewhere than the original manufacturing plant. *Abbrev.* CKD.

CARSE Cargo Automation Research. Subcommittee in connection with the development and integrated airfreight handling and control systems.

cart. Cartage, *q.v.*

CART Cargo Automation Research Team (International Air Transport Association).

C.A.R.T. Collision Avoidance Radar Trainer *aircraft*.

Cartage The Carriage, *q.v.*, or charge due to Carriers, *q.v.*, for land transport of goods carried to the docks for onward shipment destinations.

Carte du jour *French*—Bill of fare.

Cartels A merger of manufacturers to control production, marketing arrangements, prices etc.

Carte Blanche *French*—At the entire discretion; Open cheque.

Cartographer Person who draws maps and charts.

Cartography The chart and map making trade.

Carving Note A note produced and completed by a surveyor of the Department of Trade or a Consul or similar department concerned certifying that the name, official number and tonnage are carved on the ship and in exchange for which the Registrar for Shipping issues the ship's Certificate of Registry, *q.v.* These particulars are simultaneously inserted in Lloyd's and other Registers.

CAS *or* **C.A.S.** Calibrated Air Speed, *q.v.*; Certificate of Advanced Studies; Chief of Air Staff; Close Air Support *Military*; Collision Avoidance System, *q.v.*; Controlled Air Space *aviation*.

C.A.(S.A.) Member of the Accountants' Society (South Africa).

CASA Ceylon (Sri Lanka) Association of Steamer Agents.

ca.sa. Capias ad satisfaciendum *Latin*—A writ of execution.

CASE Council of Associated Stock Exchanges; Committee on Academic Science and Education, USA; Confederation for the Advancement of State Education.

Case History Past record of a company, organisation or individual setting out problems encountered and actions taken.

Casella Postale *Italian*—Post Office box. *Abbrev.* CP.

Case of Need Usually a Bill of Exchange, *q.v.*, endorsement to the effect that the person named therein ascertains to honour the bill in due time, should the drawer be unable to do so. This is signed by the drawer or the endorser.

Case Oil Liquid products shipped as cargo in small containers and packed in wooden crates or cartons. Case oil consists of petrol, paraffin, lubricating oils or other similar Petroleum, *q.v.*, products.

Case Study When encountering problems in a business organisation discussion groups are formed to study how to tackle them. It is thus a Case Study.

Cash Actual money.

CASH Cassette Aboard Ship. System for handling 40-foot cassettes or containers in the forest products trade using gantries.

Cash Bonus In life assurance annual bonuses are added to the sum assured in the policy. There are other policies, however, that give a cash bonus.

Cash Book A book-keeping accounts recorder for receipts and payments during the day. *Abbrev.* c.b. *or* C.B. *or* C/B *or* Cbk.

Cash and Carry The method of selling articles cheaply against cash where the purchaser collects from stock.

Cash Discount A discount percentage allowed from the payment of accounts if effected immediately on demand and with no absolute delays. The percentage may vary from $\frac{1}{2}$ to 5% or even more according to mutual agreement. *Abbrev.* c.d.

Cash Flow (1) The total profit in business putting aside depreciation. (2) The anticipation of the estimated income as well as expenses which would have been involved and obtained during a period of time. *See also* **Projected Cash Flow**.

Cash-in-Hand The actual money held in coin or cash by a business concern.

Cash Surrender Value *or* **Surrender Value** The amount the **Underwriters**, *q.v.*, are willing to refund for the Premiums, *q.v.*, already paid by the assured policyholder when he decides to discontinue premium payments and withdraw from the contract.

Cash with Order Payment is effected when the goods are ordered.

CASI Canadian Aeronautics and Space Institute.

Casilla postal *Spanish*—Post office box. *Abbrev.* CP.

Cask Buoy Wooden or metal cask-formed buoy.

CASO Council of American-Flag Ship Operators.

CASP Computer-Assisted Search Planning.

CASREP Casualty Report *USA*.

CASS Command Active Sonobuoy System *USA*; Cargo Account Settlement System.

Cast Anchor To let go or throw or cast the anchor of a ship. Also termed to Drop Anchor, opposite to Weigh Anchor.

Casting Vote Right of a vote allowed to the chairman of a board. The casting vote method is put into practice whenever the division of votes are equal and so an additional vote is needed to decide the Agenda, *q.v.*

Cast off Another expression to 'let go' or to 'throw off' which refers to mooring ropes, anchors, etc.

Casual Labour The engagement of a worker for a short time.

Casualty A mishap; an accident; a loss caused by death, capture, etc.

Casualty Certificate *See* **Performance Certificate**.

Casual Labourer/Worker One who works on the occasions that work arises, usually for a short time, and he may be employed on stand-by.

Casus belli *Latin*—An act which justifies war.

Casus concientiae

Casus concientiae *Latin*—A matter of conscience.

Casus fortuitus *Latin*—A matter of chance.

Casus omissus *Latin*—Omitted case. An omitted point of law not provided for in an enactment relating to the subject-matter.

Casus sentit creditor *Latin*—Borne by the creditor. This relates to the non-performance of a contract for which the creditor is liable.

CAT *or* **C.A.T.** Colleges of Advanced Technology; Clearance Air Turbulence, *q.v.*; Clear Air Terminal *aviation*; Combined Acceptance Trials *USA*, *q.v.*

Catamaran A double or treble-hulled vessel originally built of bamboo, mahogany and other wood logs tied together by the natives of the East Indies, Brazil, South Africa, etc., and used for fishing. It has now become in vogue as a yacht for racing and pleasure purposes in America and Europe. The modern catamaran is constructed in wood, aluminium or reinforced glass fibre and is also composed of two or three hulls diagonally joined together by various methods. Normally no ballast is needed since it enjoys good stability at sea. Several merchant ships and warships are now designed as catamarans.

Cataract Flow of large amount of water over a precipice. A big waterfall.

CATC *or* **C.A.T.C.** Commonwealth Air Transport Commission or Council.

Catca *or* **C.A.T.C.A.** Canadian Air Traffic Controllers Association.

Catenary From the Latin word *catena*, a chain: the curve in a rope, wire or chain when suspended between two points. Specifically used in towing where the curve described, which depends on the weight of the towline, speed and the resistance of the tow, will affect elasticity and avoidance of shock as well as the depth of water required. Also used in describing the system of overhead wires on an electrified railway.

Catenary Anchor Leg Mooring Anchorage system consisting of buoy distribution units of submarine pipelines and floating hoses with a rotating arm to which the tanker or oil vessel is moored. There is another scheme such as the ASLM (Single Anchor Leg System) which is more or less identical. Both these mooring systems are in use in many deep water ports around the world where the inner harbours are inaccessible because of shallow water. *Abbrev.* CALM.

Caterer A contractor who furnishes provisions and stores needed by a ship, aircraft, organisation. *See* **Shipsuppliers**.

Catering Act of providing food.

Catering Manager An office manager whose job is to take care of all the catering in an organisation.

Catering Officer An officer on a ship, generally a passenger ship, responsible for the catering on board, sometimes to the Purser.

Cat-head In old sailing ships a short heavy beam, fitted with sheaves, projecting from the bow to keep the anchor clear when weighing or letting-go.

Cathode The electrode at which a reduction occurs. Therefore it is the negative pole of an electrolytic cell but the positive one of a battery.

Cathodic Protection Measures to control the corrosion of ship's hulls, ballast tanks and underwater structures, caused by Electrolytic, *q.v.*, or galvanic action between dissimilar metals immersed in seawater, where one forms an anode which suffers wastage; and the other a cathode which is preserved. Such a cell is formed, for example, between a steel hull and a bronze propeller, but is also found on apparently uniform steel surfaces due to variations in the material and imperfect coatings. Protection is afforded by two methods of providing an anode which is yet more electro-negative.

a. Sacrificial. Attachment of pieces of zinc, or nowadays more usually magnesium or aluminium alloy, which will waste instead of the steelwork, and

b. Impressed Current. Introduction of an 'inert' anode which, when connected to a low voltage DC supply, dissipates a current without itself suffering significant loss.

Cat Hole Hole in a ship through which hawsers, mooring lines and springs pass.

Cat Walks Elevated walkways above a tanker's weatherdeck, to facilitate the safe movement of crew members around the vessel, thereby avoiding both the dangers of heavy seas washing over the deck of a laden tanker and avoiding the need to scramble over pipes and other obstacles.

CATIC China Aerotechnology Import Export Corporation.

CATL *or* **C.A.T.L.** Compromised and/or Arranged Total Loss.

Cattle Attendant Similar to **Cattleman**, *q.v.*

Cattle Carrier Similar to **Cattle Ship**, *q.v.*

Cattleman Another word for Cattle Attendant. A person engaged by the Consignees, *q.v.*, or Consignors, *q.v.*, on a ship which carries cattle on board to feed and care for the animals during the voyage.

Cattle Manifest Manifest made out and produced when cattle are carried on a ship.

Cattle Ship *or* **Cattle Carrier** Ship specially fitted with breast and division boards, footlocks, rumps, troughs, etc., to accommodate and carry livestock consisting of cattle, pigs, sheep, horses, asses, etc.

Cat's Paw Light air producing ripples on the surface of the sea.

Cattle Container Open-top container for the carriage of livestock, fitted with pens or boxes, and rails, to prevent the beasts falling as the ship pitches and rolls.

Catug *or* **CATUG** Short for Catamaran Tug. A rigid catamaran tug connected to a barge. When joined together they form and look like a single hull of a ship. This is more or less on the same lines as Artubar, *q.v.*, being one of the Sea Link systems or ocean-going Pusher Barge.

CATV Cable Television Systems; Community Antenna Television.

Caulk The act of inserting Oakum, *q.v.*, and tar between the planks on the decks of ships (especially near the bridge decks) to make the decks completely watertight.

Causa causans *Latin*—The immediate cause; the last link in the chain of causation; not the cause (*causa sine qua non*) of which the proximate cause is an effect but the nearest cause of the damage or effect for which relief is sought.

Causa proxima non remota spectatur *Latin*—The immediate, not the remote cause, is to be considered. This maxim was formerly used in connection with policies of marine insurance. The principle enunciated in it is now expressly set out in the Marine Insurance Act.

Causa sine qua non *Latin*—An indispensable condition. Frequently *just* Sine qua non. Indispensable condition. *See* **Causa causans**.

Caustic Soda Sodium Hydroxide.

Caution Money A sort of a deposit before proceeding with the buying negotiations. Reserve Account.

CAV *or* **C.A.V.** Construction Assistance Vehicle (a sub-aqua vehicle).

c.a.v. Curia Advisare Vult *Latin*—The court desires to consider.

Caveat *Latin*—A legal notice for the stoppage of proceedings.

Caveat emptor *Latin*—A legal interpretation for the buyer to be aware of what he is buying. He is to acquaint himself with the obvious defects in any of the goods he intends to buy. In other words 'Let the buyer beware'. *Abbrev.* c.e.

Caveat subscriptor *Latin*—A legal term. 'Let the signor or endorser beware.' If a contract is signed the people involved should have known what they have signed for. *See also* **Caveat emptor**.

Cavendo tutus *Latin*—Safe by taking heed.

CAVU Ceiling and Visibility Unlimited *aviation*.

CAXB Composite Auxiliary Boiler.

CAXBS Composite Auxiliary Boiler Survey.

Cay A sandbank, reef, or small low island found in the Caribbean.

Cb. Carboy, *q.v.*; Cumulonimbus Clouds, *q.v.*

CB *or* **C.B.** Cash Book; Chirurgiae Baccalaureus *Latin*—Bachelor of Surgery, *also Abbrev.* Ch.B.; Companion of the Bath; Compass Bearing; Container Base.

c.b. *or* **C.B.** *or* **C/B** Cash Book; Centre of Buoyancy, *q.v.*, Compass Bearing; Confined to Barracks (*military expression*); Continuous Breakdown; Country Bell; Container Base.

CBA *or* **C.B.A.** Caribbean Atlantic Airways; Collective Bargaining Agreement; Commercial Bank of Australia; Cost Benefit Analysis.

CBAA Canadian Business Aircraft Association.

C/B/B Container/Break/Bulk.

CBC *or* **C.B.C.** Canadian Broadcasting Company; China Business Conference.

C&BCISA Cardiff and Bristol Channel Incorporated Shipowners' Association.

CBD *or* **C.B.D.** Cash Before Delivery.

CBE *or* **C.B.E.** Central Bank of Egypt; Commander of the British Empire.

CBF Cubic Feet.

C.B.&H. Continent between Bordeaux and Hamburg. (Loading/discharging ports range)

CBI *or* **C.B.I.** Cape Beton Island; Central Bureau of Identification *USA*; China, Burma and India; Computer Based Information: Confederation of British Industries; Cumulative Blood Index.

c.b.i. Complete Background Investigation.

CBIS *or* **C.B.I.S.** Computer Based Information System.

cbk. Check Book, Cheque Book, *q.v.*.

cb/l *or* **CB/L** Commercial Bill of Lading

cbl Cable.

cbm *or* **CBM** *or* **C.B.M.** Conventional Buoy Mooring; Cubic Metre.

CBMIS *or* **C.B.M.I.S.** Computer-Based Management Information System.

CBMPE *or* **C.B.M.P.E.** Council of British Manufacturers of Petroleum Equipment.

CBMS *or* **C.B.M.S.** Counter-Based Message System(s). (The most sophisticated form of electronic mail.)

CBMPE *or* **C.B.M.P.E.** Council of British Manufacturers of Petroleum Equipment.

CBMU Construction Battalion Maintenance Unit *USA*.

CBO Congressional Budget Office *USA*.

CB&PGNCS Circuit Breaker and Primary Guidance Navigation Control System.

CBRT Canadian Brotherhood of Railway, Transport and General Workers.

CB's Sea Bees. *See* **Seabee**.

CBS *or* **C.B.S.** Columbia Broadcasting System *USA*.

CBST Clean Ballast Segregated Tank.

CBT Clean Ballast Tank, *q.v.*; Combined Ballast Tank(s).

CBW Chemical and Biological Warfare.

CBYRA Chesapeake Bay Yacht Racing Association.

cc Cubic Centimetre.

c.c. *or* **C/C** *or* **C/c** Cancellation Clause, *q.v.*; Cash Credit; Chamber of Commerce; Change Course *nautical*; Chronometer Correction; Civil Court; Civil

Commotions; Close Control; Colour Code; Collecting Commission; Compte Courant *French*—Current Account, *q.v.*; Conto copied to; Corrente *Italian*—Current Account; Contra Credit; Cubic Centimetre; County Council; Courant Continu *French*—Direct Current.

(c.c.) *Lloyd's Register notation* for Corrosion Control Installation in tanks; Cubic Capacity; Cubic Content; Current Cost.

c.c. *or* **C.cl.** Continuation Clause, *q.v.*

Cc Capita *Latin*—Chapter *typography*.

Cc. Cirrocumulus Clouds, *q.v.*

CC *or* **C.C.** Cape Colony; Caribbean Commission; Central Committee; Chamber of Commerce; Charity Commission; Check Clearance; Chronometer Correction *nautical*; Civil Code; Civil Commotion(s), *q.v.*; Civil Court; Common Council; Common Councilman; Compass Course *nautical*; Confined to Camp; Control Computer; Consular Clerk; Continuation Clause, *q.v.*; Close Control; Colour Code; Collecting Commission; Conto Corrente *Italian*—Current Account; Cushion Craft (Hovercraft); Corriente Continua *Spanish*—Custom Document form produced to the master in the coasting trade before he is authorised to sail. *Similar to* **A.A. document**, *q.v.*, given to foreign-going vessels in UK.

C/C Cement Carrier.

CCA Carrier Controlled Approach *USA*; Chemical Carriers Association *international body of owners trading to US ports*; Clean Channel Association, Houston *USA*.

CCA *or* **C.C.A.** Chief Clerk of Admiralty; Commission for Conventional Armaments; Court of Criminal Appeal.

CCAF *or* **C.C.A.F.** Combined Currency Adjustment Factor (Adjustment factor or inflation factor in relation to currency); Comité Centrale des Armateurs de France *French*—Central Committee of French Shipowners.

CCAI Calculated Carbon Atomaticity Index, measure of ignition quality of fuel oil.

CCB *or* **c.c.b.** Cubic Capacity of Bunkers; Controlled Carrier Bill, *q.v.*

CCBT Common Carrier Barge Transporter. *See* **Common Carrier**.

CCC Roman numeral for 300; California Coastal Commission; Commodity Credit Corporation; Commercial Court Committee; Customs Container Convention; Customs Co-operative Council.

CCC *or* **C.C.C.** Canadian Chamber of Commerce; Cargo Carrying Capacity, *q.v.*; Central Criminal Court; Commodity Credit Corporation; Commercial Court Committee.

CCCC Roman numeral for 400. Also shown as CD.

CCCP *or* **C.C.C.P.** Symbol of Soviet Union, Soyuz Sovietskikh Sotsialisticheskikh Respublik—*Russian*, Union of Soviet Socialist Republics.

CCD *or* **C.C.D.** Charge-Coupled Device; Conseil de Coopération Douannière *French*—Customs Co-operation Council.

CCDA Commercial Chemical Development Association *USA*.

CCF *or* **C.C.F.** Capital Construction Fund, *q.v.*; Co-operative Commonwealth Federation *Canada*.

CCG Canadian Coast Guard.

CCGT Combined Cycle Gas Turbine.

C.C.H. *or* **c.c.h.** Commercial Clearing House; Cubic Capacity of Holds.

cc. hr. Cubic Centimetre per Hour.

CCI Chamber of Commerce and Industry; Chambre de Commerce International *French*—International Chamber of Commerce.

CCIR International Radio Consultative Committee.

CCISG Convention Contracts of International Sale of Goods.

CCITT International Telegraphic and Telephonic Consultative Committee.

CCIW Canada Centre for Inland Waterways.

C.C.L.N. Consignment Control Label Number.

c.cm. Cubic Centimetre.

cc/min Cubic Centimetre per Minute.

CCP *or* **C.C.P.** *or* **c.c.p.** Chinese Communist Party; Code of Civil Procedure; Commonwealth Centre Part, *Australia*; Committee on Commodity Problems *FAO*; Court of Common Pleas; Credit Card Purchase; Critical Compression Pressure; Conto Corrente Postale *Italian*—Current Postal Account.

C.C.P.E. Canadian Council of Professional Engineers.

C.C.P.I.T. China Committee for the Promotion of International Trade.

CCR Conradson Carbon Residue; Critical Compression Ratio.

C.Cr.P. Code of Criminal Procedure.

CCS *or* **C.C.S.** Casualty Clearing Station; Child Care Service; Collective Call Sign; Combined Chiefs of Staff *USA*; Consolidated Cargo Service *containers*; Controlled Combustion System; Ceylon (now Sri Lanka) Civil Service; Customs Clearance Status.

CCS China Classification Society.

CCSA Certified Container Securing Arrangements; Collective Company Signing Agreement.

CCS Contract Cattle Cash Settlement Contract *livestock trade*.

CCSI China Corporation of Shipbuilding Industry.

cct Circuit.

CCT *or* **C.C.T.** Common Community Tariff or Common Customs Tariff *EEC*.

CCTA Central Computer and Telecommunications Agency *UK Govt.*

c.c.tks Cubic Capacity of Tanks.

CCTV Closed Circuit Television.

C.C.U. *or* **c.c.u.** Coronary Chart Comparison Unit *chart reading*; Coronary Care Unit (Λ clinic for the treatment of heart disease *USA*); Civil Contingencies Unit *UK.*

CCUS Chamber of Commerce of the United States.

ccy Convertible currency.

CCY *or* **CCy** Convertible Currency.

CCW *or* **C.C.W.** Counter Clockwise.

cd *or* **c.d.** *or* **C/D** Customs Duty.

cd *or* **Cd** Caddesi *Turkish*—Street; Candela, *q.v.*; Ciudad *Spanish*—city.

c/d Carried Down (Accountancy or Letter Writing).

c.d. Cash Discount, *q.v.*; Country Damage, *q.v.*; Cum Dividend *Latin*—With Dividend.

Cd Cadmium; cancelled.

CD Certificate of Deposit; Chart Datum *navigation*; Compact Disc.

CD *or* **C.D.** Carried Down; Charge D'Affaires or Corps Diplomatique; Certificate of Deposit, *q.v.*; Commercial Dock; Country Damage, *q.v.*; Consular Declaration; Cum Dividend *Latin*—With dividend; Cofferdam; Symbol for Cadmium; Roman numeral for 400, also written CCCC.

C/D Commercial Dock; Completion of Discharge; Consular Declaration; Customs Declaration.

C & D Collected & Delivered; Collection & Delivery.

C.D.A. Canadian Dental Association; College Diploma in Agriculture.

CDB *or* **C.D.B.** Cellular Double Bottom. *See* **Ballast** and **Deep Tanks**; Continuous Discharge Book. *Similar to* **Discharge Book**, *q.v.*

Cdbd. Cardboard.

CDC Centre for Disease Control and Prevention *USA.*

CDC *or* **C.D.C.** Canada Development Corporation; Combat Development Command *USA*; Commissioners of the District of Columbia; Common Development Cycle; Commonwealth Development Corporation; Command and Data Handling Console *computer*; Caribbean Defence Command.

CDCP Construction and Development Corporation of Philippines.

cd/ft Candela per square foot.

cd/fwd Carried Forward.

CDG *or* **C of DG** Carriage of Dangerous Goods at Sea. New **IMDG**, *q.v.*, code of 1 January 1991.

CDH *or* **C.D.H.** College Diploma in Horticulture; Constant Data Height.

CDI Chemical Distribution Institute; Compact Disc Interactive *computers*; Control Data Institute.

C.DK. Containers Carried on Deck *Lloyd's Register notation.*

CDM Control Distribution Module *electronics.*

CDN Compensated Density Neutron *oil drilling tool.*

CDP Concrete Drilling Platform; Controlled Depletion Polymer. The alternative paint coating material to SPC (Self Polishing Copolymers). Paint manufacturers are researching this type of paint, since SPC is said to contain Tri-organotin compounds.

C.D.P.E. Continental daily parcels express.

CDR Carbon Drag Reducer. Commander; Compensated Dual Resistivity *oil drilling tool.*

CDS *or* **C.D.S.** Construction Differential Subsidy, *q.v.*

CDSB Cargo Data Standards Board *IATA.*

CD. Sh. Commissioned Shipwright.

CDSO Companion of the Distinguished Service Order.

CDT Controlled Deep Towing *oil drilling*; Clearance Diving Team *naval.*

CDU Conical Drilling Unit, *q.v.*

C.D.V. *or* **c.d.v.** Carte-de-visite, *French*—Visiting Card; Current Domestic Value, *q.v.*

Cdw. Cum Dividend *Latin*—With Dividend.

CDW Collision Damage Water *insurance*; Collision Damage Waiver, *motor insurance.*

C.D.W. *or* **c.d.w.** Cold Drinking Water.

ce *or* **c.e** *or* **C/E** Customs Entry.

c.e. Critical Examination; Caveat emptor *Latin*—Let the Buyer Beware. *See* **Caveat emptor**.

CE *or* **C.E.** Canada East; Carbon Equivalent; Centre of Effort *sailing*; Chancellor of the Exchequer; Chemical Engineer; Chief Engineer; Chronometer Error *nautical*; Church of England; Circular Error; Common Era; Compass Error *nautical*; Council of Europe; Compression Engine.

C & E Customs and Excise

CEA *or* **C.E.A.** Canadian Electrical Association; Central Electricity Authority; Confédération Européenne de l'Agriculture *French*—European Confederation of Agriculture, Geneva; Council of Economic Advisers *US Treasury and White House Administration.*

C.E.A.C. Combustion Engineering Association; Control Electronics Assembly Commission; Commission Européenne de l'Aviation Civile *French*—European Civil Aviation Commission.

CEB Central Electricity Board.

CEBOSINE Centrale Bond Van Scheepsbouwmeesters *Dutch*—Dutch Shipbuilders' Association.

CEC *or* **C.E.C.** Canadian Electrical Code *standardisation*; Clothing Export Council of Great Britian.

CECA Communauté Européenne du Charbon et de l'Acier *French*—European Coal and Steel Community.

CED *or* **C.E.D.** Committee for Economic Development; Carbon Equivalent Difference; Council of European Developments *USA.*

CEDA Central Dredging Association.

C.E.D.A. Committee for Economic Development of Australia.

Cedi A unit of currency used in Ghana.

CEE *or* **C.E.E.** Commission économique pour l'Europe *French*—Economic Commission for Europe; Certificate of External Education; Common Entrance Examination.

CEEC *or* **C.E.E.C.** Commission of the European Economic Community.

CEFIC *or* **C.E.F.I.C.** Council of Europe Federation of Chemical Industries.

CEGB Central Electricity Generating Board.

CEI *or* **C.E.I.** Centre d'Etudes Industrielles *French*—Centre for Industrial Studies, *Switzerland*; Council of Engineers' Institute; Council of European Industries.

CEIF Council of European Industrial Federations, Paris.

Ceiling Wooden covering or flooring placed on top of double-bottom tanks as a protection to the tank tops, *q.v.*

Ceiling Price Maximum price of an article that can be disposed of under the control regulations of a state.

Ceilometer A device to measure the ceiling of the clouds.

CEJSA Council of European and Japanese Shipowners' Association.

Cel Celsius scale of degrees *temperature*.

Celà va san dire *French*—That goes without saying.

Celescope Celestial Telescope.

Celestial Navigation The science of navigation by observations of celestial or heavenly bodies, such as the sun, moon, stars and planets.

Cell *Similar to* Slot.

Cell Guides Frames built into the cargo compartments of a container ship into which containers are slotted for safe stowage.

Cellarage Storing charges for the use of a cellar.

Cellular The 'cell guides' on a container ship into which cellular container boxes fit for security and quick loading and discharging. It is well known that this method is quite a safe stowage as no shifting occurs during the voyage.

Cellular Double Bottom The inner part of the bottom of a ship where the Deep Tanks, *q.v.*, are situated and which is divided into rectangular compartments which are often used as Ballast Tanks, *q.v. Abbrev.* CDB *or* C.D.B. *Also termed* Double Bottom *or* Ballast Tanks.

CEM *or* **C.E.M.** Cost and Effectiveness Method.

CEMA Catering Equipment Manufacturers Association.

Cement Box When a ship has a leak below the waterline this may be remedied by securing a ready-made or fabricated steel box to the inside of the hull in way of the hole or split and filling it with cement. The box itself, or the box when containing the cement, is called a Cement Box.

CEMENTVOY Standard Voyage Charter Party for the Transportation of Bulk Cement and Cement Clinker in Bulk.

CEMF *or* **C.E.M.F.** Counter-Electro-Motive Force *electricity*.

Cen. Centigrade scale of degrees *temperature*.

CEN European Standards Organisation.

CENE Commission on Energy and Environment.

C/E *or* **C/Eng.** Chief Engineer; Chartered Engineer.

Census To take or obtain a detailed account of the population, production and distribution of a country.

Cent Centum *Latin*—Hundred; Per cent; %; Unit of money.

Centigramme 1/100th part of a gram equivalent to 0.15432 of an ounce.

Centilitre 1/100th part of a litre equivalent to 0.017598 of a pint.

Centimetre–Gram–Second A system of measurement units so based, derived from the metric system, for example the Dyne, *q.v.*, Erg, *q.v.*, Poundal, *q.v.*, Calorie, and Atmosphere. It is not well adapted to thermal and electrical quantities and for scientific purposes the units have been replaced by SI Units, *q.v.*

CentiStokes Units of kinematic viscosity, *marine oils. Abbrev.* cSt.

Central American Common Market Reciprocal customs union of Costa Rica, El Salvador, Guatemala, and Nicaragua *Abbrev.* CACM.

Central Atmosphere Monitor System A detector which monitors hydrogen, carbon monoxide, oxygen, carbon dioxide and refrigerant gases. *Abbrev.* CAMS.

Central Bank A rather common name in many countries from where all foreign and local financial policies are entered and transacted. As the name itself implies all business of importance, especially to the state, converges through this bank for the responsible authorities to be in a position to undertake financial dealings as a national body. Some countries have a different or specific name such as the United Kingdom which has the Bank of England.

Central Planning State planning.

Centre-Castle The raised centre and amidships part of a ship where in most cases the navigation and officers' accommodation are situated.

Centre Line The vertical plane of a ship from fore to aft at its centre.

Centre of Buoyancy The centre of gravity of the water displaced by a ship, or its underwater form through which an upward force is assumed to act. *Abbrev.* CB or C.B.

Centre of Flotation The centre of gravity of the waterplane area of a floating vessel.

Centre of Gravity The point through which the total weight of a vessel is assumed to exert a downward force. *Abbrev.* G.

Centre of Immersion The mean centre of that part of a vessel which is immersed in water, and of such displaced water. Synonymous with Centre of Gravity, Centre of Displacement, and the more usual term Centre of Buoyancy, *q.v.*

CENTROCON *or* **Centrocon** Code name for Chamber of Shipping River Plate Charterparty 1914 Homewards, amended 1 January 1937 and 14 December 1949 respectively.

CEO Chief Executive Officer.

CEOA Central European Operating Agency *NATO*

CEP Circular Error Possibility (Missile accuracy estimation about 100 metres after thousands of flights).

Cepi corpus *Latin*—I have taken the body. A sheriff's report that he has complied with a writ for an arrest.

CEPT Conférence Européenne des Administrations des Postes et de Télécomunications *French*—European Conference of Postal and Telecommunications Administrations.

CERC Coastal Engineering Research Center *US Army*.

CERCA *or* **C.E.R.C.A.** Commonwealth and Empire Radio for Civil Aviation.

CERCLA Comprehensive Environmental Response, Compensation and Liability Act. *US Oil Pollution Act 1990 under sect. 4115 of Federal Law.* Specifies design requirements for Tanker hulls.

CERES Coalition for Environmentally Responsible Economies.

CERN (1) Centre Européen de Recherches Nucléaires *French*—European Organisation for Nuclear Research. (2) Comite Economico Parala Coordinacion Latino Americano *Spanish*—Latin-American Economics Co-ordination Committee.

Cert or cert. Certificate; Certified; Certify.

Cert, A.I.B. Certificate of the Association of the Institute of Bankers.

Cert. CAM Certificate in Communication Advertising and Marketing.

Certificate A written document. Declaration, testimony or attestation of certain facts; Statement of professional qualification, or success under examination, in some cases showing marks or grades achieved; Document of Authority; statement of competence, service, efficiency or conduct.

Certificated Referring to an officer or person authorised to act or work in the grade or capacity named in the Certificate, *q.v.*

Certificate of Competency Certificate issued by the appropriate authorities confirming the competency or the efficiency of an officer of a ship in his respective grade and department.

Certificate of Damage A printed form of questionaire to be answered either by the port authorities or by the representatives of the insurance underwriters. The questions set out enquire about the cause of damage, the day of arrival of the importing ship, date of unloading, whether the damage was recorded immediately or later, the date and place of survey, the extent of the damage (including the surveyors' fees) and whether the carriers accepted liability. It is very rare that carriers accept any responsibility. The information is needed by underwriters to see whether any action for recovery can be taken. *Similar to* Certificate of Survey, *q.v.*

Certificate of Deletion An official document obtained from the Registrar of Shipping of the Port of Registry of the ship to certify the removal or deletion of the registry or of the flag. This will enable the owner to register the ship under another registry.

Certificate of Deposit An ordinary certificate known as a CD which is a negotiable bearer instrument issued by a bank as evidence of a fixed term deposit. *Abbrev.* CD *or* C.D.

Certificate of Fitness Certificate produced by a state or government after an initial or periodical survey of a chemical tanker for the carriage of dangerous chemicals in bulk. *Abbrev.* C of F.

Certificate of Incorporation An official document or certificate certifying the actual legal existence in records of company.

Certificate of Origin An invoice made out and endorsed by the original shippers and manufacturers. It shows the name and place of the manufacturers, the on-carrying vessel and full details of the merchandise, including the name and address of the importers or consignees. *Abbrev.* C of O *or* C/O.

Certificate of Ownership The Certificate of Registry, *q.v.*, showing ownership together with all the essential details of the vessel.

Certificate of Posting Post Office receipt for a packet showing the date and place of posting.

Certificate of Protection US Government certificate issued to the American owner of a yacht built outside the United States according certain legal rights.

Certificate of Record A certificate issued by the US customs authorities to the owner of an American built vessel which is owned by a non-American citizen. The vessel is obliged to carry a Sea Letter, *q.v.*

Certificate of Registry One of the essential documents which every ship has to possess before she can operate and sail from a port. This certificate is always requested to be produced in every port of call by customs and emigration authorities before clearance is given. The certificate is issued by the customs authority where the ship is registered. The details therein are mainly (1) The name of the ship. (2) Date of construction. (3) Gross and Net Tons. (4) Length Overall—Length between perpendiculars, breadth moulded, draught (draft). (5) Registry and flag. (6) Call sign or code number. (7) Names of owners and master. In selling and/or buying transactions the seller is requested to delete her by obtaining the Certificate of Deletion, *q.v.*, to enable the new owner to re-register her in his own name either with the same flag or another registry flag.

Certificate of Survey (1) Certificate issued by an appointed surveyor either by the insurance underwriters or the port authorities to conduct a survey of suspected damage to merchandise off-loaded from a ship. (2) This could also mean a Classification Survey Certificate of a ship while she was undergoing repairs or doing her periodical or Special Survey. *See* **Certificate of Damage**.

Certification Certificates issued by recognised or authorised officials or inspectors at ports where the protection of human safety is concerned. This certification generally refers to oil tank cleaning and gas freeing, thus rendering any Hot Work, *q.v.*, safe on oil tankers and the like.

Certified Cheque A **Bank Draft**, *q.v.*

Certified Copy A copy made out or photocopied from an original and attested or confirmed by two parties. It is sometimes called Attested Copy. *See* **Attestation**.

Cert. inv. *or* **cert. inv.** Certified Invoice.

Cert/s. Certificate(s).

Certiorari *Latin*—To be more fully informed. An order to remove proceedings to a higher court.

CES Coast Earth Station *satellite communications*.

CES *or* **C.E.S.** Cranial Electrotherapy Stimulation.

CESA *or* **Cesa** Committee of EU Shipbuilders' Associations.

C'est à dire *French*—That is to say; Namely.

Cestui que trust *Norman French*—He who trusts. The beneficiary under a trust or a person for whose benefit the trust is administered.

C'estui que use *Old French*—For whom is the use. For all intent and purposes, the legal owner.

C'estui que vie *Old French*—A person entrusted with property during the lifetime of another person. The property thus entrusted is called 'Pour autre vie'.

CET *or* **C.E.T.** Central European Time; College d'Enseignement Technique *French*—College of Technical Education; Common External Tariff, *q.v.*

C & ET Customs and Excise Tariff.

CETA Comprehensive Employment Training Act *USA*.

Cetane Number The ignition point number at which diesel fuel ignites when it is being pressed by the engine cylinder while in motion.

ceteris paribus *Latin*—Other things being equal. *Abbrev.* cet. par.

cet. par. ceteris paribus, *q.v.*

CF *or* **C.F.** Centre of Floatation or Flotation; Chaplain of the Forces; Charges Forward, *q.v.*; Commonwealth Fund; Compensation Fee *Post Office parcels*; Copper Fastened *Lloyd's Register notation*; Corresponding Fellow; Cost and Freight, *q.v.*

c.f. *or* **c/f** *or* **C/F** Carried Forward; Charges Forward, *q.v.*; claim form *insurance*.

C & F Cost and Freight, *q.v.*

c.f. Cubic foot/feet; Confer; Cost and Freight, *q.v.*

CFA Canadian Federation of Agriculture; Chartered Financial Analyst *USA*; France Communauté Financière Africaine *French*—Council of Foreign Affairs.

CFA franc Colonies Française d'Afrique *French*—French African Colonies Franc (A unit of currency used in Cameroon, Central African Republic, Republic of Congo, Gabon, Upper Volta, Niger, Senegal, Chad).

CFB Creative Fares Board, *q.v.*

CFC Chlorofluorocarbon, *q.v.*; Customers Foreign Currency, *banking*.

cf/d Cubic feet per day, *gas production*.

cfdt *or* **C.F.D.T.** Carrying Forward Dispatch Time, *q.v.*

CFE Contractor Furnished Equipment. *See* **Government Furnished Equipment**.

CFF Compensatory Financing Facility *finance*; Continuous Forms Feeder *computer*.

c.f.g. *or* **C.F.G.** Cubic Feet of Gas.

c.f.g.d. *or* **C.F.G.D.** Cubic Feet of Gas per Day.

c.f.g.h. *or* **C.F.G.H.** Cubic Feet of Gas per Hour.

c.f.g.m. *or* **C.F.G.M.** Cubic Feet of Gas per Minute.

c.f.h. *or* **C.F.H.** Cubic Feet per Hour.

c.f.i. *or* **C.F.I.** Chief Flying Instructor; Cost, Freight and Insurance. *Similar to* **C.I.F.**, *q.v.*

c.f.m. *or* **C.F.M.** Cubic Feet per Minute.

CFMU Central Flow Management Unit *Safety of air navigation*.

c.f.o. *or* **C.F.O.** Calling for Orders; Central Forecasting Office *meteorological office*; Channel for Orders; Chief Fire Officer; Coast for Orders.

CFP Common Fisheries Policy *EC*.

C.Fr. Cant Frame *Lloyd's Register notation*.

CFR *or* **C.F.R.** Council of Foreign Relations; Code of Federal Regulations *US Department of Transportation also Abbrev.* USCFR, *q.v.*; Cost and Freight.

CFRP Carbon-Fibre Reinforced Plastic.

CFS *or* **C.F.S.** Container Freight Station, *q.v.*; Control Flying School; Cubic feet per Second.

c.f.s. Cubic feet per second.

CFSP Common Foreign and Security Policy, *European Union.*

C.Ft *or* **cf** *or* **cft** Cubic Foot/Feet.

CFT Concrete Foundation Template *oil drilling.*

CFTC Commodity Futures Trading Commission, *q.v.*

Cft. L. Craft Loss *marine insurance.*

Cg Centigramme.

c.g. Consul General; Centre of Gravity.

CG *or* **C.G.** Captain of the Guard; Commerce en Gros *French*—Wholesale Trade; Consul General; Commissioner General; Centre of Gravity; Coast-guard Stations; Commanding General.

c.g.a. Cargo Proportion of General Average. *See* **Cargo General Average**; Certified General Accountant; Coast Guard Academy *USA;* Coast Guard Auxiliary; Compressed Gas Association *USA.*

CGA *or* **C.G.A.** Cargo's Proportion of General Average. *See* **Cargo General Average**; Certified General Accountant; Coast Guard Academy *USA;* Coast Guard Auxiliary; Compressed Gas Association *USA.*

CGC *or* **C.G.C.** Coast Guard Cutter *USA.*

CGCC Conventional General Cargo Carrier.

CGDS Canadian Government Department of Supply and Services.

cge Carriage.

cge. fwd. Carriage Forward, *q.v.*

C.G.F. Proceeds Credit Given for Proceeds *referring to allowances in insurance claim adjustments.*

C.G.F. Rec. Credit Given for Recovery *referring to allowances given in insurance claim adjustments.*

C.G.F. Recy. Credit Given for Recovery. *See* **C.G.F. REC.**

CGH *or* **C.G.H.** Cape of Good Hope.

c.g.i. Corrugated Galvanised Iron.

CGI Computer Generated Image/Imagery, as used in shiphandling simulators.

C.G.I. City and Guilds Institute; Corrugated Galvanised Iron.

C.G.I.A. City and Guilds of London Insignia Award.

CGIAR Consultative Group on International Agricultural Research.

CGIL Confederazione Generale Italiana Lavora. Confederation of Italian Trade Unions.

CGL *or* **C.G.L.** Comprehensive General Liabilities, *insurance.*

C.G.L.I. City and Guilds of London Institute.

cgm. Centigramme.

CGM *or* **C.G.M.** Conspicuous Gallantry Medal.

C.G.M.E. China, Glass, Marble and Earthenware.

cgo or Cgo Cargo, *q.v.*

CGOU Coast Guard Oceanographic Unit *USA.*

cgs *or* **CGS** *or* **C.G.S.** Centimetre-gram-second, *q.v.*; Canadian Geographical Society; Chief of General Staff.

c.g.s. cum grano salis *Latin*—With a grain or pinch of salt.

CGS *or* **C.G.S.** *or* **cgs** Carriage of Goods by Sea.

CGSA *or* **C.G.S.A.** Carriage of Goods by Sea Act. *See* **Carriage of Goods** and **Carriage of Goods by Sea Act 1924** also **Carriage of Goods by Sea Act 1936.**

C.G.S.B. Canadian Government Specification Board.

c.g.s.e. Centimetre Gramme-Second Electromagnetic.

C.G.Sta. Coast Guard Station *USA.*

cgt Capital Gains Tax.

C & GS Coast & Geodesic Survey *US nautical publication.*

CGT Compensated Gross Tons *shipbuilding calculation.*

CGT *or* **C.G.T.** Capital Gains Tax; Compensated Gross Ton. Confédération Général du Travail *French*—Confederation of Labour Organisations.

CGTB *or* **C.G.T.B.** Canadian Government Tourist Board.

c.h. Candle Hour; Central Heating; Compass Heading *nautical.*

Ch Chain; Chapter; Companion of Honour; Chirurgial *Latin*—Surgery; Chief.

CH Carrier Haulage; Compass Heading *nautical;* Customs House.

Ch. China.

C.H. Clearing Houses, *q.v.*; Companion of Honour; Customs House; Court House.

C₂H₄ Ethylene.

C_2H_4 Ethylene.

C_2H_6 Ethane.

C_3H_8 Propane.

CH_4 Methane.

CHABE Charterers Both Ends: refers to the nomination by the charterers of agents at the ports of loading and discharging.

Chacun à son goût *French*—Everyone to his taste.

Chaf Chafage; Chafing, *q.v.*

Chafing The act of rubbing together making friction and wear. *Ex.* The anchor's chafing with the bow of the ship. *Abbrev.* Chaf.

C.H.& H. Continent Between Havre and Hamburg.

Chain A measure of 22 yards or 20.108 metres consisting of 100 chainlinks.

Chain Hooks Hooks attached to the end of the chain lifting gears to hook barrels or casks for loading and discharging purposes. *See also* **Can Hooks** *and* **Barrel Hooks**.

Chain Locker Compartment located near the bow of a ship to store the Anchor Chains, *q.v.*

Chain of Command *or* **Line of Command** Definition of the relationship between commanders or managers at the various levels of an organisation which sets out the limits of their authority and responsibilities for directing and reporting.

Chain Plates Metal fittings bolted to the outside of a vessel to which the shrouds are secured.

Chain Purchase Geared device for lifting heavy weights and worked to raise or lower by an endless chain.

Chains Anchor Cable, *q.v.*

Chain Stopper A short length of chain, one end of which is shackled to a deck fitting while the other is hitched and passed around a wire hawser and used to take the strain while the wire is manipulated.

Chain Stores A store or a supermarket company organisation having a chain of stores in various places or countries.

Chair To conduct a meeting; short for chairman, *q.v.*

Chairman (1) Person elected or appointed to take charge of a meeting or conference. (2) Position held by the senior officer in a Company. His is the final decision in all matters. *See* **Chair**. (3) One of those carrying a sedan chair.

Chairman of a Company *or* **Company Chairman** Generally he is an executive or one of the senior men in a company who is given the chairmanship to preside at board meetings. Some chairmen are not active partners in the company. There are cases when a chairman also acts as a managing director.

Chaland *French*—A barge.

Chaldron 25 cwts or 1295.4 kilogrammes or 53 cwts Imperial or 2692.4 of Newcastle loads of coal or coke.

Chamber of Commerce An organisation of businessmen emanating from various companies of a particular country and/or town. Some of its functions are to unify all business concerns and to issue standard regulations to its members and subsequent commercial groups dealing in similar business. In some countries not all businessmen or companies are members of the chamber of commerce but they are indirectly involved in their way of effecting business. Bourse, *q.v.*, is another word for Chamber of Commerce. *Abbrev.* Ch of C. *or* COC *or* C.O.C.

Chamber of Shipping The Chamber of Shipping, formerly the General Council of British Shipping, is the trade association and employer's organisation for British shipowners and ship managers. It promotes and protects the interests of its member companies, both nationally and internationally, and represents British shipping to the Government, Parliament, international organisations, unions and the general public. The Chamber covers all issues which have a bearing on British shipping, ranging from fiscal policy and freedom to trade through to recruitment and training, maritime safety and the environment, navaids and pilotage. It represents six very different commercial sectors trading at sea: deep-sea bulk, short-sea bulk, deep-sea liner, ferry, cruise and offshore support. It has

121 member companies which own or manage 606 trading ships totalling 21.5 million deadweight tonnes, including some managed for foreign owners. It is governed by a President and Council, supported by sections representing the six shipping sectors. Its Director-General and staff are located in the City of London. *Abbrev.* COS.

Chamf Chamfered.

Chanc. Chancellor.

Chandler A person who deals in the selling of provisions, dried stores, etc. *Similar to* **Shipsuppliers**, *q.v.*, or ship chandlers.

Chandler's Shop A slang term usually referred to a ship carrying general cargo against a quantity of Bills of Lading, *q.v.*, with individual small items or quantities. *See* **Heterogeneous Cargo.**

Change Short for Exchange, *q.v.*

Channel Money Cash for ship and crew's port expenses.

Channel Navigation Information Service (CNIS) An organisation which provides a 24-hour radio safety service for all shipping in the Dover Strait, its Traffic Separation Scheme and associated Inshore Traffic Zones. Introduced in 1972, it is operated by HM Coastguard from the Maritime Rescue Co-ordination Centre at Langdon Battery Centre near Dover. CNIS broadcasts on VHF Channel 11 every 60 minutes (every 30 minutes in poor visibility) to give warnings of navigational difficulties and unfavourable conditions likely to be encountered in the Strait. These include adverse weather conditions, exceptional tides, misplaced or defective navigational aids and 'hampered' vessels such as oil-rigs, deep draught tankers and surveying ships. Ships using the MAREP voluntary Ship Movement Reporting Scheme introduced in 1979 are tracked, as are those contravening the International Collision Regulations, and their positions, courses and speeds are broadcast. Ships are identified by themselves, other ships, and/or by Coastguard aerial patrol. Offenders are reported to their Flag States for action to be taken in accordance to the IMO Resolution A432 (XI). The increase in speed, size and frequency of Dover Strait traffic has led in recent years to the introduction of a number of safeguards, and a steady decline in the number of recorded shipping incidents in the Dover Strait in 1994. Traffic separation became mandatory for all vessels in 1977. CNIS has the co-operation of the French MRSC (OROSSMA) at Cap Gris Nez which also maintains radar surveillance.

Channels of Distribution Various routes by which products or published information leaves the makers or originator and reaches the end user.

CHAPS Clearing House Automated Payments System *London Clearing Bank' System. See* **CHIPS**.

Charge Account Credit Account *USA*.

Charges Collect Charges enumerated on the **Waybill** or Bill of Lading, *q.v.*, which are payable by the receiver or Consignee, *q.v.*

Charges Forward An accounting word meaning that the carriage and other charges are paid on receipt of the goods by the consignee or buyer. The buyer only pays for the goods when he is in receipt of same. This may also be attributed to Freight Charges, i.e. Freight Paid at Destination or Freight At Risk or Collect Freight or Freight Forward, *q.v. Abbrev.* C/F or c.f. *or* c/f.

Charging Order A court order authorising a creditor to sell the goods or property of the debtor if the debt is not paid within the appointed time.

Charlie International Radio Telephone phonetic alphabet for 'C'.

Charlie's Wain Navigator's nickname for the **Great Bear** or **Ursa Major,** *q.v.*

Chart A map which is most useful in the navigation and manoeuvring of a ship. It depicts the depth of the sea, detailed information to the approaches of the coasts, harbours, rivers, etc.; position of light ships, lighthouses, etc.

CHARTAC *or* **Chartac** Chartered Accountant.

Charta partita *Latin*—A divided document, ie., duplicated, each half being a copy of the other. **Charterparty,** *q.v.*

Charter (1) To charter a ship for a period of time known as Time Charter, *q.v.,* or for one or other voyages known as Voyage Charter or for the management known as Bareboat Charter or Demise Charter, *q.v.* (2) A special grant given by a state or government on certain conditions to be adhered to and issued to institutions that are allowed certain rights and privileges.

Chart Correction Card A card showing corrections to charts in relation to Notices to Mariners.

Chartered (1) An adjective used in connection with the profession of a person and/or any association and/or institute where a special charter has been granted and by which they are entitled to use the word 'Chartered'. *Ex* Chartered Accountant, Chartered Bank, Chartered Institute, etc. (2) Hired. A chartered vessel is a hired vessel.

Chartered Shipbroker Professional status accorded to Fellows of the Institute of Chartered Shipbrokers, *q.v.,* which distinguishes a person from those who can describe themselves only as shipbroker.

Charterer A person or firm hiring a vessel for the carriage of goods or passengers or both.

Charterers Both Ends Charterparty, *q.v.,* expression. The charterers have the right of appointing their own agent at loading and discharging ports: *Abbrev.* CHABE *or* CABE *or* C.A.B.E. *or* c.a.b.e.

Charterers Pay Dues Charterparty term where the charterers are to pay all the dues charged on the ship. *Abbrev.* C.P.D. *or* c.p.d.

Charter Hire *or* **Charter Freight** The hiring of a ship for one or more voyages or for a period of time.

See **Charterparty** *or* **Charterer**. Freight due to the Carrier, *q.v.,* by the Charterer, *q.v.*

Chartering Agents Brokers who undertake the import and/or export of commodities in large quantities. They act as intermediaries between owners and consignors of the cargoes. *See* **Agents.**

Chartering Brokers Brokers who act as intermediaries between owners and charterers or shippers and receivers. They are mostly responsible for the drafting or signing of the Charterparty, *q.v.,* contract. *See* **Brokers.**

Charterparty A Contract of Affreightment, *q.v.,* signed between the shipowner and the charterer whereby the former hires the vessel to the latter for the carriage of goods, etc., either for a period of time known as Time Charter or Demise or Bareboat Charter or Voyage Charter. In the Demise or Bareboat Charter the whole management and running expenses of the ship are automatically handed over to the charterers until the expiry date of the charter. In some charters the insurance is the owner's responsibility, in others the payment of premiums is equally shared. The charterers have a free hand as to where and how to do business, including limitation of voyages so long as they honour the contract. Charter rate is calculated on a lump sum basis, or on a calendar month or annual basis depending on the length of the charter time. The vessel is to be returned to her owners in the same condition as she was when delivered to the charterers. Therefore surveys are effected on delivery and redelivery. In Time Charters the ship's cargo capacity is temporarily allotted to the charterers for a fixed time and committed to limited ports of call while trading. Insurance remains the responsibility of the owners. Food, stores, bunkers, port expenses, etc., are to be borne by the charterers. Again, the ship will have to be redelivered to the owners in the same condition as originally delivered to the charterers. Surveys are done before delivery and immediately after redelivery to assess any repairs which may be necessary. Charter rate is calculated on a lump sum basis per month or so much per deadweight ton per calendar month. In a Voyage Charter all expenses are provided and paid for by the owners unless otherwise agreed upon by the parties. There are numerous code forms of charterparties and each one is used in relation to the particular trade and area allocated. *Abbrev.* C/P.

Charterparty by Demise *Similar to* **Bareboat Charter,** *q.v.*

Charterparty Deletions Deletions of one or more clauses in a charterparty which are considered legal and binding. The deletion effect is considered binding in the eyes of the law and that the previous intent never existed.

Charthouse *or* **Chartroom** The living quarters of a navigating officer when on duty or on watch while a ship is on voyage. It is a room on the bridge occupying an unobstructed view of the ship itself and the sea all around her. The forepart contains the helm, navigational aids, control units, wireless/telephones and direct communication to the master's cabin which is generally situated under the charthouse. Adjacent and behind this section there is a large table on which navigational calculations can take place when charts and books are consulted. Very near to this section there may be a sleeping bunk for the officer on duty to rest.

Chartist (1) One who analyses and forecasts Stock Exchange and commodity prices. (2) Member of a workers' political party active in the nineteenth century.

Charts Maps showing in detail port approaches, depth of outer and/or inner harbours, depth of seas, distances, peculiarities, beacons, buoys, obstacles, etc. Charts can also show the general details of a harbour, river, lake, etc.

Chattel Mortgage The grant of lien on personal property by the borrower to the person or firm advancing finance as a security or guarantee for the fullfilment of the obligation of repayment of the money advanced. As soon as this debt is fully paid the lien is then removed.

Ch.B Chief of the Bureau *USA*.

Ch. B. Chirurgiae Baccalaureus *Latin*—Bachelor of Surgery.

CHC Cargo Handling Charges; Colour Hard Copy *photocopying*.

Ch. Clk. Chief Clerk.

Ch. E. Chemical Engineer; Chief Engineer.

Cheap Jack A travelling hawker of (usually) poor quality goods.

Cheap Money Money that can be obtained at low interest rates and without difficulty because of the existence of a plentiful supply of Excess Resources among Banks. Also termed Easy Money.

CHEC Commonwealth Human Ecology Council.

Checker *See* **Tally Clerk**.

Check Measuring *See* **Check Weighing**.

Check-weigher A man who, on behalf of the coalminers, checks the amounts sent up to the pit mouth.

Check Weighing When cargo weight and/or measurement disputes arise a second check will be carried out. If all is found correct the party alleging the mistake will have to bear the expenses.

Chem Chemical.

Chemical Carrier A ship having specially constructed tanks capable of containing and withstanding extremely volatile or poisonous or corrosive liquids.

Chemical Hazards Response Information System A US Coast Guard publication covering 1,250 of the most dangerous chemicals commonly transported. *Abbrev.* CHRIS.

Chem. E. Chemical Engineer.

Cheque A Bill of Exchange, *q.v.*, drawn and payable on demand. An order by a person to his banker requesting him to pay to a named person or firm a certain sum of money on demand. The person making out the cheque is called the drawer and the person to whom it is payable is the payee. Also called sometimes Bank Money. *See* **Cheque Book**.

Cheque Book Every bank customer having a current account is provided with a cheque book. It is a booklet containing a number of blank cheques which are signed by the customer requesting the bank to pay the bearer the requested cash. *See* **Cheque**.

Cheque Card or Banker's Card A card issued by a Bank to its customer which guarantees a cheque presented in payment for goods, or to obtain cash at a bank, within specified limits. Some Cheque Cards serve a dual role as Credit Cards. *See* **Banker's Card**.

Cherry Picker A road vehicle with a hydraulic arm used to lift a cradle to enable workmen to reach otherwise inaccessible heights. Used in ships' holds to enable workmen to clean or repair the upper areas of cargo spaces.

Chert One of the hardest and toughest rock formations found in drilling for oil. *See* **Chert Clause**.

Chert Clause A drilling contract clause stipulating that any Chert, *q.v.*, composition found during the time of drilling does not form part of the drilling.

Ches. Cheshire *UK*.

Chewing Tobacco *Similar to* **Quid**.

ch. fwd Charges Forward.

CHG or **C.H.G.** Cargo Handling Gear.

CH & H Continent between Le Havre and Hamburg. A charterparty clause limiting the range of ports for loading and/or discharging.

CH₄ Symbol for Methane, *q.v.*

Chicano *Spanish*—A Mexican-American.

Chief Cabin Passenger First Class Passenger.

Chief Clerk (1) A person in an office in charge of other clerks. (2) A person engaged in clerical work with the purser on board a ship. (3) An American expression for a member of a shipping group who is in charge of the clerical organisation department.

Chief Engineer The head of the engineering department. The most senior engineer officer in a ship or a shore enterprise.

Chief Executive The person in a company who has the entire management or the running of the business. He may also be the Chairman or the Managing Director.

Chief Mate *See* **Chief Officer**.

Chief Officer The second in command of a ship. He is next to the master, most especially in the navigation and as far as the deck department is concerned. The chief officer assumes the position of the Master, *q.v.*, in his absence. He is also called Mate *or* Chief Mate *or* First Mate *or* First Officer.

Chief Steward The officer in charge of all the victualling, including the cooking done by the catering staff serving on a ship. On a large cargo/passenger or passenger ship the corresponding officer is the Purser, *q.v.*, but he is of a higher officer rank since he assumes more responsibilities with regard to cuisine, smartness, inspection routines, accounts, passenger accommodation, currency exchange, etc. *See* **Steward**.

Chilled Cargo Refrigerated cargo cooled to a temperature from 2° to 6°C. This type of refrigeration does not make the cargo hard but provides a cool temperature to prevent deterioration.

Chilled Meat Meat which is stored in chambers of 4°C.

Chimb *or* **Chime** The extreme ends of a barrel or cask protruding away from each head from which Barrel Hooks, *q.v.*, are secured for lifting purposes.

Chime *See* **Chimb**.

CHIMPAC China Sea, India and Pacific Ocean *chartering*.

Chin. China.

Chinse To stop a small seam with oakum using a sharp blade as a temporary expedient where the opening in the planking will not bear the full force required to Caulk, *q.v.*, it.

CHIPFA Chartered Institute of Public Finance and Accounting, formerly knwon as Municipal Treasurers and Accountants.

Chipping Hammer A hammer with two sharp ends used to remove rust from metal. Also called Boiler Pick *or* Scaling Hammer. There is a quicker method of removing rust by means of electrical or pneumatic hammers, or scaling machines, which will do the job more efficiently with fewer men.

Chips or CHIPS California Highway Patrols *USA*; Clearing House Interbank Clearing System, New York. *See* **CHAPS**.

Chips or Chippy Slang for ship's carpenter. *See* **Shipwright**.

Chippy *See* **Shipwright**.

Chlorinated Rubber An impervious kind of paint mainly used on steel as primer and finishing coats. The effect is far superior to conventional paints.

Chlorinity The relative amount of chloride content in water.

Chlorofluorocarbon A halogenated compound, used for aerosols and as a refrigerant, the most commonly known being R12, which was banned by the Montreal Convention 1992 because its molecular content destroyed ozone in the stratosphere. Its ultimate replacement will be hydro-fluro-carbon (HFC) but for an interim substitute in refrigeration plants Hydrochlorofluorocarbon, *q.v.*, is being used.

Ch. M. Chirurgiae Magister *Latin*—Master of Surgery.

Chmn Chairman.

C. Ho. Containers carried in holds *Lloyds' Register notation*.

Chock-a-Block When the cargo hold is completely full and no more merchandise can be stowed.

Chocks Pieces of wood used as dunnage, *q.v.*, between uneven merchandise when stacked in the cargo hold of a ship.

Ch. of C. Chamber of Commerce.

Ch. of S. Chamber of Shipping.

Chopper Slang word for Helicopter, *q.v.*

CHOPT *or* **Chopt** Charterers' Option. A charter-party term. This may refer to the loading of a percentage of cargo more or less.

Chose in Action A thing which is not in a person's possession but which he has a right to recover, if necessary by legal action, or to assign, e.g. money at a bank.

Chose in Possession A legal term attributed to the power of right of possession of chattels and other properties. *Ex.* The possession of materials such as tables, chairs, houses etc.

CHP Combined Heat and Power (describes a gas turbine plant generating electricity and steam).

Ch. ppd. Charges Prepaid.

Chq. Cheque.

C. Hr. Candle Hour.

CHRIS *or* **Chris** Chemical Hazardous Response Information System.

Christiana The old city name for Oslo.

Christian Era Period of time dating from the birth of Christ.

Christmas Tree *or* **Xmas Tree** Assembly of valves on top of a well which has been organised for oil production.

Chromium A very hard white metal which is resistant to corrosion. It is mostly used in alloys for stainless steel and electroplating purposes.

Chronometer A very precise clock or watch which is used to take exact times in relation to bearings simultaneously with the intention of obtaining the nearest approximate position of the ship during navigation.

Chronometer Correction Application of advance or retard in seconds of time of a chronometer, *q.v.* Also phrased Chronometer Error.

Chronometer Error The daily loss or gain of a Chronometer, *q.v.* on Greenwich Time Signal or Greenwich Mean Time, *q.v.*

CHS Canadian Hydrographic Service.

CHSC *or* **C.H.S.C.** Central Health Services Council.

CHT *or* **c.h.t.** Cylinder Head Temperature.

Cht Chest.

Chtrd. Frt. Chartered Freight.

CHU *or* **C.H.U.** Centigrade Heat Unit.

Chuck The part of a drilling tool that holds the bit.

Chum Fish bait thrown into the sea to attract fish; a dog-salmon.

Chummer A crewmember of a fishing boat who scatters Chum, *q.v.*, to attract the fish before or during fishing.

Churning (1) Agitation as throwing up of water by a ship's propeller. (2) Short-term share dealings to maximise maritime brokerage fees.

ch.v. Check Valve.

c.h.w. Constant Hot Water.

Ci Curie, *q.v.*; Cirrus Clouds, *q.v.*; Cast Iron; Compression Ignition.

CI *or* **C.I.** Cast Iron; Certificate of Insurance; Channel Islands; China Airlines *airline designator*; Compression Ignition; Consular Invoice; Cost and Insurance.

C & I Cost and Insurance, *q.v.*

CIA Canadian Institute of Accounts/Arbitrators; Central Intelligence Agency *USA*; Culinary Institute of America.

CIA *or* **C.I.A.** Chartered Institute of Arbitrators; Chemical Industries Association; Corporation of Insurance Agents.

Cia Compagnia *Italian*—Company; Companhia *Portuguese*—Company; Compania *Spanish*—Company.

CIAA Co-ordinator of Inter-American Affairs.

CIB *or* **C.I.B.** Central Intelligence Bureau *USA*; Conseil Internationale de Blé *French*—International Wheat Council; Corporation of Insurance Brokers; The Chartered Institute of Bankers; Criminal Investigation Branch.

CIC Clean Islands Council (Honolulu); Combat Information Center *USA*; Counter Intelligence Corps *USA*.

CIC-85 May 1971 Cargo Information Card, *q.v.*

CICA *or* **C.I.C.A.** Canadian Institute of Chartered Accountants; Confederation for Agricultural Credit.

CICILS *or* **Cicils** Confédération Internationale du Commerce et des Industries des Légumes Secs *French*—International Commerce and Industry Federation of Dried Vegetables.

CICP *or* **cicp** Cast Iron Cargo Piping.

CICR *or* **C.I.C.R.** Comité International de la Croix-Rouge *French*—International Red Cross Committee.

CICT *or* **C.I.C.T.** Commission on International Commodity Trade, *in UK and USA*.

CID *or* **C.I.D.** Centre for Industrial Development *EEC*; Committee for Imperial Defence; Council of Industrial Design; Criminal Investigation Department.

ci-devant *French*—Before this; Former.

CIDS Concrete Island Drilling System, *q.v.*

CIE *or* **C.I.E.** Captain's Imperfect Entry *customs*.

Cie Compagnie *French*—Company *Abbrev.* Co or Coy.

CIEM Conseil International pour l'Exploration de La Mer *French*—International Council for the Exploration of the Sea.

CIETAC China International Economic Trade Commission.

CIF *or* **C.I.F.** Cost Insurance & Freight, *q.v.*

CIF & C *or* **C.I.F. & C.** Cost Insurance Freight and Commission. *See* **Cost Insurance and Freight**.

CIFC&I *or* **C.I.F.C.&I.** Cost Insurance Freight Commission and Interest. *See* **Cost Insurance and Freight**.

CIF & E *or* **C.I.F. & E.** *or* **c.i.f. & e.** Cost Insurance Freight and Exchange. *See* **Cost Insurance and Freight**.

CIFE&I *or* **C.I.F.E.&I.** *or* **c.i.f.e.&i.** Cost Insurance Freight Exchange and Interest. *See* **Cost Insurance and Freight**.

CIFFA Canadian International Freight Forwarders' Association Inc., Montreal.

CIF&I *or* **C.I.F.&I.** *or* **c.i.f.&i.** Cost Insurance Freight and Interest. *See* **Cost Insurance and Freight**.

CIFLT *or* **C.I.F.L.T.** *or* **c.i.f.l.t.** Cost Insurance Freight London Terms. *Similar to* CIF (Cost Insurance and Freight) but subject to London Terms which exonerates the owners of the ship from liability as soon as the cargo is discharged.

CIFW *or* **C.I.F.W.** *or* **c.i.f.w.** Cost, Insurance, Freight, War and Insurance.

CIG *or* **C.I.G.** Cassa Intergrazione Guadagni *Italian*—Redundancy Fund.

C.I.G.A. Compagnia Italiana dei Grandi Alberghi *Italian*—Italian Hotel Group.

CIHBB Cast Iron Hub Non-Ferrous Blades *propeller*.

CIHSB Cast Iron Hub Steel Blades *propeller*.

C.I.I. Centre for Industrial Innovation (At the University of Strathclyde, Scotland); Chartered Insurance Institute (A.C.I.I.—Associate; F.C.I.I.—Fellow).

C.I.I.A. Canadian Institute of International Affairs; Commission Internationale des Industries Agricole *French*—International Commission for Agriculture Industries.

CIIR Central Institute for Industrial Research *USA*.

C.I.J. Commission Internationale de Juristes *French*—International Commission of Jurists.

CILA *or* **C.I.L.A.** Chartered Institute of Loss Adjusters.

CIM *or* **C.I.M.** Canadian Institute of Mining; Comité Internationale Maritime *French*—International Maritime Committee; Commission for Industry and Manpower; Convention Internationale de Merchandises *French*—International Cargo Convention (Referring to the International Carriage of Goods by Railway).

CIME Canadian Institute of Marine Engineers.

CIMOS Computer Integrated Marine Operations.

CIMP *or* **C.I.M.P.** *or* **Cargo I.M.P.** Cargo Information Message Procedures *IATA*.

C in C Commander in Chief.

CIO Congress of Industrial Organisations *USA*, which has merged with AFL (American Federation of Labor).

CIP *or* **C.I.P.** Calling-In-Point, *q.v.*; Carriage and Insurance Paid, *q.v.*; Carriage and Insurance Paid to (place); Commercially Important Person; Council of International Programmes *USA*; Class Improvement Plan *US Military Class*.

C.I.P.A. Chartered Institute of Patent Agents.

CIPC Coastal Interdiction Patrol Craft *Air cushion vessel USA*.

Cipher *or* **Cypher** A secret code used by countries, governments and organisations. These codes are also used by businessmen transmitting messages by telex or cables to abridge phrases and sentences to save time and money and in some instances for confidentiality.

CIR *or* **C.I.R.** Commission on Industrial Relations; Commissioner of Inland Revenue; Cost Information Report.

cir, circ circa, circiter, circum *Latin*—About. *Alternative* Ca.

cir, Circ. Circumference.

Circa *Italian/Latin*—About.

Circular Notes Travellers' cheques. A written demand by a bank to its correspondents in another place to pay a named sum of money to a named person or firm.

Circulating Capital This comprises the actual stock debts and cash in hand.

Circulation Fluid *Similar to* **Mud**, *q.v.*

CIRIA Construction Industry Research and Information Association.

CIRM *or* **C.I.R.M.** Centre International Radio Medical *French*—International Medical Radio Centre.

Cirrocumulus Clouds A layer or composed patches of high clouds having a mottled appearance similar to the ripples formed by sea shore sand or ice crystals. Also called 'Mackerel Sky'. *Abbrev.* Cc.

Cirrostratus Clouds High whitish veil of fibrous clouds composed of ice crystals which can be seen forming a halo around the sun and moon. Some people forecast wind approaching. *Abbrev.* Cs.

Cirrus Clouds Ice crystal clouds seen above 20,000 ft. or 5500 metres (or in the Troposphere, *q.v.*, with a silky and fibrous whitish appearance. When these clouds appear it is generally fair weather. In some instances these clouds are called 'Mares Tails'. *Abbrev.* Ci.

CIS Commonwealth of Independent States, comprising the larger republics of the former Soviet Union.

CIS *or* **C.I.S.** Canadian International Ship Register; Cataloguing in Source; Catholic Information Society; Center for Information Studies *USA*; Central Information Service; Chartered Institute of Secretaries; Coal Industry Society; Conference of Internationally-minded Schools.

CISA Canadian Industrial Safety Association; Centro Italiano Studi Aziendale *Italian*—Italian Centre for Business Studies.

Cisco San Francisco, *USA*.

Cislunar nav. Earth to moon distance measurement.

cit *or* **c.i.t.** Citation; Cited; Citizen; Compression in Transit.

CIT *or* **C.I.T.** Chartered Institute of Transport; Compagnia Italiana Turistica *Italian Tourist Company*; Collector of Income Tax.

CITES Convention on International Trade of Endangered Species of Fauna and Flora *USA*.

Citrus Fruits Oranges, lemons, grapefruits, tangerines and limes.

C.I.V. Convention Internationale de Voyageurs *French*—International Convention Concerning the Carriage of Passengers and Luggage by Rail.

CIVC Comité Interprofessional de Vin de Champagne *French*—Interprofessional Committee of Champagne Wine.

Civ. E. Civil Engineer; Civil Engineering.

Civil Day The beginning of the day calculated from midnight. An alternative day is Calendar Day.

Civil Penalty This phrase has been inserted in Marine Insurance, *q.v.*, to draw attention to the fact that fines are often so referred to, particularly in the United States. *See* **Tovalop**.

CIWS Close-in weapons system *naval*.

C.J. Chief Judge; Chief Justice; Corpus Juris *Latin*—Body of Law.

C & J *or* **C &/or J** China and/or Japan.

CK Cask; Countersunk.

C.K. Cape Kennedy.

CKD *or* **c.k.d.** Case(s) Knocked Down; Completely Knocked Down; Cars Knocked Down.

Ck. O.S. Countersunk Other Side.

Ck. T.S. Countersunk This Side.

ckw Clockwise.

CL *or* **C.L.** Corporation of Lloyd's.

CL *or* **C.L.** *or* **c.l.** *or* **cl** Lloyd's Register indication that the propeller shaft is fitted with Continuous Liner; Common Law; Computer Language; Cut Lengths; Cutter Location; Carload; Centre Line; Civil Law; Claim; Class; Classification; Clause; Clearance; Clergymen; Clerk; Cloth; Centilitre; Collar; Calendar Line; Civil Lord; Critical List; Craft Loss; Container Load, *q.v.*; Cash Letter.

Cl. Clause; Chlorine.

C/l Completion of Loading.

C/L Craft Loss.

CLA or **C.L.A.** Cargo Landing Adaptability; Centre Line Averages.

Clad To sheathe, or sheathed, with material for protection or strength.

Cladding A covering of one material with another to protect the inner parts from exposure to corrosion, impact or damage. Hence, for example, Ironclad, a former type of battleship.

CLASS Capacity Loading and Scheduling System; Customs Local Area Signalling Services *USA*.

Class cl. Classification Clause, *q.v.*

Class Expunged If defects occur in the hull or machinery or any other item is deficient so that the ship is rendered unseaworthy or there is any infringement of the Free Board Rules, the Classification Society may strike her out from her present classification and this will be noted in the record book (*see* **Classification Records**) as Class Expunged. This remark will also be clearly shown in the Register of Shipping books as a reference to all who may be interested.

Class Maintained *See* **Conditions of Class**.

Classification *See* **Classification Societies**.

Classification Clause The marine insurance policy is subject to a classification clause which gives the underwriters access to the particulars of the classification of the ship carrying the cargo on which the premium is then charged. *Abbrev.* Class cl.

Classification Records Classification Societies hold records of the class maintenance of each ship registered or classed with them. These records are deposited for the personal and confidential information and guidance of the owner of the ship. *See also* **Conditions of Class**.

Classification Societies Non profit-making bodies directed by committees of persons representing shipowners, shipbuilders, engine-builders and underwriters for the purpose of ensuring ships are properly constructed and maintained in a seaworthy and safe condition. They make rules governing ship construction, arrange and carry out surveys during the building of a ship and throughout the vessel's trading life. They also conduct research into forms of construction, efficiency and safety of sea-going vessels, offshore equipment such as oil-rigs and shore plant.

Classification Survey Certificate *Similar to* **Certificate of Survey**, *q.v.*

Clause 40/40/20 or **40/40/20 Clause** A convention drafted under the auspices of the United Nations Conference on Trade and Development *(UNCTAD)* in which it provides that all shipping traffic between two foreign countries is to be regulated as far as the quantities of shipments are concerned on the following percentages: (1) Owners of the country of origin 40% (2) Owners of the country of destination 40%. (3) Owners of the country which is neither the origin nor the destination 20%. Since revised.

Claused Bill of Lading *Similar to* **Dirty Bill** or **Foul Bill of lading**, *q.v.*

Clauses A word used mostly in sentences or phrases appearing on insurance policies indicating what the underwriters are liable to or exonerated from in any accident. This word is also widely used in Charter-parties, *q.v.*, and Bills of Lading, *q.v.*, clarifying the rights, liabilities, immunities and procedures of the carriers, shippers and charterers.

CLB Continuous Line Bucket *underwater sea surveying*.

CLC Cargo Loading Certificate.

CLC or **C.L.C.** Canadian Labor Congress; Chartered Life Underwriter(s) of Canada; Civil Liability Certificate. P & I Club, *q.v.*, Certificate. For pollution damage. *See* **CLC 1969**; Commonwealth Liaison Committee.

C/L.C. Car/Lumber Carrier(s).

CLC 1969 International Convention on Civil Liability for Oil Pollution Damage 1969 amended by the Protocols of 1976 and 1984.

C.L.C.B. Committee of London Clearing Banks.

cld. Called; Cancelled; Cleared (referring to clearance from customs); Cooled.

Cld Cloud; Cloved.

C.L.D. Doctor of Civil Law.

CLE Council of Legal Education.

Clean (1) Relates to documents which have no detrimental effect on those directly and indirectly concerned in the business contents, such as Bills of lading, without any remarks as to the condition of the cargo. Some shippers in fact insist on having the phrase 'Clean on Board' on the Bills of Lading. This phrase covers the shippers in clearing the shipping documents through the banks. (2) Clean Bill of Health, *q.v.* (3) A berth or anchorage free from any obstructions. (4) The underwater part of a ship which is free from seaweed, barnacles, etc. *See* **Antifouling**. (5) A generic word in aviation to confirm that the aircraft is ready in all respects to take off.

Clean Bill of Health Every ship leaving port or harbour is entitled to be furnished with a Clean Bill of Health. This is an official certificate to the effect that the place or port of departure enjoys good public health and no infectious diseases are present. This is the opposite to Foul Bill of Health, *q.v.*

Clean Bill of Lading A Bill of Lading, *q.v.*, which is given to shippers without any notations as to the condition of the merchandise carried. There are shippers who specifically request the phrase 'Clean on Board' to be inserted clearly on the Bill of Lading, chiefly for two reasons: (1) To avoid misunderstanding in clearing the shipping documents through the bank. (2) Ultimate condition of the merchandise at the port of delivery and consequent claims in case of any irregularity or damage on delivery. *Abbrev.* Clean B/L. *See also* **Clean**.

Clean B/L Clean Bill of Lading.

Clean Charter (1) In concluding a Charterparty, *q.v.*, it is assumed that no underhand business has been negotiated to the detriment of the shipowner, i.e. no ulterior motives are involved. (2) A charterparty where the freight due to the owner is net, without any deductions of commission of any kind.

Clean Credit *or* **Open Credit** Instructions given by a firm or an individual to a bank to pay another firm or an individual a sum of money abroad on release of goods against the presentation of shipping documents under certain conditions.

Clean-Full When sails are kept full and bellying; off the wind. *See also* **Rap-Full**.

Cleaning of Cargo Compartment Clause A standard clause for Time Charters approved by the Documentary Council of Bimco in May 1981. The crew of the ship are to clean all cargo compartments for the necessary cargoes intended to be loaded and if required by the charterers. Such cleaning is to be performed during the voyage to prevent a possible fine on arrival at the port of loading. The owners are not responsible for the holds to be sufficiently clean in order to be acceptable to the port authorities. In fact the owners are not in any way responsible for any consequences. The charterers are to pay a negotiable lump sum each time cleaning is done.

Clear (1) Common reference to the state of the atmosphere. (2) To clear documents or ships from customs. (3) No obstructions to navigation, manoeuvring, etc. (4) To cash a bill or an account.

Clear a Bill To cash a Bill of Exchange, *q.v.*, or to obtain cash money for a bill of exchange.

Clear an Account To effect payment.

Clear a Ship To obtain the authorisation from customs for a ship to sail or for Free Pratique. *See* **Clearance**.

Clearance Another word for **Customs Clearance** which is the formality needed to obtain authorisation from the customs to allow a ship to leave port or to permit her to discharge her cargo. *See* **Clearance Inwards** and **Clearance Outwards**.

Clearance Air Turbulence National Committee for Clear Air Turbulence *US Department of Commerce 1966* has defined this term as follows: 'All turbulence in the free atmosphere in aerospace operations, i.e., not in or adjacent to visible convective activity'. The detection of CAT *or* C.A.T. is mostly done by aircraft although attempts have been made from the ground by radar techniques.

Clearance Certificate *Similar to* **Clearance Label**, *q.v.*

Clearance Inwards Customs authorisation for a ship to start commercial operations in the port of destination and/or to allow officers, crew and passengers to go ashore. *See* **Clearance** and also **Clear a Ship**.

Clearance Label A customs clearance certificate currently made use of in *UK* ports and issued to the master of a vessel authorising him to sail after having followed the necessary requirements. This is also called Cocket Card or Clearance Certificate.

Clearance Outwards Customs authorisation for a ship in port to leave for her next destination. *See* **Clear a Ship** and **Clearance** *also* **Clear**.

Clearance Terminals Another term for Port Terminals, being well equipped with all kinds of mobile, gantry and other types of cranes engaged in container traffic as opposed to any of the conventional cargo lifting gear. In most cases customs facilities are found nearby. This combines both the discharging/loading operations and customs clearance and enables the whole operation to be undertaken efficiently and quickly.

Clear Days Calculated full days. The first and last days are excluded when calculations are taken into account. *See* **Laydays**.

Clearing Banks Commercial or joint stock banks.

Clearing & Forwarding The act of clearing merchandise from customs and delivering it to the consignees.

Clearing & Forwarding Agent An agent who clears merchandise from customs sheds and delivers it to the consignees. *See* **Clearing & Forwarding**.

Clearing House An institution of bankers in the UK which handles the numerous daily cheques drawn by clients at various banks and their branches. They settle the accounts between them as well as the balance of the cheques paid to each other. These transactions are controlled by the Bank of England. The method saves formalities and valuable time for those involved by avoiding the passing of cheques between a string of brokers, dealers, etc. All the transactions are settled on an appointed account day.

Clear Sky A meteorological term for cloudless or practically cloudless sky.

Clear Total Loss Only A marine insurance clause which is not commonly in use. Sue and Labour, *q.v.*, as well as Salvage Charges are excluded from the policy. *Abbrev.* CTLO *or* C.T.L.O.

Clearway That area of ground under an aircraft from which it took off.

Clear Working Days This expression relates to Laydays, *q.v.* Sundays and Holidays are excluded in the calculations.

Cleats Pieces of wood or metal attached to the decks or sides of a ship having two projecting ends around which ropes are belayed or fastened.

Clingage Oils that remain clinging to the sides and other parts of the tanks in an oil tanker after discharge. *See* **Crude Oil Washing**.

cl. gt. Cloth Gilt *book-binding*.

CLI *or* **C.L.I.** Cost of Living Index.

CLIA Cruise Line International Association.

93

Clinker Boat A boat constructed with horizontal boards with each one overlapping in sequence. *See also* **Clinker Plates**.

Clinker Plates A system of placing wooden or steel plates overlapping each other to form a watertight coverage. *See also* **Clinker Boat**.

Clinometer *or* **Inclinometer** An instrument for establishing the angles of rolling and pitching of a ship.

CLIO Compilation of Input–Output.

clk Caulk or calk.

clld Cancelled.

clm/s Claim(s).

Clock Cards Cards used in factories and large business organisations by the employees to record the time of their presence while on duty and to establish their working hours per day, week, etc.

Close (1) To conclude a business deal or Transaction. (2) To steer towards, approach.

Close aboard Very close to the ship; almost alongside.

Close Company A company which must be at least 65% privately controlled by about six directors. A minimum of 50% dividends are distributed out of the profits to shareholders. In some countries legislation has been enacted discouraging the system of the Close Company by charging surtax on the dividends received and also imposing certain restrictions. The American term used is 'Closed Company' or 'Closed Corporation'.

Close Corporation Similar to Close Company, *q.v.*

Closed Economy State economy where no international trade is assumed to take place either in imports or exports.

Closed-end Trust An American Investment Trust.

Closed Shop An industry or any other business in which all the employees are obliged to be members of a specific trade union.

Closed Union American Trade Union with membership restricted to maintain the former's prestige and standards.

Close-Hauled Nautical term. When sailing along close to the wind.

Close Price Competitive price.

Closest Point of Approach The nearest approach of two vessels passing on different courses. *Abbrev.* CPA *or* C.P.A.

Closing a Sale The seller's action in bringing negotiations to a conclusion with the obtaining of an order from the buyer, thus securing his custom.

Closing Date (1) The latest date for cargo to be accepted for shipment by the carrier or charterer or his agent. If cargo is late in arriving it is refused and is termed Shut Out or Shut Out Cargo. (2) This can also mean the latest date for accepting stocks and shares.

Closing Prices Stock Exchange prices for stocks and shares at the end of the day's trading.

Closing Stock Goods that remain unsold at the end of the financial period. This is a credit or an asset carried forward to the next year.

Cloud Burst A sudden and very heavy continuous fall of rain.

C.L.P.A. Common Law Procedure Acts.

C.L.P.C. Cargo Loss Prevention Committee *an international union of marine insurance.*

CLR *or* **C.L.R.** Centre of Lateral Resistance *sailing*; Commercial Law Reports.

Cls Clause/s.

C.L.U. Chartered Life Underwriter.

Club Protection (or Protecting) and Indemnity Club, *q.v.*, Club Mutuelle *or* Mutual Club; P. & I. *or* Pandi Club.

Club Call A practice of Protection and Indemnity Clubs, *q.v.* A club may from time to time call upon its members to make an extra contribution to cover expenses which had not been foreseen to keep the club functioning.

Club Terms Terms and conditions relating to the Protection and Indemnity Clubs, *q.v.*

Clydeport Clyde Port Authority.

cm Centimetre, equivalent to 0.3937 inches.

CM *or* **C.M.** Carrying Modules; Certificated Master; Certificate of Merit; Chirurgiae Magister *Latin—*Master of Surgery; Circulation Manager; Commodity Modules; Computer Module *computer control system*; Condition Monitoring, *q.v.*; Corresponding Member; Corporate Member; Corporate Membership; Court Martial; Currency Movement.

c.m. Causa Mortis *Latin*—By reason of death.

C/m Call (of) More, *q.v.*

CMA *or* **C.M.A.** Catering Managers' Association; Chemical Manufacturers' Association; Civil Military Affairs.

CMB *or* **C.M.B.** Chase Manhattan Bank; Coastal Motor Boat.

C.M.B.I. Caribbean Marine Biological Institute.

CMC Carboxymethyl Cellulose; Caribbean Maritime Council; Council of Marine Carriers.

Cmd Command; Commissioned.

Cmdr. Commander.

CME *or* **C.M.E.** Chartered Mechanical Engineer.

CMEA Council for Mutual Economic Assistance *similar to* **Comecon,** *q.v.*

CMG Captain's Medical Guide.

CMI *or* **C.M.I.** Comité Maritime International *French*—International Maritime Committee.

CMIA *or* **C.M.I.A.** Coal Mining Institute of America.

cml Commercial.

CMLA *or* **C.M.L.A.** Canadian Maritime Law Association.

CMMC City Master Mariners Club, London.

C.mn.Stl. Carbon Manganese Steel, *Lloyd's Register notation.*

CMP Command Module Pilot; Comprehensive Master Plan *USA*; Cost of Maintaining Project.

Cmpl *or* **cmpl** Completed, *Lloyd's Register notation.*

cm.p.s. Centimetres per second.

C.M.R. Contrat de Transport International de Marchandises par Route *French*—International Carriage of Goods by Road. *See* **C.M.R. Convention**.

C.M.R. Convention Convention Relative au Contrat de Transport International de Marchandise par Route *French*—The Convention on the Contract for the International Carriage of Goods by Road 1965. This refers to the carriage of goods by road.

cms Centimetres.

CMS Coastal Minesweeper.

CMS *or* **C.M.S.** College of Marine Studies; Combined Marine Surcharge, *q.v.*

CMT *or* **C.M.T.** Chicago Mean Time *USA*.

CMTC Canadian Marine Transportation Center, *Halifax, Nova Scotia.*

CMUT Confederation of Trade Unions.

c/n Consignment Note; Contract Note; Cover Note; Credit Note.

C/N *or* **C.N.** Circular Note; Consignment Note, *q.v.*; Contract Note; Cover Note; Credit Note.

CNAA *or* **C.N.A.A.** Council for National Academy Awards.

CNAD Conference of National Armaments Directors *NATO.*

CNAS *or* **C.N.A.S.** Chief of Naval Air Services.

CNASA *or* **C.N.A.S.A.** Council of North Atlantic Shipping Association.

CNATCS Controller of National Air Traffic Control Services *UK*.

CNC Centralised Navigation Control; one man bridge operation classification.

CNC *or* **C.N.C.** Compagnie Nationale de Containers *French*—National Container Company; Computer-Numerical-Control.

CNEXO A French government organisation concerned with the exploitation of the oceans under the inspiration of Jaques-Yves Cousteau, the French marine explorer of worldwide fame.

CNIS Channel Navigation Information Service *UK, q.v.*

CNM *or* **C.N.M.** Comité de Normalisation de la Mecanique *French*—Standards for metric thread.

CNN Cable News Network *USA*.

CNO Chief of Naval Operations *USA*.

CNOOC China National Offshore Oil Corporation.

C-NOPB Canada-Newfoundland Offshore Petroleum Board.

CNR Certificate of National Registry.

CNRS Centre National de la Recherche Scientifiques *French*—National Centre of Scientific Research.

CNS Central Nervous System; Chief of Naval Staff.

CNSG Consolidated Nuclear Steam Generator.

CNT Canadian National Telegraph.

CNTIC China National Technical Import Corporation.

CNUCED Conférence des Nations Unies sur le Commerce et Developpement *French*—United Nations Conference on Trade Development.

c/o Care of.

Co Cobalt; Company; Course, *nautical.*

CO Colorado, *USA*.

CO *or* **c/o** *or* **C/O** Care of; Case Oil, *q.v.*; Chief Officer, *q.v.*; Carried Over; Cash Order; Certificate of Origin, *q.v.*

CO *or* **C.O.** Cash Order; Cabinet Office; Central Office; Certificate of Origin, *q.v.*; Chief Officer, *q.v.*; Clerical Officer; Colonial Office; Commonwealth Office; Criminal Office; Crown Office.

CO$_2$ Carbon Dioxide.

COA Cash on Arrival.

COA *or* **C.O.A.** Contract of Affreightment, *q.v.*; Course of Advance, *q.v.*; Condition of Admission.

Coal Hard opaque black or blackish mineral or carbonised vegetable matter found in seams or strata below earth's surface and used as fuel and in the manafacture of gas, tar, etc. Formed over aeons of time from remains of vegetation, etc., under pressure.

Coalfield Area where coal deposits are found, or extracted.

Coastal State The country where the coast is adjacent to the zone of use (Extended Zone, *q.v.* or Contiguous, *q.v.*) of a ship. It may have legal jurisdiction on the operation of the ship while entering that zone or within the territorial waters.

Coaster A small vessel, in some cases under 499 tons gross, which trades from coast to coast or on relatively short distances from one port to another. Her small tonnage exempts her from certain port charges. Despite the word 'coaster', some of these vessels are engaged on long voyages.

COASTGUARD Short for (The) Coastguard Agency, *q.v.*, of the United Kingdom.

Coastguard Agency, The The coordination of civil maritime search and rescue around the UK coastline has long been the responsibility of HM Coastguard, *q.v.* The response to oil and other pollution in UK waters has been organised by the Marine Pollution Control Unit (MPCU), *q.v.* HM Coastguard and MPCU have together formed the Marine Emergencies Organisation of the Department of Transport. On 1 April 1994, the Marine Emergencies Organisation became an Executive Agency, named The Coastguard Agency. This new agency, known as COASTGUARD for short, now discharges the Department of Transport's responsibility for both marine search and rescue and counter-pollution operations.

Coastguard, HM Part of the Coastguard Agency, *q.v.* HM Coastguard coordinates operations around the UK and to 30 degrees west in the North Atlantic. It co-operates with search and rescue organisations in north-west Europe and North America. In a maritime emergency, the Coastguard calls on and coordinates all available facilities—including Royal National Lifeboat Institution (RNLI) lifeboats, Royal Air Force and Royal Navy helicopters, other aircraft and ships, as well as merchant ships, commercial aircraft and ferries who are placed to render assistance. It also coordinates the support provided by the other emergency services when these are involved in a maritime incident. The Coastguard also has its own search and rescue resources, besides three helicopters—one at Sumburgh in Shetland, one at Stornoway in the Outer Hebrides and one at Lee-on-Solent. It maintains cliff rescue teams and a fleet of general purpose boats for inshore patrols as well as 200 vehicles.

The UK is divided into six Search and Rescue Regions: North and East Scotland, Eastern, South Eastern, South Western, Western, West of Scotland and Northern Ireland—each under a Regional Controller. Each Region has a large *Maritime Rescue Coordination Centre* (MRCC). The Regions are subdivided into three or four Districts, each under a District Controller, either in a *Maritime Rescue Sub-Centre* (MRSC) or in the MRCC itself.

Each MRCC and MRSC has a fully fitted operations room and storage for rescue equipment, vehicles and sometimes boats. The Districts are divided into a number of Sectors, Sector Officers being responsible for strategically sited Auxiliary Coastguard stations which have rescue vehicles and equipment; for patrolling their coastlines; liaison with local rescue organisations such as police, clubs and volunteers; training Auxiliaries; and promoting public safety awareness. There are 480 regular officers and some 3,500 volunteer Auxiliary Coastguards who staff 370 Coastal Response Teams and assist in operations rooms. Training of the regular Coastguards is carried out at the Coastguard Training School in Christchurch.

The Coastguard keeps a 24-hour watch, by radio on the international maritime distress frequencies and through the '999' service, from 21 strategic sites around the UK coastline. A high point system of remotely controlled radio sites provides VHF coverage all around the coast to a range of 30 miles or more.

VHF D/F and radar are fitted at certain MRCCs and MRSCs. Telex, telephones and fax are used for communications ashore and there are direct lines for the most important links. All MRCCs and MRSCs are fitted with a computerised Data Retrieval System (ADAS) to process information and compute search plans.

MRCCs for North and East Scotland and Eastern Regions maintain contact with the oil and gas offshore industries. MRCC South West communicates with search and rescue centres worldwide using INMARSAT via the British Telecom Earth Station, Goonhilly. MRCC South Eastern Region operates the Channel Navigation Information Service, *q.v.* MRCC North East Scotland broadcasts a 3-day weather forecast for fishermen in the North Sea between October and March.

HM Coastguard's Yacht and Boat Safety Scheme enables small craft sailors to give full details to the Coastguard for use if they are in distress or overdue. The Coastguard advises them on radio and lifesaving equipment.

As soon as alerted, by radio message, distress signal or telephone of ships, fishing vessels or yachts in difficulties from shifting cargo, collision, fire, stranding, breakdown, sinking, capsizing, sickness, injury, man overboard, etc. the Coastguard can take appropriate action using its communications network and resources in the area and on call. In civil maritime search and rescue operations, the Coastguard is the focal point of all action, providing the link between the various rescue services and organisations and co-ordinating their efforts in the speediest, most efficient manner possible.

HM COASTGUARD MARITIME RESCUE CENTRES

MRCC Aberdeen	Aberdeen
MRSC Shetland	Lerwick
MRSC Pentland	Kirkwall
MRSC Forth	Crail
MRCC Yarmouth	Great Yarmouth
MRSC Tyne/Tees	Tynemouth
MRSC Humber	Bridlington
MRCC Dover	Dover
MRSC Thames	Walton-on-Naze
MRSC Solent	Lee-on-Solent
MRCC Falmouth	Falmouth
MRSC Portland	Weymouth
MRSC Brixham	Brixham
MRCC Swansea	Swansea
MRSC Milford Haven	Milford Haven
MRSC Holyhead	Holyhead

MRSC Liverpool	Liverpool
MRCC Clyde	Greenock
MRSC Belfast	Bangor
MRSC Oban	Oban
MRSC Stornoway	Stornoway

Coasting Trade Relates to the carriage of cargo by vessels from one or more ports to another or others around the coasts of the same country.

c.o.b. Close of Business.

COB *or* **C.O.B.** Cargo on Board; Close of Business; Container/Oil/Bulk.

COBE Cosmic Background Explorer (*USA* 1978)

COBOL Common Business-Orientated Language *computer*.

COBRA Continental Britain Asia.

COC *or* **C.O.C.** Chamber of Commerce, *q.v.*

Coch *or* **Cochl** Cochlear *or* Cochleare *Latin* a spoon *medical prescribing*.

Cocket Card Another name for Clearance Label, *q.v.*

COCOM Co-ordinating Committee *USA*; Communications Committee.

COD Chemical Oxygen Demand, a measure of the pollutants in water discharges.

COD *or* **C.O.D.** *or* **c.o.d.** Cash on Delivery; Cargo on Deck; Chamber of Deputies; Collect on Delivery; Courier on Deck.

CODAG *or* **C.O.D.A.G.** Combined Diesel and Gas Turbine *in ship propulsion*.

Code Combination of letters which form sentences and used in telexes or telegrams to save time and expense and for security reasons.

Code Flag Small flag used in the International Code of Signals.

CODELAG Combined diesel electric and gas turbine *power configuration*.

CODELOG Combined diesel electric or gas turbine *power configuration*.

Codicil An adjunct to a will altering or cancelling it. Derived from the Latin Codicillus.

CODOG *or* **C.O.D.O.G.** Combined Diesel or Gas Turbine.

CoE Council of Europe.

COED Computer Operated Electronic Display.

Coef *or* **Coeff** Coefficient.

Coefficient of Loading *See* **Loading Coefficient**.

Coemption An act by a businessman which monopolises the market in a commodity by buying the entire available stock in order to be in an advantageous position to fix the price. This term is similar to 'Corner the Market'.

COEX Chain Operators Exchange Conference.

C of A Certificate of Airworthiness.

C of B Confirmation of Balance.

C of C Chamber of Commerce; Code of Conduct; Coefficient of Correction *nautical*.

C of E Church of England; Coefficient of Elasticity; Council of Europe.

C of F Chief of Finance; Certificate of Fitness, *q.v.*; Coefficient of Friction.

Cofferdam (1) A narrow compartment between two bulkheads extending across a ship, usually to separate oil tanks from machinery spaces in case of leaks. (2) A temporary structure within a ship to contain a leak from hull damage, shot holes, etc.

C of G Carriage of Goods, *q.v.*; Centre of Gravity.

C of G Act Carriage of Goods by Sea Act 1924, *q.v.*

C of G H Cape of Good Hope.

COFI Committee of Fisheries *In the Food and Agricultural Organisation* (*FAO*).

C of L City of London; Coefficient of Loading. *See* **Loading Coefficient**.

C of O *or* **C/O** Certificate of Origin, *q.v.*

C of P Centre of Pressure; Code of Procedure.

C of S Certificate of Seaworthiness; Chief of Staff; Church of Scotland; Conditions of Service.

COG *or* **C.O.G.** *or* **c.o.g.** Course Over Ground *nautical*; Carriage of Goods.

COGAS *or* **C.O.G.A.S.** Combined Gas and Steam *in ship propulsion*—referring to the combined gas and steam turbine.

Cognovit *Latin*—he has acknowledged or admitted. An admission of liability.

COGS *or* **C.O.G.S.** Carriage of Goods by Sea Act 1936. The American equivalent of the Hague Rules, *q.v.*; Cost of Goods Sold.

COGSA *or* **C.O.G.S.A.** *or* **Cogsa** Carriage of Goods by Sea Act—referring to the Carriage of Goods by Sea Act 1936 *USA*.

COH *or* **C.O.H.** *or* **c.o.h.** Cash on Hand.

COHSE *or* **Cohse** Confederation of Health Service Employees.

COI *or* **C.O.I.** Central Office of Information.

COIE Committee on Invisible Exports *UK*.

Coil A reel of rope of a whole length of 120 fathoms. Ropes are either twisted in a clockwise direction, known as right-handed rope, or twisted anticlockwise, known as left-handed rope. Coils of wire ropes may be of various lengths.

Co-Ins Company Insurance.

Co-Insurance Insurance risk taken by two or more underwriters together and independently of re-insurers, thus sharing the risk between them. *Abbrev.* Co-Ins.

Co. inv. Consular Invoice.

Coke The residue in a solid state remaining when the volatile elements of carbon have been extracted from coal. It is used as fuel. *See* **Coal**.

Col. Colombia; Colonel; Colorado *USA*; column.

COL *or* **C.O.L.** Cost of Living; Council of Lloyd's.

COLB *or* **C.O.L.B.** Cost of Living Bonus.

Col. Bk. Collision Bulkhead, *q.v.*

Cold Call Salesmen's speculative visit without appointment to a potential customer.

Cold Fog Suspended water droplets with a temperature below freezing point. This prevails mostly in northern Europe during winter.

Cold Move A movement by a vessel without the use of her engine(s). The move could be done by means of tugs, by Warping or Kedging.

Cold Pilot A Pilot, *q.v.*, whose duties are confined to assisting the movement of a ship within a harbour when shifting berth or docking.

Cold Ship *or* **Cold Vessel** Ship with engine(s) not available.

Cold Wall Cold water current moving from the Arctic Ocean southerly to the Gulf Stream.

COLIDAR Coherent Light Detection and Ranging.

Coll *or* **Colln** Collision

Collapsible Boat Craft made of rubber, fabric or plywood, which can be folded to save space.

Collapsible Container *See* **Container**. *Abbrev.* Coltainer.

Collapsible Davits The Davits, *q.v.* of a ship which when not in use can be folded into a position which allows a space clear of obstructions.

Collateral Security Guarantee deposited against borrowed money as a security. This could be in the form of immovables, life assurance policies, credit dividends, etc. It is invariably sought by bankers when lending money.

Coll. Com. Collection Commission.

Coll. Dam. Collision Damage.

Coll. Dge. Collision Damage.

Collect Freight *Similar to* **Freight Paid at Destination**, *q.v.*, and **Freight at Risk**, *q.v.*

Collection and Delivery Procedure of collecting goods from the premises of the Consignor, *q.v.*, and delivering them to the Consignee, *q.v.*

Collection Fees Whenever freight is paid at destination, collective fees are added to the freight. These additional fees go to the agents of the vessel at the port of destination unless agreements to the contrary exist between owners and agents.

Collective Agreement *or* **Collective Bargaining** A mutual agreement between trade unions and employers on all matters and conditions of rates of wages, employment, leave and general procedures.

Collective Bargaining *See* **Collective Agreement**.

Collective Bargaining Agreement Crew agreement on ships, in relation to wages and conditions, between the owners and their unions, generally through the ITF, *q.v. Abbrev.* CBA.

Collective Bill of Lading A Bill of Lading, *q.v.*, covering a quantity of small parcels which are distributed by the agent at the port of destination to each individual consignee. This method naturally saves a lot of minimum freights which would have to be paid separately apart from extra stamps, copies of bills of lading and charges. The corresponding American term is Omnibus Bill of Lading. *See* **House Bill of Lading**.

Collectivism Form of socialism applying the economic theory that industrial capital should be collective.

Collective-Pitch Control Similar to Cycle-Pitch.

Colliery Working Day Normal working days of a colliery, including those in a strike or other actions.

Collision Avoidance System Electronic system commonly used to prevent collisions in USA inland navigable waterways. *Abbrev.* CAS.

Collision Bulkhead The forward transverse Bulkhead, *q.v.*, ranging from the bottom part towards the bulkhead decks of a ship which are so constructed that they prevent water leaking into the forepart of the ship in the event of a collision.

Collision Clause Similar to Running Down Clause, *q.v.*

Collision Course The course of a ship which, if maintained, will bring her into contact with another in the vicinity, if the latter also maintains her course.

Collision Regulations *See* **International Regulations for Preventing Collision at Sea**.

Colln Collision.

Collusion Secret arrangement between two or more parties for gain by deceit or fraud.

Colo. Colorado *USA*.

Colog Cologarithm.

Colon (1) A unit of currency used in Costa Rica and El Salvador. (2) The Spanish word for Columbus.

C.O.L.O.S.A. Committee of Liner Operation in South America.

Colour Blindness A defect in the eyesight making it difficult to distinguish colours. Inability to recognise red and green would be very serious for a navigator.

Colours A vessel's flag or national Ensign, *q.v.*

Colporteur A person who sells bibles, religious books, etc. Derived from the French meaning hawker.

COLREG 1972 International Conference on Revision of the International Regulations for Preventing Collisions at Sea *IMO.*

COLS Communications for One Line Systems *computer.*

Coltainer Collapsible Container. *See* **Container.**

Co. Ltd. *or* **Coy. Ltd.** Company Limited, a Limited Company, *q.v.*

Column A round pillar or a Derrick Post, *q.v.*

Colza Oil Crushed rape seed oil poured on rough seas as a calming substance. It is also a lubricant and can be used to light lamps.

Com *or* **Comm** Commission; Committee.

Comb Combined.

Combi A system of adding and/or reducing air passenger seats in relation to the cargo carried so as to produce the most economical commercial income.

Combi *or* **Combi Ship** *or* **Combi Carrier** *or* **Combi Vessel** A combined ship. A ship specifically designed to carry both containers and conventional cargoes.

COMBICON Bill of lading code name for Standard Bills of Lading issued, adopted or recommended by BIMCO in conjunction with Combined Transport Bill of Lading 1971.

COMBICONBILL Combined Transport Bill of Lading 1971. An alternative document is COMBIDOC. *See* **Combined Transport Document.**

COMBIDOC Combined Transport Document, *q.v.*, issued subject to International Chamber of Commerce Rules, adopted or recommended by BIMCO in 1977.

Combination Carrier *See* **Ore/Bulk/Oil** and **Ore/Oiler.**

Combination Rate Freight rate or charges covering some or all of several forms of transport used for a shipment.

Combined Acceptance Trials Ship's trials of all kinds taken together. *Abbrev.* CAT.

Combined Marine Surcharge In the United States it is expressed as 'Marine Extension'. Underwriters charge extra premiums relative to extra hazards expected during the risk period. *Abbrev.* CMS *or* C.M.S.

Combined Transport Document A document as evidence of receipt of goods for their transportation throughout the whole combined system starting from the time the goods are taken in hand until delivery. This is covered by a convention agreed by 25 states and incorporating 17 articles. *Abbrev.* CTD *or* CT Document. *See* **Combidoc.** The title in French is Titre de Transport Combiné, *q.v.*

Combustible *or* **Combustible Liquids** Liquids or other substances giving off inflammable vapours at flash points from 80° up to 150°F. These liquids/substances are rated as Hazardous Commodities. *See* **Dangerous Cargo.** *See also* **Combustion.**

Combustion Chemical action followed by heat and explosion.

COMDAC Command Display and Control System.

COMDEV Commonwealth Development Finance Company.

COMECON (1) Council for Mutual Economic Assistance *East Europe.* (2) Communist countries comprising USSR, German Democratic Republic, Poland, Czechoslovakia, Bulgaria, Hungary and Romania.

COMET Committee of Middle East Trade.

Comet A hazy-looking object, occasionally with a star-like nucleus and occasionally with a tail, moving in an elliptical or nearly parabolic path about the sun.

Come to (1) To steer up into the wind usually with the intention of stopping the vessel. (2) Steer more to port or to starboard as indicated.

COMEX Commodity Exchange.

COMITEXTIL Comité de Co-ordination des Industries Textiles de la CEÉ *French*—Co-ordinating Committee for Textile Industries in the EEC.

COMLA Commonwealth Library Association.

COMLOSA *or* **C.O.M.L.O.S.A.** Committee of Liner Operators *South America.*

Comm Commission.

COMMAD Commission on the Merchant Marine and Defense *USA.*

Command Economy Planned Economy.

Commander (1) Naval rank below Captain. (2) Captain or Master of a ship. (3) Heavy wooden mallet; a hammer for releasing slips or knocking out pins.

Commandite *French*—A partnership where some partners provide capital but are not involved in the management.

Comme il faut *French*—Proper; as it should be; Correct.

Commencement of Hire *Similar to* Delivery Clause, *q.v.*

Commerce The various functions of buying, selling, exchanging, or distributing merchandise, especially on a large scale, either on a local or overseas basis. Traders who deal in overseas or international commerce are usually involved with banking, insurance, advertising, shipping, forwarding agents, brokers, etc.

Commercial Banks Privately owned banks providing extensive services to their clients locally and abroad on money transfers, general loans and credits, credit card facilities, deposit accounts, trusteeships and many other commercial activities.

Commercial Bill *See* **Bill of Exchange.**

Commercial Distribution The system used in the distribution of goods or products from their origin to general consumers. There are various methods of distribution, all depending on the type of material. There are items which have to pass through various phases before they are sold and consumed. Others may have to pass through necessary procedures. *Ex.* Milk in some countries is a government affair. The milk is collected from farmers and sent to a government milk centre where it is processed into different dairy products. The government then distributes these products

Commercial Port

on a wholesale basis to dealers. There are systems which are undertaken by private commercial concerns instead of the state. Manufactured goods are either sold abroad through agents or are destined for home markets. The manufacturers sell to wholesalers who in turn sell to retailers and the goods are then bought and consumed by the public through retail outlets in the form of shops, etc. Fruit and vegetables are sold to dealers who may sell in bulk or deal through wholesalers. Meat producers send their animals to abattoirs for slaughter and official inspection. The meat is then sold to bulk buying agents who sell either to other wholesalers or retailers. There are countries that have different systems of commercial distribution.

Commercial Port Ships call at a commercial port to load or unload merchandise. There may also be a certain amount of passenger trade.

Commercial Speed The prescribed speed of a vessel whereby her engines are not strained when she is loaded with cargo or in Ballast, *q.v.* Also known as Service Speed and Economical Speed, *q.v.*

COMMET Council of Mechanical and Metal Trades *UK*.

Commissary (1) Term relating to provisions; American armed forces food store; an officer providing stores to an army. (2) A deputy; a judge in a Scottish Commissary court; an officer representing a Bishop.

Commission Brokerage A remuneration due to a broker or agent for work done in the successful conclusion of business. *See* **Brokerage**.

Commission Agent An agent whose method of work is to obtain orders on behalf of various clients and then consign them direct to the individuals concerned. The agent obtains his commission on a percentage against the net cost of the order. *See* **Agency Fee**.

Commission Passed Us An extra commission over and above the normal or usual brokerage or commission for the work done by a broker. There is a general practice where all the extra commissions are ultimately distributed among those closely connected in their mutual business circle.

Commitment Fee A pledged fee due to the lender of the money as from the time of legal negotiations for a loan up to the actual time of the money lent.

Commixture The mixing of 'Homogeneous Cargo', *q.v.*, on a ship to an impossible degree of identification. In the event of any damage or shortage the general practice is that each receiver of cargo will have to bear the expense of the same proportion of the cargo which has the discrepancy.

Commodity An article or item of trade or commerce.

Commodity Exchange A building or marketplace where meetings between sellers and buyers of commodities take place and where samples to be sold are exhibited for the buyers to inspect. *See also* **Produce Exchange**.

Commodity Futures Trading Commission One of the USA's federal regulatory agencies. *Abbrev.* CFTC.

Commodity Rate American term for Freight, *q.v.*, applied to certain bulk cargoes.

Commodore (1) The senior captain of a merchant fleet. (2) Royal Navy and American Navy Senior Captain immediately under the rank of an admiral. (3) President of a Yacht Club.

Common Carrier A person or firm who undertakes to carry everyone's goods provided space is available on board the vessel. If the goods for shipment are refused, the shipper is legally covered to sue the common carrier. There are certain goods which due to their nature make the vessel unseaworthy if they are carried and therefore the carrier or agent is free to refuse them. The American corresponding phrase is General Ship.

Common External Tariff Uniform list of import duties adhered to by members of the EC and of EFTA. *Abbrev.* CET.

Common Law The general, non-statutory law of a country.

Common Law Exceptions The general exceptions covering and exempting the owner from liability of his obligations in accordance with the Carriage of Goods by Sea Act 1924, *q.v.* These consist of (1) Act of God, *Ex.* Lightning, earthquake, very rough weather. (2) Queen's or King's enemies, *Ex.* Act of war. (3) Inherent Vice, *Ex.* Deterioration of fruits, meats, vegetables, etc. (4) Jettison. *Ex.* Throwing cargo overboard in time of peril for the safety of men and ship. (5) Perils of the sea. *Ex.* Collision, stranding, sinking. (6) Barratry. *Ex.* Master scuttling ship, smuggling and (7) Restraints. *Ex.* Government embargo on landing of cargo.

Common Lien Possessory legal right over an object until the payment of outstanding accounts is settled. When a lien is exercised on goods and payment is not forthcoming the goods or the cargoes are sold by tender to obtain the necessary funds. *See* **Maritime Lien**.

Common Market A group of countries joined by treaty with a view to working in harmony in free trade, tariffs, imports/exports, training, employment, etc.

Common Notoriety The state of being well known by everyone in a disreputable way. Public exposure.

Common Stock The American equivalent for Ordinary Shares.

Common Year A year of 365 days.

commune bonum *Latin*—the common good.

communi concensu *Latin*—By common consent.

Community Transit Official document used by EC countries for clearing merchandise through customs.

COMNAVSURFLANT Commander, Naval Surface, US Atlantic Fleet.

COMO Commodore, *q.v.*

Comp. Compartment; Compass; Composite Construction *or* Composite Boiler, *Lloyd's Register notation*; Comprehensive.

Compagnia Anonima *Italian*—Limited Company. *Abbrev.* C.A.

Compagnie Anonyme *French*—Limited Company. *Abbrev.* C.A.

Comp. Air Compressed Air.

Compania Anonima *Spanish*—Limited Company. *Abbrev.* C.A.

Compania Limitada *Spanish*—Limited Company. *Abbrev.* C.L.

Companion Skylight on the upper deck of a ship allowing the light to pass through to cabins and inner accommodation.

Companion Hatch *or* **Companion Way** A small hatch on the deck of a ship just big enough to allow a seaman to pass through at ease to his accommodation below.

Companion Ladder *or* **Companion Way Ladder** (1) Narrow ladder connected to the Companion Way, *q.v.*, down to seamen's accommodation. (2) Ladder fixed on the side of a ship in a slanting position to enable the ship's crew to embark or disembark. (3) Steps connected from one deck of a ship to another.

Company Persons forming an incorporated or joint association to work on general business by exercising specific trading functions. *See* **Articles of Association**.

Company Chairman *See* **Chairman of a Company**.

Company Director A person who is chosen by the shareholders of a company to act with another or other directors to manage the company. He will have his own share of responsibility.

Company Store American term for a store under a company's control where its own employees make their purchases.

Company Structure Organisation of a company by groups, divisions and departments.

Compartment Enclosed space in a ship usually with watertight Bulkheads, *q.v.*, and door(s) or a hatch.

Compass Instrument used for observing direction on land and sea, i.e. in steering and in obtaining bearings of visual objects to establish position. It may be a gyro compass, driven by a gyroscope, which aligns itself to true North, or a magnetic compass, which aligns itself to magnetic North. The magnetic compass consists of a round bowl containing a card showing all the Cardinal Points, *q.v.*, in a horizontal position with a compass needle, *q.v.*, at the centre and floating in a liquid. The pointer mounted in the card obtains its direction according to the earth's magnetic field. It is also known as Mariner's Compass. *See* **Magnetic Meridian**; **Compass Error**. *Abbrev.* Comp.

Compass Course The course the helmsman is ordered to steer. The direction in which a ship moves, measured as the angle between Compass, *q.v.*, North and the compass heading on which she is steered. The ship's true course is found by applying variation and deviation (when using a magnetic compass) or compass error (when using a gyro compass). The direction achieved or 'course made good' will be influenced by the wind, sea and currents.

Compass Error The difference and deviation between the angles of true Meridian, *q.v.*, and the direction of the magnetic Compass Needle, *q.v.* It is the difference between magnetic and true angle.

Compass Needle A suspended steel pointer which is made to pivot in a horizontal plane and point to the magnetic North.

COMPAT Computer-Aided Trade.

Compatibility (1) The property of a fuel oil to mix with others so as not to cause problems for combustion through separation, sludging or blocking of filters. If a suitable mixture cannot be obtained the fuel is Non-Compatible. The problem can be alleviated by chemical treatment. (2) Quality of homogeneous items of any cargo that can safely be stowed together without risk of contamination, explosion corrosion or impact damage. For example butter is Incompatible, *q.v.*, with tar.

Compd. Compromised.

Compen. Compensation.

Compl. Complement; Completed *chartering*.

Complaint An American expression for Libel, *q.v.*

Complement The master, officers and crew of a ship who constitute the full crew employed.

Composite Motion A single move at suggestion put forward in a business covering several points to be discussed and resolved as one.

Composite Office An insurance company which apart from effecting insurance against fire, burglary, accidents, etc., also indemnifies on life assurance.

Composite Policy *or* **Combined Policy** An insurance policy which has been subscribed by more than one underwriter.

Composite Vessel *or* **Composite Ship** A steel-framed ship with a wooden hull and decks.

compos mentis *Latin*—Sound of mind.

Compound Arbitrage Several different arbitrages or judgments.

Compound Interest Accrued interest. *Ex.* 100 pounds are deposited at 6% compound interest. For the first year the interest will be 100 + 6 pounds. Interest on the second year will be 106 plus 6%

totalling 112.36 pounds. The next recurrent year will again be calculated at 6% on 112.36 pounds.

Compravendita *Navi Italian*—Ship Sale and Purchase.

Comprehensive Insurance In general terms an insurance which covers the whole or most of the risks in the adventure. Opposite to Limited Terms, *q.v.*

Compromised *and/or* **Arranged Total Loss** A common term used in re-insurance contracts. An Underwriter, *q.v.*, effecting a re-insurance will only recover his claims in the event of Actual Total Loss, *q.v.*, against that part re-insured after it is confirmed that he had actually settled this same claim with his clients. *Abbrev.* CATL *or* C.A.T.L. It is also expressed as Compromised Total Loss.

Compte rendu *French*—Account rendered; Report.

Compulsory Pilot A pilot who is employed at a port since the master of a ship is obliged to take a pilot for the port by law.

Compulsory Purchase The act of a governmental body in buying property by force of law for the use of itself or the general public.

Compulsory Sale (1) Sale of a property under requisition by government or official authorities. (2) Sale of a property enforced by a court of law due to insufficient funds for payment. Any balance left over after the sale is the property of the original owner of the goods or ship.

Compurgation Evidence by witnesses in favour of the accused which was accepted as proof of innocence. Abolished in England 1833.

Computerised Risk Assessment of Ship Hazards Since 1981 the UK Department of Energy has studied the avoidance of risk of collision between passing merchant ships and fixed offshore installations in UK waters. Computers are used to model shipping traffic lanes for passing platforms safely. *Abbrev.* CRASH.

Comr Commissioner.

COMRATS Commuted Rations *USA*.

COMSAT Communications Satellite.

COMSER Commission on Marine Science and Engineering Research *UN*.

Comune bonum *Latin*—Common good.

COMTAPP Computer Ticketing and Automated Passenger Processing *aviation*.

Con *or* **Contra** Against. Contra is *Latin* and *Abbrev.* Con.

CONASA Council of North Atlantic Shipping Association *USA*.

CONAT Concrete Articulated Tower.

Conbulker A ship able to carry bulk cargoes combined with containers in her holds or on deck.

Concealed Discount *See* **Trade Discount**.

Concealment of Information There is an implied condition when effecting an insurance on a ship or any other object. The insured is legally bound

to provide any material circumstance concerning the subject-matter to the Underwriters, *q.v.* Non-disclosure of any material fact will render the contract automatically null and void.

Con Cls A.R. Container Clauses—All Risks *insurance*.

Con Cls T.L. Container Clauses—Total Loss *insurance*.

Conclusive Evidence Indisputable; Clear evidence and answer.

Conclusive Evidence Clause In the trade of transporting wood this clause is inserted in the Bills of Lading, *q.v.* It is a mutual understanding between the shipper and the Carrier, *q.v.*, that the quantity of wood shown on the bills of lading is unaltered and final in case of fraud, pilferage, etc.

Concrete Island Drilling System A new concept in oil/gas drilling operations in providing a simple, endurable, and portable drilling structure very similar to the land type. This structure can be towed up to, *e.g.* 100 miles offshore and can be ballasted on the scene within 48 hours. After the operation the structure can be de-ballasted and towed back. *Abbrev.* CIDS.

Concurrent with Discharge A Voyage Charter-party clause regaining freight due to the owners by the charterers, to be settled during the course of discharging.

Condensate When natural gas is extracted from the well it contains hydrocarbons also in a form of gas but when processed they are reduced to a liquid state and eliminated. The gas becomes known as Stripped Gas, *q.v.*

Condensation *Similar to* Sweat.

Conditional Sale Agreement Sale of goods effected on a Hire Purchase, *q.v.*, basis.

Condition Monitoring Continuous measurement of the condition and performance of machinery by instrumentation and remote observation. This will indicate how far the intervals between successive maintenance inspections or overhauls can safely be extended to avoid the costs and Down Time, *q.v.*, involved in opening up machinery unnecessarily. *Abbrev.* CM *or* C.M.

Condition Order An order given by a client to pay a certain amount of money in the form of a cheque.

Condition Precedent A basic condition in a Charterparty, *q.v.* A charterparty is to be signed in good faith by the contracting parties. If it is discovered that there has been mispresentation the aggrieved party has the right to cancel the contract, unless the intended work has started. Otherwise it would be illegal to cancel. In such a case a 'Breach of Contract', *q.v.*, would arise.

Conditions and Exceptions including Negligence Clause as per Charterparty A clause commonly applied in Bills of Lading, *q.v.*, on a chartered ship protecting the Carrier, *q.v.*, from all terms, conditions, exceptions and contracts during the charter period.

Conditions of Class Recommendations which have to be attended to by the shipowner on the orders of the Classification Society, *q.v.*, before the vessel is confirmed as 'Class Maintained' *q v*, or 'Classed as per Records Book of Classification'. When all the recommended deficiencies have been dealt with the vessel's records will be marked as 'Free of Recommendations'.

Condition Sine Qua Non Absolute; Unconditional.

Condition Subsequent A condition that has to be fulfilled, otherwise the contract is automatically cancelled. An implied condition.

CONDOCK Container-Dock, *q.v.*

Conds. Conditions.

Conductivity Temperature and Depth This often refers to the reading through an instrument currently used by oceanographers to assess the conductivity, temperature and depth of seawater. *Abbrev.* CTD.

Conelrad Control of Electromagnetic Radiation.

Cone of Silence (1) An area beyond radio range when the volume of signals is either faintly heard or not heard at all. (2) Area where the detection of a submarine is very difficult, if at all.

Conference Line Very much in vogue in the past. Shipowners agreed to have a uniform list of freights which were to be kept unchanged by any one company and a special rebate was allowed to their clients after the lapse of one year. *See* **Liner Conference**.

Conference Ship *or* **Conference Line Ship** A ship of a company which is a member of a Conference Line, *q.v.*

Confidence-trickster A man whose aim is to sell as little of the goods as possible and obtain as much money as he possibly can. This is just one explanation among many for such a person; a crook or swindler. *Abbrev.* Con-man.

Confirmed Credit *or* **Confirmed Letter of Credit** Documentary bank note which goes together with an Irrevocable Letter of Credit, *q.v.* Both the negotiating and the receiving banks agree that no cancellations or alterations to the clauses appearing therein are in any way to take effect from the original instructions. The person who opens the letter of credit, however, has the authority only to change some of the clauses generally with the knowledge of the receiver. This originator of the letter of credit cannot withdraw it until the expiry date.

Confirming Bank The bank which undertakes to pay the holder of a letter of credit, *q.v.*, or other paying document.

CONFITARMA Confederazione Italiana Armatori: the Italian Shipowners' Association.

Conf.L/C Confirmed Letter of Credit, *q.v.*

CONGENBILL *or* **Congenbill** Code name for Chamber of Shipping Bill of Lading used in conjunction with Charterparty 1976 for the carriage of timber cargoes from Germany to the UK.

Conglomerate A large company or corporation formed by merger of several smaller firms having different interests.

Conical Buoy A buoy having a conical pointed top showing above the water.

Conical Drilling Unit A circular designed mobile structure originally developed as an exploration rig for operations in ice-bound waters of down to −50°C. It is approximately 85 metres in diameter and about 30,000 tons d.w. capacity. *Abbrev.* CDU.

conjuntis viribus *Latin*—By united powers.

CONLINEBILL *or* **Conlinebill** Code name approved by the Baltic and International Maritime Council in 1976 for Liner Terms Bill of Lading.

Con-man Short for 'Confidence-trickster', *q.v.*

Conn Connecticut, *USA*.

Connaissance des Temps *French*—Nautical Almanack.

Connecting Carrier That ship which loads merchandise at the Transhipment or connecting base or quay.

Conning Position *or* **Conning Station** Position with an unobstructed view where the Master or Pilot stands to control the vessel when Underway, *q.v.*, or Manoeuvring.

Conning-Tower Armoured protective superstructure on a ship, usually submarines, used as a reconnaissance and defensive station, especially during wartime.

Conrail Consolidated Rail *USA*.

Con-Ro Container Roll-on/Roll-off Ship. *See* Ro-Ro Vessel *or* Ro-Ro Ship.

Conscience Money Money paid out by an anonymous person to a particular concern. This person must have on his conscience some wrong act such as robbing, cheating or bribing, and wants to repay without the knowledge of anyone.

Consecutive Days A term referring to Laydays, *q.v.*, which are calculated irrespective of holidays, Saturdays and Sundays. *See* **Laydays**.

Consensus Ad Idem *Latin*—A complete and unqualified agreement reached between the parties and leaving nothing to be settled in future.

consensus facit legem *Latin*—Consent makes the Law.

Conservancy Charge In some countries this charge is retained for the upkeep of the approaches to waterways and canals.

Consgt Consignment, *q.v.*

Consideration

Consideration A Stock Exchange, *q.v.*, term for the sum paid by the buyer of stock for its cost plus broker's commission and transfer stamp.

Consign To deliver cargo; To entrust the husbandry or agency of the vessel; Consignment, *q.v.*

Consignee The receiver of the merchandise.

Consignment Merchandise or cargo; when goods are sent from one place to another on a consignment basis. The receiver accepts the goods to be kept in his stores until they are disposed of. There are many conditions followed but the most common ones are that (1) Once the consignment is sold then the sender is paid by the receiver. There may also be a limit as to when payment is due (2) The consignment may also be left in the receivers' stores and every sale effected is recorded by the receivers who pay the senders for the respective sale, less their agreed commission. This procedure is followed until the whole consignment is about to be exhausted and to be replenished by the same sender. *Abbrev.* Consgt.

Consignment Clause A Charterparty, *q.v.*, clause stipulating whether the owners or charterers are to appoint agents in the loading/discharging ports. Normally, the charterers prefer to appoint their agents for both functions. The shipowners occasionally appoint their own agents as Husbandry to look after their interests whenever their ship is under repair or when disputes arise.

Consignment Note A shipping document similar to a Bill of Lading, *q.v.*, supplied by the carrier for goods carried from one place to another. It contains particulars of freight payment and conditions. On delivery of the goods this note is presented to the carriers for endorsement which is a proof of receipt of the goods on board. *Abbrev.* C/N *or* C.N. *or* c.n.

Consignor The sender or the shipper of the merchandise.

consilio et animis *Latin*—By wisdom and courage.

Consolidated Rate The rate of expenses and charges relating to Consolidation, *q.v.*

Consolidated Stock Government Stock as a consolidation of The National Debt. A shortened form of word is Consols.

Consolidation Grouping several packages into one whole consignment freighted to a destination where every single parcel is delivered to its individual consignee.

Consolidator An air forwarding agent who specialises in collecting individual parcels and airfreights them in a group in one shipment. Also known as Secondary Carrier.

Consols A contraction for Consolidated Funds or Consolidated Stock.

Consorcio *Spanish*—Association.

Consortium The amalgamation of two/three or even more countries or private companies to form one whole society or business concern on one or more particular projects which would seem rather complicated or far too expensive to be undertaken singly.

Consorzio *Italian*—Association.

Constant Weights A vessel's 'fixed' allowance for crew's effects, stores, spare parts and freshwater.

Constr. Construction; Construed.

Construction Differential Subsidy A subsidy granted by the US Maritime Administration to entrepreneurs under the Merchant Marine Acts.

Construction Policy A marine insurance policy often known as Builders' Policy which makes good for expenses against risks incidental to the building of a ship. This covers the entire period the vessel is in the shipbuilding yard up to the handing over to the new owner. *See* **F.C.V.** *or* **Delivery Trip**.

Constructive Total Loss A term used in the law of marine insurance to denote a loss which entitles the insured to claim the whole amount of his insurance, on giving to the insurers Notice of Abandonment, *q.v.* Generally there is a constructive total loss when the subject-matter insured has not actually perished or lost its form or species, but has, by one of the perils insured against, been reduced to such a state or placed in such a position as to make its total destruction, though not inevitable, yet highly imminent, or its ultimate arrival under the terms of the Policy, *q.v.*, though not utterly hopeless, yet exceedingly doubtful. In such a case the insured, by giving notice within a reasonable time to the insurers of abandonment, i.e., the relinquishment of all his right to whatever may be saved, is entitled to recover against them as for a total loss. If notice is not given, the loss is treated as a partial loss unless the ship in fact has become a total loss or if there would be no possibility of benefit to the insurer if notice were given to him.

One test is that a wrecked ship has become a constructive total loss if the cost of repairing her would exceed her value when repaired. The other is that she will become so when a prudent uninsured owner would not repair her having regard to all the circumstances: in the case of a wreck the owner is entitled to take into account the break-up value, as well as the estimated cost of repairs, in reckoning whether there has been a constructive total loss. *Abbrev.* CTL *or* C.T.L.

consuetudo pro lege servatur *Latin*—Custom is observed as law.

Consul Corps Diplomatique. A Chargé d'Affaires. A person engaged by a government to reside in another place as its representative. A consul may also be engaged on an honorary basis and be a native of the same country. When a vessel arrives at a foreign port the master appoints a day to see the consul of the country where the vessel is registered in order to present the officers' and crew's passports, together with the ship's registry. These documents are returned after perusal and payment of fees due. The ship's registry is stamped against the names of the officers and crew.

Consulage Consular charges similar to Consular Fees, *q.v.*

Consular That which concerns the work of a Consul, *q.v.*

Consular Fees Charges levied by the Consulates or Consuls for services rendered. *See* **Consul**.

Consular Invoices Particular forms obtained from Consuls, *q.v.*, of the importing country. The importer is under an obligation by the importing country to present these forms before customs clearance is given.

Consular Passenger A person who is booked to travel on a ship by a Consul, *q.v.*, especially when he is repatriated and when his passage will be paid for by the state involved.

Consumer Goods Goods which are in demand by consumers. *Similar to* Consumption Goods.

Consumption Goods *Similar to* Consumer Goods, *q.v.*

Cont. Constant; Contents; Continent or mainland of Europe; Continue; Continuous.

Contact Clause A marine insurance clause under the full policy covering claims from contact against any objects including ice, irrespective of the Free of Particular, *q.v.*, or Total Loss, *q.v.*

Contact Damage Damage sustained through impact or collision with other objects. Sometimes referred to as Insurance Collision. *See* **Contact Clause**. *Abbrev.* Cont. Dge.

Cont. A.H. *or* **Cont. A/H.** Continent Antwerp Range. System of loading/discharging ports in a Charterparty, *q.v.*

Container As the name implies this is a large rectangular or square container/box of a strong structure (sometimes made of corrugated steel) that can withstand continuous rough handling from ship to shore and back. It opens from one side to allow cargo to be stacked and stowed into it. This transportation system has been in operation for some time and is considered in shipping circles to be a sophisticated way of carrying cargo. Another method of carrying cargo is by Roll-on/Roll-off or Ro/Ro, *q.v.* The container is fitted with lugs, hooks or brackets at each corner so that it can be lifted up and shifted from ship to shore and/or placed on trailers or trucks. Some of the containers in use are:—

(1) *Amphibious*. Constructed of light strong metal which when filled with merchandise floats on the water and can be floated down rivers or waterways and lifted on to either road or rail vehicles for onward despatch.

(2) *Collapsible*. This container can be dismantled after it is made use of and can be returned to its original place of shipment or where needed. It is a very economical way of transporting cargo as the dismantled container occupies little space and is great saving in freight charges.

(3) *Demountable*. A container having a hydraulic leg at each corner for the purpose of lifting and positioning it from one place to another.

(4) *Expendable*. A simple form of large case resembling a container, which, due to its comparative value *vis-à-vis* its subsequent use and expense for its return, can be disposed of or destroyed.

(5) *Insulated and Refrigerated*. A container used for provisions needing low temperatures for preservation. A large refrigerator for fresh meat, fish, fruit, vegetables, etc., needing cold or cool chambers during the voyage.

(6) *Special Type Containers*. Containers filled with bulk or liquid cargoes. This method has been found advantageous in the sense that it is quickly filled and emptied, thus saving time as well as expense.

(7) *Submersible*. This has been introduced with apparent success. It is generally constructed of a strong synthetic skin material and is used for various liquid cargoes. When emptied it can be folded or rolled and can be returned to its original place of shipment or wherever needed. The method of transportation by containers has become very common and although on the face of it the cost seems to be higher than the conventional way, it is balanced by other advantages such as (1) Pilferage is restricted to the bare minimum, if at all. (2) Lower insurance premiums. (3) Quicker discharge of goods. (4) Less breakages and (5) Easier checking. Most containers today are carried on trailers to facilitate loading and unloading operations. In many cases liners or other ships which carry containers (Container Ships, *q.v.*) normally leave the trailers behind and collect them on their return voyage.

Container Base *Similar to* Container Terminal, *q.v.*

Container Clause A standard clause for Time Charter, *q.v.*, approved by the Documentary Council of BIMCO in May 1981 whereby the charterers, *q.v.*, are wholly responsible for container security and any damage resulting from the inadequate securement of containers, tackle, etc.

Container Dock An open well deck ship designed to carry Containers, *q.v.* She can also be used for Lash, *q.v.*, barges instead of containers. When the dock is submerged the barges are steered into the well and then the Container Dock can be refloated, thus emptying the water so that the barges rest in the dry well hold. *Abbrev.* CONDOCK.

Container Freight Service Cargo freight service with Containers, *q.v.*

Container Freight Station A place or a station operated by container carriers, where the containers are assembled or collected. *Abbrev.* CFS *or* C.F.S.

Containerisation

Containerisation The act of stowing merchandise in various **Containers**, *q.v.*, at a specified place, e.g. at the premises or stores of the Consignor, *q.v.* Generally this is not done at the dock or quay area so as to avoid unnecessary congestions.

Container Load Cargo shipment which would fill up a container. *Abbrev.* CL *or* C.L. *or* c.l. *or* cl.

Container Port *Similar to* Container Terminal, *q.v.*

Container Ship *or* **Container Vessel** A ship constructed in such a way that she can easily stack Containers, *q.v.*, near and on top of each other as well as on deck. She has cell guides for the appropriate and easy handling of the containers which are mostly of a cellular construction for strength. One of the main peculiarities is the special lifting gear which will be able to lift and shift containers from one place to another for stowage purposes. This ship may not be suitable to load general or bulk or conventional cargoes unless same are carried in containers. *See* **Cellular**.

Container Stuffing The act of stacking the merchandise in a Container, *q.v.* Opposite to Container Unstuffing, *q.v.*

Container Terminal A terminal purposely built for Containers, *q.v.*, to be handled and stored on a fairly large scale. Most of the common terminals have quays with mechanical berthing facilities. Loading and unloading can take place simultaneously with the obvious unhindered departure of the ships. Also called Container Port *or* Container Base.

Container Traffic The transportation of Ro/Ro or Roll On/Roll Off.

Container Unstuffing The act of unloading the merchandise from a Container, *q.v.* Opposite to Container Stuffing, *q.v.*

Contam. Contamination, *q.v.*

Contango A Stock Exchange, *q.v.*, expression; percentage paid by buyer of stock for postponement of transfer.

Cont B.H. *or* **Cont B/H** Continent, Bordeaux, Hamburg Range. The sequence of option for the Charterer, *q.v.*, in the loading and/or discharging ports in the contract of a Charterparty, *q.v. See* **Continent**.

Contcoal Code name for Baltic and International Maritime Council Charterparty 1945 for the carriage of coal and coke from Germany, the Netherlands and Belgium.

Contd. Continued.

Cont. Dge. Contact Damage, *q.v.*

Cont. H.H. *or* **Cont H/H** Continent, Havre, Hamburg Range. The option of rotation of loading and/or discharging ports given to the charterers in a Charterparty, *q.v.*, contract. *See* **Continent**.

Contiguous Zone *Similar to* **Adjacent Zone**, *q.v.*

Continent Geographical and shipping word referring to the range of ports from Hamburg to Bordeaux, both inclusive.

Continental Shelf An area of the Seabed, *q.v.*, away from the shore being outside territorial limits having a sloping gradient depth of 200 metres and over.

Contingency Liable to occur or happen or something uncertain; provisionally liable to occur. Possible or conditional. *Ex.* Back Freight, *q.v.*

Contingency Freight Freight payable at destination being at risk to the owner or Carrier, *q.v.*, and recoverable by an insurance.

Contingent Assurance *See* **Assurance**.

Contingent Liability The responsibility held by a person for another until he has discharged his obligation.

Contingent Valuation US methodology for natural resource damage assessment following an oil spill taking into account costs of restoration and compensation for loss of resources and amenity in the interim. *Abbrev.* CV.

Continuation Clause In the Time Insurance Policy or Time Policy, *q.v.*, the Continuation Clause is added. If the ship does not arrive at her destination for some reason or other the insured is held covered until she is an 'Arrived Ship', *q.v.*, and berthed safely in the port of destination. *Ex.* The ship will be still at sea or in distress during the time of the expiration of the policy. The insured is covered by this clause for a further period of say one month. An additional premium will be paid in proportion to that of the policy. *Abbrev.* CC *or* c.c. *or* C. Cl.

Continuous Discharge Book An important document in the form of a book issued by the port authorities to a seaman and serving as a sort of history of employment while at sea from his initial seafaring days up to his discharge. It contains notes about his conduct, capacity or grade of work, names of ships engaged on and periods of work. In the first part of the book there is a photograph of the seaman which serves as identity on presentation for employment or for any other official engagement. In many countries this document is also accepted as a passport. *Abbrev.* CDB.

Continuous Welding An uninterrupted welding, as opposed to spot welding.

Contr. Controllable. *Lloyd's Register notation*.

Contra *or* **Con** Against.

Contraband The act of smuggling goods into a country by evading the payment of customs duty through the proper channels.

Contraband of War Items which by international law are not to be supplied by a neutral country to another at war and which in turn are liable to seizure or confiscation by a belligerent country.

contra bonos mores *Latin*—Contrary to good manners.

Contract In Enter into a contract, scheme or union with others. Opposite to Contract Out, *q.v.*

Contract Note Document issued by a Stockbroker, *q.v.*, showing the price, consideration, charges and taxes involved in the sale or purchase of Stocks, *q.v.*, on his client's behalf.

Contract of Affreightment *See* **Affreightment**. *Abbrev.* COA *or* C.O.A.

Contract of Beneficence Similar to Gratuitous Contract, *q.v.*

Contract of Carriage This applies to Contract of Carriage under a Bill of Lading, *q.v.*, as far as such a document relates to the Carriage of Goods by Sea. However, the bill of lading does not regulate the relationship between the carrier and the holder of same if a Charterparty, *q.v.*, is in force. The bill of lading serves as a receipt for the goods on board. So long as it remains in the hands of the charterer it would not regulate the relationship but the charterparty would. *See* **Carriage of Goods by Sea Act** 1924 and also **Charterparty**.

Contract of Novation A new substituted contract in place of the old one with the consent of all the parties.

Contractor Furnished Equipment *See* **Government Furnished Equipment**.

Contractors' All Risks A comprehensive insurance covering all the Risks, *q.v.*, undertaken by a contractor, such as the quoted price of the contract, offices, stones, scaffoldings, supports, shutterings, debris clearance in case of any destruction, surveyors, architects and consultants' fees in assessing any damage sustained, allowance for inflation, etc.

Contract Out To disassociate oneself from taking part in any action. Opposite to Contract In, *q.v.*

contra proferentem *Latin*—A principle of Marine Insurance, *q.v.*, that any uncertainty or ambiguity in the clauses in the contract is construed against the party which formulated or initiated the contract.

Cont. R/I Contributory Reinsurance.

Contribution and Calls The premium and adjusted additional fees owing to the Mutual Insurance Association or P. & I. Club, *q.v.*, for the renewal insurance of a vessel. The additional fees depend much upon the previous amount of claims.

Contributory Value *or* **Contribution Value** The value of sacrifices in the General Average Contribution, *q.v.*, in the interest of all involved and the safety of the venture. *Abbrev.* C.V.

Controllable Pitch Propeller *or* **Controlled Propeller** *or* **Variable Propeller** *or* **Variable Pitch Propeller** A mechanical propeller installed in some ships and widely used during the Second World War. Its blades revolve automatically to produce a forward or backward thrust making the ship move fore or aft. It can be controlled either direct from the bridge or from the engine room as no large gearbox units are needed, just a small mechanism alongside the hollow propeller shaft and the propeller itself. *Abbrev.* CPP *or* C.P.P. *or* cpp *or* c.p.p. *or* VPP *or* V.P.P. *or* v.p.p.

Controlled Carrier Bill A Federal Maritime Commission Bill which was introduced in the *USA* during 1978 regulating the activities of state-controlled shipping lines by preventing them from undercutting Freights, *q.v.* These state shipping lines, usually from communist or socialist countries, are listed with the Federal Maritime Commission and shippers are automatically compelled to avoid using the agents representing the companies who are seeking to cut freights. *Abbrev.* CCB.

Controlled Mooring Link A Mooring system designed to provide safe and efficient mooring even in the most severe weather conditions. *Abbrev.* CML.

Controlled Ratio Relating to Exchange Control, *q.v.* Regulations limiting the basis of the Exchange Rate.

Controller A person engaged to conduct assessments of the financial state of a company when in a state of Bankruptcy.

Controlling Depth An expression relating to the maximum draught (Draft), *q.v.*, of the sea, lake, or river shown on a chart.

Controlling Hold The cargo hold of a ship which has the largest cubic capacity compared with the other holds. This hold naturally takes more time to load and/or unload. If the ship is scheduled to leave on time overtime work will have to be performed.

Controlling Interest A person or a company who owns more than 50% shares in the business.

Control Tower A building on an airfield having the necessary equipment and personnel to observe and regulate ground and air traffic.

Cont/RO-RO/B Container/Roll-On/Roll-Off/Bulk Vessel.

Cont. V. Control Valve.

CONUS Continental United States.

CONv. Converted from Conveyance.

Conv. Conveyance.

Convenience Goods *or* **Convenience Items** Foodstuffs processed and packaged in such a way that little preparation is required before consumption.

Convenient Speed A Charterparty, *q.v.*, phrase. An undertaking by the owner and his servants that the vessel has to proceed to her loading or discharging berths at her normal reasonable speed without undue delay.

Conventional Cargo Cargo carried in the holds of conventional vessels and not in Container Ships, *q.v.*

Convention de Merchandises par Route *French*—The Convention on the Contract for the International Carriage of Goods by Road 1965. This relates to the carriage of goods by road. *Abbrev.* CMR Convention.

Convertible Currency Money currency which is freely exchanged with other currencies in the world market without any restrictions whatsoever. *See* **Hard Currency**, *q.v.*

107

Convertible Securities Securities, *q.v.*, that are easily or readily exchanged or sold in the international market.

Conveyance Order An offical order issued by a Consul, *q.v.*, or a shipmaster or a Shipping Master, *q.v.*, through the request of a shipowner or shipmanager to return a seaman or a citizen to his native land. The owner is under an obligation to return or Repatriate, *q.v.*, a seaman left behind because of sickness or failure to join his ship or on any compassionate grounds. *See* **Deserter**.

CONWARTIME 1993 Code name for BIMCO Standard War Risks Clause for Time Charters.

COO Chief Operations Officer; Country of Origin.

Cooling-off Period (1) Period allowed to a client during which a signed agreement may be revoked, after which it becomes final. (2) During a hot dispute between two parties, a mutually agreed period for reconsideration before further action.

Cooper The person who manufactures and repairs casks and barrels used for wine, spirits, tobacco, etc.

COP Cash on Presentation *of documents*; Coefficient of Performance.

COP *or* **C.O.P.** Certificate of Proficiency; Custom of Port. *See* **As Customary**.

COPA Council of Post-Secondary Accreditation *USA*.

Co-Pilot An assistant to the pilot of an aircraft while in flight.

COPM Committee on Petroleum Measurements.

Copper A red ductile metal mostly used in the alloy of bronze and brass.

Copper Paint An antifouling paint applied as a coating for underwater protection.

COPR Centre for Overseas Pest Research.

COPT Captain of the Port.

Copyright The sole right of produced or reproduced work in books, plays or any important undertaking either published or unpublished. The originator or author of such work is legally protected where there has been a breach by imitating or copying. This could be work in books, maps, charts, plans, dramas, films, novels, etc.

Copyright Reserved *Same as* 'All Rights Reserved', *q.v.*

Copy-writer Specialist in an advertising agency whose job is to influence the selling of goods or services in a skilful way to would-be clients.

Cord Wood freight measure of 128 cu.ft. out of 4ft × 4ft × 8ft.

Cordoba A unit of currency used in Nicaragua.

Core The central strand of a rope surrounded by the other strands,

CORE Canadian Offshore Resources Exposition.

C ore 7 Iron Ore Form C. Charterparty, *q.v.*, form for the carriage of iron ore from Mediterranean ports, chiefly North Africa.

C ore 8 Iron Ore Form C. Charterparty form for the carriage of iron ore from North Spain.

C and/or J China and/or Japan.

Co-respondent Either joint plaintiff or joint defender.

Corn. Cornwall, *UK*.

Cornering the Market *Similar to* **Coemption**, *q.v.*

Corn Exchange An important cereal business Bourse, *q.v.*, which has existed for over 150 years. It is situated in Mark Lane, London. Business is done every day, most particularly on Mondays. Samples of various corns are exhibited while business is going on. It has numerous members consisting of brewers, merchants and others interested in the market. There also exists a 'Corn' Exchange in Liverpool and corn exchanges once existed in most English towns and cities.

Corn Trade Clause Standard Clauses used in the corn trade.

Corp. Corporation, *q.v.*

Corporation A group of authorised persons acting by law as a Corporation. *Abbrev.* Corp. Also known as Body Corporate.

Corporation of Lloyd's *See* **Council of Lloyd's**.

Corps *French*—Hull.

corpus delicti *Latin*—The body of the crime. The essential facts of the crime charged.

corpus juris civilis *Latin*—The body of the civil law.

corr Corrected.

Correct Used in radio telephone conversation as an acknowledgement meaning 'Agreed' or 'Correct'.

Corresponding Bank *See* **Originating Bank**.

corrigenda *Latin*—To be corrected. As errors in a book.

corr.n Correction.

Cors. Corsica.

CORT Carry-On Receiver/Transmitter.

Cos. Cosecant; Cosine.

COS Chamber of Shipping, *q.v.*; Chief of Staff.

C.O.S. *or* **cos** Cash on Shipment.

Cosag *or* **C.O.S.A.G.** Combined Steam Turbine and Gas Turbine *in ship propulsion*.

COSAL Coordinated Ship's Allowance List *USA*.

Cosec. Cosecant.

Coseismal Relating to points on the earth's surface of the simultaneous arrival of the shock from an earthquake wave. A coseismal curve joining such points can be shown on a map.

COSH Committee on shipping Hydrography.

COSINE Co-operation for Open Systems Interconnection Network in Europe. *EEC research project.*

Cosmic Relating to the cosmos, the universe as a systematic whole.

Cosmoline Trade name for a thick grease used to coat weapons.

COSMOS *or* **C.O.S.M.O.S.** Coast Survey Marine Observation System; Computer Optimisation and Simulation Modelling for Operating Supermarkets. *Computer.*

COSPAR Committee on Space Research.

COST Continental Offshore Stratigraphic Test.

Cost Accountant A person employed in a company or other large business concern whose responsibility is to find the best possible means of maintaining and providing profits. Sometimes called Management Accountant.

Cost Accounts The presentation of expenditure in relation to productivity.

Cost and Freight The quotation of the merchandise includes the cost and freight. Insurance, if any, is for the account of the consignee or receiver. *Abbrev.* C & F *or* CNF.

Cost-Effectiveness The determining factor in obtaining or deciding upon the best alternative services and facilities in relation to the achievement of the maximum economic benefit.

COSTHA Conference on Safe Transportation of Hazardous Articles.

Cost Insurance and Freight *or* **Cost Freight and Insurance** Of common commercial usage in export and import work. The manufacturers, merchants or shippers, when quoting to their clients or the receivers of the merchandise, are referring to the cost of the item including insurance and freight up to the point of destination. Unloading expenses are to be borne by the consignees or receivers. This type of quotation is looked upon as advantageous to the receiver or consignees in the sense that the exact gross amount of money involved in the overall business transaction can be calculated easily. *Abbrev.* CIF *or* C.I.F. *Abbrev.* for Cost Freight and Insurance CFI *or* C.F.I. The French term is coût, assurances, fret. *Abbrev.* CAF.

Cost-per-Passenger Kilometre *Similar to* **Cost-per-Passenger Mile**, *q.v.*

Cost-per-Passenger Mile An aviation term. The number of air passengers divided by the distance covered. *Abbrev.* CPM *or* C.P.M. Sometimes also expressed as Cost-per-Passenger Kilometre. *Abbrev.* CPK *or* C.P.K.

Cost-plus The estimated cost of actual work plus an agreed percentage of profit when submitting quotations. This method is also practised in the sale of products in general.

COSTPRO Canadian Organisation for the simplification of Trade Procedures.

Costs and Expenses An insurance expression referring to the costs incurred when insurance lawyers, surveyors and others are engaged in their professional capacities.

COSUP Co-ordination of Supplies.

COT *or* **C.O.T.** Customer's Own Transport.

Cot. Cotangent.

COT Card on Tape Reader *computer*; Customs of Trade.

COU *or* **C.O.U.** Clip-on Unit.

couleur de rose *French*—rose-coloured. Attractive outlook. The bright side.

Coulisse A Paris unofficial Bourse, *q.v.*, or Chamber of Commerce, *q.v.*, made up of respected firms and Arbitrage, *q.v.*, houses.

Coulomb Unit of electric charge transferred by a current of one ampere in one second. Symbol C. named after Charles de Coulomb (1736–1806). *Abbrev.* C.

Coumarin Odorous crystalline substance ($C_9H_6O_2$) found in the seeds of the tonka beam and from which coumaric acid ($C_9H_8O_3$) can be obtained.

Coumarone Colourless liquid (C_8H_6O) distilled from coal tar or synthesised. From it resins are produced by polymerisation and used in varnish, ink, rubber compositions and chewing gum.

Council of American Flag Ship Operators An organisation mainly engaged in the maintainance and promotion of the United States Merchant Marine. All shipowners and managers of US flag ships are entitled to enrol as members. *Abbrev.* CAFSO.

Council of European and Japanese National Shipowners' Associations From its origin as an ad hoc group of shipowners in 1958, the Council was formally constituted in 1974 as an association of the national shipowners' associations of 11 nations in Europe and of Japan. There is a separate membership section of individual shipping companies trading to the USA. Known as CENSA, it concerns itself with Bulk and Tanker Trade shipping policy from the international standpoint. Its main objectives are (1) to promote and protect free market shipping policies in all sectors of shipping to meet market requirements under self-regulatory regimes in co-operation with the shipper community (2) to coordinate and present the views of its members primarily to inter-governmental fora such as the Consultative Shipping Group of governments and the OECD, and also to individual governments either directly or in support of national or regional associations such as the Ocean Common Carrier Coalition in Washington and the European Community Shipowners' Associations in Brussels. It has consultative status with UNCTAD and works closely with the International Chamber of Commerce and Intertanko.

109

Council of Lloyd's. *See* **Lloyd's.**

Counter The protruding upper part of the stern of a ship.

Counterfoil Detachable part of a document such as a ticket, cheque, certificate of posting and many others on to which essential details can be copied for retention.

Counter Guarantee The insured cargo owner's undertaking to the Underwriter, *q.v.*, in a general average distribution to reimburse any excess payments resulting in the payment of the General Average Contribution, *q.v.*

Counterpart Person in one company having similar responsibility to someone in another. The opposite business party or 'opposite number'.

Counter Trading Barter, *q.v.*, system promoted between countries to generate bilateral trade in goods without the use of money.

Countervailing Duty System of reciprocal tax or duty observed by agreement between two or more countries when goods are imported.

Country Damage The damage or deterioration resulting in bagged or baled merchandise due to dampness or through polluted material before loading operations take place. *Abbrev.* CD or C.D.

coup d' état *French*—violent change in government. Seizure of power.

Cour Courant *French*—Of the current month.

Courier Person travelling from one place to another carrying documents or packets for urgent and secure delivery; travel, company employee meeting and assisting tourists and travellers.

Courrier des Marins French Seamen's Mail.

Course of Advance In reference to navigation it indicates the anticipated route of advance over the ground. *Abbrev.* COA.

Course of Exchange Foreign Bills, *q.v.*, quoted in circulation.

Courtesy Flag A flag displaying the national colours of the port where the ship is entering. It is hoisted either at the fore-yard or any other place separate from the vessel's flag of registry.

Coût et fret *French*—Cost and Freight, *q.v.*

Coûte que coûte *French*—At any cost.

Cov Covenant; Coventry.

Cover To indemnify. To guarantee an indemnity. A word commonly used in insurance referring to the insurers or Underwriters, *q.v.*, who guarantee specific indemnities in case of loss or damage suffered by insured clients.

Cover Note *or* **Slip** A cover note is a document forwarded by a broker to his principal confirming the placing of an insurance. The cover note repeats the conditions of the policy in brief and states the rates and securities obtained. *See* **Slip**, Also **Original Slip**.

COVINCA Corporacion Venezolana de la Industria Naval C.A. An Industrial and Naval Venezuelan Corporation consisting of a group of Venezuelan businessmen.

COW Crude Oil Washing, *q.v.*

Cox Coxswain, *q.v.*

Coxswain Helmsman of a boat; person on board ship permanently in charge of and commanding boat and crew unless a superior officer is present. *Abbrev.* Cox.

Coy. Company, *q.v.*

c.p. Candle Power; Carriage Paid; Centre of Pressure; Chemically Pure; Common Pleas; Condensation Product; Constant Pressure; Court of Probate.

CP *or* **C.P.** Canadian Press; Candle Power; Cape Province, *South Africa*; Cardinal Point, *q.v.*; Carriage Paid, Casella Postale, *Italian Post Office Box*; Cathodic Protection, *q.v.*; Charterparty *or* C/P *q.v.*; Chemically Pure; Chief Patriarch; Chief of Police; Clerk of the Peace; Coaling Port; Code of Procedure; Codice Penale, *Italian Communist Party*; Command Post *military*; Conference Paper; Conference Press; Constant Pressure; Controllable Pitch *propeller*; Court of Probate; Court Procedure/Proceedings; Coût et Fret *French*—Cost and Freight; Current Paper; Custom of Port *in grain trade*; Cycloidal Propeller.

C/P Charterparty, *q.v.*; Custom of Port, *see* **C.O.P.** and **As Customary**; Cyclodial Propeller, *See* **VSP**.

C & P Carriage and Packing; Collated and Perfect.

CPA Curaçao Ports Authority.

CPA *or* **C.P.A.** Calico Printers Association; Canadian Pacific Airlines; Canadian Pharmaceutical Association; Certified Public Accountant *USA*; Chartered Patent Agent; Clyde Port Authority; Communist Party of Australia; Commonwealth Parliamentary Association; Correct Price Adjustment; Critical Path Analysis; Cost Planning and Appraisal; Claims Payable Abroad, *similar to* Settlement of Claims Abroad, *Abbrev.* SCA *or* S.C.A.; Closest Point of Approach, *q.v.*

CPCR Customs of the Port at Current Rate *chartering.*

C.P.C.U. Chartered Property and Casualty Underwriter.

cp. cycle Constant Pressure Cycle.

CPE Chinese Produce Exchange. The principal commodities dealt with are coconut, copra, oil, pepper and coffee.

C.P.F. Contributory Pension Fund.

CPF Canadian Patrol Frigate.

CPFF Cost Plus Fixed Fee.

CPG Compact Planetary Gears.

Cpge or cpgc. Course per Gyro Compass *nautical*.

C.P.H. *or* **c.p.h.** Certificate in Public Health; Cycles per Hour.

CPHA *or* **C.P.H.A.** Canadian Port and Harbour Association.

CPI *or* **C.P.I.** Chief Pilot Instructor *aviation*; Consumer Price Index.

CPIC Coastal Patrol and Intersection Craft, *USA*.

CPIR Commercial Pilot with Instrument Rating.

C.P.J.I. Cour Permanent de Justice Internationale *French*—Permanent Court of International Justice.

CPK *or* **C.P.K.** Cost-Per-Passenger Kilometre. *See* **Cost-per-Passenger Mile**.

cpm *or* **c.p.m.** Cards per Minute *computer*; Cycle per Minute.

CPL Commercial Pilot Licence; Classification, Packaging and Labelling, UK Carriage of Dangerous Goods by Road and Rail Regulations, April 1994.

CPM *or* **C.P.M.** Certified Property Manager *USA*; Cost-Per-Passenger Mile, *q.v.*

CPM *or* **CPM Policy** Contractors' Plant and Machinery Policy *insurance*.

C.P.O. Country Planning Office; Compulsory Purchase Order; Chief Petty Officer, *UK Royal Navy*; Commonwealth Producers Organisation.

CPP Central Production Platform; Controllable Pitch Propeller, *q.v.*

C.P.R. Canadian Pacific Railways; Cardiopulmonary Resuscitation.

cpr. Copper.

C.P.R.C. Central Price Regulation Council.

C.P.S. *or* **cps** *or* **c.p.s.** Custos Privati Sigillis *Latin*—Keeper of the Privy Seal; Cycles per Second *nautical*; Coupons.

CPS Contingency Planning Service. *Lloyd's*.

CPSA *or* **C.P.S.A.** Civil and Public Services Association *UK*.

CPSC *or* **C.P.S.C.** Certificate of Proficiency in Survival Craft.

cpsc Course per Standard Compass *nautical*.

Cpt. Captain.

CPT *or* **C.P.T.** Carriage Paid To, *q.v.*

C.P.T. Canadian Pacific Telegraphs.

CPTP *or* **C.P.T.P.** Civil Pilot Training Programme *aviation*.

CPU Central Processing Unit *electronics*.

C.P.U. Collective Protection Unit; Commonwealth Union.

CPV *or* **C.P.V.** Concrete Pressure,

CQ Radio signal requesting a reply.

C.Q. Change of Quarters; Conditionally Qualified.

CQD *or* **C.Q.D.** *or* **c.q.d.** Customary Quick Despatch. *Similar to* **Customary Despatch**, *q.v.*

Cr. Credit; Creditor.

CR *or* **C.R.** Carriage Risk; Carriers Risk; Central Railway; Central Registry; Chief Ranger; Coin Return *amusement machine*; Commendation Ribbon; Company's Risk; Compression Ratio; Conditional Reflex; Conference Report; Corrosion Resistant Material *Lloyd's Register notation*; Costa Rica; Currency Rate; Current Rate; Curtain Rod; China Corporation Registry *Taiwan*; Symbol for Chromium.

C & R Construction and Repair.

c/r Company's Risk.

c.r. Con Riserva *Italian*—With reservations; Cum Rights, with rights issue.

CRA *or* **C.R.A.** California Redwood Association; Canada Rheumatism Association; Commander Royal Artillery; Composite Research Aircraft; Corporation of Registered Accountants; Corrosion-Resistant Alloy.

Cradle Bed or frame adapted to the shape of a vessel's or a boat's bottom to hold her upright when she is out of the water. It can be fitted with slings for hoisting by crane, or with wheels for hauling-out on a shipway. Also for transporting and stowing machinary or awkward-shaped loads.

CRAF Civil Reserve Air Fleet *USA*.

Craft A ship or a vessel individually or collectively. This word is very commonly used in the insurance world, having a wide range of meaning of any boat or vessel engaged in the transportation of cargoes. It may also refer to barges, lighters, river boats, yachts, motor launches, etc.

Craft Clause A marine insurance clause whereby underwriters indemnify the assured while the goods are in transit by road transport and lighters, including to and from the vessel. The equivalent US insurance term is 'Risk of Boats Clause'.

Craft Port *or* **Surf Port** *or* **Overside Port** A discharging port where the cargo is not delivered direct alongside the quay. This occurs where the port has shallow water and the vessel is unable to enter. The cargo is discharged into lighters or any other similar craft and is then unloaded at that anchorage or the lighters are taken to the port to continue unloading alongside an inner port berth.

Cranky A ship which tends to become easily listed or heeled over.

CRASH Computerised Risk Assessment of Shipping Hazards, *q.v.*

Cratering This word mostly refers to offshore drilling and is encountered by oil drill platforms when the seabed caves in because of the release of underwater gas.

CRC Chemical Rubber Company; Commercial Research Committee of the International Air Transport Association.

C.R.C.C. Canadian Red Cross Committee.

CRD *or* **C.R.D.** *or* **c.r.d.** Current Rate Discharge, *q.v.*

CRDF Cathode Ray Direction Finder *Lloyd's Register notation*.

CRE *or* **C.R.E.** Commercial Relations and Exports Division *of DTI*.

Creaming *or* **Cream Digging** Early phase of discharging grain by grab unloader when the grab operates at maximum capacity.

Creative Fares Board A branch of the International Air Transport Association dealing with fixtures of air rates and fares. *Abbrev.* CFB.

CREDIOP Consorzio di Credito per le Opere Publiche *Italian*—Credit Consortium of Public Money *banking*.

Credit Account Out-going goods are against credit payments which will have to be settled later.

Credit Agency Company whose function is to investigate and report confidentially on the financial condition or credit-worthiness of other firms on request.

Credit Bank Commercial or Joint-Stock Bank.

Credit Card A handy-sized card with the name and the register number of the holder as well as the name of the issuing creditor. This is a very useful card for the tourist, businessman and the man in the street as one need not carry much money. The credit card is similar to the Banker's Card, *q.v.* Some of the organisations issuing this card are American Express, Diner's Card, Eurocard, etc. *See* **Barclay Card** *or* **Barclaycard**.

Creditor Opposite to Debtor, *q.v.*

Credit Slip Document acknowledging a payment to the credit of an account.

Credit–Squeeze Government restrictions on bank loans and credits.

Credit Status A bank or trade statement giving an estimation regarding the financial status of the firm or person who may apply for credit.

Credit Transfer Transfers are mostly done between banks. The client's instructions are followed regarding the method of sending money abroad, i.e. either by airmail, cable or telex. This is one of the new ways, especially by 'telex transfer', which have accelerated commercial transactions and procedures from one place to another. The issuing bank instructs the corresponding receiving bank with full particulars of the person or company which is called upon for payment of the money. Both banks charge a nominal fee for this service.

Creek (1) A small tributary of a river or a small stream. (2) A small inlet or bay. (3) A narrow and somewhat sheltered port.

Creeping Inflation Rising trend, since the Second World War, in commodity prices and labour costs leading to higher consumer prices in all countries.

CREFAA Convention on the Recognition and Enforcement of Foreign Arbitral Awards, New York Convention 1958.

Crew The personnel engaged on board ship, excluding the master and officers and the passengers on passenger ships.

Crew Boat A fast strongly built boat for the transportation of the crew of offshore oil rigs.

Crew List List prepared by the master of a ship showing the full names, nationally, passport or discharge book number, rank and age of every officer and crew member engaged on board that ship. This serves as one of the essential ship's documents which is always requested to be presented and handed over to the customs and immigration authorities when they board her on arrival to give a Free Pratique. The crew list is also a tool for 'Clearance Outwards', *q.v.*

Crew Manning Agency A company undertaking to provide officers and crew members for employment on board ships and aircraft.

Crew's Customs Declaration On arrival of a vessel in port, the master has to present a declaration to the customs officers apart from other declarations showing a list of dutiable articles belonging to each officer and crew member and signed by each one.

CRF *or* **C.R.F.** Capital Recovery Factor.

Crg. Carriage.

C.R.I. Caribbean Research Institute; Croce Rossa Italiana, *Italian Red Cross*.

Cribbing Heavy blocks and timbers to support a ship during her construction or to support her during drydocking.

C. Rica Costa Rica.

Cringle Rope worked round a thimble into the boltrope of a sail, through which a rope can be secured or to which running rigging may be shackled.

CRISTAL *or* **Cristal** Contract Regarding an Interim Supplement to Tanker Liability for Oil Pollution. A cargo owners' plan which is a supplementary one to Tovalop, *q.v.* This deals with compensation for oil pollution damage in connection with cargoes carried in vessels covered by the Tovalop scheme.

Critical American terminology for Urgent.

Critical Docking Draft (Draught) Draft (draught) limitation for a ship's safe entry into drydock.

Critical Moment If the centre of gravity rises above the metacentre of a ship when she is loading or discharging she will lose her stability and will develop a list.

Critical Point The point at which a vessel in distress or in an emergency requires fuel to return to her original departure point.

Critical Speed The excessive range of the engine revolutions per minute which may cause a break-down.

Cr. L. Craft Loss *marine insurance.*

CRM Counter Radar Measures.

C/Rm Control Room.

CRN Convention on Rhine Navigation in 1868; Customs Registered Number.

CRO *or* **C.R.O.** Cancelling Returns Only. *See* **Cancelling Returns**; Companies Registration Office *UK.*

CROSSMA Centre Régional De Surveillance et Sauvetage pour la Manche *French*—Channel Region Maritime Search and Rescue Centre, at Cap Gris Nez.

Cross Sea A sea running contrary to the wind, caused by rapidly changing wind direction such as results from a Cyclone, *q.v.* Ships may be endangered by the confused wave patterns.

Cross Springs A Mooring, *q.v.*, term. When the forward and after springs ropes are secured so as to cross.

Cross Swell A sea swell running against the force of the wind. *See* **Cross Sea**.

Cross the Line Said when a ship or a person crosses the Equator, *q.v.*

Crown of a Tank The highest level of a tank top of a ship.

Crow's Nest A look-out post positioned on the foremast of a ship under the masthead light. The crow's nest is used as a reconnaissance post by having a seaman on the look-out to see any land or other objects which may be of assistance in the navigation of the vessel.

CRP Contra Rotating Propeller; Controllable and Reversible Propeller.

CRS Computer Reservations System *air ticketing*; Croation Register of Shipping *classification society.*

CRS *or* **C.R.S.** Compagnies Republicaines de Sécurité, *French State Security Police*; Co-operative Retail Society or Service; Corrosion-Resisting Steel; Coast Radio Station.

Crs. Credits; Creditors.

Crt. Crate.

CRT Cathode Ray Tube *electronics.*

CRU *or* **C.R.U.** Civil Resettlement Unit; Composite Reserve Unit; Control Register User.

Crude (1) When referring to population this is the birth and death rate as per 1000. (2) Rough or unfinished or unrefined oil.

Crude Carrier An oil tanker of approximately 70,000 tons deadweight for relatively short trades. There are also larger tankers and even super and mammoth tankers of this type with Segregated Ballast Tanks meant for worldwide trading.

Crude Oil Washing A technique of cleaning tanks in oil tankers. It involves the washing of oil cargo tanks during the discharge of the crude oil, thus overcoming the main cause of pollution. This system gives a better cargo out-turn although there may be some oil sediment which could be transferred from ship to shore installations. *Abbrev.* **Cow**. *See also* **Inert Gas System**.

Cruise Climb The series of climbs in steps by an aircraft while taking off and also the period of cruising in the air.

Cruiser Stern A more or less semi-round shaped stern of a ship which is believed to improve the efficiency of the propeller and speed.

Cruising Radius *or* **Cruising Range** The maximum commercial distance a ship is liable to travel without refuelling. *See* **Steaming Distance** *or* **Steaming Range**.

Cr. U.S. US Supreme Court Reports.

Cruzeiro A unit of currency used in Brazil.

Crystal (1) Colourless and transparent Quartz, *q.v.* (2) Reception detector of crystalline formation to rectify oscillating currents.

Cryptography The art of cipher or cypher writing.

cs Cargo Segregation; Congestion Surcharge.

Cs Cirrostratus Clouds, *q.v.*

CS *or* **C.S.** *or* **C/S** Capital Stock; Carbon Steel; Case(s); Cast Steel; Certificate in Statistics; Chemical Society; Chief of Staff; Christian Science; Civil Service; Clerk of Session; Clerk of the Signet; College of Science; Colliery Screened *in coal*; Confederate States; Continental Shelf; Cotton Seed; Court of Session; Credit Sale; Come Sopra *Italian*—As above; Custos Sigilli *Latin*—Keeper of the seal; Continuous Hull Survey, *Lloyd's Register notation*; Cycle per second.

CSA *or* **C.S.A.** Canada Shipping Act; Canadian Shipowners' Association; Canadian Standards Association; Civilian Supply Association; Commonwealth Sugar Association/Agreement; Computer Science Association; Ceskoslovenske Aerolinie *Czechoslovak State Air Line.* Chlorosulphonic Acid *smoke generator*; Confederate States of America; Credit Sum(s) Assured *insurance.*

csc. Cosecant *trigonometry.*

CSC Cyprus Shipping Council.

CSC *or* **C.S.C.** Civil Service Commission; Certificate of Safety Construction; Container Service Charge; Container(s) Said to Contain; Commonwealth Scientific Committee; Conspicuous Service Cross; Convention on Safe Containers 1972 *IMO.*

CSCC *or* **C.S.C.C.** Cargo Ship Construction Certificate, *q.v.*

CSCE Conference on Security and Co-operation in Europe. Re-named the Organisation for Security and Co-operation in Europe (OSCE) in 1995.

CSD *or* **C.S.D.** Closed Shelter Deck/er, *q.v.*; Constant Speed Drive.

CSDS Combat System Development Site, *in Moorestown NJ, USA.*

CSE Certificate of Secondary Education.

C.S.E. Council of the Stock Exchange, *in London.*

CSEU Confederation of Shipbuilding and Engineering Union.

CSG *or* **C.S.G.** Consultative Shipping Group, consisting of 12 European countries and Japan.

CSH *or* **C.S.H.** Continuous Survey of Hull *Lloyd's Register notation.* When a vessel's hull is under continuous survey.

CSHBB Cast Steel Hub Non-Ferrous Blades *propeller.*

CSI Council for Security Industry.

C.S.I. Chartered Surveyors' Institution; Construction Specification Institute.

CSIR Council for Scientific and Industrial Research.

CSIRO Council of Scientific and Industrial Research Organisations; Commonwealth Scientific and Industrial Research Organisation.

CSISC Cargo Ship Safety Certificate, Interim.

Csk Cask; Countersink; Countersunk.

Csk.O. Countersunk Holes Over.

Csk. O.S. Countersunk Other Side.

Csk. T.S. Countersunk This Side.

CSL *or* **C.S.L.** Combined Single Limit.

CSM *or* **C.S.M.** Company Sergeant Major *army*; Continuous Survey on Machinery *Lloyd's Register notation* to show that a ship's machinery is undergoing continuous tests for the attainment of Special Survey classification.

CSMP Current Ship's Maintenance Project *USA.*

CSN *or* **C.S.N.** Container Safety Convention.

CSNAME Chinese Society of Naval Architecture and Marine Engineering.

C. so. Corso *Italian*—Street.

CSO Combined Sewers' Overflow.

CSO *or* **C.S.O.** Central Selling Organisation *USA*; Central Statistical Office; Chief Signal Officer; Chief Staff Officer.

C.S.P. Council of Scientific Policy.

CSP&IA *or* **C.S.P.&I.A.** Chartered Shipbrokers Protection and Indemnity Association.

CSR Certificate of Safety Radio; Cargo Systems Research.

CSR/CD Cargo Systems Research/Consulting Division.

CSS *or* **C.S.S.** Code Shipmanagement Standards; Commodity Stabilisation Service; Continuous Special Survey *Lloyd's Register notation* signifying that a vessel is undergoing periodical and continuous inspection to attain the special survey classification; Computer System Simulator; Council of Social Services.

CSSC China State Shipbuilding Corporation.

CSSEC *or* **C.S.S.E.C.** Cargo Ship Safety Equipment Certificate.

CSSRA Canadian Shipbuilding and Ship Repairing Association, *Vancouver.*

cSt CentiStokes, *q.v.*

CST *or* **C.S.T.** Central Standard Time which is between Pacific Standard Time and Eastern Standard Time; College of Science and Technology.

cstg Casting.

C.S.T.I. Council of Science and Technology Institutes.

C.S.T.V. Control System Test Vehicle.

CSU *or* **C.S.U.** Central Statistical Unit; Constant Speed Unit.

CSV Concentric Standing Valve *oil drilling.*

CSW *or* **C.S.W.** Code of Safe Working Practice; Continuous Seismic Wave.

CSWIP Certification Scheme Welding Inspection Personnel *for underwater work.*

CSWP Code of Safe Working Practice *ILO.*

CT Coiled Tubing *oil drilling.*

CT *or* **C.T.** Cable Transfer; Candidate in Theology; Cape Town; Central Time; Certificated Teacher; Civic Trust; Californian Terms *in grain trade*; Code Telegrams/Telexes; Combined Transport *document*; Community Transit System; Connecticut, *USA*: Conning Tower; College of Theology; Commercial Terms.

C/T *or* **CT** California Terms *grain terms*; Community Transit, *q.v.*

CTA *or* **C.T.A.** Canadian Tuberculosis Association; Caribbean Tourist Association; Catering Teachers Association; Chain Testers Association *UK*; Chicago Transit Authority; Commercial Travellers Association; Corn Trade Association; Combined Terminals Amsterdam; Customs Transaction Code.

CTAC Chemical Transportation Advisory Committee *USA.*

CTC *or* **C.T.C.** Canadian Transport Commission; Combined Transport Convention; Centralised Traffic Control; Carbontetra Chloride Tetrachloromethane which is widely used in drycleaning and for other industrial uses; Congress du Travail du Canada *French* —Canadian Labour Congress; Container Terminal Charge; Corn Trade Clauses.

CTCC Central Transport Users' Consultative Committee *UK.* Secretary of State Transport Section.

CTD Combined Transport Document, *q.v.*; Conductivity Temperature and Depth, *q.v.*

CT Document Combined Transport Document, *q.v.*

C.T.F. Canadian Trust Fund, *q.v.*

CTH Concentric Tubing Hanger *oil drilling.*

C.T.H. Corporation of Trinity House in London. *See* **Trinity House**.

CTI Colour Transient Improvement *television*; Container Transport International; Customs Temporary Importation.

CTIAC Chemical Transportation Industry Advisory Committee *USA*.

ct.j. Circuit Judge.

CTK Capacity Ton Kilometre, *q.v.*

CTL *or* **C.T.L.** Compromised Total Loss. *See* **Arranged Total Loss**; Constructive Total Loss, *q.v.*

CTLO *or* **C.T.L.O.** Constructive Total Loss Only. The subject-matter is expressly insured for CTL or C.T.L., *q.v.*, Clean Total Loss Only.

CTM *or* **C.T.M.** Capacity Ton Mile. *Similar to* **Capacity Tonne Kilometre**.

Ctn Carton; Cotangent.

CTO *or* **C.T.O.** Clean Total Loss Only, *q.v.*; Combined Transport Operation or Operator *USA*, Containerised and other forms of cargo transportation.

C to C *or* **C.C.** Centre to Centre.

CTOL Conventional Take Off and Landing *aviation*.

CTP Coiled-Tubing-assisted Pumpdown *oil drilling*.

CTR *or* **C.T.R.** Certified Test Record; Controlled Thermonuclear Reaction; Controlled Thermonuclear Research.

Ctr. Centre; Cutter.

Cts Cartons.

CTS *or* **CT System** Community Transit System *European Union*.

CTT Correios e Telecommunicacoes de Portugal, *Portuguese Post and Telegraph Services*; Capital Transfer Tax.

C'ttee *or* **Cttee** Committee.

CTU Coiled-Tubing Unit *oil drilling*.

CTV *or* **C.T.V.** Commodity Transporting Vessel.

CTW Cargo Tank Wing.

Cu Cuprum *Latin*—Copper.

cu Cubic; Cumulus Clouds, *q.v.*

CU *or* **C.U.** Cambridge University; Cuba; Customs Union; Charge Utile *French*—Useful Load in containerisation.

C.U.A. Canadian Underwriters' Association.

CUAP Chart Users' Advisory Panel.

Cubana Cubana de Aviacion, *Cuba State Airline*.

Cube Out Before a container is loaded the cubic space is to be taken into account as well as the weight of the merchandise.

Cubic Capacity The number of cubic feet or cubic metres in the cargo holds or spaces of a ship. This is usually given in grain and bale cubic capacities, the former being always the larger since no Broken Stowage, *q.v.*, is lost or taken into account. *See* **Bale Capacity**.

CUDOS Customs Documentation Systems.

CUFT *or* **Cu.ft.** Cubic Feet.

CUG Closed User Group *computer*.

cu.gr. Cubic Grain. *See* **Grain Capacity**.

CUKCC *or* **C.U.K.C.C.** Canada-United Kingdom Chamber of Commerce.

Culpa *Latin*—Fault, error.

culpa lata *Latin*—Slight negligence.

cum *Latin*—With; Cumulative.

Cumb. Cumberland, *UK*.

cum distribution *Latin*—With distribution.

cumd *or* **cum div.** Cum Dividend *Latin*—With dividend. The buyer of shares benefits by being credited with the dividends along with the immediate dividends that may be forthcoming. Opposite to Ex dividend or Exall.

cum gr. bl. Grain/Bale capacity in cubic metres.

cum pref. Cumulative Preference. *See* **Cumulative Preference Shares**.

cum privilegio *Latin*—with privilege.

cum rights A stocks and shares term. When quoting the purchaser has the rights' issue even although already paid.

Cumulative Hours *See* **Accumulative Hours**.

Cumulative Preference Shares Dividends of shares that have not been paid but the holders will be entitled to them in the following year at a fixed rate of interest. *Abbrev.* cum pref.

Cumulonimbus Clouds Heavy low clouds showing in a form of towers or mountains under 8000 ft from the ground with dark grey colour and fibrous texture. *Abbrev.* Cb.

Cumulus Clouds Low dense dark clouds extending vertically to about 20,000 ft or 6000 metres which normally bring thunderstorms. *Abbrev.* cu.

Cupro-nickel Metal alloy of copper with a lesser amount of nickel. This is generally used in old 'silver' coins.

CUR Currency Unit Rate(s).

Cur. Currency; Current, *q.v.*

Cur. *or* **Curt.** Current, *q.v.*

Cur. adv. vult *or* **C.A.V.** *Abbrev.* for Curia advisari vult *Latin*—When court judgment is deferred.

Curb Market *or* **Street Market** A regular meeting place in a popular and well frequented street where stock exchange business used to be transacted in the open air. In some places the old method of Curb or Street Market is still in vogue on a small scale. The London Street Market used to be in 'Change Alley' for many years. The New York Curb Exchange which was renamed Amercian Stock Exchange on 5 January 1953 used to operate in various streets in New York, such as Wall Street. *See* **American Stock Exchange**.

C.U.R.E. Care Understanding Research Organisation for the Welfare of Drug Addicts.

Curia advisari vult *Latin*—The court desires to consider. *Abbrev.* C.A.V. *or* Cur. adv. vult.

Curie A radioactive measure of the rate of radiation emitted by materials. A technical formula is used to find the disintegration. *Abbrev.* Ci.

Curing Agent A phrase mainly used in painting work where two pack components are mixed to form the so-called sophisticated paint. One pack, which has less quantity than the other pack, is the reactor which forms the special quality paint composition.

Currency Speculator Person or firms taking risks in buying and selling foreign exchange for profit.

Current (1) Present month of the year or present time, now. (2) Wind or water which tends to blow or move in one direction.

Current Account Money deposited in a bank to be used against the issue of cheques. Cheque books are supplied to the banks' customers free of charge except for government stamp duty, if any, levied on each cheque.

Current Assets Assets used in the day to day running of a business e.g., stock, debtors, bank balances and cash.

Current Domestic Value The actual value of the goods imported into a country for the purpose of assessing the customs import duty.

currente calamo *Latin*—with a running pen. Fluently. Offhand.

Current Rate Discharge The carrier in a Charterparty, *q.v.*, is to pay the unloading expenses at the customary or current rate ruling. *Abbrev.* CRD *or* C.R.D.

Current Ratio The ratio of Current Assets, *q.v.*, to Current Liabilities. Also called Working Capital Ratio. *See also* **Acid Test Ratio**.

Current Yield Percentage obtainable for the price paid for a share.

Curriculum Vitae Document, reciting essential details of a person's education, qualifications and work history which it is useful, and often essential, to produce when applying for employment. *Abbrev.* C.V.

Curtilage Land adjoining a dwelling house.

CURV Cable-Controlled Underwater Research Vessel. An unmanned robot.

Cusdec Customs Declaration.

Cusec *or* **Cu Sec.** Cubic feet per second.

Custody Bill of Lading *Similar to* **Received for Shipment Bill of Lading**.

Customary Deductions The deductions made by the average adjuster by taking into account the difference in cost of new for old ship's spares and items repaired or replaced in the General Average, *q.v.*, calculations on ships over 15 years old.

Customary Despatch *or* **Customary Dispatch** To clear the ship as quickly as possible in accordance with the customs of the port. Charterers, *q.v.*, have therefore a wide latitude as they are excused from delays provided they have done their best under

prevailing conditions. *Also termed* With All Despatch *or* Customary Quick Despatch.

Customary Freight Unit An American expression referring to limited liability per package in case of non-delivery of the cargo under the Limitation of Responsibility Clauses of the Hague Rules covering the transportation of cargoes. The aim of this liability clause is to prevent excessive claims on valuable articles and to restrain the carriers from escaping liability.

Customary Groundings Clause A marine insurance term under which the insurer is not liable for the cost of inspecting a ship's bottom after she has taken the ground in certain geographical areas, e.g. the Suez Canal, which is then not deemed as a stranding.

Custom Built Specially made to suit the customer's requirement. Bespoke.

Custom of the Port *See* **Customary Despatch** *or* **Customary Dispatch** *and also* **Custom of Trade**.

Custom of Trade Similar in procedure to Custom of the Port and Customary Despatch. It refers to general trade's customary practice. Where this term is inserted in a contract all the laws and customary processes of trade of that country supersede any clauses in the contract that may run counter to it.

Customs Broker An expert in the clearance of cargo who is employed where the importers of merchandise in certain countries are obliged by law to engage his service.

Customs Clearance *Similar to* **Clearance**, *q.v.*

Custom Co-operation Council An association of customs on an international level having its headquarters in Brussels and composed of over 50 member states. Their main object is to facilitate the international agreed tariffs imposed on all goods and to offer many other mutual advantages to its members.

Customs Duty Duty paid to the customs authorities to allow any dutiable merchandise to enter the country for home use and consumption.

Customs Tariff A detailed list of merchandise showing the rate of tax or duty enforced against each individual item when imported into a country.

Customs Union Association of states agreeing to abolish customs duty on a mutual understanding. Identical to the European Community.

Cut Oil Water contaminated oil in oil exploration. Also called Wet Oil.

Cut-throat Competition When business competition becomes so intense as to be compared to warlike conflict in the price quoting process. Sometimes the result of this type of competition will bring losses and hardship to the competitors themselves.

Cutwater The fore part of a ship which touches the sea when in motion.

CUV Construction Unit Value.

c.v. Cheval-Vapeur *French*—Horse Power, *abbrev.* HP *or* H.P.; Chief Value; Contributory Value, *q.v.*; Curriculum Vitae *Latin*—Testimonials.

CV Contingent Valuation, *q.v.*

C.V. Calorific Value; Cavallo Vapore *Italian*—Horse Power; Cavalos Vapor *Portuguese*—Horse Power; Cheque Value; Chief Value; Common Value; Common Version; Contributory Value, *q.v.*; Curriculum Vitae *Latin*—Testimonials.

CVA Certified Verification Agent *USA*.

CVD Common Value Development.

c.v.d. Come Volevasi Dimostrare *Italian*—Which had to be shown or demonstrated.

CVK *or* **C.V.K.** Centre Vertical Keel.

CVMO Commercial Value Movement Order.

CVO *or* **C.V.O.** Certificate of Value and Origin *for customs purposes*; Commander of the Royal Victorian Order.

CVP Corporación Venezolana del Petroleo, *Venezuelan Petroleum Corporation*.

CVR Cockpit Voice Recorder *aviation*; Controlled Visual Rules *aviation*.

CVS Consecutive Voyages, in the Voyage Charterparty.

cvt. Convertible.

CW Clockwise.

CW *or* **C.W.** Canada West; Carrier Wave *USA*; Chemical Warfare; Clerk of Works; Commercial Weight; Continuous Waves *nautical*.

C–W Chronometer time minus watch time *nautical*.

C+W Chronometer time plus watch time *nautical*.

CWA Clean Water Act. *See* **Combined Sewers Overflow**.

C.W.B. Canadian Wheat Board; Central Wages Board.

CWC *or* **C.W.C.** Circulating Water Channel; Country from where Consigned.

C.W.Ck. Caution and Warning System Check *computer*.

CWDE Centre for World Development Education.

C.W.E. Cleared Without Examination. Referring to the customs inspection exemption in containerisation.

CWL *or* **C.W.L.** Calm Water Line *in air cushion vessels*.

c.w.o. *or* **C.W.O.** Cash With Order; Chief Warrant Officer.

CWR *or* **C.W.R.** Cancelling War Risk, *q.v.*

CWRU Case Western Reserve University, *Cleveland, Ohio, USA*.

CWS *or* **C.W.S.** Company Welfare Services; Co-operative Wholesale Society.

cwt Hundredweight equivalent to 112 lbs.

CWTB Cylindrical Water Tube Boiler.

cx Convex.

CXT Common External Tariff of the European Community.

Cy *or* **CY** Container Yard; Currency; Capacity.

CY/CY Container Yard/Container Yard. Container is delivered from one yard to another or other yards.

C.Y.C.A. Clyde Yacht Club Association.

Cyclol Cycloidal Propeller, *q.v.*

Cyclical Fluctuation Periodical events cause economic upheavals in the commercial life of a country. A boom helps business to thrive. There are other times, however, when the business activity is so low that the nation is compelled to suffer unemployment. Thus the derivation of the word 'Cyclical'.

Cyclic–Pitch Control The movements of helicopter propeller blades by individual tilting in accordance with the cycle of rotation. *Abbrev.* CPC. *Also termed* Collective Pitch Control.

Cyclical Unemployment *See* **Cyclical Fluctuation**.

Cycloidal Propeller Propeller consisting of a revolving disc set flush with the underside of a vessel and having a number of projecting, rotating, vertical blades. Used where high manoeuvrability is required. *Abbrev.* Cyclol. *See* **VSP**.

Cyl. Cylinder.

Cyclone A low pressure system of weather, or depression, generally associated with stormy or wet weather.

Cyo *or* **CYO** Catholic Youth Organisation.

Cy-près *Old French*—Near thereto; As near as may be to a testator's intention, generally referred to charitable trusts.

CYS Cyprus Organisation for Standards and Control; Chief Yeoman of Signals, former rating in RN.

CZ Contiguous Zone.

C.Z. Canal Zone.

Cz *or* **Czech** Czechoslovakia.

C.Z.D. Calculated Zenith Distance *nautical*.

C.Zn Compass Azimuth *nautical*.

CZRV Coastal Zone Research Vessel.

D

d *or* **D** Days; Deci (10¹); Deuterium *Latin*—Five hundred; Dextro *chemistry*; December; Depth; Diameter; Drizzling Rain *weather observation*; Dyne, *q.v.*

D Date; Deeds; Declination; Declination change in hour *nautical*; Destination; Deviation; Deus *Latin*—God; Difference; Dime; Dinar, *q.v.*; Dip, short of horizon *nautical*; Dividend; Doctor; Drachma, *q.v.*; Dust *weather observation*.

DA *or* **D/A** *or* **D.A.** Deductible Average, *see* **Deductible**; Deed of Arrangement; Delayed Action; Dental Apprentice; Deposit Account; Disbursement Accounts; Discharge Afloat, *see* **AFLOAT** *also* **Always Afloat** *or* **Always Safely Afloat**; District Attorney; Documents Attached; Days After Acceptance; Documents Against Acceptance; Days After Date, *q.v.*; Draft (Draught) Aft; Double Acting *Lloyd's Register notation* re machinery.

d.a.a. *or* **D.A.A.** Discharge Always Afloat, *similar to* **Always Afloat**, *q.v.*; Direct Air Cycle; Documents Against Acceptance; Days After Acceptance.

DAB Dictionary of American Biography; digital audio broadcasting.

DAC Deductible Average Clause. *See* **Deductible**; Direct Air Cycle; Development Assistance Committee *EEC*.

DAD Documents Against Discretion *of the collecting bank*.

DAD *or* **D.A.D.** Days After Date, *q.v.*

DADS Digital Assisted Despatch System *electronics*.

DAE Dictionary of American English.

DAF *or* **D.A.F.** Delivered at Frontier, *q.v.*

d.a.f. *or* **D.A.F.** Describe as follows; Description as follows; Dry Ash Free, the basis of analysis of coal and coke.

DAFS Department of Agriculture and Fisheries of Scotland.

DAGAS Dangerous Goods Advisory Service *Laboratory of the Government Chemist*.

dag/s *or* **Dkg/s** Decagramme(s).

D. Agr. Doctor of Agriculture.

D. Agr. Sc. Doctor of Agricultural Science.

d.a.i. *or* **D.A.I.** Death from Accidental Injuries *insurance*.

DAILY *or* **daily** Daily Domestic Servant; Maid.

Daily Estimate Position Summary Wireless broadcast sent out by US maritime authority giving positions of shipping. This daily information also includes courses and speeds. *Abbrev.* Depsum.

Daily Memorandum *See* **Defence Mapping Agency Hydrography Center**.

Daily Operating Cost Daily running expenses of a vessel or any other business organisation. *Abbrev.* DOC *or* D.O.C.

Dak. Dakota, *USA*.

dal *or* **dkl** Decalitre.

Dalasi Currency unit of Gambia.

dam/s Damage(s), *q.v.*

dam *or* **dkl** Decametre.

Damage Done Phrase usually referring to collision liability or damage to piers, wharves, buoys or other harbour equipment in marine insurance. *Abbrev.* D/d *or* D/D.

Damage Feasant *Old French*—Doing damage *Ex.* When cattle trespass upon land not belonging to their owner.

Damages Legal compensation paid to an aggrieved party in a deed or contract. Damages may be classified in various categories such as (1) Nominal Damages are damages to such a small amount as to show that they are not intended as any equivalent or satisfaction to the party recovering them. (2) Exemplary Damages are the cause of ill-feeling or loss of influence resulting from past achievements, reputation or libel. (3) Liquidated Damages are payable in relation to the estimated and fair amount of money arising from the breach of contract. (4) Remoteness of Damages are those anticipated or contemplated from a breach of contract which could be gathered from the contract itself. (5) Substantial Damages are damages which represent compensation for the loss actually sustained by the plaintiff.

Damages for Detention In the absence of any special provisions such as Laytime, *q.v.*, in a Charterparty, *q.v.*, the carrier or owner of the ship has the legal right in the form of damages for detention. This could be the cost of the actual loss suffered resulting from the neglect or unrealistic detention by the Charterers, *q.v.* The damages sustained are generally charged against the running expenses of the ship plus the anticipated loss suffered because of the undue delay. *See* **Demurrage**.

Damnosa Hereditas *Latin*—Insolvent inheritance.

Damnum absque injuria *Latin*—Loss without a wrongful act—loss which does not give rise to an action for damages against the person causing it. In contrast to Injuria sine damno, *q.v.*

Damnum Infactum *Latin*—Damage not done: a threatened loss which does not occur.

Dampfschiff *German*—Steamship.

Dampfschiffahrtsgesellschaft *German*—Steam Navigation Company.

Dampfschiffgesellschaft *German*—Shipping or Steamship Company.

Dampskip *Norwegian*—Steamship.

Dampskipsaksjeselskap *Norwegian*—Joint Stock Steamship Company Limited.

Dampskipsinteressentskap *Norwegian*—Steamship Company.

Dampskipsrederi *Norwegian*—Steamship Owning Company *Abbrev.* D/R.

Dan Buoy A navigational aid in the form of a ballasted floating mark having a flag as a signal.

Dancon Code name for Chamber of Shipping Charterparty 1911 for the carriage of grain from the Romanian ports of the River Danube.

Dancroft Code name for Chamber of Shipping Charterparty 1911.

Dandy Note A customs house authority note used for the withdrawal of goods from the Warehouse, *q.v.*, which are earmarked for export.

Danger Buoy A floating mark showing a danger area for navigation.

Danger Money Additional money paid to workers engaged in hazardous or dangerous work.

Dangerous Liquids Liquids giving off inflammable vapours. Dangerous packages of this nature are to be colour labelled according to their classes. *See* **Inflammable** *or* **Flammable Liquids**.

Dangerous Oils Oils with flash points below 73°F.

Dangerous Wreck A wreck which obstructs navigation. Generally a wreck must be submerged over 10 fathoms to be considered safe.

Danmarks Rederiforening Danish Shipowners Association.

Danrsaoclonl *or* **DANRSAOCLONL** Discountless And Are Not Returnable Ship and/or Cargo Lost or Not Lost.

Dansk Skibshandler Forening Danish Shipsuppliers Association. Members of the International Shipsuppliers Association.

D.A.O. District Advisory Officer.

DAP *or* **D.A.P.** Data Automation Proposal; Direction of Administrative Planning; Do Anything Possible; Documents Against Payment.

DAQPAK Data Aquisition Package *Lloyd's Computer Program*.

DAR *or* **D.A.R.** Day After Recall, Time spent listening; Defence Aquisition Regulation *USA*; Developed Area Ratio *technical*.

D. Arch. Doctor of Architecture.

D.A.R.D. Directorate of Aircraft Research and Development.

dare pondus fumo *Latin*—To give weight to smoke. To give importance to trifles.

DARPA Defence Advance Research Projects Agency *USA*.

Darsena *Italian*—Dock.

DARTC Deep Access Reconnaissance Television Camera, *q.v.*

DAS *or* **D.A.S.** Development Advisory Ship; Delivered Alongside.

DASH Drone Antisubmarine Helicopter *USA*.

Dastard Destroyer Anti-Submarine Transportable Away Detector *USA*.

DAT Dental Admission Test *USA*.

DATA Defence Air Transportation Administration.

Data Encryption Standard A standard adopted by the National Bureau of Standards as a computer security product.

data et accepta *Latin*—Expenditures and receipts. The **data et accepta Day** is normally considered to be of 24 hours. In a Charter Party, *q.v.*, this could have various meanings in that the day might be calculated from midnight to midnight or otherwise according to the terms of the contract.

Date Line *or* **International Date Line** An imaginary line passing from North to South of the world in the Pacific Ocean on meridian 180° from Greenwich. When a ship or aircraft crosses this imaginary line to the West a day is gained. Conversely, a day is lost when crossing to the East. For every 15° longitude one hour is either advanced or retarded if calculated Westward or Eastward respectively. The Date Line is also known as the International Date Line. *See* **Antipodean Day**.

dau. Daughter.

Davits Two radial cranes on a ship which hold the lifeboats. They are constructed in such a way as to lower and lift the lifeboats the easiest way possible and are also unobstructed in case of an emergency.

Daylight Saving Achieved by advancing clocks by one or two hours on the Zone Time, *q.v.*

Daylight Saving Time When advancing one or two hours ahead of the local standard time. *Abbrev.* DST *or* D.S.T.

Daybeacon An unlighted signal beacon used as a reference point for daylight navigation. *See* **Daymark**.

Day Book A book-keeping term. Detailed purchases and sales done during the day are recorded in this book. *Abbrev.* d.b. *or* D.B.

Daymark An aid to navigation only visible by day. *See* **Daybeacon**.

Daymen Seamen who are not included in the day and night watches.

Days After Date The days after the sight of a draft, *q.v. See* **Day's Sight**. *Abbrev.* d/d *or* D/D.

Days After Sight *Similar to* Day's Sight, *q.v. See also* **Days After Date**. *Abbrev.* d/s *or* D/S.

Day's Date Referring to the day after the date of a Bill of Exchange, *q.v.*

Day Shift Normal working day. Opposite to Night Shift.

Days of Grace Days allowed for making a payment or doing some other act after the time limited for that purpose has expired. Insurance companies commonly allow a certain time for the payment of overdue premiums before forfeiting the policy. A person was formerly allowed three days of grace beyond the day named in the writ for entering an appearance. Days of grace were formerly granted in actions at the prayer of the plaintiff.

In the law of bills of exchange, days of grace are a period allowed to the drawee or acceptor of a bill of exchange to pay the bill after the due date, originally as a favour, but now as a matter of right. The number of these days varies in different countries: in England it is three, except that where the last of the three days falls on Sunday, Christmas Day or Good Friday, or a day of public fast or thanksgiving, the bill is payable on the preceding business day; but where the last day of grace is a bank holiday (other than Christmas Day or Good Friday), or where the last day of grace is a Sunday and the second day of grace is a bank holiday, the bill is payable on the succeeding business day; hence, if a bill falls due on June 11, and none of the exceptions apply, the acceptor is not bound to pay it until June 14. Days of grace are not allowed in the case of bills, notes or drafts payable on demand or at sight. *Abbrev.* D.G.

Days on Demurrage Laydays, *q.v.*, expire, demurrage commences. The period of the vessel on demurrage is called Days on Demurrage. Some shipowners enforce Damages for Detention, *q.v.*, when the days of the demurrage have elapsed and when more delays are anticipated.

Days Purposes Refers to Laydays, *q.v.*, of a charterparty calculated per day. *Abbrev.* DP *or* D.P.

Days' Sight The days after the sight of a Bill of Exchange, *q.v.*, is presented for acceptance.

Day-to-Day Loan A short loan usually for 24 hours duration given by a bank to a client Bill Broker, *q.v.*, to help him repay the Bills of Exchange, *q.v.*, which fall due on that day. Sometimes expressed as Overnight Loan or Day-to-Day Money.

Day-to-Day Money *Similar to* **Day-to-Day Loan**, *q.v.*

Day Four Oil drilling crew shift day work which normally starts at 8 a.m. and ends at 4 p.m.

Days These are referred to in the charterparty under the heading of Laydays or Demurrage, *q.v.* Days may be interpreted as Working Days, Weather Working Days, Running Days or Consecutive Days, *q.v.*

db Domestic Boiler *Lloyd's Register notation*; Decibel, *q.v.*

DB Double-ended Boiler *Lloyd's Register notation*; Dry Bulb Temperature.

D.B. Day Book, *q.v.*; Deals and Battens; Divinitatis Baccalaureus *Latin*—Doctor of Divinity; Donkey Boiler *Lloyd's Register notation*; Double Bottom. *See* **Cellular Double Bottom** *also* **Ballast Tanks**.

d.b. Deals and Battens *timber trade*.

D.B.A. Doctor of Business Administration.

d.b.a. *or* **D.B.A.** Doing Business As or At.

d.B(A) Decibel, *q.v.*

DBB *or* **D.B.B.** Deals, Boards and Battens or Deals Battens and Boards, *q.v.*

DBC Double Bottom Centre.

DBE *or* **D.B.E.** *or* **d.b.e.** Despatch or Dispatch Both Ends, *q.v.*

DBEATS *or* **D.E.B.E.A.T.S.** Despatch Payable Both Ends All Time Saved. *Similar to* **Despatch** *or* **Dispatch Money Both Ends**, *q.v.*

DBENDS *or* **D.B.E.N.D.S.** Despatch or Dispatch Both Ends, *q.v.*

DBEWTS *or* **D.B.E.W.T.S.** *Same as* **DBWTS** *or* **D.B.W.T.S.**, *q.v.*

DBLTS *or* **D.B.L.T.S.** Despatch Payable Both Ends on Laytime Saved. Charterparty term. *See* **Despatch Days**.

DBH *or* **D.B.H.** *or* **d.b.h.** Diameter at Breast Height *timber measurements*.

D.B.I.U. Dominion Board of Insurance Underwriters, *Canada*.

dbk Drawback, *q.v.*

dbl. Double.

dblr Doubler.

D.B.M. Diploma in Business Management.

Dbn. Durban, *South Africa*.

D.B.R. Double Book Rock.

db. rig Drilling Rig Barge.

DBS *or* **D.B.S.** Development Bank of Singapore; Distressed British Seaman, *q.v.*; Domestic Boiler Survey *Lloyd's Register notation*; Direct Broadcast Satellite.

DBST *or* **D.B.S.T.** Double British Summer Time.

DBWTS *or* **D.B.W.T.S.** Despatch Payable Both Ends on Working Time Saved—Voyage Charterparty term.

DC *or* D.C. Death Certificate; Decimal Classification; Deck Covering; Dental Corps; Deviation Clause, *q.v.*; Direct Current; District Commissioner; District Court; District of Columbia, *USA*; Doctor of Chiropractics; Douglas Commercial Aircraft *designation of builders*; Deputy Consul.

d.c. Da capo *Italian*—To repeat from the beginning; Direct Current *electrical*; Dead Centre; Double Column; Drift Correction *nautical*.

D/C *or* D/Cl Deviation Clause, *q.v.*

DCA Defence Communications Agency *USA*; Department of Civil Aviation; Dual-Capable Aircraft: one that can be armed with nuclear or conventional armament.

D.C.Ae. Diploma of the College of Aeronautics.

d.cap. Double Foolscap *paper measure*.

DCBT Dedicated Clean Ballast Tank.

DCC or D.C.C. Deadweight Cargo Capacity; Double Cotton Covered.

DCCD *or* D.C.C.D. Delivered at Container Collection Depot *containerisation*.

D.C.D. Diploma of Chest Diseases.

D.C.E. Diploma of Chemical Engineering; Doctor of Civil Engineering; Domestic Credit Expansion *customs*.

DCF *or* D.C.F. Discounted Cash Flow, *q.v.*

DCFM Dry Cargo Freight Market.

D.Cgo. Damaged by Other Cargo.

D.C.H. Diploma in Child Health.

D.Ch.O. Diploma in Ophthalmic Surgery.

DCJ District Court Judge *USA*.

D. Cl. *or* D/Cl Deletion Clause *chartering*.

D.C.L. Doctor of Civil Law; Doctor of Canon Law.

D.Cl.S. Doctor of Clerical Science.

DCM *or* D.C.M. Distinguished Conduct Medal.

D.Cn.L. Doctor of Canon Law.

D.C.N.S. Deputy Chief of Naval Staff.

DCO Debt Collection Order.

D.Com.L. Doctor of Commercial Law; Doctor of Comparative Law.

D.Comm. Doctor of Commerce.

DCOP *or* D.C.O.P. During Currency of Policy *insurance*.

DCP *or* D.C.P. Development Cost Plan; Diploma in Clinical Pathology.

DCPD Direct Calculation Procedural Document, *Lloyd's Register documentary system*.

DCR *or* D.C.R. Discharging at current rate *voyage chartering term*.

dcred *or* docred Documentary Credit.

DCS Departure Control System *IATA*; Distribution Contract System.

D.C.S. Deputy Clerk of Sessions; Doctor of Christian Science; Doctor of Commercial Science.

dct. Direct; Document.

D.C.T.D. Diploma in Chest and Tuberculous Diseases.

DCU Distribution Control Unit.

DCW Dead Carcase Weight.

dd *or* Dd. Delivered; Dated; Due Date, *q.v.*; Delayed Delivery; Delivered Docks, *q.v.*

d/d *or* D/D Days After Date, *q.v.*; Delivered Docks, *q.v.*; Demand Draft (Draught); Damage Done, *q.v.*

DD *or* D.D. Damage Done; Delivered at Docks; Demand Draft (Draught); Design Draft (Draught), *q.v.*; Diviniatis Doctor *Latin*—Doctor of Divinity; Despatch Days, *q.v.*; Deputy Director; Diploma in Dermatology; Direttissimo *Italian*—Fast Train; Discharge Dead *cattle trade term*; Dishonourable Discharge; Direct Debit.

DD Discharged Dead. Notation referring to termination of service on a naval rating's documents.

DDA *or* D.D.A. Dangerous Drugs Act; Duty Deposit Account; Duty Deferment Account.

DDC Deck Decompression Chamber. *See* **Decompression Sickness *or* Staggers *or* Bends**.

DDC *or* D.D.C. Damage Done in Collision *insurance*.

DDD *or* D.D.D. Deadline Delivery Date; Direct Distance Dialling, long distance telephone by dialling direct; Direct Drive Diesel.

D.½.D. Despatch Half Demurrage, *q.v.*

D½ DATSBE *or* D.½.D.A.T.S.B.E. Despatch Half Demurrage At All Time Saved Both Ends, *Similar to* D½DWTSBE *or* D.½.D.W.T.S.B.E.

DDE Direct Date Entry.

DDG Deputy Director General.

D.D.H. Diploma in Dental Health.

DDI Developing Defence Industry *NATO*.

DDL Danish Air Lines; Deputy Director of Labour.

DDM Difference in Depth of Modulation *radio bearings*.

D.D.M. Doctor of Dental Medicine.

DDO *or* D.D.O. *or* d.d.o. Dispatch Discharging Only.

D.D.O. Diploma in Dental Orthopaedics.

DDP *or* D.D.P. Delivered Duty Paid, *q.v.*

DDR Direct Debit.

d.d.s. *or* dd.s *or* Dd/s Delivered Sound *in grain trade*.

DDS *or* D.D.S. Deep Diving System; Deputy Director of Science; Director of Dental Services; Doctor of Dental Surgery, *USA*.

DDS *or* DD & SHIP *or* D.D.S. *or* D.D. & Ship Dock Dues and Shipping.

D.D.Sc. Doctor of Dental Science.

DDT *or* D.D.T. Dichloro-Diphenyl-Trichloro-Ethane *insecticide liquid*.

DDU *or* D.D.U. Delivered Duty Unpaid.

D½DWTSBE *or* D.½.D.W.T.S.B.E. Despatch Half Demurrage on Working Time Saved Both Ends.

DE *or* D.E. Deflection Error *nautical*; Delaware, *USA*; Department of Employment; Destroyer Escort

USA; Doctor of Engineering; Double Elephant. Paper 40 × 26½ inches.

D-E Diesel Electric *Lloyd's Register notation.*

d.e. Diesel Engine; Diesel Electric.

DEA *or* **D.E.A.** Department of Economic Affairs. This was established in 1964 in the UK but is now defunct; Drug Enforcement Administration *USA*.

Dead Account An account closed against further purchases because the customer has closed down or is unable to pay, and any debts have been written off.

Dead Ahead Refers to something observed in front of the ship directly in line with the pointing of her bow.

Dead Astern Opposite to Dead Ahead. Directly in line with the ship's stern.

Dead Book Register of written-off companies.

Deadeye *or* **Dead Eye** A round flat wooden block having three recesses for the Lanyards, *q.v.*, to pass through. Commonly used to fasten or extend Shrouds, *q.v.*, on a ship.

Deadfreight *or* **Dead Freight** The space booked by a Broker, *q.v.*, or Charterer, *q.v.*, to load cargo on a ship and for some reason or other it is not used. Although the booked space is unused freight will still have to be paid. *Abbrev.* df *or* D.F. *or* D/F.

Deadhead Rough wooden anchor or buoy. Also an American expression for a working passage passenger.

Dead Heading The returning of containers empty.

Dead Hours Out of Office Hours. *Also called* 'Silent Hours'.

Dead Letter An undelivered and unclaimed letter at the Post Office. *See* **Dead Letter Office**.

Dead Letter Office A department in the Post Office where undelivered letters are taken care of. *See* **Dead Letter**.

Deadlight *or* **Dead Light** Steel plate coverings fitted in addition to the portholes (*see* **Portlights**) for protection against winds, storms and waves.

Dead Loans Loans to be honoured at no fixed time, or those loans which have not been honoured at the specified time.

Deadly Dozen *See Multi-Launch Rocket System.*

Deadman A suspended deadweight on a line—e.g. used in the swinging derrick system.

Dead Money When the floating supply of gold is difficult to obtain, notwithstanding a high rate of interest.

Dead-Reckoning (1) The exact calculation in the navigation of a ship to find her position by means of compass and log line obtained from Longitude, *q.v.*, and Latitude, *q.v.*, when last determined. (2) The estimated time of arrival calculated by the navigator of an aircraft by plotting his track on maps and charts prior to take-off in conjunction with the use of the compass while flying in a straight line. *Abbrev.* Dedr. *or* d.r.

Dead-Ship Ship having no power of propulsion or officers and crew to man her. Also termed as a Cold Ship.

Dead Stock Unsold merchandise or idle capital.

Dead Water (1) Slow-moving eddy under a vessel's stern which causes drag. (2) Calm and still water between the least ebb and beginning of the tidal current. (3) Still motionless water.

Deadweight Charter A charterparty where the contract freight rates are calculated on the deadweight capacity of the vessel.

Deadweight Debt The national debt of a state to cover the current expenditure.

Deadweight Efficiency The deadweight capacity of a vessel in relation to her displacement.

Deadweight Tonnage The total cargo, plus bunkers, stores, etc., which a ship can carry up to her Plimsoll Line or Marks. *Similar to* Deadweight *or* Deadweight Capacity *or* Weight All Told.

Deadworks The upper part of the hull of a ship from the waterline when she is loaded. Upperworks is another word sometimes used.

Dealing Rings Professional traders and investors who exchange or contribute information on specific exchange market stocks.

Deals Timber lengths of not less than 5 feet equivalent to 1.5240 metres in length and not less than 2/9 inches thick, 5.644 m/m.

Deals, Boards and Battens *or* **Deals, Battens and Boards** Timber loaded from the Baltic Sea is said to have various sizes and different names. The following sizes are generally loaded with approximate lengths of 5 feet up to 9 feet:

Deals: Planks of firewood averaging 3 inches in width.

Boards: Strips of wood averaging 1 inch in thickness and 8 inches in width.

Battens: A spar of timber averaging 4 inches in thickness and 7 inches in width.

Abbrev. DBB *or* D.B.B.

Dear Money The situation when the interest rate is exceptionally high. This may be the result of a deliberate dear money policy.

Deb. Debenture, *q.v.*

De bene esse *Latin*—For what it is worth. Evidence accepted conditionally and in the absence of a witness, e.g. a dying man.

Debenture (1) Customs form certificate given to the exporter of the goods authorising him to withdraw payment of Drawback, *q.v.*, on exported or re-exported goods. (2) Joint certificate given by a company together with the guaranteed interest on the sum raised.

Debentures *Similar to* **Debenture Stock**, *q.v.*

Debenture Stock Additional finance promoted apart from and in addition to the shares forming the capital. The additional money is borrowed and debentures are issued to make up the loan capital. The holders are given a low rate of fixed interest. Debentures are redeemable after an agreed fixed time though they can also be irredeemable.

Debit Item entered on the debtor side of an account due by the receiver. *Abbrev.* D/N *or* D.N.

De bonis asportatis *Latin*—Of goods carried away.

De bonis non administratis *Latin*—Of the goods not yet administered *law*.

De bonne grâce *French*— Willingly; With good grace.

Debtor One who owes money to another. Opposite to Creditor.

Debt Service The act of payment of interest which is due against the payment of sums of money to the person or firm lending such money.

Dec. December; Declaration; Declination, *q.v.*

DEC *or* **D.E.C.** Department of Conservation *USA*; Deratting Exemption Certificate. *See* **Deratting Certificate**.

D.Ec *or* **D.Econ.** Doctor of Economics.

Deca *Greek*—Tenfold.

deceptio visus *Latin*—Optical illusion.

Decibel A standard unit of sound or noise level. *Abbrev.* d.B(A). The letter A is the adjusted measurement of normal scale to the sensibility of the ear.

Decision Speed *or* **Velocity One** Prior to take-off the pilot of an aircraft decides at a point along the runway the amount of acceleration needed. *Abbrev.* V1. for Velocity One.

Deck Boy An apprentice engaged as a seaman on a ship who has had less than nine months seafaring experience.

Deck Cargo Certificate As soon as a ship arrives in port the master or the chief officer makes out a statement giving particulars of the deck cargo carried for payment of dues. Not in common use today.

Deck Drains *Similar to* **Scuppers**, *q.v.*

Deck Gear Machinery fixed on the deck of a ship to lift cargoes and anchors.

Deck Hand Seaman who works on the deck of a ship or remains in the wheelhouse attending to the orders of the duty officers during navigation and manoeuvring. He also comes under the direct orders of the boatswain or Bo'sun.

Deckhead The underneath part of a ship's deck.

Deckhouse Small superstructure on the top deck of a vessel which contains the helm and other navigational instruments.

Deck Load (1) *Similar to* **Deck Cargo**. (2) Cargo carried on the deck of a ship.

Deck Plating Steel and/or iron plates positioned on the deck of a ship.

Deck Superintendent *See* **Superintendent**.

decl *or* **Decl** Declination.

Decl. Declaration; Declination, *q.v.*

Declaration of Compliance An official statement required confirming compliance in connection with the formation of the company.

Declaration of Ownership In general the owner of a ship is obliged to give full personal details to the Registrar of Shipping in order to qualify as a shipowner. The particulars are given in a Declaration of Ownership Form.

Declination (1) The angular distance North or South of the equator of a heavenly body. (2) Obsolete term for magnetic compass variations. *Abbrev.* decl *or* Decl.

Declivity Downward slope or gradual descent in the launching of vessels from shipyards.

Decompression Sickness *See* **Staggers** *or* **Bends**.

De-consolidation The act of separating or selecting the consolidated cargo. *See* **Consolidator** and **Consolidation**.

Decr. Decreto *Italian*—Decree; Law; Ordinance.

DECR Development and Economic Conditions Report, World Economic Outlook.

Ded. Dead Reckoning, *q.v.*; Deductible, *q.v.*, Deduction.

D. Ed. Doctor of Education.

De die in diem *Latin*—Daily; From day to day.

Deductible The percentage or the limited sum of money uninsured. The assured is to bear the expenses up to this agreed sum. In other words, if the sum insured is £500 and there is a deductible or deductible franchise of £50, or 10%, the insurers are only liable for the sum over that limit, namely £450. The sum of £50 is to be borne by the assured in case of a Total Loss, *q.v.* The underwriters will pay £500, less £50 or 10% deductible franchise. This may also refer to the Deductible Average Clause for General Average. *Abbrev.* Ded. *See* **Franchise**.

D.E.E. Diploma in Electrical Engineering.

Deed Contract under seal; Legal transaction.

Deed of Arrangement *or* **Deed of Assignment** Agreement made out in case of bankruptcy in which the debtor guarantees the debts as a surety to the creditors against the value of his properties in order to continue with the business and not be dishonoured.

Deed of Covenant Legal agreement in which one agrees to pay or donate sums of money for a fixed time, which will be assessed in his favour in his income tax returns at a later date.

Deed of Inspectorship If a company or other concern becomes insolvent those responsible hand over the whole of their financial affairs to an inspector on behalf of the creditors. The inspector may be known as a Controller, *q.v.*

Deed of Partnership A contract drawn up by the partners of a company setting out all the detailed formalities and procedures in a form of rules for the systematic running of the business as a future guideline for those concerned in the enterprise as a whole.

Deed Poll A deed where there is only one party.

Deep Access Reconnaissance Television Camera A specially built camera which is used to take photographs underwater for military and/or scientific purposes. Also used to inspect oil drilling platforms and other submerged objects. *Abbrev.* DARTC.

Deep Diver A small two-man submarine specially used for laying pipelines and studying marine life.

Deep Load Line The immersion mark on the sides of a ship when she is fully loaded with cargo. *Abbrev.* DLL *or* D.L.L.

Deep Quest Sophisticated 40-ft. underwater craft with pressure spheres and a diver support module mostly used for deep ocean exploration, particularly around oil rigs and pipeline installations.

Deep Sea Trades The traffic routes of both cargo and passenger vessels which are regularly engaged on the high seas or on long voyages.

Deepstar Three-man submarine capable of performing undersea construction projects most especially in oil drilling up to a depth of 4000 feet. Also employed for underwater photography, mapping, reconnaissance and other operations at sea.

Deep Stowage Any bulk, bagged or other type of cargo stowed in single hold ships.

Deep Tanks Steel tanks constructed at the bottom of a ship and filled either with water or other liquids such as oil, etc. These tanks can serve as ballast but the term derived from cargo tanks in general cargo vessels. The fore and aft peak tanks also serve as ballast tanks and in many cases are used to trim the vessel evenly from fore to aft. There are vessels that have side tanks on each side of the vessel. *See* **Ballast** *and also* **Cellular Double Bottom**. *Abbrev.* DT *or* D.T. *or* Dt *or* d.t.

Deep Water Berth *or* **Deep Water Quay** Berth or Quay where a vessel of deep draught or draft can moor or lie alongside without touching the ground.

Deep Water Route Specified route officially declared for the clearance of any submerged objects at a stated depth.

def defunctus *Latin*—deceased.

Def. Defendant; Deferred; Deferred Stocks and Shares; Definition.

Def a/c Deferred Accounts.

De facto *Latin*—Actually; In fact; In reality.

Default in Agreement It is stipulated in a Charterparty, *q.v.*, that if the Charterers, *q.v.*, fail to pay the owners, the latter are free to withdraw the vessel from the service, without prejudice to their rights. The owner may further claim damages through legal proceedings if no sort of agreement is reached in the meantime.

Def.cl. Deficit Clause *reinsurance.*

Defeasible Interest Insurance that automatically ceases to cover while the cargo is in transit.

Defence Bonds UK Government securities originally issued for the public in late November 1939.

Defence Clubs Clubs either within the P. & I. (Protection & Indemnity Clubs, *q.v.*) or in a separate form of Club. They indemnify individual and collective interested persons such as employees and/or their principals in the form of owners, charterers, brokers, agents, etc., against freight, demurrage and legal problems.

Defence Mapping Agency Hydrographic Center Essential radio messages transmitted to naval and merchant vessels through the Hydrographic Center in *Suitland, Maryland, USA*, where an around-the-clock watch is maintained to obtain information from any source. This in turn is relayed to ships at sea and at the same time recorded in the so-called Daily Memorandum. *Abbrev.* DMAHC *or* D.M.A.H.C.

Deferred Account Payment of annual premium by instalments instead of a down payment.

Deferred Bonds Bonds which yield agreed gradual increases of interest. Eventually transferred to fixed rates.

Deferred Credits Credits which become applicable or payable at the next period of settlement.

Deferred Ordinary Shares Holders of these shares are entitled to take any profits left after all the payments of claimants are finalised. They do not get any fixed rate of dividend or interest.

Deferred Payments *Similar to* **Hire Purchase**, *q.v.*

Deferred Rebate System A conference line system of paying out rebates of freight earnings covered during a specific period of time. This has been an old practice as an incentive to the lines' regular clients but is now being dispensed with.

Deficiency Bills Claims Bank of England loans on short period terms to the British Government.

Deficiency Clause Shipping clause commonly found in the grain trade in respect of the payment of freight which is calculated against the actual weight on delivery at the place of destination.

Deflation A decrease in the availability of money relative to buying power, resulting in a fall in demand, fewer jobs and lower living standards.

Deformation The shape of a material formed when under stress. Deformation may be elastic in which case the original shape will be resumed on removal of the stress, or only a slight deformation remains.

Defunct Company Firm or company which ceases to exist or function. A wound-up Company which ceases to operate and is written-off from the Company Register.

Deg. Degree(s) or ° chart symbol.

Degree (1) Unit of temperature in Centigrade *or* Celsius *or* Fahreinheit. (2) Unit of Latitude *or* Longitude, *q.v.* A measurement which is often used in navigation to locate the position of a ship at sea in conjunction with the use of a sextant and compass. A degree of latitude is 1/360 of a Circle. *See* **Minute**.

Degree of Fault Amount assessed to be at fault in a legal action concerning an accident.

De haut en bas *French*—From top to bottom.

DEI *or* **D.E.I.** Dutch East Indies.

Dei gratia *Latin*—By the Grace of God.

De jure *Latin*—By right; The rightful occupant of an office; In law.

del delineavit *Latin*—He, or she, drew it; Delete.

Del. Delaware, *USA*; Delivered.

DEL Direct Exchange Line.

Del Credere Agent *Italian*—An accredited agent. An agent who is given the authority to act on behalf of his principals and who guarantees the solvency of his clients.

Del Credere Commission Additional commission charged by the agent who acts as guarantor for the solvency of his clients. *See* **Del Credere Agent**.

Deld. Delivered.

Delegate Power of agreement while acting on behalf of one's principal.

Delegatus non protest delegare *Latin*—A person who is given authority by his principal cannot commit it to others without specific authorisation. *Ex.* The appointment of an agent cannot be transferred in any way without the written consent of his principal.

Deletion of Charterparty Clauses *See* **Charterparty Deletions**.

Delivered at Frontier An INCOTERM 1990 expression. The seller of the goods shipped is responsible for the payment of transport and is to provide all documentation necessary up to the named frontier. He is also responsible for the insurance risk until the arrival at the frontier. *Abbrev.* DAF *or* D.A.F.

Delivered Docks Once merchandise for shipment is delivered to the docks or loading quay the shippers or consignors are exonerated from all liabilities as far as damage or short deliveries are concerned. *Abbrev.* D.d *or* D/D.

Delivered Duty Paid An INCOTERM 1990 phrase. The seller is liable at his own expense to deliver the goods at the named port and is to bear all the transportation charges, together with duty and taxes involved on arrival. *Abbrev.* DDP *or* D.D.P.

Delivered Duty Unpaid An INCOTERM 1990 expression where the seller of the merchandise is to deliver it to the consignee/buyer up to the named place/port at his own expense, inclusive of the insurance risk. Import duty and tax are for account of the receiver or consignee. *Abbrev.* DDU *or* D.D.U.

Delivered Ex Quay An INCOTERM 1990 phrase. The seller is to deliver the goods at his own risk and expense to a named quay. *Abbrev.* DEQ.

Delivered Ex Ship An INCOTERM 1990 expression applied to sea or inland waterway transport meaning that the seller of goods meets all costs and risks in bringing the goods to the named port of destination and makes them available to the buyer onboard the vessel. The buyer bears the costs of landing and import. *Abbrev.* DES.

Delivered Horsepower The power transmitted from the main engine of a ship to the propeller shaft when in motion. *Abbrev.* DHP.

Delivered in Charge The expenses involved in transporting the container by the Carrier, *q.v.*, from the Container Service to be for the account of the consignor/consignee. *Abbrev.* DIC.

Delivery and Re-Delivery Clause A standard clause for a Time Charter approved by the Documentary Council of BIMCO in May 1981. Two separate surveyors are appointed by the owners and the charterers, who are to check the condition of the ship before and at the termination of the charter period to assess any damage that might have been caused during the time of the charter. Time to count in the delivery in favour of the charterers and alternatively time counts in favour of the owners in the re-delivery. *Abbrev.* Dely and Redely.

Delivery Cargo Release of cargo. *See* **Delivery Order** *and* **Delivery of Cargo**.

Delivery Charges The expenses for the delivery of merchandise as agreed between the Consignor, *q.v.*, and the Consignee, *q.v.*; It could also mean expenses charged to obtain the Delivery Order, *q.v.*, from the Agents of the discharging ship.

Delivery Clause A Charterparty, *q.v.*, clause regulating the place, days and time the ship will be ready for the Charterers, *q.v.*, Also known as **Commencement of Hire**.

Delivery Clerk A person engaged on the ship or on the quay who checks every single item of merchandise. The 'checker', *q.v.*, or 'Tally Clerk' as he is also called, is employed by the owners to verify the total number of packages loaded or unloaded. The checker who is engaged on the quay is generally employed by the consignors or consignees to confirm with the checker on board.

Delivery Date The date when the newly constructed ship is to be ready and delivered to her owners. *See* **Builders' Certificate**.

Delivery Note A document accompanying the goods which are earmarked to be loaded on the ship to be presented to the Chief Officer or the Tally Clerk, *q.v.*, or Checker, *q.v.*, confirming the booking of the cargo to be loaded. This document contains the total packages, type and details of the contents including weights/measurements to give a clear idea of the merchandise to be shipped on board. There are some delivery notes that only give the particulars and total packages. *Abbrev.* D.N. *or* DN. *See* **Delivery Order**.

Delivery of Cargo Release of merchandise. *See* **Delivery Order**.

Delivery Order A written order issued by the owners or agents of the ship authorising the Delivery Clerk and/or the customs officers to release the goods

125

discharged from the vessel. The relative particulars of the goods are inserted on the delivery order to be selected from other heterogeneous cargo. There is also another document known as the Shipping Order, *q.v.*, which enables the shipper or consignor to load cargo on a vessel. The Shipping Order is to be presented to the shipping clerk or delivery clerk and to the Chief Officer responsible for the receipt of the loaded goods as he should be well aware of this for the stowage position of all the merchandise to stabilise the ship. *Abbrev.* DO *or* D.O. *or* D/O. *See* **Backed Note** *also* **Delivery Note**.

Delivery Terms Used in relation to merchandise. There are various typical expressions, i.e.:
Forward Delivery: Goods to be delivered at a later date. Say after three months.
Near Delivery: Delivery is to be effected in a short time.
Prompt Delivery: Immediate delivery.
Ready Delivery: Goods ready to be despatched as soon as they are ordered or bought.
To Arrive: Delivery to be effected on arrival of the vessel where the goods are to be exported or carried.
Delivered in Charge: The expenses involved in transporting the container by the carrier from the container Freight Service. *Abbrev.* DIC.

Delivery Terms of Sale Similar to INCOTERMS. *See also* **Delivery Terms**.

Delivery Trip The first voyage taken by a newly-built ship from her builders' yard to the delivery port. The passage risk is the insurable interest of the new owners of the ship. *See also* **Construction Policy**.

Delivery Waybill Receipt notes in the road transport business recording deliveries effected by each driver.

Delta (1) The earth that is left by the force of a river which flows into the sea or ocean. The estuary thus formed looks very much like a triangle. (2) International Radio Telephone phonetic alphabet for 'D'.

Delta Wing An aircraft wing in the form of a triangle.

Dely and Redely Delivery and Re-Delivery, *q.v.*

Dem. Demurrage.

DEMA Diesel Engine Manufacturers Association *USA*.

De mal en pis *French*—From bad to worse.

Demand (1) Requirement of a purchaser or a consumer of products, where the demand of a certain commodity is in the market. This is normally, though not always, the result of relatively cheap prices against quality. (2) Request for payment or to supply. (3) A legal claim. (4) An authoritative request or claim, i.e. 'A note payable on Demand'. (5) An urgent or pressing request or requirement. (6) Call or desire for a commodity.

Demand Deposits American banking expression relating to current account deposits which are withdrawn by cheques.

Demand Draft Bill of Exchange, *q.v.*, payable at sight and which does not need an acceptance.

Demand Price Price which the buyers are prepared to pay for a specific item or service.

Demarcation Dispute Disagreement in a factory or yard between labourers engaged in different trades, such as metalworkers, shipwrights, boilermakers, fitters and electricians, as to responsibility for, and entitlement to perform, certain operations or stages in the work. Unless resolved by negotiation it may lead to a strike by one section or the whole.

DEM/DES *or* **DEM-DES** *or* **dem-des** Demurrage and Despatch. *See* **Despatch**.

Demersal Fish Fish that live on the seabed, such as Haddock, Cod, Halibut, etc.

Demijohn Large bottle in which acid liquids are carried. It is protected by a wickerwork or steel grill. Due to the corrosive or explosive nature of the contents it is carried on deck at shippers' risk.

De minimis non curat lex *Latin*—The court ignores trivial cases.

Demise To lease; To let; To charter. *See* **Bareboat Charter** *or* **Demise**.

Demography The art or study of the statistics of births, deaths and general health of the population of a country.

Demolition The process of dismantling a ship by cutting it into small pieces which are ultimately sold as steel scrap. *Also known as* Scrapping.

Demonetisation Official withdrawal of coins in circulation and consequently rendering such money illegal.

De mortuis nisi bonum *Latin*—Speak nothing except good of the dead.

Demote To lower a person's rank or grade.

Demurrage Clause Voyage Charterparty clause specifying the time and number of days allowed for loading and/or discharging.

Den. Denmark.

DEN Department of Energy.

D.en D. Docteur en Droit *French*—Doctor of Laws.

D. Eng. Doctor of Engineering.

Deniz *Turkish*—Sea.

Denizcilik *Turkish*—Maritime.

DENLA Drag Embedment Normal Load Anchor *oil drilling*.

De novo *Latin*—Anew.

Dense Fog Classified as 0 on the fog and visibility scale with objects not visible at 50 yards.

Densitometer Instrument to measure the density of fresh and sea water.

Density A mass per unit of volume. It is measured in grams per cubic centimetre in the metric system and in grams per cubic foot in the English system. To obtain the density divide the mass of the object by its volume.

Deo date *Latin*—Give to God.

Deo favente *Latin*—With God's favour.

Deo juvante *Latin*—With God's help. (*The motto of Monaco.*)

Deo volente *Latin*—God willing.

Dep Departure; Departing; Department; Deputy.

Dep. Departure; Deputy.

Department Store A large building having departments dealing in various items for sale. *Similar to* **Supermarket**.

Deponent A person making an Affidavit, *q.v.*

Deportation Mutual agreement by foreign countries wherein the judicial authorities can order an immigrant to be escorted out of a country and returned to his native land.

Deposit (1) To invest or secure money in a bank. (2) A certain amount of money (a percentage) paid by the purchaser to the seller as a security on either side. (3) In the hire-purchase system a sum of money is paid at the beginning of the hire-purchase period which is called the deposit or down payment and small amounts are then paid periodically until the full settlement is arrived at. Interest is charged over and above the outstanding amount until such time as the full settlement is effected.

Deposit Account A banking term. This is a system whereby money is deposited at a bank against a percentage interest paid yearly or half yearly. Money can be withdrawn at any time. There is also the Fixed Deposit Account—withdrawal in this case will have to be after the expiration of a year from the date of deposit. As an incentive the interest given on this account is rather higher than that of a Current Account, *q.v.*

Deposit Interest A compulsory state system in America obliging the Commercial Banks to insure against a run on the bank by the depositors.

Deposit Rate The rate of interest allowed by bankers or merchants to their clients who invest money with them. *See* **Deposit Account**.

Deposit Receipt A receipt presented to the payer or consignee of the cargo against the payment of the proportion of the General Average Bond. *Abbrev.* DR *or* D.R. *or* Dr. *See* **Average Bond**.

Depository A firm or person receiving valued articles for safe keeping.

Depot Store or warehouse or railway terminal. Place where goods are deposited for safe custody.

Depreciation Any object loses its original value immediately it is made use of. The depreciation value is calculated at a percentage value of decline. Opposite to Appreciation, *q.v.*

Depression (1) That period when commercial business suffers a great loss and in consequence of which heavy unemployment occurs. Similar to recession. (2) A meteorological word signifying an area with low atmospheric pressure where both warm and cold air currents meet together resulting in clouds and subsequent rain.

de profundis *Latin*—Out of the depths.

DEPS Departmental Entry Processing System; Data Entry Processing System.

DEPSUM. Daily Estimated Position Summary, *q.v.*

Dept Department.

Depth Charge Underwater ordnance launched by ships or aircraft to attack submarines. Sometimes called Depth Bomb.

Depth Moulded The measured distance taken vertically at the middle side of the ship from keel to freeboard deckside. The approximate measurement of the moulded side.

Depth of Hold The vertical measure or height from the top of the floor at the centre of the hold to the main deck beam amidships.

DEQ *or* **D.E.Q.** Delivered Ex Quay, *q.v.*

Der *or* **Derr** Derrick, *q.v.*

Derat Certificate *See* **Deratting Certificate**.

Deratting Certificate Issued in confirmation that a vessel has been inspected and is free of rodent infestation, or after action has been taken to successfully rid a vessel of rodent infestation.

De rigueur *French*—Compulsory; Indispensable.

Derby Derbyshire, *UK*.

Derelict An abandoned object at sea which finds its way with the current, like a boat or jettisoned cargo. The master of a vessel is duty bound to report any derelicts to the proper authorities at his first port of call for the safety of others.

Derrick Post The corresponding American term for Samson Post, *q.v.*

Derricks The booms of a vessel particularly connected with the lifting of cargoes. *Abbrev.* Der, Derr.

Derry Londonderry, *Northern Ireland*.

DERV *or* **D.E.R.V.** (1) Diesel-engined Road Vehicle. (2) Diesel fuel oil used in such vehicles.

DES Data Encryption Standard; Delivered Ex Ship, *q.v.*

Deserter A seaman who deserts his ship or who fails to be on board before she is about to leave. The master logs him (or reports him) as missing and a deserter. On the day of desertion he forfeits: (1) All or the remaining part of his wages. (2) All his wages earned on another ship until such time as the seaman returns to the place of repatriation. (3) Extra amount of wages for seaman substituted in his stead. Apart from this the deserter may also be liable for imprisonment. *See* **Conveyance Order**.

Desid. Desideratum *Latin*—Desired.

Desig Designate.

Design Draft (Draught) Draft (draught) at which the ship is designed to carry her contract deadweight and attain her design speed. *Abrrev.* DD *or* D.D.

Design Weight The weight of cargo a ship is designed to carry while attaining her designed speed.

D. ès L. Docteur ès Lettres *French*—Doctor of Letters.

Desp. Despatch or Dispatch, *q.v.*

Despatch *or* **Dispatch** Clearance Outwards, *q.v.* *Abbrev.* Desp. *See also* **Clearance**.

Despatch Bay Part of a depot where the actual shifting of cargo in the loading and discharge of ships, sail waggons, and road vehicles is done.

Despatch Both Ends *or* **Dispatch Both Ends** Money given as a remuneration in a Voyage Charterparty, *q.v.*, by the owners to the Charterers, *q.v.*, or their respective agents for time saved both while loading and/or discharging. *Abbrev.* DBENDS *or* D.B.E.N.D.S. *or* DBE *or* D.B.E.

Despatch Days *or* **Dispatch Days** In a Voyage Charterparty these are days saved while loading and/or discharging in relation to the time specified in the contract. Most of the time saved is attributed to the Charterers', *q.v.*, efforts with the result that money is saved in favour of the owners. On such occasions the Charterers, *q.v.*, or their agents are compensated by a remuneration which also serves as an incentive. This is called Despatch Money or Dispatch Money. *Abbrev.* DD *or* D.D.

Despatch Discharging Only *or* **Dispatch Discharging Only** In a Voyage Charterparty the remuneration is only allowed in reference to the Loading. *See* **Despatch Days**. *Abbrev.* DDO *or* D.D.O.

Despatch Half Demurrage *or* **Dispatch Half Demurrage** Despatch money due to the charterers or their agents and allowed by the owners at half of that of the Demurrage, *q.v.*, set in the Charterparty, *q.v.* *Abbrev.* DHD *or* D.H.D. *or* D½D.

Despatch Half Demurrage on Working Time Saved Both Ends A Charterparty, *q.v.*, term meaning Despatch Money, *q.v.* It is allowed by the owners to the charterers at half that of demurrage on all time saved, namely in loading and discharging during working time. *Abbrev.* D½DWTSBE *or* D.½.D.W.T.S.B.E.

Despatch Loading Only *or* **Dispatch Loading Only** The remuneration which is credited by the owners to the Charterers, *q.v.*, or their agents for Laydays, *q.v.*, saved during the loading operation only. *Abbrev.* DLO *or* D.L.O. *See* **Despatch Days**.

Despatch Money *or* **Dispatch Money** Money due as a remuneration to the Charterers, *q.v.*, by the owners for the quick turn-out of a cargo vessel as far as the Charterparty, *q.v.*, stipulation for the fixed Laydays, *q.v.*, is concerned. The charterers are given an agreed sum or equivalent as a remuneration for their efforts in saving laydays during the time of loading or discharging. This is not a very common term in the charterparty but it is sometimes inserted as a sort of incentive to hasten the date of departure and as opposed to the expression 'Demurrage', *q.v.* This advancement might have involved extra expense by the charterers by employing extra labour or having men working on overtime to finish the work. *Abbrev.* DM *or* D.M. *See* **Despatch Money**.

Despatch Note Printed postal form used in the forwarding of parcels by post. *Abbrev.* DN *or* D.N.

D. ès S. Docteur ès Sciences *French*—Docteur of Sciences.

Destination The ultimate place or port.

Desunt cetera *Latin*—The rest is missing.

Det. Detail; Detained; Detained at; Detention, *q.v.*

Detention Claim by the Carrier of a ship as unliquidated damages against the Charterers, *q.v.*, or receivers of the cargo after the period of Demurrage, *q.v.*, expires. Also to keep in custody by law. *Abbrev.* Det.

Detergents A surface agent for cleaning, sometimes used as a solvent but not necessarily for that purpose. Applied to various soaps in liquid or solid forms. Today mainly used in liquid form as chemicals mixed with either fresh or sea water.

Detinue A legal action taken by an owner of goods or property to recover wrongfully detained property. The claim refers to the return of the goods or their value, including damages for detention.

De trop *French*—Superfluous; Too much; Too many.

Deus ex machina *Latin*—A person who puts things in the right perspective in critical moments.

Deuterium An isotope of hydrogen symbolised by the letter 'D'.

Deutsche Mark Unit of currency used in Germany.

Deutsches Institut für Normung German Standards Institute. *Abbrev.* DIN.

Dev. Deviation.

Devaluation Reducing the value of. In international business circles a country may be compelled to reduce the official face value of its currency *vis-à-vis* those of other countries. Devaluation can be an asset in a way if the country is a successful exporter and also caters for tourism on a large scale or when the level of production from factories is unaffected by serious labour problems. On the other hand, if the country is a major importer then devaluation may be very detrimental. There is no doubt that a very difficult situation arises if repeated devaluations take place. Other nations lose confidence in the currency of the devaluating country, principally from the viewpoint of investment.

Devanning The act of taking out the merchandise from a Container, *q.v.* Also termed Unstuffing, Stripping, Unpacking or Disseminating.

Dev. Devonshire *UK*.

Dev. Cl. Deviation Clause, *q.v.*

Deviation Clause In the marine insurance policy a clause is inserted covering the assured, at an extra premium, in case of a deviation due to *force majeure* (an event beyond one's control). *Abbrev.* D/C *or* D.Cl. *or* Dev. Cl. *See* **Continuation Clause**.

Devil's Advocate One who plausibly urges the contrary or wrong side in a cause, thereby suggesting another view that might be taken.

Devils Claws A steel hand or claw shaped device bent at right angles once used as an anchor chain stopper, which has been replaced by a chain shackle.

Devise A gift or disposition by will. This word is properly only applied to real property.

Devisee A person who is bequeathed real property.

Devisor The person who bequeaths real property.

DEW Distant Early Warning *radar*. *See* **Distant Early Warning Line**.

DF *or* **D.F.** Dean of Faculty; Defensor Fidei *Latin*—Defender of the Faith; Doctor of Forestry.

D/F *or* **DF** *or* **d.f.** Deadfreight, *q.v.*; Decontamination Factor; Direction Finder; Drinking Fountain.

DFA *or* **D.F.A.** Doctor of Fine Arts.

DFC Distinguished Flying Cross.

DFD Design for Disassembly, *manufacturing*.

D.fl. Dutch Florin. A unit of currency used in Holland.

DFM Distinguished Flying Medal.

DFR Decreasing Failure Rate.

DFS Duty Free Supply.

D.F.Sc. Doctor of Financial Science.

Dft. Draft *or* Draught, *q.v.*

DFT *or* **D.F.T.** Dry Film Thickness, *q.v.*

dg Decigramme; 1/10th of a gramme.

DG *or* **D.G.** Deo Gratias *Latin*—By the Grace of God; De-gausing; Directional Gyro *nautical*; Director General.

D.G. Days of Grace, *q.v.*

DGB Dangerous Goods Board *IATA*.

DGCA *or* **D.G.C.A.** Director General of Civil Aviation.

DGD&M *or* **D.G.D.&M.** Director General Drydocks and Maintenance.

DGE *or* **D.G.E.** Director General of Equipment.

DGFT *or* **D.G.F.T.** Director General of Fair Trading *UK*.

Dghajsa A Maltese rowing boat used for the transportation of passengers or ship's crew from or to ships at anchor or on buoys.

DGI *or* **D.G.I.** Director General of Information.

D.G.M. Diploma in General Medicine.

DGN Dangerous Goods Note.

DGPS Differential Global Positioning System.

DH *or* **D.H.** Decision Height *aviation*.

d.h. Das heisst *German*—That is to say.

D&HAA *or* **D.&H.A.A.** Dock and Harbour Authorities Association.

DHD Demurrage, Half Despatch.

DHD *or* **D.H.D.** *or* **D$\frac{1}{2}$D** Despatch Half Demurrage, *q.v.*

DHDOASBE *or* **D.H.D.O.A.S.B.E.** Despatch Half Demurrage on All Time Saved Both Ends. *See* **Despatch Money**.

DHDOATSBE *or* **D.H.D.O.A.T.S.B.E.** Despatch Half Demurrage on all Working Time Saved Both Ends. *See* **Despatch Money**.

D.H.I. Deutches Hydrographisches Institut *German*—Hydrographic Institute in Hamburg, Germany.

Dhl. Dholl, *q.v.*

DHL *or* **D.H.L.** Doctor of Hebrew Letters.

Dholl Small bundle rolled together for the manufacture of ropes such as sisal.

DHP *or* **D.H.P.** Developed Horse Power; Diploma in Public Health.

DHSS *or* **D.H.S.S.** Department of Health and Social Security.

DHSV Downhole Safety Valve *oil drilling*.

DHU *or* **D.H.U.** Deckhand Uncertified.

D.Hy. Doctor of Hygiene.

D.I. *or* **Di.** Dinar. Unit of Currency.

D&I Disbursement and Interest.

D.I.A. Diploma in International Affairs.

Dia *or* **Diam.** Diameter.

Dia L. Diagonal Line.

DIC *or* **D.I.C.** Difference in Condition; Delivered in Charge, *q.v.*; Diploma of Imperial College.

Dic *or* **Dict.** Dictionary.

DICE Direct-Course-Error *aviation*.

Dictum An observation by a judge on a legal question suggested by a case before him, but not arising in such a manner as to require decision by him. It is therefore not binding as a precedent on other judges, although it may be entitled to more or less respect. 'Dictum' is an abbreviation of *obiter dictum,* 'a remark by the way.' In early times dictum meant the award of an arbitrator, who himself was known as the *dictor*. *See* **Obiter dictum**.

Dictum de dicto *Latin*—Report upon hearsay. Secondhand Story.

Dictum factum *Latin*—Said and done—No sooner said than done.

Dictum meum pactum *Latin*—My word is my bond. Maxim of the Institute of Shipbrokers and of the Baltic Exchange, London.

Dictum sapienti sat est *Latin*—A word to the wise is enough.

DIDC Depository Institutions Deregulation Committee *American financial market*.

D.I.E. Diploma of Industrial Engineer.

Die intestate To die without leaving a Will. *See* **Intestacy**.

Dies irae *Latin*—Day of Judgment.

Diesel The word has a dual purpose, meaning either diesel oil or diesel engine. It is derived from Rudolf Diesel, 1858–1913 who was the inventor of the Diesel engine in 1897. This was a 25-HP four-stroke engine which was the forerunner of the engines widely used today throughout the world in the marine and industrial fields.

Diesel Clause Charterparty, *q.v.*, clause which reads and which is self-explanatory:— The vessel shall have liberty as part of the contract to proceed to any port or ports at which bunker oil is available for the purpose of bunkering at any stage of the voyage whatsoever and whether such port or ports are on or off direct and/or customary route or routes between any of the ports of loading and discharging named in the charter, and may there take oil bunkers in any quantity at the discretion of owners even to the full capacity of fuel tanks and deep tanks, any other compartments in which oil can be carried, whether such amount is or is not required for the chartered voyage.

Dies non *Latin*—Due to some particular reason no business is done on a specified day. *Ex.* Bank or court or public holidays.

Dieu et mon droit *French*—God and my right, the motto used on the royal coat of arms of England since the time of Richard I.

Diff. Difference.

Differences A Stock Exchange, *q.v.*, term. When a speculator fails to attend to the shares bought or sold on the appropriate day, the account is carried over to the next statement, on which day he is to settle the difference in the respective fluctuation of the prices ruling on that day.

Digital Computer Device to calculate on numbers digitally represented.

Digest Collection of legal rules or maxims.

D.I.H. Deputy Inspector of Hospitals; Diploma of Industrial Health.

D.I.M. Diploma of Industrial Manufacture.

Dim. Dimension.

Dime An American currency unit equivalent to 10 cents.

Dim sum *Chinese*—snack; small item of refreshment.

Din or **D.I.N.** Das ist Norm *German*—This is normal; Deutsche Industrie-Norm *German*—German Industrial Standards. *See* **Din Rating**; Dinar, *q.v.*

Dinar A unit of currency used in Algeria, Iraq, Jordan, Kuwait, Libya, Tunisia and Yugoslavia.

D. Ing. Doctor Ingeniareae *Latin*—Doctor of Engineering.

Din Rating German Industrial Standards. *Abbrev.* Din or D.I.N.

Dioptic Lights Refracted prisms and lenses which produce concentrated parallel beams. They are for searchlights and coastal lights.

Dip (1) The angle between the horizontal plane through the observer's eye and the direction of the observed horizon, which depends on the observer's height above the surface. This angle in minutes and the distance of the horizon in nautical miles are roughly equal to the square root of the height of eye in feet. When corrected for refraction using nautical tables, dip is applied to an observed altitude (by sextant) to obtain true altitude. (2) The angle between the earth's surface and the line of its magnetic force. This is 90 degrees at the magnetic poles and zero at the magnetic equator. (3) A candle. (4) The action of an object dropping below the visible horizon. (5) The quick sequence of lowering and rehoisting an ensign in salute. (6) The half-hoisted position of signal flag(s). In naval practice, an 'answer pendant' in this position acknowledges that another ship's signal has been observed but not understood. A flag officer may display a ship's identifying pendants in this position as a sign of disapprobation.

Dip. Diploma.

DIP Direct Iron Pellet. *See* **Direct Reduced Iron**; Document Image Processing *document filing*.

Diplomatic Privilege Diplomatic immunity enjoyed by or accredited to certain diplomatic personnel in foreign countries.

Dip MSO Diploma in Management of Ship Operations.

Dip Needle A magnetised needle, similar to a compass needle but mounted on a horizontal axis, in an instrument to measure Dip, *q.v.*

Dipper The seaman's slang word for Ursa Minor, *q.v.*

Dip. Tech. Diploma in Technology.

Direct Bill of Lading A Bill of Lading, *q.v.*, which is used when direct loading and discharging ports are involved. *Abbrev.* Direct B/L.

Direct B/L Direct Bill of Lading, *q.v.*

Direct Drive This concerns the engine power transmitted direct to the main driving shaft or the propeller shaft and no gear box is installed to reduce the revolutions before the engine power is transmitted. An engine drive without a gear-box.

Direct Expenditure *Similar to* **Prime Cost**, *q.v.*

Direct Insurance Insurance but with no re-insurance. Insurance effected directly between the insured and the Underwriters, *q.v.*, or through their brokers.

Direction Finder *See* **Radar**.

Direct Lift Control Automatic equipment in an aircraft which adjusts the lift and pitch positions during the landing approach. *Abbrev.* DLC.

Director Person who has a special position in a company to manage it and who may virtually be called agent of the company.

Directorate The position of the officer acting as a director of a company; A group of directors or body of directors; Office of a director.

Directors and Officers' Liability Insurance Lloyd's underwriters carry most of the business in covering the liabilities of directors and their office executives as regards damages and court cases brought against them in the course of their office duty. Some American underwriting companies are also in the market to insure presidents and other officials in their private corporations. *Abbrev.* D & O Insurance.

Direct Port A Charterparty, *q.v.*, phrase for a ship to sail direct without stopping or deviating anywhere. *Abbrev.* DP or D.P. or d.p.

Direct Reduced Iron Metallic manufacturing system in reducing iron oxide at below the fusion point of iron. This compressed form of iron for scrap purposes consists of pellets, lumps, and cold moulded briquettes which are loaded in bulk. *Abbrev.* DRI.

Direct Selling The system of selling without a Middleman, *q.v.*, or Broker, *q.v.*

Direct Tax The tax the consumer of goods pays when he buys food and other articles and when tax is levied before being released from the customs.

Dirham Unit of currency used in Morocco.

Dirty Ballast The oil polluted water remaining after tank cleaning of crude, diesel and other oils.

Dirty Bill *or* **Dirty Bill of Lading** *or* **Claused Bill of Lading** *or* **Unclean Bill of Lading** *or* **Foul Bill of Lading** If the Bill of Lading, *q.v.*, has any remarks inserted on it as to condition or quality, it is said to be a Dirty Bill or Dirty Bill of Lading or as above. This could be a remark that the cargo was wet with rain. The shippers are opposed to accepting a Dirty Bill, etc., as in due course claims may turn against them or the bank may refuse to accept the shipping documents. Therefore, the Dirty Bill will hamper the smooth running of the banks. In some cases the shippers obtain a Letter of Indemnity, *q.v.*, which in effect is not legally acceptable. Example of a remark inserted on a Dirty Bill 'Bags received wet and torn, contents leaking'. Because of this remark one can experience difficulties as invariably no bank is bound to accept it. This is the opposite meaning of Clean Bill of Lading, *q.v.*

Dirty Oils *or* **Black Oils** In regard to petroleum bulk cargoes the following are termed black or dirty oils: Crude, furnace and diesel oils.

Dirty Money Extra money allowed to those port workers engaged in the unloading of grains in bulk by pneumatic means. During the discharge a lot of dust is created in and outside the holds. This term is also called Dust Allowance *or* Danger Money, *q.v.*

Dirty Ship A tanker which has been carrying dirty or unrefined oils.

DIS *or* **D.I.S.** Dairy Industry Society; Defence Intelligence School *USA.*

Dis *or* **Disct.** Discount, *q.v.*

Dis. A. Seaman's Discharge Book.

Disabled A ship which is unable to start or proceed on her voyage either through lack of officers or crew or through damage. Also termed Disabled Ship *or* Disabled Vessel, *q.v.*

Disabled Ship *or* **Disabled Vessel** When a ship is unable to sail efficiently or in a seaworthy state. This could be the result of engine trouble, or damage to the hull or ship's gear or even through lack of officers and crew. *See also* **Disabled**.

Disbursement Clause A marine insurance clause covering expenses incurred by the ship supplementary to the main insurances on hull and machinery. This may be charged on a certain percentage against the main insurances and is part and parcel of the Policy Proof of Interest (P.P.I.) or Free Interest Admitted (F.I.A.) or Interest or No Interest.

Disbts. Disbursements, *q.v.*

DISC *or* **D.I.S.C.** Domestic International Sales Corporation *USA.*

Disc. Discharge, *q.v.*; Discount.

Disch. Discharge, *q.v.*; Discharging.

Discharge (1) Discharge of a Bill of Exchange, *q.v.*, puts an end to the right and subsequent payments effected in the normal way. (2) To unload cargo from the ship's holds on to the quay, craft or lighter. *Abbrev.* Disc. *or* Disch. (3) To relieve, or relief, from obligation or responsibility. (4) To let off or shoot. (5) To pay off, dismiss; dismissal. (6) Outflow from a pump.

Discharge Book Booklet issued by the Registrar of Shipping or Shipping Master of the country where the seaman is registered and where records are kept regarding his credentials. An essential document for officers and seamen since it serves as an official certificate confirming sea experience in the employment for which he was engaged. Whenever an officer or a seaman is discharged he is to be in the presence of the Registrar of Shipping or Consul of the registered ship's country together with the master who signs and rubber stamps the discharge book with the name of the ship. *Also called* Continuous Discharge Book *or* Seaman's Book.

Disclosure A word closely associated with the Latin phrase 'Uberrimae fidei', *q.v.* when a person seeks to insure something he is in duty bound to give all pertinent information in utmost good faith, failing which the contract would become null and void. *See* **Utmost Good Faith**.

Discount Deduction from original face value to serve as an incentive. There are various methods of discounts in business and trade activities such as a discount for cash payment or for payment within a fixed time limit; a discount on wholesale or retail goods; a discount shown in shop windows known as 'Sale'. *Abbrev.* Dis. *or* Disct.

Discount Bills of Exchange Purchase of Bills of Exchange through a third party at a discount price. This method may even gain advantage by receiving the money before it is due.

Discount Broker Broker acting as an intermediary between buyers and sellers.

Discount Cash Flow Calculations or techniques used by entrepreneurs in the investment of a proposed project. If the interest rate reduces the anticipated capital to a balance it is known as Cash Flow. *See* **Net Present Value**. *Abbrev.* DCF *or* D.C.F.

Discount Houses Those engaged in the business of Bank Bills, Treasury Bills, Discounting Bills of Exchange, *q.v.*, etc, on a large scale. To obtain a low rate of interest, business is done in large amounts of money borrowed against the guarantee of the Bills of Exchange not yet matured.

Discountless And Are Not Returnable *and/or* **Cargo Lost or Not Lost** A Charterparty, *q.v.*, phrase referring to the chartered freight agreed and paid for which will not be returnable or refunded so long as no breach is committed by the carrier or owner. This generally refers to that freight which is 'paid in advance' or 'prepaid' which system is similar to 'Advance Freight or Freight Paid in Advance or Freight Prepaid', *q.v. Abbrev.* Danrsaoclonl *or* DANRSAOCLONL.

Discount Rate Referring to Discount Interest.

Discount Stores Stores similar to supermarkets negotiating in bulk buying and trading in relatively large amounts of groceries and various commodities, thus being able to give discounts against the retail buying by their customers.

Discretionary Income Net income; Income remaining after a person has paid for his living expenses and all taxes.

Discriminating Duty An import duty charged on merchandise depending on the country from which it originated. Generally this import duty is levied *vis-à-vis* their mutual relationship, such as the special preferential tariffs which are practised between the Common Market (EC) and Commonwealth countries or between Commonwealth states. *See* **Preferential Duty**.

Discus *or* **DISCUS** Distilled Spirits Council of the United States.

Disembarkation When passengers or officers and crew of a ship or an aircraft leave.

DISERO Disembarkation Resettlement Offer. *UN procedure.*

Disgorging The act of removing wine sediment from bottles.

DISH Data Interchange for Shipping.

Dishonour When the holder of a Bill of Exchange, *q.v.*, is unable to discharge on its Due Date, *q.v.* Alternatively, when a Drawer, *q.v.*, is unable to meet the money demanded for payment. *See* **Dishonoured Bills**.

Dishonoured Bills Bills of Exchange that are unacceptable when presented. *See* **Dishonour** *and* **Bill of Exchange Dishonoured**.

disjecta membra *Latin*—Scattered remains.

Dismasted When a vessel loses a mast(s) because of an accident such as a collision, a storm or enemy action.

Dispatch *See* **Despatch**.

Dispatch Days *See* **Despatch Days**.

Dispatch Note *See* **Despatch Note**.

DISPL *or* **Displ.** Displacement Tonnage or Displacement, *q.v.*

Displacement Light or loaded displacement. The light displacement is the weight of the vessel either in long tons of 2240 lbs or in metric tons of 1000 kilogrammes. The loaded displacement is the weight of the vessel plus stores, cargo, the weight of officers and crew members as well as passengers if any. *Abbrev.* Displ.

Displacement Scale Scale used to find the individual inch immersion of various ships' drafts (draughts).

Disponent Owner e.g. the time charterer of a ship. An operator entitled through a charter arrangement to exploit the ship commercially.

Disposable Income Income remaining after various deductions for tax, national insurance contributions, etc., have been made.

Disposal Letter A request by the shipper to the owner, agent, or master of a ship to be given a letter showing or confirming that all the formalities for the shipment of the goods have been executed. This letter may serve a good purpose as far as the Letter of Credit, *q.v.*, is concerned in order to make sure that the respective documents are in good order. This will avoid any delays and consequent litigation.

DISPS Diverless Sub-sea Production System *oil drilling*

Disseise To debar or dispossess the ownership of an estate or freehold. *See* **Disseisin**.

Disseisin Act of debarring a person's possession of the ownership of an estate or a freehold. Also called Disseise.

Disseminate Act of unloading merchandise from a container, after which every individual consignment is separated before being delivered to the respective consignees. *See also* **Devanning**.

Dissolution of Partnership Discontinuation of the firm's or company's usual business connections or functions.

Dissolved Gas Alternative for Natural Gas, *q.v.*

Dist. Distance; District.

Distance In the marine field distance is taken as Cables, *q.v.*, i.e. tenths of a mile or nautical/statute miles or metres or kilometres.

Distance Endurance Maximum distance that can be covered with the whole bunkers in a ship. Also known as Maximum Distance Range and also Cruising Radius, *q.v.*

Distance Freight If a ship is unable to arrive at her port of destination to unload her cargo due to some unforeseen reason her master has the right to call at an alternative safe port. In so doing some extra expenses may be incurred chiefly because of the deviation. The owner charges an extra freight in the ratio to the distance covered, so called 'distance freight'.

Distance Tables A book containing nautical distances between ports—used as a ready reckoner for various purposes such as voyage estimating.

Distant Early Warning Line An advanced distant radar warning chain for the detection of aircraft approaching North American shores. *Abbrev.* DEW Line.

Dist. Atty. District Attorney.

Distillate Fuel Extract from crude oil which is commonly known as gas or diesel oil. Used in auxiliary engines, mainly generator sets. For marine use grades have been established by ISO Fuel Standard (DIS) 1994 using the terms DMX, DMA, DMB and DMC, *q.v.*

Dist. M.G. *or* **DMG** *or* **D.M.G.** Distance made good *nautical*.

Distrain To seize property or goods against debt.

Distrainor The person who legally seizes goods or properties of another for rent or other debts due.

Distress In shipping circles this term refers to a ship which may require assistance at sea for various reasons.

Distress Cargo When cargo is not offered as expected, vacant space may be avoided by completion of loading with cargo that is available at a very low rate. *See* **Distress Freight**.

Distress Freight When the holds of a ship are not fully loaded the Carrier, *q.v.*, will accept further Cargo, *q.v.*, at reduced freight rates to fill up the vacant space.

Distress Signals 1. The following signals, used or exhibited either together or separately, indicate distress and need of assistance:

(a) a gun or other explosive signal fired at intervals of about a minute;

(b) a continuous sounding with any fog-signalling apparatus;

(c) rockets or shells, throwing red stars fired one at a time at short intervals;

(d) a signal made by radiotelegraphy or by any other signalling method consisting of the group ···−−−··· (SOS) in the Morse Code

(e) a signal sent by radio telephony consisting of the spoken word 'Mayday';

(f) the International Code Signal of distress indicated by N.C.;

(g) a signal consisting of a square flag having above or below it a ball or anything resembling a ball;

(h) flames on the vessel (as from a burning tar barrel, oil barrel, etc.);

(i) a rocket parachute flare or a hand flare showing a red light;

(j) a smoke signal giving off orange-coloured smoke;

(k) slowly and repeatedly raising and lowering arms outstretched to each side;

(l) the radiotelegraph alarm signal;

(m) the radiotelephone alarm signal;

(n) signals transmitted by emergency position-indicating radio beacons.

2. The use or exhibition of any of the foregoing signals except for the purpose of indicating distress and need of assistance and the use of other signals which may be confused with any of the above signals is prohibited.

3. Attention is drawn to the relevant sections of the International Code of Signals, the Merchant Ship Search and Rescue Manual and the following signals:

(a) a piece of orange-coloured canvas with either a black square and circle or other appropriate symbol (for identifications from the air);

(b) a dye marker.

Distressed British Seaman A British seaman who is left behind because of sickness or any other mishap beyond his control. The Shipping Master or British Consulate undertakes to send him back to the place of repatriation at the first opportunity. This has nothing to do with desertion or a Deserter, *q.v. Abbrev.* DBS *or* D.B.S. *See* **Repatriate**.

Distributive Cost Cost of the commodity estimated from the place of manufacture up to the time of consumption or use. The rise or fall of the cost mostly depends upon the current cost of living, advertising, etc. *Also called* Distribution Cost.

Distributor Person appointed by an agent or his principal to take charge of the distribution of the items in the market. In general a distributor works on exclusivity.

DIT *or* **D.I.T.** Detroit Institute of Technology.

DITB Department of Industry Training Board *UK*.

Ditto *or* **Do** *Italian*—The same.

Diurnal Wind A sea breeze on land and alternatively a breeze on the sea from land. These occur due to convection currents.

Div. Diversification, *q.v.*; Dividend; Division.

Diversification This can refer to either a country or a factory. Diversification of certain industries helps to create full employment. Producing a wide range of products saves bad repercussions if any one product were to disappear from the market. Through the range of products some by-products can sometimes be obtained. *Abbrev.* Div.

Divest To eliminate or take away vested rights such as interests or estates.

Dividend Rate of interest set by a company to be distributed among its shareholders, usually every year, in relation to the number of shares each shareholder possesses. This calculation is decided upon by the directors after the net profit is finally arrived at by the auditors.

Dividend Mandate Dividend paid into the bank on instructions given by the shareholder.

Dividend Warrant The holders of stocks and shares are notified by circular letters giving full details, together with enclosed respective cheques for the amount of interest approved by authorised persons.

Dividers Instrument used in the Chartroom, *q.v.*, or on the Bridge to measure distances on a Chart.

Dixi *Latin*—I have spoken.

DIY Do it yourself.

Diz. Dizianario *Italian*—Dictionary.

D.J. Dinner Jacket; Diploma in Journalism.

D.J.S. Doctor of Judicial Science.

D. Jur. Doctor of Jurisprudence.

D.K. Disbursing Clerk; Dust Keel or Box Keel.

Dk. Deck; Dock.

Dkg *or* **Dag** Dekagramme.

Dkl *or* **Dal** Dekalitre.

Dkm *or* **Dam** Dekametre.

DL *or* **D. Lat.** Difference of Latitude.

D.L. Doctor Legum *Latin*—Doctor of Law.

D/L *or* **DL** Dead Line; Demand Loan; Double Ledger.

DLA Defence Logistics Agency *USA*.

DLC Direct Lift Control, *q.v.*

DLD Difference in longitude *nautical also abbrev.* D. Long.

Dld *or* **DLD** Deadline Date Delivered.

DLH *or* **D.L.H.** Deutsche Lufthansa, *West German State Airline.*

D. Lit. Doctor of Literature.

DLL *or* **D.L.L.** Deep Load Line, *q.v.*

d.l.m. Des laufenden monats *German*—Of the current month.

DLO *or* **D.L.O.** Dead Letter Office or Returned Letter Office; Despatch Loading Only, *q.v.*

D. Long. Difference in longitude *nautical.*

Dls Dollars.

D.L.S. Doctor of Library Science.

DLSS Deep Submergence Diving System *US Navy.*

Dly *or* **DLY** Delivery.

DM Deutsche Mark-German Mark; Draftsman; Despatch Money, *q.v.*; Dock Master, *q.v.*

D.M. Deputy Master; Despatch Money, *q.v.*; Dock Master, *q.v.*; Doctor of Medicine; Doctor of Music.

d.m. Dieses Monats *German*—Instant.

dm Decimetre.

DMA Defence Manufacturers Association UK; Defence Mapping Agency *USA*; Dominion Marine Association, *Canada.*

DMA A pure general purpose distillate fuel, which may be referred to as Marine Gasoil. Bright and clear in appearance.

DMAHC *or* **D.M.A.H.C.** Defence Mapping Agency Hydrography Center.

D. Maths Doctor of Mathematics.

DMB A distillate fuel, similar to DMA, *q.v.*, but with a small trace of carbon residue from which it can appear black.

DMC A distillate fuel with more residue than DMB, *q.v.*, and limited sediment.

DMC *or* **D.M.C.** Direct Manufacturing Cost; Development Member Countries; Dutch Maritime Code.

DMD *or* **D.M.D.** Dentariae Medicinae *Latin*—Doctor of Dental Medicine *USA*; Doctor of Medical Dentistry.

DME *or* **D.M.E.** Diploma in Mechanical Engineering; Distance Measuring Equipment *aviation.*

D. Med. Doctor of Medicine.

d.mer. parts Difference of Meridian Parts *nautical.*

DMF *or* **D.M.F.** Le Droit Maritime Français. The monthly published leading Maritime Law Reports in France.

DMG Distress Message Generator.

DMG *or* **D.M.G.** *Similar to* **Dist. MG** *or* **Dist. M.G.**, *q.v.*

Dmge Damage.

DMIX Digital Multi-media Information Exchange.

d.m.m.f. Dry Mineral Matter Free. The analysis of coals and cokes.

DMO *or* **D.M.O.** Marine Diesel Oil; District Medical Officer.

DMP *or* **D.M.P.** Difference in Meridional Parts *nautical.*

D.M.P. Diploma in Medical Psychology.

DMPI *or* **D.M.P.I.** Desired Mean Point of Impact.

D.M.R. Diploma in Medical Radiology.

D.M.R.(T) Diploma in Radiotherapy.

DMS *or* **D.M.S.** Data Management System; Diploma in Management Studies; Directorate of Military Studies; Doctor of Medical Science.

DMU Distance Measuring Unit.

DMX A pure distillate fuel used for emergency and auxiliary engines and stored outside main machinery spaces because of its low flashpoint.

DMZ Dimilitarized Zone.

DN *or* **D.N.** Despatch Note, *q.v.*

D.N. *or* **D/N** Debit Note. Dominus Noster *Latin*—Our Lord.

Dn. Down.

DNA Deoxyribonucleic Acid.

D.N.A. Deutscher Normen Ausschuss *German*—West German Standards Institute.

DNB *or* **D.N.B.** Dictionary of National Biography, *UK.*

dne Douane *French*—Customs.

D.N.E.E. Deltion Nafticou Epimelitiriou Ellados. The Bulletin of the Greek Chamber of Shipping.

D.N.I. Dana Normalisasi Indonesia. The Indonesian Standards Institute.

D.N.J. Det Norske Justervesen. The Norwegian Weights and Measures Bureau.

DNL Det Norske Luftarsselkap. Norwegian State Airlines.

D.N.O. Debit Note Only.

D Note United States currency banknote of 500 dollars.

DnV *or* **DNV** *or* **D.N.V.** Det Norske Veritas *Norwegian classification society.*

DNVPS Det Norske Veritas Petroleum Services.

DO *or* **D.O.** Deck Officer; Delivery Order, *q.v.*; Diesel Oil; Distribution Office.

D.O. Doctor of Optometry; Doctor of Osteopathy; Diploma in Ophthalmology.

D/O Delivery Order, *q.v.*; Direct Order.

do Diesel Oil; Ditto *Italian*—The same.

DOA *or* **D.O.A.** Date of Arrival; Date of Availability; Dead on Arrival; Department of the Army *USA*; Dissolved Oxygen Analyser.

d.o.b. Date of Birth.

DOC *or* **D.O.C.** Daily Operating Cost, *q.v.*; Department of Commerce; Department of Communications; Direct Operation Cost; Document of Compliance, a requirement of the ISM, *q.v.*, code for ship operators.

Doc. Document; Doctor.

D.o.c.a. Date of Current Appointment.

DOCD Development Operators Co-ordination Documents *USA*.

D.o.c.e. Date of Current Enlistment.

Doc. Eng. Doctor of Engineering.

Dockage Dock facilities; Charges for the use of a dock.

Dock Boss An American expression referring to the chief tally clerk or checker responsible for the rest of the tally clerks and to the information relating to the loading and discharging as a whole to the Harbour Master.

Dock Charter *or* **Berth Charter** Charterparty, *q.v.*, wherein it is stipulated that the ship has to be at the loading and/or discharging berth before she can be said to be an Arrived Ship, *q.v.*, for the Laydays, *q.v.*, to start counting. If, on the other hand, a ship is Chartered, *q.v.*, on the berth she is hired to load at the sole risk of the charterers insofar as the whole payment of freight is concerned. If no full cargo is available the dead freight due lies as the sole responsibility of the charterers.

Docker Registered transport or port worker.

Docket Act of registering letters, telexes and documents by inserting dates when received and replied to.

Docking Plan Detailed plan of the outer part of a ship specifically needed when she has to be docked or drydocked for underwater maintenance or repairs. This generally gives the profile of the ship and all the underwater sections including details such as machinery spaces, bulkheads, bilge keels, longitudinal bottom lines and other pertinent measurements.

Docking Survey Period of time when the ship is in the dry dock to undergo her periodical classification survey.

Dock Master *or* **Dockmaster** Person who is in charge of operations in a dry dock or floating dock. *Abbrev.* DM.

Dock Pass (1) Written authority which can be requested by the Dock Master, *q.v.*, before the ship sails. (2) A document furnished by the harbour or dock authorities after the Dock Dues are settled.

Dock Rent *See* **Dock Dues**.

Docks (1) An enclosed or sheltered place near or in a harbour or river where ships can berth to load or unload cargoes or embark and disembark passengers. (2) *See* **Dry Docks**. (3) Encircled or enclosed space in a courtroom where an accused stands trial.

Dock Warrant A receipt given in return for goods stored in a dock warehouse. *Abbrev.* D/W.

Dock Weight Note When merchandise is imported the dock authorities issue documents showing the detailed goods off-loaded from the ship. Particulars include the number of packages, type of merchandise, tare and net weights, marks or numbers, names of importers/exporters and dates of entry. *Abbrev.* DWN *or* D.W.N.

Doct. Doctor *Latin*—Doctor.

Doctus cum libro *Latin*—Learned with a book.

Documents Against Payment *or* **Payment Against Documents** Before withdrawal of documents from a bank the consignee has to pay the respective cost and expenses of the merchandise. This type of transaction is advantageous to shipper or consignor as the goods are safeguarded insofar as solvency is concerned. *See* **COD** *or* **C.O.D.** *and* **Letter of Credit**. These documents are needed for the withdrawal of goods from customs sheds or quays. Shipping documents mainly consist of Bills of Lading, *q.v.*, Marine Insurance Policy, Certificate of Origin Forms, Specification of Goods, etc., and are mainly withdrawn or collected from the banks. *Abbrev.* D/P.

Documentary Bill Bill of Exchange, *q.v.*, attached to the shipping documents showing the authorised name and address of Consignees, *q.v.* Alternatively, a Bill of Lading, *q.v.*, attached to a Bill of Exchange, *q.v.*, Insurance Policy, *q.v.*, invoices, cargo specifications and other relative shipping documents.

Documentary Credit An order by the importer of goods to his bankers to remit a sum of money and credit his exporter abroad through other corresponding bankers against the goods to be shippped and so avoid delays. *Similar to* **Letter of Credit**, *q.v.*

Document of Title A legal document concerning the right of ownership of the goods withdrawn or negotiated. There are many such titles. *Ex.* Bill of

Documents of Title

Lading, Dock Warrant, Delivery Order, Letter of Credit, Ships Register, all *q.v.*, etc.

Documents of Title For goods in transit: Bills of Lading and Export Invoices which the consignee must present to the carrier as evidence of title and the right to delivery of the merchandise.

Documents Against Payment Documents of the goods or merchandise will be released to the importer against the signature of the Bill of Exchange, *q.v.* The bill can be presented either by the bank or the agent of the ship, but generally by the former.

DOD Department of Defense *USA.*

Dodecagon Flat or plain figure having 12 equal joined sides and angles.

Dodger Canvas or other material screen to protect against sea spray and wind while a ship is steaming in the open seas.

DODOWTSBE *or* **D.O.D.O.W.T.S.B.E.** Despatch Half Demurrage on All Working Time Saved. *See* **Despatch Money**.

DoE Department of the Environment.

DOE Department of Energy *USA.*

D.O.E. Department of Employment.

DofE Department of Education.

Dog Hooks Cargo lift gear in the form of linked hooks to handle cargo, particularly barrels. *See* **Barrel Hooks**.

Dog Star Nickname for Sirius, or Canis Majoris, the brightest star in the heavens, with a magnitude of 1.4.

Dog Watch *See* **Bell** *and* **Watch Duties**.

D.oh.C. Double or Dual Overhead Cam.

DOI Department of Information; Department of Interior *USA.*

D & O Insurance Directors' and Officers' Liability Insurance.

D.O.L. Dock Owner's Liability; Daily Official List *Stock Exchange*; Doctor of Oriental Learning.

Dolce vita *Latin*—The sweet life.

Doldrums *See* **Variables**.

Dole Unemployment benefit given as a social service by a government.

Dollar Unit of currency used in Australia, Bahamas, Barbados, Belize, Bermuda, Canada, Guyana, Hong Kong, Jamaica, Liberia, New Zealand, Singapore, Taiwan, Trinidad and Tobago, United States and Zimbabwe.

Dollar Stocks Canadian and United States Stocks.

Dolly (1) Low platform on rollers or wheels used to move heavy objects including machinery, aircraft and boats. (2) Rolling platform on which moving picture television and video cameras are moved into position.

Dolly shop A marine store or pawnshop.

Dolly-wagon A wagon used to take away the dirt from a mine.

Dols Dollars, $.

Dolus *Latin*—Fraud; Wilful injury; Deceit.

Dom. Dominica, *West Indies*.

D.O.M. Deo Optimo Maximo *Latin*—To God the best and the greatest; Dominus Omnium Magister *Latin*—God the Master, or Lord of all.

DOMES Deep Ocean Mining Environment Study *USA.*

Domestic Bill of Exchange *Similar to* **Inland Bill of Exchange**, *q.v.*

Domestic Shipping American meaning of waterborne movement of commodities in the domestic oceans, inland waterways and the Great Lakes.

Domik *or* **Umak** A large open boat of Eskimo origin.

Dominium *Latin*—Lordship; Ownership.

Dominus illuminato mea *Latin*—The Lord is my light. Motto of Oxford University.

D.O.M.O. Dispensing Opticians Manufacturing Organisation.

Dom. Proc. Domus Procerum *Latin*—House of Lords.

D.O.M.S. Diploma in Ophthalmic Medicine and Surgery.

DON Department of the Navy *USA.*

Don *or* **don** Donec *Latin*—Until.

Donatio *Latin*—Gift; present.

Donatio mortis causa *Latin*—Gift on account of death. A gift given to a Donee, *q.v.*, or third party after the death of a donor, *q.v.*

Donation Duty Tax levy on donations received. *Similar to* **Gift Tax**, *q.v.*

Donee Person who receives the donation.

DONG Dansk Olie & Naturgas—Danish Oil and Natural Gas.

Dong Unit of currency used in North and South Vietnam.

Donkey Boiler Auxiliary boiler mostly used to supply power to generate electricity, heating, air conditioning and other services when a ship is in port.

Donor Person who gives the donation.

Door to Door *or* **Door/Door** A Container, *q.v.*, shipment term meaning that the Consignor, *q.v.*, assumes responsibility for the cargo until it reaches the Consignee, *q.v.*, from warehouse to warehouse or from door to door inclusive of all expenses except customs taxes and duties. It is considered as one whole conveyance operation from beginning to end. Despite some extra expense being involved for this facility, it has proved convenient and popular since it saves time on clearance procedures needed from ship and customs.

DOP *or* **D.O.P.** Director of Ports; Dropping Outward Pilot, *q.v.*

Doppler effect Apparent change of frequency when source of vibrations is approaching or receding *physics*.

D.Opt. Diploma in Ophthalmics.

Dormant Balance Money to the credit of a bank's customer lying idle for a considerable time.

Dormant Partner *or* **Sleeping Partner** One of the shareholding partners of a company who does not work or engage in the business affairs.

Dors. Dorsetshire, *UK.*

dos à dos *French*—Back to back. *See* **Back To Back Credit**.

DOSV *or* **D.O.S.V.** Deep Ocean Survey Vessel.

DOT *or* **D.O.T.** Department of Trade *USA*; Department of Transport *Canada*; Department of Transportation *USA*; Department of Treasury *USA*; Direct Optical Tuning *electronics*; Dropping Out or Outward Pilot, *q.v.*

D.o.T. Department of Telecommunications.

DOTI Department of Trade and Industry.

DOTIPOS *or* **D.O.T.I.P.O.S.** Deep Ocean Test Instrument Placement and Observation System.

Dott. Dottore *Italian*—Doctor.

Doub. Sleek 15/18 ft. underwater craft which can easily manoeuvre in deep waters and stay submerged for up to about 70 hours.

Double Berth (1) A mooring or berth alongside another ship. On this berth the ship awaits another berth before she starts loading and/or discharging unless the operation is carried out by lighters. (2) Cabin of a ship with two bunks or beds.

Double Bottom *See* **Ballast** *and also* **CDB** *or* **C.D.B.**

Double entendre *French*—Play of words; Double meaning.

Double Insurance Object insured twice at the same time. In the event of double insurance the insured may claim from any or all the underwriters concerned but is not entitled to recover in excess of the defined indemnity.

Double Option Option or chance to negotiate up to a fixed time. Agreement permitting the holder to sell or buy.

Double Pricing Displaying two prices at one time, cancelling the higher price and showing the lower one. Commonly seen when shops display 'Sale' or 'Sales' in the windows and show two prices indicating the marked difference of the normal *vis-à-vis* the sale price.

Double the Cape Said when a ship sails around Cape Horn.

Doubler *or* **Doubling** An extra steel plate attached to the hull of a ship for strengthening purposes.

Double Skin Double layer construction in a ship or boat. The sides of a container ship are built of a double skin. Also the construction of the double bottom of a ship can be said to be double skin.

Double Whip Purchase, *q.v.*, or any mechanical arrangement of tackle which eases and hastens the force applied by a combination of pullies. It consists of two single Blocks, *q.v.*, of which one remains fast while the other moves and this hastens the loading or unloading of a ship's cargo. *See* **Burtoning**.

Douceur Bribe or gratuity. This is not commonly in use today.

Douglas Sea State A scale of sea conditions—e.g. wave heights.

DOW *or* **D.O.W.** Died of wounds.

DOWB *or* **D.O.W.B.** Deep Ocean Work Boat.

Dowel Wood inserted between the edges of two boards or other material to join them together.

Dow Jones Industrial Average The main American share index, equivalent to the British Footsie, *q.v.* The founder was Charles H. Dow, who was the first editor of the Wall Street Journal which published it.

Down by the Head When the Draft or Draught, *q.v.*, of a vessel at the bows exceeds that of the stern.

Down by the Keel When both the Draft or Draught, *q.v.*, fore and aft are even.

Down by the Stern When the Draft or Draught, *q.v.*, aft exceeds that at the bows.

Down on Even Keel When a vessel is loaded to her load lines evenly fore and aft.

Down Payment Money deposited. Closely related to Hire-Purchase, *q.v.*

Downpour Sudden very heavy rain.

Downstream In the direction of the current of a river or stream. Also refers to industrial processing and manufacture after raw materials have been extracted and refined.

Down the Wind With the course of the wind. Opposite meaning of By the Wind, *q.v.*

Downtime The unproductive time in a factory when work is interrupted from such causes as power failure or machinery breakdown.

Down to Her Load Line *Similar to* **Full and Down**, *q.v.*

Down to Her Marks *Similar to* **Down to Even Keel**, *q.v.*, *or* **On Even Keel**, *q.v.*

Downtown The part of a city which is the main business district.

Downtrend A falling off in business or performance.

Downturn Similar to Downtrend, *q.v.*

Downward Phase Period of time when business as a whole is experiencing a gradual fall in prices, employment, etc.

Downwind In the same direction that the wind is blowing; on the leeside or to leeward.

Doz. Dozen.

d.p. Direct Port, *q.v.*

DP *or* **D.P.** Data Processing; Data Processing Department; Days Purposes, *q.v.*; Displaced Person; Directional Propeller *Lloyd's Register notation*; Domus Procerum *Latin*—House of Lords; Diploma in Psychiatry; Doctor of Philosophy; Duty Paid; Dynamic Positioning, *Lloyd's Register notation* for Dynamic Positioning Vessel, *q.v.*

DP *or* **D/P** Drilling Platform.

D/P Degree of Polymerization; Diametric Pitch; Documents Against Payment, *q.v.*

d.p.a. *or* **D.P.A.** Deferred Payment Account.

DPA Defence Mapping Agency.

DPB *or* **D.P.B.** Deposit Pass Book.

DPD *or* **D.P.D.** Dynamically Positioned Drillship. *See* **Dynamic Positioning Vessel**.

Dpf. Deutsch Pfennig, a small German coin one-hundredth of a mark.

D.P.H. Department of Public Health; Diploma in Public Health; Doctor of Public Health.

D. Ph. Doctor of Pharmacy; Doctor of Philosophy.

D. Pharm. Doctor of Pharmacy.

D.P.H.D. Diploma in Public Health Dentistry.

D. Phil. Doctor of Philosophy.

D.P.Hy. Doctor of Public Hygiene.

D.P.O. Distributing Post Office; District Pay Office.

d.p.o.b. Date and place of birth.

D. Pol.Sc. Doctor of Political Science.

DPP Director of Public Prosecutions.

dpr *or* **d.p.r.** Daily pro rata.

DPS *or* **D.P.S.** Descent Propulsion System *aviation*; Data Processing Services; Dual Purpose Ship, *q.v.*

D.Ps.Sc. Doctor of Psychological Science.

D. Psy. Docteur en Psychologie *French*—Doctor of Psychology.

Dpt. Department.

D.pt. Distributed Profit Tax.

d.p.t. Distributed Profit Tax.

Dpty. Deputy.

DPV *or* **D.P.V.** Dynamic Positioning Vessel, *q.v.*

d.p.v. *or* **D.P.V.** Duty Paid Value, *q.v.*

D.P.W. Department of Public Works.

d.q. Direct Question.

dr. Double Reduction Gear; Drachma *Unit of Greek currency*; Drift Angle *nautical*.

d.r. Data Report; Dead Reckoning, *q.v.*

Dr Debtor; Deposit Receipt; Doctor; Door; Drawer.

DR Discharged Run: notation in a naval rating's service certificate to show his service has been terminated by desertion.

DR *or* **D.R.** Dead Reckoning *nautical*, *q.v.*; Dead Reckoning Position; Deposit Receipt, *q.v.*

D/R *or* **DR** Dead Rise; Deposit Receipt; Double Reduction.

DRA Defence Research Agency of the UK Ministry of Defence. *Farnborough, UK.*

DRA *or* **D.R.A.** Dead Reckoning Analyser *nautical*.

Drachma Unit of currency used in Greece.

Dracones Word derived from the Greek language meaning serpent. In shipping this means a flexible container designed to carry liquids other than water through inland waters and in open seas. Usually dracones are used for the transportation of oils. Mostly

constructed of strong woven nylon acrylonitrile rubber and coated with neoprene on the outside.

Draft *or* **Draught** These words, with their alternative spelling, have many meanings, some of which are: (1) An order to pay a sum of money or Bill of Exchange, *q.v.* (2) Detachment of men from larger body for special duty. (3) A sketch or rough copy of a document. (4) Depth or height of the submerged part of a ship when she has no cargo. This is termed Light Draft (Draught). The Load or Loaded Draft, *q.v.*, is the depth up to which the ship is submerged when loaded. (5) Loss in weight due to evaporation or shrinkage. *Abbrev.* Dft *or* Drft. *See* **Bulkhead Draught**, **Mean Draught and Moulded Draught**.

Draftage In the Australian Charterparty known as the 'Australian Grain Charterparty' the freight is paid per long ton of 2240 lbs or 1016 kilos net minus a draftage of 21 lbs per 2000 lbs wheat.

Draft check A check on a ship's draft—usually carried out by a chief officer—during the loading of a vessel to ensure the cargo is stowed correctly and the ship is not overloaded.

Draft-horse *or* **Draught-horse** A horse used to draw heavy wagons and in logging.

Draft Scale *See* **Immersion Scale**.

Draft Survey A survey of a vessel's draft. Used, for example, to establish cargo quantities for bill of lading purposes where no shore scales are available.

Drag (1) When a vessel drifts from her original anchorage. (2) The air or sea resistance against the sides of the ship while in motion. In terms of air resistance this also applies to aircraft and vehicles such as cars, trucks, etc.

Drag Anchor When an anchor fails to hold.

Drag Chains Chains and Weights secured to the sides of a newly-built ship to brake the speed of her slide down the slipway at launching. Very necessary where there is a restricted area of water.

Drag Marks *or* **Drag Markings** The numerical figures painted white on both sides of the bow and stern of a ship starting from the bottom up to the maximum draft (draught) allowed.

Drag, to To draw a rope, chain, net Grapnel, *q.v.*, along the seabed to locate or recover something.

Dramatis personae *Latin*—Characters of the play.

Draught Marks Graduations in feet or metres marked on either side of the ship at the bow and stern which serve to show at the waterline the depth of immersion from the keel.

Draw The position or state of the sails when they are filled with wind.

Drawback Refund of money paid by customs for duty already paid on merchandise which was meant for export. *Abbrev.* dbk. *See* **Bonded Stores** *or* **Bonded Goods**.

Drawee Person or firm named on a Bill of Exchange, *q.v.*, or receiving a Cheque, *q.v.*, from another person or firm. Opposite to Drawer, *q.v.*

Drawer Person or firm drawing a Bill of Exchange, *q.v.*, or issuing a cheque for payment to another person or firm. Opposite to Drawee, *q.v.*

Draw-works A particular kind of operation carried out in the oil drilling industry whereby an electrically-driven rotary drilling rig is used and which is controlled by the Driller, *q.v.*

Dr Bus. Admin. Doctor of Business Administration.

DRC *or* **D.R.C.** Daily Running Cost. *See* **Daily Operating Cost**. Damaged Received in Collision *insurance phrase.*

DRC obs *or* **D.R.C. obs** Damage Received in Collision including with Objects.

DRCS Distress Radio Call System.

DRE *or* **D.R.E.** Dead Reckoning Equipment *nautical.*

Dreadage *or* **Dreading** In grain Charterparties, *q.v.*, the Charterers, *q.v.*, have the option to load general cargo, so long as the freight charged is similar to that of the grain and all the extra expense due to the nature of the cargo is borne by the Charterers, *q.v.*, or shippers.

Dredge *or* **Dredging** When the depth of water in a port or harbour is shallow due to sand, silt or wreckage, etc., the authorities clear or dredge the area by engaging a special vessel to remove the obstruction. In some ports or canals dredging is done regularly.

Dredger A vessel employed in removing sand, shingle, mud, rock, etc., for the deepening of a harbour or channel. It may be specified as a Bucket Dredger, Suction Dredger, Grab Dredger or Backhoe Dredger. Dredgers discharge spoil into Hopper Barges or pump it ashore, while there are Hopper Dredgers which transport spoil as well as dredging it.

Dredging an Anchor Dragging an anchor on the bottom to slow the movement of a ship, particularly the bow, as an aid to manoeuvring.

D.R.F.Ahd. Double Ring Full Ahead *nautical in manoeuvring.*

Drft *or* **Dft** Draught *or* Draft, *q.v.*

DRG Defence Research Group *NATO.*

DRI Direct Reduced Iron, *q.v.*

Drift (1) To move under the influence of the current, or wind, or both; the distance so moved or deviation from course. (2) A surface current. (3) Lateral movement of a projectile when fired from a gun. (4) Mass of floating debris washed out to sea by floods and appearing like a small island. (5) Tool for fairing rivet holes. (6) To punch out a locking pin.

Drifter A fishing vessel which lies to her line of buoyant nets.

Drift Ice Loose floating ice drifting from its original point. Also termed Sailing Ice.

Drill (1) To bore a hole in any material. Has particular reference to boring into the ground or seabed for the extraction of oil and gas. (2) Tool or device for boring holes. (3) Practice routinely conducted on board ship to train and exercise the crew and passengers so that they will be prepared for an emergency.

Driller Person employed on a drilling rig responsible for the drilling gear, crew and operations. He is virtually in charge of the whole operation.

Drill Ship Vessel purposely constructed to drill offshore areas for oil and/or gas exploration. It can operate on various sites in a field area by anchoring at that point, in shallow or deep water. Whenever the water does not exceed a depth of 500/700 ft. a drill ship can be replaced by a semi-submersible rig or a jack-up rig. These rigs remain stationary during rough seas since they remain fastened to the seabed.

Drilling Crew The complement of a drill ship or oil rig.

Drive on/Drive off *Similar to* Roll on/Roll off, *q.v.*

Drizzle Rain that precipitates in fine droplets.

DRL Dead Reckoning Latitude *nautical.*

DRM *or* **D.R.M.** Direction of Relative Movement *nautical.*

Drm. Drum.

DRMP Dredged Material Research Program *USA.*

Drn. Drain.

DRO *or* **D.R.O.** Deferred Repairs Outstanding.

Drogue Sea Anchor, *q.v.*

Droit *French*—Right; Legal Right.

Droit de Quai. *See* **Quay Dues**.

Drome Airdrome or Aerodrome.

Drone Pilotless aircraft normally used for practice and target purposes.

Droogdokk *Norwegian*—Dry dock.

Drop Anchor *See* **Cast Anchor**.

Dropping Outward Pilot A Time Charterparty, *q.v.*, clause. The place generally outside harbour limits where a Pilot is released. It is sometimes termed Dropping Out Pilot. *Abbrev.* DOP. This expression is also connected with the phrase 'When and Where Ready'.

Drop Shipment When, because of the substantial volume of goods ordered, the manufacturer or Consignor, *q.v.*, agrees to transport it directly to the warehouse of the Consignee, *q.v.*

Dr P.H. Doctor of Public Health.

Drs Debtors; Doctors.

Drss *or* **DRSS** Drill Rig Safety System.

DRT *or* **D.R.T.** Dead Reckoning Trace *nautical.*

dr. t. Troy dram.

Drug on the Market Unsold merchandise which has been the cause of abundance or surplus.

Drx Drachma, Greek unit of currency.

Dry Barrel A cask or barrel designed to contain dry commodities and unsuitable for liquid products.

Dry Bulk Cargo shipped in a dry state and in bulk, for example, grains, cereals or cement.

Dry Cargo Merchandise other than liquid carried in bulk.

Dry Cargo Ship *or* **Dry Cargo Vessel** Ship which carries all merchandise, excluding liquid, in bulk.

Dry Compass Compass without liquid. *See* **Compass**.

Dry Dock Large basin where all the fresh/sea water is pumped out to allow a ship to dock in order to carry out underwater cleaning and repairing. *See* **Dry Docks**.

Dry Docks These consist of three types:— (1) Graving Docks. Basins with sliding gates at the entrances which allow water to pour in so that ships can enter the dock when it is full. The gates are then closed and the water inside is pumped out until the closed basin or dock is empty and the ship(s) settle on to cradles and are fastened. The vessel is then ready for underwater cleaning, painting and/or other necessary repairs. (2) Slip Docks. Docks built on a gradient enabling a ship in a cradle to be hoisted from the sea on to a particular platform in the dock where she can be secured and serviced. (3) Floating Docks. These are docks which are immersed so that ships can enter through the vacant channel areas. When all the stagings are fixed the docks are refloated to their original levels and the vessel is then ready to be repaired, cleaned or painted in the dry.

Dry Film Thickness Generally referred to the film thickness of dried paints when applied on surfaces. Thickness of paints is calculated in microns or mils. *Abbrev.* DFT *or* D.F.T.

Dry Freight Freight on General or other cargo of a nature that does not call for special attention on voyage.

Dry Hole Non-productive oil or gas wells. Also expressed as Duster, *q.v.*

Drying Oils Natural oils and fats consisting of:— cotton seed, china wood, hempseed, linseed, atticea, mustard seed, peanut, pine, rubber seed, soya bean, tung and walnut.

Drysalter Dealer in salted and dried commodities such as meats, fish etc.

Dry Steam *See* **Steam**.

Dry Stores *or* **Dry Goods** *or* **Dry Provisions** General provisions or victuals stowed in a ship's stores and needing no refrigeration to remain in good condition until consumed. Dry stores include all canned food, and consumables in bags such as sugar, beans, peas, etc.

Dry Tank A tank under the boilers forming part of the double bottom tanks, which is always kept dry to protect it from corrosion.

Dry Tank Certificate An official certificate issued by the local port authorities to an oil or other liquid petroleum tanker both at the loading and discharging ports. The certificate is issued after completion of loading and discharging as to comply with the port safety regulations.

DS *or* **D.S.** Dal Segno *Italian*—Repeat from the Sign *music*; Danske Standardiseringsraad, Danish Standards Institute; Debenture Stocks; Dental Surgeon; Department of State; Deputy Secretary; Director of Science; Director of Surgery; Dock Survey; Doctor Scientiae *Latin*—Doctor of Science.

DS *or* **D/S** Days After Sight. *See* **Day's Sight**.

D&S Demand and Safety.

D/S Days After, Sight. *See* **Day's Sight**.

d.s. Date of Survey; Daylight Saving; Days After Sight, *q.v.*; Day's Sight, *q.v.*; Document Signed.

DSA *or* **D.S.A.** Danish Ship Suppliers Association.

D.S.A.O. Diplomatic Service Administration Office *UK*.

DSARC Defence System Aquisition Review Committee *USA*.

d.s.b. *or* **D.S.B.** Discharged to Sick Bay. *Similar to* **d.s.q.** *or* **D.S.Q.**, *q.v.* Double Side Bank *radio frequency*.

DSC Danish Shipping Council; Digital Selective Calling *communications*; Distinguished Service Cross; Doctor of Surgical Chiropody.

DSC Dynamically Supported Craft.

D. Sc. Doctor of Science.

Dsf. Dusseldorf, *Germany*.

D.S.I. Dairy Society International.

D.S.I.R. Department of Scientific and Industrial Research *UK*.

dsl. elec. Diesel Electric.

DSM Distinguished Service Medal.

DSN Deep Space Network.

DSO Distinguished Service Order; District Staff Officer.

d.s.p. Decessit sine prole *Latin*—Died without issue.

d.s.q. *or* **D.S.Q.** Discharged to Sick Quarters. *Similar to* **d.s.b.** *or* **D.S.B.**, *q.v.*

DSRD *or* **D.S.R.D.** Department of Scientific Research and Development.

dr. rig. Drill Rig or Drilling Rig. *See* **Drill Ship**.

DSRV Deep Submergence Rescue Vehicle, a contrivance which can be carried on the casing of a submarine.

DSSV Deep Submergence Search Vehicle.

DST Drill Steam Test *oil drilling*; Double Stack Train.

DST *or* **D.S.T.** Double Summer Time.

DSV Diving Support Vessel.

d.t. Double Time; Deep Tank. *See* **Deep Tanks**.

Dt. Deep Tank. *See* **Deep Tanks**.

DT *or* **D.T.** Deep Tank; Dental Technician; Double Time; Dust Tight.

D.T. Doctor of Theology.

D.T.a. or **DTA** or **D.T.A.** Deep Tank Aft.

d.t.b.a. or **D.T.B.A.** Date to be advised *insurance*.

dtd. Dated.

DTE Date Terminal Equipment *computer*.

D. Tech. Doctor of Technology.

D.T.f. or **DTF** or **D.T.F.** Deep Tank Forward.

DTG Date Time Group.

D.T.H. Diploma in Tropical Hygiene.

D.Th. Doctor of Theology.

DTI or **D.T.I.** Department of Trade and Industry; Direct Trader Input.

D.T.M. Diploma in Tropical Medicine.

DTMA Deep Tank Midship/Aft.

DTM or **D.T.M.** Deep Tank Midship.

D.T.m.f. or **DTMF** Deep Tank Midship/Forward.

DTN Date Terminal Network *computer*.

DTO District Training Office/Officer.

DTp Department of Transport *UK*.

DTS District Traffic Superintendent.

DTV Drillstring Torsional Vibration *oil drilling*.

Dual–Purpose Ship Specially constructed ship able to carry different types of cargoes such as ore and/or oil. Sometimes called Ore/Oil ships. *Abbrev.* DPS. *See* **Ore/Bulk/Oil** or **Oil/Bulk Oil Carrier**.

Dual Rate Contract A form of Loyalty Contract between a shipper and a courier or conference line.

Dual Valuation or **Dual Valuation Clause** Hull marine insurance clause. Refers to the mutually arrived at premiums which are to be paid for insurance in the case of either accidental damage with Total Loss or for Total Loss Only, *q.v.* Normally the insured insures the object after taking into account depreciation while the Underwriters, *q.v.*, have a foreseeable interest in the daily increment of labour and repair charges in case of a future claim. In this case an agreement between them is reached under the subtitle of Dual Valuation Clause where premiums are shown for dual coverage, i.e. a premium for Total Loss Only or for both Total Loss and eventual repair. The former premium will be cheaper than the latter. Another explanation is for two valuations on a ship in the hull policies to be considered as Actual Total Loss, *q.v.* It is provided that the insured has two options in this coverage. He may choose to repair up to the extent of the damage to the Total Loss, *q.v.*, or Constructive Total Loss, *q.v.* Alternatively, he may abandon the ship to the underwriters and claim for the repair cost. If, however, the insured decides not to repair the ship then underwriters will automatically be liable only to the limitation of the lower repair cost, namely for the total loss valuation. *Abbrev.* D/V.

Dub. Doubtful; Dubious.

DUC Danske Undergrunds Consortium, Danish Underground Consortium.

Duct Tube or channel from which fluids pass.

Dud Cheque An overdrawn cheque or a false cheque.

Due Date The date when Bills of Exchange, *q.v.*, are due for payment. Three days are added to the time for which the bills of exchange have been drawn. Saturdays and Sundays are excluded. *Abbrev.* d/d or D/d.

Due Diligence The owners or carriers are legally bound to make the ship seaworthy. Usually termed as follows: 'The owners are to use due diligence to make the ship seaworthy throughout'.

Dumb Barge Barge without engines, sails or other propulsion.

Dumb Compass or **Dummy Compass** *Similar to* **Pelorus**, *q.v.*

Dum bene se gesserit *Latin*—In the course of good manners or behaviour.

Dumping Merchandise marketed in a foreign country on a cut-throat basis. Locally produced goods can be so much in abundance that the traders prefer to sell abroad at attractive prices and in some instances even at a slight loss rather than being compelled to dispose of the goods at a much greater loss in their own country. Such business has given vent to logical complaints by the traders' counterparts in the importing country and in turn customs authorities can provide protection by imposing relatively high import tariffs to combat the 'dumping'.

Dum spiro spero *Latin*—While I breath I hope. Motto of South Carolina, *USA*.

Dum vita est, spes est *Latin*—While there is life, there is hope.

Dun Repeated calls for payment; Urgent demand for payment.

Dung Surplus fodder and litter as well as fittings and utensils collected from a cattleship after the cattle have disembarked.

Dungaree Heavy blue cotton cloth mostly used for working.

Dunnage Various pieces of wood, logs or planks used for wedging between Broken Stowage, *q.v.*, in the holds of a ship to prevent the cargo shifting and to steady the whole group of packages. Dunnage is additionally useful as the spaces in between enable air to circulate around the cargo. *See* **Broken Stowage**.

Duopoly Two producers of goods who begin a cut-throat competition with a view to overpowering each other in business. Sometimes they find it more sensible to come to some amicable agreement and so they may merge and arrange to have stable prices on a profit-sharing basis.

Duopsony A market situation monopolised by two producers or buyers. *See* **Duopoly**.

Duplicate Original A copy direct from the original document.

Dur. Durham, *UK*.

dur. Duration.

Duralumin An alloy of aluminum, copper, silicon and manganese used in aircraft manufacture due to its strength and lightness.

Durable Goods Items which when possessed and made use of remain usable for a long period of time. These comprise furniture, cookers, dishwashers, washing machines, refrigerators, etc. The price of these items fluctuates with resultant activity in the trade cycle. *See* **Depression**; **Cyclical Fluctuation**.

durante vita *Latin*—During life.

Duration Clause Refers to the Time Policy, *q.v.*, covered for 12 months, and both the first and the last days mentioned in the policy are covered in the insurance.

Dust Allowance *Similar to* **Dirty Money**, *q.v.*

Dust Counter An instrument enumerating the quantity of dust in a volume of air.

Duster An oil drilling term for a non-productive oil well. Also expressed as Dry Hole. *See* **Dry Hole**.

Dusty Money Extra payment for port workers in the grain trades due to the presence of dust during loading/discharging.

Dut. Dutch.

Dutch Auction An unusual auction where all bids start from the highest and continue to the lowest. The price is reduced by the auctioneer until a purchaser is found.

Dutch Roll An aviation term. The aircraft's tendency of oscillating or yawing, *q.v.*, while in the process of rolling.

Dutiable Merchandise *or* **Dutiable Stores** Imported merchandise or stores bound by the law of the country to be subject to duty before customs clearance and used for home consumption.

Duty Free Shop A shop selling a variety of commodities such as cigarettes and spirits and other expensive items at a price which excludes import duty. The goods are offered to bona fide air and sea passengers either at the airport or on board the aircraft or ship. At the departure lounge passengers have to produce their passports and/or embarkation cards. While on board no identification is needed once the passengers are automatically part of the complement of the aircraft or ship.

Duty Paid Contract A contract whereby the seller of the goods undertakes to pay the import duties, customs and warehousing expenses apart from transport and labour costs. Not a very common contract but some government tenders are awarded as such.

Duty Paid Value The overall cost of the merchandise, after calculating the entire expenses, including the duty imposed. *Abbrev.* d.p.v. *or* D.P.V.

DUVM *or* **D.U.V.M.** Doctor of Veterinary Medicine *USA*.

D.V. Deo Volente *Latin*—God Willing.

D/V Dual Valuation, *q.v.*

DVLC Driver and Vehicle Licensing Centre *Swansea UK*.

D.V.M. Doctor of Veterinary Medicine *USA*.

D.V.M.S. Doctor of Veterinary Medicine and Surgery.

DVOR *or* **D.V.O.R.** Doppler Very High Frequency Ommi-Range Beacon *aviation*.

d.v.p. Decessit vita patris *Latin*—Died during his or her father's life.

DVS Doctor of Veterinary Science (*or* Surgery).

d.w. Delivered Weight.

D/W Dock Warrant, *q.v.*; Drainwater; Deadweight, *q.v.*

d.w. *or* **D/W** *or* **DWT** *or* **D.W.T.** (1) Deadweight, i.e. The weight capacity of a ship in tons. *See* **Deadweight Tonnage**. (2) Dock Warrants, *q.v.*

DWA *or* **D.W.A.** Dock Water Allowance.

Dwarf Planets Mercury, Earth, Venus and Mars.

DWAT *or* **D.W.A.T.** Deadweight All Told.

DWB Dual Walking Beam *electronics*.

dwc *or* **d.w.c.** *or* **DWC** *or* **DWCC** *or* **D.W.C.C.** Deadweight Cargo Capacity.

dwct *or* **d.w.c.t.** *or* **DWCT** *or* **D.W.C.T.** Deadweight Cargo Tons or Tonnage.

DWL *or* **D.W.L.** Displacement Water Line *in aircushion (hovercraft) vessels*.

DWN *or* **D.W.N.** Dock Weight Note, *q.v.*

DWP Design Waterplane or Load Waterplane; Dual Wall Packer *oil drilling*.

D.W.S. Department of Water Supply.

dwt. Deadweight Ton; Pennyweight, Troy.

d.w.z. Dat wil zeggen *Dutch*—That is to say.

DX *or* **D.X.** Distance; Distant.

Dy Delivery.

Dyd or D.Y. Dockyard.

dyn. Dynamics; Dyne, *q.v.*

Dynamic Peg Rate of Exchange.

Dynamic Positioning Vessel Specialised vessel employed in conjunction with oil exploration at sea. The special instruments on board are capable of tracing underwater geology. One of her main jobs is to find the blocked holes where previous oil drilling took place. *Abbrev.* DPV. *Lloyd's Register notation. Abbrev.* DP.

Dyne A unit of force producing an acceleration of 1cm per sec² when acting on a mass of 1 gramme. *Abbrev.* Dyn.

D.Z. Doctor of Zoology.

DZ. Der Zeit *German*—Of this time.

Dz. Dozen.

E

E Earl; East, *q.v.*, Eire; English; Refitted Engine *Lloyd's Register notation*: Sleet *weather observation*. E Base of Naperion Logarithms *nautical*; Electricity *nautical*; Electromotive *force of cell*; Engine/Engineering, Error(s).

E3 European-Economical-Ecological. A 1992 design for a double hulled VLCC, *q.v.*, created by a group of European shipyards.

EA Each.

E & A Exploration and Appraisal *oil drilling*.

EAA *or* **E.A.A.** Engineers' and Architects' Association *USA*; Ethylene Acrylic Acid; Experimental Aircraft Association.

EAAFRO East African Agriculture and Forestry Research Organisation.

EAC East African Community; European Agency for Co-operation.

EAD Entry Acceptance Data.

EADP European Association of Directory Publishers.

EAEC European Atomic Energy Commission.

EAF Emergency Action File.

EAG Economists Advisory Group.

EAN European Article Number.

e.a.o.n. *or* **EAON** Except as otherwise noted.

EAR *or* **E.A.R.** Expanded Area Ratio *technical*.

EARB European Airlines Research Bureau.

Early Bird Discount Discount on the price of an air passage booked well in advance of departure, e.g. three months before departure.

Early Closing Usually refers to the practice in some places where shops are required to close in the afternoon on one day a week.

Early Production and Testing System Procedure followed by oil drilling enterprises before they venture to start developing an oilfield. Sufficient information is gathered to determine the potential commercial return on exploiting the expected oil yield. *Abbrev.* EPTS.

Earmarked (1) Money reserved for a particular use. (2) Reserved item or cargo meant for a particular purpose or destination.

Earned Premium The premium which is totally earned by the underwriters while the policy has, as yet, not been terminated. *Ex.* An insured ship suffers a Constructive Total Loss, *q.v.*, and settlement is effected by the underwriters before the termination of the existing insurance policy.

Earnest *or* **Earnest Money** Money deposit or deposit of any other valuable object to serve as a bond or security between the buyer and the seller. It is often called Earnest. Also called Token Money. *See* **Deposit**.

Earnings Amount of money one obtains for a week or a period of time in return for work carried out. May refer to freight earnings in which case it is the collection of the freight by the carrier or agents of the ship.

Earnings Yield The ratio of annual profits net of tax and preference dividend to the market value of the ordinary shares in issue of a limited company. In other words the annual earnings of a company expressed as a percentage of the current share price, earnings in this context being the annual profit net of prior charges and tax.

EAS *or* **E.A.S.** Equivalent Air Speed *aviation*; Estimated Air Speed *aviation*.

Easement A legal right over the land of another. Right of way.

Ease Off To slacken a line or a mooring rope.

EASI European Association for Shipping Informatics.

East Cardinal point of the compass, directly opposite to West.

Eastern Hemisphere *or* **Old World** *See* **New World**.

Eastern and Northern Canada Refers to certain Canadian ports consisting of Halifax (Nova Scotia); St. John (New Brunswick); St. Lawrence River ports Baie Comeau, Montreal, Quebec, Sorel, Three Rivers and Churchill (Manitoba). *Abbrev.* E & NC.

Easting The distance gained in sailing eastwards from a given meridian, opposite to Westing.

EASTROPAC Eastern Tropical Pacific.

Eastward Movement of a vessel or aircraft proceeding to a destination further east.

Easy Area Refers to ice in the sense that navigation is possible in an area where there is prevailing ice.

Easy Money Policy Policy taken by a bank in lowering the rate of interest. Especially practised during a recession. An American term similar to Cheap Money, *q.v.*

Easy Sail Sailing with sails filled but not sheeted hard. Progress is slower but the motion generally more comfortable.

Easy Tank Cleaning Det Norske Veritas notation for vessels fitted with tanks of stainless or coated steel having smooth surfaces. *Abbrev.* ETC.

Easy Trimmer A term describing the trimming facilities of a bulk carrier where the owner is not saying that the vessel is capable of thoroughly self-trimming a bulk cargo, but can easy trim thereby reducing the need for manual trimming.

e.a.t. *or* **EAT** *or* **E.A.T.** Earliest Arrival Time; Expected Approach Time *aviation.*

EATCHIP European Air Traffic Control Harmonisation and Integration Programme.

EAX Electric Automatic Exchange.

EB *or* **E.B.** Elektrisk Bureau (Norway); Encyclopaedia Britannica; Existing Business *insurance.*

Ebb The fall or the reflex of the tide.

Ebb Current. *See* **Ebb.**

Ebb Side *See* **Ebb.**

Ebb Stream *See* **Ebb.**

Ebb Tide The flowing back or receding of tidal water.

EBC *or* **E.B.C.** European Brewery Convention.

EBCS *or* **E.B.C.S.** European Barge Carrier System. *See* **Bacat, LSD, Lash, Valmet, Trimariner** and **B.V.C.**

EBI European Investment Bank established in 1958 by the European Economic Community.

EBI *or* **E.B.I.** Electronic Bearing Indicator *radar.*

EBIC European Banks International Corporation.

EBL *or* **E.B.L.** Electronic Bearing Line *radar.*

EBM *or* **E.B.M.** Electronic Bearing Marker *radar.*

E.B.N. *or* **E.BY.N.** East by North.

EBR Excess Baggage Receipt. An air excess baggage ticket receipt.

EBS *or* **E.BY.S.** East by South.

E.B.U. European Banking Union.

ec Earth Closet; Enamel Coated; Enamel Covered; Error Correction *nautical*; Extension Course *nautical.*

Ec *or* **Ecua** Ecuador.

EC *or* **E.C.** East Coast; Eastern Command; Educational Committee; Electric Council; Electronic Computer; Emergency Commission; Engineer Captain; Episcopal Church; Established Church; Eurocheque; European Community, in Brussels; Executive Committee.

ECA Electronic closing and accounting *insurance.*

ECA *or* **E.C.A.** Early Closing Association; East Coast of Africa; Economic Commission for Africa *United Nations;* Economic Co-operation Administration; European Congress of Accountants; Emergency Controlling Authority *aviation.*

ECAC *or* **E.C.A.C.** European Civil Aviation Conference.

ECAFE Economic Commissions for Asia and Far East *United Nations.*

ECALO Energy Conserving Automatic Light Output.

ECAREG Eastern Canada Traffic System *Shipping Control System.*

ECC *or* **E.C.C.** European Cultural Centre; Engineroom Control Centre; Exchange Control Copy.

Ecc Eccellenza *Italian*—Excellency.

ECCM *or* **E.C.C.M.** East Caribbean Common Market.

E.C.C.P. East Coast Coal Port.

ECD *or* **E.C.D.** Early Closing Day; Estimated Completion Date.

ECDC Economic co-operation in Developing Countries.

ECDIS Electronic Chart Display and Information System.

ECE Early Childhood Education; Engine Condition Evaluator.

ECE *or* **E.C.E.** Economic Commission for Europe *USA in Geneva.*

E & CE Engineering and Construction Engineering.

ECEPA Environmental Challenges for European Port Authorities.

ECF *or* **E.C.F.** Export Cargo Form *in containerisation.*

ECFMGE Educational Commission for Foreign Medical Graduate Education *USA.*

ECFTUC European Confederation of Free Trade Unions in the Community.

ECFVG Educational Commission for Foreign Veterinary Graduates *USA.*

ECG *or* **E.C.G.** Electrocardiograph; European Co-operation Grouping; Export Credits Guarantee.

E.C.G.B. East Coast of Great Britain.

E.C.G.C. Empire Cotton Growing Corporation.

E.C.G.D. Export Credits Guarantee Department.

Ech Echelon.

Echo International Radio Telephone phonetic alphabet for 'E'.

ECHR European Court of Human Rights.

E.C.I. East Coast of Ireland; Export Consignment Identification Number.

ECICS *or* **E.C.I.C.S.** Export Credit Insurance Corporation of Singapore.

ECIR Euro-Currency Interest Rates.

E.C.I.T.O. European Central Inland Transport Organisation.

ECJ *or* **E.C.J.** European Court of Justice *Strasbourg.*

ECLA *or* **E.C.L.A.** Economic Commission for Latin America *United Nations.*

Eclipse The eclipse of the sun happens when the moon passes between the earth and the sun. The sun therefore becomes completely or partly hidden. The eclipse of the moon is when the earth passes between the moon and the sun resulting in obscuring of the sun from the moon.

ECM *or* **E.C.M.** Electric Coding Machine; Electro-Chemical Machining; Electronic Counter Measure *aviation*; European Common Market.

ECMA European Computer Manufacturers' Association.

ECME *or* **E.C.M.E.** Economic Commission for the Middle East *United Nations.*

E.C.M.T. European Conference of Ministers of Transport.

E.C.N.A. East Coast of North America.

ECNR *or* **E.C.N.R.** European Council for Nuclear Research.

E-COM Electronic Computer Originated Mail.

e.con. E contrario *Latin*—On the contrary.

Economics The science or study of the sources of supply or processes of production alongside the respective expenditures in relation to the needs of men. The governing of the distribution of wealth.

Economic Speed Average calculated speed of a ship, an aircraft or vehicle taking into consideration the most economical way of using fuel, lubricating oils, etc., in relation to hourly or daily operating costs. Sometimes also expressed as Commercial Speed or Service Speed.

ECOR *or* **E.C.O.R.** Engineering Committee on Oceanic Resources *USA.*

ECOSOC Economic and Social Committee of United Nations.

ECOWAS Economic Community of West African States.

ECP External Casing Packer.

ECPD Engineer's Council for Professional Development *USA*; Export Cargo Packing Declaration.

ECPS Environment and Consumer Protection Service.

ECR *or* **E.C.R.** Estimated Cost of Repairs; Exchange Cross Rate.

ECS Electronic Chart System; European Communication Satellite; Environmental Control System.

E.C.S.A. East Coast of South America.

ECSC European Coal and Steel Community.

ECSI *or* **E.C.S.I.** Export Cargo Shipping Instructions.

ECT *or* **E.c.T.** Electroconvulsive Therapy; Equivalent Chill Temperature, *q.v.*

ECTUA European Council of Telecommunications Users Associations.

ECU European Currency Unit.

Ecua *or* **Ec.** Ecuador.

E.C.U.K. East Coast of the United Kingdom.

ECUSA East Coast of the United States of America.

ECV *or* **E.C.V.** Each Cargo Voyage *insurance.*

E.C.W.A. Economic Commission for Western Asia.

Ed. B. Bachelor of Education.

ed Editor; Edited; Edition.

e.d. Error Detecting *nautical;* Expected Departure; Extra Duty.

ED *or* **E.D.** Education Department; Election District; Electronic Device; Employment Department; Entertainment Duty; Estate Duty; Ex. Dividend, *q.v.*; Existence Doubtful *chart abbreviation*; Extra Dividend; Emergency Distance, *q.v.*

e.d.a. *or* **EDA** Early Departure Authorised.

e.d.c. *or* **E.D.C.** Error Detection and Correction *nautical*; Extra Dark Colour; Export Development Corporation *USA.*

EDC *or* **E.D.C.** Economic Development Committee; European Defence Committee; Expected Date of Confinement; Express Dairy Company.

Ed. D. Doctor of Education.

EDD *or* **E.D.D.** Expected Date of Delivery; English Dialect Dictionary

Eddy Air or water current moving against the main stream; currrent moving in a curved or circular fashion.

EDF Electricité de France. *French*—Electricity of France.

EDF *or* **E.D.F.** European Development Fund.

Edged Securities Public securities and stocks backed by the government of a state.

Edge Water An oil reservoir having an outer rim of water which causes oil to flow due to the existent water pressure. An oil term. *See* **Flowing Well**.

EDH *or* **E.D.H.** Efficient Deck Hand. Second in rank to Able Seaman, *q.v.*

EDI Electronic Data Interchange.

EDIA Electronic Data Interchange Association.

EDIFACT Electronic Data Interchange for Administration, Commerce and Transport.

EDIG European Defence Industry Group.

EDIMAR Electronic Data Interchange for Maritime Traffic.

Ed. in. Ch. Editor in Chief.

EDIS Electronic Data Interchange Systems.

EDIT *or* **E.D.I.T.** Estate Duties Investment Trust.

editio princeps *Latin*—The first printed edition *of a book.*

Ed.M. Master of Education.

Edn. Edition.

EDO Engineering Duty Officer *USA.*

e.d.o.c. *or* **EDOC** Effective Date of Change.

EDP *or* **E.D.P.** Electronic Data Processing.

EDR European Data and Research.

EDS Electrodynamic Suspension.

e.d.t. *or* **E.D.T.** Eastern Daylight Time.

EDTC European Diving Technology Committee.

EE *or* **E.E.** Electrical Engineer; Electrical Engineering; Electronic Engineer; Electronic Engineering; Employment Exchange; Envoy Extraordinary; Estado Español *Spanish*—The Spanish State; Early English; A UK inward customs form for Home Trade vessels from 200 tons gross upwards.

e.e. *or* **E.E.** Errors Excepted; Ear and Eye.

E.E. *or* **E.O.E.** Errors Excepted or Errors and Omissions Excepted. *See* **Errors Excepted.**

e.&e.a. *or* **E.&E.A.** Each and Every Accident *insurance.*

EEA Electronic Engineering Association *UK*; European Economic Area.

EEAIE Electrical, Electronic and Allied Industries, Europe.

EEB Exports to Europe Branch.

EEBS Emergency Excape Breathing Support System.

EEC *or* **E.E.C.** English Electric Company; European Economic Community (Common Market); Electric Engine Control.

EEDC Economic Employment and Development Commission.

EEF Employers' Educational Foundation *USA.*

EEG Electroencephalograph.

EEIG European Economic Interest Group, established in Paris 27 April 1993.

EEIG Unitas A grouping of European Classification Societies, *q.v.*, based in Brussels.

EEL Emergency Exposure Limit; Energy Exposure Level.

e.&e.l. *or* **E.&E.L.** Each and Every Loss *insurance.*

EEMA European Electronic Mail Association.

E.E. & M.P. Envoy Extraordinary and Minister Plenipotentiary.

e.&e.o. *or* **E.&E.O.** Each and Every Occurrence *insurance.*

EER Evacuation, Escape and Rescue.

EERI Earthquake Engineering Research Institute *USA.*

EES *or* **E.E.S.** European Economic Space; European Exchange System.

E.E.T. Eastern European Time.

EETO Eligible Exchange Traded Options *payment contract. See* **Futures Contract**.

e (ex.g.) Exhaust Gas Economiser *Lloyd's Register notation.*

EEZ *or* **E.E.Z.** Exclusive Economic Zone, *q.v.*

EF *or* **E.F.** European Foundation.

eff. Effetto *Italian*—Bill Note or Promissory Note; Efficiency; Effigy.

Effecting an Insurance Concluding an insurance between the Underwriters, *q.v.*, and the assured, in most cases through a broker.

Effecting a Policy The issuing of an insurance policy after effecting insurance with the underwriters, generally through the medium of a Broker, *q.v.*

Effective Date Binding date. Generally referred to a contract to take effect legally.

Effective HP *or* **Effective Horse Power** The power actually produced to the propeller of a ship which is roughly 50% of the Indicated Horse Power, *q.v. Abbrev.* EHP *or* E.H.P.

Effectively Closed. Term used to confirm that a liquid container is hermetically sealed.

Effective Perceived Noise Decibel The effective unit of noise in relation to the quantity of annoyance at various frequencies. *Abbrev.* EPNdB.

Effective Power The towing power needed to pull a vessel's hull without rudder, propeller, shaft, etc. *Also called* Towrope Horsepower.

E.F.P.H. Equivalent Full Power Hour *nuclear energy.*

EFTA *or* **E.F.T.A.** European Free Trade Association.

EFTPOS Electronic Funds Transfer at Point of Sale.

E.F.T.S. Elementary Flying Training School.

EFU Energetic Feed Unit.

EFZ Exclusive Fishing Zone.

Eg *or* **Egy** Egypt; Egyptian.

e.g. *or* **E.G.** Ejusdem Generis, *q.v.*; Exempli Gratia *Latin*—for example or by way of example.

EG *or* **E.G.** Engineers' Guide; Ejusdem Generis.

EGE Exhaust Gas Economiser *Lloyd's Register notation.*

e.g.e. Eau, Gaz, Electricité *French*—Water, Gas, Electricity.

E.Ger East Germany.

EGES Exhaust Gas Economiser Survey *Lloyd's Register notation.*

Eggbeater *See* **Helicopter**.

EGM Extraordinary General Meeting, *q.v.*

e.g.m.b.h. *or* **e.G.m.b.H.** Eingetragene Gesellschaft mit Beschrankter Haftung *German*—Registered Liability Company.

ego The self which is conscious and thinks.

EGP External Gravel Pack *oil drilling.*

EGR Exhaust Gas Recirculation.

e.g.t. *or* **E.G.T.** Exhaust Gas Temperature.

Egyptair Egyptian National airline.

EHA *or* **E.H.A.** Equipment Handover Agreement.

EHC *or* **E.H.C.** European Hotel Corporation (A joint enterprise by European airlines and bankers.)

EHF *or* **E.H.F.** Extreme High Frequency.

EHMA European Harbourmasters' Association.

e.h.p. *or* **EHP** *or* **E.H.P** Effective Horse Power, *q.v.*; Electric Horse Power; Equivalent Horse Power.

EHS *or* **E.H.S.** Extra high strength.

E.H.W.S. Extreme High Water Level Spring Tides.

E I East India; East Indian; East Indies; Electrical Insulation; Endorsement Irregular.

EI *or* **E.I.** Each Incident *insurance.*

E.I.A. East Indian Association; Electronic Industries Association *USA*: Engineering Industries Association; Environmental Impact Assessment.

E.I.B. Economisch Instituut Voor de Bouwnijverheid, *Dutch*—Building Institute of Economics; European Investment Bank; Export-Import Bank.

EIBA *or* **E.I.B.A.** European Institute of Business Administration.

E.I.C. Electrical Industries Club; Engineering Institute of Canada.

EID *or* **E.I.D.** East India Dock *in London*; Electrical Inspection Directorate; Employment Income Distribution *in the International Labour Office*.

EIF *or* **E.I.F.** European Investment Fund.

Eighteenmo A book based on 18 leaves, 36 pages to the sheet; also called octo-decimo, decimo-octavo *typography. Abbrev.* 18mo.

E in C Engineer-in-Chief.

E.Ind. East Indies.

einschl Einschliesslich *German*—Including; Inclusive.

EIR Equipment Interchange Receipt *intermodal transport.*

EIR *or* **E.I.R.** Earned Income Relief *income tax*; Environmental Impact Report; European Income Relief.

EIS *or* **E.I.S.** Education Institute of Scotland; European Intelligence Service; Executive Information Service; Export Intelligence Service.

E.I.T.B. Engineering Industry Training Board.

EIU Economist Intelligence Unit; Even if Used, *q.v.*

EJC Engineers' Joint Council *USA*.

Ejection Seat Equipment designed to catapult the pilot and his seat from an aircraft in case of an emergency.

Ejectment Recovery of land.

EJU Export Japan Unit.

ejusd. Ejusdem *Latin*—similar; of the same. *See* **Ejusdem Generis**.

Ejusdem Generis *Latin*—Of the same nature. This expression applies to a clause in conjunction with others. For instance, any damage to cargo arising without the actual fault or privity of the carriers or their agents. *Ex*: A vessel loaded with grain passes through heavy weather and, to avoid water getting into the holds through the vessel's ventilators, the master orders them to be closed. In consequence of this the grains may suffer damage by Sweat and for which the owners are held responsible. If, on the other hand, the ventilators remained open then the whole cargo would have been damaged by sea water. The master's action in this particular case was to save the cargo and possibly the ship and lives. Therefore both the owners and master are exonerated from blame. On arrival at the port of destination the master should lodge a Sea Protest, *q.v.*, to this effect to protect the carriers and himself from any eventual claim by the receivers. *Abbrev.* e.g. *or* E.G. *or* Ejusd. Gen. *See* **Proximate Cause**.

Ekranoplane A land/sea/air vehicle using air cushion, foils and wings developed in Russia and reputed to be capable of 500 km/h.

EL Equipment Letter *Lloyd's Register notation*; Elastic Limit, *q.v.*

el Electrically Heated *Lloyd's Register notation*; Elevation.

E.L. Electrical Laboratory; Electronics Laboratory; Employer's Liability, *q.v.*, Elastic Limit, *q.v.*

ELA Equipment Leasing Association *UK*.

Elastic Demand Demand for a product which is sensitive to price.

Elastic Limit The maximum amount of force applied to a material before it can take another form. When force is withdrawn the material returns to its original state. This could apply to ropes in general and steel plates. *Abbrev.* EL *or* E.L.

Elasticity of Demand The price of the demand or order is adjusted in accordance with the quantities given. The products in general are all the same kind. *Ex.* For an order consisting of 100 cases of beer the buyer will be charged 10% less than the price of 50 cases.

Elastic Supply Production linked to Elastic Demand, *q.v.*, and therefore also sensitive to price.

ELB Export Licensing Branch.

ELCAS Elevated Causeway (Modular) System: a transportable pier that can be used anywhere in amphibious operations.

ELDO European Launching and Development Organisation.

ELEC European League for Economic Co-operation.

elec *or* **elect** Electric: Electrical; Electricity.

ELECTRA Trademark for London Electricity Board.

Electra Electrical, Electronics and Communications Trades Association.

Electrical Chipping Hammer *or* **Pneumatic Chipping Hammer** *See* **Chipping Hammer**.

Electrode Conductor through which electricity enters or leaves an electrolyte, gas, vacuum or other medium. *See* **Electrolytic Action**.

Electrolysis Process of chemical decomposition by electric action. *Ex.* The electrolysis of a solution of copper chloride will produce copper at the cathode and chlorine at the anode.

Electrolytic Action. Electrolysis, *q.v. See* **Electrode**.

Electromagnetic Having both electrical and magnetic character or effects (applied esp. to waves or radiations which travel with the same velocity as light). *Abbrev.* EMP.

Electromotive Force Electrical potential voltage difference. *Abbrev.* emf.

Electronic Relative Motion Analyzer The US Maritime Administration instituted regulations for vessels of any nationality carrying hazardous cargoes and entering US waters to install Anti Computer Radars, eventually named Electronic Relative Motion Analyzer. This is in the 1978 American Port and Tanker Safety Act, *Abbrev.* ERMA.

Elements There are many meanings of this word, one of which (in chemistry) means any of the substances that cannot be resolved by chemical means into simpler substances.

Elev. Elevation.

Elevated Compass Magnetic compass positioned in such a way as not to be influenced by metal parts of the ship. Generally it is put above the bridge.

Elevating Transfer Vehicle Fork-lift vehicle used in the lifting and transporting of containers, pallets and merchandise. *Abbrev.* ETV.

e.l.f. Early Lunar Flare; Extra or Extremely Low Frequency.

ell Eller *Swedish*—Or.

ellipt Elliptical.

ELMA Electro-Mechanical-Acid.

ELOISE European Large Orbiting Instrumentation for Solar Experiments.

E.Long East Longitude.

E.L.R. Export Licence Regulations.

ELSA Emergency Life Support Apparatus.

El Salv. El Salvador.

ELSBM *or* **E.L.S.B.M.** Exposed Location Single Buoy Mooring.

ELT *or* **E.L.T.** Emergency Locator Transmitter, *q.v.*; European Letter Telegram.

e.l.v. *or* **E.L.V.** Extra Low Voltage.

E.L.W.S. Extra Low Water-Level Spring, *tides*.

Ely Easterly.

e.m. Electro Magnetic; Emergency Maintenance; Expanded Metal; Emergency Memorandum.

EM Electromagnetic; Electronic Mail; End of Medium.

EMA European Monetary Agreement; Engineers' and Managers' Association.

E.M.B. Empire Marketing Board.

emb Embargo, *q.v.*; Embossed.

Embargo A government order to stop movements of ships and cargoes in or out of ports to safeguard the interests of the country. This is practised mostly during a time of insurrection, war, riots, etc. *Abbrev.* emb.

Embarkation When officers, crew members and passengers are boarding a ship. It is the opposite of Disembarkation, *q.v.*

Embarkation Deck A place on a passenger ship used for embarkation procedures.

EMBO European Molecular Biology Organisation.

EMC Engineering Manpower Commission.

emcee M.C.; Master of Ceremonies.

EMCF European Monetary Co-operation Fund.

emer Emergency.

Emergency Boat Life Boat, *q.v.*

Emergency Distance The length of runway available should an aircraft need to abort its take-off.

Emergency Locator Transmitter Transmitter which is obliged to be used by law, especially by aircraft in North America, in order to emit distress signals. *Abbrev.* ELT.

Emergency Position Indicating Radio Beacon Device carried aboard a vessel which radiates a distress signal in emergency. It will float clear if the vessel sinks or should be attached by line to a survival craft. Life saving equipment in terms of SOLAS, *q.v. Abbrev.* EPIRB.

EMF European Motel Federation.

emf Electromotive force, *q.v.*

EMI Electrical and Musical Industries (Ltd.).

Emigrant A person who leaves his or her country for whatever reason in order to settle in another country.

Emigrant Ship A passenger ship specially equipped to carry emigrants. *See* **Emigrant.**

EMIP Equivalent Mean Investment Period.

E.M.K. Elektro-Motorische Kraft *German*—Electromotive Force.

EML *or* **E.M.P.** Estimated Maximum Loss *insurance*.

e.m.o.s. Earth's Mean Orbital Speed.

EMPA *or* **E.M.P.A.** European Maritime Pilots' Association.

EMPL *or* **E.M.P.L.** Estimated Maximum Probable Loss *in re-insurance*.

Employer's Liability In almost all cases an employer is considered to be liable for expenses incurred which may result from injuries sustained by his employees during the course of their working hours. If not enforced by law the employer is bound to insure his employees to cover himself against any unexpected expense due to accidents or injuries.

Emporium Derived from the Greek *emporion*. A place, building or shop which is a centre of commerce.

EMR *or* **E.M.R.** Eastern Mediterranean Region; Electromagnetic Response.

EMS *or* **E.M.S.** Electronic Mail Service; European Monetary System; Electromagnetic Suspension.

EmS Emergency procedures supplement to the IMDG Code, *q.v.*, *IMO.*

EMU *or* **E.M.U.** European Monetary Unit.

emu *or* **e.m.u.** Electromagnetic Unit.

EMW Equivalent Mud Weight *oil drilling*.

Emy Emergency.

EN (1) European Standard, from CEN, *q.v.* The initials are followed by digits for each standard. (2) Equipment Number.

ENAB Export to North America Branch.

Enamel In shipping this refers to paints of hard gloss applied in addition to the flat paint or undercoat to produce a good finish.

En autre droit *Old French*—A trustee and executor.

Enc Enclosed.

E&NC Eastern and Northern Canada, *q.v.*

Encash The act of exchanging cheques for cash.

Encashment Credit An authority for a person who has an account with one bank to draw cheques on his name at another and at a different place for a limited sum.

encl *or* **Enc** *or* **Enclo** Enclosure.

Encroachment (1) Recovery of land. (2) Intruding on the rights of others. The use of unauthorised extended boundary.

Encumbrance Charges; Debts; Liability; Responsibility.

Encumbrance Free of Liability referring to lien, claims, charges etc if a mortgage or debts incurred are discharged for the buyer of the ship or other articles. In other words the buyer of any item is freed from liability held previously against such item.

E.N.D. Epitheorisis Naftiliakou Dikeou. Greek Shipping Law Review.

End-Door Ship A drive on/drive off ship carrying vehicles. Equipped with two hinged doors, one at the stern and the other at the bow. Vehicles can easily enter or leave the ship via the hinged doors which are lowered in line with the quay. Similar to RO/RO *or* Roll On/Roll Off ship. *See* **Ro-Ro Vessel**.

Endorse To sign and/or write one's name at the back of a Cheque, *q.v.*, or a bill or a Bill of Lading, *q.v.*, or any other document.

Endorsee Person who is assigned to a Bill of Exchange, *q.v.*, or a Cheque, *q.v.*, Bills or any other document which is signed by the endorsee when accepted or cashed.

Endorsement of a Bill of Lading On presentation of an original Bill of Lading, *q.v.*, the consignee must sign it before a Delivery Order, *q.v.*, is issued by the owners or agents to enable him to obtain the merchandise. *See* **Endorsee** *also* **Endorser**.

Endorser The person who signs or writes one's name on the back of a Cheque, *q.v.*, or on a Bill, *q.v.*, or any other document.

Endowment Policy Insurance policy covering the assured and providing a lump sum of money with or without bonuses, after a period of years or to his or her estate if death occurs earlier. A combined investment and life assurance. *See* **Assurance.**

End. Tel. Endereço Telegrafico *Portuguese*—Cable Address.

ENE *or* **E.N.E.** East North East.

ENEA *or* **E.N.E.A.** European Nuclear Energy Agency.

ENEL Ente Nazionale per l'Energia Elettrica *Italian*—National Electrical Energy Agency.

Enemies Hostile persons; opponents.

ENF *or* **E.N.F.** European Nuclear Force.

Enfaced Foreign bond term for assented.

Enfranchise To grant the privilege; To set free; To confer some liberty or freedom upon.

eng. Engine.

Eng Engineer; Engineering; Engraved; Engraving; England; English.

ENG Electronic News Gathering.

engg. Engineering *USA.*

Eng. Fdn Engine Foundation.

Engineer Superintendent *See* **Superintendent**.

Engine Transplant Replacement of a power unit in situ.

Engr. Engineer; Engineering.

Eng. Rm. Engine Room.

ENI Ente Nazionale Idrocarburi *Italian*—National Hydrocarbon Corporation.

E.N.I.A.C. *or* **ENIAC** Electronic numerical Integrator and calculator or computer; an automatic mathematical calculator used for fast work calculations.

ENIT Ente Nazionale Industrie Turistiche, *Italian*—State Tourist Office.

En masse *French*—All together; In a body.

En passant *French*—By the way; Passing.

En plein jour *French*—In broad daylight.

Enquête du Pavillon A warship of a nation authorised by international law to follow and stop a suspected ship outside territorial waters. The warship is obliged to show her colours before she can proceed with this request.

En rapport *French*—In sympathy with; In agreement.

En règle *French*—According to rules; In Order.

Enrolled Ship A US ship engaged in the coasting trade or in fishing which is not necessarily required to be registered so long as she is not engaged in foreign trade. Also, yachts and pleasure craft are not obliged to be registered.

En route *French*—On the way.

Ens Ensign, *q.v.*

ENSC European National Shippers' Council.

Ensign *See* **National Flag**.

En.S.L. Engineer Sub-Lieutenant.

ENSO El Niño/Southern Oscillation: Phenomenon of the currents in the South Pacific producing marked changed in sea surface temperature which affect the climate in many countries.

ENSPM École Nationale Supérieure du Pétrole et des Moteurs *French*—Higher national school of petroleum and engines, i.e., fuel and power.

Entd Entered.

Entente cordiale *French*—Cordial understanding between states.

Entered In *or* **Entered Inwards** Customs clearance of a ship to enable her to start discharging. Opposite to Entered Out *or* Entered Outwards. *See* **Clearance.**

Entered Out *or* **Entered Outwards** Customs clearance of a ship to enable her to start loading. Opposite to Entered In *or* Entered Inwards. *See* **Clearance.**

Entered Tonnage The total gross registered tonnage entered for insurance with the mutual insurance association or P. & I. Club, *q.v.*

Enterprise *Same as* **Business,** *q.v.*

En train *French*—In progress.

Entre nous *French*—Confidentially.

Entrepreneur A French word literally meaning contractor. A business contractor or organiser who undertakes a business venture and anticipates a successful outcome.

Entry A customs form used for the clearance of ships or merchandise. There are various forms used for merchandise. These may be where duty is levied, for warehousing, for withdrawal without duty, for transhipment, for ship's use, etc. *See* **Clearance.** *See also* **Entry for Duty, Entry for Free Goods** and **Entry for Warehousing.**

Entry for Duty Customs form used to clear goods liable for duty. *See* **Entry.**

Entry for Free Goods A customs form used to clear goods not liable for duty. *See* **Entry.**

Entry for Warehousing A customs form used to clear goods and warehouse them until such time as they may be required to be re-exported or supplied as stores for ships or transhipped on to another port or for home consumption. *See* **Bonded Stores.**

Enure To take effect.

Environmental Quality Objectives The requirements or goals or general objectives of a given environment which is practised in many parts of the USA and EEC countries. This can refer to a factory or any other object for the protection of human health and the safeguarding of the natural environment, especially as far as animals and plant life are concerned. *Abbrev.* EQO *or* E.Q.O.

Environmental Quality Standards Similar to Environmental Quality Objectives but the objectives are concentrated on the standards of materials. *Abbrev.* EQS *or* E.Q.S.

e.n.z. En zoo voort *Dutch*—And so on.

E.O. Education Officer; Emergency Operation; Employers' Organisation; Engineer Officer; Executive Officer, Experimental Officer; Executive Order.

e.o. Ex Officio *Latin*—By virtue of office.

EOA *or* **E.O.A.** Essential Oil Association; Examination Opinion and Advise.

EOD *or* **E.O.D.** Explosive Ordinance Disposal; Every other day; Entry on duty.

e.o.d. Entry on duty.

E.&O.E. *or* **EOE** Errors and/or Omissions Excepted.

e.o.h.p. Except otherwise herein provided *insurance.*

eo instanti *Latin*—Instantly; At that time; At that instant.

EOLM Electro-Optical Light Modulator.

e.o.m *or* **E.O.M.** End of month; Every other month.

eo nomine *Latin*—Under this name. On that account.

EONR *or* **E.O.N.R.** European Organisation for Nuclear Research.

e.o.o.e *or* **E.O.O.E.** Erreur ou Omission Excepté *French*—Error or Omission Excepted. *See* **Errors Excepted.**

EOOW Engineering Officer of the Watch.

EOP Economic Order Point; End of Passage: completion of sea passage.

EOQ Economic Order Quality.

EOQC *or* **E.O.Q.C.** European Organisation for Quality Control. *See* **Environmental Quality Objectives.**

e.o.t. End of Transmission.

EOTC European Organisation for Testing and Certification.

EOTP *or* **E.O.T.P.** European Organisation for Trade Promotion.

e.p. Easy projection; Electrically polarised; End paper; En passant *French*—In passing; Estimated position *nautical*; Expanded polystyrene; Extreme pressure; Editio princeps *Latin*—First edition.

Ep Electro-plate; Electro-plated.

EP *or* **E.P.** Extra-ordinary and Plenipotentiary *diplomatic*; Estimated Position *nautical*; European Parliament; Extreme Pressure.

EPA *or* **E.P.A.** Environment Protection Agency, in connection with the US Federal Water Pollution Center; European Productivity Agency.

EPC Engineering Procurement Contract *oil drilling.*

EPCA European Petro-Chemical Association.

EPCC Environment Programme of the Community Commission of EEC.

EPCI Engineering Procurement, Construction and Installation.

Epcot Experimental Prototype Community of Tommorow.

EPD *or* **E.P.D.** Excess Profits Duty, *customs duty.*

EPDA Emergency Powers Defence Act.

EPF European Packing Federation, The Hague.

EPFZ Ecole Polytechnique Féderale, Zurich *French*—The Federal Polytechnic School, Zurich.

EPI Energy Performance Index *refinery operations.*

EPI *or* **E.P.I.** Electronic Position Indicator; Earned Premium Income; Estimated Premium Income.

EPIRB Emergency Positioning Indicating Radio Beacon, *q.v.*

EPLT European Part Load Tariff.

EPNdB Effective Perceived Noise Decibel, *q.v.*

EPNS *or* **E.P.N.S.** Electro-Plated Nickel Silver.

E pluribus unum *Latin*—One from many. Motto of the US.

EPO European Patent Office.

EPOS Electronic Point-of-Sale Service *bar codes.*

Epoxy Type of synthetic resin blended in sophisticated paint coatings and considered to be a long-life rust preservative with service endurance.

EPRI *or* **E.P.R.I.** Electric Power Research Institute.

e.p.s. Earnings per share.

EPS Electronic Placing Support *insurance.*

EPSRC Engineering and Physical Science Research Council.

E.P.T. Ethylene Propylene Terpolymer *synthetic rubber*; Excess Profits Tax.

EPTA *or* **E.P.T.A.** Expanded Programme of Technical Assistance *United Nations.*

EPTO European Permanent Traffic Observatory: European Commission project for the monitoring and analysis of traffic in VTS, *q.v.*, areas.

EPTS Early Production and Testing System, *q.v.*

EPU *or* **E.P.U.** European Political Union; Empire Press Union.

EPUL Ecole D'Université de Lausanne *French*—Institute of Technology of the University of Lausanne.

Epus Episcopus *Latin*—Bishop.

eq Equal; Equator; Equatorial; Equipment; Equivalent.

Eq.T. Equation of Time *nautical.*

Equal Altitudes When two or more heavenly bodies are recorded on the same altitude on either side of the Meridian, *q.v.*

Equated Time Accounts that fall due and averaging the convenient time. *Similar to* **Average Due Date,** *q.v.*

Equation of Time The difference between the Mean Time, *q.v.*, and the Apparent Time, *q.v. Abbrev.* ET *or* E.T. *or* Eq. of T. *nautical.*

Equator An imaginary line circumscribing the globe midway between the poles and at its greatest circumference (24,901.96 miles). It constitutes the zero from which latitudes North and South are calculated.

Equilibrium The balance position of an object.

Equilibrium Price Price at which regular demand and supply of a manufactured product will remain constant.

Equilibrium Relative Humidity The average moisture content of an object. *Abbrev.* ERH.

Equin Equinoctial *nautical. See* **Equinox.**

Equinox Derived from the Latin *Aequus* meaning equal and *Nox* meaning night. Equinox refers to when the sun is overhead at the Equator, *q.v.*, resulting in equal days and nights. Vernal equinox about 20 March, autumnal equinox about 22 September.

EQUIP *or* **E.Q.U.I.P.** Equipment Usage Information Programme.

Equitable Mortgage Money lent without any formal deed drawn up but simply depositing the title deeds of the property mortgaged with the lender. *See* **Mortgage.**

Equities Shares which are issued by associations or companies.

Equity (1) Shareholding of stock in a company or corporation. (2) Value of property, i.e. excess of charges upon it.

Equity Capital (1) The capital that is formed by the ordinary shareholders of the business. (2) That part of the capital resulting from the sale of the stock.

Equity Crude The crude oil remaining as the property of the oil company after tax and royalties have been paid.

equiv. Equivalent.

Equivalent Chill Temperature Wind chill cooling effect considered to be in the range -7 to -20 degrees C. *Abbrev.* ECT.

Equivalents Standard equivalent sizes and capacity of containers which are 20, 30, 35 and 40 feet long. All these have been used in the past but it is known that the two most commonly in use are the 20 ft. and the 40 ft. *See* **Container.**

e.r. Echo Ranging; Electronic Reconnaissance; Emergency Rescue; External Resistance; Emergency Request.

E.R. Eastern Region of British Rail; Engine Room.

ERA *or* **E.R.A.** Electrical Research Association; Electronic Reading Automation; Emergency Relief Administration; Engine Room Artificer; Equal Rights Amendments *USA.*

ERAB Energy Research Advisory Board.

E.R.Bh. Engine Room Bulkhead *Lloyd's Register notation.*

E.R.C. Economic Research Council; Electronic Research Centre.

ERC European Registry of Commerce.

ERD Extended Reach Drilling *oil.*

ERDA Energy Research and Development Administration *USA.*

ERDF European Regional Development Fund.

Erector American expression for workman in a shipyard employed to join together ships' structures.

e re nata *Latin*—Under the present circumstances. As matters stand.

Erg Unit of work equal to one dyne-centimetre.

E.R.G. Electrical Resistance Gauge.

Ergo *Latin*—Therefore; Consequently.

ERGOM European Research Group on Management.

Ergonomics System of study in order to determine the efficiency of physical effort.

ERH Equilibrium Relative Humidity, *q.v.*

ERIC Educational Resources Information Center *USA*; Energy Rate Input Information.

Erload Expected Ready to Load. *See* **Erload Clause**.

Erload Clause *or* **Erload 1957** A Charterparty, *q.v.*, expression for Expected Ready for loading or shipment.

ERM Exchange Rate Mechanism, *q.v.*

E/Rm Engine Room.

ERMS *or* **Erms** European Radio Message Service.

ERNIE Electronic Random Number Indicator Equipment *electronics*.

ERO *or* **E.R.O.** European Regional Organisation, a United Nations branch of the International Confederation of Free Trade Unions.

Erosion The abrasive wearing away effect of materials by fluids, such as erosion of the ship's side by sea water.

e.r.p. Effected Radiated Power.

ERP *or* **E.R.P.** European Recovery Programme.

err Errate; Erratum *Latin*—Mistake(s).

Errare est humanum *Latin*—To err is human.

Erron Erroneous.

Errors Excepted *or* **Errors and Omissions Excepted** Very common business accounts term abbreviated E&OE *or* E.O.E. and shown at the bottom of bills. This is a statement by the company or person presenting the bill to the effect that any errors and/or omissions therein shown are accepted for correction.

ERS *or* **E.R.S.** Earth Resources Satellite; Engine Repair Section; Experimental Research Society *USA*.

ERV *or* **E.R.V.** Each or Every Round Voyage.

e.r.w. Electric Resistance Welding.

e.s. Electrical Sounding; Electric Starting; Electrostatic.

ES Engine Survey *Lloyd's Register notation*.

ESA Endangered Species Act *USA*; European Space Agency; European System of Integrated National Accounts; Experimental Stress Analysis (the testing of scale structures).

E.S.A. European Savings Account; Economic Stabilisation Administration; European Space Agency.

ES. adv. Engine Survey Advanced *Lloyd's Register notation*.

Esc *or* **ESC** *or* **esc** Escudo, Portuguese currency, *see* **Escudo**; Economic and Social Council; Escompte *French*—Discount; European Shippers' Council, The Hague.

Escalation *or* **Escalation Clause** Marine insurance policy covering any increments in the building costs of ships due to the Inflation, *q.v.*, of the day. Also

inserted in other industrial contracts enabling the contractor to adjust his prices in accordance with the cost of labour and materials.

ESCAP *or* **E.S.C.A.P.** Economic and Social Commission for Asia and Pacific, *Bangkok*.

Escape Hatch A small Hatch, *q.v.*, wide enough for a man to pass through in an emergency. Such a hatch is positioned in the tweendecks of a ship to allow men to get out of a hold or compartment when cargo is trimmed.

Escheat An old legal process which is being done away with whereby tenure of land used as property will be taken over by the state if the tenant leaves no heirs and no one to claim it.

ESCOM Electricity Supply Commission of South Africa.

Escrow Sealed document held by a third party to be presented as soon as all the conditions of a deed or a contract are fulfilled.

Escudo Unit of currency used in Portugal, Chile, and Cape Verde Islands. *Abbrev.* Esc, ESC or esc.

ESD *or* **E.S.D.** *or* **e.s.d.** Echo Sounding Device *Lloyd's register notation* to indicate that the ship is fitted with an Echo Sounding Device, (Fathometer, Echo Sounder and Acoustic Sounding Device); Emergency Shut-Down *oil drilling*; Epitheorrisis Synkinomiakou Dikeou, Greek Transport Law Review.

ESDAC European Space Data Centre.

ESE *or* **E.S.E.** East-South-East; Engineers Stores Establishment.

ESFJC Eastern Sea Fisheries Joint Committee.

ESFSWR *or* **E.S.F.S.W.R.** Extra Stainless Flexible Steel Wire Rope.

ESG *or* **E.S.G.** English Standards Gauge.

ESGM Electrostatically Supported Gyro Monitor.

ESH *or* **E.S.H.** Equivalent Standard Hours.

Esk Eskimo.

ESKA Ekranolytny Spasatyelny Kater Amphibiya, *USSR*—air-cushion vessel.

ESN *or* **E.S.N.** Engine Serial Number.

ESOCO Economic and Social Council.

ESOMAR *or* **E.S.O.M.A.R.** European Society for Opinion and Market Research.

esp Especially.

ESP Electrical Submersible Pump; Extrasensory Perception.

Espac European Space Data Centre.

ESPO European Community Sea Ports Organisation, based in Brussels.

ESPRIT European Strategic Programme for Research in Information Technology. Established 1984.

Esprit de corps *French*—Regard for honour and interests of body or organisation one belongs to.

ESQ *or* **E.S.Q.** Extra Special Quality (generally referring to the quality of steel wire ropes).

Esq Esquire, a courtesy title.

ESR Equivalent Service Rounds.

ESRO European Satellite Research Organisation.

Essence Similar to Gasoline, *q.v.*

EST *or* **E.S.T.** Earliest Start Time; Eastern Standard Time *Eastern Coast of USA*: Eastern Summer Time *USA.*

Est. Estonia; Estimated; Estuary.

Established Brands Refers to products that are very well known to the general public, and therefore should be easy to sell.

Established Charges A book-keeping term for administrative or indirect expenses.

Estate Agent Person who acts as an agent for the buying or selling or valuation on collected rents, immovable lands, houses, etc.

Estate Duty Duty levied on inherited property.

E.S.T.I. European Space Technology Institute.

Estimated Position *nautical.*

estm. Estimation.

Estoppel A rule of evidence whereby a party is precluded from denying the existence of some statement of facts which he has previously asserted. An action cannot be founded on an estoppel. Unlike other evidence, an estoppel must be pleaded. An estoppel may be waived. Estoppels are generally divided into three kinds—of record, by deed, and *in pais* (by conduct). Estoppel by record is based on the principle that a record imports such absolute verity that no person against whom it is producible shall be permitted to aver against it. The most important kind of estoppel by record occurs in the case of judgments: thus, where A wrongfully signed judgment against B, it was held that so long as the judgment stood B could not dispute it, and that his remedy was to apply to the court to set it aside. A record concludes the parties thereto, and their privies, whether in blood, in law, or by estate, upon the point adjudged, but not upon any matter collateral or adjudged by inference. A judgment in an action *in rem* is absolutely binding upon all the world. A conviction on the same facts is no estoppel in a civil action because the parties are not the same. Estoppel by deed or specialty is based on the rule that no man is allowed to dispute his own solemn deed: hence, as a general rule, if a person executes a deed containing a recital or statement, he cannot afterwards deny the truth of it, or show it to be incorrect. The general rule is that an indenture estops all who are parties to it, while a deed-poll only estops the party who executes it, since it is his sole language and act. Estoppel *in pais* (by conduct) occurs, *e.g.*, in the case of estoppel by entry, by acceptance of rent, by partition or by acceptance of an estate.

Estray A lost animal of value found astray.

Estuarial Service *or* **Operation** Shipping traffic on a river estuary.

Estuary Wide, usually navigable, mouth of a river where the tides meet the river flow.

e.s.u. *or* **E.S.U.** Electrostatic Unit.

E.S.V. Earth Satellite Vehicle; Emergency Support Vessel (additional service rig to actual oil rigs or a support); Emergency Safety Valve; Emergency Shutdown Valve; Emergency Slide Valve.

ESWR *or* **E.S.W.R.** Extra Steel Wire Rope.

ET *or* **E.T.** Eastern Time; Electric Telegraph; Electronic Technician; English Translation; Entertainment Tax; Equation of Time; Estimated Time or Expected Time; Exchange Telegraph.

ETA *or* **eta** *or* **E.T.A.** Expected Time of Arrival (referring to the time of arrival of a ship, train, aircraft, etc.)

étab Etablissement *French*—Business place.

et al Et alia *Latin*—And others.

ETAP *or* **E.T.A.P.** United Nations Expanded Assistance Programme. *See* **World Health Organisation.**

ETB English Tourist Board.

ETB *or* **E.T.B.** *or* **e.t.b.** Expected Time Berthing.

etc. et cetera *Latin*—And so on *or* and so forth.

e.t.c. En tout cas *French*—In any case.

ETC Electronic Trade Confirmation *stock exchange facility.*

ETC *or* **E.T.C.** Easy Tank Cleaning, *q.v.*; Energy Transportation Corporation; Entrepreneur de Transport Combiné *French*—Combined Transport Co-operator; European Trade Committee; European Travellers Cheques; Expected Time of Commencement or Completion.

Etch Primer *Similar to* **Wash Primer,** *q.v.*

ETD *or* **etd** *or* **E.T.D.** Estimated/Expected Time of Departure; Extension Trunk Dialling.

ETE *or* **ete** *or* **E.T.E.** Expected/Estimated Time en Route.

Eth Ethiopia; Ethiopian.

Ethanol Short for Ethyl Alcohol. It can be extracted from sugar cane, waste plant material or whole grains.

Ethylene Oxide Industrial chemical product used as a sterilizing gas. *Abbrev.* Eto.

Ethylene Vinyl Acetate Copolymer Polymer containing aluminium hydroxide, magnesium carbonate and zinc borate used in fire retardant fillers.

ETI *or* **e.t.i.** *or* **E.T.I.** Elapsed Time Indicator; Estimated Time of Interception.

ETL *or* **e.t.l.** *or* **E.T.L.** Estimated/Expected Time of Loading; Estimated/Expected Time Leaving; European Transport Law.

Eto Ethylene Oxide, *q.v.*

ETO *or* **eto** *or* **E.T.O.** European Theatre of Operations; Estimated Time Off; Estimated Take-off *aviation.*

ETP *or* **etp** *or* **E.T.P.** Estimated Turning Point.

ETPO *or* **E.T.P.O.** European Trade Promotion Organisation.

E.T.P.S. Empire Test Pilots' School.

ETR *or* **etr** *or* **E.T.R.** Estimated/Expected Time of Readiness; Estimated/Expected Time of Return.

ETS *or* **ets** *or* **E.T.S.** Estimated/Expected Time Sailing; Expiration Time of Service; Energy Transportation Security *USA*.

ETS European Telecommunications Standards.

ETSI European Telecommunications Standards Institute.

e.t.s.p. Entitled to severance pay.

et seq. Et sequens *Latin*—And those that follow.

et sequens *Abbrev.* et seq., *q.v.*

et sequentia *Latin*—And what follows.

ETSU Energy Technology Support Unit *UK*.

ETU *or* **E.T.U.** Electrical Trades Union; Expected Time Unloading.

ETUC European Trade Unions Confederation.

ETUI European Trade Unions Institute.

ETV *or* **E.T.V.** Educational Television; Elevating Transfer Vehicle, *q.v.*

etw Etwas *German*—Something.

EU Ecuatoreana, National Airline of Ecuador *airline designator*; European Union.

E.U. Estados Unidos *Spanish*—United States; Etats–Unis *French*—United States; Experimental Union.

E.U.A. Estados Unitos da America *Portuguese*—United States of America; Estados Unidos de America *Spanish*—United States of America; Etats-Unis d'Amerique *French*—United States of America; European Unit of Account, *q.v.*

EUGOES European Group of Energy Suppliers.

EUI *or* **E.U.I.** European Committee on Insurance.

Eur Europe.

EurACS European Association of Classification Societies, *q.v.*

EURAM European Research in Advancced Materials.

Euratom *or* **EURATOM** European Atomic Energy Community, established 1958. For the development of nuclear research and industries.

Eureka A copper nickel wire used on electrical resistance coils for low temperature coefficient.

Eur. Ing. European Engineer.

Euro European Regional Office (a branch office in the Food and Agriculture Department of the United Nations).

Eurobonds Securities issued by companies or countries outside their own country's markets to raise funding in a variety of currencies and usually not transferrable to local bearers.

Eurobudget Air fares offered at standard seating but with no basic catering service. Symbolised by the letters YB.

Eurocard Credit Card, *q.v.*

Eurocheques Cheques provided by banks in some forty countries which enable customers to draw cash to a specified limit or pay for purchases in the local currencies of European countries.

Eurocontrol *or* **EUROCONTROL** European Organisation for the Safety of Air Navigation.

Euro Currency *See* **European Unit of Account.**

Euro Dollars American dollars which are being held by persons and countries outside the United States. The counterpart to External Sterling, *q.v. See also* **European Unit of Account.**

EUROFINAS European Finance Houses Association.

Euromarket European Common Market *Abbrev.* E.C.M.

European Association of Classification Societies, The. Formed in 1979 and comprising Bureau Veritas, Germanischer Lloyd, Lloyd's Register of Shipping, and Registro Italiano Navale. These societies are all members of IACS, *q.v.* Together they class over 24,000 seagoing ships as well as many inland waterway ships, a large proportion of which navigate the Rhine. The Association's objects are: to improve safety standards and prevent maritime pollution; to co-ordinate members' technical standards; and to promote a uniform interpretation of international conventions. *Abbrev.* EurACS.

European Unit of Account Unit of account used in borrowing systems between some European countries such as Switzerland, France, Germany, Italy, etc. Superseded by the E.C.U. (European Currency Unit). The abbreviation for European Unit of Account is E.U.A. The currency mostly used internationally is the Eurodollar, *q.v.*

EUROS European Register of Shipping.

E.U.V. Extreme Ultra Violet.

EV Exposure Value *photography.*

eV *or* **e.V.** Efficient Vulcanisation; Eingetragener Verein *German*—Registered Association; Electronic Volt.

EVA Extra Vehicular Activity *aviation.*

EVAC Ethylene Vinyl Acetate Copolymer, *q.v.*

Evaporator A distillation unit on board a vessel used to produce fresh water from sea water. In the past ships were compelled to replenish their fresh water supply at ports on the way to their destination.

evce. Evidence.

Evening Star *See* **Morning Star**.

Even if Used Voyage Charterparty expression, referring to Laydays, *q.v.*, and Demurrage, *q.v.*, which are not to be accounted for even if used outside normal hours. *Abbrev.* EIU.

Even Keel When the Draft (draught) *q.v.*, of a ship fore and aft are the same.

Eviction To expel a person, especially a tenant, from property or land, etc.

evy Every.

Ew Electric Welding.

EW *or* **E.W.** England and Wales; Electrically Welded *Lloyd's Register notation.*

EWA Europäische Währungsabkommen. *German* —European Monetary Agreement.

E.W.O. Essential Work Order.

E.W.R. Early Warning Radar.

E.W.S. Emergency Water Supply.

Ex all When dividends or bonuses or any other right for claims are due these are credited and also paid to the sellers. *Abbrev.* x-all or x-a.

ex animo *Latin*—from the heart; sincerely.

ex ante *Latin*—from before. The position before an occurrence has happened.

Ex aq. Ex aqua *Latin*—Out of water.

Exc Excellency; Excellent; Except; Exception; Exchange;

Ex C Extra Master.

Ex Capitalisation Stock Exchange, *q.v.*, term. The quoted price of the share offered excludes any capital distribution or business be it past or present. *Abbrev.* Xep *or* X.ep.

ex capite *Latin*—out of the head; from memory.

Ex cathedra Official statement.

exempli gratia *Abbrev.* e.g., *q.v.*

Excepted Days of the week that are not to be counted as lay days or laytime, *q.v.*, in a Voyage Charter Party are said to be excepted.

Exception Clauses Clauses used in Contracts of Affreightment, *q.v.*, and in Marine Insurance, *q.v.* In contracts of affreightment these clauses relieve carriers from the named perils as well as all accidents on the part of their servants while on duty. In marine insurance these clauses generally follow the Running Down Clauses, *q.v.*, or Collision Clauses, thus limiting protection afforded by underwriters in certain risks. Alternatively, the risks may be covered by insurance with a Protection and Indemnity Association. *See* **Protection and Indemnity**.

Excess An insurance word meaning the amount of money that is deducted from a partial loss before the claim is applied to the policy. In other words, for each and every claim Underwriters, *q.v.*, will only be liable to pay claims over the amount of money or Excess mentioned in the policy *See* **Franchise**.

Excess Demand When the price of the commodity is below the cost price, the demand will exceed the supply.

Excess Insurance That excessive part or balance of the whole insurance which Underwriters, *q.v.*, refrain from covering but which can be placed and accepted on the same conditions by other underwriters.

Excess Liability Clause A set of Institute clauses to be found in marine insurance policies in relation to Collision, Sue and Labour, *q.v.*; General Average, *q.v.*; Salvage Charges, etc. *Ex.* Normally marine insurance covers only three-quarters of collision claims. The remainder is for the Mutual Clubs or P. & I. Clubs, *q.v.*, to settle. In the Sue and Labour Clause, *q.v.*,

underwriters indemnify the insured against the expenses incurred by the owners' representatives who attend to repairs locally or abroad with a view to minimising the repair charges as much as possible. Underwriters also cover against the risk of claims for General Average, Salvage and Salvage charges.

Excess Point Insurance term signifying the limit at which the insurance risk is covered.

Excess Supply When the stock of a commodity exceeds demand.

Excess Value Insurance A supplementary insurance covering the excess contribution in the General Average, *q.v.*

Exch Exchange, *q.v.* Exchequer.

Exchange One of the three divisions of economics (production, exchange and distribution). Exchange has various meanings, such as the rate of exchange of foreign currencies; a place where commercial business is transacted by its members, etc. *Abbrev.* Exch. *or* Xchange. The corresponding French word is Bourse, *q.v.*

Exchange Broker An intermediary who arranges foreign exchange business for his clients.

Exchange Control A system imposed by a government to control the movement of money to or from abroad in order to protect its country's economy.

Exchange Rate Mechanism Part of the European Monetary System adopted by the European Council in 1978. The currencies of member states are linked to the European Currency Unit, *abbrev.* ECU, and by cross rates to each other. The mechanism establishes margins within which these rates may fluctuate before compulsory bilateral intervention. *Abbrev.* ERM.

Exchange Time Charter Form 1946 United States Government form of Time Charter approved by the New York Produce Exchange on 6 November 1913 and amended on 20 October 1921, 6 August 1931 and 3 October 1946.

Excise Tax or duty imposed on locally manufactured or produced goods.

Excise Duty Duty levied by the Customs House Authority on home-made commodities, while customs or import duty tax refers to local consumption. Commonly termed Customs Duty, *q.v.*

Excisemen Inland Revenue Officers.

Excluded Clause Time Charter, *q.v.*, clause warning the Charterer, *q.v.*, to avoid named ports for some reason or other.

Excl Exclude; Exclusive.

Excluded Losses Losses in marine insurance not recoverable from Underwriters, *q.v.* These can include war perils, frustration, loss of life, personal injuries, part of the expenses in collision cases, customs fines, wilful misconduct, etc.

Exclusive Economic Zone Certain coastal states of the developing countries are accorded sovereign rights over the 200-mile limit as regards sea pollution. *Abbrev.* EEZ *or* E.E.Z.

Ex contractu *Latin*—Resulting from a contract; In consequence of a contract.

Ex Coupon Without interest. *Abbrev.* X.Cor, x.cor, x.cp *or* X.cp.

excp. Excluding coupon similar to Ex Coupon, *q.v.*

Ex. C.R. Exchequer Exchequer Court Reports (referring to Canadian Admiralty Reports).

Exct *or* **Exec** Executor, *q.v.*

Excursion Fare Comprehensive fare discounted at an attractive rate for air, land or sea passengers which includes certain amenities such as free incidental transportation, guide, hotel accommodation.

exd Examined.

Exd *or* **ex div** Ex dividend; Excluding the subsequent dividend; Ex Dock, *q.v.*

Ex delicto *Latin*—In consequence of or arising out of a wrong or fault.

Ex dividend *Latin*—Without dividend which is due. *Similar to* Ex all. Opposite to Cum Dividend. *Abbrev.* Ex div *or* XD *or* xd *or* x.d. *or* x.div.

Ex Dock Merchandise which is sold from ship's side excluding loading costs. *Abbrev.* Exd.

Ex drawings When bonds are sold the benefit of interest is not included. *Abbrev.* X.dr. *See* **Ex Dividend**.

Executive Director A member of a Company's Board of Directors who has a management or executive function. The Managing Director is one of them. *See also* **Company Director**.

Executor A person appointed by a testator to execute his will. An executor can also be a bank nominated to act likewise. *Abbrev.* exor. *or* Exct *or* Exec.

Executor de son tort *Norman French*—One who without authority meddles with the goods of a deceased person as if he had been appointed an executor.

Execx Executrix.

Exemplary Damages *See* **Damages**.

Exempt Company Private Company.

Exequatur *Latin*—When a foreign government recognises a Consul, *q.v.*

exes *or* **exps** Expenses.

Ex facia *Latin*—In accordance with documents.

Ex factory Price or quotation for goods from the factory. Excluding the cost of transportation from the factory to the place of embarkation or destination.

Exfoliation The wear and tear of copper sheeting on wooden ships.

Ex g Exhaust Gas *Lloyd's Register notation.*

Ex g *or* **ex gr.** Ex gratia *Latin*—Free of charge; As a favour.

Ex gratia *Latin*—By favour *Abbrev.* ex g. *or* ex gr., *q.v.*

Ex Gratia Payment Payment effected without the force of the law; As a sign of goodwill; As a favour.

Ex hypothesi *Latin*—By hypothesis. From a supposition, proposition, conjecture. Assumed to be true without proof.

Eximbank Export-Import Bank.

Ex. Imp *or* **Eximbank** Export-Import Bank, *q.v.*

Ex. In *or* **Ex. Int** Excluding Interest.

Ex Interest Without interest. *Abbrev.* X.in *or* X.I. *See* **Ex Dividend**.

Exit Interview Interview between a company executive and an employee who is leaving to agree the terms of his retirement and perhaps to ease the problems and financial difficulties he will encounter.

Ex-l Ex libris *Latin*—From the library of; From the books of.

Ex lib Ex libris *Latin*—From the books of; From the library of.

Ex libris *Abbrev.* ex-l, *q.v. or* Ex lib, *q.v.*

Ex Mer Ex Meridian.

Ex mero motu *Latin*—Of one's own accord.

Ex n *or* **Ex New Shares** Excluding new issues of shares or having no rights for the new shares. Also *Abbrev.* x.n. *or* x.new.

Ex mudo pacto non oritur actio *Latin*—No action can arise out of a bare promise.

Ex off Ex Officio *Latin*—From office; By virtue of office.

Ex officio *Abbrev.* Ex off., *q.v.*

Exor/s Executor(s), *q.v.*

Ex p Ex parte *Latin*—On one side only.

Exp Export; Exported; Exporter; Expand; Expansion; Expedition; Expense; Experience; Experiment; Expiration; Expire; Explosion; Exportation; Express; Expunged.

Expan Expanded; Expansion.

Expansion Hatch Small hatch on the deck of an oil tanker which provides access to the oil tanks. It has high Coamings to allow for the volume expansion which takes place at high temperatures.

Ex parte *Latin*—Court application made by a person who is not a party to the proceedings, in the absence of the other party; one sided; one side only. *Abbrev.* Ex p.

Expatriate To leave, and renounce the citizenship of one's country to emigrate and live in another country; to send out of the country, banish, exile; a person residing in a foreign country while contracted to work there.

Expendable Pallet Pallet of cheap material which is non-returnable and may be destroyed when it has served its original purpose.

Expenditure *or* **Extraordinary Sacrifice** The first factor to be observed and taken into consideration in General Average, *q.v.*, is that any action undertaken is to be extraordinary sacrifice or expenditure.

exper Experiment.

Experientia docet *Latin*—Experience teaches.

Expertise Expert knowledge.

Experto crede *Latin*—Believe the one who speaks from experience.

Exploitation Well An oil drilling term meaning an oil well exploration area.

Explosimeter An instrument by means of which the atmosphere can be tested and where concentrated inflammable gases and vapour are traced.

Exp. M. Expanded Metal.

Expo Exposition, Exhibition; Business Fair.

Exp. O. Experimental Order.

Export Bill of Lading US shipping term meaning the issue of an Export Bill of Lading by the carriers signifying that the cargo carried is meant for foreign ports only.

Export Declaration. In normal cases when goods are exported the exporter has to fill up a customs form, inserting all the details of the goods about to be shipped. This also serves as statistical data. *See* **Export Licence.**

Export House An experienced firm concentrating its business in export markets throughout the world. Generally speaking this entails the obtaining of export licences, booking for shipments, effecting insurances, arranging packing and transportation, etc. Its duties may also include a guarantee of payment for the merchandise by directing the shipping documents through banks and thus ensuring payment on the delivery of these shipping documents.

Export Licence An application form authorised by the Trade Department of a country to export goods. *See* **Export Declaration.**

Export-Import Bank Government agency of the United States established in 1934. Its main functions are to encourage US trade by aiding and supplying credits to countries outside the United States. *Abbrev.* Ex-Imp or Eximbank.

Export Quotation *or* **Export Price** This includes the specific price of the commodity meant for export plus all the charges incurred up to the final destination of the goods, namely CIF *or* C.I.F., *q.v.*

Exports Business transaction undertaken by a person, a company or a country with others overseas. Exportation can be effected by land, sea or air but most of it is done by sea. There are two classifications of export; visible and invisible. Visible exports are all those things which are exported abroad and in return money is paid. Invisible exports are those that bring payments for financial services, foreign investments, general shipping charges, etc.

Exposed Location Single Buoy Mooring A new concept in use by oil explorers around oil field areas. It consists of a huge buoyancy chamber with ballast compartments which can be used if and when needed. *Abbrev.* ELSBM or E.L.S.B.M.

Ex post facto *Latin*—By a subsequent act; Retrospective. *Ex.* a law having a retrospective effect.

Express Authority Written or verbal authority.

Express Delivery A Post Office word meaning the urgency of the letter. An extra amount of money is charged for Express Delivery Letter.

Express Liner A fast ship used on a scheduled service between specified ports.

Express Warranties Extra clauses to the Implied Warranty Clauses in marine insurance exempting the underwriters from some particular risk such as in the case of territorial waters where wars, riots, etc., are imminent. Express Warranties, whether in marine insurance or other contracts, must be included in the contract form in order to be binding as opposed to Implied Warranties, *q.v.*, which are automatically construed as being understood by both parties. *See* **Warranty.**

Exp.T. Expansion Tank.

Expunged A word commonly seen in Classification Societies' remarks. A vessel is deleted or expunged from her class when the rules have not been strictly adhered to.

Ex rights A stock exchange term. The quoted price of the shares exclude any rights which exist or which are to be issued later.

Exs Examples; Expenses; Excesses.

ext. Extended, e.g. of a time hire agreement.

Ext. Exterior; External; Extra; Extreme.

External Deficit A country's deficit in the balance of payments. *Opposite to* **External Surplus,** *q.v.*

External Sterling Sterling money possessed by people of non-sterling areas. The counterpart to Euro Dollars, *q.v.*

External Surplus A country's credit in the balance of payments. **Opposite to External Deficit,** *q.v.*

Extortion Obtaining money or a promise, etc., by violence or intimidation. It can also mean civil offence committed by public officers of a country in taking advantage of their positions.

Extra Charges Charges incurred in proving a claim, e.g. expenses of an inspection, which may be recoverable if the claim is paid, depending upon agreement in the policy.

Extra Charges Follow the Claim An expression in insurance signifying that the expenses for investigating and adjusting a claim are not accepted for settlement by Underwriters, *q.v.*, unless actually authorised by them.

Extra Cost Marginal cost, *q.v.*

Extractive Industry An industry which is involved in extracting or producing raw materials from land and sea and possibly even from air. *Ex.* Rough diamonds; fertilisers; crude oil, etc.

Extradition A request made by the authorities of one country to another to surrender a person or persons to be tried at the place where the crime had been committed and where he, she or they were involved.

Extra Insurance Clause A Charterparty, *q.v.*, clause giving the relative remuneration of the owners to the charterers, *q.v.*, in case extra insurance is charged on merchandise shipped because the vessel is not entirely accepted as satisfactory as far as classification or age or flag is concerned.

Extra judicium *Latin*—Out of court.

Extraordinary General Meeting A meeting convened by the members of a company or an association to resolve urgent matters that cannot wait for an Annual General Meeting. *Abbrev.* EGM.

Extraordinary Sacrifice *Similar to* Expenditure, *q.v.*

Extreme Draught *or* **Extreme Draft** The vertical height from the waterline to the lowest part of a ship immersed in the water at any point extending from fore to aft. Keel Draft (Draught) is an identical expression.

Extremity Extreme point, very end; extreme adversity.

Ex turpi causa non oritur actio *Latin*—No legal rights flow from a contract which has an unlawful basis.

Ex voto *Latin*—By vow.

EXW Ex Works.

Ex Warehouse The buyer is to provide means for the payment of land transport from the warehouse apart from the other expenses relating to the shipment of the merchandise.

Eye A loop at the end of a rope. This eye is very useful in mooring ropes where quick action is needed when the ship is being tied up to the quay, moles, etc. *See* **Eyebolt**.

Eyebolt An eye-shaped form of a bolt used as a recess for a hook. *See* **Eye**.

Eyes The two sides of a ship's bows near to the Hawse, *q.v.*, holes.

Eye Tackle A Tackle, *q.v.*, with a loop or an Eye, *q.v.*, on every block.

Eyrir A coin of Iceland, one-hundredth of a Krona—plural is aurar.

EZC European Zone Charge.

F

F Fahrenheit; Family; Far; Farad *electric capacitance*; Fast; February; Fellow; Ferrovia *Italian*—Train; Finance; Fine; Fixed Light *chart symbol*; Fluorine; Fog; Folio; Force; Forecastle *Lloyd's Register notation*; Founder; Franc; France; Frauen *German*—Women; Freddo *Italian*—Cold; Longitude Factor *nautical*; Phase Correction *nautical*; Altitude Factor; Friday; Symbol for First Class passenger *aviation*.

F *or* **f** Following; Female; Fathom; Fiat *Latin*—Let it be done; Fog; Foot/Feet; Forecastle *Lloyd's Register notation*; Forte *Italian*—Strong; Forward; Frequency; Force.

F.A. Factory Act; Family Allowance; Field Ambulance; Finance Act; Financial Adviser; Fireman Apprentice; Food Administration; Football Association.

Fa Firma *German*—Company; Florida, *USA*.

F.a. Fire Alarm; Free Alongside, *q.v.*, *see* **F.a.s**; Freight Advance, *see* **Freight Paid in Advance**; Freight Agent.

f.a. First Aid; Fire Alarm; First Attack; Free Alongside, *q.v.*; Free Aperture; Friendly Aircraft; Full Air.

F.a *or* **F.a.s.** Free Alongside *or* Free Alongside Ship.

f & a *or* **F & A** Fore and Aft.

FAA *or* **F.A.A.** *or* **F.a.a.** Federal Aviation Administration *USA*; Federal Aviation Agency *USA*; Federal Aviation Authority *USA*; Fellow of the Association of Anaesthetists; Fellow of the American Association for the Advancement of Science; Fellow of the Australian Academy of Science; Fleet Air Arm; Free of All Average, *q.v.*

f.a.a. Free of All Average, *q.v.*

F.A.A.A.S. Fellow of the American Academy of Arts and Sciences; Fellow of the American Association for the Advancement of Science.

FAACA Federal Aviation Administration Certificate of Approval *USA*.

F.A.A.O.S. Fellow of the American Academy of Orthopaedic Surgeons.

FAAP Federal Aid to Airport Program *USA*.

FAB Flown as booked *post flight confirmation*.

F.A.B. *or* **f.a.b.** Franco à Bord *French*—Free on Board, *q.v.*

Fabricated or Pre-fabricated Vessel A vessel built by assembling large sectional parts constructed in various departments, in a drydock or on a slipway. When the ship is complete, it can be undocked or launched. Mass production was achieved in this manner during the Second World War.

fac Facsimile; Factor; Factory; Faculty; Factum simile *Latin*—Facsimile, *q.v.*

Fac Facultative, *q.v.*

FAC *or* **F.A.C.** *or* **f.a.c.** Fast As Can, *q.v.*; Federal Appeals Court *USA*; Freight Abandonment Clause, *q.v.*; Fast Attack Craft *USA*; Forwarding Agency's Commission.

F.A.C.A. Fellow of the American College of Anaesthetists.

F.A.C.A.M. Fédération des Agents Consignataires de Navires et d'Agents Maritimes de France *French*—The federation of shipping and forwarding agents in France.

F.A.C.C.A. Fellow of the Association of Corporate and Certified Accountants.

FACCOP *or* **F.A.C.C.O.P.** *or* **faccop** *or* **f.a.c.c.o.p.** As Fast as Can, Custom of the Port. *Similar to* **Fast As Can**, *q.v.* A Charter Party, *q.v.*, term meaning that loading and discharging operations are to be executed as quickly as possible.

F.A.C.D. Fellow of the American College of Dentists.

Face to Face Selling Direct personal contact between seller and buyer without the intervention of an intermediary in the negotiation of the sale.

Face Value Nominal or par value of stocks, shares, bonds, debentures, etc. Value shown on the face of the bank note, bonds, debentures, etc. Does not necessarily mean that the up-to-date value of such notes is in conformity with the market value, as the latter fluctuates according to the rate of the purchase and the sale offering at that time.

F.A.C.I.A.A. Fellow of the Commercial and Industrial Artists' Association.

facil Facility.

Facile princeps *Latin*—Easily first.

Facilitation of International Maritime Traffic 1965 A meeting convened by IMO who designed a documentation to come into force on 2 June 1984 with a view to expedite international maritime traffic concerning the procedures for crew change, passenger and cargo clearance and to save substantial costs. *Abbrev.* F.A.L.

Facsimile The exact copy of reproduction. *Abbrev.* fac, facs *or* facsm.

FACMTA Federal Advisory Council of Medical Training *USA*.

F.A.C.O.G. Fellow of the American College of Obstetricians and Gynaecologists.

Façon de parler *French*—Way of speaking.

Fac/oblig Facultative/obligatory, *q.v.*

F.A.C.P. Fellow of the American College of Physicians.

FACS *or* **F.A.C.S.** Federation of American Controlled Shipping (FACS members are mostly the crews of Liberian and Panamanian vessels) formerly known as American Committee for Flags of Necessity. *Abbrev.* ACFN; Fellow of the American College of Surgeons; Fellow of the Association of Certified Secretaries of South Africa.

facs *or* **facsm** Facsimile, *q.v.*

Facsimile Exact copy of a picture or message transmitted instantly from one place to others by radio and/or landline. *Abbrev.* FAX or Fax. Also used to describe the transmitting/receiving instrument itself.

Facsimile Transmission Link between two or more companies in the courier services group. *Abbrev.* Fax.

fact Factura *Spanish*—Invoice, *q.v.*

FACT Federation against Copyright Theft.

Factor Agent who buys and sells merchandise on behalf of others; a wholesaler in a specific trade such as in shirts, boots, etc.

Factor Markets Demands on which the elements of production are negotiated in buying and selling.

Factor Payments Payments effected against the goods and services used in production.

Factors of Production Items used and services rendered in production. Similar to Agents of Production, *q.v.*

Factory Cost The main cost plus on-cost.

Factory Ship Special ships used in the whaling and fishing industry. They can undertake long voyages of several months before returning to their home bases. They are equipped with plant to process their catches ready for market, before returning to base.

Facts *or* **FACTS** Fully Automated Cargo Tracking System *electronics*.

Factum *Latin*—Anything that is made a fact; Deed; Contract.

Facultative Right of option in the sense that Underwriters, *q.v.*, have the right to decide whether they accept an insurance risk or not. *Abbrev.* Fac.

Facultative Insurance *or* **Placing** Single Voyage Marine Insurance, *q.v.*

Facultative/Obligatory Similar to Facultative, *q.v.*, but the procedure is between the re-insurer and re-assured. *Abbrev.* Fac/Oblig.

Facultative Re-Insurance Underwriter's, *q.v.*, procedure in presenting every risk individually in the re-insurance market. When all the re-insurances are placed the Underwriter will then accept the insurance risks.

Faculty A University Department.

FAD Fleet Air Detachment *USA*.

F.A.D. *or* **f.a.d.** Free Air Delivered.

FAEB Federation of Associations of Periodical Publishers.

Faellesbank Danish and Norwegian Central Bank which also performs the services of a savings bank and other duties.

Faer Faeroe Islands.

FAF Free at Field *aviation*: when chartering an aircraft delivery is taken at a specified airfield; Fuel Adjustment Factor, *q.v.*

Faggot A bundle of sticks or twigs bound together as fuel; a bundle of steel rods; 120 lbs of steel.

FAGO *or* **F.A.G.O.** Fellow of the American Guild of Organists.

FAGS *or* **F.A.G.S.** Federation of Astronomical and Geophysical Services; Fellow of the American Geographical Society.

F. Ahd Full Ahead.

Fahr Fahrenheit.

FAI *or* **F.A.I.** Federation Aeronautique Internationale *French*—International Aeronautical Federation; Fellow of the Auctioneers' and Estate Agents' Institute; Fresh Air Inlet.

FAIA *or* **F.A.I.A.** Fellow of the American Institute of Architects; Fellow of the Association of International Accountants; Fellow of the Australian Institute of Advertising.

F.A.I. ex. Fellow of the Australian Institute of Export.

F.A.I.H.A. Fellow of the Australian Institute of Hospital Administration.

F.A.I.I. Fellow of the Australian Insurance Institute.

F.A.I.M. Fellow of the Australian Institute of Management.

F.A.I.P. Fellow of the Australian Institute of Physics.

F.A.I.P.M. Fellow of the Australian Institute of Personnel Management.

FAIR *or* **F.A.I.R.** Federation of Afro-Asian Insurers and Re-Insurers.

Fair Average Quality In the grain trade the purchaser is entitled to have a 'fair average quality' which bears comparison with similar shipments of grain during the same month. Grains also cover any kind of cereals such as peas, beans, etc. *Abbrev.* FAQ or F.A.Q. *or* F.a.q.

Faire suivre s'il vous plâit *French*—Please forward.

Fairleads Small openings in a vessel's Bulwarks, *q.v.*, through which mooring ropes can be led inboard directly to mooring winches or via bollards. Can be of various designs, some little more than holes with strengthened edges, others fitted with rollers to help prevent chafing of the rope.

Fair Holes Round holes on metal sheets made exactly against each other which enable rivets to pass through easily. Opposite to Blind Holes, *q.v.*

Fair Tide A tidal current which increases the speed of the vessel.

Fair Trade Policy A policy recognised by all governments in granting reciprocal concessions or restrictions in trade between themselves. Also known as 'Beggar-my-neighbour policy', *q.v.*

Fairway An area of deep water or a channel outside the entrance to a harbour. Similar to Mid-Channel. *See* **Fairway Buoy**.

Fairway Buoy A buoy stationed outside the entrance to a harbour or in a channel, used as a marker for navigation. It is generally painted Black and White or White with vertical stripes in Red and White. A Fairway Buoy is also occasionally called a Mid-Channel Buoy.

Fairway Speed Speed to be observed by all traffic navigating within a Fairway, *q.v.*

Fair Wear and Tear Excepted A Time Charter-party, *q.v.*, expression referring to the condition of a ship as a whole upon re-delivery. The normal wear and tear which would have taken place while the vessel was on hire is considered part of the contract. It is used with the following clauses: 'Vessel to be delivered (fair wear and tear excepted)'; 'Vessel to be re-delivered in like good order and condition (fair wear and tear excepted)'. The vessel is to be given back on the expiration of the charter in the same good order as when delivered to the charterers but fair wear and tear which occurred during the time of the charter is to be ignored or excepted.

Fair Wind A favourable wind which assists a ship in the course of her voyage.

F.A.I.S. Fellow of the Amalgamated Institute of Secretaries.

Fait accompli *French*—An accomplished fact; a thing already done.

fak Faktura *German*—Invoice.

fak *or* **FAK** First Aid Kit; Freights All Kind, *q.v.*

Fake One turn in a coil of rope.

FAKS Fault Avoidance Knowledge System *diesel engine monitoring*.

FAL 1965 Facilitation of International Maritime Traffic 1965, *q.v.*

FAL Forms Short for Standardized Facilitation Forms, issued by IMO in pads of 50. *See* **Facilitation of International Marine Traffic 1965**.

Falk I. *or* **Falk Is** Falkland Islands.

Fall The Whip or tackle with which a boat or raft is lowered from a ship or hoisted out of the water. When a boat is hooked on to a forward and an after fall, these are collectively 'the falls'.

Fall Astern When a ship is passed by another and remains behind or astern.

Falling Market Declining market; when the value of goods falls in the market. Generally, dealers hold on to their stocks rather than sell at a loss.

Falling Tide Decreasing depth of the water *nautical*.

Fall-Out A radioactive dust following a nuclear explosion.

Fall-pipe Ship A medium-size vessel which carries bulk aggregates, stone or riprap and is fitted with conveyors and a telescopic pipe by which the material is precisely deposited, with the aid of dynamic positioning, to bury and protect seabed pipelines.

Falls The ropes and blocks used to lower the lifeboats.

F.A.L.P.A. Fellow of the Incorporated Society of Auctioneers and Landed Property Agents.

Falsa demonstratio non noget *Latin*—A false statement does not alter or cancel a contract or a document.

False Bay In Cape Province, South Africa, at the Cape of Good Hope.

Falsus in uno, falsus in omnibus *Latin*—False in one thing, false in everything.

fam Family; Familiar.

f.a.m. *or* **F.A.M.** Foreign Airmail; Free at Mill *in grain trade*.

FAMA *or* **F.A.M.A.** Fellow of the American Medical Association.

F.A.M.E. Filippino Association of Marine Employers.

F.A.M.E.M.E. Federation of Associations of Mining, Electrical and Mechanical Engineers.

F.A.M.H.E.M. Federation of Associations of Materials, Handling Equipment Manufacturers.

Famosus libellus *Latin*—A scandalous libel.

F.A.N.Y. First Aid Nursing Yeomanry.

FANZAAS *or* **F.A.N.Z.A.A.S.** Fellow of the Australian and New Zealand Association for Advancement of Science.

FAO Food and Agriculture Organisation *United Nations special agency*; Fleet Accountant Officer.

FAO *or* **F.A.O.** *or* **f.a.o.** Finish All Over; For the attention of.

FAP Future Annual Premium *insurance*.

F.A.P. First Aid Post; Força Aerea Portuguesa *Portuguese Air Force*.

F&AP *or* **F.&A.P.** Fire and Allied Perils *insurance*.

FAPHA *or* **F.A.P.H.A.** Fellow of the American Public Health Association.

FAPS *or* **F.A.P.S.** Fellow of the American Physical Society.

FAQ *or* **F.A.Q.** *or* **F.a.q.** Fair Average Quality, *q.v.*; Free Alongside Quay, *q.v.*

FAQS *or* **F.A.Q.S.** *or* **F.a.q.s.** Fair Average Quality of Season. *See* **Fair Average Quality**.

FAR *or* **F.A.R.** False Alarm Rate; Federation Aviation Regulations *USA*; Federation of Arab Republics; Free of Accident Reported *insurance*.

Farad Unit of electrical capacitance, being the capacitance of a capacitor which has a potential of 1 volt between its plates when charged with one coulomb.

Far East Regional Investigation Team A team of experts comprising special officers of the Salvage Association, an Average Adjuster, *q.v.*, a London maritime lawyer and a regional insurance claims manager formed to investigate information involving cargo frauds and mysterious sinkings which had arisen under cargo policies, most especially in the Far East. *Abbrev.* FERIT.

Fare Price paid by, or charged to, a traveller for his passage by road, rail, sea or air, inclusive of baggage to a certain limit, from one place to another.

Farewell Whistle When the master of a vessel wishes to send a farewell signal when leaving harbour, he gives three prolonged blasts on the ship's whistle. If the vessel wishes to salute another vessel a similar signal is sounded. The other party acknowledges the salute with a reciprocating three blasts and in turn another single whistle signal is given by the original saluting vessel.

Farm In To take over the financial part of an enterprise which would have been disrupted or halted because of the continued capitalisation needed for the venture. *Ex.* The costs incurred in oil exploration are such that considerable finance is needed in order to undertake the venture successfully. *See* **Farm Out**.

Farm Out In a large enterprise where unlimited sums of money are needed a consortium comprising companies and bankers is formed in order to obtain ultimate commercial success. One of the companies of the group thus formed farms out or shares in the venture with others, otherwise it would be unable to supply the enormous amount of finance required to continue with the venture. *See* **Farm In**.

FARP Fully Automated Radar Plotting.

FAS Flexible Accounting Scheme *customs*; Free Arrival Station *containers*; Free Alongside Ship *INCOTERM*; Firsts and Seconds *timber*; Free Alongside, *q.v.*

f.a.s. First and Seconds; Free Alongside Ship, *q.v.*

F.A.S. Faculty of Architects and Surveyors; Federation of American Scientists *USA*; Fellow of the Anthropological Society; Fellow of the Antiquarian Society.

FASA or **F.A.S.A.** Federation of ASEAN Shipowners Associations; Fellow of the Acoustical Society of America; Fellow of the Australian Society of Accountants.

F.A.S.A.P. Fellow of the Australian Society of Animal Production.

FASB Financial Accounting Standards Board, *q.v.*

FASC Federation of Asian Shippers' Councils.

fasc Fasciculus *Latin*—Bundle.

F.A.S.C. Fellow of the American Society of Civil Engineers.

FASRON Fleet Aircraft Service Squadron *USA*.

Fast Fixed, attached, secured.

FAST or **F.A.S.T.** First Automatic Ship Transport; Fixed Activated Sludge Treatment (for sewage plants); Formula for Assisting the Specification of Trains; Factor Analysis System; Fast Automatic Shuttle Transfer.

Fast As Can Charterparty, *q.v.*, expression. An obligation on the part of the Charterers, *q.v.*, or agents in a charterparty to despatch the ship as soon as possible and without undue delay. This, of course, is on the understanding that it is in accordance with the Custom of the Port, *q.v. Abbrev.* FAC or F.A.C. or f.a.c. *Similar to* **As fast as can**, *q.v.*

Fast Ice Sea ice attached to the shore, to an ice wall, to an ice front or between shoals and grounded icebergs. It may extend for a few metres or hundreds of miles. If more than two metres above sea level, it becomes an ice shelf.

Fast Lane (1) The outer traffic lane on a motorway used for overtaking. (2) Slang for a style of high living.

F. Astn Full Astern.

FAT Final Acceptance Trial *USA*.

F&T Fire and Theft *insurance*.

fat or **fath.** Fathom, *q.v.*

Fathom (1) Nautical measure of 6 feet or 1.8288 metres. (2) Timber measure of 216 cubic feet (6'×6'×6').

Fathometer *Similar to* **Echo Sounder** and **Acoustic Sounding**. *See* **ESD**.

FATIS or **F.A.T.I.S.** Food and Agriculture Technical Information Service.

F.A.U. Friend's Ambulance Unit.

Fault Free Analysis The system carried out by Lloyd's Register of Shipping in determining the risk analysis in the three phases of gas freeing in chemical tankers in the process of loading, discharging and the voyage. *Abbrev.* FFA.

Faute de mieux *French*—For want of better.

Faux pas *French*—A slip in behaviour; a false step.

fav Favour, *q.v.*

FAV or **Fav.** First Available Vessel.

FAVO Fleet Aviation Officer.

Favour A word sometimes used in commercial correspondence. *Abbrev.* fav.

Favourable Balance or **Favourable Payments** or **Favourable Trade** *See* **Active Balance of Payments**.

Favourable Current *Similar to* **Fair Tide**, *q.v.*

Favourable Wind *Similar to* **Fair Wind**.

FAX Facts; Fuel Air Explosion; Telefax.

Fax or **FAX** Facsimile, *q.v.*; Facsimile Transmission, *q.v.*; Telefax.

FAXCOM Facsimile Communications.

Fazendas Large estates in the coffee growing areas of Brazil.

f.b. Flat Bar; Fog Bell; Freight Bill.

FB Fog Bell; Flinder's Bar.

FB *or* **F.B.** Fishery/Fisheries Board; Fire Brigade; Flat Bar; Flat Bottom; Flying Boat.

FBA *or* **F.B.A.** Federal Bar Association *USA*; Federation of British Astrologers; Fellow of the British Academy; Freshwater Biological Association.

F.B.A.A. Fellow of the British Association of Accountants and Auditors.

FBC Fluidised Bed Combustion *steam generators*.

F.B.C.M. Federation of British Carpet Manufacturers.

F.B.C.P. Fellow of the British College of Physiotherapists.

F.B.C.S. Fellow of the British Computer Society.

fbd *or* **Fbd** Freeboard, *q.v.*

F.B.E.A. Fellow of the British Esperanto Association.

F.B.G. Federation of British Growers.

FBH Fire Brigade Hydrant; Free on Board Harbour.

FBI Federal Bureau of Investigation *USA*; Federation of British Industries, now amalgamated with CBI *Confederation of British Industries*.

F.B.I.A. Fellow of the Bankers' Institute of Australasia.

F.B.I.M. Fellow of the British Institute of Management. Obsolete. *See* **FIMgt.**

FBM *or* **fbm** Feet Board Measure, *see* **Board Measure**; Fleet Ballistic Missile.

F'bm *or* **fbm** Feet Board Measure, *see* **Board Measure**.

F.B.O.A. Fellow of the British Optical Association.

FBP Federation of British Ports; Final Boiling Point; Foreign Buyers Program.

F.B.Ps.S. Fellow of the British Psychological Society.

FBR Fast Breeder Reactor; forward Buying Rate.

F.B.R.A.M. Federation of British Rubber and Allied Manufacturers.

F.B.S. Fellow of the Botanic Society.

F.B.S.C. Fellow of the British Society of Commerce.

FC *or* **F.C.** Federal Cabinet; Fire Control; Fisheries Convention; Full Claim *insurance*.

f.c. File Cabinet; Fire Clay; Follow Copy *printing*; Foot Candle; For Cash; Fuel Cell.

F&C Full and Change.

FCA Farm Credit Administration *USA*; Federation of Canadian Artists; Free Carrier, *q.v.*

F.C.A. Fellow of the Institute of Chartered Accountants.

F.C.A. (Aust) Fellow of the Institute of Chartered Accountants in Australia.

F.C.A.A. Fellow of the Australasian Institute of Cost Accountants.

F.Cant Forward Cant Frames.

fcap Foolscap.

FCAR *or* **F.C.A.R.** *or* **f.c.a.r.** Free of Claim for Accident Reported, *q.v.*

F.C.B.A. Fellow of the Canadian Bankers' Association.

FCC Fully Cellular Containership; Freight Connectors Coalition, USA, formed in 1995.

FCC *or* **F.C.C.** Federal Communications Commission *USA*; Federal Council of Churches; First Class Certificates; Food Control Committee; Freight Container Certificate; French Chamber of Commerce.

F-CCA Florida–Caribbean Cruise Association.

F.C.C.S. Fellow of the Corporation of Certified Secretaries.

FCDA Federal Civil Defense Administration *USA*.

F.C.E. Fellow and Chartered Engineer.

fcg Facing.

F.C.G.I. *or* **FCGI** Fellow of the City and Guilds of London Institute.

F. chgs Forwarding Charges (charges due to Forwarding Agents).

fchse Franchise.

FCI *or* **F.C.I.** Fellow of the Clothing Institute of Commerce; Finance Corporation for Industry; Fluid Control Institute *USA*; Food Corporation of India.

F.C.I.A. Fellow of the Corporation of Insurance Agents; Foreign Credit Insurance Association.

F.C.I. Arb. Fellow of the Chartered Institute of Arbitrators.

FCIB Fellow of the Chartered Institute of Bankers; Fellow of the Corporation of Insurance Brokers.

FCIC *or* **F.C.I.C.** Fellow of the Chemical Institute of Canada.

F.C.I.I. Fellow of the Chartered Insurance Institute.

F.C.I.L.A. Fellow of the Institute of Loss Adjusters.

F.C.I.P.A. Fellow of the Chartered Institute of Patent Agents.

F.C.I.S. Fellow of the Chartered Institute of Secretaries.

F.C.I.T. Fellow of the Chartered Institute of Transport.

F.C.I.V. Fellow of the Commonwealth Institute of Valuers.

FCL *or* **F.C.L.** Full Container Load, *q.v.*

FCL/LCL Full Container Load/Less than Container Load. Generic term referring to a container service offering both possibilities.

FCM Fire Control Module *electronics*.

FCMA Fisheries Conservation Management Act *USA*.

F.C.M.A. Fellow of the Institute of Cost and Management Accountants.

F.C.M.I.E. Fellow of the Colleges of Management and Industrial Engineering.

F.C.M.S. Fellow of the Society of Consulting Marine Engineers and Ship Surveyors.

F.C.O. Foreign and Commonwealth Office.

Fco *or* **fco** Franco Abordo *or* Franco *Italian*—Free of Expenses to the ship. *Similar to* **Free In & Out**, *q.v.*; Fair Copy.

FCOD Fire, Collision, Overturning and Derailment *insurance*.

F.C.O.G. Fellow of the College of Obstetricians and Gynaecologists. *Similar to* **F.R.C.O.G.**, *q.v.*

F.Comm. A. Fellow of the Society of Commercial Accountants.

F.C.P. Fellow of the College of Preceptors.

FCPO *or* **F.C.P.O.** Fleet Chief Petty Officer.

FCR *or* **F.C.R.** Forwarders Certificate of Receipt *or* Forwarding Agent's Certificate of Receipt, *q.v.*

FCRA *or* **F.C.R.A.** Fellow of the College of Radiologists of Australasia; Fellow of the Corporation of Registered Accountants.

F.C.S. Fellow of the Chemical Society.

f.c.s. *or* **f.c.&s.** *or* **F.C.&S.** Free of Capture and Seizure, *q.v.*

FCSA Fellow of the Institute of Chartered Secretaries and Administrators.

FCSI Foodservice Consultants Society International.

fcsle Forecastle, *q.v.*

f.c.s.s.r.c. *or* **F.C.S.S.R.C.C.** Free of Capture and Seizure etc and of Strikes, Riots and Civil Commotions, *q.v.*

FCST Fellow of the College of Speech Therapists.

FCT Forwarders Certificate of Transport; Forwarding Agent's Certificate of Transport; Frequency Changing Transformer, *q.v.*

f.c(t) Fin courant *French*—At the end of the month.

FCU Fighter Control Unit.

FCV *or* **F.C.V.** Full Contract Value *or* Full Completed Value (referring to the insurance of new-building ships).

F.C.W.A. Fellow of the Institute of Cost and Works Accountants. *Similar to* **F.C.M.A.**, *q.v.*

FD Floating Dock; Forced Draft (Draught); Frequent Diversity.

fd Fjord; Forward; Founded; Fund.

F.D. Fidei defensor *Latin*—Defender of the Faith, *also abbrev.* fid def; Fire Department; Forced Draught *Lloyd's Register notation*; Free Discharge; Free Docks.

f.d. *or* **F/d** Free Delivery, *q.v.*; Free Despatch *or* Dispatch, *q.v.*; Free Docks; Flight Deck (of a ship); Focal Distance.

f&d *or* **F&D** Fill and Drain; Freight and Demurrage.

FDA Food and Drugs Administration; Food Distribution Administration.

FDC Federal District Court *USA*; Fire Direction Center *USA*; First Day Cover (relating to stamps).

FD&C Food Drug and Color Regulations *USA*.

f.d.c. Fleur de coin *French*—Mint (in stamps).

FDD *or* **F.D.D.** Francs de droits *French*—Free of charges.

F.D.&D. Freight, Demurrage & Defence, *in P.& I. insurance*, *q.v.*

FDDI Fibre distributed data interface.

FDIC *or* **F.D.I.C.** Federal Deposit Insurance Corporation *USA*.

FDL Fast Development Logistic Ship.

Fdn Foundation (in connection with shipbuilding).

FDO *or* **F.D.O.** For Declaration Purposes Only (used in connection with contracts and insurance).

FDR Flight Data Recorder *aviation*.

Fdr Founder.

F.dr/s Fluid dram(s), equivalent to 60 grains avoirdupois.

FDS Fleet Data Service, Houston, *USA*.

F.D.S. Fellow in Dental Surgery.

F.D.S.R.C. Fellow in Dental Surgery of the Royal College of Surgeons.

F.D.S.R.C.Ed. Fellow in Dental Surgery of the Royal College of Surgeons of Edinburgh.

FD Zug Fernschnellzug *German*—Long Distance Express.

Fe Ferrum *Latin*—Iron.

FE *or* **F.E.** Far East; Foreign Edition; Forward End; Free of Encumbrance, *q.v.*; symbol for iron.

f.e. First Edition; For Example.

Fe₂O₃ Fe_2O_3 Oxide of iron, rust. The result of corrosion which occurs only in the presence of both water and oxygen. Rusting is accelerated by impurities in the iron and by the presence of acids and other electrolytes in the water.

FEA Federal Executive Association; Federal Energy Authority; *Similar to* **USFEA**, *q.v.*; Foreign Economic Administration.

FEANI Fédération Européenne d'Associations Nationales d'Ingénieurs. *French*—European Federation of National Engineers Associations, *in Netherlands*.

Featherbedding The employment of unnecessary extra workers on a job because of trade union insistence.

Feather the Oar To position the blade of an oar horizontally.

Feathering Screw Propeller *See* **Controllable Propeller**.

Feb February.

feb *or* **febb** Febbraio *Italian*—February.

FEB *or* **F.E.B.** Foreign Exchange Broker.

F.E.B.A. Foreign Exchange Brokers' Association.

Fe.Bd. Free Board.

FEC Forward Error Correcting (radio broadcasting method); Far Eastern Command.

F.E.C.B. Foreign Exchange Control Board.

FECE Far Eastern Conference.

FECT or **F.E.C.T.** Far East Conference on Tariffs.

Fed Federal; Federation.

F.E.D.C. Federation of Engineering Design Consultants.

FEDERAGENTI Federazione Nazionale Agenti Raccomandatori Marittimi Agenti Aerei e Pubblici Mediatori Marittimi. The Italian Federation of the National Association of Shipbrokers and Agents who also act as mediators in mercantile and aviation matters.

Federal Courts Courts of law in the United States.

Federal Reserve Bank A US Federal or Central Bank established in 1913 for the purpose of regulating finance with its headquarters in Washington. It has 12 Federal Reserve Districts. *Abbrev.* F.R.B.

Federal Water Pollution Control Act The principal United States domestic statute enacted for the strict control of marine pollution around the coasts. These regulations form part of United States Coast Guard activities. *Abbrev.* FWPCA.

Federation Union Union of professional bodies or states or provinces under a common head or government as is commonly practised in e.g. the *USA* and Australia.

FEDIS FIATA EDI Systems Co-operative Society.

Feedback Form of interference when transmitting by radio telephone.

Feeder Lash or **Feeder for Lighter Aboard Ship** A Lash ship (or Lighter Aboard Ship, *q.v.*). Identical to a small dry dock which when submerged can accommodate eight to 20 barges and when refloated can sail for her destination. The barges can be refloated and taken out by again submerging the FLASH, as it is abbreviated.

Feeder Services Shipping traffic employed moving cargoes which are unloaded by large vessels in major ports for re-shipment or transhipment to their destinations in smaller vessels. The purpose of this is to economise on the operation of the larger vessels by avoiding delays and expense at the smaller ports, apart from the fact that there could be a Bar Draft, *q.v.*, or other difficulties for the entry and berthing there of the large ships. *See* **Feeder Vessels**.

Feeder Vessels Ships used to carry transhipment cargoes to various destinations after having been discharged from larger vessels at a single port. *See* **Feeder Services**.

Feed Water The replacement water pumped into a boiler system to make good for that lost in operation.

Feel the Bottom Nautical term which refers to difficult steering and the reduction of normal speed when a ship is navigating in shallow waters.

Fee Simple Inherited estate without limitation.

Feet In metric system one foot equals 30.48 centimetres; *abbrev.* ft.

F.E.F.C. Far Eastern Freight Conference.

F.E.F.C.O. Fédération Européenne des Fabricants de Carton Ondulé *French*—Federation of Corrugated Container Manufacturers.

FEIA or **F.E.I.A.** Flight Engineers' International Association.

fel or **Fell** Fellow.

Fel.AIEE or **Fel.A.I.E.E.** Fellow of the American Institute of Electrical Engineers.

Felix culpa *Latin*—A fortunate mistake.

Fellow Subsidiaries Sister companies or firms.

Fellurometer An infra-red distance measuring instrument to obtain accurate differential measuring techniques.

Felo de se *Latin*—Suicide.

Felony A major crime.

Fe_2O_3 White Rust or Red Oxide of Iron.

fem or **f.e.m.** Forza elettromotorica *Italian*—Electromotive Force.

Feme covert Married woman.

Feme sole A spinster or widow.

Fender A sort of protection used by a ship's crew to prevent her from chafing or rubbing against another ship, quay or rocks. A fender may be of various materials and shapes including tyres and/or oblong or round objects made of rubber, plastic, canvas or fibrous material.

FENEX Federatie Van Nederlandse Expedteursorganisaties, *Dutch* Organisation of Nederland's Forwarding Agents.

F.E.O. Fleet Engineer Officer.

FEPEM or **F.E.P.E.M.** Federation of European Petroleum Equipment Manufacturers.

Ferae naturae *Latin*—Of wild nature or disposition.

FERC Federal Energy Regulatory Commission *USA*.

FERIT Far East Regional Investigation Team.

Ferm. Fermentation.

Ferr. Ferrovia *Italian*—Railway.

Ferry Ship or boat used to convey passengers, with their baggage, road vehicles, or railway carriages and wagons on a regular schedule between two or more points. May refer to an aircraft, car, bus or lorry similarly engaged.

Ferry Port *Similar to* **Packet Port**, *q.v.*

Ferrous Metals *See* **Non-Ferrous Metals**.

FERTICON Code name for Chamber of Shipping Charterparty 1942, for the carriage of fertilizers.

FERTIVOX 88 Revised North American Fertilizer Charter Party 1988.

Fertility Rate Birth rate.

Fertilizers Soil manure in form of salt obtained from the earth/rocks. The following are some: ammonium phosphate; ammonium sulphate; basic slag also known as sodium nitrate; ammonium nitrate; chili nitrate also known as sodium nitrate; ground rock phosphate; kainit(e), kalisalt; nitrate of soda; potash or muriate of potash; potassium sulphate; superphosphate and urea.

F.E.S. Federation of Engineering Societies.

FESCO Far East Shipping Conference Company.

Festina lente *Latin*—Hasten slowly.

FET Field Effective Transistor; Federal Excise Tax *USA*.

FETS Field Effective Transistors.

FEU Forty-foot equivalent unit container.

fev. Fevereiro *Portuguese*—February; Février *French*—February.

FEWAC Far East/West Africa Conference.

FEX Fleet Exercise *USA*.

ff. Thick fog *nautical*; Folios; Following; Fortissimo (very loud) *music*.

f.f. Fixed focus.

F.f. Fortsetzung folgt *German*—To be continued.

f&f Fittings and fixtures.

f.f.a. *or* **F.F.A.** Free From Alongside, *q.v.*; Free of Foreign Agency; Free Fatty Acid.

FFA Fault Free Analysis, *q.v.*

FFA *or* **F.F.A.** Fire Fighting Appliances; Fellow of the Faculty of Actuaries.

FFAPC *or* **F.F.A. per cent.** Free Fatty Acidity per cent.

FFAS *or* **F.F.A.S.** Fellow of the Faculty of Architects and Surveyors.

FFC Foreign Funds Control *USA*.

FFCC Forward Facing Cockpit Crew *aviation*.

FFF Air Ticket *or* **F.F.F. Air Ticket** Fly Full Fare. Also expressed as Triple F. Economy Class Ticket. IATA air ticket of economy class.

F.F. Hom. Fellow of the Faculty of Homoeopathy.

F.F.I. Fellow of the Faculty of Insurance; Finance for Industry; For further instructions.

f.f.i. *or* **F.F.I.** Free from Infection.

FFIA Federal Food Indemnity Administration *USA*.

F.Fl. Fixed & Flashing light, Visual Lights *nautical*.

F.F.O. Fixed and Floating Objects *insurance*.

F.F.P.S. Fellow of the Faculty of Physicians and Surgeons.

F.F.P.S.G. Fellow of the Faculty of Physicians and Surgeons Glasgow.

F.F.R. Fellow of the Faculty of Radiologists.

FFSA Firth of Forth Shipowners' Association, Leith, *UK*.

FF.SS. Ferrovie dello State *Italian State Railways*.

ffss Full Frequency Stereophonic Sound.

fly *or* **ffly** Faithfully.

FG Ariana Afghan Airlines *airline designator*.

FG *or* **F.G.** Federal Government; Fire Guard; Fog Over Land *meteorology*; Friction Glazed *paper*; Foreign Going, *q.v.*; Fully Good.

fg Fine Grain *in leather*; Fog Over Land; Friction Glazed *paper*.

f.g. Fully good.

f.g.a. *or* **F.g.a.** *or* **F.G.A.** Free of General Average, *q.v.*; Foreign General Average, *q.v. See also* **Foreign General Average Clause**.

FGC Fifth Generation *computer*.

FGD Flue Gas Desulphurisation, a measure to reduce air pollution.

f.g.f. Fully Good Fair.

F.G.I. For Guidance and Information.

FGMDSS Future Global Maritime Distress and Safety System.

Fgn Foreign; Foreigner.

F.G.S. Fellow of the Geological Society.

FGSA *or* **F.G.S.A.** Fellow of the Geological Society of America.

fgt. Freight.

F/H *or* **F.H.** Fire Hydrant; Free hold; Field Hospital; Fore Hatch; Fog Horn, *q.v.*

f.h. *or* **F.h.** First half of the month; Fog Horn, *q.v.* Fore Hatch; Free Hatch.

FHA Federal Housing Administration *USA*; Finance Housing Administration; Finance Houses Association.

F.H.C.I. Fellow of the Hotel and Catering Institute.

FHEX *or* **F.H.E.X.** Fridays and Holidays Excepted *or* Excluded.

FHINC *or* **fhinc** Fridays and Holidays Included, *q.v.*

FHLBB Federal Home Loan Bank Board *USA*.

Fhld *or* **f'hold** Freehold.

fhp *or* **FHP** *or* **F.H.P.** Friction Horse Power.

FHR Flame and Heat Resistant.

FHR *or* **F.H.R.** Federal House of Representatives *Australia*.

FHWAC *or* **F.H.W.A.C.** First High Water After Completion (in statement of facts—Charterparty, *q.v.*).

FI Icelandair *airline designator*.

F.I. Falkland *or* Farne *or* Fiji Islands; Fire Insurance.

f.i. For instance; Free In, *q.v.*

FIA *or* **F.I.A.** Fédération Intérnationale de l'Automobile *French*—International Federation of Automobiles; Fellow of the Institute of Actuaries.

f.i.a. *or* **P.P.I.** *or* **Interest** *or* **No Interest** Full Interest Admitted *or* Policy Proof of Interest. *See* **Interest** *or* **No Interest**.

FIAC *or* **F.I.A.C.** Fellow of the Institute of Company Accountants.

FIAI *or* **F.I.A.I.** Fellow of the Institute of Industrial and Commercial Accountants.

F.I.A.I.I. Fellow of the Incorporated Australian Insurance Institute.

F.I.A.M. Fellow of the International Academy of Management.

F.I.A.M.A. Fellow of the Incorporated Advertising Managers' Association.

F.I.A.N.Z. Fellow of the Institute of Actuaries of New Zealand.

FIAS or **F.I.A.S.** Fellow Surveyor Member of the Incorporated Association of Architects and Surveyors.

FIAT Fabbrica Italiana Automobile Torino *Italian—motor car, company, and factory.*

FIATA Fédération Internationale des Associations de Transitaires et d'Assimilés. *French*—International Federation of Freight Fowarders Associations.

Fiat Money Inconvertible paper money made legal tender.

FIAV or **F.I.A.V.** Fédération Internationale des Agences de Voyage *French*—International Federation of Travel Agents.

F.I.B. Fellow of the Institute of Building.

FIB or **F.I.B.** or **f.i.b.** Free into Barge; Free into Bond; Free into Bunkers, *q.v.*; First Aid Box.

FIBC Flexible Intermediate Bulk Container: A container in the form of a bag or sack which can replace drums for the transport of substances in bulk.

Fibre or **Fibre Ropes** Ropes which are manufactured from vegetable fibres such as sisal, cotton, manilla, hemp, coir, flax, etc.

FIBV World Federation of Stock Exchanges.

FIC Frequency Interference Control.

F.I.C. Fellow of the Institute of Chemistry.

FICA Federation Insurance Contributions Act *USA*; Food Industries Credit Association.

F.I.C.A. Fellow of the Commonwealth Institute of Accountants; Fellow of the Institute of Chartered Accountants.

F.I.C.C.I. Federation of Indian Chambers of Commerce and Industry.

F.I.C.E.A. Fellow of the Association of Industrial and Commercial Executive Accountants.

F.I.Chem.E. Fellow of the Institute of Chemical Engineers.

F.I.C.M. Fellow of the Institute of Credit Management.

F.I.C.S. Fellow of the Institute of Chartered Shipbrokers, *q.v.*; Fellow of the International College of Surgeons.

F.I.C.S.A. Fellow of the International Civil Servants' Associations.

FICT or **F.I.C.T.** Fédération Internationale de Centres Touristiques *French*—International Federation of Tourist Centres.

F.I.C.W.A. Fellow of the Institute of Cost and Works Accountants.

Fid A tapered and pointed piece of round wood used for rope splicing or for separating the strands of a rope.

FID Flame-Ionization Detection; Photo-Ionization Detection.

FID or **F.I.D.** Falkland Islands Dependencies; Fédération Internationale de Documentation *French*—International Federation for Documentation; Fellow of the Institute of Directors; Field Intelligence Department.

Fiddle Static or adjustable framework around dining tables on board a ship which prevent dishes and other utensils from sliding off during rough seas.

Fiddley Casing or grating above the engine room or boiler room; the upper part of a boiler room below the funnel.

Fidei defensor *Latin*—Defender of the Faith.

Fidelity Guarantee Guarantee based on the integrity and loyalty of the person doing it.

Fidelity Guarantee Insurance An insurance taken out by employers to indemnify themselves against any financial loss suffered due to theft by employees.

FIDENAVIS Spanish Ship Classification Society, having its main office in Madrid, Spain.

Fides et justitia *Latin*—Fidelity and justice.

FIDH or **F.I.D.H.** Fédération Internationale des Droits de l'Homme. *French*—International Federation for Human Rights.

FIDIC or **F.I.D.I.C.** Fédération Internationale des Ingénieurs-Conseils. *French*—International Federation of Consulting Engineers.

FIDO Fog Investigation Dispersal Operation.

Fiduciaria *Italian*—Society.

Fiduciary Good Faith; In Trust; A Trustee.

Fiduciary Issue A Bank of England security issue backed by government.

Fiduciary Loan Loan given without any guarantee or security, but simply on the good faith of the receiver of the loan.

F.I.E.D. Fellow of the Institution of Engineering Designers.

F.I.E.E. Fellow of the Institution of Electrical Engineers.

Field Used in reference to exploration and development. An area where oil or gas has been found in commercially profitable quantities.

F.I.E.M. Fédération Internationale de l'Enseignement Ménager *French*—International Federation of Home Economies.

F.I.E.R.E. Fellow of the Institute of Electronic and Radio Engineers.

Fieri Facias *Latin*—Cause it to be done. A writ of execution to levy a judgment debt. It commands the authorities to recover the sum due out of the goods of the debtor, including accrued interest. *Abbrev.* fi.fa.

FIESTA Finite Element Structural Analysis *Lloyd's Register.*

F.I.F. Fellow of the Institute of Fuel.

FIFA or **F.I.F.A.** Fellow of the Institute of Shipping and Forwarding Agents; Fédération Internationale de Football Association *French*—International Football

167

Federation; Fellow of the International Faculty of Arts.

fi.fa. *Latin*—Fieri Facias, *q.v.*

F.I.F.E. Fellow of the Institution of Fire Engineers.

Fiferail Rail which takes a number of Belaying Pins, *q.v.*

FIFO *or* **fifo** First In First Out, *q.v.*

Fifth Freedom The right to transport passengers between two airports which are foreign to the carrying aircraft's nationality.

Fighting Ship A shipping term for a Conference Line Ship, *q.v.*, which is engaged to compete on a cutthroat competition basis with other tramp ships. Any loss sustained by this vessel is covered by the conference organisation.

F.I.G.M. Fellow of the Institute of General Managers.

F.I.H.E. Fellow of the Institute of Health Education.

F.I.I. Fellow of the Imperial Institute. *Also abbrev.* F.I.Inst.

F.I.I.A. Fellow of the Indian Institute of Architects; Fellow of the Institute of Industrial Administration.

F.I.I.M. Fellow of the Institute of Industrial Management.

F.I.Inst. Fellow of the Imperial Institute. *Also abbrev.* F.I.I.

F.I.I. Tech. Fellow of the Institute of Industrial Technicians.

F.I.J. Fellow of the Institute of Journalists.

F.I.L. Foreign Insurance Legislation.

F.I.L.A. Fellow of the Institute of Loss Adjusters.

FILD *or* **F.I.L.D.** Free In Liner Terms Discharge *chartering.*

F.I.L.E. Fellow of the Institute of Legal Executives.

Filler Currency unit of Hungary, equal to one hundredth of a Forint.

FILO *or* **F.I.L.O.** First In Last Out, *q.v.*; Free In Liner Out, *q.v.*

filos Filosofia *Italian*—Philosophy.

Filtage Profit margin in the coal trade.

F.I.Mar.E. Fellow of the Institute of Marine Engineers.

FIMBRA Financial Intermediaries, Managers and Brokers Regulatory Association.

F.I.Mech.E. Fellow of the Institute of Mechanical Engineers.

FIMgt Fellow of the Institute of Management.

FIMTA *or* **F.I.M.T.A.** Fellow of the Municipal Treasurers and Accountants. Formerly CIPFA or C.I.P.F.A.

Fin Finland; Finnish; conning tower of a submarine.

fin Final; Finance; Financial; Financier; Finish.

f.i.n.a. Following items not available.

Final Course Final heading taken while navigating in order to arrive at the ultimate destination.

Final Landings Last day on which all the cargo has been discharged from the ship.

Final Port Refers to the last port of unloading of a ship.

Final Sailing The time when the ship leaves her berth and has reached the precincts of the port on the understanding that she is to proceed on her voyage. This is considered commercially to be her final sailing.

Finance The general meaning of regulating money matters. If referring to a state, this word has a connection with the management of the revenues of the state.

Finance Bill Bill of Exchange, *q.v.*, drawn for the loan of a company. Governments generally use this system and they are known as Treasury Bills, *q.v.*

Finance Company *or* **Finance House** A company whose business is in financial speculation by lending money, guaranteeing finance against properties and so on. A merchant bank can also be called a Finance House or Finance Company.

Finance House Finance Company, *q.v.*

Financial Accounting Standards Board An organisation set up by the American Institute of Certified Public Accountants and financed by the industry and professional accountants who regulate and ensure that financial statements comply with the principles practised or accepted by the board. *Abbrev.* FASB.

Financial Guarantee *or* **Financial Warranty** An undertaking to the effect that a stipulated sum of money will be covered and/or paid in full as agreed between the contracting parties.

Financial Year The Statement of Accounts, *q.v.*, period or date of the year.

Financier A person expert in finance, whose chief business is concerned with supplying or obtaining money on a commercial basis.

Fin.Co Final Course, *q.v.*

Fine Bill Bill of Exchange, *q.v.*, easily discountable due to its backing by well-established Finance Houses, *q.v.*, or Commercial Banks.

Fineness Measure of the amount of fine or pure metal in a coin or in alloys used in minting coin.

Fine Paper First quality paper.

Fines (1) Type of iron ore. (2) Deleterious solid particles in fuel and lubricating oils.

Finest Rate of Discount Lowest rate of discount.

FINIDA Finnish International Development Agency.

Finis coronat opus *Latin*—The end crowns the work.

Finished Goods Products that are prompt or ready for sale and use.

Finlands Skeppshandlareförening r.f. *Similar to* Suoman Laivakauppiaitten Yhdistys r.y. *Finnish Ship-suppliers Association.*

FINMARE Societa Finanziaria Marittima *Italy.*

Finnair Finnair Aero O/Y. *Finnish State Airlines.*

FINPRO Finnish Committee on Trade Procedure.

Fin. Sec. Financial Secretary.

FINSIDER Società Finanziaria Siderurgica *Italian* —Iron and Steel Financial Society.

Fin. Co. Final Course *nautical.*

F.Inst.A.M. Fellow of the Institute of Administrative Management.

F.Inst.F. Fellow of the Institute of Fuel.

F.Inst.F.F. Fellow of the Institute of Freight Forwarders.

F.Inst.P. Fellow of the Institute of Physics.

F.Inst.Pet. Fellow of the Institute of Petroleum.

f.i.o. For Information Only; Free In and Out, *q.v.*

FIO *or* **F.I.O** *or* **f.i.o.** Fellow of the Institute of Opticians; Free In and Out, *q.v.* Fleet Information Office.

FIOL *or* **F.I.O.L.** *or* **f.i.o.l.** Free In and Out and Lashed.

FIOP *or* **F.I.O.P.** Fellow of the Institute of Printing.

FIOS *or* **F.I.O.S.** *or* **f.i.o.s.** Free In and Out Stowed. *See* **Free In and Out**.

FIOSPT *or* **F.I.O.S.P.T.** *or* **f.i.o.s.p.t.** Free In and Out Spout Trimmed, *q.v.*

FIOST *or* **F.I.O.S.T.** *or* **f.i.o.s.t.** Free In and Out Stowed and Trimmed. *See* **Free In and Out**. *Also* **F.I.O. Stowed & Trimmed & Discharged**.

FIOT *or* **F.I.O.T.** *or* **f.i.o.t.** Free In and Out Truck.

FIPA *or* **F.I.P.A.** Fellow of the Institute of Practitioners in Advertising.

F.I.Q.S. Fellow of the Institute of Quantity Surveyors.

f.l.r Floating-in Rate *fuel indicator reading.*

FIR *or* **F.I.R.** Flight Information Region, *q.v.*

Fir Firkin, *q.v.*

FIRAV *or* **Firav** First Available. *See* **First Carrier**.

Fire One of the excepted perils for which carriers are not liable, provided this is not caused through the fault of the carriers themselves.

Fire Insurance Policy An insurance policy covering the holder against any accidental fire. Some consequent effects such as loss of profits, etc., may also be covered under the terms of the contract.

Fire Point The continual ignition temperature of fuel or the lowest temperature at which oil will ignite.

That temperature at which a substance, normally in the form of a liquid, starts burning after ignition. Naturally it is higher than Flash Point, *q.v. See also* **Ignition Point**.

Fire Policy An insurance policy covering damage caused by fire. *See* **Fire Insurance Policy**.

Fire Triangle A triangular form representing the elements of fire and explosion. These are Hydrocarbon Gas, *q.v.*, Oxygen and any source of ignition such as a naked flame, sparks, etc.

Firkin (1) A wooden cask containing about 56 lbs of butter, fish, etc. (2) Ale or beer measure of nine imperial gallons.

Firm A company or partnership.

Firm and Ready to Trade The phrase expressed by a Shipbroker or other businessman indicating a firm offer right away in order to hasten and finalise the contract or business.

First Available This refers to the first availability of an aircraft, ship, train or any other transport facility. *Abbrev.* FIRAV or Firav.

First Carrier (1) The first aircraft undertaking the initial flight of a cargo. The second and third transhipment operations are undertaken by the second and third Carriers respectively. (2) The first of the ships involved in a continuous series of transhipments.

First Class Mail Mail considered to be of a priority nature.

First Class Paper Refers to a Bill of Exchange, *q.v.*, or Government Bill which has as its heading a popular name.

First Dog Watch Seaman's watch at sea from 4 p.m. to 6 p.m. *See* **Bell** and **Watch Duties**.

First In First Out Accountancy method in calculating the valuation of assets as per original price obtained. *Abbrev.* FIFO or fifo.

First In Last Out The method used when discharging employees or workers from a company, especially during redundancy. The first employees to join the company will be the last to be discharged and those who were the last to be employed will be the first to be discharged. *Abbrev.* FILO or filo.

First Mate *See* **Chief Officer**.

First Officer *See* **Chief Officer**.

First Open Water A voyage Charterparty phrase denoting that the ship is to report for loading or discharging on the first day of the port being free from ice. It enables the ship to navigate in ice-free water, though in certain instances a day has to be fixed. *Abbrev.* FOW *or* F.O.W. *or* f.o.w.

First Refusal First preference; Priority (referring to decisions on offers and bids).

First Three The first three Leading Underwriters, *q.v.*, whose names are shown on the Slip, *q.v.*, when the subject-matter is put on to the insurance market. *See* **Leading Underwriters** *Also called* Leaders.

First Watch Seaman's watch duty at sea from 8 p.m. to midnight. *See* **Bell** and **Watch Duties**.

FIRTO Fire Insurers' Research and Testing Organisation.

F.I.S. Family Income Supplement; Flight Information System.

F.I.S.A. Fellow of the Incorporated Secretaries Association.

FISC Freight, Insurance and Shipping Charges.

Fiscal Drag Restraint on growth and demand due to increasing rates of taxation under conditions of inflation, *q.v.*, aggravated when increased wages and salaries bring people into higher tax brackets.

Fiscal Policy This refers to the country's spending and taxation policy. When governments increase taxation and cut spending, it is said to be a 'tight fiscal policy'. On the other hand, if spending is increased and taxation is eased or cut, it is termed a 'loose fiscal policy'.

Fiscal Quarter *See* **Fiscal Year**.

Fiscal Year The closing up of financial books at a fixed date each year concerning public revenue. Also known as Financial Year.

Fish To lift the flukes of an anchor.

Fishing a Boom Additional strong wood tightened around the boom of a Derrick, *q.v.*, to protect it against excessive stress when heavy loads are lifted.

Fishing Port As the name implies it is a port used mainly by fishermen and which on the whole does not necessarily have to be as large as a commercial port, though a few of them are in existence around the world.

Fishing Vessel Insurance Plan The Federal Canadian programme designed to insure fishing boats.

Fish Meal The remains of fish after being used for human consumption which is specially treated and turned into animal fodder.

Fishmonger Person selling or dealing in fish.

F.I.S.M. Fellow of the Institute of Supervisory Management.

f.i.t. *or* **FIT** *or* **F.I.T.** Free of Income Tax.

f.i.t. *or* **F.I.T.** Fabrication in Transit; Free Into Truck; Free in Transit.

Fit *or* **F.I.T.** Federal Income Tax *USA*; Fluid Injection Test.

Fittage Sometimes referred to as commission or brokerage.

Fitted for Oil Fuel *Lloyd's Register notation* for a ship which is equipped to use oil fuel.

f.i.t.w. *or* **F.I.T.W.** Federal Income Tax Withholding *USA*.

Five Freedoms *See* **Freedom**.

FIW *or* **f.i.w.** Free Into Wagon, *q.v.*

fix Fixture, *q.v.*

Fix To secure or obtain cargo for a ship. The ship is fixed when she is confirmed as being chartered or contracted to load cargo at agreed freight rates. Fixing is the act of chartering a ship. *See* **Charter**. A fix is also the determination of a ship's navigational position by

observation of terrestrial or heavenly bodies or with the assistance of radio aids.

Fixed Assets Tangible assets like premises, furniture, land, etc. *See* **Fixtures And Fittings**.

Fixed Capital *Similar to* **Fixed Assets**, *q.v.*

Fixed Light A constant luminous light *nautical*. *Abbrev.* F.

Fixed Objects Structures to which ships can be moored such as quays, piers, etc., or objects at sea which are aids to navigation such as buoys, lighthouses, etc.

Fixed Oils Oils extracted from olives, aniseed, palm kernels, nuts, etc.

Fixing a Cargo To book a cargo. *See* **Fix**.

Fixing a Ship To charter a ship. *See* **Charter** and **Charterparty**.

Fixing Broker A shipbroker who fixes ships for the transportation of cargo. He generally deducts the ship's port expenses from the freight collected if any.

Fixing Letter A form of letter produced by the chartering broker after negotiation of the Charter, *q.v.*, of a ship is concluded, showing the terms and conditions agreed upon. This provisional contract is signed by both parties (owner and charterer) in preparation and as a sign of a finalised charterparty.

Fixture When a Charter, *q.v.*, or contract for a ship to be drydocked or to undertake repairs is concluded. *Abbrev.* fix.

Fixtures and Fittings In book-keeping this phrase commonly refers to buildings, furniture and other concrete objects. *See* **Fixed Assets**.

FIYTO Federation of Youth Travel Organisations.

FJ Air Pacific *airline designator*.

F.J. *or* **Fjd** Fjord.

F.J.I. Fellow of the Institute of Journalists.

Fjord *Norwegian*—An inlet from the sea.

F.K. Flat Keel, *Lloyd's Register notation* denoting that the ship's keel is flat.

F.K.C. Fellow of King's College, London.

FKSU Federation of Korean Seafarers.

Fl Flawless *diamonds*.

Fl. Flashing Light, *q.v.*; Flemish.

FL *or* **F.L.** Flag Lieutenant; Freight Liner, container trains of British Rail; Flashing Light, *q.v.*, Florida, *USA*; Flight Level *aviation*; Floor; Florin; Fürstentum Liechtenstein, Principality of Liechtenstein.

FLA Fluid Loss Additive.

F.L.A. Fellow of the Library Association.

F.L.A.A. Fellow of the Library Association of Australia; Fellow of the London Association of Accountants.

Flag A piece of cloth, generally Bunting, *q.v.*, made in one of various shapes and sizes, designs and colours to serve as a sign, symbol, ensign or for signalling.

Flag Clause A clause generally applied to Bills of Lading, *q.v.*, conditions where the carriers' liabilities

Flight Bag

are to be referred to the laws of the country printed or shown on the Bill of Lading.

Flagman A person with a flag for signalling warnings.

Flag Officer Senior naval officer, i.e. admiral, vice-admiral or rear-admiral. A Flag Lieutenant is an admiral's A.D.C.

Flag Out *or* **Flagging Out** To cancel the national flag registration of a ship and re-register her under a Flag of Convenience, *q.v.*

Flagrante Delicto *Latin*—In the very act of committing the crime.

Flag of Convenience Generally describing the flag of registry of a ship owned and operated by foreign nationals and/or corporations who derive financial benefit from so doing.

FLAGS Far North Liquids and Associated Gas System.

Flags of Necessity *Similar to* Flags of Convenience, *q.v.*

Flagstaff That pole of a ship on which a flag is hoisted. Also called Flagpost.

Flag State The nation in which a ship is registered and which holds legal jurisdiction as regards operation of the ship whether at home or abroad.

Flag Surtax Surcharge imposed on foreign ships by a nation. A sort of protective tax for the country's own vessels.

Flag, Ownership *or* **Management** Certain Underwriters, *q.v.*, request details of the ship's flag, owners or management whenever cargo is to be carried on Flag of Convenience, *q.v.*, ships before they accept the insurance risk. *Abbrev.* F.O.M. *or* f.o.m.

Flake To lay out a rope in a regular pattern.

Flamboy A kerosene-type torch emitting light and used as a warning.

Flanking Buoy A small float positioned between the towing lines from the Towboat, *q.v.*, to the vessel which is being towed, used on the Mississippi River. It gives a clear indication of the position of the vessel being towed, especially in sharp bends of the river.

Flanking Rudder Rudders which are also carried forward. They are backing rudders and are often used on riverboats carrying passengers and vehicles. Flanking rudders are installed on large craft engaged on the Mississippi River. *See* **Bow Thruster**.

Flap Moving device at the rear edge of the wing of an aircraft which is used during take-off and landing.

F.L.A.S. Fellow of the Land Agents' Society.

FLASH Feeder LASH or Feeder for Lighter Aboard Ship, *q.v.*

Flashing Lamp Lamp used at sea for sending messages in Morse Code, *q.v.* Also Morse Lamp, *q.v.*

Flashing Light Light emitted during dark hours at regular intervals. *Abbrev.* FL *or* F/L *or* Fl.

Flashlight American for electric torch.

Flash Point The degree of heat at which an explosion will occur in certain inflammable items such as paint, oil, petrol, etc. The bigger the flash point the less dangerous the item is. Such cargoes are always carried on deck at shippers' risk and are not covered under the Carriage of Goods by Sea Acts. Also known as Ignition Point.

Flat Container A container with a floor but no roof. It may have panels at each end, with or without half-height panels at the sides, or corner posts only.

Flat Cost The American expression for Prime Cost, *q.v.* The cost of manufactured goods, including wages, expenses, and cost of material up to the time of its finished state.

Flat Rate Constant rate of freight. Flat rates are given by Carriers to Charterers, *q.v.*, or shippers to ensure that the freights will not fluctuate during the voyage in relation to the prevailing freight market. This is of advantage to the receivers of the cargo as the ultimate price calculations can be readily assessed.

Flats Generally square-shaped wooden platforms where cargo is stacked and packed in such a way as to be carried or lifted in one operation. This method is extensively used both on conventional and sophisticated ships. The flats are so constructed that the fork-lift vehicle and the ship's cargo handling gear can easily handle them. It gives quicker handling, better stacking and less pilferage. Quite often called Pallets, *q.v.*

Flat Top Barge A barge of a standard deadweight tonnage which is used for offshore activities but which can also be adapted to many other forms of commercial work.

FLB Federal Land Bank *USA*.

FLE *or* **F.L.E.** Fire, Lightning and Explosion *insurance*.

Flea Market An outside market where cheap and second-hand items are sold.

Fleet Group or number of ships owned or managed by the same organisation; Unit comprising several warships; A nation's seagoing naval force as a whole.

Fleet Policy An insurance policy covering the fleet or group of ships of a shipowning company as a whole while having some Limit per Bottom, *q.v.*, or limit of liability for each vessel.

FLEX Federation Licensing Examination *USA*.

Flg Flange; Following.

Flg. Flag.

Flg. Plt. Flange Plate.

F.L.I.A. Fellow of the Life Insurance Association.

FLIC Forwarders Local Importers Control *UK*.

Fli/Flo *or* **Flo/Fli** Float In/Float Out or Float Out/Float In, *q.v.*

Flight Bag A container or a bag or an envelope containing copies of manifests, load sheets, stowage plans and other documents of an aircraft. The information is invariably needed for customs clearance and other official reasons when calling at overseas airports. The Flight Bag corresponds to the Ship's Bag, *q.v.*

Flight Coupon Air Ticket.

Flight Crew All the crew members engaged in an aircraft during a flight.

Flight Data Recorder *See* **Flight Recorder** and **Black Box**.

Flight Deck Deck or part of a deck high up in the ship clear of all obstructions for the operation of fixed-wing aircraft or helicopters.

Flight Director An aviation instrument showing the pilot the direction the aircraft is taking. It shows the movements of the aircraft to the left, right, up or down, level, etc.

Flight Envelope The limited acceleration or speed of an aircraft.

Flight from Cash Substituting a Barter, *q.v.*, system for cash transactions.

Flight Information Details of aircraft arrivals and departures issued at airports.

Flight Information Region Air traffic control above and within the territorial air space of a state. All aircraft passing through this space are obliged to report to Air Traffic Control. *Abbrev.* FIR *or* F.I.R. *See* **Air Control**.

Flight Number Number allocated to the flight of each aircraft. For scheduled flights these numbers provide references for the information needed by passengers, cargo carriers and receivers or anyone needing to know the aircrafts' whereabouts.

Flight Recorder A device carried by commercial aircraft and strong enough to withstand impact in case the aircraft crashes. It records information relating to the state of the aircraft, including speed, controls, engines, height, etc., during the course of the flight. It is also known as the Black Box or Flight Data Recorder. *See* **Black Box**.

Flight Stage An aircraft's single flight from one airport to another. A long-haul flight may overall involve two or more such stages or sections.

Flight Time Time spent by an aircraft between taking off and landing.

Flint Flintshire, *UK*.

FLIP Floating Instrument Platform. *Similar to* Flip Ship, *q.v.*

Flip Ship A stationary or stable tubular vessel in the form of a ship which can be used as a research laboratory. Its stern section is flooded so that it remains steady and stationary underwater. It is motorless and has to be towed to different locations. Sometimes it is called FLIP or Floating Instrument Platform.

FLIR Forward-Looking Infra-Red *aviation*.

Flitch Sawn timber of about 4 inches thick and 12 inches wide.

Flli Fratelli *Italian*—Brothers.

F.L.N *or* **f.l.n** Following Landing Numbers, *q.v.*

Float-board One of the boards of a paddle wheel.

Floatels *or* **Floatinns** Floating Hotels.

Float In/Float Out *or* **Float Out/Float In** A system of loading/unloading or handling small barges by means of being towed by small tugs and directed in or out of a barge carrying ship. *See* **Barge Carrying Vessels**.

Float-inns *or* **Floatels** Floating Hotel.

Floating The nation's debt with regard to Treasury Bills and Exchequer Bonds.

Floating Assets Current Assets. Items which are easily converted into money.

Floating Beacons Beacons used as markers in hydrographic and marine surveying works.

Floating Charge A business loan against the security of assets, in which the lender has first preference on the property secured in case of bankruptcy or liquidation.

Floating Clause A Charterparty, *q.v.*, clause which provides that the vessel sails to a safe port or berth where she has to remain always afloat. *Similar to* Always Afloat, *q.v.*

Floating Debt Debt incurred which is of a temporary nature as opposed to Funded Debt, *q.v.*

Floating Dock A floating pontoon with two side walls. When a ship is to be docked the tanks inside the pontoon and the sides are filled with water so that the whole unit submerges. The ship is then moved into the channel formed by the two sides and when made fast on either side the water in the tanks is pumped out again so that the pontoon unit, or floating dock, is refloated with the ship which can be worked on in dry conditions. The floating dock is sometimes called Pontoon Dock. *See also* **Dry Dock**.

Floating Money Currency which is allowed to fluctuate freely in foreign exchange transactions according to demand and not subject to exchange control.

Floating Oil Storage Oil stored on floating vessels. It has been the practice for oil to be stored in large laid-up oil tankers in order to offset the loss involved while the tankers are inactive. *Abbrev.* FOS *or* F.O.S. *See* **Floating Storage Unit**.

Floating Policy *or* **Open Policy** A marine insurance policy which is commonly used to cover merchandise which is partly carried in various parcels and on different ships whose names are not known in advance. The Underwriters, *q.v.*, issue the Floating or Open Policy showing the lump sum value of the whole consignment. The gross value is correspondingly reduced with every individual shipment effected until the whole sum insured becomes exhausted. *See* **Open Cover**. *Abbrev.* F/P *or* F.P.

Floating Production Ship *or* **Floating Production Vessel** *or* **Floating Production Platform** *or* **Floating Production Unit** Ship or rig for combined oil drilling and production. *Abbrev.* respectively FP, FPV, FPP, FPU.

Floating Production, Storage and Off-Loading Ship or unit. A tanker designed or converted to perform these three functions when moored in an offshore oilfield.

Floating Storage Unit A floating unit at sea which is connected to a moored oil tanker whereby oil is transferred and directed to the specific tank sections of the unit, according to the grades of oils. It is used as an oil storage unit so that oil can be pumped out whenever needed. *Abbrev.* FSU *or* F.S.U. Sometimes called Floating Oil Storage & Off-loading.

Floating Territory Another word for ship.

Float On/Float Off A specially designed vessel which can partially submerge on the principle similar to that of a Floating Dock, *q.v.*, used to transport barges or pontoons, laden or otherwise, floating equipment such as cranes and rigs, or smaller vessels, to any destination. *Abbrev.* FO/FO. *See also* **Barge Aboard Catamaran** *and* **Articulated Tug Barge**.

Float Plane Seaplane,*q.v.*

Floats, Orange *See* **Marine Pyrotechnics** and **Orange Smoke Flares**.

Flo.B. Floor Beams.

Flo.Con. Floating Containers. *See* **Container**.

Flo/Fli *or* **Fli/Flo** Float Out/Float In or Float In/Float Out. *See* **Float in/Float out**.

Floe A large block of floating ice.

FLOGWING *or* **FLAW** Fleet Logistic Air Wing *USA*.

FLOOD Fleet Observation of Oceanographic Data *USA*.

Floodable Holds Holds of a bulk carrier or tweendecker that can be flooded to provide ballast facilities.

Floodable Length The lengthwise area of a ship from the Margin Line, *q.v.*, downwards which can be flooded without her sinking.

Flood Tide After the low tide and before the high tide. Also called Rising *or* Advancing Tide.

Floor Ceiling Wood covering on the tops of tanks to protect the metal from accidental impact.

Floor Load Limit The maximum weight allowed to be the stored on the upper floors of warehouses, or the maximum weight per foot which can be utilised in an aircraft for cargo and passengers.

Floor Man *or* **Rotary Helper** *or* **Rough Neck** He is generally called a rotary helper or rough neck and works on the derrick floor which lies above the main deck of the rig. His main duty is to assemble or dismantle the heavy drilling pipe.

Floor Price The lowest possible quoted price in the market for a commodity or anything else.

Floors Vertical steel plates between the inner and outer bottoms of a ship.

Flotation The issue on a stock exchange of stocks in a new public company.

Flotsam Goods lost by shipwreck and found floating on the sea.

Flow Meter A device to measure fluids flowing in enclosed pipes, ducts and conduits.

Flow Moisture Point Concentrated moisture that tends to liquefy. *Abbrev.* FMP.

Flowing Well An easily producing gas or oil well which needs no artifical means to lift the raw material. *See* **Edge Water**.

Fl.oz. Fluid Ounce.

f.l.p. Fault Location Panel.

Flr. Florin.

fl.rt. Flow Rate.

FLT Flight; Fork Lift Truck.

Fluctuation The rise and fall of the price rates of money and goods.

Fluctuat nec mergitur *Latin*—She is tossed by the waves and does not sink. Motto of Paris, France.

Flue Gas Mixture of nitrogen and carbon dioxide which can be produced either from a boiler plant or from a generator. It is used in conjunction with Crude Oil Washing, *q.v.* It replaces the air (oxygen) in an oil tank so as to avoid an explosion and is one of the Inert Gases used for COW (Crude Oil Washing).

FLUG Flugfelag Island *Icelandic State Airlines*.

Fluid In reference to a liquid form.

Flush Deck Hatches Hatches which have no Coamings. They are positioned on top of the lower Holds, *q.v.*, and other 'tween decks.

Flush Deck Ship A ship having the upper deck extending from bow to stern and having no well decks.

Flukes The projecting pointed ends of an anchor which hook into the seabed to keep the ship from moving or drifting.

Fly-by-Wire Electrical flight control signal system between the control tower and air pilots.

Fly Dragging A seine net fishing method used in deep water.

Flying Boat An aeroplane whose fuselage is in the form of a buoyant hull for landing on water, with wing floats to aid stability when afloat.

Flying Bridge (1) Raised narrow fore-and-aft gangway connecting the poop to the forecastle in a sailing ship. (2) Narrow platform with a standard compass above the navigating bridge, similar to Monkey's Island. (3) Ferryboat plying across a river under the combined force of the stream and the tension of a hawser.

Flywheel A heavy revolving wheel which balances and equalises the driving force on its connected shaft.

FM Field Marshal; Fire Main; Frequency Modulation.

Fm *or* **Fthm** Fathom, *q.v.*

FMB Federal Maritime Board *USA*; Federation of Master Builders.

FMC Federal Maritime Commission, Washington, *USA*.

FMCG Fast-Moving Consumer Goods.

FMCS Federal Mediation and Conciliation Service *USA*.

FMCW Frequency Modulated Continuous Wave *electronics.*

FMEA Failure Mode and Effect Analysis *electronics.*

FMF Fleet Marine Force; Food Manufacturers' Federation.

FMGEMS Foreign Medical Graduate Examination in Medical Science *USA.*

FMP Flow Moisture Point, *q.v.*

F.M.R.S. Foreign Member of the Royal Society.

FMS Formation Microscanner *geophysical.*

FMVSS Federal Motor Vehicle Safety Standards *USA.*

FN Spanish Ship Register.

F/N Flight Number.

F.N.A. *or* **f.n.a.** For necessary action(s).

FNARM Federazione Nazionale Agenti Raccomandatori Marittimi—*Italian Shipbrokers' Association.*

FNASA *or* **F.N.A.S.A.** Federation of National Associations of Shipping Agents.

FNASBA *or* **F.N.A.S.B.A.** Federation of National Associations of Shipbrokers and Agents.

FNBC First National Bank of Chicago *USA.*

FNC File(d) No Claim *insurance.*

FNCB First National City Bank *USA.*

F.N.I. Fellow of the Nautical Institute.

FNMA Federal National Mortgage Association *USA.*

FNMM Friends of the National Maritime Museum.

fnp. Fusion Point.

FNPP Floating Nuclear Power Plant.

FNWC Fleet Numerical Weather Center, US Navy Meteorological Office Center.

FNZIM *or* **F.N.Z.I.M.** Fellow of the New Zealand Institute of Management.

FNZSA *or* **F.N.Z.S.A.** Fellow of the New Zealand Society of Accountants.

FO *or* **F.O.** Field Officer; Firm Offer; Flying Officer; Foreign Office; Free Overside, *q.v.*; Freight Officer; Fuel Oil.

F/O *or* **f/o** Firm Offer; For Orders; Free Overside, *q.v.*, Folio; Full Out Terms *grain trade similar to* **Full Out Rye Terms**, *q.v.*

F.O. *or* **f.o.** Firm Offer; For Orders; Full Out Terms *grain trade similar to* **Full Out Rye Terms**, *q.v.*

F & O Freight and Demurrage.

FOA & F.O.A. Foreign Operators Administration *USA*; Free On Aircraft or Airport *airfreight.*

FOB *or* **F.O.B.** *or* **f.o.b.** Free On Board, *q.v.*

FO&B *or* **F.O.& B.** Fuel Oil and Ballast.

FOBA *or* **F.O.B.A.** Free On Board Aircraft *airfreight.*

FOBAS Fuel Oil Bunkering Analysis and Advisory Service, *q.v.*

FOBFO *or* **F.O.B.F.O.** Federation of British Fire Organisations.

FOBS *or* **F.O.B.S.** Free On Board and Stowed *airfreight.*

FOBT *or* **F.O.B.T.** Free On Board and Trimmed.

FOC *or* **F.O.C.** Flag of Convenience, *q.v.*; Fire Offices Committee *insurance*; Free on Car; Free of Charge; Free of Cost; Free of Claim.

Focl. Forecastle, *q.v.*

Focl. Dk. Forecastle Deck.

FOD *or* **F.O.D.** Free of Damage. *See* **Free of Damage Absolutely.**

FOD. abs *or* **F.O.D. abs.** Free of Damage Absolutely, *q.v.*

FOD *or* **F.O.D.** Flags of Discrimination.

FOE *or* **F.O.E.** *or* **FE** *or* **F.E.** Friends of the Earth; Free of Encumbrances, *q.v.*

FO/FO Float-On/Float-Off.

Fog Horn *or* **Fog Signal** A sound signal warning used regularly during fog. *Abbrev.* FH or F.H.

Fog Patch Isolated area of fog.

F of F Firth of Forth,*UK.*

Föhn Wind Hot southerly wind in the Alps.

FOIL *or* **F.O.** *or* **f/o** Fuel Oil.

Following Landing Numbers In the 'Average Clause', *q.v.*, the cargo being landed has to be computed on a landing number system known as 'following landing numbers'. The items of cargo are therefore numbered in the order they are landed, including damaged ones. By this method it may be possible to trace the extent and cause of damage. *Abbrev.* F.L.N. or f.l.n.

F.O.M. *or* **f.o.m.** Flag, Ownership and Management, *q.v.*

FOMC Federal Open Market Committee *USA Federal Reserve.*

FON *or* **F.O.N.** Flag of Necessity. *Similar to* **Flag of Convenience**, *q.v.*

FONASBA *or* **F.O.N.A.S.B.A.** Federation of National Associations of Shipbrokers and Agents.

Foot of Sail The lower edge of a sail.

Footsie Colloquialism for FT-SE 100, the London Stock Exchange/Financial Times equity index of the one hundred largest UK companies. *See* **FT-SE.**

FOP *or* **F.O.P.** Forepeak; Friendship Oil Pipeline *Eastern Europe*; Fuel Oil Port.

Fo.P. Forepeak.

FOPI *or* **F.O.P.I.** Federal Office of Private Insurance *Swiss.*

FOQ *or* **F.O.Q.** *or* **f.o.q.** Free on Quay, *q.v.*

FOR *or* **F.O.R.** *or* **f.o.r.** Free On Rail.

for. *or* **For.** Foreign.

For *or* **For'd** Forward.

FORATOM Forum Atomique Européen *French* —European Atomic Forum.

Forbund *Swedish*—Federation.

Force de Cheval *French*—Horse Power. French metric system corresponding to 0.9863 H.P.

Fore Before.

Fore-and-Aft (1) Relating to both ends, or the entire length of a ship. (2) In line with or parallel to the longer dimension, as opposed to crosswise or Athwartships, *q.v.* (3) Describes a sailor's naval uniform or 'rig', with a buttoned jacket, as opposed to a jumper.

Fore-and-After A chain span connecting a type of disengaging gear for the falls in the bow and stern of a boat designed to be lowered at sea.

Forebody That part of a ship which lies in front of amidship (midship). *Abbrev.* For'd B.

Fore Cabin Passengers Generally applicable to Second Class Passengers.

Forecast The prediction of the weather to come within a stated period.

Forecastle Head The deck over the forecastle.

Foreclose *or* **Foreclosure** To bar the Mortgagor, *q.v.*, from enjoyment of mortgaged property. This happens when no more money can be expected to be paid in full settlement of the mortgage.

Foredate To date a note before its actual time; as a post-dated cheque.

Foredeck The foremost part of a ship in the main or upper deck.

Forefoot The lower part of the ship's stem that meets the keel.

Foreign Bills Foreign Bills of Exchange drawn or negotiated from one place to another.

Foreign Exchange The transfer of money or funds from one country to another foreign country in order to settle pending accounts. Also extensive trade carried on between various countries on an international scale which results in a balance of payments due among the countries. The exchange rates of money, which are daily and officially published, are geared to the financial standing of any particular country on the day. *Abbrev.* Forex.

Foreign General Average Marine insurance term in which the Underwriters, *q.v.*, undertake to pay the adjustments arrived at by foreign adjusters at the port of destination or where the voyage ended in the General Average. *q.v. Abbrev.* F.G.A. *or* f.g.a.

Foreign General Average Clause In the case of General Average, *q.v.*, the port or country of destination or where the voyage ended must govern in the general average assessment unless any special stipulation to the contrary exists in the contract of affreightment. *See* **Foreign General Average**.

Foreign Going A general nautical expression for a ship's master who has long service coupled with experience and who has passed the respective examinations. A master in this category can take a ship of any size and navigate her in the oceans of the world. *Abbrev.* FG *or* F.G.

Foreign Going Vessel Any vessel trading outside the hometrade limits. Its voyages involve calls on foreign ports as well as home ports.

Foreign Trade Commercial business or trade transacted and calculated as a whole between one country and others or vice versa.

Foreign Traders Index A file of reputable overseas firms interested in informing or advertising American made goods and which is kept at the US Department of Commerce in Washington D.C.

Foreland A Cape, *q.v.*, or headland.

Forelock A piece of metal in the form of a stud or split pin which is inserted through a hole in the Shackle, *q.v.* At the far end is a hole where the pin is also passed through to keep the stud locked in its position.

Foremast The first mast nearest to the bow of a ship.

Foreningen *Danish*—Union.

Fore Noon Watch Seaman's watch at sea from 8 a.m. up to mid-day. *See* **Bell** and **Watch Duties**.

Fore Peak Tank This is more or less identical to the Aft Peak Tank, *q.v.*, but the water tank is positioned in the fore part of the ship instead of the aft part. The Fore and Aft tanks are important in the trimming of the ship.

Foresail The lowest of the square sails set on the Foremast, *q.v.*, of a Sailing Ship.

Foreshore Level between high and low water marks.

Forest Product Carrier A timber carrier having very large hatchways so that large logs or pieces of timber can be stowed in the holds without hindrance. The operation is handled by the ship's own Gantry, *q.v.*, cranes.

Forestay A rope or wire from the fore masthead to the stem, *q.v.*, to keep it steady.

Foretop A platform situated high on the Foremast, *q.v.*

Forex Foreign Exchange.

Forfeited Shares Shares taken back by a company or firm which the holder is unable to pay for.

Forint Unit of currency used in Hungary.

Fork Lifter Alternatively known as Fork Lift Truck and Elevating Transfer Vehicle. *Abbrev.* ETV *or* E.T.V. It is used to lift, shift and stack containers, pallets and other merchandise, using its two protruding mechanically operated prongs.

Formaldehyde A pungent colourless gas which is soluble in water and is used in plastic manufacture and disinfectants.

Form O Code term for American charterparty for the carriage of cotton where freight is paid on the net tonnage of the ship.

for.rts. Foreign Rights.

for'sl Foresail.

FORT *or* **F.O.R.T.** *or* **f.o.r.t.** Full Out Rye Terms, *q.v.*

Fortes fortuna adjuvat *Latin*—Fortune favours the brave.

175

Fortiter in re suaviter in modo *Latin*—Bravely in action, gently in manner.

For the Account Stock exchange term meaning to settle account days in the following days allowed.

FORTRAN Formula translation, computer programming algebraic language.

Forward Toward the bow of the ship *nautical*.

Forward Contract Contract made for commodities which are to be supplied at a future date.

Forward Dating Dating of Cheques and other commercial documents in advance.

Forward Delivery *See* **Delivery Terms**.

Forwarders Certificate of Receipts *or* **Forwarding Agent's Certificate Receipt** Receipt made out by a forwarding agent for the goods received for shipment. *Abbrev.* FCR *or* F.C.R.

Forwarding The act of forwarding merchandise from one place to another on behalf of other persons or clients. *Abbrev.* Fwdg. *See* **Forwarding Agents**.

Forwarding Agents Persons engaged in taking care of cargo or personal effects to be despatched from one country to another by sea, land or air. A similar expression is Freight Forwarders, *q.v.*

Forwarding Agent's Certificate Receipt *Similar to* Forwarders Certificate Receipt, *q.v.*

Forward Sundays and Holidays to Count An expression found in a Charterparty, *q.v.*, which gives the right of the ship to claim despatch money on all time saved.

FOS *or* **F.O.S.** Free of Stamp, *q.v.*; Floating Oil Storage, *q.v.*; Free On Steamer. *Similar to* **Free On Board.**

FOSFA Federation of Oils, Seeds and Fats Association.

FOST Flag Officer Sea Training.

FOT *or* **F.O.T.** Fuel Oil Treatment; Free On Truck; Free of Tax.

Foul *or* **Fouling** (1) To be in the way or to be an obstruction. *Ex.* A ship while manoeuvring in a restricted area or port may foul or collide or touch another ship. (2) The anchor and/or chain cables if entangled with other submerged objects are said to be foul. (3) The species of marine organisms and vegetable matter which grow on the bottom of a ship. *See* **Antifouling** and **Antifouling Paints**.

Foul Anchor *See* **Foul** *or* **Fouling**.

Foul Bill of Health When a vessel sails from a place infected by a contagious disease she will be furnished with a Foul Bill of Health. Also expressed as a Suspended Bill of Health, *q.v.*

Foul Bill of Lading *Similar to* **Dirty Bill** *or* **Dirty Bill of Lading**, *q.v.*

Foul Bottom The state of a ship's underwater hull when marine growths such as weed or barnacles are attached, which reduces her speed.

Foul Ground An area of the seabed to be avoided when anchoring because of obstructions such as wreckage and outcrops of rock or coral. Designated on charts by the word 'foul'. Also fouled by old anchors such as off Bhavnagar (India).

Foul Propeller When the propeller of a vessel cannot turn because of obstruction by ropes or debris around it.

Founder When a ship is filled with water to the extent that she sinks.

FOW *or* **F.O.W.** *or* **f.o.w.** First Open Water, *q.v.*; Free On Wagon; Free On Wharf.

Foxes Short left-handed laid yarns.

Foxtrot International Radio Telephone phonetic alphabet for 'F'.

fp Free Piston.

f.p. Fine paper; Fireplace; Fixed Price; Flame Proof; Flashproof; Foolscap; Foot Path; Foot Pound; Freezing Point; Full Point.

FP *or* **F.P.** Fully Pressurised *gas carriers*; Fire Point.

FP *or* **F/P** *or* **F.P.** Federal Parliament; Fire Policy *insurance*; First Preference, *similar to* **Last Refusal**; *q.v.*; Flash Point, *q.v.*; Floating Policy, *q.v.*; Forward Perpendicular; Free Pratique.

FP *or* **F.p.** *or* **f.p.** Freezing Point; Fully Paid.

FPA *or* **F.P.A.** Fire Protection Association; Foreign Press Association; Food Products Administration; Forth Ports Authority; Free of Particular Average, *q.v.*

f.p.a. Free of Particular Average, *q.v*

F.P.A. abs. Free of Particular Average Absolutely, *q.v. Also abbrev.* FPA. abs.

FPAAC *or* **F.P.A.A.C.** Free of Particular Average American Conditions.

FPAEC *or* **F.P.A.E.C.** Free of Particular Average English Conditions.

F.P.A.N.Z. Fellow Public Accountant of the New Zealand Society of Accountants.

F.P.A u.c.b.s. Free of Particular Average unless caused by stranding.

F.P.A. unl. Free of Particular Average unless.

f.p.c. For private circulation.

FPC Federal Power Commission *USA*; Forest Protection Service.

f.pd. Fully paid.

FPF Floating Production Facility *oil drilling*.

F.Ph.s. Fellow of the Philosophical Society.

FPIL *or* **F.P.I.L.** *or* **F.p.i.l.** Full Premium if Lost, *q.v.*

FPILIP *or* **F.P.I.L.I.P.** Full Premium if Lost from an Insured Peril. *See* **Full Premium if Lost**.

FPILIPIA *or* **F.P.I.L.I.P.I.A.** Full Premium If Lost by a Peril Insured Against. *Similar to* Full Premium if Lost from An Insured Peril.

FPK Flat Plate Keel.

F.Plt. Face Plate.

fpm *or* **f.p.m.** Feet per Minute.

FPO *or* **F.P.O.** Field Post Office; Fire Prevention Office; Fleet Post Office.

FPP *or* **F.P.P.** Fixed Pitch Propeller.

Freeboard

FPS *or* **F.P.S.** Fellow of the Pharmaceutical Society; Financial Planning Simulator *electronics*.

fps. *or* **f.p.s.** Feet or Foot Pound per Second.

FPSO *or* **F.P.S.O.** Floating Production, Storage & Offloading, *q.v.*

fpsps *or* **f.p.s.p.s.** Feet per second per second.

FPT *or* **F.P.T.** Fast Patrol Boat; Fixed Price Tender; Fore Peak Tank, *q.v.*

FPV Floating Production Vessel, *q.v.*

f.q. Fiscal Quarter. *See* **Fiscal Year**.

F.Q.A.W.T. Flush Quick Acting Water Type.

Fr. France; Fratelli *Italian*—brothers; Frau *German*—Mrs or Wife.

FR Fully Refrigerated: referring to LPG ships.

FR *or* **F.R.** Federal Republic; Freight Release, *q.v.*

F/R Freight Release, *q.v.* Full Release.

fr. Franc, *q.v.*; Free; Frei *German*—Free; Frequent; Front; From.

fr *or* **f.r.** Folio recto *Latin*—Right hand pages.

FRAM *or* **F.R.A.M.** Fellow of the Royal Academy of Music; Fine Resolution Antarctic Model *oceanography*; Fleet Rehabilitation and Modernization *USA*.

Franco le long du navire *French*—Free Alongside Ship, *q.v.*

Franco transporteur *French*—Free Carrier, *q.v.*

Franco Wagon *French*—Free on Wagon.

F.R.A.S. Fellow of the Royal Aeronautical Society.

F.R.A.G. Fellow of the Society of Australian Gynaecologists.

Franc Unit of currency used in France, Belgium, Benin, Mali, Rwanda, Switzerland.

Franchise The part or percentage of the uninsured sum of money of the adventure. Franchise may either be deductible or non-deductible. Deductible franchise is warranted free from particular average say 10% or £10 or from the first 10% of £100 or £10. In case of accident the Underwriters, *q.v.*, are liable for the whole claim, less £10. Non-deductible franchise is Warranted free from Particular Average, *q.v.*, under say 10% or £10. Claims under the insured value of 10% or £10 are not met by the underwriters. However, if the whole amount is inclusive of the Particular Average, *q.v.*, the underwriters will be liable.

Franco Domicile An airline term meaning that the consignor or seller of the goods is to bear the cost of freight and handling charges inclusive of transportation from the airport up to the consignee's place of receipt. In other words, this means the expenses incurred include freight costs, clearance and duty at the place of destination.

Franco *or* **Rendu** *or* **Free Contract** The exporter pays all the expenses to the importer up to the government or customs warehouse. In short, the price quotation of the goods includes costs, freight, insurance and all other charges up to the point of delivery at customs sheds. This clause is not common as the

exporter is at risk in the sense that the expenses may be altering every day and are not known to him.

Frank To impress a mark on a letter or parcel indicating that postage due has been or will be paid by affixing postage stamps. But *see* **Freepost**.

F.R.A.N.Z. Fellow Registered Accountant Member of the New Zealand Society of Accountants.

Frap The act of passing ropes around a Tackle, *q.v.*, to increase tension for hoisting purposes.

F.R.A.S. Fellow of the Royal Astronomical Society.

Fraud Deceit, imposture, deceptive trick, cheating, a swindle with a view to unlawful gain or advantage.

Fraus Omnia Vitiat *Latin*—Fraud impairs; Fraud debases everything.

FRB Federal Reserve Bank, *q.v.*, *or* Board *USA*; Fisheries Research Board *Canada*.

FRC Fast Rescue Craft.

FRC *or* **F.R.C.** *or* **f.r.c.** Free of Reported Casualty *insurance*; Free revolving crane; Fast Rescue Craft.

Fr.Can. French Canadian.

FR & CC *or* **F.R. & C.C.** *or* **f.r.c.c.** *or* **fr & c.c.** Free of Riots and Civil Commotion, *q.v.*

FRCM *or* **F.R.C.M.** Fellow of the Royal College of Music.

FRCO *or* **F.R.C.O.** Fellow of the Royal College of Organists.

F.R.C.O.G. Fellow of the Royal College of Obstetricians and Gynaecologists.

FRCP *or* **F.R.C.P.** Federal Rules of Civil Procedures *USA*; Fellow of the Royal College of Physicians, London.

F.R.C.P. Ed. Fellow of the Royal College of Physicians of Edinburgh.

FRCS *or* **F.R.C.S.** Fellow of the Royal College of Surgeons.

FRCVS *or* **F.R.C.V.S.** Fellow of the Royal College of Veterinary Surgeons, London.

Fre Facture *French*—Invoice.

F.R.Econ.S. Fellow of the Royal Economic Society (*now ceased*).

F.R.E.D. Figure Reading Electronic Device.

Free Freeway; Sailing with the wind *in sailing*.

Free Alongside *or* **Free Alongside Ship** *or* **Free Alongside Quay** An INCOTERM 1990 expression. Buyer is to pay all expenses as from the time of shipment of the merchandise up to the point of destination. These include loading and landing expenses, freight, insurance and all other charges involved in the process of carrying the goods. *Abbrev.* FAS *or* F.A.S. for Free Alongside Ship and FAQ *or* F.A.Q. for Free Alongside Quay. The French equivalent is Franco le long du navire.

Freeboard The vertical measurement from the ship's side amidships from the Load Waterline, *q.v.*, up to the upperside of the Freeboard Deck, *q.v.*, or Main Deck. *Abbrev.* fbd *or* Fbd.

177

Freeboard Certificate Certificate of approval regarding the position of the Loading Disc, *q.v.*, and Loadline Markings. *See* **Annual Loadline Certificate**.

Freeboard Deck Upper main deck with permanent enclosure or with weather openings capable of being hermetically closed. In flush deck and detached superstructure vessels, the Freeboard, *q.v.*, is taken to be in the upper deck. When the vessel has a complete superstructure deck then the shelter deck will be below that superstructure.

Freeboard Marks Similar to Plimsoll Mark *or* Annual Load Line Certificate, *q.v.*

Free Capital Cash in hand available against an opportunity.

Free Carrier An INCOTERM 1990, *q.v.*, expression. This is similar to FOB, *q.v.*, but meets the exigencies of multimodal transport methods in containers, trailers, etc. The seller is responsible for delivering the goods at a named place to the Carrier, *q.v.*, or to a Freight Forwarder, *q.v.*, appointed by the buyer. The seller is responsible for export formalities. All other charges are for the buyer, though he may request the seller to arrange carriage on his behalf. Certain rules apply where goods are loaded on a lorry, or in a rail wagon, or onboard inland water transport at the seller's premises. *Abbrev.* FCA.

Free Contract *See* **Franco** and **Rendu**.

Free Currency The currency which enjoys a relatively stable international exchange value and tends to maintain its face value *vis-à-vis* other currencies.

Free Delivery Shipper or consignor is to deliver the merchandise free of expenses up to the place of destination agreed in the contract. Opposite to Free Alongside Ship, *q.v. Abbrev.* F.D.

Free Despatch *or* **Free of Despatch** Owners do not pay or give remuneration or despatch money to Charterers, *q.v.*, if loading and/or discharging are finished before the prescribed time in the Charterparty, *q.v.* Opposite to Despatch Days *or* Dispatch Days, *q.v.*, and also Despatch Money *or* Despatch Money, *q.v. Abbrev.* F/D *or* f.d.

Free Domicile The Consignor, *q.v.*, bears all delivery costs until goods are with the Consignee, *q.v.* Also termed Free House.

Freedom of the Seas The rule of all maritime nations in declaring that in the Open Seas, *q.v.*, or High Seas, *q.v.*, and where these areas are also termed outside Territorial Waters, *q.v.*, all vessels of any nationality enjoy freedom of navigation and no nation has any right to deny them such without good and legal reasons; or unless any areas are declared as War Zones, *q.v.*, by specific nations in time of war. The Latin term is Mare Liberum, *q.v.*

Freedoms Commonly known as the Five Freedoms concerning the liberties enjoyed by aircraft of all nations. These liberties of course can only exist when there is a peaceful state of affairs and on the basis of mutual understanding. They consist of (1) Freedom to fly over the territory of another nation but without landing. (2) Freedom to land due to technical faults or for medical assistance. (3) Freedom to collect passengers, their belongings and cargo under whose flag the aircraft is flying. (4) Freedom to land passengers, their belongings and cargo. (5) Freedom to collect and land passengers, their belongings and cargo from one place to another or vice versa.

Freedom Ship A Tweendeck Ship of 15,000 tons deadweight capacity built in Japan to carry bulk and general cargo economically at a normal speed of 15 knots.

Free Entry Customs form for the clearance of imported merchandise to be exempted from customs duty.

Free Enterprise A system where the individual is free to undertake commercial activity as long as it does not conflict with the law, as opposed to being under government control.

Free fall Describes a liferaft or liferaft which can be let go from a ship using a slip rather than ropes or wires for lowering it.

Free from Alongside Buyers of merchandise are exempted from paying lighterage expenses at the port of discharge. *Abbrev.* FFA *or* F.F.A. *or* f.f.a.

Free Goods The merchandise that is excluded in the payment of import or local tax duty. This could be goods in transit or of a special category or for re-exportation or for warehousing.

Freehold Absolute ownership; an estate in real estate.

Free House Same as Free Domicile, *q.v.*

Free In An insertion in a Bill of Lading, *q.v.*, to show that all charges for loading including stevedoring are for the account of the shippers or consignees and not for the ship. *Abbrev.* f.i.

Freeing Port An opening in the Bulwark, *q.v.*, of a ship which allows water washing on deck to fall back into the sea.

Free In Out and Spout Trimmed Cargo is loaded and spout trimmed as well as unloaded at the expense of the shippers or receivers. *Abbrev.* FIO&ST *or* F.I.O&S.T. *or* f.i.o&s.t.

Free In Liner Discharge *or* **Free In Liner Out** Cargo is loaded free of all loading charges to owners of the ship but on Liner Terms, *q.v.*, namely stevedoring for owners account. *Abbrev.* FILD *or* F.I.L.D. for Free In Liner Discharge and FILO *or* F.I.L.O. for Free In Liner Out.

Free Into Bunkers Price of merchandise includes the expenses up to loading into barge or bunkers. Full charges are to be for the buyers. *Abbrev.* FIB *or* F.I.B. *or* f.i.b.

Free Into Wagons The quotation includes expenses up to the loading on to wagons and all charges are the responsibility of the buyers. *Abbrev.* FIW *or* F.I.W. *or* f.i.w.

Free List List of goods that are allowed to be imported free of duty.

Free Of An insurance term excluding liability to the Underwriters, *q.v.*, up to (free of) that amount or action. *See* **Franchise** *also* **Free of Particular Average**.

Free of Address Voyage Charterparty expression. When no remuneration is allowed by the owners to the Charterers, *q.v.*, it is termed Free of Address. This is the opposite to Address Commission, *q.v.*

Free of All Average This refers to Total Loss or Total Loss Only. *See* **Total Loss** *also* **Total Loss Only**. *Abbrev.* FAA *or* F.A.A. *or* F.a.a.

Free of Capture and Seizure This is one of the clauses in insurance where Underwriters, *q.v.*, are not liable for any cause resulting from wartime activities and hostilities. In general it is termed 'Warranted Free of Capture and Seizure'. *Abbrev* FCS *or* F.C.S. *or* FC&S *or* F.C.& S.

Free of Capture and Seizure etc and of Strikes, Riots and Civil Commotions Similar to Free of Capture and Seizure, *q.v.*, and the exceptions of liability to the underwriters extend to Strikes, Riots and Civil Commotions. *Abbrev.* FCS & SRCC *or* F.C.S. & S.R.C.C.

Free of Claim for Accident Reported An insurance clause generally expressed 'Warranted Free of Claim for Accident Reported'. A warranty that underwriters, *q.v.*, accept the risk, though not necessarily on every reported accident. *Abbrev.* FCAR *or* F.C.A.R. *or* f.c.a.r.

Free of Damage *or* **Free of Damage Absolutely** Marine Insurance, *q.v.*, clause covering Total or Constructive Total Loss of General Average (Excluding General Average Damage to Hull and Machinery), Salvage Charges, *q.v.*, and the three-quarters of the Running Down Clause, *q.v.*

Free of Encumbrances This expression is invariably found in the sale contracts of ships and other items. It is one of the most important factors for the purchaser. It certifies and guarantees that there exist no burdens on the ship or articles in any way. Namely no debts are incurred or mortgages enforced that would automatically be transferred from the sellers to the purchasers in the eyes of the law. *Abbrev.* F.E. *or* F.O.E.

Free of Freshwater Damage By this clause insurance Underwriters, *q.v.*, are not liable for damage caused by fresh water to the cargo.

Free of General Average Marine Insurance, *q.v.*, term where the assured is not covered by General Average, *q.v. Abbrev.* F.G.A. *or* f.g.a.

Free of Income Tax On most dividends or interest due, especially from government shares, no income tax is paid provided the holder is a non-resident and in some instances a declaration is to be signed to confirm same.

Free of Particular Average *Similar to* **Total Loss Only**, *q.v.*, or **Free of Interest** though the Underwriters, *q.v.*, still remain liable for losses arising from stranding, sinking, burning and other expenses resulting from these mishaps. The conditions generally refer to the Institute Cargo Clauses, Hull *q.v. Abbrev.* FPA *or* F.P.A.

Free of Particular Average Absolutely The conditions laid down in the Institute Clauses are more rigid than the Free of Particular Average, *q.v.* These conditions, as opposed to the normal Free of Particular Average, exclude General Average Damage to the hull of the ship only, except if such damage has been the result of extinguishing fire or by fire contact in the course of the salvage operation only. *Abbrev.* FPA abs. *or* F.P.A. abs.

Free of Recommendations *See* **Conditions of Class**.

Free of Riots and Civil Commotions Underwriters, *q.v.*, are not liable for damages emanating from strikes and labour disturbances. This clause comes under the heading of the Institute Cargo Clauses W.A. *See* **Institute Warranties**. *Abbrev.* FR&CC *or* f.r.c.c.

Free of S.R.C.C. Free of Strikes, Riot and Civil Commotion *insurance*.

Free of Stamp When new securities are issued in many cases no stamp duties are charged. *Abbrev.* FOS *or* F.O.S.

Free of Stem *See* **Subject to Stem**.

Free of Turn *or* **Free Turn & Berth** Charterparty, *q.v.*, clause meaning that Laydays, *q.v.*, start to count on the vessel's arrival at a port whether a loading/unloading berth is available or not, subject of course to the vessel being ready to work (any time waiting for berth is for charterers' account). *Abbrev.* FT *or* F.T *or* f.t for Free of Turn and FT&B *or* F.T. & B. *or* f.t.&b. for Free Turn & Berth.

Free on Quay *See* **Free Alongside** *or* **Free Alongside Quay**. *Abbrev.* FOQ *or* f.o.q.

Free Overside *Similar to* **Ex-Ship**. *Abbrev.* FO *or* F.O. *or* f.o. *or* F/O *or* f/o.

Free Port A port or section of it where ships can unload goods without the payment of customs duty. Ports of this nature are ideal as places for the transhipment of cargoes. *See* **Entrepôt**.

Freepost Postal arrangement where the franking of a letter is paid by the receiver and free to the sender.

Free Stem Charterparty, *q.v.*, phrase implying that the loading is to start as soon as the vessel has arrived at her loading berth.

Free Time *or* **Turn Time** Period of time in a Charterparty, *q.v.*, between the handing over of the Notice or Letter of Readiness, *q.v.*, to the Charterers, *q.v.*, and the beginning of the Lay Days, *q.v. Abbrev.* FT *or* F.T. *or* f.t.

Free Trade Policy practised by a state on an international level permitting the import of merchandise freely and where no tax is imposed. *Abbrev.* F.T. *or* FT *or* f.t.

Free Turn & Berth *or* **Free of Turn** Charterparty, *q.v.*, expression meaning that the ship is not to wait for loading and/or discharging. *See* **Free of Turn**.

Free Zone Secluded place or port where no customs duties are imposed on imported merchandise which will be warehoused under customs supervision pending re-exportation as ships' stores or as cargo for onward transmission to another named destination. *See* **Entrepôt**. *Abbrev.* FZ *or* F.Z. Also called Free Port.

F.R.E.I. Fellow of the Australian Real Estate and Stock Institute.

Freight Transportation charges for cargo carried by a ship. There are different methods in the payment of freights. It may be Freight Prepaid, *similar to* **Advance Freight**, *q.v.*, *or* **Freight Payable Destination** *or* **Freight Paid at Destination**, *q.v.*, or **Flat Rate** *or* **Lumpsum**, *q.v. or* **Dead Freight**, *q.v.*

Freight Abandonment Clause Marine Insurance, *q.v.*, clause where the underwriters are not liable for freight in case of Total or Constructive Total Loss, *q.v.*, whether Notice of Abondonment, *q.v.*, is produced or not.

Freight Account The statement of freight earned by a vessel over a voyage or a period of time.

Freight All Kinds Uniform tariff of freight practised in the container business irrespective of the distance covered and weight of containers as opposed to freight charged on conventional cargo which is calculated on weight or measurement, whichever is greater. *Abbrev.* FAK *or* F.A.K. *or* fak.

Freightage The entire cargo, either volume or weight capacity, available on a ship.

Freight At Risk *Similar to* Collect Freight *or* Freight Payable *or* **Paid at Destination**, *q.v.*

Freight Car American for railway wagon.

Freight Clause *Similar to* **Hire and Payment Clause**, *q.v.*

Freight Clerk (1) Person employed in a shipping company who does clerical work and collects the freight earnings against the respective Bills of Lading, *q.v.*, and Manifests of ships where the office acts as agents. (2) American shipping expression. A shipping clerk employed on board a merchant ship and doing the work of a Supercargo.

Freight Collect *Similar to* **Freight at Risk**, *q.v.*

Freight Collision Clause A Marine Insurance, *q.v.*, clause covering the recovery of freight paid in the event that the damage is caused by collision. *See* **Freight Abandonment Clause.**

Freight Container A receptacle with a minimum internal volume of 1 cubic metre constructed for practically permanent commercial usage. It must be accessible to receive cargo and be fitted with devices that can be readily handled for transportation without difficulty. It has no connection with vehicles or conventional or loose general cargoes. *See* **Container**.

Freight Contingency The consignee's insurable interest in the remaining cargo still undelivered from a vessel until such time as the cargo is received by the consignee at port of destination.

Freight Earnings *See* **Earnings**.

Freighter A cargo ship. A Charterer, *q.v.*, of a ship engaged in carrying cargo for freight inclusive of chartered vessels.

Freight Forward *Similar to* **Freight Paid at Destination**, *q.v.*

Freight Forwarder/s Firm engaged in arranging to forward parcels of cargo by sea, land or air from one place to another on behalf of its clients. Also termed Shipping and Forwarding Agents *or* Forwarding Agents, *q.v.*

Freight Futures *See* BIFFEX.

Freight In Full A Charterparty, *q.v.*, provision where all the port charges are to be paid by the owner of the vessel. This usually relates to a Voyage Charter contract.

Freight in Full of All Port Charges Pilotages, Consulages, Light Dues, Trimming, Lighterage at Discharging Port Chartering term meaning that the owners of a ship are to pay the entire charges on cargo carried.

Freight Notes Bills or notes sent by ship agents and shippers, showing the freight charges due for the cargo meant for shipment.

Freight Paid at Destination *or* **Freight at Risk** *or* **Collect Freight** *or* **Freight Forward** Freight is paid on safe arrival of the merchandise. This method is naturally much preferred by shippers and receivers in the sense that should the ship be unable to reach her port of destination through some mishap or other, the owners or carriers are not legally covered to receive payment for the freight. This is the opposite to Advance Freight, *q.v.*, *or* Freight Paid in Advance.

Freight Paid in Advance *Similar to* **Advance Freight**, *q.v.*

Freight Policy The obligation by the Underwriters, *q.v.*, in the Institute Time Clauses Freight for the amount or part of lost freight to be made good whenever the following occurs: Accidents in loading, discharging, handling, bunkering, contact with aircraft, explosions, bursting of boilers, breakage of shafts or any latent defect in the machinery or hull, negligence of the master, seamen, engineers, pilots etc. Marine Insurance, *q.v.*, policy protecting the assured for the freight charged. *See* **Marine Insurance** *and* **Maritime Policy**.

Freight Per Revenue Ton Freight which gives the total maximum amount on merchandise on either weight or measurement (W/M) whichever the greater. *See* **Revenue Ton**. This freight is commonly observed by Liner vessels.

Freight Rate The rate charge due to the carrier for cargo carried. Also termed Rate of Freight.

Freight Release *Similar to* **Advance Freight**, *q.v. Abbrev.* FR *or* F.R. This could also mean the Delivery Order, *q.v.*

Freight Tonnage Freight charged when a ship carries cargo either against a ton of weight or a ton of volume. The ton weight is either short or long. A short ton (also referred to as a net ton) is calculated as 907.18 kilos or 2000 lbs. weight whereas a long ton is 1016.06 kilos or 2240 lbs. The volume ton (termed cubic or metric ton) is either 40 cubic feet or 35.3 cubic feet to the ton respectively. *See* **Freight Ton** *or* **Shipping Ton**.

Freight Tonne Weight calculated as 1,000 kilos per cubic metre of volume. *See also* **Freight Tonnage**.

Freight Tonne Kilometre *or* **Tonne Kilometre** The actual freight payload in an aircraft used in conjunction with the distance covered. *See* **Seat Kilometre**, and **Capacity Tonne Kilometre**.

French Canada The Province of Quebec.

French Leave Departure without notice or permission or suspiciously.

Freon A trademark referring to odourless refrigerants based on Chlorofluorocarbon, *q.v.*

Freq. Frequency.

Frequency Changing Transformer An electrical unit generally used on a ship to allow 50 and 60 cycles to be mixed together when shore power is communicated to the ship. *Abbrev.* FCT.

F.R.E.S. Fellow of the Royal Entomological Society.

Fresh Freshwater.

FRESH *or* **F.R.E.S.H.** Foil Research Hydrofoil.

Fresh Water Allowance An allowance for additional draft/reduced freeboard when navigating in fresh water conditions—e.g. along a river.

Fresh Water Damage *Abbrev.* FWD *or* F.W.D. *or* f.w.d. *See* **Bilge Water Damage**.

Fresh Water Mark Load Line, *q.v.*, mark on each side of the ship limiting the immersion she is allowed to be at when navigating in fresh waters.

Fret To be eaten up or pitted due to heavy corrosion. Also called Frettage; *French*—Freight.

Fret ou Port Payé *French* Freight or Carriage Paid by the Consignor *INCOTERMS*.

Frettage *See* **Fret**.

F.R.F.P.S. Fellow of the Royal Faculty of Physicians and Surgeons, Glasgow.

FRG Federal Republic of Germany, *West Germany*.

Fr.G. French Guiana.

F.R.G.S. Fellow of the Royal Geographical Society.

Frgt Freight, *q.v.*

F.R.H.B. Federation of Registered House Builders.

F.R.Hist.S. Fellow of the Historical Society.

F.R.H.S. Fellow of the Royal Horticultural Society.

Fri. Fribourg, *Switzerland*; Friday.

FRI *or* **F.R.I.** Food Research Institute.

F.R.I.B.A. Fellow of the Royal Institute of British Architects.

F.R.I.C. Fellow of the Royal Institute of Chemistry, now F.R.S.C.

F.R.I.C.S. Fellow of the Royal Institution of Chartered Surveyors.

Friction Wake Current The water surrounding a vessel which is dragged while the ship is in motion. The current increases in force from fore to aft while the water accumulates at the stern.

Fridays and Holidays Included Charterparty, *q.v.*, term referring to the Laydays, *q.v.*, which are to be accounted for on Fridays and Holidays. *Abbrev.* FHINC or fhinc.

Fridays and Holidays Excepted *or* **Fridays and Holidays Excluded** Charterparty, *q.v.*, term regarding the Laydays, *q.v.*, which are not to be counted in case work is done on Fridays and holidays. *Abbrev.* FHEX or F.H.E.X.

Fridays, Sundays and Holiday to Count This clause refers to the voyage Charterparty, *q.v.*, meaning that the Laydays, *q.v.*, are to count in favour of the Charterers, *q.v.* Alternatively, if the above days are actually used by the charterers, the owners are not entitled to count them in the overall laydays. *Abbrev.* FSHC or F.S.H.C.

Fridge Refrigerator.

Fridge Ship *or* **Refrigerated Ship** A ship suitably fitted to receive frozen merchandise. These vessels (often referred to as 'reefers') can carry the following commodities vis-à-vis the type of ship used: Chilled Meat Ship—frozen meats and fish; Frozen and Chilled Cargo Ship—meats, fish and chilled merchandise; Fruit or Air-Cooled Ship—for the carriage of fresh fruits.

FRIN Fellow of the Royal Institute of Navigation.

Fringe Benefits Extra benefits enjoyed by employees over and above normal remuneration. These may include pension arrangements, holidays with pay, free meals, transport, medical assistance, etc.

F.R.I.Q.S. Fellow of the Institute of Quantity Surveyors.

Frisco San Francisco, *USA*.

FRISCO Fast Reaction Integrated Submarine Control *USA*.

Frld Freehold.

frm. From.

FRMetS *or* **F.R.Met.S.** Fellow of the Royal Meteorological Society.

F.R.M.S. Fellow of the Royal Microscopical Society.

FRN Floating Rate Note.

F.R.N.S. Fellow of the Royal Numismatic Society.

FRO *or* **F.R.O.** Fire Research Organisation; Fire Risk Only *insurance*.

f.r.o. Fire Risk Only *insurance*.

FROF *or* **F.R.O.F.** *or* **frof** *or* **f.r.o.f.** Fire Risk On Freight *insurance*.

From Stem to Stern A phrase meaning the whole length of the ship.

From the Loading Thereof An insurance expression. The risk attaches only as soon as the merchandise is actually on board. Underwriters, *q.v.*, are not liable while the goods are in transit.

Front Boundary between areas of relatively warm and cold air masses *meteorology*.

Front Door A last resort usually practised by the Discount Houses, *q.v.*, when they are short of funds by selling their Bills, *q.v.*, to the Bank of England or Central Bank at a special reduced bank rate. This is the opposite to Back Door, *q.v.*

Frost Smoke *See* **Sea Smoke**.

Froz. Food Cls. Frozen Food Clauses.

Frozen Assets Money or capital which is dormant and cannot be utilised. A person's or company's money in a foreign bank could be frozen due to the country's regulations. The most usual way in which assets are frozen is when countries are at war with each other.

Frozen Meat Meat which is carried and stored at a temperature of 9.5°.

FRP Fibreglass Reinforced Plastic.

FRS *or* **F.R.S.** Fellow of the Royal Society; Fire Rescue and Safety; Federal Reserve System.

F.R.S.A. Fellow of the Royal Society of Arts.

F.R.S.C. Fellow of the Royal Society of Chemistry.

F.R.S.E. Fellow of the Royal Society of Edinburgh.

F.R.S.L. Fellow of the Royal Society of Literature.

FRSM Fellow of the Royal Society of Medicine.

FRSS *or* **F.R.S.S.** Fellow of the Royal Statistical Society.

Frt. Freight, *q.v.*

Frt. fwd. Freight Forward. *Similar to* **Freight Paid at Destination**, *q.v.*

Frt/ppd Freight Prepaid. *Similar to* **Advance Freight**, *q.v.*

frust Frustrato *Italian*—Disappointment.

Frustration When unforeseen circumstances would compel the contracting parties to cancel the contract. These circumstances must be based on facts. *Ex.* Declaration of War; Outbreak of Epidemics; Undue Delay.

Frustration Clause A marine insurance clause which indemnifies the Underwriters, *q.v.*, in case of loss or frustration of the enterprise due to declaration of war during the voyage.

Frustration of Adventure Charterparty, *q.v.*, term. Though not very often used may save complications when inserted in the contract and applied.

Frwk. Framework.

FS *or* **F.S.** Ferrovie dello Stato *Italian*—State Railways; Field Security; Financial Secretary; Financial Statement; Finlands Skeppshandlareförening r.f. *See* **FSA** *or* **F.S.A.**

fs. Facsimile, *q.v.*

f.s. Factor of Safety; Faire Suivre *French*—Please Forward; Far Side; Fire Station; Flight Service; Fog over the sun; Foot Second.

FSA *or* **F.S.A.** Farm Society Administration; Federal Security Agency *USA*; Financial Services Act *UK*; Finnish Ship Suppliers Association, *similar to* FS *or* F.S. and SLY *or* S.L.Y.; Foreign Service Allowance; French Suppliers Association; Finnish Shipowners' Association.

FSA *or* **f.s.a.** Foreign Storage Area.

F.S.A.A. Fellow of the Society of Incorporated Accountants and Auditors.

FSC Federal Supreme Court *USA*; Fellow of the Society of Chiropodists; Field Studies Council; Finnish Shippers Council; French Shippers Council.

FSE Field Support Equipment; Free Surface Effect *ship's stability*.

F.S.E. Fellow of the Society of Engineers.

FSH Follicles-Stimulating Hormone.

FSHC *or* **F.S.H.C.** Fridays, Sundays and Holidays to Count, *q.v.*

F.S.I. Fellow of the Surveyors' Institute; Federal Stock Item.

FSIA Foreign Sovereign Immunities Act *USA*.

FSL *or* **F.S.L.** First Sea Lord; Full Signed Line, *q.v.*

f.s.l. Full Signed Line, *q.v.*

FSM *or* **F.S.M.** Federation Syndicate Mondiale *French*—World Federation of Trade Unions.

F.S.M.A. Fellow of the Incorporated Sales Managers' Association.

FS Method Federal Standard Method.

F.S.N.A. Fellow of the Society of Naval Architects.

FSO *or* **F.S.O.** Field Security Officer; Fleet Signals Officer; Fleet Supply Order; Fraud Squad Officer.

FSP *or* **F.S.P.** Field Security Police; Foreign Service Pay.

FSPB *or* **F.S.P.B.** Field Service Pocket Book.

FSQS Food Safety and Quality Service, Branch of US Department of Agriculture.

F.S.R. Field Service Regulations.

F.S.S. Fellow of the Royal Statistical Society.

FSSU *or* **F.S.S.U.** Federated Superannuation Scheme for Universities.

FST Flat Square Tube *television*.

FSU *or* **F.S.U.** Family Service Unit; Floating Storage Unit, *q.v.*

F.S.V.A. Fellow of the Incorporated Society of Valuers and Auctioneers.

FSWR *or* **F.S.W.R.** Flexible Steel Wire Rope.

F&T *or* **f&t** Fire and Theft *insurance*.

FT *or* **F.T.** *or* **f.t.** Financial Times; Free Trade; Free Trader; Free of Turn, *q.v.*, Full Terms, *q.v.,*; Formal Training; Free Time, *q.v.*

ft. Foot; Feet; Flat paper.

FTA *or* **F.T.A.** Free Trade Agreement; Freight Transport Association.

FTAC Foreign Trade Arbitration Commission.

F.T.A. Index Financial Times Actuaries Index.

FT&B *or* **F.T.&B.** *or* **f.t.&b.** Free Turn and Berth. *Similar to* **Free of Turn**, *q.v.*

FTC *or* **F.T.C.** Flying Training Command; Full Technical Certificate; Federal Trade Commission; Fast Time Constant; Flight Test Center *USA*.

ft.C. Foot-Candle.

FTD Foreign Technology Division.

ftg. Fitting.

fth. *or* **fthm.** Fathom, *q.v.*

ft/hr Foot per Hour.

F.T.I. Fellow of the Textile Institute.

F.T.I.I. Fellow of the Taxation Institute Incorporated.

F.T.I.T. Fellow of the Institute of Taxation.

ft.-L. Foot-Lambert.

ft.lb. Foot Pound Force.

f to f Face to Face.

FTHP Flow Tubing Head Pressure *oil drilling*.

FTK *or* **LTK** *or* **RTK** Freight Tonne Kilometre *or* Load Tonne Kilometre *or* Revenue Tonne Kilometre. *See* **Freight Tonne Kilometre.**

FTO Foreign Trade Organisation; Fruit Trade Organisation.

FTP Field Terminal Platform *oil drilling*.

F.T.S. Flying Training School.

ft/s Feet or Foot per Second.

ft/s² Feet or Foot per Second Square.

ft³/s Cubic Feet per Second.

FT-SE Joint trade mark of the London Stock Exchange and the Financial Times used to denote a range of indices or benchmarks for the performance of stockmarkets, introduced from 1984. Thus FT-SE 100 (nicknamed Footsie) consists of the 100 largest UK companies with a capital market value above £1bn, representing three-quarters of the total share market; FT-SE Mid 250 consists of the next 250 medium size or 'second line' companies, with a total share of one-

fifth of the market; FT-SE SmallCap, which measures some 500 small or 'third line' companies; FT-SE Actuaries, a real-time index which combines the 100 and Mid 250 by what are called 'industry baskets'; and FT-SE Actuaries All-Share, FT-SE All-Share for short, a composite end-of-day index covering all three segments, large, medium, and small, i.e. the whole UK market. There are also FT-SE Eurotrack 100 for Europe excluding the UK, and FT-SE Eurotrack 200 for all Europe. *See also* **Hang Seng** *and* **Dow Jones** *and* **Nikkei**.

ft. sec. Foot Second.

f.t.t.b. For the time being *insurance*.

fttr. Fitter.

FTU *or* **F.T.U.** Federation of Trade Unions, Hong Kong.

FTW *or* **F.T.W.** Free Trade Wharf.

FTZ Free Trade Zone. Similar to Free Zone, *q.v.*

FU Feed Unit.

Fuel Combustible liquids or other material which is burnt in engines to provide energy to put them in motion.

Fuel Adjustment Factor Same as Bunker Adjustment Factor, *q.v.*

Fuel Oil The heaviest grades of residual fuel used for marine and industrial purposes. The alternative word is Boiler Oil.

Fuel Oil Bunkering Analysis and Advisory Service A Lloyd's Register Service since 1982 for shipping and industry which analyses fuel samples obtained from the bunker manifold and makes reports to the owners.

Fuel P Fuel Port.

FUEN *or* **F.U.E.N.** Federal Union of European Nationalities.

Full and Bye Sailing close to the wind while keeping the sails filled.

Full and Complete Cargo The full amount of cargo needed to immerse a ship to the Load Lines *q.v.* A margin of 10% more or less is normally allowed to be a reasonable figure.

Full and Down When a ship is fully loaded and down to her marks.

Full Container Load A containerised shipment which is a single complete shipment, as distinct from a Less than Container Load shipment that may be consolidated or grouped with others to load a container. *Abbrev.* FCL.

Full Freight The total amount of freight collected or to be collected as per Charterparty, *q.v.*, or Bills of Lading, *q.v.*, calculated freights.

Full Funding The Government action to raise money through various methods of taxation to balance its budget rather than by borrowing.

Full Hole An oil/gas drilled hole being to the maximum diameter. An oil drilling term.

Full Interest The insured has complete interest in the whole subject-matter insured.

Full Interest Admitted *Similar to* **Interest or no Interest,** *q.v.*

Full Out Rye Terms The sellers indemnify themselves to the buyers as to the state and condition of the grain even up to the port of destination. *Abbrev.* FORT *or* F.O.R.T. *or* F.o.r.t.

Full Premium if Lost As the name implies, the assured is paid the full claim of the insurable loss including the amount of the original premium paid. *Abbrev.* FPIL *or* F.P.I.L.

Full Reach and Burden The full cargo space in the holds of a ship, plus the lawful deck capacity, which is offered to the Charterers, *q.v.* The passenger accommodation remains reserved for the owners unless otherwise agreed.

Full Rudder The usual extreme angle of the rudder's movements in a ship.

Full Scantling Ships *See* **Ship with Freeboard**.

Full Signed Line When the Slip, *q.v.*, is circulated in the usual manner at Lloyd's Underwriters, *q.v.*, start accepting or signing lines. Each line accepted or Signed, *q.v.*, is termed 'written line' and abbreviated WL *or* W.L. *or* w.l. When the full amount of insurance is closed or in some instances exceeded the amounts signed for are reduced and a proportion is taken against lines signed by each underwriter to adjust the exact sum of insurance required. This calculation is called 'Full Signed Line' and abbreviated FSL *or* F.S.L. *or* f.s.l. In reinsurance the corresponding word used is 'Full Written Line' and abbreviated FWL *or* F.W.L. *or* f.w.l. *See* **Being Full Signed Line**.

Full Terms Despatch money due for loading and discharging. *See* **Despatch Money**. In the freight market business the owners are to grant certain allowances from the freight to the agents/charterers apart from the usual commission. This is called 'full terms'. *Abbrev.* FT *or* F.T. *or* f.t.

Full Weight Capacity The greatest net weight a container, *q.v.*, will hold, the load being limited by weight rather than volume. *Abbrev.* FWC. See also Fully Loaded Weight and Capacity.

Fully Loaded Weight and Capacity This is referred to when a Container, *q.v.*, is fully stowed to its maximum weight or capacity. *Abbrev.* FWC *or* F.W.C. *Similar to* **Full Container Load**, *q.v.*

Fully Paid Shares *or* **Fully Paid Up Shares** Shares of a company which have actually been paid fully, up to the normal amount.

Fully Rigged Ship Sailing ship which is fully square rigged.

Fully Subscribed The extra number of new shares offered over the needed quota and which are in due time unacceptable.

Fully Written Line *See* **Full Signed Line** and **Being Full Signed Line**.

Fumigation The act of disinfecting the cabins and all compartments of a ship to destroy any rats, mice, cockroaches, insects and other pests. This is an essential precaution taken by shipowners to protect their vessels and to ensure the health of persons living on board.

Functus Officio *Latin*—Having discharged his duty.

Fund International Monetary Fund, *q.v.*

FUND 1971 International Convention on the Establishment of an International Fund for Compensation for Oil Pollution Damage 1971 supplemented by the Protocol of 1984 *IMO.*

Funded Debt A debt which was originally of a temporary nature and which has become permanent. *Opposite to Floating Debt, q.v.*

Funding The method of converting a short term debt into a long term debt.

Funds Accumulated money in hand or capital in stock; government securities; Incomes due to a government.

Funnel *or* **Chimney** In most cases a streamlined cylindrical steel casing which is part of the superstructure of a ship. All the exhaust pipes from the main and auxiliary engines are connected to it and fumes are expelled high up and away from the decks of the vessel. Fumes from the Galley, *q.v.*, are also redirected to pass through the funnel. It is sometimes called a smokestack, though this is not a common word.

Fur. Furlong, *q.v.*; Further.

Furl (1) Furlong, *q.v.* (2) To gather or roll up a sail.

Furlong An English linear measurement equal to 220 yards or one-eighth of a mile. *Abbrev.* Fur or furl.

Furn. Furnace.

Future Markets *or* **Terminal Markets** Transactions for purchase against an anticipated future delivery. This occurs in trades dealing with commodities such as grain, sugar, wool, tea, coffee. The delivery date may have some connection with the harvest. *See* **Forward Contract**.

Futures Commodities or stocks bought and sold at an agreed price for the buyer to take at a future appointed time.

Futures Contract Forward contract for the purchase or sale of goods or commodities to be delivered or acquired at a future date at a price agreed when the contract is made.

FV *or* **F.V.** *or* **f.v.** Fire Vent; Fishing Vessel; Flush Value; Folio Verso *Latin*—Back of the pack or sheet; the left-hand page, usually having an even page number.

F.V.A. Fellow of the Valuers' Association.

FVC or **F.V.C.** Fishing Vessel Clauses, *marine insurance covering fishing vessels.*

F.V.I. Fellow of the Valuers Institution.

FVIP Fishing Vessel Insurance Plan, *q.v.*

FW or **F.W.** Fresh Water.

FWA or **F.W.A.** Federal Works Agency *USA*; Fresh Water Allowance, *q.v.*

FWAD or **F.W.A.D.** Fresh Water Arrival Draft (Draught).

FWC or **F.W.C.** Full Weight Capacity *q.v.*; Fully Loaded Weight and Capacity, *q.v.*

FWD or **F.W.D.** or **f.w.d.** Four-Wheel Drive; Free of Fresh Water Damage, *See* **Bilge Water Damage**; Front Wheel Drive.

fwd Forward.

Fwdg Forwarding, *q.v.*

Fwdr Forwarder. Similar to Forwarding Agent, *q.v.*

FWE or **F.W.E.** Finished With Engines.

FWI or **F.W.I.** Federation of West Indies; French West Indies.

FWL or **F.W.L.** or **f.w.l.** Foilborne Water Line, referring to waterline of hydrofoil; Fully Written Line.

FWO Federation of Wholesale Organisations; Federation of Workers' Organisations, Merchant shipping in Rotterdam, Holland.

FWP or **F.W.P.** Fresh Water Pump.

FWPCA Federal Water Pollution Control Administration *USA*, or Federal Water Pollution Control Act.

FWR Forwarder's Warehousing Receipt.

FWT or **F.W.T.** Fair, Wear and Tear. *See* **Fair Wear and Tear Excepted**. Feed Water Tank; Fresh Water Tight.

Fwt. Featherweight.

FWT & GD or **F.W.T. & G.D.** Fair, Wear, Tear and Gradual Deterioration, *insurance policy exemption.*

FWTH or **F.W.T.H.** Flush Water Tight Hatch.

f.x. Foreign Exchange.

FX Forecastle *nautical.*

fxd. Fixed.

Fxle Forecastle, *q.v.*

FY or **F.Y.** Fiscal Year, *q.v.*

FYG or **F.Y.G.** For your guidance.

FYG&I For your guidance and information.

FYI For your information.

FYLC or **F.Y.L.C.** Five Year Loadline Certificate.

FYP or **F.Y.P.** Five Year Plan *or* Four Year Plan.

FZ Fishing Zone.

FZ or **F.Z.** Free Zone, *q.v.*

F.Z.S. Fellow of the Zoological Society.

G

g Gale force 8 or 9 *Beaufort notation*; Gallon, *q.v.*; Gauche *French*—Left; Gauge; General Factor; Gloomy; Astronomy *nautical*; Acceleration due to gravity; Gold; Gourde *Haitian currency*; Grain; Gramme; Grammes; Gravity; Gros *French*—Large; Guilder *Dutch currency*; Gulf.

G Gauge; German; Giga (10⁹); Grand (US$1000); Centre of Gravity; Grain; Grammar; Greek; Grid; Greenwich Meridian, upper branch; Storm force 10 *Beaufort notation*.

GA *or* **Ga** Georgia, *USA*.

G.A. *or* **GA** Garuda Indonesian Airways *airline designator*; General Average, *q.v.*; Go Ahead *in telex communications*.

G/A *or* **g/a** General Agent(s), *q.v.* General Assembly; General Assignment; General Average, *q.v.*; Geographical Association; Geologist's Association; Gone Away; Government Actuary; Greenwich Meridian, lower branch; Ground to Air *aviation*.

GAACCIA *or* **G.A.A.C.C.I.A.** General Arab Association of Chambers of Commerce, Industry & Agriculture.

G.A. Act General Average Act, *q.v.*

G.A. in Full Clause An insurance term in the cargo policy whereby Underwriters, *q.v.*, undertake to pay General Average Contribution, *q.v.*, in full irrespective of any excess due in the Contributory Value, *q.v.*

GAA General Agency Agreement, between the National Shipping Authority and US Shipowners.

GAAP Generally Accepted Accounting Principles, *q.v.*

Gab. Gabon, *West Africa.*

GAB General Arrangement to Borrow, *q.v.*

GAC *or* **G.A.C.** General Average Certificate; General Average Claim. *See* **General Average**: General Average Contribution, *q.v.*

G/A Con General Average Contribution, *q.v.*

GAD *or* **G.A.D.** *or* **G.A. Dep.** *or* **G/A Dep** General Average Deposit, *q.v.*

GA. Dis. *or* **G.A. Dis.** General Average Disbursements.

GADV *or* **G.A.D.V.** Gross Arrived Damaged Value. Opposite to Gross Arrived Sound Value, *q.v.*

GAFTA Grain and Feed Trade Association.

GAG *or* **G.A.G.** General Average Guarantee, *q.v.*

GAI Guaranteed Annual Income *USA.*

Gaining Rate The daily error recorded for a chronometer when running fast. Opposite to Losing Rate, *q.v.*

GAL *or* **G.A.L.** General Average Loss. *See* **General Average**.

Gal *or* **gall** Gallon, *q.v.*

Gal. cap. Gallon capacity.

Gale Winds blowing force 8 and over measured on the Beaufort Scale, *q.v.*

Galena Lead sulphide.

Galley Cooking place or kitchen of a ship.

Galley Range The kind of installation the galley, *q.v.*, has for cooking, namely whether fuel burning, electric or any other system.

Galley Wireless Sailor's slang for rumour. From the supposed source of reliable information about the ship's programme being the cook.

Galliot An old Dutch trading or fishing boat; small Mediterranean galley.

Gallon A measure of capacity. The imperial gallon contains ten imperial standard pounds' weight of distilled water weighed in air against brass weights, with the water and the air at 62 degrees Fahrenheit, and with the barometer at thirty inches. This gallon contains 277.274 cubic inches: it is to be distinguished from the American gallon, which contains 231 cubic inches. A gallon equals four quarts or 4.54596 litres. *Abbrev.* Gall.

Galls *or* **Gls.** Gallon(s), *q.v.*

Galofaro Whirlpool in the Strait of Messina.

Galv. Galvanic; Galvanised; Galvanometer, *q.v.*

Galvanic Action Electrolytic reaction produced by sea water on various metals which will in time result in their corrosion.

Galvanized Iron Zinc coated iron sheet generally plated by hot dipping. *Abbrev.* GI or G.I. or g.i.

Galvanometer A very sensitive and delicate instrument by which current passing through it is detected and measured. *Abbrev.* Galv.

GAM Guided Aircraft Missile.

Gangway A narrow platform of wood or metal having wooden bars or protective ropes either side which is used as a passageway for people boarding or leaving a ship. Used generally when the vessel goes alongside a quay. *See* **Manropes**. Also a passageway for people within a ship.

Gantline A line or rope passing through a Block, *q.v.*, from the top of the mast of a ship to help crew members and ship's rigging to be carried aloft. *Also* Girtline.

Gantry *or* **Gauntry** High level structure on the deck of a ship or on shore to support a moving crane on rails.

Gantry Cranes *See* **Gantry** *or* **Gauntry.**

GAO General Accounting Officer, *US Navy*; General Accounting Office *USA*.

GAP General Arrangement Plan; General Assembly Programme; Great American Public; Gross Agriculture Product.

GAPAN *or* **G.A.P.A.N.** Guild of Air Pilots and Air Navigators.

GAR *or* **G.A.R.** General Average Refund; Guided Aircraft Rocket.

Garble To unintentionally distort or confuse. This word is often used when telex communications become illegible.

Garbling Removing damaged tobacco from a full bale to avoid further damage to the whole.

Garboard Strake Ship's shell plating near to the keel plate. *Abbrev.* Gar. Str.

Garnishee A debtor who has been warned to pay his debt not to his own creditor but to some third party who has obtained a final judgment against the creditor. *See* **Garnishee Orders.**

Garnishee Orders Court orders to persons forcing them to refrain from using or paying out any funds in their possession belonging to other persons. *See* **Garnishee.**

Garnishment The court notices or orders stopping the payment of money and goods. *See* **Garnishee** and **Garnishee Orders.**

GARP Global Atmospheric Research Programme. An international meteorological research programme.

Gar. Str. Garboard Strake, *q.v.*

Garuda Indonesia State Airlines.

Gas. Gasoline, *q.v.*

GAS *or* **G.A.S.** General Average Sacrifice.

GASC German-American Securities Corporation.

G.A. & S. *or* **G/A and S** General Average and Salvage.

Gas Carrier A tanker intended to carry liquefied petroleum gases (*abbrev.* LPG) in the region of about 75,000 cu. metres and which is capable of handling two different kinds of cargo.

Gas Certificate A certificate produced by the port authorities after the empty tanks of an oil tanker have been found free of any dangerous or explosive airs or gases. The certificate is needed before a tanker is allowed to enter a commercial port or before entering a dry dock for necessary repairs or other work. *See* **Gas Freeing.**

Gas Free *See* **Gas Freeing.**

Gas Free Certificate *Similar to* **Gas Certificate,** *q.v.*

Gas Freeing The process of clearing dangerous and explosive airs or gases from empty petroleum tanks. The tanks are cleaned by special detergents before the port chemist boards the ship to test and certify that she is gas free prior to her entering port. *See* **Gas Certificate.**

Gas Line The alternative US term is Vent Line. This is a system installed in oil tankers in order to allow the intake of fresh air and to expel dangerous gases at the same time.

Gas Oil Fuel Oil or Diesel Oil.

Gasoline Distillate of crude petroleum; alternatively known as petrol or benzine or essence.

Gastank Waybill *or* **GASTANK WAYBILL** Non-negotiable Gas Tank Waybill used in the LPG trade.

Gastime *or* **GASTIME** Code name for BIMCO Time Charter Party for the carriage of liquefied gas.

GASV *or* **G.A.S.V.** Gross Arrived Sound Value, *q.v.*

GASVOY Gas Voyage Charter Party to be used for Liquid Gas except LNG, *q.v.*

Gas Well A well which produces natural gas.

GAT *or* **G.A.T.** Greenwich Apparent Time *nautical.*

GATCO *or* **G.A.T.C.O.** Guild of Air Traffic Control Officers.

Gate Number Number of the gate at which passengers assemble and which they pass through to embark on an aeroplane.

GATT *or* **G.A.T.T.** General Agreement on Tariffs and Trade.

Gauge Glass A thin glass tube used for measuring the contents of oil or water in tanks.

Gauntry *Similar to* **Gantry,** *q.v.*

GAV *or* **G.A.V.** Gross Annual Value *income tax.*

Gavel Payment of Rent; Payment or Tribute to a superior.

GAW *or* **G.A.W.** Guaranteed Annual Wage.

Gaz. Gazette; Gazetteer.

GB *or* **G.B.** Great Britain; Grid Bearing *nautical*; Guide Book; Gunboat.

g.b.e. Gilt Bevelled Edge *in glass, mirrors, etc.*

G.B. & I. Great Britain and Ireland.

GBL Government Bill of Lading *UK*.

GBO *or* **G.B.O.** *or* **g.b.o.** Goods in Bad Order.

GBRMPA Great Barrier Reef Marine Park Authority, established to recommend areas for inclusion in the marine park zoning scheme.

GBS Gravity-based Structure *oil drilling.*

GBSBP Gravity Base Structure Production Platform *oil drilling.*

Gc. *Lloyd's Register notation* showing that the ship is fitted with Gyro Compass, *q.v.*

GC Gas Chromatography; General Cargo.

GC *or* G.C. Garage Combined *insurance*; George Cross; Grand Chancellor; Grade Chapter; Grain Certificate; Grid Course *nautical*; Great Circle *navigation*; Gyro Compass.

GC *or* Gencargo General Cargo.

GCA *or* G.C.A. Gold Clause Agreement; Greek Collective Agreement; Ground Controlled Approach *aviation*.

g.cal. Gramme Calorie.

G/Captain *or* G.Capt. Group Captain.

GCBS Now defunct and reconstituted as the Chamber of Shipping, *q.v.*

GCC *or* G.C.C. Gulf Co-operation Council; Grain Cargo Certificate. *See* **Grain Certificate**. Gas Carrier Code, now International Bulk Code. *Abbrev.* IBC.

GCD *or* G.C.D. General Certificate of Distribution.

g.c.d. *or* **gcd** Greatest Common Denominator *mathematics*.

G.C.E. General Certificate of Education.

gcf *or* **g.c.f.** Greatest Common Factor *mathematics*.

GCI *or* G.C.I. Ground Controlled Interception *radar*.

GCIC German Chamber of Industry and Commerce.

G.C.M. General Court Martial; Good Conduct Medal.

gcm *or* **g.c.m.** Greatest Common Measure or Multiple.

GCR *or* G.C.R. Gas Cooled Reactor; General Cargo Rate; General Community Rate; Ground Controlled Radar.

GCS Greek Chamber of Shipping.

G & CSA Glasgow and Clyde Shipping Association.

GCT *or* G.C.T. Greenwich Civil Time, *q.v.*

G.D. Geared Diesel; Graduate of Divinity.

g.d. General Duties; Good Delivery.

g-d Gravimetric Density.

GDBA Guide Dogs for the Blind Association.

GDH Goods on Hand.

Gdk. Gdansk, *Poland*.

GDP *or* G.D.P. Gross Domestic Product, *q.v.*

Gd.R. Guard Rail.

GDR *or* G.D.r. German Democratic Republic, *East Germany*.

Gds. Goods.

GDWQ Guidelines for Drinking Water Quality, *issued by World Health Organisation*.

GE *or* g.e. General Electric; Gilt Edged; Gross Energy; Gyro Error *nautical*.

G.E. General Electric; Gross Energy.

Gear Equipment or apparatus appurtenant to a ship. Also called Cargo Gear.

Gearless Meaning a 'gearless vessel' which is a ship having no gear for lifting cargo. In many cases a vessel of this type is specfically used for bulk cargoes where loading/discharging is effected by chutes, grabs or shore cranes. *Abbrev.* GLESS.

GEBCO General Bathymetric Chart of the Oceans.

G.E.C. General Electric Company.

GECREF World Geographic Reference System.

GEF Global Environmental Facility *World Bank*.

Gegluga *Polish*—Shipping.

GEICO Government Employees Insurance Company *USA*.

GEIS General Electric Information System.

Gel. Gelatin.

GEM Grain Elevator Mij BV, at Rotterdam, Holland; Ground Effect Machine, also known as Hovercraft; Guidance Evaluation Missile.

GEMM Gilt-edged Market Maker *London Stock Exchange*.

GEMS Global Environmental Monitoring System *UN*.

Gen. General; Generator; Genesis; Geneva, Switzerland; Genoa, Italy.

GENEM Groupement d'Exportation de Navires et d'Engines de Mer en Acier *French*—General Maritime Exporters Association.

General Agent An officially appointed agent who has a wide scope of authority in binding his principals during the course of his business. *Abbrev.* G.A. *or* g.a.

General Agreement on Tariffs and Trade International trade organisation with headquarters in Geneva, Switzerland. The aim is to co-ordinate import duties and tariffs for the benefit of the member countries. *Abbrev.* GATT *or* G.A.T.T.

General Arrangements to Borrow Agreement of ten central banks from UK, USA, France, Italy, The Netherlands, Germany, Belgium, Sweden, Canada and Japan, to lend to the International Monetary Fund in order to cover loans to borrowing nations. *Abbrev.* GAB.

General Average An internationally accepted rule of the sea. When a ship is in danger of total loss the master has the right to sacrifice property and/or incur reasonable expenditure to prevent the total loss. Measures taken for the sole benefit of any particular interest are not general average. On successful completion of the adventure, or if it is abandoned in a place of safety, on successful attainment of such place the ship is declared as entering 'under average'. Security in the shape of deposits or guarantees is taken from each cargo interest and an average adjuster is appointed. The adjuster calculates the value of the saved interest and each interest is required to contribute a rateable proportion to make good the general average loss. Underwriters are liable only if the peril leading to the general average act was an insured peril, and their liability is limited to the insured value of sacrificed insured property and to their proportion of the general average contribution payable by the assured. *Abbrev.* G/A *or* g/a *or* GA *or* G.A.

General Average Account An open account for the purpose of receiving contributions from the cargo owners in a **General Average**, *q.v.* act. *See* **Average Bond**.

General Average Contribution The proportion payable by one of the parties involved in a general average act to make good the loss suffered in that act. *Abbrev.* GAC *or* G.A.C. *or* GA. Con. *See* **General Average Account**.

General Average Deposit Deposit demanded by carriers or their agents from cargo owners when a General Average, *q.v.*, loss is encountered prior to the release of the cargo. This is similar to Average Bond, *q.v. Abbrev.* GAD *or* G.A.D. *or* G.A Dep. *See* **Average Bond**.

General Average Disbursements Shipowner's expenses as part of a General Average Act, *q.v.*, which are in turn recoverable from a General Average Fund or Refund, *q.v. Abbrev.* G.A. Disb. *See* **Average Bond**.

General Average Fund The total arrived at by adding together general average expenditure and the value of property sacrificed in a general average act, plus costs of its adjustment. *See* **Average Bond**.

General Average Guarantee An undertaking by a financial house or an underwriter to pay the contribution due towards a general average fund. *Abbrev.* GAG *or* G.A.G. *See* **Average Bond**.

General Cargo Mixed cargo consisting of various types of merchandise, such as cases of tinned fruit, vegetables, meats, fish, cartons of cigarettes, biscuits, bags of sugar, dried peas, etc. The most adaptable ship for this type of general cargo is a Tweendecker, *q.v.*, where the cargo can easily be sorted out in different layers and sections.

General Council of British Shipping Now defunct and replaced by the Chamber of Shipping, *q.v.*

General Lien *See* **Lien**.

General Meeting Meeting of all the members of a company or association held in accordance with the Memorandum of Association. *Abbrev.* GM.

General Partner An active partner with unlimited liability.

General Purpose Rating A seaman whose duties on board include work on deck, in the engine room or other departments. *Abbrev.* GPR *or* G.P.R. *or* G.P. Rating.

General Purpose Tanker *or* **GP Tanker** A notional term for an oil carrier in the range 16,500 to 25,000 dwt.

General Ship The American expression for Common Carrier, *q.v.*

General Strike A strike in which all the unions of a country are involved.

General Trader A vessel employed in tramping.

General Transfer *Similar to* **Block Transfer**, *q.v.*

General Transire Coasters engaged on regular ports of call give notice of the cargo carried to facilitate the procedure of clearing from customs inwards and outwards during a period of one year. In certain ports this system also saves money in port dues as compared with ships which pay dues on each call. In the General Transire system the ships pay a uniform and economical port tariff for one whole year irrespective of the number of the calls made.

General Union Industrial Union.

Generalised System of Preferences A system of customs duty preference which is used by the EC to cut tariffs on imports from the developing world. *Abbrev.* G.S.P. *or* GSP.

Generally Accepted Accounting Principles An organisation set up by the American Institute of Certified Public Accountants. It obviates the need for government control. *Abbrev.* GAAP.

Gen'l General *USA*.

Genn. Gennaio *Italian*—January.

Genoa/Savona/Spezia/Leghorn Range of loading/discharging ports in Italy. *Abbrev.* GSSL *or* G.S.S.L.

Genoa/Savona/Spezia/Leghorn/Naples *or* **Civitavecchia** Range of loading/discharging ports in Italy. *Abbrev.* GSSLNCV *or* G.S.S.L.N.C.V.

Geo Georgia, *USA*.

GEO Geostationery *satellite configuration*.

Geodesy Branch of mathematics dealing with figure and area of the earth's surface.

Geographical Mile *or* **Nautical Mile** One minute of great circle of earth, fixed by British Admiralty at 6080 feet or 1853.18 metres. *See* **Nautical Mile**.

Geographical Rotation A shipping term currently used in Charterparties, *q.v.*, regarding the regular sequence of the geographical positions of the ports of loading and discharging. *Abbrev.* GR *or* G.R. *Similar to* **In Geographical Rotation**, *q.v.*

Geolograph An instrument used to find out the speed and penetration effects of the drill during drilling operations.

Geology Science of the earth's crust, its strata and their relations and changes.

Geometric On-Line Definition A shipbuilding system used for the outlining of the steelwork by means of electronic machines. *Abbrev.* GOLD.

GEON *or* **geon** Gyro Erected Optical Navigation.

Geophysics Physics pertaining to the earth including, geodesy, seismology and others.

GEOREF Geographical Reference *ships/aircraft reporting system*.

GEOS Geostationary Operational Environmental Satellites.

Ger. German; Germany.

GES *or* **ges** Gesellschaft *German*—Firm or Society.

GESAMP *or* **G.E.S.A.M.P.** Group of Experts on the Scientific Aspects of Marine Pollution, made up of experts nominated by eight United Nations bodies, including the IMO.

Gesellschaft mit beschränkter Haftung *German*—Limited Liability Company. *Abbrev.* G.m.b.H. *or* GmbH.

GEV Ground Effect Vehicle.

GeV Thousand million electron volt.

Gez. Gezeichnet *German*—Signed.

GF *or* **G.F.** General Foods; Government Form, referring to charterparty in UK; Ground Fog or Shallow Fog *weather observation.*

GFA *or* **G.F.A.** *or* **g.f.a.** Good Fair Average; Goods Freight Agent; Government Freight Agent *appointed by the Ministry of Defence on behalf of the UK Government.*

GFC *or* **G.F.C.** Gas Free Certificate. *See* **Gas Certificate**.

GFCM *or* **G.F.C.M.** General Fisheries Council for the Mediterranean, Fisheries and Agriculture Organisation Branch of UN.

GFG Good Food Guide.

GFR *or* **G.F.R.** German Federal Republic, West Germany.

GFTU General Federation of Trade Unions.

GG Glasgow.

GH General Hospital; Ghana; Ghana Airways *airline designator*; Grid Heading *nautical.*

G–H Gibraltar-Hamburg Range.

GHA *or* **G.H.A.** Greenwich Hour Angle *nautical.*

GHAMS Greenwich Hour Angle Mean Sun *nautical.*

GHE *or* **ghe** Ground Handling Equipment.

GHI *or* **G.H.I.** Good Housekeeping Institute.

GHOST Global Horizontal Sounding Technique.

GHQ *or* **G.H.Q.** General Headquarters.

GI *or* **G.I.** Air Guinea *airline designator*; American soldier. The expression GI is derived from the US Army term 'Government Issue' which means kit or equipment; Government of India.

gi. Gill.

g.i. Galvanised Iron.

GIA Garuda Indonesian Airlines.

Giant Planets Jupiter, Uranus, Neptune and Saturn.

Gib. Gibraltar.

GIB Guy in the Back Seat *US Navy aviation slang.*

Gibair Gibraltar Airways.

Gibbous Moon The moon when the shape of its illuminated face is gibbous, that is, between half and full.

GIDEP Government Industry Data Exchange Program, *US.*

GIDI Guidelines for Data Exchange and Interbridge.

GIE Groupement d'Interêts Économiques.

Giffen Goods Goods of inferior qualities.

GIFT *or* **G.I.F.T.** Glasgow International Freight Terminal.

Gift Tax A state tax on gifts received *USA; Similar to* **Donation Duty**, *q.v.*

Gig Light two-wheeled one-horsed carriage; light narrow clinker-built ship's boat for oars or sails; rowing-boat chiefly used for racing.

giga Prefix meaning one thousand million (10^9).

GIH Governmental Industrial Hygienists *USA.*

GIGO Garbage In, Garbage Out, referring to computers, erratic or inaccurate output of data in consequence of the initial inaccurate input.

Gill One quarter of a measured pint, *Abbrev.* gl.

Gilt-edged Securities Fixed-interest of securities. As government stocks they are not difficult to sell.

Gimbals. A set of rings and pivots attached and arranged in such a manner as to hold a compass always in a horizontal position.

Gimlet *or* **Grimlet** A small instrument with a wooden crosspiece handle attached to a pointed screw-type piece of metal used to bore holes in wood.

G.I Mech.E. Graduate of the Institute of Mechanical Engineers.

Gin Block Metal pulley in a skeleton frame.

Ginko *Japanese*—Bank.

Ginny Virginia, *USA.*

G. Inst. T. Graduate of the Institute of Transportation.

G in T Goods in Transit.

G.I.O. Guild of Insurance Officials.

Giorni Correnti *Italian—Similar to* **Consecutive Days**, *q.v.*

Gird To bind; Application of a transverse heeling force to a tug when the tow-rope is allowed to grow on the beam under strain, a potentially dangerous situation.

Girders (1) The principal longitudinal and transverse structural members built into a ship for strength. (2) Main beams or supporting members in any other structure, such as the steel beams in the frame of a building. *Abbrev.* Gir(s).

Giro A Post Office system of money transfer whereby a person can order the Giro to pay any amount of money to another person or organisation.

Gir/s Girder(s), *q.v.*

GIS German International Ship Register.

Giu. Giugno *Italian*—June.

Giudice togato *Latin*—Gowned judge; Official/Real Judge.

GIUK Greenland, Iceland and United Kingdom.

Giver-On A Stock Exchange, *q.v.*, expression referring to the Broker, *q.v.*, who hands over stocks plus interest to another broker. The latter broker is called Taker-In.

Give Way (1) To keep clear; when in proximity to another vessel, to alter course so that the other may maintain her course. This action will be governed by the Regulations to Prevent Collisions at Sea (COLREGS). (2) An order to pull on the oars.

Gk *or* **Gr.** Greek.

GK Gomei Kaisha, *q.v.*; Goshi Kaisha, *q.v.*

gl Gill, *q.v.*

g/l Grams per litre.

GL *or* **G.L.** Germanischer Lloyd, German classification society; Ground Level.

Gl. *or* **G.L.** Galvanized Iron; Gill, *q.v.*

GLA General Lighthouse Authority. There are three: Trinity House, Northern Lighthouse Board and the Commissioners for Irish Lights.

Glam. Glamorganshire, *UK*.

Gland Packing Material used in the propeller shaft of a ship to prevent oil leakage from the inside (oil which is used as lubricant) or sea water penetrating from outside. *See* **Shaft Tunnel**.

Glas. Glasgow, *Scotland*.

Glasnost *Russian*—Publicity; Openness to criticism. Word introduced into USSR government policy by M. Gorbachev.

GLASS Global Authority For Ship Standards.

GLB General Laybarge *oil drilling*.

G.L.C. Greater London Council (now abolished).

GLC/MS Gas Liquid Chromatography/Mass Spectrography.

GKS Geographical Kernel System.

GLCM Ground-Launched Cruise Missile *US Air Force*.

Gld *or* **gldr.** Guilder, of currency used in Holland.

Glen A narrow valley having steep sides usually with a lake, stream or river at the bottom.

GLESS Gearless, *q.v.*

Glidepath The descending path of an aircraft when landing.

Global Maritime Distress and Safety System Concept of satellite and terrestrial radiocommunication services developed by IMO for worldwide navigational warnings and search and rescue operations. *Abbrev.* GMDSS.

Global Positioning System Satellite navigation service under development by US Department of Defence and the Soviet Union providing continuous world-wide all-weather position fixes for land, sea and air. *Abbrev.* GPS.

Globe Another word for the world depicted on a sphere.

GLOMEX Global Oceanographical and Meteorological Experiment.

Glos. Gloucester; Gloucestershire.

Glovays Astronomicheskaya Observatoriya, astronomical observatory in USSR.

GLP Great Lakes Ports, *q.v.*

GLT Greetings Letter Telegram.

Glut This word can be used to describe a situation when there are insufficient buyers for a particular commodity; a surplus.

GM *or* **G.M.** General Manager; General Merchandise; General Mortgage; Grand Master; George Medal; Good Middling; Guided Missile; Gyro-Magnetic Compass.

GM General Meeting; Geometric Mean; Gunner's Mate, *q.v.*

GMAA German Maritime Arbitration Association.

G-man Federal Criminal Investigating Officer in America.

GMAT *or* **G.M.A.T.** Greenwich Mean Astronomical Time.

GMB The UK's general trades union and second largest with membership in a wide variety of trades and craft, formerly the General, Municipal and Boilermakers.

GMB *or* **G.M.B.** *or* **g.m.b.** Good Merchandise Brand; Good Merchantable Brand.

GmbH *or* **G.M.b.h** *or* **G.m.b.H.** Gesellschaft mit beschränkter Haftung, German Limited Liability Company.

GMC *or* **G.M.C.** General Management Committee; General Medical Council; Ground Movement Controller *aviation*.

GMDSS Global Maritime Distress and Safety System, *q.v.*

GMLS *or* **G.M.L.S.** General Maritime Law of Salvage.

GMQ *or* **G.M.Q.** *or* **g.m.q.** Good Merchantable Quality.

GMR Ground Mapping Radar.

GMRS Global Maritime Radiotelephone Service, an association of major telecommunications companies.

GMS *or* **G.M.S.** Geostationary Meteorological Satellite.

GMT *or* **G.M.T.** Greenwich Mean or Meridian Time, *q.v.*

GMWU General and Municipal Workers' Union *UK*.

GN *or* **G.N.** Air Gabon *airline designator*; Guidance Notes.

GNC Global Navigation Chart.

GNEPI *or* **G.N.E.P.I.** Gross Net Earned Premium Income.

GNP *or* **G.N.P.** *or* **g.n.p.** Gross National Product, *q.v.*

GNPC Ghana National Petroleum Corporation.

GNSS Global Navigation Satellite Systems.

GO or G.O. Gas Oil; Gas Operated; General Office.

GOB *or* **G.O.B.** *or* **g.o.b.** Good Ordinary Brand.

GOC *or* **G.O.C.** General Officer Commanding; General Operator Certificate: with regard to the Global Maritime Distress and Safety System, *q.v.*

Godown Far Eastern expression for a Warehouse, *q.v.*, or cargo shed.

GOES Geostationary Operational Environmental Satellites.

GOFAR Global Geological and Geophysical Ocean Floor Analysis and Research.

Going concern Flourishing business.

Going price Current ruling price.

Going public When a private company switches to being a public one and markets its shares.

GOLD Geometric On-Line Definition, *q.v.*

Gold Clause Agreement An agreement between parties concerned with the carriage of goods by sea, whereby they agree to increase the limits imposed by law for shipowners' liability in respect of cargo.

Golden Handshake Compensation offered in the form of money or other credits to employees of a company who are unable to continue to work for some reason, such as redundancy. Also called Terminal Benefits or Redundancy Payment, *q.v.*

Gold Franc A franc containing 65 milligrams of gold.

Gold Fixing The fixing of the price of gold on the London and New York gold markets.

Gold Market Referring to London and New York gold markets.

Gold Premium The difference of the face value of paper currency when it is less than the actual corresponding gold value.

Gold Standard System under which a national currency has a precise value in gold.

Golf International Radio Telephone phonetic alphabet for 'G'.

Gomei Kaisha *Japanese*—Family partnership with unlimited liability. *Abbrev.* GK.

Goniometer Device or instrument for measuring angles.

Good Faith In Bona Fide, *Latin*—Good and Honest intention. One of the most essential characteristics in an insurance contract and if the utmost good faith (as it is accepted in the legal aspect) is not observed by either party the contract becomes automatically null and void.

Good Merchantable Quality & Condition In reference to the sale of merchandise which should be of a sound customary standard. *Abbrev.* GMQ&C or G.M.Q.&C. *or* g.m.q&c.

Good Order & Condition As a rule this clause is to be found inserted on Bills of Lading, *q.v.*, in regard to the receipt of the merchandise on board 'in apparent good order and condition'.

Goods In marine insurance this refers to merchandise but excluding personal effects as well as ship's stores and provisions. However, in a Charterparty, *q.v.*, and on Bills of Lading, *q.v.*, goods refer to all kinds of merchandise except livestock.

Goods and Chattels Material moveable belongings *law.*

Good Ship Seaworthy Ship. There must be an implied warranty of seaworthiness or a good ship for the Carrier, *q.v.*, to undertake a contract for the Carriage of Goods, *q.v.* The word 'good' is usually put before the word 'ship' in Marine Insurance Policies, Contract of Affreightments, Charterparties, Bills of Lading (all *q.v.*) and suchlike documents. *See* **Seaworthiness**.

Goods On Approval Goods, *q.v.*, are finally accepted on condition that the purchaser is allowed a certain number of days to inspect them.

Goods on Consignment *Similar to* **Consignment**, *q.v.*

Good Sound Merchandise *Similar to* **Good Merchantable Quality** and **Condition**, *q.v. Abbrev.* G.S.M. *or* GSM *or* g.s.m.

Goodwill The intangible reputation and business valuation obtained after a certain number of years.

GOP *or* **G.O.P.** Gross Operating Profit.

GOR *or* **G.O.R.** Gross Original Rate. *Similar to* **Original Gross Rate**, *q.v.*

GOS *or* **G.O.S.** Global Observation Station.

Goshi Kaisha *Japanese*—Partnership. *Abbrev.* GK.

GOSIP Government Open Systems Interconnection Profile.

Go-Slow In reference to an industrial dispute where the workers adopt a slow method of working or work strictly according to the rule book. *Similar to* Slowdown. *See also* **Work to Rule** *or* **Working to Rule**.

Gosplan Russian State Planning Commission.

Gosud. Gosudarstvo *Russian*—State.

Gosudarstvenni Universalni Magazin Russian Universal State Store. *Abbrev.* GUM.

Gosudarstvo *Russian*—State *Abbrev.* Gosud.

Gothenburg Standard Pit prop standards and sleepers calculated at 180 cu.ft.

Gou. Gourde, *q.v.*

Gourde Unit of currency used in Haiti. The US dollar is also in use. *Abbrev.* Gou.

Gov. Governor.

Government Furnished Equipment On awarding a contract for the construction of a US naval vessel the authorities pinpoint the systems, items or equipment to be furnished to the privately-owned shipyard who are also known as 'prime contractors' (*abbrev.* PC). Any remaining equipment necessary is to be supplied by the private yard or prime contractor or by those sub-contracted. The latter is called Contractor Furnished Equipment and *abbrev.* CFE.

Government Securities Treasury Bills and Funded Stocks.

Government Stocks Gilt-edged Stocks.

Gov.Gen. Governor General.

Govt. Government.

GOX *or* **Gox** Gaseous Oxygen.

GP *or* **G.P.** *or* **g.p.** General Practitioner; General Purpose. *See* **General Purpose Ratings**; Geographical Position *nautical.*

GPA Global Program on Aids *WHO.*

G.P.A. General Practitioners' Association.

GPC *or* **G.P.C.** General Policy Committee, General Council of British Shipping; General Purpose Committee; Guaranteed Post Card

Gp Comdr. *or* **Gp. Cmdr.** Group Commander.

Gpd *or* **g.p.d.** Gallons per day.

GPE Guided Projectile Establishment.

Gp. Fl. Group Flashing Light *visual aid to navigation.*

G.Ph. Graduate in Pharmacy.

GPH *or* **G.P.H.** *or* **g.p.h.** Gallons per Hour.

GPL *or* **G.P.L.** *or* **g.p.l.** Gallons per Litre.

GPM *or* **G.P.M.** *or* **g.p.m.** Gallons per Minute.

Gp. Occ. Group Occulting Light, *visual aid to navigation.*

GPO *or* **G.P.O.** General Post Office; Government Printing Office *USA.*

GPR General Purpose Rating(s), *q.v.*

G.P. Rating General Purpose Rating(s), *q.v.*

GPS *or* **G.P.S.** General Purpose Ship; Global Positioning System, *q.v.*

GPTS General Purpose Automatic Test System.

GPU Ground Power Unit. Airport equipment providing heat and ventilation for aircraft before take-off, as well as starting engines.

GPW *or* **G.P.W.** Gross Plated Weight.

GQ *or* **G.Q.** General Quarters.

GR *or* **G.R.** General Reserve; Geographical Rotation, *q.v.*; Ground Rent.

Gr. Grain, cubic; Greece; Greek; Greenwich.

gr. Grade; Grain; Gravity, *q.v.*; Gross, i.e. 144.

Grab Lifting equipment mostly used in loading/unloading bulk cargoes. It lifts cargo by gripping. There is another grab system which uses large magnets to lift scrap metal.

Grab Damage A Charterparty, *q.v.*, clause covering the owner of the ship for any consequent damage which may occur by the use of a Grab, *q.v.*, while loading /unloading bulk cargoes. Surveys are effected before and after a Charter, *q.v.*, to assess such damage if any.

Grab Discharge Clause A Charterparty, *q.v.*, clause used in connection with bulk cargoes and mostly concerns ships not really suitable for bulk cargoes, resulting in extra loading/unloading

expenses. The extra time and cost involved are borne by the carriers.

GRACE Group Routing and Charging Equipment.

grad. Gradient; Grading; Graduate.

Graft An American slang word for 'fraud'.

Grain In the Merchant Shipping Act 1894 the word grain indicates all types of grain such as barley, corn, maize, nut kernels, nuts, oats, pulses, rice, rye, seeds, wheat, etc.

Grain Capacity The whole underdeck cargo volume or cubic capacity of a ship and ignoring Broken Stowage, *q.v. Similar to* **Grain Space Capacity**, *q.v.*

Grain Cargo Certificate *Similar to* **Grain Certificate**, *q.v.*

Grain Certificate A certificate issued by the customs authorities confirming that the ship has loaded grain with the necessary precautions to avoid possible shifting. Very essential in the port of loading and in the port of discharging. It is in fact compulsory to produce it to the proper authorities before departure for outward clearance and before discharging at the port of destination. *Abbrev.* GC or G.C. and GCC or G.C.C. for Grain Cargo Certificate. *See* **Grain Fittings** and **Shifting Boards.**

Grain Contract Forms Contract forms issued by the London Corn Trade Association commonly used to cover the grain trades between various ports of the world.

Grain Cubic Capacity A measurement of a vessel's cargo compartments as an indication of the space available for the carriage of any bulk commodity—not only grain.

Grain Fittings There is an international obligation for ships carrying full cargoes of grain to fit shifting boards and feeders to stop the grain from shifting. The shifting boards and feeders are to be provided by the carrier or owner. Before a grain ship is allowed to sail from the port of departure it is general practice for a general survey to be carried out by the competent authorities. If all is well she is confirmed to be fully equipped in accordance with the grain fitting directives and given a Grain Certificate, *q.v.*, or Grain Cargo Certificate to this effect. This certificate is an important document to be presented to the customs at the discharging port before discharging can take place.

Grain Laden When a ship is partly loaded with grain exceeding more than a third of her registered tonnage, she is said to be Grain Laden.

Grain Space *or* **Grain Space Capacity** Cubic or volume capacity in feet or metres of the entire cargo holds, including the hatchways and other uneven spaces. A ready reference for the owner or his servants to estimate the amount of bulk cargo a ship can take once the stowage factor is known. In normal cases the volume capacity is given in cubic feet. *See* **Bale Capacity** *or* **Grain Capacity.**

Granary Relatively large warehouse where grain in bulk is stored.

Grande Vitesse *French*—Fastest international freight train covering 300 km per day. Now obsolete in France but still referred to in some documents. *Abbrev.* GV. Train Grande Vitesse, *abbrev.* TGV, is a super express passenger train.

Grants-in-Aid Funds or grants allowed by a government to local departments.

GRAV Symbol for API Gravity.

Grapnel A small anchor with several Flukes, *q.v.*

Gratuitous Contract One-sided contract: a contract which would only benefit one party and allows no compensation to the other party. *Also called* Contract of Beneficence.

Gratuity Tip or gift as a sign of gratitude; Sum, in addition to pay, in recognition of service rendered.

Graving Act of scraping, cleansing, washing and painting the underwater parts of a ship.

Graving Dock A Dry Dock, *q.v.*, originally for graving a ship's bottom, i.e. burning or scraping off growths and tarring or painting them. Now virtually synonymous with a dry dock and used in addition to bottom cleaning for carrying out hull repairs or for the building of ships which have their keels laid on the dock bottom and are floated out on completion.

Gravity Volume weight ratio of a substance calculated against the same volume of water at normal temperature and pressure.

Gravity Tank Fresh water tank placed in a high position in a ship and in such a way that the water is transferred to low storage tanks and/or conveyed through pipes to cabins and lavatories.

Gr.Br. Grande Bretagne *French*—Great Britain.

Gr. Brit. Great Britain.

GRC *or* **G.R.C.** Glass Fibre Reinforced Concrete.

GRCM Graduate of the Royal College of Music.

GRE Graduate Record Examination *USA*.

Grease Pencil A type of pencil in various colours which writes on glass, porcelain and plastic surfaces, very useful for plotting and for marking radar screens.

Great Circle A circle on the surface of the earth whose plane passes through the earth's centre.

Great Lakes Ports Ports in the lakes of Canada and/or USA popular for grain shipments. In Canada —Port Arthur and Fort William in Lake Superior; Hamilton, Kingston, Toronto and Prescott in Lake Ontario. In USA—Chicago, Milwaukee in Lake Michigan; Duluth and Superior in Lake Superior and Toledo in Lake Erie.

Greek Shipsuppliers Greek Members of the International Shipsuppliers Association.

Greenbacks Slang term for US legal-tender banknotes.

Green Card Certificate of comprehensive insurance cover for a vehicle being used abroad.

Greengrocer Retailer in fruit and vegetables.

Greenheart Wood generally used for underwater purposes. It is supposed to be the heaviest, strongest and most decay resistant of all woods.

Green Light Green Light System (for traffic); clear signal to proceed with a job or to take action.

Green Light *or* **Red Light** *See* **Red** *or* **Green Light System**.

Green Seas Waves sweeping over a ship encountering rough weather.

Greenwich *Similar to* **Greenwich Mean Time**, *q.v.*

Greenwich Civil Time *Similar to* **Greenwich Mean Time**, *q.v.*

Greenwich Mean Time An imaginary line or Meridian, *q.v.*, taken as the universal basis and passing at 0°C through Greenwich, UK, where the mean solar time is internationally calculated for the commencement of the day starting from midnight. Other words used are Greenwich Time, Greenwich Civil Time, Standard Time, Universal Time, Zero Meridian and Prime Meridian (all *q.v.*).

Gremlo Associaçao Portuguesa dos Armadores da Marinha Merchanti Portuguese Shipowners Association, Lisbon.

Grey Area An intermediate area of uncertainty.

Greyslick *Similar to* **Slick**, *q.v.*

Gridiron Framework of baulks of timber laid flat on the foreshore or alongside a wall upon which the keel of a vessel may rest on a falling tide to expose her bottom, rudder and propeller for inspection, cleaning and repairs.

Grisola The bronze/brass cover of the magnetic compass comprising lights and navigational bearing aids.

GRN *or* **G.R.N.** Goods Received Note.

Gro. Gross, *q.v.*

GRO *or* **G.R.O.** General Register Office; Greenwich Royal Observatory.

Grocer A retailer in provisions and general foodstuffs.

Grog Rum diluted with water; strong alcoholic drink.

Gross 144 in number. In the investment field gross means before deduction of income tax.

Gross Arrived Sound Value The insurance assessment on the remaining sound valuation of the insured damage. *Abbrev.* GASV or G.A.S.V.

Gross Charter *or* **Gross Form Charterparty** *or* **Gross Terms** A Contract of Affreightment, *q.v.*, or a Charterparty, *q.v.*, whereby the owner or carrier is to pay for all port expenses including loading, discharging, tallying, etc.

Gross Combination Weight The entire weight of a vehicle including its load, occupants and fuel.

Gross Domestic Product The entire value of the goods and services produced by a country. As GDP rises, there is growth, but when it falls the economy is said to be in recession. *Abbrev.* GDP.

Gross for Weight It is usually the case that due to the insignificant weight of the outside wrapping the weight of the wrapping is not taken into account. *Ex.* The weight of bags compared with their contents, for example, cement, is negligible.

Gross Form Charterparty *See* **Gross Charter** *or* **Gross Terms**.

Gross Freight Freight money collected or to be collected without calculating the expenses relating to the running cost of the ship for the voyage undertaken.

Gross National Product This refers to the national income of a state for the total output of services rendered and goods produced in a given period of a year including overseas earnings, but excluding capital depreciation. *Abbrev.* GNP or G.N.P. or g.n.p.

Gross Profit Amount of gain arrived at without calculating the respective expenses which will lead to the net profit.

Gross Terms *See* **Gross Charter**.

Gross Ton/s Unit of 100 cubic feet or 2.831 cubic metres used in arriving at the calculation of gross tonnage.

Gross Weight The weight of the contents and the container. *Abbrev.* Gr. wt.

Ground When a vessel contacts the bottom of the sea or the ground; if, by doing so, she is immobilised, she is Aground. *See* **Stranding**.

Groundage Another word for Harbour Dues, *q.v.*, or Port Dues.

Ground Crew Technicians stationed on the ground for handling and servicing aircraft.

Ground Effect The extra attraction to the ground an aircraft experiences when approaching the ground.

Ground Swell Large waves resulting from a distant storm; Public feeling or opinion.

Ground Tackle Tackle, *q.v.*, or appliances to make fast and secure to the ground such as anchors, chain cables, windlasses and other fittings.

Grounding Clause A marine insurance clause exempting the Underwriters, *q.v.*, from liability in case of Grounding, *q.v.*, in certain waterways like canals, lakes, rivers, etc.

Group Taken commercially this word refers to a group of companies.

Groupage (1) Port Dues charged to a ship for the use of the port and quay. (2) The method of collecting a number of consignors and their cargoes, all shipped under one consignment and covered under one Bill of Lading, *q.v.*, known as Groupage or Collective Bill of Lading. All the merchandise is consigned to an agent who will in turn deliver to each individual receiver.

Groupement The seven Swiss Banks based in Geneva dealing with international accounts.

Growth Positive development in some aspect of the economy.

Grp. Group, *q.v.*

GRP *or* **G.R.P.** Glass Reinforced Plastic.

G.R.S.M. Graduate of the Royal Schools of Music (the Royal Academy and the Royal College).

Grs. t. Gross Tons or Gross Tonnage.

GRT *or* **G.R.T.** *or* **g.r.t.** Gross Registered Ton/Tonnage.

G.R.T.M. *or* **gr.t.m.** Gross Ton-Mile.

Gr.Tons *or* **GrT.** Gross Tons.

Gr. wt. Gross weight, *q.v.*

GRY *or* **g.r.y.** Gross Redemption Yields *stocks and shares.*

GS *or* **G.S.** Grand Secretary; General Staff; Grate Surface *Lloyd's Register notation*; General Secretary; Ground Speed *navigation.*

Gs. Guaranis, a unit of currency used in Paraguay.

g.s. Good Safety.

GSA *or* **G.S.A.** General Sales Agents; General Services Administration *USA*; German Shipsuppliers Association; Greek Shipsuppliers Association.

GSB *or* **G.S.B.** Graduate School of Business, Stanford University.

GSC German Shippers' Council; Group Switching Centre *electronics.*

GSCC Greek Shipping Co-operation Committee.

GSK *or* **G.S.K.** General Ship Knowledge.

GSM *or* **G.S.M.** *or* **g.s.m.** Grams per Square Metre; Good Sound Merchandise, *q.v.*; Good Sound Merchantable, *q.v.*

GSO *or* **G.S.O.** General Staff Officer.

GSP *or* **G.S.P.** Generalised System of Preferences, *q.v.*; Good Service Pension.

GSPB Good Safe Port Berth *charter party clause.*

GSPBAAAE Good Safe Port Berth Always Afloat at All Ends *charter party clause.*

GSPBENDS Good Safe Port Both Ends *charter party clause.*

GSS Global Surveillance System.

GSSL *or* **G.S.S.L.** Genoa/Savona/Spezia/Leghorn. Range of ports in Italy.

GSSLNCV *or* **G.S.S.L.N.C.V.** Genoa/Savona/Spezia/Leghorn/Naples and Civitavecchia. Range of ports in Italy.

GST *or* **G.S.T.** Goods and Service Tax *Canada*; Greenwich Siderial Time *nautical.*

G.S. & W.R. Great Southern and Western Railway In Ireland.

GSTP General System of Tariffs and Preferences *UNCTAD.*

gt. Gilt.

g.t. Gas Light; Gilt Top *book binding*; Gross Tonnage, *q.v.*; Gross Tons, *q.v.*

GT Gas Turbine.

GT *or* **G.T.** Grand Touring; Greeting Telegram; Gross Ton(s), *q.v.*; Gross Tonnage, *q.v.*; Globe Temperature.

Gt. Br. *or* **Gr. Brit.** Great Britain.

GTC *or* **G.T.C.** Good Till Cancelled; Government Training Centres.

Gtd Guaranteed.

g.t.e. Gilt Top Edge.

GT-E Gas Turbo-Electric.

Gtee Guarantee, *q.v.*

GTM *or* **G.T.M.** General Traffic Manager.

gtm *or* **g.t.m.** good this month.

GTMA Galvanised Tank Manufacturers' Association; Gauge and Tool Makers' Association *UK*.

gtr. Greater.

GTS *or* **G.T.S.** Gas Turbine Ship/Steamer; Geared Twin Screw *in pumps*; Global Telecommunication System; Government Telecommunications Service *UK*; Greenwich Time Signal.

GTV *or* **G.T.V.** Gas Turbine Vessel.

gtw *or* **g.t.w.** Good this week.

GTW Gross Train Weight.

GU Guam, an island in the Pacific Ocean (USA).

Guano The droppings of seabirds. Exported notably by Peru.

Guar. Guarantee, *q.v.*; Guaranteed.

Guarami Currency unit of Paraguay. *Abbrev.* G.

Guarantee Pledge made by a bank or a person or a firm in favour of another for a particular sum of money or for a specific responsibility; a warrant or formal assurance stating specifications or undertaking obligations to be met; to indemnify or protect. *Abbrev.* Gtee *or* Guar. *See* **Guarantor**.

Guarantee Engineer Expert marine engineer temporarily employed on a newly built ship by the builders and paid by the owners. His duties are to supervise the satisfactory performance of the engines on board, while still on guarantee by the manufacturers. *See* **Guarantee Period**.

Guarantee Period A term used when a newly built vessel is taken over by the contracting owner. This guarantee is generally observed for a period of one year as from the delivery time and covers all defects arising in respect of the ship's performance. Any accidents due to navigation are excluded. *See also* **Guarantee Engineer**.

Guarantee Space Per Ton In the Centrocon Charterparty the owners guarantee a margin of deadweight cargo of grain to be carried according to the capacity of the ship. Failing this, the charterers, *q.v.*, have the right to deduct proportionately from the freight earned.

Guaranteed Deadweight Owners' guarantee which is verified by the builders' deadweight scale and/or log books. This is often seen in the Contract of Affreightment, *q.v. See* **Guaranteed Deadweight**.

Guaranteed Deadweight and Bale Space A Charterparty, *q.v.*, term where the freight rate is calculated on the agreed rate per ton of the vessel's Deadweight, *q.v.*, against the Balespace, *q.v.* In this case the owners guarantee on figures for deadweight and bale space and if during the loading any reductions occur through any fault of the owners the charterers are allowed a pro rata reduction of the guaranteed figures.

Guaranteed Freight Freight offered to cargo owners on a guaranteed basis.

Guarantor A person who indemnifies or makes a surety for another or others. Also called Surety. *See* **Guarantee**.

Guat. Guatemala.

Guay Guayaquil, *Ecuador*.

Guernsey A Channel Island.

Gui *or* **Guin.** Guiana.

GUI Graphical User Interface *computers*.

Guil. Guilder, *q.v.*

Guilder Unit of currency used in the Netherlands, Surinam, Netherlands Antilles. Also called Florin.

Gulf Gulf of Mexico; Arabian Gulf.

Gulf Ports Comprising Gulf of Mexico, Port Arthur to Tampa both inclusive.

Gulf States American States bordering the Gulf of Mexico: Texas, Louisiana, Alabama and Florida; Nations bordering the Persian or Arabian Gulf.

Gulf Stream Warm current of water moving from the Gulf of Mexico up to the coasts of North West Europe. Because of this warm area no ice is formed during the cold winter season.

GUM Gosudarstvenni Universalni Magazin, *q.v.*

Gumi *Japanese* Corporation or Company.

Gunmetal Type of bronze usually of 88–90% copper, 8–10% tin and 2–4% zinc, having a dark grey or black finish, formerly used for cannon and now for bearings and other parts including small anchors requiring high resistance to corrosion and wear.

Gun Runner A person illegally transporting weapons and ammunition.

Gunship Helicopter used for war purposes.

Gunsmith A firearms manufacturer or repairer.

Gunwale Timber covering the uppermost strake in a boat. Plank covering the heads of the timbers in a wooden ship. Pronounced 'gunnel'. The name is possibly derived from its having served to support small guns. It has come to mean the projection above the upper deck on each side of small vessels. Water will come onboard if a craft heels 'gunwale under'.

Gust An abrupt blast of wind happening within a limited duration.

Guys Ropes or wires or chains secured to the head of a derrick or boom of a ship. They serve to guide the position or shifting of cargoes to and from the ship and quay. The word Guys is in fact derived from guide. *See* **Burtoning**.

g.v. Gravimetric Volume; Gross Valuation.

GV Grande Vitesse, *q.v.*

Gvt. Government.

GVW *or* **G.V.W.** Gross Vehicle Weight *containerisation*.

Gwh Gigawatt-hour.

GWP Globe Warming Potential.

GWP *or* **G.W.P.** Government White Paper.

Gy. C. *Lloyd's Register notation* signifying that the ship is equipped with a Gyro Compass, *q.v.*

Gypsy The rotating drum of a Winch, *q.v.*, or Windlass, *q.v.*

Gyro Compass An automatic compass having one or more Gyroscopes, *q.v.*, which point or direct to True North. *Abbrev.* Gy.C.

Gyroplane Helicopter, *q.v.*

Gyroscope A rapidly spinning wheel fixed in something, e.g. a compass, to keep it in equilibrium.

Gyro Pilot An instrument which automatically steers a ship very accurately compared with human navigation. A heading is set and the gyro pilot will direct the vessel accordingly.

Gyroplane *See* **Helicopter**.

Gyro Repeater An instrument used to take bearings, steering, obtain Azimuths, *q.v.*, etc. It is in direct contact with the Gyro Compass, *q.v.*

Gyuaji *Japanese*—Referee.

GZ *or* **G.Z.** The figure representing the righting arm between forces acting through the Centre of Gravity and the Centre of Buoyancy of a ship. A GZ Curve illustrates the stability of a ship at various angles of heel.

H

h Altitude *astronomical*; Height above sea level; Harbour; Heat; Height; Horizontal; *Lloyd's Register notation* for Hours; Hull; Hundred.

H Symbol for Hydrogen; Half Page, Harbour; Hard Pencil; Hatch; Haven; Hazardous Cargo; Hail *weather observation*; Heavy Sea; Henry *unit of inductance*; High Loren, *PRR horizontal of earth's magnetic field*; Houses of Parliament; Hull; Hundred High Pressure Area; Hydrant; Hydraulic(s); International flag coloured Red/White (vertical) which when hoisted denotes a pilot is on board ship.

HA *or* **H.A.** Hospital Assistant; Hour Angle *nautical*.

h.a. Hoc anno *Latin*—This year; Hujus Anni *Latin*—This year's.

Ha. Haiti; Haitian; Hawaii; Approximate Altitude.; Hatch.

ha Hectare; Hektar *German*—Hectare.

HAA Helicopter Association of America; Hotel Accountants' Association.

Haar A sea-mist which occurs on the east coast of England and Scotland during the summer season.

hab Habitat. *Latin*—inhabits.

Hab. corp Habeas corpus. *Latin*—May you take the body. In most cases it refers to a writ to present a person before the court.

Habeas corpus *Abbrev.* Hab. Corp., *q.v.*

habt Habeat *Latin*—Let him have.

HACC Hellenic American Chamber of Commerce in New York.

Hachures Short lines on a map representing the flow from high to low ground or to show mountains.

Hack. Hackney, taxicab.

Hackney A legal word for a cab or taxicab.

HAD *or* **H.A.D.** Havre/Antwerp/Dunkirk. Range of ports of call for loading and/or unloading.

h.a.d. Head Acceleration Device; Hereinafter Described.

HADS Hypersonic Air Data Sensor.

HAF *or* **H.A.F.** High Abrasive Furnace, *rubber compound filler*; High Altitude Fluorescence.

HAF Black *or* **H.A.F. Black** High Abrasive Furnace Black, black rubber compound filler.

HAFMED Headquarters Allied Forces Mediterranean.

HAFO Home Accounting and Finance Office *US*.

H.A.G.B. Helicopter Association of Great Britain.

Haggle To wrangle or argue over the price.

Hague Conventions Agreements signed by the powers at The Hague as to the rules of international law binding on them. Those of 1899 relate to the pacific settlement of international disputes, the laws and customs of war on land, and the adaptation of the principles of the Geneva Convention to maritime war; that of 1904 relates to hospital ships; and those of 1907 relate to the pacific settlements of international disputes, the employment of force for the recovery of contract debts, the opening of hostilities, the laws and customs of war on land, the rights and duties of neutral powers in war on land, the status of enemy merchant ships on the outbreak of hostilities, the conversion of merchant ships into warships, the laying of submarine contact mines, the bombardment by naval forces in time of war, the adaptation of the principles of the Geneva Convention to maritime war, the restriction of the exercise of the right of capture in maritime war, the establishment of the International Prize Court, the rights and duties of neutral powers in maritime war and a declaration prohibiting the discharge of projectiles and explosives from balloons.

H.Ahd Half Ahead *navigation*.

Hajozasi *Hungarian*—Shipping and Navigation.

Half Commission Man Stock Exchange term which referred to a person who was neither a Broker, *q.v.*, nor a Stockjobber but who introduced business via the broker, thereby sharing the commission.

Half-High *or* **Half-Height** A container having an open top and half the height of the standard one.

Half-Mast When a flag is hoisted at half-mast it signifies a ship in distress or in mourning.

Half Tide A tide which is between the lowest and the highest.

Halliard *or* **Halyard** *See* **Halyard**.

Hall-Mark *or* **Hallmark** Mark inscribed on gold or silver objects to indicate that they have been tested to the requirements of the authorities. *Abbrev.* hm. or h.m.

Halon A member of the ozone-destroying halo-carbon family of chemicals. Once used extensively in fire and explosion suppressant systems, halons are being phased out under United Nations agreements (Montreal Protocol) and banned from installation in ships since 1991.

Halsey System Generally a bonus given to workers equal to half the hours' normal rate for the time saved on some work.

HALTATA or **H.A.L.T.A.T.A.** High and Low Temperature Accuracy Testing Apparatus.

Halyard or **Halliard** A rope or a line to hoist flags, sails or pennants.

HAM Hardware Association Memory *computer.*

Ham Hamburg, *Germany.*

Hammered If a member of a Stock Exchange, *q.v.,* is unable to pay his debts, the hammer is used to announce the defaulter.

Hampered Ship A ship which due to its inability to manoeuvre is restricted in effecting any kind of avoiding action.

HAMS or **H.A.M.S** Hour Angle Mean Sun *astronomy.*

Han Hanover, *Germany.*

h and c Hot and Cold *in plumbing.*

Handels *Dutch*—Commercial.

Handelsaktiebolag *Swedish*—Trading Company Limited.

Handelsbolag *Swedish*—Trading Company.

Handelsgesellschaft *German*—Trading Company.

Handelsselskab *Danish*—Trading Company.

Hand Flag Small flag used for hand signalling the semaphore alphabet.

Hand Flares *See* **Marine Pyrotechnics.**

Handling Charges Charges due to the shipping agents or forwarding agents for loading and/or unloading cargoes. In many cases calculated per cubic ton or cubic metre and in other instances per ton weight of 2240 lbs. or 2000 lbs. or 1000 kilos.

Hands Crew members of a ship.

Handspike A bar lever.

Handy Billy A seaman's expression for (1) Lightweight portable fire-fighting pump which is easily handled by a single man in the engine room. (2) For the deck hand it refers to a Small Purchase or Tackle, *q.v.,* for setting up Halyards, *q.v.,* or slackening Mooring Lines, *q.v.,* or mooring ropes when mechanical means are not available.

Handymax Dry bulk carrier of 35/50,000 tons d.w. which is popular for full efficiency, flexibility and low Draft (Draught), *q.v.,* drawing less than 12 metres or 39 feet.

Hangar A large shed in or near to an airport or airfield where aircraft are serviced and stored.

Hang Seng The Hong Kong Stockmarket Index.

Hank (1) Metal ring or spring clip sewn or shackled to the Luff, *q.v.,* of a sail which secures it to a stay while free to slide up and down the same. (2) Coil of cotton yarn measuring 840 yards or worsted measuring 560 yards.

Hansard Official report of proceedings in British Parliament.

Hanse Medieval league; guild of merchants; political and commercial league of Germanic towns.

Hants Hampshire, *UK.*

H/A or **D** Havre and Antwerp or Dunkirk, Loading/unloading ports of call.

HAP High Attitude Platform.

h. app. Heir Apparent.

Har. Harbour, *q.v.*

HAR Householder's All Risks *insurance.*

Harbour Sometimes called Haven, *q.v.,* but not so common commercially. It is an enclosed safe water area deep enough for ships to enter and where they can anchor, berth and shelter for various purposes such as loading, discharging, repairing and other requirements. As the name implies it must be protected in such a way as to give complete security in case of storms, apart from the fact that it must be fully equipped and have the necessary customs sheds and port organisation. *Abbrev.* Har or Hbr.

Harbour Dues or **Port Dues** Customs levy due in respect of ships entering and/or using the harbour, docks, basin, locks, quays, etc. Charges are paid according to the number of days the ship remains in port and in relation to her size as well as the Net Registered Tons/Tonnage, *q.v.*

Harbour Master A person usually having the experience of a certificated master mariner and having a good knowledge of the characteristics of the port and its whole area. He administers the entire shipping movements that take place in and within reach of the port he is responsible for.

Harbour Stores Provisions and stores supplied to the ship during her stay in port. Most of the harbour stores provided are for the daily need of the crew for fresh food, and other urgent items.

Hard A sloping platform on the quayside for the door ramp of the Roll On/Roll Off, *q.v.,* ship to rest upon when lowered. Place at the water's edge hardened to permit the hauling out or beaching, and launching, of small vessels and boats for bottom cleaning and repairs.

Hard Aground A vessel which has gone aground and is incapable of refloating under her own power.

Hard and Fast Said of a ship very firmly ashore or aground.

Hard Cash The actual cash money available as opposed to other credits like cheques.

Hard Currency That currency which enjoys international exchange value and tends to maintain its value.

Hard Traders Difficult people to deal with in business.

Hardware Metalware and other articles like brass, copper, iron, nails, screws, padlocks, etc.

HARM High Radiation Anti-Radar Missile *military.*

Harmattan Dry desert wind in W. Africa.

Harness Cask Cask meant to hold meat.

HARP High Altitude Relay Point.

Hartal *Indian*—A stoppage of work; strike.

Harter Act (1893) This US statute refers to merchandise or property transported from or between ports of the USA and foreign ports. Now partially superseded by the Carriage of Goods by Sea Act 1936.

Harv. Harvard University *USA.*

Harvest Moon Full Moon nearest the autumnal equinox.

HAS Helicopter Air Service; High Altitude Sample.

H.A.S. Hellenic Astronautical Society in Greece; Hydrographic Automated System.

Hash Hashish.

HASP Hardware Assisted Software Polling *computer*; High Altitude Sampling Programme; High Altitude Space Platform.

H.Astn Half Astern *navigation.*

HASWA Health and Safety at Work Act, *UK.*

HAT *or* **H.A.T.** Highest Astronomical Tide.

Hat. Hatch, *q.v.*

Hatch Beams *or* **Hatch Webs** The transverse steel structure on the inner side of the Hatch Coamings, *q.v.*, and so fitted as to support the wooden hatch covers and cover the whole hatchway before it is overlaid by tarpaulins to prevent water getting into the holds. New ships are constructed with steel pontoons or steel covers and the wooden hatch covers are done away with.

Hatch/Hold Refers to the minimum fixed quantities in tons of cargo a ship can load or unload per day per hatch. *Abbrev.* Ha/Ho *or* HA/HO.

Hatch Webs *Similar to* **Hatch Beams,** *q.v.*

HATS *or* **H.A.T.S.** Hour Angle of the True Sea *nautical.*

Haul To shift from one place to another; to transport; to pull on a rope or wire. To haul off is to sail closer to the wind to avoid danger. To haul to the wind, or haul her wind, is to bring a vessel closer to the wind.

Haulage (1) Transport of goods by road. (2) Charge for transport by road or rail.

Haulage Contractor One who engages to carry goods in land vehicles.

Haul Out (1) To bring a craft out of the water on a Slipway, *q.v.*, or Patent Ship or by crane. Pleasure yachts are frequently hauled out for the winter season. (2) To stretch the foot or a sail by its clew(s).

Haute mer *French*—High Seas, *q.v.*

Hav. Haversine, *q.v.*

Haven Another word for Port or Harbour, *q.v.* This generally refers to an inlet of the sea or the mouth of a river having adequate protection for ships.

Haversine Half of versine, i.e. (1-cos)÷ 2 *navigation. Abbrev.* Hav.

HAWB *or* **H.A.W.B.** House Airwaybill, *q.v.*

Hawker Seller disposing of his goods in the street.

Hawse (1) The space of water between the bow of a ship at anchor and the position of her anchor. (2) The deck space in the bows of a ship where the Hawse Pipes, *q.v.*, are situated.

Hawse Pipe The tube which connects a hole in the bow of a ship, the Hawse hose, to the foredeck and through which the anchor cable passes.

Hawser Large strong rope used for towing purposes and for securing or Mooring ships. Hawsers are now mostly made of steel. *See* **Towing Hawser**.

Hawser Laid Ropes that are usually composed of three to four Strands laid right handed. The manufacturers' expression is 'Shroud Laid'. *See* **Cable Laid**.

Haz. Hazardous, *q.v.*; HAZ—Heat Affected Zone.

HAZCHEM Hazardous Chemical.

HAZMAT Hazardous Material.

HAZOP Hazard and Operability. In hazardous cargo studies.

HB *or* **H.B.** *or* **H/B** Half Bound *book-binding*; Hard Back, referring to book cover; Hard Pencil; Holy Bible; House Boat.

Hba. Habana *Spanish*—Havana, the capital city of Cuba.

HBC *or* **H.B.C.** High Breaking Capacity; Hudson's Bay Company.

Hbf Hauptbahnhof *German*—Central or Main Station.

H.B.M. Her/His Britannic Majesty or Her/His Britannic Majesty's Service.

HBO Home Box Office.

Hbr. Harbour, *q.v.*

Hbt. Hobart, *Tasmania.*

Hby Hereby.

HC Calculated Altitude *aviation*; Heating Coil *q.v.*; Hydraulic Coupling

Hc *or* **H.c.** *or* **H.C.** Habitual Criminal; Hague Convention; Held Covered, *q.v.*; High Court; Higher Certificate; Highly Commended; Highway Code, *q.v.*; House of Commons; Housing Centre; Hatch Coamings; Home Consumption *customs forms*; House of Clergy; Home Counties.

h.c. Hand Control; Held Covered; High Capacity; High Carbon; Honoris Causa *Latin*—For Honour's Sake; Hot and Cold.

H/C Held Covered, *q.v.*

H.C.A. High Court of Admiralty.

H. Cap *or* **Hcap** *or* **Hcp** Handicap or Handicapped.

H.C.B. House of Commons Bill.

HCC *or* **hcc.** Hull Construction Certificate *Lloyd's classification*; Hydraulic Cement Concrete.

HCD *or* **hcd** High Current Density.

HCE *or* **hce** Human-Caused Error.

HCF *or* **H.C.F.** High Cost Factor; Highest Common Factor; 100 Cubic Feet.

HCFC Halon fluorocarbon *refrigerant*.

HCG *or* **H.C.G.** Horizontal Location of the Centre of Gravity.

HCH *or* **H.C.H.** Hexachlorocylohexane *insecticide*.

HCI *or* **H.C.I.** Hotel Catering Institute.

HCIL Hague Conference on International Law.

H.C.J. High Court of Justice.

HCL Hydrochloric Acid.

HCL *or* **H.C.L.** *or* **h.c.l.** Horizontal Centre Line.

H/C. L.U *or* **h/c l/u** Held Covered Leading Underwriters, *q.v.*

HCM *or* **H.C.M.** His/Her Catholic Majesty.

H.C.M.M. Honorable Company of Master Mariners, London, UK.

HCMS Hull Condition Monitoring System.

H Com. High Commissioner.

hcptr Helicopter.

HCS *or* **H.C.S.** *or* **h.c.s.** High Carbon Steel.

HCSA *or* **H.C.S.A.** Hospital Consultants and Specialists' Association.

HCWM High Capacity Washing Machine, *q.v.*

h.d. Heavy Duty; Hochdruck *German*—High Pressure; Hook Damage *in cargo handling*.

Hd *or* **hd** Head *nautical*; Hogshead, *q.v.*

HD *or* **H.D.** *or* **H/D.** Hatch Door; Havre-Dunkirk; High Density; Honourable Discharge; Honorary Degree; Hour Difference *nautical*; Hydrographic Department.

HDA *or* **H.D.A.** High Duty Alloy; Horizontal Danger Angle *aviation*.

HDATZ High Density Air Traffic Zone.

hdbk. Handbook.

Hd.cr. Hard Chromium.

H.D.D. Higher Dental Diploma.

Hdg. Heading *nautical*.

HDI Household Disposable Income, *q.v.*

H.Dip.E. Higher Diploma in Education.

Hdk Hurricane Deck.

Hdl. Handle.

HDL High Density Lipoprotein.

Hdlg Handling.

HDM *or* **H.D.M.** *or* **h.d.m.** High Duty Metal.

HDMR High Density Moderated Reactor.

HDR High Dry Rock.

Hdqrs *or* **H.Q.** *or* **H.Qrs.** Head Quarters/Head-quarters.

HDST *or* **H.D.S.T.** *or* **h.d.s.t.** High Density Shock Tube.

HDTV High Definition Television.

HDV *or* **H.D.V.** Heavy Duty Vehicle.

HDWC *or* **H.D.W.C.** Heavy Deadweight Cargo. *See* **Heavy Cargo**.

hdwd Hardwood.

Hdwnd Headwind.

He Helium.

HE *or* **H.E.** Heading Error *nautical*; Heeling Error; Height of Eye *nautical*; High Explosive; His Eminence; His Excellency; Home Establishment; Hydraulic Engineer.

Head The bow or fore part of a ship.

Headfast Rope positioned at the ship's head or bow for the sake of mooring or making her fast to the quay or wharf, etc.

Headgate The gate that allows water to go into the Lock, *q.v.*

Heading (1) The direction in which a vessel's bow is pointing. (2) Steaming or moving in a certain direction. (3) Passage in a mine or underground working for drainage or ventilation. (4) The level direction in which the horizontal axis of an aircraft in flight is pointing.

Head of Navigation The farthest part of the river where merchant ships are able to navigate freely.

Head-On Head to head collision.

Headrope (1) The bolt rope sewn onto the head of a sail. (2) A mooring rope or wire leading from the bow of a ship to the shore.

Heads Seaman's word for lavatory from its former position in the head of the ship outside the bows proper. Americans say Head.

Headsails The sails that are set forward of the Foremast, *q.v.*

Head Sea That sea that rolls or hits directly against the movement of the ship.

Head-Up-Display Flight information displayed on the windshield of an aircraft. *Abbrev.* HuD.

Headsman A man in command of a whaling boat.

Headswell A sea swell heading against the direction of the ship's forward movement.

Headwater The upper part of a stream which forms the river.

Headway A ship's forward movement.

Headwind A wind which is blowing against the direction of the ship's forward movement.

Heart When referring to rope this is the core used in the centre of three Strands, *q.v.*

Heat High Explosive Anti-Tank.

Heat Barrier Air flowing past an aircraft which is travelling faster than the speed of sound can result in metal reaching melting point. Therefore such metals must be made of special materials.

Heave Upward movement of the sea; the act of lifting or raising a weight; to haul cable or rope by hand or Capstan, *q.v.*, or Windlass, *q.v.*

Heave the Lead To take soundings with lead and line.

Heave-to To bring a vessel's head near to the direction of the wind and/or sea and keep her stationary by trimming sails or the use of engines. This manoeuvre may be to assist boarding or lowering a boat, or to maintain a waiting position.

Heave Up Anchor To weigh anchor.

Heaving Line A thin rope with a heavy knot, 'monkey's fist', or small sandbag at one end for throwing from ship to shore or vice versa. The other end is hitched to the heavier rope, wire or hawser which is hauled across for Mooring, q.v., purposes.

Heaviside Layer That part high up in the atmosphere which reflects wireless waves back to earth.

Heavy Address This refers to the heavy 'Address Commission' in a Charterparty, q.v., which virtually amounts to reduction of the charter rate payable by the owners to agents for the quick despatch of a ship. When the rate is say 5% or over, generally due to more than one broker/agent being involved in the contract, the commission would have to be distributed between them. See **Address Commission.**

Heavy Cargo Cargo which is heavy in relation to its cubic volume. Alternatively it is that cargo which weighs over a ton or 20 cwts against its cubic measurement of 40 cubic feet or over a ton of 1000 kilos against the measurement of 35 cubic metres.

Heavy Goods Vehicle Special licence for Heavy Goods-carrying vehicles. *Abbrev.* HGV *or* H.G.V.

Heavy Grain Corn, Maize, Rye, Wheat. *Abbrev.* HG *or* H.G. *See also* **Heavy Grain Soyas, Sorghum**.

Heavy Grain Soyas, Sorghum According to the Merchant Shipping Act of 1894 this term covers all kinds of grain. Wheat, rye and maize are considered to be 'heavy grains' while barley and oats are said to be 'light grains'. Other commodities termed as grains are rice, paddy, pulses, seeds, nuts and kernels. *Abbrev* H.S.S. or h.s.s.

Heavy Industry Enterprises involved in manufacturing heavy materials, such as steel, or machines, such as engines and plant vehicles. It includes shipbuilding and drydocking.

Heavy Lift The bulk lift or package which weighs over 2000 lbs. is considered to be a heavy lift in shipping. It is generally the case that Carriers allow a higher weight limit when considering a heavy lift. Normally extra charges are added to the freight rates when heavy lifts are carried.

Heavy Lift Derrick *or* **Heavy Lift Ship/Vessel** Derrick or ship having special equipment to carry heavy weights. May vary from five tons up to 600-ton lifts.

Heavy Sea Extremely high wave action which makes the movements of a ship difficult.

Heavy Water Water with a density about 10% greater than that of ordinary water; the oxide of deuterium or heavy hydrogen.

HEC Health Education Council; Hydroxyethyl Cellulose *polymer.*

He. Cls. B. Heating Coils in Bunkers *Lloyd's Register notation.*

He. Cls. C. Heating Coils in Cargo Tanks *Lloyd's Register notation.*

HECP Harbour Entrance Control Port *USA.*

Hectare Metric system measurement of area equal to 2.471 acres.

Hecto. One hundred times.

Hectog. Hectogram.

Hectol. Hectolitre.

HECTOR *or* **H.E.C.T.O.R.** Heated Experimental Carbon Thermal Oscillator Reactor.

Heel (1) For a ship to lean to one side. To lean over. (2) Lowest end of a mast. (3) Where the Sternpost meets the keel.

Heel Block Block at the lower part of a Derrick, q.v.

HEF *or* **H.E.F.** High Energy Fuel.

HEG High Efficiency Gun *oil drilling.*

H.E.H. His/Her Exalted Highness.

HEI *or* **H.E.I.** Heat Exchange Institute.

Height Above the Hull Height from the uppermost continuous deck of a ship up to the highest mast. This is useful information especially if the ship has to pass under bridges.

Heir A beneficiary under a will.

Heirloom Chattel that follows devolution of real estate; piece of personal property that has been in family for generations.

Hel *or* **Heli** *or* **Heptr** Helicopter, q.v.

Helco E. Cyprus Ore Charterparty form.

Held Covered An agreement by underwriters to extend the terms of the insurance in specified circumstances subject to an additional premium. *Abbrev.* H/C *or* H.C.

Held Covered Leading Underwriters When an insurance risk is covered at a Premium, q.v., pending the acknowledgement or confirmation from leading Underwriters, q.v. *Abbrev.* H/C. L/U *or* h/c. l/u *See also* **Held Covered**.

Helicopter An aircraft used by civilians and the armed forces alike for the transportation of personnel, stores, medical services from ship to shore and vice versa, as well as for use in restricted areas where normal aircraft are unable to operate and manoeuvre. It is also used for reconnaissance and as an air ambulance. It has a large propeller rotating overhead and another small one at the rear for manoeuvrability. In colloquial terms it is called Chopper, Eggbeater or Whirlybird. *Abbrev.* Hecptr *or* Hel *or* Heli *or* Helo.

Heliograph Apparatus for signalling using a mirror to flash the reflection of the sun's rays; Apparatus for photographing the sun; Instrument for measuring the intensity of sunlight.

Helioscope A telescope capable of looking at the sun without dazzling the eyes.

Heliport A landing platform or area for a Helicopter, q.v.

Helm A tiller; A wheel generally installed on the bridge or wheelhouse of a ship to turn the rudder during manoeuvring and navigation. It is in fact the steering wheel of the ship.

HELMEPA Hellenic Environment Protection Association.

Helmsman The seaman at the Helm, *q.v.*, or the wheel of a ship. *See* **Quartermaster.**

Helo Heliport.

HELP *or* **H.E.L.P.** Haulage Emergency Link Protection; Helicopter Electronic Landing Path.

Hemis Hemisphere, *q.v.*

Hemisphere The earth is divided into Northern and Southern hemispheres and these may be separately projected on a map. *Abbrev.* Hemis.

Henry Unit of inductance. *Abbrev.* H.

HEO *or* **H.E.O.** Higher Executive Officer.

HEOS High Eccentricity Orbit Satellite.

HEP Hydro-electric Power.

HEPA High Efficiency Particulate Air *recirculation.*

HEPCAT Helicopter Pilot Control and Training.

HERA High Explosive Rocket Assisted.

Heref. Herefordshire, *UK.*

HERF High Energy Rate Forming.

HERMES Heavy Element and Radioactive Material Electromagnetic Separator.

Herts Hertfordshire, *UK.*

Hertz Unit of Frequency, i.e. 1 cycle/second. *Abbrev.* Hz.

HERU Higher Education Research Unit.

HESH High Explosive Squash Heads.

HET Heavy Equipment Transporter.

Heterogeneous Cargo General or various cargoes in contrast to Homogeneous Cargo, *q.v. Ex.* Cargo consisting of cases, cartons, bales, crates, loose, etc. A slang term for this is Chandlers Ship.

HETI Helsinki Stock Exchange Trading and Information *System.*

HEW Health Education and Welfare *USA.*

Hex. Hexagon; Hexagonal.

H–Ex High Expansion Foam.

hf Half; High Frequency.

HF *or* **H.F.** Hard Lead Pencil; High Frequency, *q.v.*; Home Fleet.

h.f. Height Finding; High Frequency; Hold Fire; Hook Fast.

H/F Hlutafjelagid *Icelandic*—Limited Company.

Hf. Bh Half Breadth.

HFC House Flag Clause, *q.v.*; Hydrofluorocarbon.

HFC *or* **hfc** High Frequency Current.

HFDF *or* **H.F.D.F.** High Frequency Detecting and Finding; High Frequency Direction Finder.

HFG High Frequency Gas.

HFO *or* **H.F.O.** Heavy Fuel Oil.

hfo High Frequency Oscillator.

H.F.R.A. Honorary Fellow of the Royal Academy.

HFS *or* **H.F.S.** High Frequency Sensibility. *See* **Decibel.**

Hfx Halifax, *Canada* or *UK.*

Hg Hydrargyrum *Latin*—Mercury.

HG *or* **H.G.** Hand Generator; Haute-Garonne *France*; Heavy Grain; High German; High Grade; His/Her Grace; Horse Guards.

hg Hectogram; Heliogram.

HGB Handelsgesetzbuch *German*—Commercial Law Code.

HGCA *or* **H.G.C.A.** Home Grown Cereals Authority. *UK.*

Hgd *or* **hhd** Hogshead, *q.v.*

HG&PE Household Goods and Personal Effects *insurance.*

hgpt *or* **h.g.p.t.** Hard Gloss Paint, generally expressed as Enamel Paint.

HGSS *or* **H.G.S.S.** *or* **hgss** *or* **h.g.s.s.** Heavy Grains, Soyas, Sorghum.

Hgt Height.

HGV *or* **H.G.V.** Heavy Goods Vehicle, *q.v.*

Hgy Highway.

HH Heavy Hydrogen; Havre/Hamburg, ports of call range for loading/discharging.

H.H. Her/His Highness; His Holiness; His Honour.

H/H Havre/Hamburg, grain trade ports of call range for loading/discharging.

HHA *or* **H.H.A.** Half-Hardy Annual *in horticulture.*

HHB *or* **H.H.B.** Half-Hardy Biennial *in horticulture.*

Hhd *or* **hgd** Hogshead, *q.v.*

HHDWS *or* **H.H.D.W.S.** Heavy Handy Deadweight Scrap.

HHFA Housing and Home Finance Agency *USA.*

H.H.H. Very hard lead pencil.

H-Hour Operation Starting Time *or* Zero Hour *or* O-Hour.

HHP *or* **H.H.P.** Hard-Hardy Perennial *in horticulture.*

H.H.P. *or* **h.h.p.** High Holding Power, *anchors.*

HHT Hand-held Terminal(s).

HHW *or* **Hhw** *or* **H.H.W.** Higher High Water, *tide tables.*

HI *or* **H.I.** Hard Iron *magnesium*; Hawaiian Islands; High Intensity, Horizontal Interval.

H.I. *or* **H.J.** Hic Jacet *Latin*—Here lies.

Hi. High; Hindu; Humidity Index.

Hi. ac. High Accuracy.

Hibernia *Latin*—name for Ireland.

HIBEX High Acceleration Booster Experiment.

HIC Hydrogen Induced Cracking, *q.v.*

HICAPCOM High Capacity Communications System.

HICAT High Altitude Clear Air Turbulence *aviation.*

Hidden Tax Tax or levy indirectly influencing the cost of goods and services, or profitability in general.

Hi-Ex High Explosion Foam.

hi-fi High Fidelity.

hifor High Level Forecast.

High High Pressure Area.

High Build Paints Paints of more than 75 microns or three milligrams thickness when dried.

High Capacity Washing Machine An oil tank cleaning machine which has a cleaning rate of approximately 60 cubic metres per hour. *Abbrev.* HCWM.

High Clouds Clouds above 20,000 feet.

High Cube A container, at 9ft. 6 inches, higher than the standard, used for lightweight cargo which is voluminous or is very high.

High Density Cargo Aviation term for cargo which is heavy in relation to its volume.

High & Dry Said of a stranded vessel which becomes dry when the tide ebbs.

High Frequency Radio frequency from three up to 30 megacycles per second. *Abbrev.* HF *or* H.F. *or* hf *or* h.f.

High Noon Mid-day.

Highs & Lows Term used in the financial columns of newspapers where Stocks and Shares, *q.v.*, are shown at their highest and lowest prices.

High Seas Outside Territorial Waters, *q.v. Similar to* Open Seas, *q.v.*

High-Speed Diesel Diesel engine working at over 1000 revolutions per minute. *Abbrev.* HSD *or* H.S.D.

High-Speed Steel An alloy of vanadium tungsten and chromium making a steel with hard cutting edges. Used for drills and lathes. *Abbrev.* HSS *or* H.S.S.

High Stowage Factor Opposite to Low Stowage Factor, *q.v.*

High Sulphur Fuels Fuel oils containing in excess of 1% sulphur. *Abbrev.* HSF *or* H.S.F.

High Test Hypochlorite Powerful deodorising agent in solution or powder form widely used to eliminate odours emitted by obnoxious cargoes. *Abbrev.* HTH.

High Tide *Similar to* **High Water,** *q.v.*

High Viscosity Fuel Fuel oil having a viscosity in excess of 1500 redwood. Redwood No. 1 or thereabouts. *Abbrev.* HVF *or* H.V.F.

High Water *or* **High Water Mark** Maximum height possible for tidal waters to reach; the highest mark of the tide shown on the shore. Also known as High Tide. *Abbrev.* HW *or* H.W. *or* hw.

Highway Code Laws and regulations formulated by the authorities concerned in relation to the behaviour of traffic or vehicle drivers as well as pedestrians. *Abbrev.* Hc *or* H.C.

High Wind Boisterous wind exceeding the Beauford Force Scale 7.

Himalaya Clause *or* **Himalaya Bill of Lading Clause** A clause inserted in the Bill of Lading, *q.v.*, as a consequence of a court case in 1962 involving the vessel *Himalaya*. The basic features of a typical modern Himalaya clause extend the benefit of every right, limitation and exemption under the bill of lading to servants, agents and independent contractors of the carrier, express that the carrier acts, for the purpose of the above, as the agent or trustee of such servants, agents or independent contractors and expressly makes such servants, agents or independent contractors parties to the contract of carriage. By way of variation other Himalaya clauses extend protection by extending the definition of 'Carrier' to include servants, agents and independent contractors. English and Commonwealth cases on the clause have dealt with stevedores. There is no reason, however, why other parties satisfying the definition of servant, agent or independent contractor should not take advantage of the clause insofar as case law permits. Courts in the US have, as a matter of construction of the clause, considered claims by dry dock owners and by inland carriers. Servants, agents or independent contractors of stevedores appear not to be covered, however, as there is no privity of contract between them and the carrier, and the same fate, for the same reason, awaits servants, agents and independent contractors of the consignee.

HIMICS Hitachi Management Information Control System.

HINSIB Honduras International Naval Surveyors and Inspection Bureau.

Hinterland The industrial area behind a port; land situated behind coast or river's banks.

HIP Hot Isostatic Pressing *material technology*.

HIPAR High Power Acquisition Radar.

HIPEG High Performance External Gun.

HIPERNAS High Performance Navigation System.

HIPOE High Pressure Oceanographic Equipment.

Hipot High Potential; High Potentiality.

HIPS *or* **H.I.P.S.** High Impact Polystyrene.

Hire Alternative to Charter, *q.v.*

Hire and Payment Clause Charterparty, *q.v.*, clause also called Freight Clause stipulating the terms and conditions of the charter rates.

Hirel High Reliability.

Hire Money Time Charter Money.

Hire-Purchase A common method of buying goods on a deferred payment basis. A Down Payment, *q.v.*, or Deposit, *q.v.*, is paid to the seller and the remainder of the money owed is paid periodically, generally monthly. This system has become very popular. On the one hand, the seller is encouraging more sales and at the same time acquiring extra money on the interest charged for the money outstanding. On the other hand, the buyer finds it so convenient that he may buy other items on the same basis. *Also termed* **Deferred Payment,** *q.v.*

hi.T High Torque.

hi. temp High Temperature.

HIV Human Immunodeficiency Virus. The retrovirus responsible for AIDS, *q.v.*

hiv Hiver *French*—Winter.

HIVOS High Vacuum Orbital Simulator.

H.J.S. Hic Jacet Sepultus *Latin*—Here lies buried.

hk Hook.

H.K. Handelskammer *German*—Chamber of Commerce; Hong Kong *also abbrev.* Hg. Kg.

H.K.J. Hashemite Kingdom of Jordan.

HKMA Hong Kong Management Association.

HKSA Hong Kong Shipowners' Association.

HKTA Hong Kong Tourist Association.

HKTAG Hong Kong Trade Advisory Group.

hl Hectolitre; Hektolitre *German*—Hectolitre.

HL *or* **H.L.** *or* **h.l.** Hawser Laid, *q.v.* Horizontal Line, See **Accidental Vanishing Point**; House of Lords.

h.l. Hoc loco *Latin*—In this place.

HLBB Home Loan Bank Board.

HLDG *or* **Hldg** Holding.

HLHS Heavy Lift Helicopter System.

HLLV Heavy-Lift Launch Vehicle.

HLMC Heavy Lift Mast Crane.

HLO Helicopter Landing Officer/s.

Hlpr Helper.

H'Ls Holes.

HLS Heavy Logistics Support.

Hlutafjelagid *Icelandic*—Limited Company.

HLW Higher Low Water.

HLWN Higher Low Water Neap Tides.

hm *or* **h.m.** Hand made; Hall Mark, *q.v.*; Hoc mense *Latin*—In this month; Huius mensis *Latin*—This month's; Hectometre(s).

Hm *or* **H.M.** Harbour Master; Head Master or Head Mistress; Home Mission; Hull and Machinery.

H.M. Her/His Majesty.

H & M Hull Machinery(ies) *marine insurance*; Hull and Materials *sailing ship insurance.*

h. Mach etc. Hull Machinery etc. *marine insurance.*

HMC Hellenic Marine Consortium.

H.M.C. Head Masters' Conference; Her/His Majesty's Customs.

HMC & E His/Her Majesty's Customs and Excise.

HM Coastguard *See* **Coastguard Agency, The**.

HMEC Heavy Machinery Export Council.

H.M.F. Her/His Majesty's Forces.

HMG His/Her Majesty's Government; Heavy Machine Gun.

HMMIP Hague-Merchant Marine Industries Post *USA.*

HMO *or* **H.M.O.** Harbour Master's Office.

hmp *or* **h.m.p.** Hand-made paper.

HMPE High Module Polyethylene, synthetic rope fibre.

H.M.S. Her/His Majesty's Ship or Service.

h.m.s. Hours, minutes and seconds.

H.M.S.O. Her/His Majesty's Stationery Office.

HMTA Hazardous Materials Transportation Act of 1974 *USA.*

HMTCA Hellenic Marine Technical Consultants Association.

H.N.C. Higher National Certificate.

H.N.D. Higher National Diploma.

Hnrs Honours.

HNS Convention on the Carriage of Noxious and Hazardous Substances by Sea.

h.o. Hand Over, Hold Over.

Ho Holmium, metallic element occurring in Apatite, *q.v.*

HO. Holds *Lloyd's Register notation.*

H.O. Handelsorganisation *German*—Trade Organisation; Head Office; Home Office; Hydrographic Office; Observed height; Hostilities Only *insurance.*

H & O *or* **h & o** Hook & Oil Damage *cargo handling.*

ho *or* **hse** House.

Hoar Frost Very small ice crystals similar to frozen dew.

Hoarding The act of holding or hiding money or goods; overstock oneself with food, etc., in time of scarcity.

HOB *or* **hob** Height of Burst.

H.O.C. *or* **h.o.c.** Heavy Organic Chemical; Held on Charge.

Hoc anno *Latin*—In this year. *Abbrev.* h.a.

Hockle Kinking or twisting of a strand in a rope, rendering it unfit for service.

Hoc Loco *Latin*—In this place, *Abbrev.* h.l.

Hoc tempore *Latin*—At this time. *Abbrev.* h.t.

H.O.D. Head of Department.

Hodie mini, cras tibi *Latin*—It's my turn today, yours tomorrow.

Hodometer Instrument for measuring distance travelled by wheeled vehicle,

hof *Lloyd's Register notation* for vessels burning High Viscosity Fuel.

H of C *or* **H.O.C.** House of Commons *UK.*

H of K House of Keys.

H of L House of Lords *UK.*

H of R House of Representatives *USA.*

Hog, Hogging, Hogged These terms refer to the longitudinal bending of a ship when her centre section is raised while the bow and stern are lowered. This arises: (1) at sea where the length and angle of incidence of waves is such that a crest will support the centre while bow and stern are over the troughs, and (2) when cargo is badly distributed, with the heaviest weights at the extremes. The opposite is Sagging, *q.v.*

Hogshead A measure containing half a pipe, a fourth part of a tun, or 63 gallons. This measure is not known to the law. Often it means one containing fifty-four gallons: but the number of gallons in a hogshead varies when it is used as a measure for different kinds of wine or for brandy. A pipe contains two hogsheads. *Abbrev.* Hd. *or* hd. *or* hhd.

Ho/Ha Hold/Hatch. *See* **Hatch/Hold.**

Hoist (1) To lift something by an elevator, crane, tackle, hydraulic machinery or any other means; Any devise so used. (2) The Halyard, *q.v.*, side of a flag.

Hoisting Rope Special flexible wire rope for lifting purposes, generally being of six strands with 19 wires in each Strand and in most cases having a hemp rope at the centre.

Hol *or* **Holl** Holland.

HOLC *or* **H.O.L.C.** Home Owner's Loan Corporation *USA.*

Hold The underdeck space being below the lower deck in which cargo is placed. *See* **Hatch**. *Abbrev.* Ho.

Hold Covered *Similar to* Held Covered, *q.v.*

Holder in Due Course A holder who has taken a bill of exchange, cheque or note, complete and regular on the face of it, under the following conditions, namely: (a) that he became the holder of it before it was overdue, and without notice that it has been previously dishonoured, if such was the fact; and (b) that he took the bill, cheque or note in good faith and for value, and that at the time it was negotiated to him he had no notice of any defect in the title of the person who negotiated it.

Hold Harmless An American insurance expression. The insured agrees to refund any additional and rightful claims put forward by a third party if his own claim has been originally settled.

Holding Company A company which holds more than 50% of the shares or voting power or which may appoint the majority of the directors of another company, called the subsidiary company. **See Subsidiary Company**.

Holding Ground According to the nature of the sea bottom in a particular spot and its capacity to keep an anchor fast and not dragging, it is described as good or bad holding ground.

Holiday An area left during painting.

Holidays with Pay Apart from national or official holidays (Bank holidays, etc.) employees are normally entitled to a certain number of days off during a year while remaining on full pay. The amount of time allowed for holidays usually depends on length of service with the company concerned.

Holograph A handwritten will by the testator; a document written by one's own handwriting.

Holystone A light stone for scouring floors and decks.

Home Trade Trade or business done within short distances or limits of the country. The trade limits are declared by the countries concerned in such trades.

Home Use Entry Customs House form to clear dutiable merchandise from the Warehouse, *q.v.*, for home consumption.

Homeward B/L *See* **Homeward Bill of Lading.**

Homeward Bill of Lading Referring to a Bill of Lading., *q.v.*, where the goods are coming from another country. *Abbrev.* Homeward B/L.

Homicide par imprudence *French*—Manslaughter.

Homme d'affaires *French*—A businessman; Agent of the law.

Homme de bien *French*—A respectable or charitable man.

Homogeneous Cargo Cargo of the same like or nature. It consists of identical packages and weights in contrast to Heterogeneous Cargo, *q.v. Ex.* Full load of bagged potatoes or onions or bagged cereals or grain in bulk.

Hon. *or* **Hond.** Honduras.

Hon. Honorary; Honour.

Honi soit qui mal y pense *French*—Shamed be he who thinks evil of it. Motto of the Order of the Garter.

Hono. Honolulu.

Honorarium *Latin*—Fee paid as a remuneration for service voluntarily rendered. It is given freely and not demanded.

Honorary An unpaid position and which is held for the sake of honour, such as Honorary Secretary, Honorary Chairman, Honorary Consul, etc. *See* **Honorary Appointment** *Abbrev.* Hon.

Honorary Appointment Assuming the responsibility of a particular position or work without getting paid. *See* **Honorary.**

Honoris causa *Latin*—Honorary; For the Sake of Honour. Describes an honorary degree so bestowed by a university.

Honour Policy *Similar to* **Policy Proof of Interest** *also* **Interest or No Interest,** *q.v.*

Hons. With Honours, in graduation decorations.

Hon. Sec. Honorary Secretary.

Hook Hook Damage, *q.v.*

Hook Damage Damage caused by cargo hooks during loading or discharge. Also termed Hook, *q.v.*

Hoop A circle or a band of steel or wire around wooden buckets, barrels, kegs, etc., to hold them together and to reinforce the wooden sections.

Hopper A chute or funnel into which bulk cargo (e.g. grain) is fed which leads to another structure such as a conveyor belt, or to railway wagons, road trucks or to a bagging plant. Can be permanent or portable.

Hopper Barge A barge engaged in harbour dredging work. The contents collected are dumped outside the harbour limits by opening its bottom doors.

Hoppered Hold A hold of a bulk carrier with sloping plating to the port and starboard lower hold sides, helping to secure the stow of a bulk cargo.

Hor. Horizon; Horizontal.

Hora fugit *Latin*—The hour flies.

Horis *or* **HORIS** Hotel Reservations and Information System.

Horizon Imaginary line where the sky seems to meet the earth or the sea. *Abbrev.* Hor *or* Horz.

Horizontal Cost. An expression used in accountancy whereby calculations are made on the basic value of stocks and the cost of furniture and other items.

Horizontal Integration The fusion or amalgamation of two or more competitive companies into one large company. This is the general meaning of Merger, *q.v.*, which may give rise to universal strength and even sometimes, but not always,to monopoly. Also called Lateral Integration.

Horizontal Situation Indicator An instrument displaying an aircraft's position in relation to radio navigation and instrument landing beams as well as the bearing and distance in nautical miles to a selected radio beacon. *Abbrev.* HSI *or* H.S.I.

Horology The art and scientific construction of the equipment to measure the passage of time (clocks and watches, etc.) and the study of measuring time.

Hor Pax *or* **Hor/pax** Horizontal Parallax *nautical*.

Horresco referens *Latin*—I tremble or hate to say it.

Horribile dictu *Latin*—Horrible to relate.

Hors de combat *French*—Out of action; disabled.

Horse Power A unit of power to lift 33,000 footpounds per minute or a unit to measure the power or rate of work by an engine equivalent to 550 foot pounds per second. *Abbrev.* HP *or* H.P. *or* hp *or* h.p.

Horse Trade Political or commercial bargaining, or business exchange conducted with shrewdness or cunning where concessions may be offset by other advantages.

H.O.R.U. Home Office Research Unit.

Horz. Horizon, *q.v.*; Horizontal.

Hos *or* **Hose R.** Hose Rack.

Hos.C. *or* **Hose C.** Hose Connection.

Hosier A person dealing in socks and stockings.

Hospt. Hospital.

Host Country A country accepting or providing for foreign visitors, tourists, conferences, and exhibitions, and international organisations.

HO/TA Holds/Tanks *Lloyd's Register notation*.

HOTEL International Radio Telephone phonetic alphabet for 'H'.

Hotline *or* **Hot-line** Telephone communications constantly maintained between heads of state and others concerned with urgent contact.

Hot Money (1) Foreign currencies or funds coming into a country to obtain better rates of interest. This improves the financial state of the importing country. (2) Extra money paid to port workers who are engaged in the holds of ships where the temperature becomes uncomfortable. Applicable to the grain trade and others.

Hotol *or* **HOTOL** Horizontal take-off and landing *aviation*.

Hot Pursuit The law permits the chasing of a foreign vessel smuggling or fishing in Territorial Waters, *q.v.*, and to arrest her before she reaches the High Seas, *q.v.* In a few cases the continual pursuit into the high seas is also permitted. This is known as 'The right of hot pursuit'.

Hot Work Work done by using acetylene torches and similar work connected with extreme and dangerous heat.

Hours Accumulative *See* **Accumulative Hours**.

Hours Purposes Time allowed to the Charterers, *q.v.*, in a Time Charterparty for the total loading and discharging of the merchandise. *Similar to* Reversible Laydays.

House Airwaybill Groupage Airwaybill. *Abbrev.* HAWB *or* H.A.W.B. *See* **Groupage.**

House Bill of Lading *See* **Groupage Bill of Lading.**

House Flag Clause Some Time charterparties allow the Charterers, *q.v.*, to have the option of painting the ship's funnel and topside markings with their own colours at their own expense. The ship is to be redelivered with her original colours. The Charterers have also the option to hoist their own house flags during the charter period. Time taken for painting and repainting is to count as charter time. *Abbrev.* HFC Clause *or* H.F.C. Clause.

Household Disposable Income A person's net income remaining after normal living expenses and taxes have been paid. *Abbrev.* HDI.

House Organ An internal company publication for the information of employees.

House-to-House Container traffic term. The container is provided by the carrier at the consignor's or exporter's premises where it is stuffed (*see* **Stuffing**) by the consignor himself and is carried to the consignee's premises who unpacks it himself.

Hovercraft A vessel used for the transportation of passengers and cargo riding on a cushion of air formed under it. It is very manoeuvrable and is also amphibious. This craft is very commonly seen in service crossing sea channels and can travel at speeds of up to 50 m.p.h. and over with a load capacity of about 60 cars and 180 passengers or about 600 passengers. There are always improvements being carried out to this new concept of transportation. Sometimes called Air Cushion Craft. *See* **Hydrofoil**.

Hoy Former small coastal sailing vessel usually sloop-rigged in England, with two masts in Holland.

Hoyrylaiva *Finnish*—Steamship.

H.p. Precomputed Altitude *nautical*.

H.P. *or* **h.p.** Half pay; High Power; High Pressure; Hire-Purchase, *q.v.* Horse Power, *q.v.*

HP *or* **H.P.** Harbour Patrol *USA*; Hardy Perennial *horticulture*; Horizontal Parallax *nautical*; Horse Power, *q.v.*; House Physician; Houses of Parliament; Hawse Pipe; High Pressure.

hPa hectoPascal *SI unit of pressure equivalent to 1 millibar.*

HPA Houston Port Authority.

hp.cyl. High Pressure Cylinder.

HPD Hydraulic Pump Drives.

HPF *or* **H.P.F.** Highest Possible Frequency.

Hpgc. Heading per Gyro Compass *nautical.*

hp.hr. Horse Power-Hour.

HP/HT High Pressure/High Temperature *oil drilling.*

HPI Hire Purchase Information; Hire Purchase Interest.

HPLC High Performance Liquid Chromatography.

HPN *or* **H.P.N.** Horse Power Nominal. Normally *abbrev.* NHP *or* N.H.P. for Nominal Horse Power, *q.v.*

HPNS High Pressure Nervous Syndrome *oil drilling.*

H.P.R. Hungarian People's Republic.

h.pres. Heir Presumptive.

HPS *or* **H.P.S.** Hazardous Pollution Substances; High Pressure Sodium *lights;* High Presure Steam; Hire-Purchase System, *see* **Hire-Purchase**.

Hpt. High Point; Highest Point; High Pressure Test.

HPTA *or* **H.P.T.A.** Hire-Purchase Trade Association.

Hpu. Hydraulic Pumping Unit.

HPW *or* **H.P.W.** High Pressure Water; High Pressure Watts.

HQ *or* **H.Q.** *or* **H.q.** Headquarters; Hoc Quaere *Latin*—Watch for this.

HR *or* **H.R.** Hellenic Register of Shipping *classification society*; Home Rule; House of Representatives *USA*; House of Representatives Bill.

hr. Rectified Altitude *nautical*; Hour.

Hra. Herra *Finnish*—Mr

H.rd. Half Round.

Hrd. Hard.

HRIP *or* **H.R.I.P.** Hic requiescit in pace *Latin*—Here rests in peace.

H.R.R. Higher Reduced Rate *income tax.*

HRS Hellenic Register of Shipping. *Symbol* HR.

H.R.S. Hydraulic Research Station.

hr/s. Hour(s).

HRT Honolulu Rapid Transit; Hormone Replacement Therapy.

HRU Hydrostatic Release Unit *lifesaving apparatus.*

hs. Heating Surface *Lloyd's Register notation.*

Hs. Handschrift *German*—Manuscript.

HS *or* **H.S.** Heating Surface *Lloyd's Register notation;* Hic sepultus *Latin*—Here is buried; Hic situs *Latin*—Here is placed; Hinged Shelf; High Sheriff; Home Secretary; Honorary Secretary; Hospital Ship; House of Surgeons; House Surgeon.

h.s. Hoc Sensu *Latin*—In this sense; Sextant Altitude *nautical.*

H₂S Hydrogen Sulphide.

HSA *or* **H.S.A.** Horizontal Sextant Angle *nautical.*

HSAA Health Sciences Advancement Award, *USA.*

HSBC Hong Kong and Shanghai Banking Corporation.

HSC Health and Safety Commission.

HSD *or* **H.S.D.** Half Shelter Decker; High Shot Density *oil drilling*; High Speed Diesel, *q.v.*

HSE Health, Safety and Environmental.

HSE *or* **H.S.E.** Health and Safety Executive.

HSF *or* **H.S.F.** High Sulphur Fuel, *q.v.*

HSHT Hydro Sonic Hull Tender.

HSI *or* **H.S.I.** Horizontal Situation Indicator.

HSM His/Her Serene Majesty; Hull Stress Monitoring.

HSMS High-Strength Microsphere.

HSP *or* **H.S.P.** *or* **h.s.p.** High Speed Printer *computer.*

HSR *or* **H.S.R.** *or* **h.s.r.** High Speed Reader *computer*; Hotel Single Rate.

HSS *or* **H.S.S.** *or* **h.s.s.** Handy Sized Ship; Heavy Grain, Soyas, Sorghum; High Speed Steel; History of Science Society *USA.*

HSSC *or* **H.S.S.C.** Harmonised System of Survey and Certification **IMO.** Heating, Sweating, Spontaneous Combustion *regarding insurance and grade of cargo.*

HSST *or* **H.S.S.T.** High Speed Surface Transport.

HT *or* **H.T.** Half Time *Lloyd's Register notation for survey*; Hawaiian Territory; High Tension; High Tide; Home Trade.

ht *or* **h.t.** Halftone; Heat Threaded; Heat Treatment; High Tension; Highest *chart symbol*; Height; Hoc Tempore *Latin*—At the time; Hoc titulo *Latin*—Under this title; heat.

H.T. *or* **H-T** High Temperature.

h. & t. Hardened and Tempered.

h.t.a. Heavier than Air.

h.t.b. High Tension Battery *electricity.*

HTC *or* **htc** Hold till called for.

HTD High Technical Diploma.

Htg. Heating.

HTG High Temperature Gas.

H.T.G.C.R. *or* **H.T.G.R.** High Temperature Gas Cooled Reactor.

HTH High Test Hypochlorite, *q.v.*

HTLV Human T-cell Lymphotropic Virus *virus of AIDS.*

HTM *or* **H.T.M.** Heat Transfer Medium.

Htn. Hamilton, *Bermuda.*

Hto. Hereto.

HTOL *or* **H.T.O.L.** Horizontal Take Off and Landing *aviation.*

Htr or **htr.** Heater *Lloyd's Register notation.*

HTR High Temperature Reactor. A type of nuclear power station.

HTS or **H.T.S.** or **h.t.s.** Half Time Survey; Heat Threaded Steel; High Tensile Steel.

Ht/s Height(s), *chart symbol.*

HTU Heat Transfer Unit.

HTVAL Symbol for Heating Value.

HTW High Temperature Water.

H.U. Harvard University *USA.*

HUCR Highest Useful Compression Ratio.

HUD Head-Up Display, *q.v.*

HUET Helicopter Underwater Escape Trainer.

HUGO or **H.U.G.O.** Highly Unusual Geophysical Operation.

HUK Hunter Killer Force *USA.*

Hulk (1) Out of commission and dismantled ship. (2) Hull of an old ship out of commission but which is made use of as a store ship.

Hull Shell or body of a ship.

Hull and Machinery Marine Insurance, *q.v.*, expression for hull, machinery or materials.

Hull Down The position or sight of a ship when seen on the horizon, only her masts and superstructure being visible.

Hull Insurance A term in Marine Insurance, *q.v.*, regarding the insurance of the hull of a ship, including any accidents arising from impact, collisions, etc., while berthing or during navigation.

Hull Int. Hull Interest.

Hull Number *Similar to* **Yard Number**, *q.v.*

Human Capital The actual number of human beings available for employment in a business enterprise or a country.

Humanum est errare *Latin*—To make mistakes is human.

Humi Humidity, *q.v.*

Humidity A moist, damp atmospheric condition. Relative humidity is the amount of moisture in the atmosphere as compared with that of complete saturation at the given temperature.

Hummocked Ice or **Packed Ice** An extensive icefield caused by heavy blocks of ice pressed together due to strong winds which prevents navigation. *See* **Packed Ice.**

HUMS Health and Usage Monitoring System.

Hun. or **Hung.** Hungarian; Hungary.

Hund. Hundred.

Hundred Deals. *Similar to* **Standard Measure**, *q.v.*

hur. or **hurr.** Hurricane, *q.v.*

Hurricane A very strong tropical wind of 64 knots (118 mph) or over.

Hurricane Deck The uppermost deck of a ship commonly seen on passenger ships where some sort of shade is provided for the passengers.

Hurricane Lamp or **Hurricane Lantern** A Lamp burning kerosene through a wick, so designed that when exposed in the open the wind will not extinguish it. *Also* Storm Lantern.

Hurt Certificate Document attesting to an injury received in service by a naval person.

Hush Money A slang term for money paid to induce a person or persons to keep silent about a particular event.

HV or **H.V.** Health Visitor; High Vacuum; High Velocity; High Voltage; High Visibility; Hoc Verbum *Latin*—This word.

HVAC Heating, Ventilation and Air Conditioning.

Hvalfangerselskapet *Norwegian*—Whaling Company.

HVAR High Velocity Aircraft Rocket.

HVEM High Voltage Electron Microscope.

HVF or **H.V.F.** High Viscosity Fuel Oil, *q.v.*

HVSA or **hvsa.** High Voltage Slow Activity.

HVSS or **hvss.** Horizontal Volute Spring Suspension.

Hvy. Heavy.

HW or **H.W.** or **hw.** High Water *chart symbol*; Heavy Weather Damage *insurance;* Hot Water.

H/w Herewith.

HWB Hot Water Boiler(s) *Lloyd's Register notation.*

HWC or **h.w.c.** Hot Water Circulation.

HWD or **H.W.D.** or **h.w.d.** Heavy Weather Damage *insurance.*

HWDAS Heavy Weather Damage Avoidance System. A US method to facilitate the navigation of large ships in bad weather.

HWF&C or **H.W.F.&.C.** High Water at Full and Change *nautical.*

HWI or **H.W.I** or **h.w.i.** High Water Interval.

HWL or **H.W.L.** or **h.w.l.** High Water Line.

H.W.L.B. High Water London Bridge.

HWM or **H.W.M.** or **h.w.m** High Water Mark.

HWNT or **H.W.N.T.** or **h.w.n.t** High Water Neap Tide.

HWONT or **H.W.O.N.T.** High Water Ordinary Neap Tide.

HWOST or **H.W.O.S.T.** High Water Ordinary Spring Tide.

HWP High Water Pressure or High Pressure Water.

HWS or **H.W.S.** Hot Water Soluble; Hurricane Warning System.

HWST or **H.W.S.T.** High Water Spring Tide.

HWT Hot Water Tank.

Hwy. Highway.

Hy or **Hvy.** Heavy.

Hyd. Hydrant; Hydraulic; Hydrostatic.

Hydr. Hydrographer.

HYDRA Hydrographic Digital Positioning and Depth Recording System.

Hydrant A pipe to which a hose can be attached for drawing water from the main supply. *Abbrev.* Hydt.

Hydraulic Currents Currents that occur in sea channels as a consequence of two opposing currents or tides which makes manoeuvring difficult.

Hydro Hydrostatic(s).

Hydrog. Hydrography, *q.v.*; Hydrographic.

Hydrocarbon A combination of hydrogen and carbon often found in gas and oil.

Hydrocarbon Gas Gas extracted from petroleum.

HYDROCHARTER Code name for Hydrocarbon Voyage Charter (amended 1975).

Hydrochlorofluorocarbon One of the compounds used in cargo refrigeration, the most commonly known being R22, which will be phased out under the Montreal Protocol 1992 because of its ozone depletion and Global Warming potential in favour of Hydrogenfluorocarbon, but is an intermediate replacement for Chlorofluorocarbon (CFC), *q.v. Abbrev.* HCFC.

Hydrodynamics The science of the application of fluid forces, comprising hydrostatics and hydrokinetics.

Hydrofoil A craft more or less similar to the Hovercraft, *q.v.*, insofar as it flies over water and thus eliminates friction between the water and the hull. Under acceleration it rises above water but remains in contact with the surface through supporting legs. They can be quite large, carrying many passengers across narrow seas or up and down large rivers. Improvements and alterations are continuously taking place. *See* **Hovercraft**.

Hydrogen Colourless invisible odourless gas, an element, the lightest substance known, forming two-thirds in volume of water.

Hydrogen Bomb Also known as fusion or thermo-nuclear bomb. Immensely powerful bomb utilising fusion of hydrogen atomic nuclei.

Hydrogen Induced Cracking Cracking which occurs in pipes made of very good quality and of high toughness. *Abbrev.* HIC.

Hydrography The science of marine surveying and investigation of the seas, lakes and rivers and the study of tides and currents; the compilation of charts and related publications. *Abbrev.* Hydroc.

Hydrology The science concerning the laws and properties of water.

Hydrometer A device to measure the specific gravity of liquids.

Hydrophone An instrument for the detection of sound-waves in water. It is used to locate the underwater presence of submarines and other objects.

Hydroplane A fin-like device enabling a submarine to rise or fall; a light, fast, motor-boat designed to skim over the surface; a seaplane.

Hydroscope An optical device to see objects under the surface of the sea.

Hydrostatic Test A pressure test applied by means of water being poured into compartments or tanks until they are completely filled. By this action, the tightness and strength of material can be checked. Each classification society has its testing requirements and methods.

HYDROX *or* **Hydrox** Hydrogen and oxygen mixture.

Hydt. Hydrant, *q.v.*

Hygrometer An instrument for measuring humidity of air or gas.

Hygroscope An instrument indicating but not measuring humidity of air.

Hyg. Hygiene.

Hygrothermograph An instrument to register relative humidity and temperature.

Hyp. Hypothecate, *q.v.* Hypothecation, *q.v.*

Hyperbolic Aids Aids to navigation such as Loran, Decca, etc., necessitate the use of special charts whereupon hyperbolic position indicator lines are marked.

Hypermarket This is a very large enclosed market-place, bigger than a supermarket, which offers a wide range of household goods and clothing. It is generally situated outside a town or city and has its own large car parking area for the use of its customers.

Hypersonic Speed over five times faster than that of sound.

Hyp.log. Hyperbolic Logarithm.

Hypothecate To assume responsibility for a debt without transfer of the security, as when a shipper borrows from a bank but retains the goods being shipped, or an owner mortgages his ship.

Hypothecation The act of mortgaging. *Abbrev.* Hyp.

Hypothecation, Letter of Letter of Guarantee. *See* **Letter of Hypothecation.**

Hypsography That part of geography dealing with the measuring of elevations and altitudes.

Hypsometer An instrument to measure altitude by the method of determining the temperature at which water boils against a stated or needed height.

Hypsometry The act of measuring angles of elevations and altitudes at sea level.

HYSWAS Hydrofoil Small Waterplane Area Ship.

Hythergraph An instrument to measure the condition of air in order to establish the Dew Point.

Hz Hertz *radio frequency.*

hzy Hazy *weather observation.*

I

i Id. *Latin*—That; Inner *Lloyd's Register notation*; Moment of inertia *in mechanics*.

I I-Beam; Interest; International Code Flag Signal (black disc on yellow background signifying, 'I am directing my course to port'); Island; Roman numeral for the figure 1.

IA *or* **I.A.** Incorporated Accountant; Indian Army; Initial Allowance *income tax*; Institute of Actuaries; Institute of Abitrators; Iowa, *USA*; Iraqi Airways *airline designator*.

i.a. Immediately available; In absentia *Latin*—In absence; Initial Action; Indicated Altitude.

Ia. Indiana, *USA*; Iowa, *USA*.

I/A Isle of Anglesey.

i.a. Im auftrage *German*—By order of.

I.A.A. International Academy of Astronautics; International Advertising Association.

IAADFS International Association of Airport Duty Free Stores.

IAADP Institute of Administrative Accounting and Data Processing.

IAAF International Amateur Athletic Federation.

IAAI International Airports Authority of India.

IAAS Incorporated Association of Architects and Surveyors.

IAASP *or* **I.A.A.S.P.** International Association of Airport and Seaport Police.

IAB *or* **I.A.B.** Industrial Advisory Board; Inter American Bank.

IABA *or* **I.A.B.A.** International Association of Airbrokers and Agents.

IAC *or* **I.A.C.** Industrial Advisory Council *USA*; International Association of Classification Societies, *also abbrev.* I.A.C.S. *or* IACS.

IACA *or* **I.A.C.A.** Independent Air Carriers' Association; International Agreement in Civil Aviation.

IACOMS *or* **I.A.C.O.M.S.** International Advisory Committee on Marine Sciences *FAO*.

IACS *or* **I.A.C.S.** International Annealed Copper Standards; International Association of Classification Societies, *q.v.*

IADB Inter-American Defence Board; Inter-American Development Bank.

IADC International Association of Drilling Contractors *USA*.

IADS International Association of Drycargo Shipowners.

IADSO International Association of Drycargo Shipowners Organisation.

IAE *or* **I.A.E.** Institute of Automobile Engineers; International Aero Engineers.

IAEA International Atomic Energy Agency.

IAEC International Association of Environmental Co-ordinators, *Brussels*.

I.A.F. International Astronautical Federation; International Arbitration Forum.

IAG International Auditing Guidlines *accounts*.

IAGLP International Association of Great Lakes Ports.

IAHM *or* **I.A.H.M.** Incorporated Association of Head Masters.

IAIN *or* **I.A.I.N.** International Association of Institutes of Navigation.

IALA *or* **I.A.L.A.** International African Law Association; International Association of Lighthouse Authorities, Paris, France—*Founded by Trinity House UK*.

IALS *or* **I.A.L.S.** International Association of Legal Science.

IAM *or* **I.A.M.** Institute of Administrative Management; Institute of Advanced Motorists *UK*.

IAMAP *or* **I.A.M.A.P.** International Association of Meteorology and Atmospheric Physics.

IAME International Association of Maritime Economists, founded in 1992.

IAMI International Association of Maritime Institutions, *q.v.*

IAN *or* **I.A.N.** International Association of Navigation.

IANA Intermodal Association of North America.

I.A.N.C. International Airline Navigators' Council.

IAOPA International Council of Aircraft Owner and Pilot Associations.

IAPC International Auditing Practices Committee *accounts*.

IAPCO International Association of Private Container Owners, Monte Carlo, Monaco.

IAPH International Association of Ports and Harbors, Tokyo, Japan.

IAPIP International Association for the Protection of Industrial Property.

IAPSO *or* **I.A.P.S.O.** International Association for the Physical Sciences of the Oceans.

IARA International Agricultural Research Institute *UN*; Inter-Allied Reparations Agency.

I. Arb. Institute of Arbitrators.

IAS *or* **I.A.S.** Indian Administrative Services; Institute of Aerospace Sciences *USA*; Indicated Air Speed, *q.v.*; Institute of Advanced Studies; Integrated Air System, *q.v.*

i.a.s. Immediate Access Storage *computer*; Indicated Air Speed, *q.v.*; Instrument Approach System.

IASA *or* **I.A.S.A.** International Air Safety Association.

IASC International Association of Seed Crushers.

i.a.s.o.r. Ice and Snow on Runway.

IASMM International Association for the Study of Maritime Mission.

IASST International Association for Sea Survival Training.

i.a.t. Inside Air Temperature.

IATA *or* **I.A.T.A.** International Air Transport Association; International Amateur Theatre Association.

IATC Inter-American Travel Congress.

IATTC Inter-American Tropical Tuna Commission.

I.A.U.P.L. International Association of University Professors and Lecturers.

i.a.w. In accordance with.

IAWPR *or* **I.A.W.P.R.** International Association on Water Pollution Research.

ib Ibidem *Latin*—In the same place.

IB Iberia, *Spanish Airlines*.

IB *or* **I.B.** In Board; In Bond, *q.v.*; Incendiary Bomb; Industrial Business; Information Bureau; Institute of Bankers; Institute of Building; Intelligence Branch; International Bank; Invoice Book; Intelligence Book.

I/B Inland Bill/s.

IBA Independent Broadcasting Authority *UK*; Indonesian-British Association; Insurance Brokers (Registration) Act 1977; International Bauxite Association.

IBA *or* **I.B.A.** Independent Bankers' Association; Industrial Bankers' Association; Investment Bankers' Association; Independent Broadcasting Authority; Industrial Building Allowance *income tax*; International Bar Association.

IBAA *or* **I.B.A.A.** Investment Bankers' Association of America.

IBAM *or* **I.B.A.M.** Institute of Business Administration and Management, Japan.

IBC Intermediate Bulk Container.

IBC *or* **I.B.C.** Institute Builders' Risk Clauses *insurance*; Intermediate Bulk Carrier.

IBC Code International Code for the Construction and Equipment of Ships carrying Dangerous Chemicals in Bulk, *IMO*, which applies to ships built after 1 July 1986.

IBCS Integrated Bridge Control System.

IBE Institute of British Engineers.

IBEC Indo-British Economic Committee; International Basic Econmony Corporation *USA*.

IBEL Interest-bearing eligible liability.

IBF International Banking Facilities.

I.B.I. *or* **i.b.i.** Invoice Book Inwards.

ibid *or* **ib** Ibidem *Latin*—In the same place.

IBI Inter-Governmental Bureau of Information.

IBIA International Bunker Industry Association, founded in 1993.

Ibidem *Abbrev* ib *or* ibid, *q.v.*

IBIN International Banking Institute, *Hamburg*.

IBM International Brotherhood of Magicians.

IBN *or* **I.B.N.** Institut Belge de Normalisation, *Belgian Standards Institute.*

IBNR *or* **I.B.N.R.** Incurred But Not Reported Losses *insurance.*

IBO *or* **ibo** Invoice Book Outwards.

I.B.P. *or* **i.b.p.** Initial Boiling Point.

IBP International Biological Programme.

IBRD *or* **I.B.R.D.** International Bank for Reconstruction and Development, *also known as* the World Bank.

IBRO *or* **I.B.R.O.** International Bank Research Organisation.

I.B.S. *or* **I.B.(Scot)** Institute of Bankers in Scotland.

IBSFC *or* **I.B.S.F.C.** International Baltic Sea Fisheries Commission.

I.bu. Imperial Bushel.

IBWM *or* **I.B.W.M.** International Bureau of Weights and Measures.

ic *or* **IC** Inland Container.

IC Ice Crystals *weather observation*; In Commission; Indian Airlines *airline designator*; Inter-Container; Internal Combustion.

IC *or* **I.C.** Identity Card; Imperial College of Science, Technology and Medicine in London; Imperial Conference; Industrial Court; Information Centre; Institute Clauses, *q.v.*; Intelligence Corps; Internal Communications.

I/C *or* **I.C.** *or* **i/c** In charge; In command; In commission.

I–C Indo-China.

ICA Ice Cream Alliance; Ignition Control Additive; Industrial Catering Association; International Civil Aviation; International Co-operation Administration; Insurance Companies Act; International Co-operative Alliance; International Commercial Arbitration; International Court of Arbitration; International Council on Archives.

I.C.A. Institute of Chartered Accountants in England and Wales; Institute of Company Accountants; Institute of Contemporary Arts; International Commission on Accoustics.

ICAA *or* **I.C.A.A.** International Civil Airport Association; Investment Counsel Association of America.

ICAAAA *or* ICAA Intercollegiate Association of Amateur Athletics in America.

I.C.A.B. International Cargo Advisory Bureau.

ICAC Independent Commission Against Corruption *Hong Kong*. International Cotton Advisory Committee.

ICAL International Conference on Air Law.

I.C.A.M. International Civil Aircraft Markings.

I.C.A.N. International Commission for Air Navigation.

ICAO *or* I.C.A.O. International Civil Aviation Organisation.

I.C.A.S. Institute of Chartered Accountants of Scotland.

ICAS *or* I.C.A.S. Institute Clauses for Air Sendings (All Risks), *q.v.*

I.C.B. Indian Coffee Board; International Container Bureau.

I.C.B.M. Intercontinental Ballistic Missile.

ICC Inter-Conference Committee *shipping*.

I.C.C. Institute Cargo Clauses, *q.v.*; International Chamber of Commerce; International Commodity Classifications; International Congregational Council; International Correspondence Colleges; Interstate Commerce Commission, *q.v.*; International Chemical Code.

IC & C *or* I.C.& C. Invoice Cost and Charges.

ICCAS International Conference on Computer Applications in automation of Shipyard Operations and Ship Design.

ICCE *or* I.C.C.E. International Council of Commerce Employers.

ICCGB Italian Chamber of Commerce in Great Britian.

ICCH *or* I.C.C.H. International Committee Clearing House.

ICCICA Interim Co-ordinating Committee on International Commodity Arrangements. *USA*.

ICCO International Conference of Container Operators, *group of Freight Forwarders, Paris*.

ICD *or* I.C.D. Inland Clearance Depot, *q.v.*

Ice *or* Icel *or* Iclnd Iceland; Icelandic.

I.C.E. Institute of Chemical Engineers; Institute of Civil Engineers; Internal Combustion Engine; International Cultural Exchange; In Car Entertainment.

Ice Anchor An anchor used to secure a ship in icebound waters.

Ice Barrier Ice in the sea or lake which is impenetrable for navigational purposes.

Iceberg A huge floating mass of ice. Detached portion of a glacier.

Icebound A vessel is icebound when she is unable to proceed on her voyage because of hard ice. For the same reason, a port may be Icebound and traffic in or out is prevented.

Icebreaker Ship constructed for the special purpose of cutting channels in winter ice for the passage of other vessels.

Ice Class A system of classification for vessels constructed with the ability to navigate safely in ice:
Class 1★—for extreme ice
1 —severe ice
2 —intermediate ice
3 —light ice.

Ice Field Area of pack ice consisting of any size of floes which is greater than 10 km across.

Ice Jam Accumulated broken ice blocking a narrow channel or canal.

Icelandair The airline of Iceland.

ICEM Inter-Governmental Committee for European Migration.

Iceport Embayment in an ice-shelf or floating glacier, which may be temporary, where a vessel may moor alongside and discharge cargo onto the ice.

ICER International Centre of the European Railways, Rome.

ICES *or* I.C.E.S. International Council for the Exploration of the Sea.

Ice Tongue Narrow strip of ice at sea.

ICETT Industrial Council for Educational and Training Technology *UK*.

ICF *or* I.C.F. Import Cargo Form, *in containerization*; International Compensation Fund *developed by IMCO* (now the IMO).

ICFC *or* I.C.F.C. Industrial and Commercial Finance Corporation, *q.v.*

ICFTU International Confederation of Free Trade Unions.

ICFU International Confederation of Free Trade Unions.

ICGC International Cargo Gear Certificate.

ICGS International Cargo Gear Survey/Surveyor.

ICHCA *or* I.C.C.H.C.A. International Cargo Handling Co-ordination Association.

I.Chem.E. Institute of Chemical Engineers.

ICI *or* I.C.I. Imperial Chemical Industries; Investment Costing Institute *USA*.

ICIA *or* I.C.I.A. International Credit Insurance Association.

ICIE International Centre for Industry and the Environment; International Council for Industrial Editors.

Icing The formation of ice on ships or aircraft.

ICITO Interim Commission of International Trade Organisations.

ICJ *or* I.C.J. International Commission of Jurists, *q.v.*; International Court of Justice, *q.v.*

ICL International Call Letters.

ICL *or* I.C.L. International Computers Ltd.; International Confederation of Labour.

ICLL International Convention on Loadlines.

ICM *or* I.C.M. Institute of Computer Management; Institute of Credit Management.

ICMA Institute of Cost and Management Accountants; International Cable Makers' Association;

International Christian Maritime Association; International Congress of Maritime Arbitrators.

ICMES International Co-operation on Marine Engineering Systems.

ICMH International Commission for Maritime History.

I.Co. Initial Course *nautical.*

ICO Icelandic Conference Organisation; Intermediate Circular Orbit *satellite configuration*; International Coffee Organisation.

ICOM International Council of Museums, *Paris.*

Icon. Iconographic.

ICONN Island Complex Offshore, New York and New Jersey.

ICOR *or* **I.C.O.R.** Incremental Capital–Output Ratio, *q.v.*; Inter-Governmental Conference on Oceanic Research, *a branch of UNESCO.*

I.Corr.T. Institute of Corrosion Technology.

I.C.O.T. Institute of Coastal Oceanography and Tides.

ICOTAS *or* **I.C.O.T.A.S.** International Committee on the Organisation of Traffic at Sea.

ICP Impressed Current Protection *electronics*; Intelligent Cellular Peripheral *computer*; International Phonetic Association; Inventory Control Point *USA Army.*

ICPA *or* **I.C.P.A.** International Co-operative Petroleum Association.

ICPC Internal Contracts Policy Committee.

ICPEMC International Commission for Protection against Environmental Mutagens and Carcinogens *UN.*

ICPL *or* **I.C.P.L.** International Committee of Passenger Lines.

ICPO Irrevocable Corporate Purchase Order, *q.v.*

ICPO *or* **I.C.P.O.** International Criminal Police Organisation, *similar to* International Criminal Police Commission, *abbrev.* Interpol.

ICRC International Committee of Red Cross.

ICS *or* **I.C.S.** Imperial College of Science and Technology, London; Indian Chemical Society; Indian Civil Service; Instalment Credit Selling; Institute of Chartered Shipbroker, *q.v.*; International Chamber of Shipping; Integrated Container Service; International Code of Signals, *q.v.*; Institute of Chartered Surveyors *UK.*

ICSEMS *or* **I.C.S.E.M.S.** International Commission for the Scientific Exploration of the Mediterranean Sea.

ICSLS *or* **I.C.S.L.S.** International Convention for Saving Life at Sea.

ICSP Inter-Agency Committee on Standards Policy *USA.*

ICSPRO Inter-secretariat Committee on Scientific Programmes Relating to Oceanography.

ICSU *or* **I.C.S.U.** International Council of Scientific Unions *UNESCO.*

ICSW International Committee on Seafarers Welfare.

ICSW *or* **I.C.S.W.** International Conference of Social Work.

ICT *or* **I.C.T.** International Computers and Tabulators.

ICTA *or* **I.C.T.A.** Imperial College of Tropical Agriculture in Trinidad; International Council of Travel Agents.

Ictus Iurisconsultus *Latin*—Counsellor at Law.

ICU *or* **I.C.U.** International Code Use; International Credit Unions.

ICVA International Council of Voluntary Agencies.

ICW *or* **I.C.W.** In connection with; Interrupted Continuous Wave.

I.C.W.A. Indian Council of World Affairs; Institute of Cost and Works Accountants.

ICWP Inter-Conference Working Party; International Council of Women Psychologists.

I.C.Z. Isthmian Canal Zone.

id Idem *Latin*—The same.

ID Idaho, *USA.*

ID *or* **I.D.** Identification; Import Duty; Induced Draught; Industrial Dynamics; Infectious Disease; Information Department; Inside Diameter; Institute of Directors; Investigation Department; Intelligence Department.

I.D. *or* **i.d.** Inside Diameter; Internal Dimensions.

IDA *or* **Ida** *or* **I.D.A.** Industrial Diamond Association; International Development Association *United Nations.*

IDAC Import Duties Advisory Committee.

Idaresi *Turkish*—Administration.

I.D.B. Illicit Diamond Buyer; Inter-American Development Bank.

idc In due course, i.e. at a later date.

IDC Industrial Development Certificate.

I.D. Card Identity/Identification Card.

IDD *or* **I.D.D.** International Direct Dialling, *overseas direct telephone dialing.*

Idem *Abbrev.* id., *q.v.*

Ident Identification *aviation.*

Ider per idem *Latin*—The same for the same.

I.D.F. Inter-Departmental Flexibility, *q.v.*

IDI *or* **I.D.I.** Institut de Droit International *French*—Institute of International Law.

IDL *or* **I.D.L.** International Date Line, *q.v.*

Idle Capacity Labour and/or machines available in excess of that required for current production.

Idle Money Capital that is producing neither earnings nor interest nor dividends.

Idle Time Period when workmen and/or machines are not producing.

Idle Vessel *or* **Idle Tonnage** Ship(s) laid up for lack of shipping demand.

Idler A slang word for a crew member of a ship who does not take part in the watches. This refers to a boatswain, purser, carpenter, cook, steward, donkeyman, etc.

IDLH Immediately Dangerous to Life and Health.

Ido Idaho *USA*.

IDP Intelligence Data Processing System *USA*; International Driving Permit.

idr Idraulica *Italian*—Hydraulic.

IDR Import Duty Report Number; Intermediate Data Rate *telecommunications*.

IDRC International Development Research Centre *Canada*.

I.D.S. Institute of Dental Surgery.

IDSO *or* **I.D.S.O.** International Diamond Security Organisation.

IDSPS International Deep Sea Pilotage Service.

IDT *or* **I.D.T.** Industrial Design Technology.

i.e. Id est *Latin*—That is; Inside Edge.

IE *or* **I.E.** Index Error *nautical*; Initial Equipment; Institution of Electronics; Institution of Engineers; Solomon Airlines *airline designator*.

IEA *or* **I.E.A.** Institute of Economic Affairs; Institute of Engineers in Australia; International Economic Association; International Electrical Association; International Energy Agency, *US state sponsored organisation*.

IEAT Industrial Estates Authority Thailand.

IEC *or* **I.E.C.** International Economical Corporation; International Electrical Committee.

IEE *or* **I.E.E.** Institution of Electrical Engineers.

IEEE Institute of Electrical and Electronics Engineers *USA*.

IEEIE Institute of Electrical and Electronics Incorporated Engineers.

IEI *or* **I.E.I.** Industrial Education Institute; Industrial Engineering Institute.

I.E.(I). Institute of Engineers (India).

I.E.M.E. Institute of Electrical and Mechanical Engineering.

I.E.R.E. Institution of Electronic and Radio Engineers.

I.E.S. *or* **I.E.S.S.** Institution of Engineers and Shipbuilders in Scotland.

i.e.t. Initial Engine Test.

I.Ex. Institute of Export.

i.f. In full; information feedback; intermediate frequency.

IF Internally Flawless *diamonds*.

IF *or* **I.F.** Ice Fog *weather observation*; Information Feedback; In full; Intermediate Frequency; Institute of Fuel; Inspector of Fisheries.

I/F Insufficient Funds *in banking*.

IFAD International Fund for Agricultural Development *UN*.

IFALPA *or* **I.F.A.L.P.A.** International Federation of Airline Pilots' Associations.

IFAN *or* **I.F.A.N.** International Federation for the Application of Standards.

IFAP International Federation of Agricultural Producers.

IFATCA *or* **I.F.A.T.C.A.** International Federation of Air Traffic Controllers' Associations.

IFB Invitations for Bid *USA*.

IFC International Finance Corporation *United Nations*.

I.F.C. Institute Freight Clauses *marine insurance*; International Finance Corporation, *q.v.*

IFCA International Flight Catering Association.

IFCCTE *or* **I.F.C.C.T.E.** International Federation of Commercial, Clerical and Technical Employees.

IFCOPD International Fund for Compensation for Oil Pollution Damage. *See* **FUND 1971**.

IFCT International Finance Corporation Thailand.

IFCTU *or* **I.F.C.T.U.** International Federation of Christian Trade Unions.

I.F.E. Institution of Fire Engineers.

IFF *or* **I.F.F.** Institute of Freight Forwarders; International Freight Forwarders; Identification Friend or Foe *radar*.

IFFFA International Federation of Freight Forwarders' Associations. Now FIATA, *q.v.*

IFFTC *or* **I.F.F.T.C.** International Freight Forwarding Trading Council.

IFIP International Federation of Information Processing *USA*.

IFJ *or* **I.F.J.** International Federation of Journalists.

IFL Icelandic Federation of Labour; International Frequency List.

IFLA International Federation of Landscape Architects; International Federation of Library Associations and Institutes.

IFLO *or* **I.F.L.O.** International Federation of Librarians' Association.

I.F.N.A. *or* **i.f.n.a.** In full net absolutely, *re insurance claim*.

IFO *or* **I.F.O.** Intermediate Fuel Oil.

IFP *or* **I.F.P.** International Flyte Pack, *q.v.*

IFPMA *or* **I.F.P.M.A.** International Federation of Pharmaceutical Manufacturers' Associations.

IFR *or* **I.F.R.** Instrument Flight Rules, *see* **All-weather Air Stations.**

IFRB International Frequency Registration Board.

IFR&D Industry Finance Research and Development *International Air Transport Association*.

Ifremar *or* **IFREMAR** Institut Français de Recherche pour l'Exploitation de la Mer. *French*—The French Institute for Research in the Exploitation of the Sea.

IFS *or* **I.F.S.** International Finance Statistics; Irish Free State.

IFSA *or* **I.F.S.A.** In Flight Food Services Association.

I.F.S.M.A. International Federation of Ship Masters' Associations.

IFT *or* **I.F.T.** International Frequency Tables *radio frequency.*

IFTA *or* **I.F.T.A.** International Federation of Teachers' Associations; International Federation of Travel Agents/Agencies.

IFTM International Forwarding and Transport Message.

IFVC *or* **I.F.V.C.** Institute Fishing Vessel Clauses *marine insurance.*

IG *or* **I.G.** Imperial Gallon; Inspector General; Interior Gauge; Inert Gas, *see* **Inert Gas System**.

IGA *or* **I.G.A.** International Geographical Association.

I.Gas.E. Institute of Gas Engineers.

IGAT International Gold Award Training *UK*; Iranian Gas Trunk Pipeline *between Iran and USSR.*

IGC *or* **I.G.C.** International Geophysical Committee; International Geophysical Co-operation.

IGC Code International Code for the Construction and Equipment of ships carrying liquefied Gases in Bulk 1983. *IMO.*

I.G.D. Illicit Gold Dealer.

I.G.E. Imposta Generale Sull'Entrate *Italian*—Turnover Tax.

IGG Inner Gas Generator(s).

Igloo Palletised container for air freight designed to stow easily in aircraft holds.

Ign. Ignition; Incognito *or* ignotus *Latin*—Unknown.

Ignition Delay Time taken between the start of ignition and the combustion in a diesel engine.

Ignition Lag The split second of time between the opening of the injection valve to the cumbustion chamber in an engine and ignition. The time varies according to the type of engine.

Ignition Point Temperature at which point a substance will ignite or explode. Also expressed as Flash Point. *q.v.*

Ignorantia Juris Neminem Excusat *Latin*—Ignorance of the law is no excuse.

IGO *or* **I.G.O.** Inter-Governmental Organisation.

IGOR Injection Gas Oil Ration; Intercept Ground Optical Recorder.

IGORTT Intercept Ground Optical Recorder Tracking Telescope.

IGOSS *or* **I.G.O.S.S.** Integrated Global Ocean Station System.

IGP International Gravel Pack *oil drilling.*

IGPM *or* **I.G.P.M.** Imperial Gallons per Minute.

igr Igitur *Latin*—Therefore.

IGS Inert Gas System, *q.v.*

IGT *or* **I.G.T.** Institute of Gas Technology, an energy research and education centre in Chicago, *USA.*

IGY *or* **I.G.Y.** International Geophysical Year.

IH Inland Haulage *containerisation.*

IH *or* **I.H.** Issuing House *or* Issuing Houses, *q.v.*

IHA *or* **I.H.A.** International Hotel and House Association; Issuing Houses' Association, *q.v.*

IHB *or* **I.H.B.** International Hydrographic Bureau, Monaco.

IHC *or* **I.H.C.** International Health Convention; International Hotels Corporation.

IHF *or* **I.H.F.** International Helsinki Federation *for human rights.*

I.H.K. Internationale Handelskammer *German*—International Chamber of Commerce.

IHO *or* **I.H.O.** International Hydrographic Organisation. *See* **International Charts**.

I.H.O.U. Institute of Home Office Underwriters.

IHP *or* **I.H.P.** *or* **i.h.p.** Indicated Horse Power, *q.v.*

IHR *or* **I.H.R.** International Health Regulation.

IHS *or* **I.H.S.** International Hydrofoil Society; Iesous *Greek*—Jesus; in hoc signo *Latin*—by this sign; Jesus Hominum Salvator *Latin*—Jesus Saviour of Man.

IHS-NA *or* **I.H.S.-N.A.** International Hydofoil Society North America.

II Illegal Immigrant; Roman numeral for figure 2.

IIA Information Industry Association *USA.*

IIAP Insurance Institute for Asia and Pacific.

IIAS *or* **I.I.A.S.** International Institute of Administrative Sciences.

IIB *or* **I.I.B.** Incorporated Insurance Broker(s); Institut International de Brevets *French*—International Institute of Patents.

IICC *or* **I.I.C.C.** International Institute in Commercial Competition.

IICL Institute of International Container Lessors, New York.

IIE *or* **I.I.E.** Institute for International Education *USA*; International Institute of Embryology.

III Roman numeral for figure 3.

I.I.L. Insurance Institute of London.

I.I.M. Institute of Industrial Managers.

IIMS International Institute of Marine Surveyors; founded 1991.

I.I.P. Institut Internationale de la Presse *French*—International Press Institute; International Institute of Philosophy; International Ice Patrol.

IIPA *or* **I.I.P.A.** Institute of Incorporated Practitioners in Advertising.

IIR Institute for International Research.

IIR *or* **I.I.R.** International Institute of Refrigeration.

IIRS *or* **I.I.R.S.** Institute for Industrial Research and Standards in Ireland.

IISI *or* **I.I.S.I.** International Iron and Steel Institute in Brussels.

IISO *or* **I.I.S.O.** Institution of Industrial Safety Officers.

IIT Inspector of Income Tax.

IITA International Institute of Tropical Agriculture *Nigeria*.

Ijara Service contract in Islamic law.

IJC *or* **I.J.C.** International Commission of Jurists, *q.v.*

I.J.K. Internationale Juristen-Kommission *German* —International Commission of Jurists.

IK *or* **I.K.** Inner Keel of a ship.

IKAPI Indonesian Publishers' Association.

IL Illinois *USA*.

IL *or* **I.L.** Import Licence; Including Loading; Instrument Landing; In Line; Institute of Linguists.

I/L Import Licence.

ILA *or* **I.L.A.** Institute of Loss Adjusters; Instrument Landing Approach *aviation*; International Law Association; International Longshoremen's Association.

I.L.A.A. International Legal Aid Association.

ILB *or* **I.L.B.** Irish Land Boundary, *for customs purposes*.

ill Illustrissimus *Latin*—Most Distinguished.

Ill Illinois *USA*.

ILLC *or* **I.L.L.C.** International Load Line Certificate.

ILLEC *or* **I.L.L.E.C.** International Load Line Exemption Certificate.

ILMB Irish Livestock and Meat Board.

Il n'y a pas à dire *French*—There is nothing to say; The matter is settled.

ILO *or* **I.L.O.** Industrial Liaison Officer; International Labour Office; International Labour Organisation, *q.v.*

i.l.o. In lieu of.

I.L.O.A. Industrial Life Officers' Association *life assurance*.

ILOC Irrevocable Letter of Credit. *See* **Irrevocable**.

ILRAD International Laboratory for Research on Animal Diseases *Kenya*.

ILS *or* **I.L.S.** Incorporated Law Society; International Latitude Service: Instrument Landing System *aviation*; International Law of the Sea. *See* **International Maritime Law.**

i.l.t. In lieu thereof.

ILU *or* **I.L.U.** Institute of London Underwriters.

ILW International Low Water.

ILWU International Longshoremen's and Warehousemen's Union *USA*.

ILZSG International Lead and Zinc Study Group.

IM Institute of Management, London. (Formed in 1992 on merger of the British Institute of Management and the Institution of Industrial Management.)

IM *or* **I.M.** Impulse Modulation; Intermediate Modulation; Inland Marine Risk; Institute of Metals; Institution of Metallurgists; Interceptor of Missile; Isle of Man.

1ma Prima *Italian*—First; First class.

I.M.A. Industrial Marketing Association; Industrial Medical Association; International Management Association.

IMAC International Court of Arbitration for Marine and Inland Navigation. At the Chamber of Commerce in Gdynia.

I.Mar.E. Institute of Marine Engineers *UK*, *q.v.*

IMAS Integrated Monitoring and Surveillance: London Stock Exchange service which identifies abnormal or irregular trading; International Marine and Shipping.

IMB *or* **I.M.B.** Institute of Marine Biology: International Maritime Bureau *London*.

IMC *or* **I.M.C.** Institute of Management Consultants; Instrument of Meteorological Conditions *aviation*; International Maritime Committee; International Materials Conference.

IMCA International Marine Contractors Association, formed in 1995.

IMCC Integrated Mission Control Center *USA*.

IMCO *or* **I.M.C.O.** Inter-Governmental Maritime Consultative Organization, now IMO.

IMCOS International Meteorological Consultancy Service.

I.M.D. Indian Medical Department.

IMDEX International Marine Traffic Data Exchange.

IMDG *or* **I.M.D.G.** International Maritime Dangerous Goods.

IMDGC *or* **IMDG Code** International Maritime Dangerous Goods Code, *IMO.*

IME *or* **I.M.E.** Institute of Marine Engineers, *q.v.*; Institution of Mechanical Engineers; International Marine Engineers.

I.Mech.E. Institution of Mechanical Engineers.

IMEDE International Management Development Institute, Lausanne, Switzerland.

IMEF *or* **I.M.E.F.** International Maritime Employers' Federation, *q.v.*

IMEG *or* **I.M.E.G.** International Management and Engineering Group.

IMES Irish Marine Emergency Service.

I.Mets Institute of Metals.

IMEX International Marine Exhibition, *generally in Earls Court, London*.

IMF *or* **I.M.F.** International Marketing Federation; International Monetary Fund *United Nations*.

IMI Imperial Metal Industries; Improved Manned Interceptor; International Maritime Industry.

IMIB Inland Marine Insurance Bureau *USA*.

IMIF International Maritime Industry Forum.

I.Min.E. Institute of Mining Engineers.

Imit Imitation.

IML *or* **I.M.L.** International Maritime Law, *q.v.*

IMLA International Maritime Lecturers Association.

IMLI International Maritime Law Institute, set up in Malta by IMO in 1988.

IMM or **I.M.M.** Institute of Mining and Metallurgy; International Mercantile Marine.

Immature Creditor Nation An under-developed country which through its ability and resources has improved and developed its trade balance so favourably that investment abroad has yielded a surplus. *Opposite to* Immature Debtor Nation, *q.v.*

Immature Debtor Nation An under-developed country which through the borrowing of money on a large scale has created an adverse balance of payments with other nations. *Opposite to* Immature Creditor Nation., *q.v.*

Immed. Immediate.

Immersion Scale An American term for a table showing the weight in tons needed to immerse a vessel to different drafts. *Also called* Draft Scale.

Immigrant A person who enters a country from abroad in order to settle in that country.

Immigration The act of entering a country for the purpose of settling in that country. *Opposite to* emigration. *See* **Emigrant.**

IMMTS or **I.M.M.T.S.** Indian Mercantile Marine Training Ship.

Immunities In law this means exemption from taxation, jurisdiction, etc.

IMO or **I.M.O.** International Money Order, *q.v.*; International Meteorological Organisation; International Maritime Organization, *q.v.*

IMO/ILO Guidelines for Packing Cargo in Freight Containers or Vehicles *IMO.* Supplements the IMDG Code, *q.v.*

IMO-MODU *Lloyd's Register notation* for the construction and equipment of mobile offshore drilling units.

IMO Number A seven-digit number allotted to every merchant ship hull which is its internationally recognised identity number, introduced by International Maritime Organisation resolution to assist registration and combat fraud. For ships of all flags and registries, this is the same as the Lloyd's Register Number or LRN.

IMP Interplanetary Monitoring Platform.

Imp Imperial—former size of paper 22 ins. × 30 ins.; Imperator *Latin*—Emperor; Imported; Importer; Imprimatur *Latin*—Let it be printed.

IMPA or **I.M.P.A.** International Marine Purchasing Association; International Maritime Pilots' Association.

IMPACT Implementation, Planning and Control Technique.

Imperfect Market A market in which the usual forces of economic efficiency are frustrated by restrictions on production, prices, profits and distribution.

Imperial Preference Countries within the British Commonwealth who enjoyed special mutual import duty tariff.

Imperial Standard The customary units of weights and measures used in the United Kingdom. Gradual integration with the metric system is taking place since entry into the Common Market (EEC).

Imp. gal. Imperial gallon.

Implied Authority Authority granted by implication or implied by law according to the circumstances. An appointed agent is automatically acting on behalf of his principals.

Implied Warranties In Marine Insurance, *q.v.*, contracts, there are implications or rules to be observed apart from the expressed warranties. These are (1) The contract must be based on the utmost good faith. (2) The vessel insured must be seaworthy in all respects and she is to be fit to encounter the usual port and sea perils. *See* **Warranty** *and also* **Express Warranties.**

Import Berth or Quay A special berth or quay for the discharge of imported cargoes.

Import Duty Tax imposed by the local customs authority on goods imported into a country before they are withdrawn by the consignees for home consumption. *See* **Imports.**

Import Duty Report A document filled up at the loading port giving details of the cargo loaded, to be presented to the authorities on arrival at the port of discharge. *Abbrev.* IDR. Generally in use in South Africa.

Importer/s Person(s) or firm(s) engaged in the importation of goods into a country.

Import List or **Inward Manifest** A list of imported goods in alphabetical sequence for use by customs officers.

Imports The importation of goods into a country is virtually essential. Some nations are compelled to import more commodities than they export. There are two types of imports—visible and invisible. Visible imports relate to merchandise of all kinds while invisible imports concern services received and paid for.

Import Shed A government warehouse for storing imported merchandise awaiting re-export or customs clearance for local consumption.

Impossible to Perform This refers to a contract which for some reason or other has to be cancelled due to *force majeure* (circumstances beyond one's control). *Ex.* Declaration of war would prevent a company chartering a ship to sail to a war zone or, due to certain circumstances, a ship becomes prohibited by law from entering a port with her cargo.

Impost General name for import taxes or duties.

Impound (1) To confiscate; to place a suspected document in the custody of the law when it is produced at a trial. (2) Water retained in, or admitted into, an empty dock to bring the water level up to that outside the dock.

Imprest Money advanced to a person to be used in state business.

Imprest Account A provision set aside by a company to cover petty expenses.

Imprimatur *Abbrev.* Imp., *q.v.*

Imprimis *Latin*—In the first place; among the first things.

Improper Navigation Bad navigation of a ship by the servants of the owners.

Improver Apprentice; a person who works for a nominal wage or fee to improve his experience in a trade.

IMPS Institutional Meat Purchase Specifications *USA*; Integrated Marine Propulsion System.

Impunity Exemption from punishment, as when a person gives evidence for the state against a fellow accused. This is known as turning King's/Queen's Evidence.

Impx Imperatrix *Latin*—Empress.

IMRA *or* **I.M.R.A.** Industrial Marketing Research Association.

IMRAMN International Meeting on Radio Aids to Marine Navigation.

IMRAN International Marine Radio Aids to Navigation.

IMRC *or* **I.M.R.C.** International Marine Radio Company.

IMRO Investment Managment Regulatory Organisation *UK financial.*

IMRS Industrial Market Research Services in Thailand.

IMS International Marine Surveyors; International Military Staff *NATO.*

IMS *or* **I.M.S.** International Marine Science; Industrial Management Society; Industrial Methylated Spirit; Institute of Management Services; Institute of Manpower Studies.

IMSA *or* **I.M.S.A.** Institute of Marketing and Sales Association.

IMT Immediate Money Transfer.

IMTA International Marine Transit Association; International Meat Trade Association.

IMU *or* **I.M.U.** International Mathematical Union.

I.Mun.E. Institution of Municipal Engineers.

IMV *or* **I.M.V.** Internal Movement Vehicle.

In. India; Indiana, *USA*; insulated.

In *or* **Inch** Inch.

IN Indonesian Classification Society *symbol.*

INA *or* **I.N.A.** Institute of Nautical Archaeology; Integrated Navigation Aid *USA.*

In absentia *Latin*—In or during the absence.

In and/or Out Insurance abbreviation covering both deck and/or under-deck cargo at the same rate.

In and/or Over *or* **In and/or Out** Shipping term referring to cargo being carried either on or under deck. In the insurance business this phrase means that the cargo thus insured is covered whether shipped on deck or in the Holds, *q.v.*, of a ship.

In articulo mortis *Latin*—At the point of death.

In Ballast *or* **On Ballast** A ship is said to be in or on ballast when she is without cargo and is partly filled with solid or liquid ballast for stability at sea. The bottom or double bottom tanks are filled with either fresh or sea water. There are some modern ships that utilise the ballast thus carried for a dual purpose by using edible oils as cargoes. In bygone days ships used to load solid objects as ballast such as sand, stone, pig iron, soil, etc. *See* **Ballast.**

In Bd. *Similar to* I.B. In Bond, *q.v.*

In bd. Inboard, *q.v.*

In blank Trust, *similar to* Carte Blanche.

Inboard Internal side of the ship's hull as opposed to Outboard, *q.v.*

In Bond Goods liable to duty, which are kept in a government warehouse. Subsequently they are cleared by customs to be used for home consumption, ship's stores or re-exportation. When goods are in bond no duty is paid but there is a nominal fee called store rent. They can also be stored in a Private Bond, *q.v. See* **Bonded Stores** and **Bonded Warehouse.**

Inbond Association of international standing, dealing in bonds.

In Bot. Inner Bottom.

Inbound Freight Merchandise transported on a home-bound ship.

INBUCON *or* **Inbucon** International Business Consultants.

inc Include; Inclusive; Income; Incorporated, *q.v.*; Increase.

Inc. An abbreviation for 'Incorporated'. In the USA it corresponds to the English Ltd. or Limited Liability Company. Incorporated companies are called Corporations.

INCAF International Commission for the North West Atlantic Fisheries.

In camera *Latin*—An expression usually used in legal circles, meaning not in open court; evidence heard in private.

Ince. Insurance.

Incep. Inception.

Inception of the Policy The effective time of the insurance policy or when the insurance attaches.

Incharpass 1967 Charterparty, *q.v.*, form of contract issued by the Institute of Chartered Shipbrokers for the charter of a passenger ship.

Inchmaree Clause Marine Insurance, *q.v.*, clause which covers loss or damage to the hull or machinery of a ship caused during the course of loading and/or discharging operations. It also covers accidents in bunkering, explosions on board or elsewhere, bursting of boilers, breakage of shafts or any latent defect regarding the ship as a whole. The negligence of masters, seamen, engineers or any servants of the shipowner is also included. The name comes from an 1887 court judgment in the case of the *SS Inchmaree*. The clause was also known as the additional perils clause. It has been superceded by the MAR form of policy which incorporated the concept.

Inchoate Bill Incomplete endorsed Cheque which can normally, but not always, be completed by the holder; a blank cheque.

Inch-Trim-Moment The change of one inch immersion in the difference between the forward and aft draught (draft), considering a half-inch increase from one end of the ship against a half-inch decrease at the other end. *Abbrev.* ITM *or* I.T.M.

Incidental Non-Marine Insurance risks which do not fall under the category of marine, though they may be placed in the marine insurance market, such as those referring to riverboats, small craft, barges, piers, bridges, etc. *Abbrev.* INM *or* I.N.M.

Incl. Include; Included; Including.

Inclinometer *See* **Clinometer**.

Incog. Incognito *Latin*—Unknown; Unofficial.

Incognito *Abbrev.* Incog, *q.v. or* Ign, *q.v.*

Income Periodical receipts from one's business, lands, work, investments, etc.

Income and Earned Surplus Statement Double-Entry Book-keeping *USA*.

Income Tax A tax imposed by the government of a country on the income derived by every business or individual residing in that country. Every country has its own particular law as to rates of tax and procedure.

In Commission A phrase referring to a ship fitted out to sail with all tackle and in a seaworthy state. *Opposite to* Out of Commission.

Incompatible Describes fuel oils which cannot satisfactorily be mixed. Also cargoes which cannot be stowed together without risk of contamination, corrosion, explosion or other damage on voyage. *See* **Compatibility**.

Incor. Incorporated, *q.v.*; Incorporation.

Incorporate (1) To combine into one body or to form a legal business association. (2) To admit as a member of an association or corporation. *See* **Incorporated.**

Incorporated (1) A constituted business and legal corporation or association. *See* **Incorporate**. (2) *Abbrev.* Inc. The corresponding English word is Ltd. or Limited Company, *q.v.* Incorporated companies in America are called Corporations. *See* **Incorporate.**

Incorporated Insurance Broker A member of the Corporation of Insurance Brokers.

Incorporation of a Company Formation of a company in the *USA*.

Incrd. Increased.

Incrd. Val. Increased Value.

Incre. Increment *navigation*.

Increment This word is normally used to refer to an increase in wages or salaries.

Increased Value Marine insurance clause securing an additional percentage on a hull policy against loss on anticipated chartered freights, voyage charter hire, disbursements, managers' commissions, etc. Also termed Disbursements or Excess Clause. *Abbrev.* I.V. *or* I/V *or* iv *or* i.v.

Incremental Capital–Ouput Ratio The relative increase in the capital stock of a business over a stated time. *Abbrev.* ICOR.

INCT Informal Composite Negotiation Test *United Nations*.

In curia *Latin*—In open court.

In custodia legis In the hands or custody of the law; seized by the order of law.

Ind. Independent; Index; India; Indiana, *USA*; Indianapolis, *USA*; Indication.

IND *or* **I.N.D.** In nomine Dei *Latin*—In the name of God.

Indebiti solutio *Latin*—Payment of that which is not owed. *See* **Quasi-contract**.

Indemnify To compensate for loss, expenses incurred, etc.

Indemnity Clause (1) A Charterparty, *q.v.*, clause indemnifying the owner of the vessel for any irregularity of the Charterer, *q.v.*, master or officers in signing Bills of Lading, *q.v.*, contracts and other papers on behalf of the charterers, including the agents, but excluding any faults in the navigation where the responsibility lies with the owner and the master. (2) This may also refer to the exoneration of liability of the tug owner in a towage operation. If the tug becomes liable to a third party while towing she is indemnified for any consequent damage.

Indent To request officially; a written order to supply or furnish.

Indenture Any sealed agreement or contract, especially that which binds an apprentice to his master.

Independent Inclusive Tour A holiday package deal involving air travel which also includes accommodation and/or car hire. *Abbrev.* ITX.

Independent Line *Similar to* **Non-Conference Line**.

Independent Wire Rope Code An independent wire rope recommendation table. *Abbrev.* IWRC *or* I.W.R.C.

Ind Est Industrial Estate, *q.v.*

Ind. H.P. Indicated Horse Power, *q.v.*

India International Radio Telephone phonetic alphabet for 'I'.

Indicated Air Speed The speed of an aircraft through the air which is indicated by instruments. *Abbrev.* IAS *or* i.a.s.

Indicated Horse Power The power produced by an engine and shown by an indicator, a recording instrument attached to the engine. *Abbrev.* IHP *or* I.H.P. *or* Ind. H.P.

Indictment A formal accusation; a written statement of a criminal charge.

Indirect Arbitrage Commonly known as 'Compound Arbitrage'. It concerns various offices in different places buying/selling the same speculation. *See* **Arbitrage**.

Indirect Cost The cost of labour, materials and services not immediately related to each unit of production but which is allocated to unit cost or overhead.

Indirect Damage Damage not proximately caused by an insured peril.

Indirect Tax Tax paid by the consumer in the form of increased price for the taxed goods.

Indorse *Similar to* **Endorse**, *q.v.*

Indorsee *Similar to* **Endorsee**, *q.v.*

Indorsement *or* **Endorsement** Signature.

Indorser *Similar to* **Endorser**, *q.v.*

INDPRO Indian Committee/Organisation on Trade Procedures and Facilitation, *New Delhi.*

indre Indenture.

Indty Indemnity.

Industrial Action Cessation or disruption of work in the form of a strike, go-slow or work to rule by employees in pursuance of a dispute or to draw attention to a grievance. *See* **Strike**.

Industrial and Commercial Finance Corporation An organisation of the Bank of England and other UK banks who help small local industries. *Abbrev.* I.C.F.C.

Industrial Bank Institution or a bank dealing in commercial and industrial commitments as to Hire-Purchase, *q.v.*, loans, money transfers, issuing cheques, etc.

Industrial Business Life assurance on which the premiums are paid to collectors weekly or monthly.

Industrial Estate An area set aside with units to accommodate a number of industries or commercial firms. *Abbrev.* Ind Est. Other terms used are Industrial Centre/Zone/Park.

Industrialised Country A country equipped for and mainly occupied, for its prosperity, in manufacture.

Industrial Dispute Disagreement between management and employees which may result in Industrial Action, *q.v.*, or a strike, *q.v.*

Industrial Marketing The promotion and introduction of manufactured articles in both local and overseas markets.

Industrial Spirits Light refined petroleum products, generally solvents or fuels such as white spirit and methylated spirit.

Industrial Union A union representing all sections and grades of workers in industry.

Ined. Ineditus *Latin*—Unpublished.

Inert Gas System A system associated with Crude Oil Washing, *q.v.*, in which a layer of inert (non-explosive) gas is placed on the top of each cargo of crude oil and which gradually replaces the oil as it is pumped away.

Inertial Guidance Automatic gyroscopic guidance system for aircraft and missiles using data computed from acceleration and the physical properties of the earth, dispensing with the magnetic compass and independent of ground-based radio aids. Also termed Inertial Navigation (System).

Inertial Reference System An instrument used in conjunction with automatic flight control in sophisticated airliners. *Abbrev.* IRS.

Inertia Selling Leaving or sending unsolicited goods and attempting to charge for them if they are not returned. Promoting a service which is charged for in the absence of a positive refusal by the potential customer.

Inerting The act of avoiding an explosive atmosphere by replacing the oxygen in an oil tank by an Inert Gas.

In esse *Latin*—In existence; In being.

In ex In extenso *Latin*—At length. *See* **In Extenso**.

In extenso *Latin*—Entirely or extensively. This can refer to the reading of a document throughout its length, leaving nothing out.

Inf. Inferior; Infinitive; Information, Infra *Latin* —Below.

INF Irradiated Nuclear Fuel.

Inferior Good Something which has become obsolete, generally due to an improved standard of living. Air travel has largely replaced sea and train travel; the latter has therefore become 'Inferior Good'.

In flagrante delicto *Latin*—In the very act of committing a crime.

Inflammable Liquids *or* **Flammable Liquids** Liquids liable to spontaneous combustion which give off inflammable vapours at or below 80°F. These come under the category of Hazardous Commodities. Here are some inflammable substances: benzine, benzole, carbon disulphide, ether (sulphuric), ethyl (acetate), gasolene, lythene, motor spirit, naptha, nickel (carbonyl), coaltar paints, enamels, photographic printing inks, tar oil compounds, kerosene, mineral oils, etc. *See* **Dangerous Liquids** *and* **Blue Book**.

Inflatable Liferaft Modern apparatus which is immediately handy for saving life at sea in case a vessel in distress is abandoned. The raft is stowed as a folded package until released overboard when on hitting the water it will inflate itself. It contains lifeboat rations for each person according to its capacity as well as Pyrotechnics, *q.v.*, for attracting attention and a VHF transmitter or EPIRB, *q.v.*

Inflation A persistent increase in the prices of food and services generally over time.

Inflight Catering The furnishing of provisions for passengers and crew during flight.

Info Information.

Information Desk A counter at an airport, seaport or railway station where travel details can be obtained.

Infra *Latin*—Further below, beneath, after, on, in a page or a book. *Abbrev.* If *or* Inf, *q.v.*

Infra Dig. Infra Dignitatem *Latin*—Beneath one's dignity; Undignified.

Infra Dignitatem *Abbrev.* Infra Dig., *q.v.*

Infrastructure The basic items of a country with reference to roads, transport, communications, and public services such as electricity, water supply, etc. Sometimes known as Social Overhead Capital.

In futuru *Latin*—In or for the future.

Ing. Ingenieur *German*—Engineer; Ingegnere *Italian*—Engineer.

In Geographical Rotation Voyage Charterparty clause signifying the direction of a vessel's loading and/or discharging ports in a regular sequence and without deviation from her course. Any deviation expenses will be for the account of the Charterers, *q.v.*

Ingl. Inghilterra *Italian*—England.

INGO *or* **I.N.G.O.** International Non-Governmental Organisation.

In God we Trust Motto of the United States of America.

Ingot A mass, usually oblong, of cast metal, especially of gold, platinum, silver, copper, etc.

Inh. Inhaber *German*—Proprietor.

In illo tempore *Latin*—At that time.

in/in Inch per inch.

in init In initio *Latin*—In the beginning.

In irons Describes a sailing vessel which, in the course of tacking or through a shift in the wind, is unable to pay off and fill its sails on either tack. *Also termed* In Stays.

init Initio *Latin*—Starting; beginning.

In integrum *Latin*—In the original condition; to bring back to its original state.

Injunction A court's order to a person or firm to do something, known as a mandatory injunction, or to refrain from doing something, known as a restrictive injunction.

Injuria *Latin*—A legal wrong; A wrong act.

Injuria sine damno *Latin*—A legal wrong without consequent or actual damage or loss to the person who is aggrieved.

In Kind Type of remuneration for services rendered in the form of acceptable articles instead of cash.

Inkl Inklusiv *German*—Inclusive.

Inland Bill of Exchange A local Bill of Exchange, *q.v.*, only drawn or paid in that same country. Sometimes called Domestic Bill of Exchange.

Inland Clearance Depot A shipping depot conveniently positioned near to an industrial and commercial area, within reasonable reach of the sea and airports. Since maintenance charges would run to some considerable expense, most of these depots are operated by a Consortium, *q.v.*

Inland Marine Marine Insurance covering various craft and barges navigating rivers and canals as well as wharves, piers, bridges and other structures so situated.

Inland Waters Marine Insurance, *q.v.*, term referring to lakes, streams, rivers, canals, waterways, inlets, bays and the like.

Inland Waterway Water inland which is navigable by vessels which could be a river, canal, lake or tidal inlet.

Inland Water Vessels Ships authorised to navigate in restricted areas and not required to have a load line certificate.

In lieu of weight 2% This expression is occasionally inserted in coal charterparties. The Carriers or owners accept payment of freight at destination on ascertaining the exact weight, less the agreed percentage of say 2%. If a shortage is found the carriers will not be liable unless the master or officers declare that the amount discharged was actually loaded. Freight is adjusted accordingly at the port of destination.

In liq *or* **liqn.** In liquidation.

In lim In limine *Latin*—At the outset.

In loc In loco *Latin*—In its place.

In loco *Latin*—In place; In proper place. *Abbrev.* In loc.

In loco citato *Latin*—Summoned.

In loco parentis *Latin*—In the place of a parent.

INM *or* **I.N.M.** Incidental Non Marine, *q.v.*; International Nautical Mile, *q.v.*; Integrated Noise Model, for calculating noise exposure in airport operations.

INMARSAT International Mobile Satellite Organization; originally International Maritime Satellite Organization.

In medias res *Latin*—In the midst of things.

Inner Bottom The Tank Top, *q.v.*, of a ship.

Inner Harbour Innermost part of a port area which is more or less completely protected from the sea.

Innocent Misrepresentation. *See* **Misrepresentation**.

Innocent Passage *Similar to* Territorial Waters, *q.v.*

Innominate Applies to an undertaking or an obligation under a contract when there is doubt whether in law it is a warranty or a condition.

In nomine *Latin*—In the name of.

In omnia paratus *Latin*—Prepared for all things.

Inops consilii *Latin*—Without advice or suggestion.

In and/or Out (1) Goods that are carried below or above deck of a cargo ship. (2) Insurance abbreviation referring to the coverage of both on-deck and under-deck cargo.

Inp Input.

In pais *Latin*—By conduct.

In pari delicto *Latin*—Equally guilty. In an illegal contract both parties are equally guilty.

In pectore *Latin*—In the soul; Secretly.

In perpetuum *Latin*—For ever.

In personam *Latin*—A remedy against the person only. A legal action against an individual.

In posse *Latin*—Possibility of being; *Opposite to* **In esse**, *q.v.*

In praesenti *Latin*—At the moment; At the present; Presently.

In primis *Latin*—In the first place.

In principio *Latin*—In the beginning. *Abbrev.* In pr.

In pro. In proportion.

In propria persona *Latin*—In his own presence.

Inputs The goods and service involved in the outcome of the production.

Inquest (1) Inquiry into cause of some mishap or other action with a view to be rectified. (2) Judicial inquiry before a jury to ascertain facts.

In re *Latin*—Regarding; In the matter of.

In Regular Turn *Similar to* Berth Terms, *q.v.*

In rem In the civil law, the expressions *in rem* and *in personam* were always opposed to one another, an act or proceeding *in personam* being one done or directed against or with reference to a specific person, while an act or proceeding *in rem* was one done or directed with reference to no specific person, and consequently against or with reference to all whom it might concern, or all the world. The phrases were especially applied to actions, an *actio in personam* being the remedy where a claim against a specific person arose out of an obligation, whether *ex contractu* or *ex maleficio*, while an *actio in rem* was one brought for the assertion of a right of property, easement, status, etc., against one who denied or infringed it. In maritime circles an action *in rem* is a legal action against an object or the owners thereof (e.g. an action naming the ship).

In rerum natura *Latin*—In the nature of things.

In Rev. International Revenue.

INRI *or* **I.N.R.I.** Jesus Nazarenus Rex Judaeorum *Latin*—Jesus of Nazareth King of the Jews.

Ins. Insurance; Inspector.

in s. In situ *Latin*—In its (original) place.

INS Inertial Navigation System, *q.v.*; Institutional Net Settlement *stock exchange service.*

INSA Indian National Shipowners' Association in Bombay; International Shipowners Association in Gdynia.

In saecula saeculorum *Latin*—For ever and ever.

Insce Insurance, *q.v.*

Inscrutable Fault When the fault in a collision incident cannot be decided.

INSEAN Instituto Nazionale Studi e Esperienze Architettura Navale. *Italian*—National Institute of Naval Architectural Studies and Experiences.

Inscribed Stocks *or* **Inscribed Shares** Generally refers to government stocks and shares where no certificates are given, but are inscribed in the records. Therefore any transfers which may be needed have to undergo formal procedures before this act is confirmed. *See* **Registered Stock**.

In/sec Inches per second.

Ins. Gen. Inspector General.

Inshore In or in the vicinity of the shore.

Inshore Navigation Navigating within easy reach of the coastline.

In situ *Abbrev.* in s., *q.v.*

INSMAT Inspector of Naval Material *USA.*

Insolv. Insolvent, *q.v.*

Insolvency Bankruptcy.

Insolvent Insufficient money or credit to pay debts. *Abbrev.* Insolv.

In Soundings When a vessel is navigating in waters under 100 fathoms (600 feet) deep. *Opposite to* **Off Soundings**, *q.v.*

Insp. Inspector. *See* **Superintendent**.

Inst. Instance: Instant; Institute; Institution.

Installation Floater An American Insurance Policy, *q.v.*, covering machinery, equipment or structures in transit to a site where installation or construction is to be carried out. It may cover a period of time.

Instalment Part payment. *See* **Hire-Purchase**.

In Standard Atmosphere Atmospheric conditions for maximum payload take-off of an aircraft. *Abbrev.* ISA *or* I.S.A.

Instant In commercial correspondence this word means the same month of the year. *Abbrev.* Inst.

Instanter *Latin*—The act to be done presently or immediately. The act to be done within 24 hours.

Instant Purchase Excursion An international air ticket purchased at any time, generally valid for three months. The stay must include a Saturday night at the place of destination *Abbrev.* IPEX.

In statu quo *Latin*—In the same state as before.

In Stays *Similar to* **In Irons**, *q.v.*

Inst Cls. Institute Clauses, *q.v.*

Institute Cargo Clauses Clauses drafted by Marine Underwriters, *q.v.*, in connection with Marine Insurance, *q.v.*, on merchandise. *Abbrev.* ICC *or* I.C.C.

Institute Clauses Clauses in Marine Insurance, *q.v.*, which are drafted by Underwriters, *q.v.*, in connection with marine insurance on merchandise. *Abbrev.* IC *or* I.C. *See* **Institute Warranties.**

Institute Clauses for Air Sendings (All Risks) The aviation insurance coverage of cargo against all risks. *Similar to* the Institute Cargo Clauses, *q.v.*, (All Risks) by sea. These air cargo clauses, as they are also termed, have been authorised and issued since 1965. Merchandise by air postage is not covered. *Abbrev.* ICAS *or* I.C.A.S.

Institute of Chartered Shipbrokers, The This is the professional body for those engaged in shipping business which covers chartering, ship management, sale and purchase, liner trades and port agency. It was founded in 1911 for those engaged in fixing ships and cargoes. In 1920, a Royal Charter was granted and in 1984 membership was opened to all nationalities. Membership (MICS) is achieved through qualifying examinations, which are held annually, Fellowship (FICS) for those with appropriate seniority and influence in their professions involves a member also becoming a Chartered Shipbroker. *Abbrev.* ICS.

Institute of Marine Engineers (UK) The membership of this Institute includes all ranks of ship engineers from chief to cadet. Membership of IMarE is also open to any branch of the engineering profession which has an application in the design, building or operation of ships, or their machinery or contributes in any way to marine technology. According to their qualifications and responsibilities, officers may be admitted as 'Fellow', 'Member', 'Associate Member' or 'Associate'. As well as seeking to bridge the link between the various backgrounds in engineering, the Institute aims to provide a professional service to the membership. The Institute produces a wide range of technical publications covering a comprehensive framework of subjects related to engineering such as types of engine, ship and naval architecture, marine boilers, fire fighting equipment in ships, automation, steering gear, chemicals in ships, business management and various other similarly related subjects.

IMarE sponsored conferences and seminars are often held jointly with other professional Institutions, particularly the Nautical Institute, as well as such bodies as the Department of Transport or Merchant Navy Training Board. IMarE members are eligible for registration with the Engineers Registration Board according to certification, some chief engineers qualifying for 'C.Eng' registration.

Institute of Strategic Studies A London based institute internationally acknowledged regarding information on military matters throughout the world.

Institute Replacement Clause An insurance clause. In case of damage or loss to any part or parts of the insured machinery the amount recoverable from Underwriters, *q.v.*, will not exceed the cost of repairs or replacement of parts and the forwarding expenses, including the refitting of the parts. Duty involved will also be included in the recovery. *Abbrev.* I. Rep. C.

Institute Time Clauses Marine Insurance, *q.v.*, clauses covering the hull and cargo of a ship. The clauses are altered from time to time according to the exigencies then prevailing. *Abbrev.* ITC *or* I.T.C.

Institute Time Clauses, Hulls The standard United Kingdom Market clauses for the insurance on ships. These have been subject to regular revision over the years in order to keep pace with the changing requirements of the world's insurance markets and shipowning communities.

Institute Warranties A set of express warranties for use in policies covering ships. Mainly these are navigational warranties restricting the ship's navigational areas. Breach of the warranties is held covered subject to payment of an additional premium and change of policy conditions, if required by the underwriters. *Abbrev.* IW *or* I.W. *or* Inst. Wties.

Institutional Investors Large organisations such as banks, trusts, insurance companies or manufacturers who invest their income, fund contributions, premiums or money deposits in government stocks with a view to a profitable and secured investment.

Inst. N. Institute of Navigation.

Inst. T. Institute of Transport.

Inst. War etc. Institute War Clauses *insurance.*

Inst. Wties Institute Warranties.

Insul. Insulation.

Insulated and Refrigerated As the phrase itself implies it is used for provisions needing low temperatures in order to preserve them.

In sum. In summary.

Insur. Insurance, *q.v.*

Insurable Interest The interest one has in relation to property exposed to peril whereby one may lose financially by the loss of, or damage to, such property or may incur a liability in respect thereof. A person who effects a marine insurance contract without an insurable or a reasonable expectation of acquiring such interest is guilty of an offence under English law.

Insurable Value Insurance term indicating the value of the insured object including the premium of the insurance. (1) The value of a ship at the commencement of the voyage including all the expenses incurred to make her seaworthy for the voyage(s). (2) The cost of the goods when the risk commences plus delivery and shipping expenses as well as the insurance premiums. (3). The amount of freight payable in advance or at destination plus Insurance Premiums, *q.v.*

Insurance *or* **Assurance** The act of making good in money for the insurable object which has suffered a loss. The insurers or Underwriters, *q.v.*, undertake to pay relative compensation in the form of money in respect of any part or total loss of the insured object in consideration of the payment of a sum of money called a Premium, *q.v.* There are many types of insurance although the main warranties and conditions remain, but always subject to the contract of agreement known as the Policy, *q.v.* The owner of the policy is called the policyholder who is to present it if and when he submits any claims. In the group insurances (*see* **Life Insurance**) the holders of the insurance policies are called Certificate Holders. There are a wide variety of insurances in general, some of which are: Accidents, Credit, Death, Engineering, Marine, Motor Vehicle, Yacht, Personal Valuables, Riot and Civil Commotion, etc. The main two Implied Warranties, *q.v.*, are that (a) the person has to have an Insurable Interest, *q.v.*, in the matter and that (b) information or particulars given to the underwriters must be in the 'utmost good faith' based on the principle of Uberrimae Fidei, *q.v.* Failing one or both of these two implied warranties will render the contract of insurance null and void. *Abbrev.* Ins. *or* Inse.

Insurance Broker An agent or intermediary between the assured client and the Underwriter, *q.v.*, working to conclude a contract of insurance on ships, freight, merchandise, etc.

Insurance, National Insurance organised by the government of a state for the welfare of the inhabitants. National Insurance methods vary in accordance with the standard of living of each country.

Insurance Policy A document in the form of a contract whereby the Insurers, *q.v.*, agree to indemnify the assured against named risks while exempting themselves from exposure to certain events. On every claim these exemption clauses in the policy are taken into account.

Insurance Premium The sum of money which is due to the insurer or Underwriter, *q.v.*, by the Insured, *q.v.*, for the insurance Policy, *q.v.*, effected. The insurance premium is generally paid in advance annually, but there are occasions when by agreement the premium is paid half yearly, quarterly or monthly. *See* **Premium**.

Insurance Slip *See* **Original Slip** and **Held Covered**.

Insured *or* **Assured** The person who is covered or compensated for the loss or damage.

Insurers Those companies or associations or Underwriters, *q.v.*, that engage in a business to make good or Indemnify, *q.v.*, any damage that might occur in a risk or from an accident. There are many types of insurers which may be classified as follows: Lloyd's Underwriters; Proprietary Companies; Mutual Clubs or Companies; Collecting Friendly Societies; Self Insurers; Mutual Indemnity Associations.

Ins Val. Insured Value.

Int Intercept *navigation*: Interest, *q.v.*; Internal; Intermittent; Interval.

INT *or* **I.N.T.** International Charts, *q.v.*; Isaac Newton Telescope.

Intake Measure In Pitch Pine Charterparties coded Pixpinus, *q.v.*, freight on all deals, battens, boards, scantlings and ends is charged in full as customary irrespective of the woods.

Int. al. Inter alia *Latin*—Among other things.

In tandem One behind the other. Often refers to towage in tandem.

Intangible Assets Assets which have no cash value or those which are more or less of no use when a business becomes insolvent. *Ex.* Goodwill, short lease, trademarks, etc.

INTASAFCON *or* **Intasafcon** International Tanker Safety Conference.

INTASCALE International Tanker Nominal Freight Scale Association Ltd. (1962). *See* **American Tanker Rate**.

Intcl. Inter Coastal.

Int. Corps Intelligence Corps.

Integral Self-contained refrigerated container.

Integrated Air System System enabling oil engines using heavy fuel oil to economise on consumption. *Abbrev.* IAS.

Integrated Navigation Systems Combined navigational method having all electronic instruments incorporated in one single automatic navigation system. These may include radar, decca, transit satellite, automatic pilot, etc. *Abbrev.* INS.

Integrated Ship Design and Production System of computer graphics and calculations for shipbuilding. *Abbrev.* ISDP.

Integrated Tug Barge A large barge of about 600 ft. and 22,000 tons cargo capacity, integrated from the rear on to the bow of a tug purposely constructed to push the barge. The barge has a self-unloading system to transfer cargoes on to barges and/or quays at an average rate of 1,400 tons per hour. The cargo can be operated simultaneously from both sides of the ship. This is somewhat similar to the Artubar or Ocean Going Barge or Tug Combinations or Sea Link System. *Abbrev.* ITB.

Intellectual Property Owners A non-profit association representing holders of US Patents, Trademarks, *q.v.*, and Copyrights, *q.v.*

Intelsat International Telecommunications Satellite Consortium, 1971.

In tempo utile *Italian*—At the appropriate time.

Intention to Proceed A phrase in regard to the anticipated decision taken in the execution of an action, such as the intention of a state to buy an aircraft or ship.

Inter *Latin*—Between; Amid.

Inter alia *Abbrev.* int.al. *q.v.*

Inter. Arr. Internal Arrangements.

INTERBANK International Bank for Reconstruction and Development *Washington*. Commonly known as the World Bank.

Intercardinal Points Referring to North East, North West, South East and South West.

INTERCARGO The International Association of Dry Cargo Shipowners, *q.v.*

Interchange Container A container of universal dimensions that allow it to be transferred between aircraft.

Interchange Report Standard container document signed by two parties concerned in the shipment of cargo containers in relation to the outside condition or survey of the containers.

INTERCOA 80 Code name for INTERTANKO Charterparty for tanker chartering.

INTERCONSEC 76 Code name for INTER-TANKO Voyage Charterparty, *q.v.*, on consecutive voyages.

Inter-Departmental Flexibility (1) A method in the feasibility of the services of deck and engine room officers and ratings of doing interchangeable work where necessary. (2) A rating who agrees to work in a different department during a heavy workload period. *Abbrev.* I.D.F.

Interest The remuneration given for the amount of money deposited with a bank or other business concern. Interest is generally calculated every six months or annually on a percentage on the money deposited. *Abbrev.* Int.

Interest *or* **No Interest** Also called Policy Proof of Interest, *q.v.* A policy wherein the underwriter agrees to waive proof that an insurable interest is enjoyed by the assured as a condition of claim payment. In other policies the underwriter is not liable for any claim where the assured is unable to prove that his interest in the subject-matter of the insurance exists at the time of loss. P.P.I. policies are invalid in a court of law, but are not illegal except where no interest exists or where there was no reasonable expectation that it would exist at the time the policy was effected. Also called Honour Policy, *q.v.*

Interest Policies Ancillary interests appertaining to the assured in the insurance policy. *Ex.* A shipowner has an insurance interest of the full value of the vessel plus the cargo carried as well as disbursements, freights, etc.

Interest Rate The rate of interest.

Interest Warrants Payment advices of Stocks and Shares which are settled at intervals.

Interferometer An instrument to measure length by means of light waves in electricity.

INTERFUND International Monetary Fund in Washington connected with INTERBANK, *q.v.*

Interim *Latin*—In the meantime; Meanwhile, Temporary.

Interim Accounts General half-yearly unaudited account presented by a company.

Interim Certificate of Class *or* **Interim Class Certificate** Certificate issued by a Classification Society, *q.v.*, to a ship registered with it, after satisfactory repairs are effected as per recommendation of the surveyor of the society. This Interim Class Certificate will allow the ship to remain in her original class.

Interim Dividends Dividends issued by instalments before the final Dividend, *q.v.*

Interline Agreement Arrangement between airlines to exchange passenger bookings and aircraft routes.

Interlineation A correction or writing between two lines in a document which according to law will have to be initialled by the two parties signing the contract.

Interlkd. Interlocked.

Interlocking Directorates The appointment of individuals as directors in two or more separate companies.

Interlocutory A detailed written document regarding judgment or determination of a court case.

Intermediary Acting between two parties or persons. Acting as a Broker, *q.v.*, or an Agent, *q.v.*

Intermediate Goods *or* **Intermediate Products** Nearly finished goods or goods partly in a raw state.

Intermediate Port A loading and discharging port situated between the initial port of departure and the final destination.

Intermediate Shaft A shaft between and connecting the inner and outer propeller shafts of a ship.

Intermodal Pertaining to through transportation of unit loads using a combination of two or more land, sea or air systems.

Intermodal Container Container capable of transfer between systems and being carried by a combination of any land, sea or air modes of transport.

Intern. International.

Internal Debt The national debt of a country towards its inhabitants for the issue of bonds, war loans, etc.

International Air Transport Association An international organisation founded by about six airlines at a meeting at The Hague on 28 August 1919, when air transportation was in its initial stages. By 1930 the association had increased to some 23 members consisting of nearly all the European airlines. After the 1939-45 world war the membership went up to over 40 companies and now most if not all the international airlines are a party to it.

International Association of Classification Societies Eleven leading Classification Societies, *q.v.* (1994):

American Bureau of Shipping	*Abbrev.* ABS
Bureau Veritas	*Abbrev.* BV
China Classification Society	*Abbrev.* CCS
Det Norske Veritas	*Abbrev.* DNV
Germanischer Lloyd	*Abbrev.* GL
Korean Register of Shipping	*Abbrev.* KR
Lloyd's Register of Shipping	*Abbrev.* LR
Nippon Kaiji Kyokai	*Abbrev.* NK
Polski Rejestr Statkow	*Abbrev.* PRS
Registro Italiano Navale	*Abbrev.* RINA
Registry of Shipping (Russia)	*Abbrev.* RS

and there are two Associates:

| Croatian Register of Shipping | *Abbrev.* CRS |
| Indian Register of Shipping | *Abbrev.* IRS |

In addition to Classification work, the members undertake Statutory Certification connected with international conventions on behalf of 100 national governments. Classification Surveys over 40,000 ships, 90% of the world's tonnage. Several hundred thousand periodic surveys are performed by 5000 surveyors each year. *Abbrev.* IACS. The IACS Council appoints a permanent representative to the International Maritime Organisation, *q.v.* See **Classification Societies** *and* **Lloyd's Register of Shipping**.

International Association of Dry Cargo Shipowners, The Commonly known as INTER-CARGO, it was established in 1980 and is devoted to the interests of the dry bulk shipping industry, with 160 members (1994) in 28 countries controlling over a thousand large bulk carriers with a deadweight total of 70 million tonnes. Associates are from many shipping and related financial professions. It holds its sessions at main shipping centres around the world, and has consultative status with IMO and UNCTAD. Apart from being a forum and focus for the main concerns in the dry bulk trade and maritime safety, it addresses issues such as insurance, charterparties, national and international legislation and freight payments.

International Association of Maritime Institutions Formed in 1993 by a merger of the International Association of Navigation Schools and the Association of Marine Engineering Schools.

International Bank *Similar to* World Bank, *q.v.*

International Bank for Reconstruction and Development An international organisation established just after the 1939-45 World War for the reconstruction and development of war-stricken countries and most particularly for those still underdeveloped. The World Bank, also known as the International Bank, has had a tangible interest on the financial side. Some of the principal aims of the bank are to improve the standard of living of underdeveloped countries by offering financial help for reconstruction, to promote investments and also to advance loans if necessary by way of guarantees. *See*

International Finance Corporation *also* **World Bank.**

International Chamber of Shipping *Abbrev.* ICS. The shipping industry's principal international non-governmental organisation, founded in 1921 and assuming its present title in 1948. Half the world's tonnage is represented by its membership from over 30 national shipowners' associations. It promotes the shipowners' interests in technical and legal concerns as reflected in its structure of Insurance, Marine, Maritime Law and Trade Procedures Committees. The Marine Committee has separate sub-committees on chemical carriers, construction and equipment, container, gas carriers, oil tanker, radio and nautical affairs, and panels concerned with the Panama and Suez Canals. The Trade Procedures Committee has a sub-committee on EDI. The Chamber is very active in safety management and environmental protection matters and the advancement of maritime technology. It co-operates with the International Shipowners' Federation, with which it shares a secretariat, MARISEC, *q.v.*, and has consultative status with the International Maritime Organisation (IMO). The ICS is based in London.

International Charts The use and issue of charts by authorities in accordance with the systems agreed upon by the International Hydrographic Organisation. *Abbrev.* IC *or* I.C.

International Civil Aviation Organisation An international organisation established in 1944 regulating air safety, efficiency and discipline. Most of the civil or state-owned airlines are members and contribute to its active technological and professional experience. It is one of the United Nations specialised agencies, being brought by time and technology increasingly closer to its kindred agencies in the realm of sea-borne trade and transport. The headquarters are in Montreal. *Abbrev.* ICAO.

International Code of Signals A book defining methods of signalling at sea and abbreviated codes with their translations, as well as the shapes and colours of flags and pennants with their meanings. The first code used by several nations was Captain Marryat's of 1817. Others followed until that of the British Board of Trade in 1867. This was revised, agreed by a conference of maritime powers in Washington in 1889, published in 1897, and translated into several languages. It failed, however, a severe testing in the First World War and another conference was held in 1927 when major changes were agreed. A new book in two volumes, including radio signals, was published in 1932 and came into force in 1934. In 1959, responsibility was assumed by IMCO (now IMO), *q.v.* which completed a revision in nine languages in 1964. The resulting single volume code was adopted in 1965 and became effective in 1969. *Abbrev.* ICS.

International Commission of Jurists A Swiss based International Commission at Geneva comprising an independent group of lawyers campaigning for the rule of law throughout the world. *Abbrev.* ICJ or I.C.J.

International Committee on Seafarers' Welfare *Abbrev.* ICSW. The body co-ordinating welfare activities at international level which meets at the ILO offices in Geneva. Its membership includes major societies providing port welfare, government agencies, the ITF and the ISF. It is concerned with recreation, counselling, and assisting seafarers who are left stranded or in difficulties.

International Court of Justice A body created by the Charter of the United Nations consisting of 15 members of whom nine members form a quorum. Members of and suitors to the court have diplomatic immunity. It is situated at the Peace Palace, The Hague, and is an important component in the UN system. A variety of maritime cases have been before it at one time or another. *Abbrev.* ICJ or I.C.J.

International Criminal Police Organisation. *Abbreviated* ICPO or I.C.P.O. and identical to Interpol, *q.v.*

International Date Line An imaginary line passing from North to South of the world in the Pacific Ocean on meridian 180 degrees from Greenwich. When a ship or aircraft crosses this imaginary line to the West a day is gained. Conversely, a day is lost when crossing to the East. *Abbrev.* IDL or I.D.L. *See* **Date Line**.

International Federation for the Application of Standards An international forum established in 1974 for the mutual exchange of information regarding standardisation. *Abbrev.* IFAN or I.F.A.N.

International Flyte Pack *or* **Flyte Pack** *or* **Flight Pack** A fast airport-to-airport delivery service of small packets or letters which are normally enclosed in a brief-case measuring about 15 inches by 12 inches by 3 inches. There are certain Flight Pack operators that guarantee the period of time taken from the point of departure to destination. *Abbrev.* IFP or I.F.P.

International Finance Corporation An international finance institution formed in July 1956, very closely connected with the World Bank, most especially with that section of the International Bank for Reconstruction and Development, *q.v.* The aim is to increase development of productivity in its member countries' enterprises without the involvement of guarantees by their respective governments. Affiliation is open to all members of the World Bank. *Abbrev.* IFC *or* I.F.C.

International Labour Organisation *Abbrev.* ILO. Founded in 1919, and based in Geneva, it is now one of the oldest United Nations agencies with over 160 member states. Its objects are social justice for workers in all fields, and their protection from disease and injury; adequate wages; prevention of unemployment;

and the protection of children, young persons and women. Its annual General Conference is attended by representatives of governments and of employers' and workers' organisations. At longer intervals, it holds special maritime sessions. In the intervals, there are meetings of the Joint Maritime Commission (JMC) of members from the shipowners' and seafarers' sides organised by the International Shipping Federation (ISF) and the International Transport Workers' Federation (ITF). There have been a number of ILO conventions covering wages, manning, recruitment and shipboard conditions.

International Maritime Employers' Federation Formed in 1993 to represent 16 shipowners and shipmanagers, employing over 15,000 seafarers of 28 nationalities and operating ships in 20 countries. Concerned with wages, conditions, labour developments, industrial relations and the promotion of good employment practice and high training standards. It replaces the former London Committee—Asean Seaman. *Abbrev.* IMEF.

International Maritime Law Also known as International Law of the Sea. The uniform systems and rules covered by this law are acknowledged and followed carefully by most nations. *Abbrev.* IML or I.M.L.

International Maritime Organization *Abbrev.* IMO. Recognition of the need to improve safety at sea led to the United Nations Maritime Conference in 1948, which adopted a Convention to establish the Inter-Governmental Maritime Consultative Committee (IMCO) as a permanent body devoted exclusively to maritime matters and to co-ordinate action by nations. IMCO came into existence in 1958 (and in 1982 became IMO) and started work in 1959, taking over several Conventions already developed, including those for the Safety of Life at Sea of 1948 and the Prevention of the Pollution of the Sea by Oil of 1954. These two subjects have been primary objectives of IMO, though it is empowered to develop new conventions, and has done, as the need arises. The organisation is based in London with nearly 300 staff under the Secretary-General. It is governed by the Assembly, representing some 140 member states which meets every two years. 32 member governments form a Council, which governs IMO between sessions of the Assembly. The technical work is pursued by the Marine Safety Committee (MSC), the Marine Environment Protection Committee (MEPC), the Legal Committee, the Technical Co-operation Committee and the Facilitation Committee. Each has a number of sub-committees. The committees are composed of representatives from member states.

IMO's purpose is to determine acceptable standards, and to develop international treaties relating to shipping, monitoring their implementation by governments, and keeping them up to date in line with advances in technology. The treaties take the form of conventions, protocols thereto and, in some cases, agreements. Any member may put forward a proposal for an international treaty and this is referred by the Assembly or the Council to the appropriate committee. The committee, using its machinery for meetings, working groups, and consultations with expert organisations, prepares a draft convention which is submitted to a conference of all UN member states and relevant UN agencies. When the conference adopts a convention, governments are invited to ratify it within a specified time. Ratification imposes an obligation on a nation to be bound by the convention and to take national legislative action accordingly. When a specified number of countries have ratified the instrument, and in some cases if this represents a certain percentage of world tonnage, it enters into force from a date which allows sufficient time for measures to be taken to give it effect.

Additionally, the Assembly, or sometimes the Council, adopts codes and recommendations by resolution. These cover a wide variety of subjects, mainly technical and related to safety or the prevention of pollution. These are advisory rather than mandatory, in that governments are not bound by them. They are, however, very important and in practice most countries implement them and incorporate them in national law.

The first conference in 1960 adopted a new International Convention on the Safety of Life at Sea (SOLAS) which came into force in 1965. A new International Convention for the Prevention of Pollution from Ships was adopted in 1973 (later incorporated within the Convention of 1978 and known as MARPOL 73/78). These were the most important of the early achievements of IMCO/IMO. By 1994, IMO had adopted some 40 conventions and protocols and hundreds of codes and recommendations.

IMO CONVENTIONS

Maritime safety:
International Convention for the Safety of Life at Sea (SOLAS), 1960 and 1974
International Convention on Load Lines (LL), 1966
Special Trade Passenger Ships Agreement (STP), 1971
Convention on the International Regulations for Preventing Collisions at Sea (COLREG), 1972
International Convention for Safe Containers (CSC), 1972
Convention on the International Maritime Satellite Organization (INMARSAT), 1976
The Torremolinos International Convention for the Safety of Fishing Vessels (SFV), 1977

International Convention on Standards of Training, Certification and Watchkeeping for Seafarers (STCW), 1978
International Convention on Maritime Search and Rescue (SAR), 1979
Marine pullution:
International Convention for the Prevention of Pullution of the Sea by Oil (OILPOL), 1954
Convention on the Prevention of Marine Pollution by Dumping of Wastes and Other Matter (LC), 1972
International Convention for the Prevention of Pollution from Ships, 1973, as modified by the Protocol of 1978 relating thereto (MARPOL 73/78)
International Convention Relating to Intervention on the High Seas in Cases of Oil Pollution Casualties (INTERVENTION), 1969
International Convention on Oil Pollution Preparedness, Response and Co-operation (OPRC), 1990
Liability and compensation:
International Convention on Civil Liability for Oil Pollution Damage (CLC), 1969
International Convention on the Establishment of an International Fund for Compensation for Oil Pollution Damage (FUND), 1971
Convention relating to Civil Liability in the Field of Maritime Carriage of Nuclear Materials (NUCLEAR), 1971
Athens Convention relating to the Carriage of Passengers and their Luggage by Sea (PAL), 1974
Convention on Limitation of Liability for Maritime Claims (LLMC), 1976
Other subjects:
Convention on Facilitation of International Maritime Traffic (FAL), 1965
International Convention on Tonnage Measurement of Ships (TONNAGE), 1969
Convention for the Suppression of Unlawful Acts Against the Safety of Maritime Navigation (SUA), 1988
Protocol for the Suppression of Unlawful Acts Against the Safety of Fixed Platforms Located on the Continental Shelf (SUAPROT), 1988
International Convention on Salvage (SALVAGE), 1989
See also **World Maritime University**.

International Mobile Satellite Organization
Abrev. INMARSAT. From 1966, the International Maritime Organization (IMO)—IMCO as it then was—began to consider the use of space technology to overcome the congestion on maritime radiocommunication frequencies and in 1975 a conference was convened with the aim of establishing a system using satellites. A Convention and an Operating Agreement were adopted the following year. INMARSAT was

established in 1979 with headquarters in London to serve the maritime community with improved distress, safety and commercial communications. It began operating in 1981. Amendments to the Convention in 1985 and 1989 extended operations to serve aircraft and land-based vehicles. The organization comprises three bodies. The Assembly is composed of representatives of member countries, expanded to 75 countries (1994) and meets every two years. The Council, which meets at least three times a year, represents the 18 largest investor countries, with four others to ensure a geographical spread. The Directorate of 500 permanent staff carries out the day-to-day tasks.

INMARSAT has evolved as the only provider of global mobile satellite communications for commercial and distress and safety applications at sea, in the air and on land. Its network supports telephone, telex, facsimile, electronic mail and data connections for maritime use; flight-deck voice and data, automatic position and status reporting, and passenger telephones for aircraft; and two-way data communications, position reporting, electronic mail and fleet management for land transport. It is also used in cases of human and natural disasters. A number of systems were operated as of 1993. INMARSAT A, the original system first used in the mid-1970s, links some 23,000 ship-borne and transportable land terminals with their associated antennae via satellite to 40 land earth stations (LES), also called coast earth stations (CES), for connection to the telecommunications networks. INMARSAT B, the eventual replacement for INMARSAT A, was introduced in 1993 to provide a wider range of services taking advantage of digital technology and including live video and teleconferencing facilities using 13 LESs, while its terminals can be of suitcase size. INMARSAT M provides voice, facsimile and data communications with briefcase terminals for individuals on the move. It uses 13 LESs. INMARSAT C, available from 1991, uses the world's smallest terminals for data and text messaging with modem facility via 24 LESs and is installed in thousands of fishing vessels and small craft. INMARSAT-AERO is a range of three systems to serve aviation operations and flight-deck communications with aircraft anywhere in the world via 10 LESs, called ground earth stations (GES). The A, B and C systems are approved for use in the GMDSS, *q.v.*, and to pass Marine Safety Information (MSI). INMARSAT services are also being extended to Air Traffic Control and GPS, *q.v.* The system depends on four owned second series satellites stationed over the West Atlantic, Eastern Atlantic, Indian and Pacific Oceans, backed up by leased capacity on several other spacecraft. INMARSAT plans the launch of four larger third-generation satellites.

International Monetary Fund An international finance organisation of the United Nations. Its main purpose is to stabilise international money exchange rates of all its member nations in relation to their financial strength. Members of the IMF are called upon to abide by the standing rules set out in respect of the official rates of exchange allotted or regulated by this bank to each member. Members enjoy the opportunity of being helped in financial difficulties. Apart from the many duties it has, it undertakes commercial bank transactions. *Abbrev.* IMF *or* I.M.F.

International Money Order An economical method of transferring a small and limited amount of money from one place to another through the Post Office. *Abbrev.* IMO *or* I.M.O.

International Nautical Mile A mile equivalent to 1,852 metres or 6,076.12 ft. *See* **Nautical Mile**.

International Oil Pollution Compensation Fund An inter-governmental agency designed to pay compensation for oil pollution damage exceeding the shipowner's liability. It was created by an IMCO (now IMO) Convention in 1971 and started its operations in October 1978. Contributions for it come mainly from the oil companies of member states. *Abbrev.* I.O.P.C.F. *or* IOPCF.

International Parity Price Price of a tender or bid submitted when calculated on an international exchange basis. *Abbrev.* IPP.

International Regulations for Preventing Collisions at Sea Upon the invitation of the International Maritime Organization, a Conference was held in London from 4 to 20 October 1972 for the purpose of revising the International Regulations for Preventing Collisions at Sea, 1960. The revised Regulations came into effect on July 1977. At the 12th session of the Assembly in November 1981, a Resolution was adopted amending the International Regulations for Preventing Collisions at Sea, 1972. These Amendments came into force on 1 June 1983.

International Shipping Federation *Abbrev.* ISF. Founded in 1909, it has a membership of shipowners' associations of more than 30 nations and is concerned with the employment of seafarers and good employment practice. It deliberates on wages and conditions, relations with governments and with the unions, including the International Transport Workers' Federation, national and international employment law, recruiting, manning and training, regulations affecting health and safety, Port State Control, etc. It co-ordinates the employer's position in the Joint Maritime Commission (JMC) and the International Labour Organisation (ILO), *q.v.* Its activities are handled by a Labour Affairs Committee and a Manning and Training Committee. It works closely with the International Chamber of Shipping (ICS) and shares with it a secretariat, MARISEC, *q.v.* The ISF has consultative status with the International Maritime Organization (IMO), *q.v.*, on such matters as certification, the safety management code and piracy. The ISF is based in London.

International Shipping Federation Located at 30-32 St Mary Axe, London EC3A 8ET, the International Shipping Federation (ISF) was founded in 1909. Its membership comprises national shipowner associations representing 29 nations throughout the world and almost two thirds of world merchant tonnage.

The objective of the ISF is to promote the interests of its members with regard to all sea-going personnel matters.

Its functions are:
— to provide an opportunity for shipowners of different countries to exchange information and views on all social and personnel matters affecting shipping;
— to serve as a medium of preparation and co-ordination for employers' corporate policies at international conferences involving shipping personnel matters. In particular, it provides the secretariat for the shipowners' group at International Labour (Maritime) Conferences and also for the shipowners' side of the ILO Joint Maritime Commission;
— to keep its members advised of relevant international developments, both generally and in the specialised maritime international bodies; to attend meetings of, and make representations as appropriate to, such organisations;
— to liaise with other non-governmental organisations (employer and union) in pursuance of its policies.

The ISF is governed by the ISF Council, which meets annually and on which every ISF member is represented. There is no permanent committee structure, but meetings are convened as necessary to deal with specific subjects.

The office-bearers consist of a President and three Vice-Presidents chosen from leading shipowners of the member associations. The Secretariat is based in London and provided by the General Council of British Shipping.

International Standardization Organization This body is responsible for numerous works of standardization in mechanical, dental, computer, automobile, food and many other fields. *Abbrev.* ISO *or* I.S.O.

International Standards Industrial Classification. A standards classification certificate which is issued by the United Nations. *Abbrev.* ISIC *or* I.S.I.C.

International Waterways Consist of international straits, inland and interocean canals and rivers where they separate the territories of two or more nations. Provided no treaty is enforced both merchant ships and warships have the right of free and unrestricted navigation through these waterways.

Interpleader Proceedings The process whereby a person who is or expects to be sued by two or more parties, claiming adversely to each other, for a debt or goods in his hands, but in which he himself has no interest, obtains relief by procuring such parties to try their rights between or amongst themselves only. When a person is in possession of property in which he claims no interest, but to which two or more other persons lay claim, and he, not knowing to whom he may safely give it up, is sued by one or both, he can compel them to interplead, i.e., to take proceedings between themselves to determine who is entitled to it.

Interpol International Criminal Police Commission with its HQ in Paris. See also ICPO and I.C.P.O.

inter praesentes *Latin*—Between those present.

In terrorem *Latin*—As a warning.

Intersat *or* **Intelsat** International Telecommunications Satellite Consortium.

inter se *Latin*—Between themselves.

Interstate Pertaining to relations *vis-à-vis* states of federal governments, such as Canada and the USA—Interstate trade or railways.

Interstate Commerce Commission The movements of passengers and the transport of goods from one state to another within the USA. This includes the use of land, sea and air. *Abbrev.* ICC *or* I.C.C.

INTERTANKBILL 78 Code name for BIMCO approved Bill of Lading on tankers for Voyage Charter Party.

INTERTANKTIME 80 Code name for INTERTANKO time charterparty.

INTERTANKVOY 76 Code name for a Tanker Voyage Charterparty. A revised Charterparty, *q.v.*, of 1970 originally issued by BIMCO and the Japan Shipping Exchange Inc. Tokyo. This form of C/P is suitable for the carriage of both clean and dirty products and for use in the Parcel Trades.

Intertropical Convergence Zone Low atmospheric pressure close to the Equator, *q.v.*, near to the Doldrums, *q.v. Abbrev.* ITCZ *or* I.T.C.Z.

Intervention An action taken by a country's bank to sell and buy its own currency or even another member currency in the financial market to protect its value.

Intestacy The condition when a person who dies before he makes a will. The property will be distributed acording to law of the country.

Intex *or* **INTEX** Integer extraction; International Futures Exchange, *Bermuda*.

In the Bank A phrase used for Discount Houses, *q.v.*, when they borrow money from the Bank of England.

INTIS International Transport Information System.

Intl International.

In tort *Latin*—Describes an action in law for a wrong or injury not arising out of contract.

In toto *Latin*—Totally; Wholly; Entirely.

Intra muros *Latin*—Within the walls.

In trans In transit; In transitu *Latin*—In transit.

In transitu *Latin*—In transit. *Abbrev.* In trans.

INTRASTAT International Transport Statistics Working Group.

Intra vires *Latin*—Within the powers a person or a body occupies or within the authority of an agent.

Introductory Offer Presentation of a new product to a competitive market usually at an attractively discounted price and sometimes accompanied by gifts, special privileges or opportunities to win prizes.

INTUG International Telecommunications Users Group, in the Netherlands.

I.Nuc.E. Institute of Nuclear Engineers.

Inv. Invenit *Latin*—Designed this; Invent; Inventor; Inverness; Investment; Invoice.

Invar A metal composed of about 60% of iron and 40% of nickel which is mostly used in the measuring tape and balance wheels of clocks because of the negligible coefficient linear expansion.

Inventory A schedule or detailed list of items.

Inventer Electrical device to convert direct current (DC) into alternating current (AC).

Investment Banks An American phrase equivalent to the English Issuing Banks or Merchant Banks, *q.v.* They are institutions that undertake services by accepting Bills of Exchange, *q.v.*; issuing loans, securities, cheques, hire-purchase notes, providing capital for commerce, transferring business deals in foreign currencies, transacting local and foreign finance, etc.

Investment *See* **Trusts**.

Investment Trust *See* **Trust**. The American term is Closed-end Trust.

Inv. et del. Invenit et delineavit *Latin*—Designed and drawn.

Invisible Exports These are shipping services, insurance and similar items that account for the apparent excess of a country's imports over exports. Also known as Invisible Trade or 'Invisibles'.

Invitation to Treat The insertion of prices against each item in a show window or against each item in a catalogue when such items are for sale.

Invitus *Latin*—Without wishing; Without Agreement; Unwilling.

Invoice List of goods shipped or sent, with prices and charges.

Involuntary Savings Forced savings. Certain restrictions imposed by a country to save consumers' expenditure in order to force or encourage them to save money.

Involuntary Unemployment A person who is unemployed through the lack of suitable work. This is the opposite of Voluntary Unemployment.

Inward Cargo. Merchandise to be imported into the country.

Inward Charges Charges relating to a vessel's entry into a port. The charges relate to pilotage, port dues, mooring fees, tug assistance, etc.

IO *or* **I.O.** India Office; Inspecting Officer; Intelligence Officer.

I/O *or* **I &/or O** In and/or On Deck or In and/or Over Deck. Stowage of cargo to be either on deck or under deck.

I/O In and/or On Deck or in and/or Over Deck. Stowage of cargo to be either on or under deck; Inspecting Order; Input/Output; Input-In/Input Out *oil drilling*.

IOA *or* **I.O.A.** International Omega Association.

IOB *or* **I.O.B.** Institute of Bankers; Institute of Book-keepers; Institute of Builders.

IOBI *or* **I.O.B.I.** Institute of Bankers of Ireland.

IOBS *or* **I.O.B.S.** Institute of Bankers of Scotland.

IOC *or* **I.O.C.** Indirect Operating Cost; Initial Operational Capacity; Inter-Governmental Oceanographic Commission; UNESCO Paris; International Olympic Committee.

IOCC International Offshore Craft Conference.

IOCU International Office of Consumers' Union.

IOD *or* **I.O.D.** Institute of Directors.

IODE *or* **I.O.D.E.** International Oceanographic Date/Data Exchange.

IOE *or* **I.O.E.** International Organisation of Employers.

I.o.E. Institute of Export.

IOF *or* **I.O.F.** Institute of Fuel; International Oceanographic Foundation; Independent Order of Foresters.

I.o.F. Institute of Fuel.

I of Arb. Institute of Arbitrators.

IOFF Independent Ocean Freight Forwarders.

IOI *or* **I.O.I.** International Ocean Institute.

I of J Institute of Journalists.

I of M Isle of Man.

IOM *or* **I.O.M.** Institute of Office Management; Isle of Man.

IOME *or* **I.O.M.E.** Institute of Marine Engineers.

IOMMP *or* **I.O.M.M.P** International Organisation of Masters, Mates and Pilots.

Ion A particle passing through an electric field in liquid or gas and carrying an electric charge.

Ionosphere The layer of atmosphere above 60 kilometres from the earth from which radio waves are reflected back to earth. *See* **Sky Wave.**

Ionospheric Storm Disturbance caused by intense solar flares or eruption causing a flash of ultra-violet light and a stream of charged particles. These cause electrification of the lower layers in the earth's upper atmosphere blocking short radio waves and changing the behaviour of long waves to the detriment of communications. The use of satellites helps to overcome the problem.

IOOC International Olive Oil Council *USA*.

IOP Institute of Packaging; Institute of Physics; Institute of Printing; Input/Output Processor.

IOP *or* **i.o.p.** Irrespective of Percentage, *q.v.*

IOPCF International Oil Pollution Compensation Fund, *q.v.*

IOPEC International Oil Pollution Exhibition and Conference.

IOPF International Oil Pollution Fund, *similar to* International Oil Pollution Compensation Fund, *q.v.*

IOPP International Oil Pollution Prevention Certificate.

IOPPEC International Oil Pollution Exhibition and Congress.

IOS Institute of Oceanographic Sciences *UK*.

IOSC International Oil Spill Conference.

IOSCO International Organisation of Securities Commissions.

IOT *or* **I.O.T.** Institute of Transport.

IOTA *or* **I.O.T.A.** Institute of Traffic Administration.

IOT&E Initial Operational Testing and Evaluation *for air cushion vehicles* (Hovercraft, *q.v.*)

IOTS In Orbit Test System *satellite*.

IOTTSG *or* **I.O.T.T.S.G.** International Oil Tanker and Terminal Safety Group.

IOU *or* **I.O.U.** I owe you, *q.v.*

I.O.W. Isle of Wight.

I owe you A formal acknowledgement of money lent by simply inserting this sentence or the abbreviation IOU *or* I.O.U. and signed for. Normally this informal receipt is done through friendship or trust.

IP Intermediate Premium.

IP *or* **I.P.** Imperial Preference. This used to mean the taxing of imports from parts of the British Empire at lower rates than those from foreign countries. Indicated Power; Induced polarisation *oil exploration*; Institute of Petroleum; Intermediate Pressure; Instalment Plan.

IPA *or* **I.P.A.** Institute of Practitioners in Advertising; Institute of Private Affairs in Australia; International Publishers' Association.

IPAA International Petroleum Association of America.

IPAI International Primary Aluminium Institute.

IPBA India, Pakistan and Bangladesh Association.

IPBM Interplanetary Ballistic Missile.

IPC *or* **I.P.C.** Institute Port Risk Clauses, *also abbrev.* P.R.C, *see* **Port Risks**. International Publishing Corporation.

IPCC Intergovernmental Panel on Climate Change.

IPCS Institution of Professional Civil Servants.

I.P.D. In praesentia dominorum *Latin*—In the presence of the Lords while in session; Individual Package Delivery.

IPE International Petroleum Exchange of London Ltd.

IPEX *or* **Ipex** Instant Purchase Excursion, *q.v.*

I.P.F. Intaken pile fathom.

IPG *or* **I.P.G.** Inter-Governmental Preparatory Group.

IPI *or* **I.P.I.** International Press Institute.

IPIECA International Petroleum Industry Environmental Conservation Association.

IPL Interprovincial Pipe Line; International Processing Language.

IPM *or* **I.P.M.** Inches per Minute.

IPP Institute for Plasma Physics *Germany*; International Parity Price, *q.v.*; Interpulse Period.

IPPB Intermittent Positive Pressure Breathing.

IPR Initial Production Rate; Institute of Public Relations; Internal Press Release.

I.P.R.C. *or* **I.P.C.** Institute Port Risk Clauses *insurance*. *See* **Port Risks**.

I.Prod. E. Institution of Production Engineers.

IPs. Ipswich, *UK*.

IPS Institute of Purchasing and Supply.

IPS *or* **I.P.S.** *or* **ips** Inch per Second.

IPSA *or* **I.P.S.A.** International Passenger Ship Association. International Professional Security Association.

Ipse dixit *Latin*—He himself said it.

Ipsissimis verbis *Latin*—In the very words. Quoted exactly.

Ipso facto *Latin*—By actual fact; Immediately; Facts speak for themselves.

Ipso jure *Latin*—By the law itself.

IPSS International Packet Switched Service.

IPTS *or* **I.P.T.S.** *or* **i.p.t.s.** International Practice Temperature Scale.

IPU International Parliamentary Union.

IPY *or* **I.P.Y.** *or* **i.p.y.** Inches per year.

i.q. Idem quod *Latin*—The same as.

IQ Intelligence Quotient, being the percentage ratio of mental age to actual age; International Quota.

IQA Institute of Quality Assurance.

IQS *or* **I.Q.S.** Institute of Quantity Surveyors.

Ir. Ireland.

Ir *or* **IR** Infra Red; Iran Air *airline designator*.

IRAN Inspect and Repair as Necessary.

Iran Air Iran National State Airlines.

IRAS Infra-red Astronomical Satellite.

IRASA *or* **I.R.A.S.A.** International Radio Air Safety Association.

IRBM Intermediate-Range Ballistic Missile.

I.R.C. Industrial Reorganisation Commission *U.K.*; International Red Cross; International Research Council; International Rice Commission.

IRCA International Register of Certified Auditors.

IRD or **I.R.D.** Inland Rail Depot *containerisation.*

IRDA or **I.R.D.A.** Industrial Research and Development Authority.

Ire. Ireland.

IRE or **I.R.E.** Institute of Radio Engineers *USA.*

I.Rep. C. Institute Replacement Clause, *q.v.*

IRF International Road Federation, Geneva.

IRFAA International Rescue and First Aid Association.

IRFO International Road Freight Office,

I.R.I. Instituto per la Ricostruzione Industriale *Italian*—Institute of Industrial Reconstruction.

IRMC or **I.R.M.C.** International Radio Maritime Committee.

Irn Iron, *Lloyd's Register notation.*

IRN Import Release Note.

IRO or **I.R.O.** International Refugee Organisation; International Relief Organisation.

iro or **i.r.o.** In respect of.

Iron Mikes Slang term for Gyro Compass.

Ironmonger A person dealing in hardware and ironware.

Iron Pellets *Similar to* Direct Reduced Iron, *q.v.*

Iron Sponge A material used to remove sulphur and other impurities from commercial gases which are subject to spontaneous combustion when soaked thoroughly with impurities.

IRPC Industrial Relations Policy Committee.

IRR Internal Rate of Return; Iraqi Republic Airlines.

Irredeemable A stock exchange saying. Shares are not allowed to be re-purchased.

Irredeemable Debentures Debentures which cannot be re-purchased at any time.

Irrespective of Percentage An insurance expression referring to the Franchise, *q.v.*, in the Policy, *q.v.*, of Marine Insurance, *q.v.*, which is excluded in the Particular Average. In other words 'Not Subject to Franchise'. *Abbrev.* I.O.P. *or* i.o.p.

Irrevocable Generally refers to an Irrevocable Letter of Credit which cannot be revoked or withdrawn once it is opened. *See* **Letter of Credit.**

Irrevocable Corporate Purchase Order A bank instrument whereby the buyer is covered by the bank for the amount due to be paid or settled against the merchandise ordered from abroad. *Abbrev.* ICPO *or* I.C.P.O.

IRRP International Rice Research Institute, *Manila.*

IRS Indian Register of Shipping, classification society, *q.v.*; Inertial Reference System, *q.v.* Internal Revenue Service *USA.*

IRSG International Rubber Study Group, *London.*

IRTF Infra-red Telescope Facility.

IRU International Road Transport Union, Geneva.

IRVR Instrumented Runway Visual Range *aviation.*

IS or **I.S.** Irish Society; Indian Summer *Plimsoll mark.*

ISA International Standards of Auditing.

ISA or **I.S.A.** Indonesian Sawmill Association; Institute of Shipping & Forwarding Agents. A.I.S.A. (Associate) F.I.S.A. (Fellow); Instrument Society of America; International Federation of the National Standardisation Associations; International Standard Atmosphere *aviation weather observation*; International Standardization Organization; International Sugar Agreement; Israel Shipowners' Association; Italian Shipsuppliers Association.

ISAA Irish Ship Agents' Association.

ISAS or **I.S.A.S.** Institute of the South African Shipbrokers.

ISASI or **I.S.A.S.I.** International Society of Air Safety Investigators.

ISBN or **I.S.B.N.** International Standard Book Number.

ISC Integrated Ship Control *IMO nautical safety requirements*; International Sending Code *telegraphic code.*

ISC or **I.S.C.** International Statistical Classification; International Chamber of Shipping; International Sugar Council; Israel Shipping Council; Italian Shipping Council.

ISD or **I.S.D.** International Subscriber Dialling.

ISDN Integrated Services Digital Network.

ISDP Integrated Ship Design and Production, *q.v.*

ISE Institute of Shipping Economics, Bremen, West Germany; International Submarine Engineering. In Port Moody, British Columbia, Canada.

ISESA International Ship Electrical Service Association.

ISF or **I.S.F.** International Shipping Federation, *q.v.*

ISFA or **I.S.F.A.** Institute of Shipping and Forwarding Agents.

ISG or **Isg** Imperial Standard Gallon.

ISGOTT International Safety Guide for Oil Tankers and Terminals, a guide published by the International Chamber of Shipping and the Oil Companies' International Marine Forum.

ISHR International Society of Human Rights.

ISI International Statistical Institute; Iron and Steel Institute.

ISIC or **I.S.I.C.** International Standards Industrial Classification, *q.v.*

ISIS Independent Schools Information Service *UK*; In-Service Inspection System, for pressure systems *Lloyd's Register.*

Isl. Island.

Isletmesi *Turkish*—Management.

ISLW *or* **I.S.L.W.** Indian Spring Low Water.

ISLWG *or* **I.S.L.W.G.** International Shipping Legislation Working Group *UNCTAD.*

ISM International Safety Management code *IMO.*

ISM *or* **I.S.M.** Institute of Supervisory Management.

ISMA International Ship Managers' Association.

ISNAR International Service for National Agricultural Research.

ISO *or* **I.S.O.** Imperial Service Order; International Standardization Organization, *q.v.*; International Standards Organization.

ISO 9000 The ISO 9000 serially numbered documents comprise the International Standards Organisation's Standards for Quality Management, quality assurance and quality systems. They were adopted from British Standards (BS 5750) first published in 1979 from standards of the Ministry of Defence and government departments. These standards were also adopted by CEN, the European Standards organisation (In consequence, BS 5750 was in 1994 redesignated as BS EN ISO 9000.) Registration under the standards requires documentation of a quality system which defines in detail what is to be done in every department of an organisation, by whom and how, in order to satisfy customers' needs.

ISOA International Support Vessel Owners' Association.

ISOC *or* **I.S.O.C.** International Shipping and Offshore Oil Conference.

Isobar A line on a chart or map connecting places at which the atmospheric pressure is the same.

Isobaric Describes a chart having continuous lines joining points on the earth's surface where barometric pressure is the same.

Isochronous Rolling When the period of each roll of a ship is exactly the same.

Isogonic Line A line drawn on a map indicating equal angles of magnetic variation.

Isohaline A drawn line on a map passing through the oceans delineating equal salinity.

Isohel A drawn line on a map to show equal sunshine duration.

Isohyet A drawn line on a map to show points of equal rainfall.

Isoneph A drawn line on a map to show average cloudiness over a period of time.

ISOSO International Symposium on Ship Operations.

Isotherm A drawn line on a map connecting places which have the same mean annual temperature.

ISR Institute of Shipping Research, Bergen, Norway; International Sanitary Regulations.

ISRA International Ship Reporting Association.

ISS Institute of Space Sciences; Institute of Strategic Studies, *q.v.*; International Social Service, *Geneva.*

ISSA *or* **I.S.S.A.** International Sailing Schools' Association; International Ship Suppliers Association.

ISSN International Standard Serial Number.

Issued Capital Actual money paid up for the formation of a company.

Issuing Bank *Similar to* Investment Bank, *q.v.*

Issuing Broker A broker who is engaged in a new issue of securities.

Issuing Carrier The company or carrier which produces the Air Way Bill, *q.v.*, and other details of airfreighted cargo.

Issuing House *or* **Issuing Houses** Financial experts who are in the business of issuing shares for public limited companies. *Abbrev.* IH *or* I.H.

Issuing Houses Association An institution containing members of the Issuing Houses, *q.v. Abbrev.* IHA *or* I.H.A.

IST *or* **I.S.T.** International Shipping Trustees, Indian Standard Time.

Ist. Istituto *Italian*—Institute.

ISTEL Information Systems and Telecommunications *UK.*

Isth. Isthmus, *q.v.*

Isthmus A narrow strip of land connecting two larger land areas. *Abbrev.* Isth.

Istiraki *Turkish*—Partnership.

I.Struct.E. Institute of Structural Engineers.

ISU International Salvage Union, *Rotterdam*; Iowa State University.

ISWG *or* **I.S.W.G.** Imperial Standard Wire Gauge.

i.t. Internal Thread; In Transit.

it *or* **IT** *or* **I.T.** Inclusive Tour; Income Tax; Independent Tank; Industrial Tribunal, *established 1971*; Inland Transport; Insulated Tank; Intelligence Test; Interline Traffic, Internal Air Traffic.

IT Information Technology.

ITA *or* **I.T.A.** Industrial Training Act 1964 *UK*; Industrial Transport Association; Independent Television Authority; Institut du Transport Aérien *French*—Air Transport Institute; Institute of Traffic Administration; Institute of Travel Agents; International Telecommunications Administration.

Ital. Italics; Italy.

ITAWDA Integrated Tactical Amphibious Warfare Data System *USA.*

ITB Industrial Training Board *UK*; Integrated Tug Barge, *q.v.*; International Time Bureau.

ITC *or* **I.T.C.** Insulated Tank Container; Integrated Turbine Control; International Tin Council; Installation, Time and Cost; Institute Time Clause (Hull), *see* **Institute Clauses**; International Tonnage Certificate; International Transport Consortium, composed of various international forwarding agents; International Travel Catering.

ITCFOD. Abs *or* **I.T.C.F.O.D. Abs.** Institute Time Clauses (Hulls) Free of Damage Absolutely *marine insurance.*

ITCFPA Abs. *or* **I.T.C.F.P.A. Abs.** Institute Time Clauses (Hulls) Free of Particular Average Absolutely *marine insurance.*

ITCXS Abs. *or* **I.T.C.X.S. Abs.** Institute Time Clauses (Hulls) Excess All Claims Absolutely *marine insurance.*

ITCXSPA *or* **I.T.C.X.S.P.A.** Institute Time Clauses (Hulls) Excess Particular Average *marine insurance.*

ITCZ *or* **I.T.C.Z.** Intertropical Convergence Zone, *q.v.*

ITDC *or* **I.T.D.C.** Indian Tourism Development Corporation.

item *or* **itidem** *Latin*—Likewise; ditto.

ITF *or* **I.T.F.** International Transport Workers' Federation.

ITH Intermediate Tow Head *oil drilling.*

ITI Institute of Translation and Interpreting *London.*

ITIC *or* **I.T.I.C.** International Tsunami Information Centre.

ITM *or* **I.T.M.** Inch Trim Moment, *q.v.*; Institute of Travel Managers.

ITMA *or* **I.T.M.A.** Institute of Trade Mark Agents.

ITMC *or* **I.T.M.C.** International Tonnage Measurement Convention.

ITMRC *or* **I.T.M.R.C.** International Travel Market Research Council.

ITO Income Tax Office; International Trade Organisation of UN.

ITOA *or* **I.T.O.A.** International Tanker Owners' Association.

ITOPF International Tanker Owners' Pollution Federation.

ITP *or* **I.T.P.** Intercept Terminal Point *navigation.*

ITS *or* **I.T.S.** Industrial Training Service; International Trade Secretariat; International Tanker Services, Norway.

ITSA *or* **I.T.S.A.** International Tonnage Stabilization Association; International Tank Storage Association.

ITT *or* **I.T.T.** *or* **It. &t.** International Telephone and Telegraph *USA.*

ITT Institute of Travel and Tourism.

ITTC International Towing Tank Conference.

ITU *or* **I.T.U.** International Telecommunications Union; Inter-American Telecommunications Union; Intensive Treatment Unit *medical.*

ITV Independent Television.

ITW Independent Tank Wing. *Side ballast tank of a ship.*

ITWF International Transport Workers' Federation.

ITX Independent Inclusive Tour, *q.v.* Independent Tank Common *ship's ballast tank.*

IU International Unit.

IUAI *or* **I.U.A.I.** International Union of Aviation Insurers.

IUAPPA International Union of Air Pollution Prevention Associations.

IUB International Union of Biochemistry.

IUCN International Union for Conservation of Nature and Natural Resources, *Switzerland.*

IUD Intra-Uterine Device.

IUE International Ultra Violet Explorer.

IUMI *or* **I.U.M.I.** International Union of Marine Insurance, Zurich, Switzerland.

IUMU *or* **I.U.M.U.** International Union of Maritime Unions.

IUS Inertial Upper Stage.

IUOTO *or* **I.U.O.T.O.** International Union of Official Travel Organisations.

IUR International Union of Railways.

IV Roman numeral for the figure 4; In Vertretung *German*—On behalf of; Proxy.

IV *or* **I.V.** Invoice Value; Increased Value; Insured Value; Initial Velocity; In verbo *Latin*—By word.

I.V.A. Impôt sur la valeur ajoutée *French*—Value Added Tax—VAT.

I.V.C. Institute Voyage Clauses *marine insurance.*

IVM Isolation Valve Module, *oil drilling.*

I.V.R. Internationale Vereiningung des Rheinschiffsregisters *German*—International Association of Rhine Ships' Register.

IVSI Instantaneous Vertical Speed Indicator *aviation.*

IW *or* **I.W.** Innere Weite *German*—Inside Diameter; Institute Warranties, *q.v.*

i.w. Indirect Waste; Inside Width.

IWA *or* **I.W.A.** Inland Waterways Association; Institute of World Affairs; International Wheat Agreement.

IWC International Weather Code.

IWC *or* **I.W.C.** International Wheat Council; International Whaling Commission.

I.W.D. *or* **I.W. & D.** Inland Waterways and Docks.

I.W.E. Institute of Water Engineers.

I.W.F.S. International Wine and Food Society.

IWG *or* **I.W.G.** Imperial Wire Gauge.

IWL Institute Warranty Limits *insurance geographical claim limits.* See **Institute Warranties.**

IWM *or* **I.W.M.** Institute of Works Management.

IWPC *or* **I.W.P.C.** Institute of Water Pollution Control.

IWRB *or* **I.W.R.B.** Independent Water Resources Board; Indonesian Water Resources Board.

IWRC *or* **I.W.R.C.** Independent Wire Rope Core.

IWS International Wool Secretariat, *London.*

IWSA *or* **I.W.S.A.** International Water Supply Association.

IWSG International Wool Study Group, *London.*

IWSP *or* **I.W.S.P.** Institute of Works Study Practitioners.

IWT *or* **I.W.T.** Inland Water Transport.

IWTA *or* **I.W.T.A.** Inland Water Transport Authority; Inland Water Transport Association.

IWW *or* **I.W.W.** Industrial Workers of the World.

IX Roman numeral for figure 9.

IXDD International Telex Direct Dialling.

I.Y. Imperial Yeomanry.

IYC *or* **I.Y.C.** Institute Yacht Clauses *insurance.*

IYHF International Youth Hostel Federation.

Izv Izvestia, Russian newspaper.

J

J Irradiation Correction *altitude*; Jet aircraft; Joule *physics*; Justice; Judex *Latin*—Judge; Jurum *Latin*—of law; Mr Justice.

JA *or* **J.Adv.** Judge Advocate.

JA *or* **J.A.** Judge Advocate; Justice of Appeal; Junior Achievement, *q.v.*

j.a *or* **JA** *or* **J/A** Joint account, *q.v.*; Judge Advocate.

Ja *or* **Jan** January.

JA Jamaica; Judge Advocate.

JAA *or* **J.A.A.** Japan Aeronautics Association; Jewish Athletic Association.

Jac Jacobean; Jacobus *Latin*—Jacob.

JAC *or* **J.A.C.** Joint Advisory Airworthiness Committee.

Jachtwerf *Dutch*—Yacht yard.

Jack Ship's flag, smaller than ensign, especially one flown from a jack-staff at bow indicating nationality; Short form for Union Jack.

Jackass Barque A four-masted sailing vessel square-rigged on fore and main masts, fore-and-aft rigged on mizzen and jigger.

Jacket The lower part of an offshore platform fixed to the bottom of the sea by means of piles.

Jack Ladder *See* **Jacob's Ladder**.

Jack-staff A flagpole positioned at the bow of a ship. *See* **Jack**.

Jackstay A rope, wire or rod set taut between two points and acting as a traveller. On the yard of a square-rigged vessel, the head of the sail is bent to it. On a mast, it is used to hold the luff of a gaff topsail. Between the land and a wreck, it will support a Breeches Buoy, *q.v.* Between ships, it is used in Replenishment at Sea. It can be rigged above the deck as a lifeline or as a ridge rope to support an awning, and for other purposes.

Jack Tar A nickname for a British sailor probably because in the old days the sailor used to wear tarred clothing.

Jack Up A sort of platform used for offshore oil drilling. It is towed as a barge into the drilling area and its legs are hydraulically lowered until they settle well into the seabed. Usually jack-ups are used in reasonably shallow water.

JACOB Junior Achievement Corporation Business.

Jacob's Ladder A rope ladder with wooden rungs and wire or cord ropes attached to the sides. Used temporarily while slung over the side by crew members, pilots and others when accommodation ladders are not in use. Also termed Pilot Ladder, *q.v.*, Bosun's Ladder, Jack Ladder.

JACT *or* **J.A.C.T.** Joint Association of Classical Teachers.

JAD Julian Astronomical Day.

JADB Joint Air Defence Board *USA*.

JAEC Japanese Aero Engine Corporation; Joint Atomic Energy Commission; Joint Atomic Energy Committee *US Congress*.

JAF Judge Advocate of the Fleet.

JAFC Japan Atomic Fuel Corporation.

J.A.G. *or* **J. Ad. G.** Judge Advocate General, *q.v.*

JAIEG Joint Atomic Information Exchange Group.

JAL Japan Air Lines; Jet Approach and Landing Chart.

Jam Jamaica.

JAMA Journal of the American Medical Association.

Jamahiriya In Arabic language it literally means 'of the masses' or 'belonging to the people'.

JAMDA Japan Marine Machinery Development Association.

Jamming Deliberate radio or radar interference.

J. and/or l.o. Jettison and/or Loss Overboard. *See* **Jettison**.

J. and/or w.o. Jettison and/or Washed Overboard.

JANA Japanese Navigational Aids Association.

janv. Janvier *French*—January.

Jap. Japan.

Japanese Ice Clause A Marine Insurance, *q.v.*, hull clause excluding liability in the event of damage caused when in contact with ice in Japanese waters from 15 November to 30 April, the winter period. Not much in vogue today.

Jarnvagsaktiebolag *Swedish*—Railway Company Ltd.

JAS Japan Air System *airline*.

Jastop Jet Assisted Stop *aviation*.

JAT Jugoslovenski Aero Transport, Yugoslav State Airlines.

JATCC *or* **J.A.T.C.C.** Joint Aviation Telecommunication Co-ordination Committee.

JATCRU *or* **J.A.T.C.R.U.** Joint Air Traffic Control Radar Unit.

JATO *or* **J.A.T.O.** Jet Assisted Take-Off *aviation*.

JATS *or* **J.A.T.S.** Joint Air Transportation Service.

JATWS Japan Association of Temporary Work Services.

Jav Java.

Jaw (1) The hollow in a block which takes the sheave. More usually called the swallow or sheave-hole. (2) Semi-circular fitting at the end of a boom or gaff which rests against the mast.

JAWS Joint Airport Weather Studies *USA*.

Jax Jacksonville, *USA*.

Jaycee Junior Chamber of Commerce *USA*; A person belonging to that Chamber.

J.B. Jurum Baccalaureus *Latin*—Bachelor of Laws.

j.b. Junction Box.

JBAA Journal of the British Archaeological Association.

JBC Jumbo Barge Carrier.

JBCNS Joint Board of Clinical Nursing Studies.

JBCSA *or* **J.B.C.S.A.** Joint British Committee for Stress Analysis.

JBL Journal of British Law.

J.C. Jesus Christ; Juris-consultus *Latin*—Juriconsult; Justice Clerk; Jockey Club.

JCAR Joint Committee on Applied Radioactivity.

J.C.B. Juris Canonici Baccalaurcus *Latin*—Bachelor of Canon Law; Juris Civilis Baccalaureus *Latin*—Bachelor of Civil Law.

J.C.C. Joint Computer Conference; Joint Communications Centre; Joint Consultative Committee; Joint Consultative Council; Junior Chamber of Commerce.

JCCC Joint Customs Consultative Committee.

JCCIUK Japanese Chamber of Commerce and Industry in UK.

J.C.D. Juris Canonici Doctor *Latin*—Doctor of Canon Law.

J.C.D. Juris Civilis Doctor *Latin*—Doctor of Civil Law.

JCFA Japan Chemical Fibres Association.

JCI *or* **J.C.I.** Junior Chamber International.

J.C.L. Juris Canonici Licentiatus *Latin*—Licentiate of Canon Law; Juris Civilus Licentiatus *Latin*—Licentiate of Civil Law.

JCLA Joint Council of Language Associations.

JCRA Joint Common Risk Agreement.

JCS Joint Chiefs of Staff; Joint Commonwealth Societies; Joint Container Service; Journal of the Chemical Society.

JD *or* **J.D.** Junior Deacon; Junior Dean; Junior Delinquent; Diploma in Journalism; Jurum Doctor *Latin*—Doctor of Laws; Justice Department *USA*.

J.d. Jordanian Dinar, unit of currency used in Jordan.

JD Japan Air System *airline designator*.

JDA Japan Defence Agency; Japan Domestic Airlines.

JDB Japan Development Bank.

J/deg. Joule per Degree.

JDI *or* **J.D.I.** Joint Declaration of Interest.

J.Dip.M.S. Joint Diploma in Management Studies.

JDODC Japan China Oil Development Corporation.

JDS *or* **J.D.S.** Job Data Sheet; Job Description and Specification *insurance*; Job diagnosis survey.

JE Job evaluation *insurance*; Japanese Encephalitis.

JEA *or* **J.E.S.** Joint Export Agency.

JEDI Joint Electronic Data Interchange.

Jeep *or* **GP** General Purpose *referring to a vehicle*.

JERI *or* **J.E.R.I.** Japan Economic Research Institute.

Jerque Note A customs inward entry certificate.

Jerquing *or* **Rummaging** The searching of a ship by customs officers to ascertain that no smuggling is occurring. *See* **Rummage**.

Jerry Building A building or buildings which have been constructed with the use of very cheap materials in a short time. The jerry builder aims to make as much profit as possible.

JESA *or* **J.E.S.A.** Japanese Engineering Standards Association.

Jet Jet propelled or driven engine; Jetsam, *q.v.*

Jet Engine Mostly fitted to aircraft. Its fuel is burnt and the resultant gases are then discharged through its rear exhaust system, resulting in the achievement of very high speed. It was invented by Sir Frank Whittle. *See* **Jet Propulsion**.

Jet Lag A somewhat debilitating effect on the human body as the result of the rapid change in time zones through fast air travel.

Jetliner A commercial aircraft of large size propelled by jet engines.

Jet-p. Jet Propelled; Jet propulsion, *q.v.*

Jet Propulsion The driving forward of a vehicle, usually an aircraft, by the burning of fuel and the expulsion of the resulting gases. *Abbrev.* Jet-p. *See* **Jet Engine**.

JETRO Japan External Trade Organisation.

JETS Joint Enroute Terminal Systems.

Jetsam The goods thrown away by jettison and washed up on shore—alternatively goods from a wreck remaining under water. *See* **Flotsam**.

Jett. Jettison and Washing Overboard, *q.v.*

Jettison and Washing Overboard *or* **Jettison and/or Washed Overboard** Marine insurance term concerning the act of throwing or washing overboard either cargo or ship's stores/equipment in time of danger and emergency. *Abbrev.* J. and/or W.O.

Jett. Loss Jettison Loss. *See* **Jettison and Washing Overboard**.

Jetty A Mole constructed to protect a harbour. Also a Quay or landing pier.

Jetty Clause A marine insurance clause covering grain and similar merchandise up to its arrival at landing sheds where a survey is usually carried out to verify its condition. Point of destination or transit is not covered under the Policy, *q.v.*, of Insurance. This period can be covered under the Transit Clause, *q.v.*

Jeu. Jeudi *French*—Thursday.

Jeu de mots *French*—Play on words.

J/F Journal Folio; Jewellery and Furs *insurance*.

J.Form An insurance policy indemnifying the insured where no sea transportation or transit takes place.

JFPS *or* **J.F.P.S.** Japan Fire Prevention Society.

JFRCA Japan Fisheries Resources Conservation Association.

JFS *or* **J.F.S.** Japan Fishery Society.

JG *or* **j.g.** Junior Grade *US Navy*.

JHC Joint Hull Committee, *q.v.*, *insurance*.

JHF Joint Hull Formula.

JHIU Japan Hull Insurance Union.

J.I. Journalists Institute.

JIC Joint Industrial Council.

JICA Japan International Co-operation Agency.

JIFA Japan Institute for Foreign Affairs.

Jigyo *Japanese*—Achievement; Business; Enterprise; Undertaking

Jigyodan *Japanese*—Business group.

JIM *or* **J.I.M.** Japan Institute of Metals.

JIMA *or* **J.I.M.A.** Japan Industrial Management Association.

Jingle Bell American term for the signal bell used to transmit engine movement orders from the bridge.

JIS *or* **J.I.S.** Japan Industrial Standards.

Jitsugyo *Japanese*—Business generally; the business world.

JJ. Judges; Justices.

JL Japan Air Lines *airline designator*.

JLA Jamaica Library Association.

JLCD *or* **J.L.C.D.** Joint Liaison Committee on Documents.

JLRSS Joint Long-Range Strategic Studies *USA*.

JMA *or* **J.M.A.** Japanese Meteorological Agency; Japan Management Association.

JMC Joint Maritime Commission *ILO. Dealing with labour conditions for seafarers.*

JMSDE Joint Merchant Shipping Defence Committee.

JMSDF *or* **J.M.S.D.F.** Japan Maritime Self Defence Force.

J.M.S.O. Joint Meetings of Seafarers' Organisation.

JNR *or* **J.N.R.** Japan National Railways.

JNSA *or* **J.N.S.A.** Japanese National Shippers Association.

JNTA *or* **J.N.T.A.** Japan National Tourist Association.

Jnt. Joint.

JNTO *or* **J.N.T.O.** Japan National Tourist Organisation.

Jnt. Stk. Joint Stock(s), *q.v.*

Job Analysis Minute study of a job to find out factors needed to execute the work successfully in so far as the workers and the management are concerned.

Jobber A dealer in securities. Formerly a broker on the London Stock Exchange, now replaced by the Broker/Dealer.

Jobber's Turn The difference in the selling price against the buying price of shares when the former is higher.

Job Card A card used in factories, workshops, etc., showing details of work carried out or authorising repairs to be undertaken.

Job Evaluation The assessment of the cost, including wages, of a job. Can also mean the assessment by management of a person's aptitude to carry out a certain task or tasks.

Job-holder The person permanently in the job; permanent civil servant.

Job-hopping The habit of frequently changing one's job in search of a better salary and conditions.

Jobman. Job Management.

JOG Junior Offshore Group *relating to yachting*.

JOIDES Joint Oceanographic Institutions for Deep Earth Sampling. A body which coordinates the Ocean Drilling Programme for taking core samples in the oceans of the world.

Joint Account A bank account in the name of two or more persons. *Abbrev.* J/A *or* J.A.

Joint Cargo Committee A group of London company underwriters and Lloyd's underwriters who meet to discuss matters relating to cargo insurance and to make recommendations to the cargo insurance market.

Joint Holders Joint Registered Shares.

Joint Hull Committee A group of London company underwriters and Lloyd's underwriters who meet to discuss matters relating to hull insurance and to make recommendations to the hull insurance market.

Joint Owners Part Owners.

Joint Partner One of the partners in an enterprise.

Joint Stocks Stocks held in a company that are transferable by each holder without the consent of another party. *Abbrev.* Jnt. Stk(s).

Joint Stock Banks Joint stock companies for the purpose of banking.

Joint Stock Company A term originally applied to those unincorporated companies or large partnerships with transferable shares formed at the beginning of the last century (joint stock companies under the common law), and as to the legality of which doubts have been entertained. The formation of joint stock companies was legalised and facilitated by various Acts, the first of which was passed in 1825, and the law on the subject is now contained in the Companies Acts.

Joint Venture A venture or business activity undertaken by two or more people or firms in a merger or partnership.

Jolly Boat A small ship's boat or tender.

Jolly Roger (The) A black flag inscribed with a skull and two crossed bones. Also commonly known as 'The Pirate Flag'.

Jones Act The Merchant Marine Act 1920 (the 'Jones Act') has as its principal stated purpose 'to provide for the promotion and maintenance of the American merchant marine'. It requires that merchandise transported between US ports must be carried in vessels built in and registered in the United States and owned by persons who are citizens of the United States. For a corporation or partnership to own a vessel which qualifies for Jones Act status, the corporation must be at least 75% owned by US citizens. Further, all licensed officers and pilots and 75% of the remaining crew on these vessels must be US citizens. Merchandise is defined to include valueless material, such as sewage, salvage and dredged material, transported anywhere in the US Exclusive Economic Zone. Amongst other provisions, the Act authorises the Federal Maritime Commission to make rules and regulations to counter foreign laws and practices favouring foreign vessels.

JOOD Junior Officer of the Deck *USA*.

JOOW Junior Officer of the Watch *USA*.

JOS *or* **J.O.S.** Junior Ordinary Seaman, *q.v.*

JOT *or* **J.O.T.** Joint Observer Team.

Joule The unit of work or energy. It is equal to 1 Newton metre and corresponds also to the energy dissipated in one second by a current of 1 ampere through the resistance of 1 ohm. 1 joule equals .73756 foot pound Avoirdupois. *Abbrev.* J.

Jour. Journal, *q.v.*; Journey.

Journal Book or register in book-keeping containing daily transactions. *Abbrev.* Jour.

Journeyman A skilled workman who has completed his apprenticeship. A person who has gained experience in his work over several years.

J.P. Jet Pilot; Jet Propelled; Jet Propulsion; Judge President; Justice of the Peace.

JPA Jacksonville Port Authority *USA*.

JPC Jet Propulsion Center *USA*.

J.P. Fuel Jet Propulsion Fuel.

JPL Jet Propulsion Laboratory.

Jpn. Japan.

J.P.S. Jet Propulsion System.

J.P.T.O. Jet Propelled Take-Off *aviation*.

Jr. Jour *French*—Day.

JR Japan Rail; Jugoslavenski Registar Brodova, Yugoslavian Classification Society *symbol*.

JRDF Joint Rapid Deployment Force, formed in 1994 by the UK.

J.S. Japan Society in USA; Judgment Summons *legal*. Judicial Separation *legal*.

JSA *or* **J.S.A.** Japanese Shipowners' Association.

JSEA *or* **J.S.E.A.** Japanese Ship Exporters' Association.

Jsey. Jersey Island.

JSIF Japanese Shipbuilding Industry Foundation.

JSMDA *or* **J.S.M.D.A.** Japan Ship Development Association.

JSMEA *or* **J.S.M.E.A.** Japan Ship Machinery Association.

JSS *or* **J.S.S.** Joint Service Standard.

JST *or* **J.S.T.** Japanese Standard Time, being nine hours ahead of GMT.

JSU Japanese Seafarers' Union.

jt. Joint.

JTAG Japan Trade Advisory Group *DTI*.

Jt. Agent *or* **jt. agt.** Joint Agent.

JTCC Joint Technical and Clauses Committee *insurance*.

Jtly. Jointly

JTO *or* **J.T.O.** Jump Take-Off *aviation*.

Jt.r *or* **Jt.R.** Joint Rate.

JU Jeunesse Universelle *French*—World Youth.

j.u. Joint Use.

Ju *or* **Jun.** June.

Juan Spanish for John.

J.U.D. Juris Utriusque Doctor *Latin*—Doctor of Canon and Civil Law.

Jud. Judgment, *q.v.*; Judicial.

Judge Adv. Judge Advocate, the crown prosecutor at a court martial.

Judge Advocate-General He is the adviser of the Secretary of State for War in the UK in reference to courts martial and other matters of military law. *Abbrev.* J.A.G. *or* J. Ad. G.

Judgment *or* **Judgement** The sentence or order of the court in a civil or criminal proceeding. *Abbrev.* Jud. *or* Judgt.

Judgment Creditor A person or firm who wins a legal action against a debtor. The latter person or firm is called Judgment Debtor, *q.v.* See **Judgment Summons**.

Judgment Debtor A person or firm who has been ordered through legal action to pay money to another. The latter firm or person is referred to as Judgment Creditor, *q.v.* See **Judgment Summons**.

Judgment Summons The process used to procure the committal of a judgment debtor. *See* **Judgment Creditor** and **Judgment Debtor**.

Judicium Dei *Latin*—God's judgement.

Judo Modern form of ju-jitsu (Japanese wrestling).

Juil. Juillet *French*—July.

Juliet International Radio Telephone phonetic alphabet for 'J'.

Jumbo Bags Large bags of half a ton capacity or more, usually equipped with hooks for ease of lifting by ship or shore gear.

Jumbo Derricks Ship's derrick designed with a very heavy working load.

Jumper Stay A strong rope or wire or stay positioned between the foremast and the funnel.

Jump Ship American slang meaning 'To desert a ship'.

Junior Achievement US system offering school-leavers the opportunity to start a small business on their own, under the supervision of volunteer businessmen and industrialists. *Abbrev.* JA *or* J.A.

Junior Ordinary Seaman A young seaman of not less than $16\frac{1}{2}$ years of age. *Abbrev.* JOS *or* J.O.S.

Junk (1) A flat-bottomed sailing boat used in Chinese seas. (2) Pieces of old cordage very widely used for oakum. *See* **Caulking**. (3) Discard rope.

Jure divine *Latin*—By divine law or right.

Juris Utriusque Doctor *Latin*—Doctor of both civil and canon law. *Abbrev.* J.U.D.

Jurisdiction The official power to administer law and justice; the territory where authority is exercised.

Jurisprudence The science or the formal principles on which legal rules are based; The legal science; Sometimes used of a body of law.

Jurisprudent A learned lawyer.

Jurist A person versed in legal matters; lawyer.

Juror A member of a jury in a court of law.

Jurum Doctor *Latin*—Doctor of Laws.

Jury Mast A temporary mast in place of a lost or broken one.

Jury Rigged Temporarily equipped or replaced.

Jury Rudder An emergency and temporary rudder rigged when the proper rudder has been lost.

Jus canonicum *Latin*—Canon law.

Jus civile *Latin*—Civil law, usually Roman civil law.

Jus commune *Latin*—The common law.

Jus disponendi *Latin*—The right of disposing.

Jus divinum *Latin*—Divine law.

JUSE Japanese Union of Scientists and Engineers.

Jus gentium *Latin*—Law of the nations.

Jus gladii *Latin*—The law of the sword.

Jus mariti *Latin*—A husband's right to his wife's moveable property.

Jus naturae *Latin*—The law of nature.

Jus non scriptum *Latin*—The unwritten law.

Jus sanguinis *Latin*—Legal right of descent or heritage.

Jus scriptum *Latin*—The written law or legislation.

Justitia omnibus *Latin*—Justice to all. Motto of Columbia, USA.

Jute Cls. Jute Clauses *marine insurance.*

J.V. Junior Varsity.

JV *or* **jv** Joint Venture, *q.v.*

J. & W.O. *or* **J. & w.o.** Jettison and Washing Overboard, *q.v.*

JWP Joint Working Party.

JWS Japan Welding Society.

Jy. July; Jury.

JZ Jugoslovenski Zekeznice, Beograd. Yugoslavian Railways, Belgrade.

JZS *or* **J.Z.S.** Jugoslovenski Zavod Za Standadizacija *Yugoslav Standards Institute.*

K

k Cumulus Clouds, *q.v.*; Kilogramme, *q.v.*; Knot, *q.v.*; Kopek, *q.v.*; Koruna, *q.v.*; Krona, *q.v.*; Krone, *q.v.*

K Symbol representing the element of Potassium; Karat, *q.v.*; Kelvin *temperature*; Ketch; Knot, *q.v.*; Koruna, *q.v.*; Krona, *q.v.*; Krone *q.v.*

Ka Komppania *Finnish*—Company.

k/a Known as.

Kabushiki Kaisha *Japanese*—Joint Stock Company Limited. *Abbrev.* KK.

Kaffirs A South African stock exchange word for gold mining shares.

Kai *Japanese*—Sea.

Kaigun *Japanese*—Navy.

KAIIN Zen Nihon Kaain Kumiai *All Japan Seamen's Union.*

Kaiji *Japanese*—Maritime; Marine.

Kaijo *Japanese*—Marine; Maritime.

Kaiun *Japanese*—Shipping.

KAL Korean Air Lines.

Kalimantan Indonesian part of Borneo.

Kan *or* **Kans** Kansas, *USA.*

Kangaroo Australian stock and share term for mining shares; dealers in these.

Kanhosen *Turkish*—Tourist ship.

Kanuck *or* **Canuck** A French Canadian.

Kapok Fine cotton wool surrounding seeds of a certain tree, used for stuffing cushions, lifejackets, etc.

Karat *See* **Carat**.

KATT Kind attention.

Kayak A small one-man Eskimo canoe of light wooden framework covered with sealskins.

kb Kilobar; kilobyte.

KB *or* **K.B.** Knight Bachelor; Knight of the Order of the Bath.

K.B. King's Bench; Kommanditbolaget, Norwegian Limited Company.

Kbhvn Kobenhavn *Danish*—Copenhagen.

KBr Symbol for potassium bromide.

KC *or* **K.C.** *or* **K of C** Knight of Columbus.

K.C. King's Counsel; Knight Commander; Kansas City, *USA.*

KC Koruna, *q.v.*

kcal kilocalorie.

KCB *or* **K.C.B.** Knight Commander of the Bath.

KCl Potassium chloride. *From the mineral sylvite. Used as a fertiliser and in photography. Has low toxicity.*

KCLO₃ Symbol for potassium chlorate. Used in weedkillers, explosives and pyrotechnics.

KCMG *or* **K.C.M.G.** Knight Commander of the Order of St Michael and St George.

KCN Symbol for potassium cyanide.

K₂CO₃ Symbol for potassium carbonate.

KCVO *or* **K.C.V.O.** Knight Commander of the Royal Victorian Order.

KD Kuwait Dinar.

KD *or* **K.D.** Knocked Down. *See* **Knocked Down Condition**.

K.d. Kuwait Dinar, unit of currency used in Kuwait.

KDC *or* **K.D.C.** Knocked Down Condition, *q.v.*

KDLC *or* **K.D.L.C.L.** Knocked Down in less than car load. *See* **Knocked Down Condition**.

K.D.M. Kongelige Danske Marine, *Royal Danish Marine.*

Ke *or* **K.E.** Kinetic Energy.

KE Korean Air Lines *airline designator.*

KEAS *or* **K.E.A.S.** Knots Equivalent Air Speed.

Kedge (1) A small anchor to keep a ship steady or for working. (2) Standard weight of $21\frac{1}{2}$ tons of coal generally in use in the Tyne area, UK.

KEE Knowledge Engineering Environment *computer system.*

Keel A flat steel plate running along the centre-line of a vessel, to which are attached floor (or bottom) plates extending outwards to both sides, all being supported by lateral transverse frames termed Keelsons, *q.v.*

Keelage Dues paid by a ship making use of certain British ports.

Keel Block A short baulk of timber used to support a ship in drydock. The blocks are built up and spaced in such a way as to allow the drydock labourers to work freely in scraping, cleaning, painting, making repairs, inspecting, etc., under the ship's keel.

Keelboat A shallow cargo boat or barge mainly used on Western American rivers.

Keel Clearance *Similar to* **Underkeel Clearance**.

Keel Draft (Draught) *Similar to* Extreme Draft (Draught), *q.v.*

Keelson *or* **Kelson** A longitudinal member on top of the keel of a ship or boat to which the frames or timbers are attached.

Keeping House *See* **Act of Bankruptcy**.

Keg A small barrel or cask.

Keizai *Japanese*—Economy; Economic.

Kelp A species of very long and large seaweed.

KEMA Keuring Van Electrotechniche Materialen *Dutch*—Electrotechnical Materials Testing Institute.

Ken Kentucky, *USA*.

Kennedy Round Reference to various tariff advantages enjoyed in many areas overseas in the sphere of the General Agreement on Tariffs and Trade. *Abbrev. GATT.*

Kensetsu *Japanese*—Construction; constructive.

Kentledge Loose counterbalance weight of a crane. Also pig-iron or any other heavy material used to ballast a ship. Considered obsolete nowadays when liquids like water and edible oils are being used for ballasting.

keV Kilo-electron Volt.

Key currency Additional money reserve kept by the IMF, *q.v.*, in addition to gold reserves.

Keyhole Socket Slotted fitting on the deck of a Container Ship, *q.v.*, for securing the lashings of Containers, *q.v.*

Key Industry A most important industry when compared with others.

Key Industry Duty A scheduled list of items which are regarded as most essential for a country and on which no import duty is imposed or when a special concession is allowed when imported. *Abbrev.* KID or K.I.D.

Key Man An employee thought to be indispensable. *See* **Key Worker**

Key Worker A person deemed to have a skill or experience essential to the work of others in an enterprise. Similar to Key Man, *q.v.*

kfm Kaufmann *German*—Merchant.

KG Height of the centre of gravity above the keel of a ship.

kg *or* **Kilo** Kilogramme. Metric weight system equivalent to 2.2046 lbs.

kg cal Kilogramme calorie.

KGB Komitet Gosudarstvennoi Bezopasnosti *former Soviet State Security Service.*

kg.cum. Kilogramme per cubic metre.

kgf Kilogramme force.

KGFS King George's Fund for Sailors.

K.G.K. Kabuskiki Goshi Kaisha *Japanese Joint Limited Company.*

Kgl Königlich *German*—Royal.

Kg m Kilogrammetre, energy that will raise one kilogramme to the height of one metre.

Kg/m² Kilogramme per square metre.

Kg/m³ Kilogramme per cubic metre.

Kgn Kingston, capital of Jamaica.

K.H.M. King's Harbour Master *also abbrev.* Q.H.M. Queen's Harbour Master.

kHz. Kilohertz.

KI Biro Klasifikasí Indonesia. *Indonesian Classification Society.*

KIA Killed in Action.

KIAS *or* **Kias** Knots indicated airspeed.

Kick An oil drilling expression. When rocks containing high pressure oil or gas are penetrated by the drill bits the flow will immediately start from the well. This action is called Kick.

KID *or* **K.I.D.** Key Industry Duty, *q.v.*

Kil Kilometre.

Kilderkin A cask for liquids containing 16 or 18 gallons; this as measure.

Kilk Kilkenny, Republic of Ireland.

Killick *or* **Killock** A small anchor; naval slang for leading rating.

Kilo Kilogramme; International Radio Telephone phonetic alphabet for 'K'.

Kilohm Kilo-ohm (one thousand ohms) *electricity.*

Kilojoule One thousand joules. *See* **Joule**.

Kilometre One kilometre (one thousand metres) equals 0.62137 of a statute mile or 1093.613 yards or 3280.86 feet.

Kilowatt One thousand watts *electricity.*

kin Kinematic. This means motion considered abstractly without reference to force or mass.

Kina Currency unit of Papua New Guinea.

Kinds of Oils *See* **Oils**.

King George's Fund for Sailors *Abbrev.* KGFS. The was formally launched at the Mansion House in London on 5 July 1917 with the Lord Mayor in the Chair. This was the new and permanent name of the Sailor's Fund, whose inaugural meeting had been held in March of that year when the shipowners, officers of the Royal and Merchant Navies, merchants, representatives of the Marine Charities and others met in answer 'to a generally expressed desire that some organisation should be formed to benefit Marine Charities generally'. King George's Fund for Sailors is the central and the only fund which covers all maritime charities. It exists to give financial aid and advice to marine benevolent organisations and institutions. Each year around 90 such bodies receive help from the resources of KGFS.

Many institutions give regular financial assistance to former seafarers or to their dependants who are in need; some make emergency payments to those in trouble; others provide homes for those who through age, infirmity or financial incapacity can no longer look after themselves; educational grants are made to the dependants of seafarers or to those preparing for a life at sea; and the Fund assists the missions which provide physical and spiritual comfort to seafarers when they come ashore.

All who earn or have earned their living at sea—the men and woman of the Royal Navy, the Merchant Navy and the fishing fleets, together with their dependants—qualify for the assistance that KGFS and the charities it supports can give.

In 1993, grants of over £2¼ million were given to over 80 charities, funds and trusts caring for seafarers and their dependants.

Kingpost *Similar to* **Samson Post**, *q.v. Abbrev.* KP or K.P.

King's or Queen's Enemies The armed forces at a state of war with the United Kingdom, or the country to which a chartered ship belongs, whose action represents one of the exceptions in the Carriage of Goods Act 1924, exempting the Carrier, *q.v.*, from his normal liabilities.

King's or Queen's Warehouse The Customs' stores, in UK, where seized goods are retained.

Kin visc Kinetic Viscosity.

Kip (1) Unit of force equivalent to 1000 lbs. (2) Unit of currency used in Laos.

KISA *or* **K.I.S.A.** Thousand Square Inches Absolute.

Kisen *Japanese*—Ship.

Kite Slang for accommodation bill, *q.v.*, which is endorsed by a person as a guarantor to another. Should the latter fail to honour his obligation the former acceptor would become liable. The word kite may also be referred to as a British Standards Institution (BSI) mark of approval.

Kite Flying Endorsing to Kite, *q.v.*, or an accommodation bill, *q.v.*, as a security or guarantee. The word 'flying' relates to 'endorsing'.

KIVI Koninklijk Instituut Van Ingenieurs *Royal Dutch Institute of Engineers.*

kj Kilojoule, *q.v. See* **Joule**.

K.K. Kabushiki Kaisha *Japanese Limited Company.*

K/K Knock for Knock, *q.v.*

Kl Kilolitre.

KLM *or* **K.L.M.** Koninklijke Luchtvaart Maatschappij *Royal Dutch Airlines, Dutch State Airlines.*

Klondiker A specialised ship which anchors offshore and loads fish from fishing vessels. The fish is frozen or processed and when the ship has a full load she proceeds to the market at her home port. Commonly foreign fishing vessels deliver their catches to such a vessel instead of landing them at their own ports.

km Kilometre, *q.v.*

km² Square kilometre.

km³ Cubic kilometre.

K.Mess. King's Messenger. Q. Mess stands for Queen's Messenger.

Km/h Kilometres per hour.

KMnO₄ Symbol for potassium permanganate.

KMO *or* **K.M.O.** Kobe Marine Observatory, Japan.

kmwhr Kilomegawatt per hour.

Kn Knot, *q.v.*

KNAK *or* **K.N.A.K.** Kongelig Norsk Automobilklubb *Royal Norwegian Automobile Club.*

KNAN *or* **K.N.A.N.** Koninklijke Nederlandse Akademie Voor Naturwetenschappen *Royal Dutch Academy of Sciences.*

KNK *or* **K.N.K.** Kita Nippon Koku *Northern Japanese Airlines.*

KNMI *or* **K.N.M.I.** Koninklijke Nederlandse Meteorologisch Instituut *Royal Dutch Meteorological Institute.*

KNO₂ Symbol for potassium nitrate *saltpetre.*

Knocked Down Condition Dismantled vehicles, machines and other merchandise. This is done at times in order to save space in shipments. *Abbrev.* KDC or K.D.C.

Knock for Knock A common practice in insurance when two vehicles collide. Mostly used in motor insurance when each insurer bears the insured's expenses. *Abbrev.* K/K.

Knock Off Slang term meaning to finish work or duty.

Knock-Out Agreements Agreements between bidders in an auction sale who agree to abstain from bidding to buy goods for resale or for other advantageous purposes. Such agreements are rendered illegal and if proved may be prosecuted as fraudulent.

Knot Unit of speed in navigation which is the rate of nautical miles per hour. One knot equals one nautical mile (6,080 feet or 1,852 metres) per hour.

KNRV *or* **K.N.R.V.** Koninklijke Nederlandse Redersvereniging *Royal Netherlands Ship-owners' Association.*

KNSM *or* **K.N.S.M.** Koninklijke Nederlandse Stoom-Boat Maatschappij *Royal Dutch Steam Ship Company.*

KNVL *or* **K.N.V.L.** Koninklijke Nederlande Vereniging Voor Luchtvaart *Royal Dutch Aero Club.*

KNVTO *or* **K.N.V.T.O.** Koninklijke Nederlandse Van Transport Ondernemingen, Den Hague *Association of Netherlands Transport Enterprises*, The Hague.

K.O. *or* **k.o.** Keep off; Keep Out; Kick Off; Knock Out *in boxing.*

Kobe One of the most popular financial centres in Japan and a major port.

Kobo Currency unit of Nigeria equivalent to one hundredth of a naira.

Koeki *Japanese*—Trade.

Kogyo *Japanese*—Business Association/Society.

KOH Symbol for potassium hydroxide.

Kokan *Japanese*—Exchange.

Koku A Japanese Standard Measure representing 10 cubic feet, alternatively 120 board feet.

K.O.M. Knights of the Order of Malta.

Kommanditbolag *Swedish*—Limited Partnership Company.

Kommandit-gesellschaft *German*—Joint Stock Company. *Abbrev.* KG or K.G.

Kommandiitiyhtio *Finnish*—Limited Company.

Kommandittselskap *Norwegian*—Limited Company. *Abbrev.* K.S. *or* K.S.

Komp. Kompanie *German*—Company.

Kompagni *Danish*—Company.

Komurculuk *Turkish*—Coal Company.

Kon. Koninklijke, *q.v.*

Koninklijke *Dutch*—Royal. *Abbrev.* Kon.

kop Kopek, *q.v.*

Kopek Old spelling for copeck, a unit of currency used in USSR, one hundred copeks equals one rouble.

Kor Korea.

Koruna Unit of currency used in Czechoslovakia, one hundred heller equals one koruna.

Kosokusen *Japanese*—Ferry.

KOTRA Korean Trade Promotion Corporation, *Seoul.*

KP *or* **K.P.** Kingpost. *See* **Samson Post**.

kph *or* **k.p.h.** Kilometres per hour.

Kr *or* **K.R.** Keel Rider; King's Regulations.

kr Krona, *q.v.*; Krone, *q.v.*

KR Classification Society Korean Register of Shipping.

K.R & A.I. King's Regulations and Admiralty Instructions; Q.R and A.I stands for Queen's Regulations and Admiralty Instructions.

K.R.C. Knight of the Red Cross.

Krona Unit of currency used in Sweden and Iceland. *Abbrev.* kr. One hundred øre equals one krona.

Krone Unit of currency used in Norway and Denmark. *Abbrev.* kr. One hundred øre equals one krone.

Krugerrand A 22 carat gold coin minted in South Africa. The first bullion coin to contain exactly one ounce of pure gold, with a little alloy to harden it.

Ks *or* **KS** Kansas, *USA.*

K.S. *or* **K/S** Kommandittselskap, *q.v.*

K₂S Symbol for potassium sulphide.

K₂SO₃ Symbol for potassium sulphite.

K₂SO₄ Symbol for potassium sulphate.

KSC Kuwait Shareholding Company.

KSFT King Size Filter Tipped *cigarettes.*

KSM *or* **K.S.M.** Kungliga Svenska Marinen *Royal Swedish Navy.*

KSRC Kuwait Shipbuilding and Repair Company.

Kt. Karat, *q.v.*; Knight

Kt *or* **Knt** Knight.

KT *or* **K.T.** Knight of the Order of the Thistle; Knight Templar.

KTAG Korea Trade Advisory Group.

Kt.Bach. Knight Bachelor.

Kto. Konto *German*—Account.

K.U. Kuwait Airways *designator.*

Kumiai *Japanese* Society; Association; Corporation; labour union.

Kuna Currency unit of Croatia.

K.U.T.D. *or* **k.u.t.d.** Keep up to date.

Kuw. Kuwait.

Kuzbas Kuznetsk Basin in the USSR. It contains large deposits of coke and coal and is also known for its production of steel and iron.

KV Kilovolt, one thousand Volts. *Also abbrev.* k.v.

kVah Kilovolt–ampere hour.

kVp Kilovolt Peak.

kW Kilowatt, *q.v.*

K.W.A.C. *or* **k.w.a.c.** Key word and context.

Kwacha Unit of currency used in Zambia, one hundred ngwee equals one kwacha.

kWh Kilowatt-hour.

K.W.I.C. *or* **k.w.i.c.** Keyword in context.

K.W.O.C. *or* **k.w.o.c.** Keyword out of context.

K.W.O.T. *or* **k.w.o.t.** Keyword out of title.

K.W.T. *or* **k.w.t.** Keyword in title.

KY Kentucky, *USA.*

ky *or* **k** Kyat, *q.v.*

Kyat Unit of currency used in Burma, one hundred pyas equals one kyat.

Kyo Kyoto, Japan.

Kyokai *Japanese*—Association.

Kyokukai *Japanese*—Polar Sea.

L

l Difference in Latitude; League; Lightening; Link; Litre; *q.v.*; Logarithm; Logarithmic; Long.

l *or* **L** Libra; Pound Sterling; Low Pressure Area.

£ Libra; Pound Sterling.

L Drizzle *weather observation*; 'Freeboard Marking' on each side of the ship near to the load line and assigned to timber cargoes for all zones and seasons; Latitude, *q.v.*; Line; Long; Low Loran PRR; Lower Limb *astronomy*; Locus *Latin*—Place; Roman numeral for figure 50; Lira, Italian currency; Wavelength.

LA Linea Aerea Nacional *airline designator*; Los Angeles, *USA.*

La Louisiana, *USA.*

L/A *or* **L.A.** Ledger Account; Letter of Authority; Lettre d'avis *French*—Letter of Advice; Lloyd's Agent; Legislative Assembly; Local Authority; Literature in Arts.

l.a. Landing Account, *q.v.*; Law Agent; Legal Adviser; Lloyd's Agent'; Local Agent; Low Altitude *weather observation.*

LA₂ Liquid Hydrogen.

LAA *or* **L.A.A.** Libyan Arab Airlines.

L.A.A.A. London Association of Accountants and Auditors.

Lab. Label; Labour; Labour *politics*; labour; laboratory; Labrador.

LAB Laboratory Animals Bureau; Live Animal Board *IATA*; Low Altitude Bombing.

LAB 77 Title of Lloyd's Average Bond, set in 1977.

Lab. Cur. Labrador Current.

Label Clause A marine insurance clause covering the faded identification letters and marks on cases and canned products because of contact with sea water. The Underwriters, *q.v.*, reimburse for the overall cost of relabelling. *Abbrev.* L.cl.

Laborare est orare *Latin*—To work is to pray; work is worship.

Labor omnia vincit *Latin*—Labour conquers all things. Motto of Oklahoma, USA.

Labour Said of a ship when she rolls and pitches in heavy seas.

Labour Force The strength or supply of a workforce in employment.

Labouring When a ship is pitching and rolling heavily *nautical.*

Labour Legislation Parliamentary Acts regarding conditions for workers.

Labour Market The supply of labour with reference to the demand for it.

Labour Movement The working classes as a political force; also meaning the formation of unions to represent their members' interests.

Labour-Saving Invention Any system of automation which saves labour and time.

Labour Turnover The degree of movement relating to the hiring or dismissal of workers.

Labor Union American Trade Union.

Labrador Coastal territory in the northern part of the Province of Newfoundland, Canada.

LABS Low-Altitude Bombing System *aviation.*

LAC Laboratory Animals Centre; Latin-American and Caribbean; Leading Aircraftman; Lights Advisory Committee; London Approach Control *aviation*; Lunar Aeronautical Chart.

L.A.C. London Athletic Club; Licenciate of the Apothecaries' Company.

L.A.C.C. Lloyd's Aviation Claims Centre.

LACD London Accident Claims Department *insurance.*

LACE Liquid Air Cycle Engine.

LACH Lightweight Amphibious Container Handlers *USA.*

Laches Legal term for procrastination or unreasonable delay in asserting a person's rights.

LACSA Lineas Aereas Costaricenses *Costa Rica State Airlines.*

LACT Lease Automatic Custody Transfer.

LACV Lighter Air Cushion Vessel *US Army.*

LAD Light Aid Detachment.

LADE Lineas Aereas del Estado *Argentine State Airlines.*

Laden in Bulk A ship loaded with bulk cargo. Grains, coals and other commodities which are shipped loose are termed as being shipped in 'bulk'.

Laden Load Line *Similar to* **Loadline marks**, *q.v.*

Lading, Bill of *See* **Bill of Lading**.

LADS Laser Airborne Depth Sounding, *q.v.*, Lloyd's Analysis of Dynamic Systems.

Laesa majestas *Latin*—Crime against a sovereign; High treason.

LAFTA Latin American Free Trade Association.

Lag. Lagoon.

Lagan *or* **Ligan** Recoverable heavy cargo thrown overboard from a sinking ship and which is marked by a buoy. Also Ligsam.

LAGEOS Laser Geodynamics Satellite *USA*.

LAHR *or* **L.A.H.R.** London, Antwerp, Havre and Rouen. Ports used by the grain traffic trade.

LAHS Low Altitude High Speed *aviation*.

Lagoon Shallow waters bordering the sea through the natural formation of sand, pebbles, coral, etc. *See also* **Atoll**.

LAIA Latin American Integration Association.

Laid Down Describes a ship when her keel is first assembled in the shipyard and marks the beginning of her construction.

Laid Up A ship which is out of commission. The reason could be for fitting out, awaiting better markets, needing work for classification, etc. *Abbrev.* L/U *or* l/u.

Laid-up Returns *See* **Cancelling Returns**.

Laid Up Tonnage Laid up, *q.v.*, ships in general or in total.

Laissez-faire *French*—This refers to freedom of action as opposed to state planning and restriction; let well alone.

Laiva *Finnish*—Ship.

Laivanisannisto *Finnish*—Shipping.

Lake A large amount of water surrounded by land.

Lake Cls. Lake Time Clauses *marine hull insurance*.

Lake I/V & Ex Great Lakes Hull Clause—Increased Value and Excess Liabilities.

Lake SS Classed ships having Special Survey to trade in the Great Lakes service. *See* **Laker**.

Lake T.L.C. Lake Total Loss Clauses *marine insurance*.

Lake T.L.O. & Exs. Lake Total Loss Only & Excess *marine insurance*.

Laker A particular style of ship (long and narrow) designed for the carriage of cargoes about the Great Lakes System of North America.

La Manche The French meaning for English Channel.

LAMCON BILL *or* **Lamcon-bill** Bill of Lading, *q.v.*, used in conjunction with the Lamcon charterparty.

Lame Duck A member of the Stock Exchange, *q.v.*, who is unable to pay his debt in due time and is about to be Hammered, *q.v.*, and expelled from the House; A business that is not flourishing; An inefficient person.

LAMPS Light Airborne Multipurpose System *helicopter equipment system*.

LAMS Lloyd's Analysis of Mooring Systems.

LAN *or* **L.A.N.** Local Apparent Noon *nautical*; Local Area Network.

LANBY Large Automatic Navigational Buoy, *q.v.*

Lancs. Lancashire, *UK*.

Land To bring an aircraft to rest on land or sea.

Land Agent A person engaged by a land owner to collect rents, let houses, farms, etc. He is also engaged in the buying and selling of houses in some instances.

Land Bank An agricultural bank.

Land Breeze Air blowing seaward from land.

Land Bridge Rate The combined tariff rate charged when a Land Bridge operation takes place.

Land Days Landed Terms *Similar to* **Franco Domicile** *or* **Rendu** *or* **Free Contract**, *q.v.*

Land Effect Coastal refraction.

Landfall (1) First sighting of land from a ship at sea. (2) The silhouette appearance when approaching land at sea.

Land Ice (1) Floating ice that has separated from a glacier. (2) A continuous field of ice connected to the land during winter.

Landed Terms Expenses involved in the transportation of the goods to be imported, including all the port charges at the point of arrival. *Abbrev.* L.T. *or* l.t.

Landed Weight The final weight of the merchandise when unloaded.

Landeszentralbank Central German Bank, based in Frankfurt. *Abbrev.* Lzb.

Landing Account Full details of the merchandise coming off a ship showing the marks, number of packages, damage or shortages if any, etc. *Similar to* Out-Turn Report. *Abbrev.* l.a.

Landing Book Company and warehousekeepers use landing books to record the amount of merchandise stored and the commencement date for rental charges is entered in the book.

Landing Cost The arrived cost of the merchandise on the C.I.F., *q.v.*, value, plus unloading, import duty, if any, and other incidental charges.

Landing Craft A flat bottomed vessel having a Low Draft (Draught), *q.v.*, to allow her to enter into shallow and/or to run aground on sandy beaches. Some landing craft are used to carry personnel and/or heavy vehicles which can drive ashore when the bow doors are opened.

Landing Field A fairly large levelled field where aircraft can land or take off.

Landing Gear This refers to the undercarriage (wheels) of an aircraft.

Landing Officer A Customs Officer, mainly in UK, in control of both import and export operations.

Landing Order An authorised document for the Dock Company or Wharfinger to enable acceptance of cargo discharged from a ship.

Landing Ship A sea-going floating dock in the form of a ship which can be flooded while in the open seas to dock another vessel or to accommodate cargo barges. The Trimariner has also been developed, being more or less identical but on a larger scale. *See* **Capricorn** *also* **Landing Ship Dock**.

Landing Ship Dock A ship in the form of a landing craft capable of carrying conventional landing craft and also designed to carry air cushion vehicles. *Abbrev.* L.S.D. *See* **Capricorn** *also* **Landing Ship**.

Landing Stage Platform, fixed or floating, for embarking and disembarking passengers.

Landlocked *or* **Land Locked** This can refer to a port which is sheltered from all angles by land and in consequence of which provides a safe harbour. Also describes a territory without a seaboard or sea coast.

Landlord Property owner of house, land, shop, business premises, etc. Such places that yield rent.

Landlubber *See* **Landsman**.

Landmark *or* **Land Mark** One meaning of this can be a navigational indicator or warning in a conspicuous position on land as a reference or guidance for ships.

Landsick Said of a ship when her movements are impeded by the nearness of land.

Landslip The sliding down of a mass of land on a cliff or mountain.

Landsman A non-seafaring man. Sometimes also referred to as landlubber who is a person ignorant of the sea and ships.

Landswell A roll of water near the shore.

Lane Metres The described capacity of Ro/Ro vessels. The width of a lane metre although being a minimum 2.5 metres to suit standard container dimensions, plus a small space for access by stevedores, varies depending on the characteristics of each ship.

Lanyard A short rope or line attached to something to secure it; a cord hanging around the neck or looped around the shoulder to which may be attached a knife, whistle, etc.

LAP Lineas Aereas Paraguayas *Paraguay National Airlines*.

La P. La Paz, *Bolivia*.

Lapilli *Latin*—Fragments of lava erupted from a volcano.

Lapsus calami *Latin*—A slip of the pen.

Lapsus linguae *Latin*—A slip of the tongue.

Lapsus memoriae *Latin*—A lapse of memory.

LAPT *or* **L.A.P.T.** London Association for the Protection of Trade.

Lap-top A portable computer, small enough to be used on a person's lap.

LAR *or* **L.A.R.** Liquefied Argon; Live Animals Regulations.

LARAS Lower Airspace Radar Advisory Service *aviation*.

Large Automatic Navigation Buoy A fairly large buoy of approximately 50 feet in diameter and with a reasonable amount of water ballast for stability. It is fitted with electric power to generate long-range light for navigational signals. *Abbrev.* LANBY *or* L.A.N.B.Y.

Larboard Old term for the port side or left-side of a ship. Opposite to Starboard. Now replaced by Port to avoid confusion with starboard.

Larceny The unlawful taking and carrying away of things personal, with intent to deprive the rightful owner of the same.

LAS *or* **L.A.S.** League of Arab States; Legal Aid Society; Lord Advocate for Scotland; Low Altitude Satellite; Lower Air Space.

Lascar Indian seaman.

LASER *or* **Laser** Light Amplification by Stimulated Emission of Radiation.

Laser Airborne Depth Sounding System using an aircraft to direct red and green laser beams reflected respectively by the sea surface and the sea bed. *Abbrev.* **LADS**.

L.A.S.H. Lighter Aboard Ship. Barges are floated up to the stern of a mother ship then hoisted inboard by gantry crane and thereafter lifted directly to their stowage positions.

Lash To secure with wire or line. *See* **Lashings**.

Lasher or Rigger A port worker whose job is to lash the cargo on board a ship before sailing. It is generally the case that the seamen on board will attend to additional Lashing, *q.v.*, immediately after the ship sails, to ensure safety.

Lashing Eye A hook, ring or other fitting anywhere on the ship to which to Lash, *q.v.*, cargo or Containers, *q.v.*

Lashing Post A post or hook positioned on the deck of a ship to which is fastened deck cargo to prevent shifting or washing overboard.

Lashings Ropes or other tackle to fasten deck cargoes or vehicles in the holds to prevent shifting.

LASL Los Alamos Scientific Laboratory USA since 1972.

LASP Low Altitude Space/Surveillance Platform.

LASS Lighter Than Air Submarine Simulator.

Lastage The name given to sand, soil, gravel and other similar material used as ballast for ships. Since modern ships have double bottoms for liquid ballast purposes these materials are hardly being used, if at all.

Last Days *Similar to* **Laydays**, *q.v.*

Last Dog Watch *or* **Second Dog Watch** Seaman's watch at sea from 6 p.m. to 8 p.m.

Last in First Out (1) Assumption in the costings adopted by firms that goods recently bought are used first and those in stock may be old and as such cannot be appraised at the current ruling price. However, for regularity's sake, the stock is priced at a fixed value similar to its original price for accounting purposes. (2) A system when discharging employees from factories or other employment. The last to be employed are the first to be discharged in a gradual process. *Abbrev.* Life *or* l.i.f.o.

Last Trading Day Last day of dealing in the current Stock Exchange account period. *Obsolete*.

Lat. Latitude, *q.v.* Latus *Latin*—Wide.

LAT *or* **L.A.T.** Local Apparent Time *nautical*; Los Angeles Time; Lowest Astronomical Time *nautical*.

lat. Latitude Astronomical Time *nautical*.

LATAG Latin–American Trade Advisory Group.

LATCC *or* **L.A.T.C.C.** London Air Traffic Control Centre.

LATCRS *or* **L.A.T.C.R.S.** London Air Traffic Control Radio Station.

Lat/Def. Latent Defect.

Latent Defect A defect in the construction of a ship or machinery that is not readily discernible to a competent person carrying out a normal inspection. Discovery of a latent defect does not give rise to a claim on the ordinary hull policy, but damage caused thereby is usually covered. *Abbrev.* Lat/Def.

Lateen A triangular sail on a long yard at an angle of 45 degrees to the mast. Similar to those on Arabian dhows.

Lateral Integration *Similar to* **Horizonal Integration**, *q.v.*

Latex A milky liquid originally extracted from a tree grown in Brazil called *hevea Brasiliensis* and from which rubber is extracted. Malaysia and Indonesia are the chief producers of rubber.

LATF *or* **L.A.T.F.** Lloyd's American Trust Fund.

lat. ht. Latent Heat.

Latine dictum *Latin*—Spoken in Latin.

Latitude Distance which is measured in degrees North or South of the Equator. *Abbrev.* Lat.

LAUA *or* **L.A.U.A.** Lloyd's Aviation Underwriters' Association.

Laughing Gas Nitrous oxide, with an intoxicating effect when inhaled, and formerly used as an anaesthetic.

LAUK *or* **L.A.U.K.** Library Association of the United Kingdom.

Launch To cause a ship to slide into the water for the first time; To push out to sea; To cause a vessel to pass from the land into the water.

Launch Escape System A safety system used for astronauts in case they want to bail out in emergency. *Abbrev.* LES.

Launching Clause A marine insurance clause specifying the amount and limitation of Underwriters', *q.v.*, liability during the launching of a newly built ship.

Launching Ways Heavy timber structures under the bottom of a ship on which she slides down to the water at launching.

Laus Dei *Latin*—Praise to God.

LAV Lymphadenopathy–associated Virus *virus for AIDS. Similar to* HTLV.

LAW *or* **L.A.W.** League of American Writers; Light Anti-Armour Weapon; London Association of Wharfingers.

Law Division Clerk A clerk employed by the customs in the USA whose job is to open up and examine packets and/or merchandise which has either been refused or left behind and unclaimed.

Lawful Cargo *or* **Lawful Merchandise** Cargo or merchandise which is considered to be legal in accordance with the laws of a country.

Law Merchant Mercantile law: Laws relative to commerce, and bills of exchange, partnership and other matters reflecting general European usage.

Lawsuit A case before the courts of law or equity in which there is a dispute between two parties.

Lawyer A member of the legal profession. An expert in legal science; Advocate; *q.v.* Barrister; Solicitor.

Lay Barge A large sized barge of strong structure specially designed to lay pipes on the seabed in preparation for carrying gas or oil products from offshore oilwells.

Lay/Can An abbreviation for Laydays/Cancelling.

Laydays The dates within which a ship is to present for loading. If she is too late to meet the last date—the cancelling date—she may be cancelled.

Lay-off A slack time in industry or business. A vessel which anchors outside a harbour. Cessation of work or employment, out of work. To send workers off the job when production declines temporarily.

Lay-off Pay Reduced wages paid to men laid-off or sent home.

Laytime The time permitted a charterer under a voyage charterparty to perform cargo operations.

Lay-up Berth Place in a port or sheltered anchorage reserved for ships without employment and laid up. *See* **Laid Up Tonnage**.

Lay-up Return A return of part of the annual premium on a ship time policy paid back to the assured by the underwriter because the ship has been laid up and not exposed to full navigational risks for a period of not less than 30 consecutive days. Such return is not paid until the natural expiry date of the policy, and is forfeit if the vessel becomes a total loss before such date. Also called Laid-up Returns.

Lazaretto *or* **Lazaret** (1) In the old days, this was a hospital where persons with infectious diseases, principally leprosy, were isolated or quarantined until they recovered. From lazar, an old word for a leper, as was the beggar named Lazarus. *See* **Quarantine**. (2) A small compartment aft in old merchant ships used as a hospital but often as a store-room.

Lazy Guy A rope or Tackle, *q.v.*, fastened to a ship's Boom, *q.v.*, to prevent it from swinging when not in use.

L.B. Bachelor of Letters; Local Board; Lloyd's Broker.

lb. Lebra *Latin*—Pound Weight.

L.B. *or* **l.b.** Landing Barge.

LB *or* **L.B.** Baccalaureus Litterarum *Latin*—Bachelor of Letters; Bachelor of Literature.

LBA *or* **L.B.A.** Lloyd's Brokers' Association.

lb.ap. Pound Apothecaries.

lb.av. *or* **lb. avoir.** Pound Avoirdupois.

LBC Local Baggage Committee *IATA*.

LBCH *or* **L.B.C.H.** London Bankers' Clearing House.

lb. chu. Pound Centigrade Heat Unit.

lb. f. Pound Force.

lb. ft. Pound-Foot.

L.B.H. *or* **L.W.H.** Length, Breadth and Height or Length, Width and Height.

lbm Pound Mass.

LBM Linear Block Manifold *oil drilling*.

LBMA *or* **L.B.M.A.** Liberian Bureau of Maritime Affairs.

LBO *or* **L.B.O.** Leveraged Buy-out *financial*.

LBP *or* **L.B.P.** Length Between Perpendiculars, *q.v.*

LBS *or* **L.B.S.** Lifeboat Station; London School of Business Studies.

L.b.s. Lectori Benevolo Salutem *Latin*—Greetings to the kind reader.

lb.t. Pound Troy.

LBTF Land Based Test Facility *USA*.

LBV Landing Barge Vehicle.

l.b.w. Leg before wicket *cricket*.

Lc. Little Change *weather observation*.

l.c. Legal Currency; Letter Card; Low Calorie; Low Carbon; Lower Case *typography*.

l.c. *or* **Loc. cit.** Loco citato *Latin*—In the place cited.

LCA *or* **L.C.A.** London Court of Arbitration.

LCAC Landing Craft Air Cushion *US naval landing craft*.

L.C.A.O. Lloyd's Central Accounting Office.

LCB *or* **L.C.B.** Longitudinal Centre of Buoyancy; Lord Chief Baron.

l.c.b. Longitudinal Centre of Buoyancy.

LCC *or* **L.C.C.** Lancashire Cotton Corporation; London Chamber of Commerce; London County Council *now defunct*.

LCCI London Chamber of Commerce and Industry.

LCD Liquid Crystal Display, *quartz analogue watch or electronic calculator, etc.*

L.C.D. *or* **l.c.d.** Lowest Common Denominator; Lower Court Decision.

LCDR Lieutenant Commander.

LCE London Commodity Exchange, *q.v.*

L.C.E. Licentiate in Civil Engineering.

L.C.F. *or* **l.c.f.** Longitudinal Centre of Flotation; Lowest Common Factor.

LCFTA *or* **L.C.F.T.A.** London Cattle Food Trade Association.

L.C.G. *or* **l.c.g.** Longitudinal Centre of Gravity.

LCH The London Clearing Houses owned by the major UK clearing banks.

L.CH. Licentiatus Chirurgiae *Latin*—Licentiate in Surgery; Lord Chancellor.

LCI Landing Craft Infantry.

L.C.J. Lord Chief Justice.

L.C.L. *or* **l.c.l.** Lower Control limit.

L.Cl. Label Clause, *q.v.*

LCLV Liquid-Crystal Light Valve.

LCM *or* **L.C.M.** London College of Music; Least Common Multiple.

L.C.M. *or* **l.c.m.** Least Common Measure.

LCN *or* **L.C.N.** Load Classification Number *aviation*; Local Civil Noon *nautical*.

LCNR Low-Cost Non-Returnable, *q.v.*

LCO *or* **L.C.O.** Launch Control Officer.

LCP *or* **L.C.P.** Licentiate of the College of Preceptors; Low Cost Production; Liquid Crystal Polyster, synthetic rope fibre.

l/cr Lettre de Crédit *French*—Letter of Credit, *q.v.*

LCT *or* **L.C.T.** Landing Craft Tank; Local Civil Time.

LCTA *or* **L.C.T.A.** London Corn Trade Association. An association specialising in the grain grade. *See* **Grain Contract Form**.

L.C.T.F. *Similar to* **C.T.O.**, *q.v.*

LCU Landing Craft Utility.

LCY London City Airport.

Ld. Limited; Pound Weight; Load; Lord.

LD *or* **L.D.** Licentiate of Divinity; London Docks; Low Density; Doctor of Letters *in Holland*; Load Draft (Draught); Light Draft (Draught).

L/D Letter of Deposit.

L&D Loans and Discounts; Loss and Damage.

l.d. Line of Departure.

Lda *or* **lda** Limitada *Portuguese*—Limited.

L.D.A. Light Defence Aircraft.

l.d.b. Light Distribution Box.

LDC Less Developed Countries *finance*. Long Distance Call; Low Dead Centre.

L.D.C. *or* **l.d.c.** Long Distance Call; Low Dead Centre.

L.d.d. *or* **L.D.D.** Lost During Discharge.

LDDC London Docklands Development Corporation.

L.Def. Latent Defect, *q.v.*

LDEF Long Duration Exposure Facility, *q.v.*

Lderry *or* **Ldy** Londonderry, *Northern Ireland*.

Ldg. Leading; Loading.

Ldg and Dely *or* **ldg and dely** Landing and Delivery.

L.Div. Licenciate in Divinity.

L.dk. Lower deck.

LDL Low Density Lipoprotein.

L.D.L. *or* **L.d.l.** Lost During Loading.

Ld. Lmt. Load Limit.

L.D.M.A. London Discount Market Association.

Ld. May. Lord Mayor.

LDMC Livestock Development and Marketing Corporation, *Burma*.

Ldmk *or* **LDMK** Land Mark.

Ldn. London, *UK*.

LDO Limited Duty Officer *USA*.

Ldp. Lordship.

LDPE *or* **Ldpe.** Low Density Polyethylene.

L.D.P.S. Lloyd's Data Processing Services.

L.D.S. Licentiate in Dental Surgery; Laus deo semper *Latin*—Praise be to God always.

L.D.Sc. Licentiate in Dental Science.

lds *or* **Lds** Loads.

LDSTSA London Deep Sea Tramp Shipowners' Association.

L.D.T. *or* **l.d.t.** Lost During Transhipment.

LDT *or* **L.D.T.** Light Displacement Tonnage.

L.D.X. Long Distance Xerography.

Le *or* **Leb.** Lebanese or Lebanon.

L.E. Land's End; Leading Edge; Leg End.

Lea. League; Leather; Leave.

Lead Heavy easily fusible soft malleable base metal of dull pale bluish-grey colour; a lump of lead with a hollowed out base 'armed' with tallow is used by seamen to determine the nature of the seabed; also to discover the depth of water when used with a line marked off in fathoms. The user is known as the leadsman.

Leadage Transport expenses of coal or minerals from the pithead to the loading berth.

Lead *or* **Leader** Leading Underwriter, *q.v.*

Leading Question A question suggesting the answer. A question leading to the answer.

Leading Underwriters The first insurers or underwriters who sign the Slip, *q.v. Also called* Leaders. *See* **First Three**.

Lead Line *or* **Sounding Line** *or* **Sounding Tape** *See* **Lead.**

Lead Yard The shipyard which originally builds a ship of its own design and other yards build vessels of similar design on licence.

League Three nautical or geographical miles; a sea mile averages 6080 feet or 1852 metres depending on the geographical position.

Leakage The liquid which escapes or oozes through a hole or other openings from barrels, casks, pipes, etc.

LEAPS Long Term Equity Anticipation Securities. *USA Finance*.

LEAR Low-Energy Anti-Proton Ring.

Lease Contract for the possession of an object (such as land, ship, building, etc.) which is hired by the owner to another person or firm for a period of time against the payment of rent.

Leatherette Artificial leather used among other things for bookbinding.

Leb. Lebanon.

LEC Local Employment Committee; Local Export Control.

Lect. Lecture; Lecturer.

Led. Ledger.

LED Light Emitting Diode.

Ledger Principal book of the set used for recording trade transactions, containing debtor and creditor accounts.

LEEA Lifting Equipment Engineers Association.

Lee Anchor The leeward one of two anchors to which a vessel is riding when moored.

Lee Board A board lowered over the side of a ship to reduce the drift to leeward.

Leeside *or* **Leeward** Sheltered side; the side away from the wind.

Lee Tide *or* **Leeward Tide** Tide which runs in the same direction as the wind.

LEFO *or* **L.E.F.O.** Land's End for Orders.

Left-Hand Lay This refers to the twist direction of the Strands, *q.v.*, of a rope.

Left Out Short Shipped.

Leg Equivalent Continuous Sound Level. Measure of total sound energy monitored over a set period and used in establishing environmental noise contours.

Legacy Property or moneys gifted or bequeathed at the death of a person, by way of a will.

Legacy Tax A tax on inheritance. In England Legacy Duty was abolished by the Finance Act 1949. It is now replaced by Inheritance Tax.

Legal Battle Arguments that may eventually lead to court cases for settlement.

Legal Holiday National holiday of a state.

Legal Quay A wharf licensed by the customs authority for bonded stores.

Legal Reserve Amounts of money set aside by various business organisations as security.

Legal Tender Money in the form of coins and notes which are legally authorised by the state to be accepted for payment when presented.

Legatee Recipient of a Legacy, *q.v.*

Leg.Chgs. Legal Charges.

Legist An expert in the law.

Legitimus contradictor *Latin*—Legitimate defendant.

Leics Leicestershire, *UK*; Leicester, *UK*.

Leip. Leipzig, *Germany*.

Lek Unit of currency used in Albania, one hundred quindarka equals one lek.

LEL Lower Explosive Limit. Also known as Lower Flammable Limit.

LEM *or* **Lem** Lower Explosive Mixture; Lunar Excursion Module.

Lempira Unit of currency used in Honduras, one hundred centavos equals one lempira.

Len. Length; Lengthened, *Lloyd's Register notation*.

Length Between Perpendiculars The distance along the length of a vessel's summer load waterline from the fore-side of the stern to the after side of the rudder post or, if there is no rudder post, to the centre of the rudder stock. *Abbrev.* LBP.

Length of Cable *Similar to* **Shackle**, *q.v.*

Length Overall The length of a vessel measured horizontally between lines vertical to the extremes of the bow and the stern, above or below the waterline. *Abbrev.* LOA.

Lentoposi *Finnish*—Airmail.

LEO Low earth orbit *configuration of a satellite.*

LEPORE Long Term and Expanded Programme of Oceanic Research and Exploration.

LES Land Earth Station *satellite communications*; Launch Escape System, *q.v.*

Lèse-Majesté An offence committed against a sovereign power or state; treason.

L.ès L. Licencié ès Lettres *French*—Licentiate in Letters.

Lessee Holder of a tenancy of a house, etc., under a Lease, *q.v.*

Lessor The person who leases a property. *See* **Lease.**

Less Than Carload Freight American equivalent of Less Than Container Load.

Letter Carrier Postman; mail-carrier.

Letterhead Details as appropriate of name and address(es), registered and/or otherwise, telephone, telex and fax numbers, VAT registration, directors names, and logo or trade mark printed on the writing paper used in a firm's correspondence. Also a sheet of paper so printed.

Letter of Assignment A document by which the appointer or assignor allots or transfers his rights to another party.

Letter of Attorney *Similar to* **Power of Attorney**, *q.v.*

Letter(s) of Credence Diplomatic credentials presented by an envoy, ambassador, or diplomat appointed to a foreign country showing his authority to represent his own country.

Letter of Credit An authority by one person (A) to another (B) to draw cheques or bills of exchange (with or without a limit as to amount) upon him (A) with an undertaking by A to honour the drafts on presentation. An ordinary letter of credit also contains the name of the person (A's correspondent) by whom the drafts are to be negotiated or cashed: when it does not do so, it is called an open letter of credit. A letter of credit is in fact a proposal or request to the person named therein, or (in the case of an open letter) to persons generally, to advance money on the faith of it, and the advance constitutes an acceptance of the proposal, thus making a contract between the giver of the letter of credit and the person cashing or negotiating the draft, by which the former is bound to honour the draft. Letters of credit are sometimes used in conjunction with circular notes, in which case the letter of credit is called a letter of indication. Circular notes are forms of drafts, generally for some specific amount, given with the letter and requiring to be signed by the bearer. Circular notes are chiefly used by persons travelling abroad, and may be obtained from almost any banker. Confirmed credit, a type of commercial credit, in which a bank confirms that it will honour drafts drawn and presented in conformity with the terms of credit. *Abbrev.* L.C. or L/C or l.o.c.

Letter of Hypothecation A document given by the owners of goods in a Bill of Lading, *q.v.*, to bankers for money advanced. The bankers are in a position to exercise a lien on the goods if the money is not forthcoming.

Letter of Indication *or* **Letter of Identification** An identification form used by the payee to identify himself.

Letter of Introduction A letter whereby the writer introduces the bearer of it to the addressee to make them acquainted and at the same time requests a favour to be done. In normal cases the addressee is a personal friend of the sender.

Letter of Marque *or* **Mark** A commission issued by the admiralty or an admiral of a belligerent state licensing the commander of a privately owned ship to cruise in search of, attack and capture enemy vessels or to make reprisals for damage sustained by enemy action. A Letter of Reprisal was similar. A vessel with such a commission was also called a Letter of Marque and commonly known as a privateer. Ships captured were condemned by an Admiralty Court for disposal and valuation for prize money. The practice, originating in the 13th century, was abolished by the signatories of the Congress of Paris in 1856 and universally by the Hague Convention of 1907.

Letter of Readiness *See* **Notice of Readiness**.

Letter of Subrogation The right enjoyed by an underwriter to take over the rights and remedies available to an assured, following payment of a claim on the policy, in order to recover up to the amount of the claim from another party who was responsible for the loss. *See* **Subrogation**.

Letter Rogatory A commission from one judge to another requesting him to examine a witness. This procedure can operate between foreign countries.

Letters of Administration A legal authority given to one or more individuals or a trust corporation to administer a deceased person's estate.

Letters Patent An official document granting exclusive rights to the manufacturers of items, or originally invented objects, and which nobody can legally imitate without permission.

Letter Testamentary 'Testamentary power' in the power of making a valid will, either generally, or with reference to particular kinds or dispositions of property.

Let the Buyer Beware *See* **Caveat Emptor** and also **Caveat Subscriptor**.

Lettre de Cachet *French*—Sealed Letter; Royal Letter of arrest or imprisonment.

Lettres Rogatoires *French*—Written evidence heard abroad. *Similar to* **Letter Rogatory**, *q.v.*

Leu Unit of currency used in Romania, one hundred bani equals one leu.

LEV *or* **Lev** Lunar Excursion Vessel; Unit of currency used in Bulgaria, one hundred stotinki equals one lev.

Levant Eastern Mediterranean countries.

Level of Living American expression for Standard of Living.

Level Premium Premium, *q.v.*, paid at regular intervals by equal instalments to cover long-term insurance.

Level Temporary Insurance Similar to Term Insurance, *q.v.*

LEX Law.

Lex. Lexicon.

Lex fori *Latin*—The law of the place of action.

Lexicographer Author of a dictionary.

Lex loci actus *Latin*—The law of the place where a legal act takes place.

Lex mercatoria *Latin*—Law Merchant, *q.v.*.

Lex non cogit ad impossibilia *Latin*—The law does not compel the impossible.

Lex non scripta *Latin*—Common law; the unwritten law.

Lex scripta *Latin*—The written law; statute law.

Lex situs *Latin*—The law of the place where the property is situated.

Lex talionis *Latin*—Law of retaliation; 'an eye for an eye, a tooth for a tooth'.

LF *or* **L.F.** Summer fresh water load line markings on each side of a ship assigned to carry timber; Life Guards; Low Frequency.

Lf *or* **lf** Ledger Folio; Lifeboat, *q.v.*; Low Frequency *electricity*.

L.F.A. Local Freight Agent.

LFAB *or* **L.F.A.B.** Lloyd's Form of Average Bond.

LFB *or* **L.F.B.** London Fire Brigade.

l.f.c. Low Frequency Current.

L.F.G. *or* **LFG** Liquefied Flammable Gases. *See* **Inflammable Liquids** *or* **Flammable Liquids**.

LFL *or* **L.F.L.** Lower Flammable Limit. *Similar to* **Lower Explosive Limit**.

LFO *or* **L.F.O.** Light Fuel Oil.

L.F.P.S. Licentiate of the Faculty of Physicians and Surgeons.

LFSA *or* **L.F.S.A.** Lloyd's Form of Salvage Agreement.

l.ft Linear Foot; Linear Feet.

LGB *or* **L.G.B.** Local Government Board.

LGC Laboratory of the Government Chemist.

L. Ger. Low German.

LGFM *or* **L.G.F.M.** London Grains Futures Market.

Lgn. Leghorn—Italy. *In Italian* Livorno.

L.Grk. Low Greek; Late Greek.

LGSM Licentiate of the Guildhall School of Music.

Lgth. Length.

lg.tn Long ton of 2240 lbs.

LH Lufthansa, German Airlines *airline designator*.

LH *or* **L.H.** Licentiate of Hygiene; Lightening Hole; Lighthouse(s), *q.v.*

l.h. Left Hand.

LHA *or* **L.H.A.** Landing Helicopter Assault; Licentiate of the Australian Institute of Hospital Administration; Local Hour Angle *nautical*; Lord High Admiral.

LHAR *or* **L.H.A.R.** London, Hull, Antwerp and Rotterdam—rotation of ports for loading/discharging of grain.

L.H.C. Lord High Chancellor.

L.H.D. Literarum Humaniorum Doctor *Latin*—Doctor of Humanities; Left Hand Drive.

LHF Liquid Hydrogen Fuel.

LHO *or* **L.H.O.** Livestock Husbandry Officer.

L.H.S. Left Hand Side.

L.H.T. Lord High Treasurer.

LHW *or* **L.H.W.** Lower HighWater, *q.v.*

LHWC Longshore and Harbour Workers' Compensation Act 1984.

LHWN *or* **L.H.W.N.** Lowest High Water Neap Tides.

L.I. Licentiate of Instruction *USA*; Long Island.

l.i. Letter of Introduction; Longitudinal Interval.

LIA *or* **L.I.A.** Life Insurance Association; Lebanese International Airlines.

Liabilities Being liable for debts or claims by an individual, company or any other business concern. *Opposite to* **Assets**, *q.v.*

Liability for Oil Pollution Legislation for the prevention of oil pollution on an international level was passed in 1971 and over the years a scheme was introduced by Tankers Owners' Voluntary Agreement Concerning Liability for Oil Pollution (*abbrev.* TOVALOP, *q.v.*), and Contract Regarding an Interim Supplement to Tanker Liability for Oil Pollution (*abbrev.* CRISTAL, *q.v.*)

LIARS Lloyd's Instantaneous Accounting Record System.

Lib. Liberia.

LIB *or* **L.I.B.** Lloyd's Insurance Brokers.

LIBA *or* **L.I.B.A.** Lloyd's Insurance Brokers' Association.

LIBC *or* **L.I.B.C.** Lloyd's Insurance Brokers' Committee.

Libellant An American expression for plaintiff.

Libel Defamation in a permanent form. False defamatory words, if written and published, constitute a libel.

Libelling American term for the Arrest or Seizure, *q.v.*, of a ship.

Liberty Replacement Various designs of tween-deckers constructed from the 1960s onwards to replace the then ageing war-built liberty fleet—e.g. SD14s, freedoms.

Liberty Ship *or* **Liberty Type Ship** Ship of about 7,000 Gross Tonnage mass-produced in the USA during the Second World War between 1942 and 1945 and very rapidly constructed to meet the need for sea transport. A large number of these ships were still trading commercially and profitably for many years after the War.

LIBO London Inter-Bank Offering *rate of interest*.

LIBOR *or* **L.I.B.O.R.** London Inter-Bank Offering Rate, *q.v.*

Libra *or* **L** Pound Weight; Pound money.

LIC Local Import Control.

Licence A document or permission obtained from the respective authorities of a state permitting individuals or firms to act or use the services requested. Licences are issued for a fixed period of time after which an extension of authority has to be granted. There are many types of licences.

Licentiate A holder of a university licence or attestation of competence from collegiate or examining body.

Lic. Med. Licenciate in Medicine.

LICOS *or* **Licos** Liquid Cargo Operations Similator *electronics*

Lic. S. Licentiate in Surgery.

Lic. Theo. Licentiate in Theology.

LID *or* **L.I.D.** Lloyd's Intelligence Department; Lloyd's Information Department.

LIDAR Light Detection and Ranging.

Liecht. Liechtenstein.

Lien A legal right by which a person is entitled to obtain satisfaction of a debt by means of property belonging to the person indebted to him. Liens are classified as Maritime, Possessory or General. *See* **Maritime Lien**.

Lien Clause A Charterparty, *q.v.*, clause protecting the owners or carrier by imposing a lien on all cargoes and freights due against the Charter Rate or Charter Freight, *q.v.*

Lieut. Lieutenant.

Life Annuity A sum of money advanced every year as income to a person during his/her lifetime. Payments cease as soon as the person dies. *See* **Life Insurance** *or* **Assurance**.

Life Belt A belt made of buoyant material to support a person in water. Also a belt of strong fabric to hold a person by the waist while working high up in a building and elsewhere.

Lifeboat A specially constructed double ended boat which can withstand heavy, rough seas. It is used either by lifesavers who go out to sea in any weather to save people from drowning or by crew members and passengers in case a ship is in danger of foundering. Lifeboats are fully equipped with survival gear. *Abbrev.* L.f. *or* lf. *See also* **Royal National Lifeboat Institution**.

Lifeboat Drill *or* **Boat Drill** The master of every vessel is bound by international law to make the officers, crew and passengers (if carried) adequately acquainted with the procedures of lowering and the use of lifeboats in case of emergency. Lifeboat exercise or drill, as it is called, takes place fairly regularly.

Lifebuoy A device for supporting a person in water. It is usually in the shape of a ring and made of cork or other approved material. It is enveloped by hard canvas and painted for identification. It can easily be put over the head and supported under the arms. Lifebuoys are easily accessible to those on board ship in case of emergency. A self-igniting light is attached by means of a short cord to the lifebuoy which can be used as a distress signal by the person using it. *See* **Distress Signals** and **Marine Pyrotechnics**.

Lifefloat Another word for Liferaft, *q.v.*

Lifejacket *See* **Lifebuoy**. *Abbrev.* l.j.

Life Insurance *or* **Assurance** Life insurance is a contract by which the insurer, in consideration of a certain premium, either in a gross sum or by annual payments, undertakes to pay to the person for whose benefit the insurance is made a certain sum of money or annuity on the death of the person whose life is insured; the latter is sometimes called the *cestui que vie* or 'the life.' If the insurance is for the whole life, the insurer undertakes to make the payment whenever the death happens; if otherwise, he undertakes to make it in case the death should happen within a certain period, for which period the insurance is said to be made. The utility of this contract is obvious. A creditor is enabled thereby to secure his debt; and annuitant, the continuance of his income after the grantor's decease; a father, a provision for his family, available in case of his own death. Policies may also form the subject of settlements and mortgages.

Life insurance is very ordinarily called life assurance in distinction to other kinds of insurance; but the distinction has no legal significance, and is by no means strictly adhered to.

Life Interest An interest for one's own life, or the life of another.

Lifeline Rope or line rigged on an exposed deck for the safety of the crew in heavy weather; Line thrown from shore or a ship to save someone at risk of drowning. Rope between a wreck and the shore to assist the rescue of survivors; Colloquially can mean social or commercial assistance offered to an individual or organisation in trouble.

Life Preserver Life jacket, belt or buoy to prevent someone from drowning if he is washed or falls overboard or in the event of the ship sinking.

Liferaft There are two types of rafts which have been added to lifeboats in recent years as a means of lifesaving at sea. The wooden liferaft is designed to float well above the surface of the sea. The inflatable liferaft automatically inflates. When inflated it is completely covered and protects the occupants, it being protected on all sides. Liferafts are fully equipped with essential items to cover a limited number of days at sea. See **Lifeboat**.

Lifesaving The organised endeavours of specially trained persons to help distressed people at sea. Their activities include rendering assistance by means of lifeboats, line throwing (see **Marine Pyrotechnics**) first-aid assistance, and transfers to hospital. In the past many people have lost their lives in their attempts to save the lives of others in distress. In recent years air/sea rescue methods have come into existence. The lifeboat is gradually being replaced by the use of helicopters and ship-to-shore communications have also made lifesaving operations easier.

LIFFE London International Financial Futures Exchange.

Lifo or **Li/fo** or **L.I.F.O.** Last in, First Out, q.v.; Liner in Free Out.

LIFT or **L.I.F.T.** London International Freight Terminal.

Lift One parcel of merchandise; Elevator; raising or lifting one grab by a crane or ship's derrick.

Lifting Gear The lifting equipment of a ship for loading and discharging operations.

Lift On/Lift Off Where goods are lifted on and off a vessel by means of ship's gear—as opposed to roll on/roll off, q.v.

Lift Unit Frame Lift trucks for handling all types of cargo containers. Abbrev. LUF.

Ligan Similar to **Lagan**, q.v.

Light Buoys They are anchored, floating buoys which exhibit a fixed or flashing light. As with other types of buoys, they are used to mark a channel or to alert shipping to dangers, wrecks or other obstructions.

Light Cargo Cargo which exceeds the standard 40 cubic feet measurement against a weight ton of 20 cwts or 35 cubic metres against a ton of 1000 kilos. Light cargo fills up the cargo holds.

Light Draft (Draught) The Draft (Draught), q.v., of a ship when she is completely empty of cargo. Abbrev. LD or L.D. See **Draft (Draught)**.

Light Load Line or **Light Line** The position of the line of immersion when a vessel is on ballast only. Abbrev. LLL or L.L.L.

Lighter A large open boat used in loading and unloading ships—known also as a barge, q.v.

Lighter Aboard Ship or **Lash Ship** See **L.A.S.H.**.

Lighterman Port worker engaged in loading and discharging from or on to a Lighter, q.v.

Light Grain See **Heavy Grain, Soya, Sorghum**.

Lighthouse A building from which lights are shown to guide ships at sea. See **Trinity House Lighthouse Service**.

Lightning Rod or **Lightning Conductor** Metal rod attached at the highest point of a building or a ship's mast to divert the electrical discharge accompanying a thunderstorm into the ground or the sea.

Lightship or **Lightvessel** A vessel mounting a powerful lantern, which is moored on a permanent station in the vicinity of a port approach channel or of a dangerous shoal as a guide to safe navigation. She is equipped with a radio beacon and fog signals. She may or may not be manned.

Lightweight The actual weight, or displacement, of an unladen vessel. See **Light Displacement Tonnage**.

Light Wood Wood consisting of deals, battens or other similar lightwood.

Lignite A sort of dark brown coal.

Lignum Vitae Brownish green-coloured heavy wood known for its extreme hardness and durability. Small fittings and furniture requiring these qualities, such as wooden balls and bowls, deadeyes, parrel beads, pulley blocks and fairleads are made from it.

Like Kind Similar to **Ejusdem Generis**, q.v.

LILO or **L.I.L.O.** Last In Last Out.

Lim. Limerick, Republic of Ireland; Limit.

LIM Licentiate of the Institute of Metallurgists; London Insurance Market.

Lima International Radio Telephone phonetic alphabet for 'L'.

Liman Russian—Shallow lagoon or saltmarsh at the mouth of a river; Turkish—Port or harbour.

Limber Hole Hole in a floor frame or deck plate allowing the flow of loose water to a bilge whence it can be pumped.

LIMDSM London Insurance Data Standards Manual.

Limit A stock exchange term. The limit or maximum price the Broker, q.v., is authorised to bid for the buyer of stocks; alternatively the lowest amount of bid given to sell the shares.

Limitada Portuguese—Limited. Abbrev. Ltda.

Limitation of Shipowner's Liability A shipowner is not liable at all for loss of or damage to goods on the ship in certain cases, e.g., when caused by fire and is not liable for loss of life or injury to persons or things caused by or on board the ship in other cases beyond a certain amount for each ton of the ship's tonnage, except in certain cases mentioned. If the damages actually sustained exceed the amount thus arrived at, they are paid into court, and distributed among the claimants in proportion to their claims. The claimants are said to prove against the fund in court just as creditors prove against an insolvent estate. These limitations of liability only apply to cases where the loss or injury has not been caused by the shipowner's actual fault or privity. Where two vessels are injured by collision, and both are to blame, so that the enactment as to apportioning the liability for the damage applies, and one of them obtains a limitation of his liability under the Merchant Shipping Act, a set-off is allowed between the amounts of the respective damages.

Limited When referring to a company this has certain advantages in regard to the margin of liability applicable to the shareholders. The company is restricted to the amount of liability or fixed sum of money declared in the Articles of Association, *q.v.*, or Memorandum of Association, known as Capital. The management will have to be directly answerable for any debts or faults caused by improper management. *Abbrev.* Ltd. *See* **Limited Company**.

Limited Company The members are not liable beyond the unpaid-up part (if any) of the nominal amount of the shares in respect of which they are registered in the books of the company. When a share has been fully paid up, no further liability exists.

Limited Partner A limited partnership is one in which, although there must be one or more partners responsible for all the liabilities of the partnership, there may be one or more partners who are under no liability if they contribute an agreed sum for partnership purposes provided that they take no part in the management and that the partnership is registered as a limited partnership. A limited partnership is not dissolved by the death of a limited partner, and in the case of his becoming of unsound mind the court will not consider this a ground for dissolution of the partnership unless his share cannot be otherwise ascertained and realised. The bankruptcy law applies to limited partnerships as if they were ordinary partnerships, and if all the general partners become bankrupt the assets of the partnership vest in the trustee.

Limited Terms An insurance expression meaning either Total Loss Only, *q.v.*, or Free of Particular Average *q.v.*, or any other conditions not covering partial losses in the risk. In fact it is the opposite in a way to Comprehensive Insurance, *q.v.*

Limiting Sector The route of an aircraft starting from loading up to the landing area.

Limit per Bottom This refers to a Fleet Policy, *q.v.*, where the Underwriters, *q.v.*, insure a fleet of ships with a limit of liability per vessel.

LIMNET London Insurance Market Network.

LIMNMG *or* **L.I.M.N.M.G.** London Insurance Market Network Management Group.

Limnology The physical, chemical and biological study of lakes and ponds.

LIMRV Linear Induction Motor Research Vehicle *air-cushion vehicle.*

LIMTCG *or* **L.I.M.T.C.G.** London Insurance Market Technical Co-ordination Group.

LIN *or* **L.I.N.** Liquefied Nitrogen.

Lincs. Lincolnshire, *UK.*

Line There are many meanings of this word, some of which are: (1) Collective name for ships trading regularly on schedule. (2) A rope of small circumference generally used to secure a small boat. (3) A railway track. (4) A formation of ships.

Line, The Sailor's term for the Equator, as in 'Crossing the Line'.

Liner Conference An association of shipping lines running to schedules in the same trade at uniform rates of freight and giving advantages to regular clients. They provide what is called a liner service. *See* **Conference Lines**.

Liner Terms The responsibility and cost of loading, carrying and discharging cargo is that of the carrier, from the moment the goods are placed alongside the vessel in readiness for loading until discharged alongside at their destination. Time spent cargo-handling is also at the carrier's risk.

Line Throwing Apparatus *See* **Marine Pyrotechnics**.

Lin.ft. Linear feet/foot.

Lingua Franca An international language, especially a mixture of Italian, French, Greek and Spanish used in the Levant. Can also mean a common language, understood and spoken by many people.

Linie *German, Danish*—Shipping Line.

Linje *Norwegian, Swedish*—Shipping Line.

LINS Loran Inertial Navigation System.

LIP *or* **L.I.P.** Life Insurance Policy.

Liq. Liquid; Liquor.

Liquid This can mean capital in the form of money.

Liquid Assets Cash or property of a readily realisable nature.

Liquid Cargo Cargo of a liquid nature which in most cases is carried in bulk. This may consist of vegetable, mineral or whale oil, molasses, gas oil, etc.

Liquid Compass A Compass, *q.v.*, having its card floating in alcohol or other liquid with a low freezing point.

Liquid Petroleum Gas Carrier A tanker designed for the carriage of petroleum gas in a form of butane or propane and/or both. *Abbrev.* LPGC *or* L.P.G.C.

Liquidated Damages The amount agreed upon by a party to a contract to be paid as compensation for the breach of it, and intended to be recovered, whether the actual damages sustained by the breach are more or less, in contradistinction to a penalty.

Liquidation Under the Bankruptcy Act, 1869, a debtor unable to pay his debts might present a petition stating that he was insolvent, and thereupon summon a general meeting of his creditors. If the meeting passed a special resolution declaring that the affairs of the debtor were to be liquidated by arrangement, and not in bankruptcy, the creditors appointed a trustee, with or without a committee of inspection, and the property of the debtor vested in the trustee and became divisible among his creditors, in the same way as if he had been made bankrupt. The subsequent proceedings generally followed the same course as those in an ordinary bankruptcy, except that the close of the liquidation and the discharge of the trustee were fixed by the creditors. The theory of the proceeding was that the affairs of the estate were brought under the immediate control of the creditors, without the delays and expenses caused by the supervision of the court as in bankruptcy. Liquidation is not known to the bankruptcy law, the only alternative to bankruptcy being an arrangement effected either privately or through the court. The liquidation of a company is now known as winding up: but the person who carries out the winding up is known as a liquidator.

Liquidator A person appointed to carry out the winding up of a company. In the case of a member's voluntary winding up, with or without supervision, one or more liquidators are appointed by the company. In the case of a creditor's voluntary winding up, the creditors and the members each nominate a liquidator, the liquidator of the creditors having preference. Where a company is being wound up subject to supervision, the court may appoint additional liquidators. In the case of a compulsory winding up by the court a provisional liquidator may be appointed by the court as soon as a petition for winding up has been presented. After the winding up order is made, the official receiver is *ex officio* the provisional liquidator until he or some other person is appointed as liquidator. The duties of a liquidator are to get in and realise the property of the company, to pay its debts, and to distribute the surplus (if any) among the members. The chief difference between a liquidator in a winding up by the court and a liquidator appointed in a voluntary winding up is that the former cannot as a rule take any important step in the winding up without the sanction of the court, while the latter is not so restricted, and also does as of course various things which, in a winding up by the court, are done either by the court or by the liquidator under a general authority of the court, *e.g.*, settling the list of contributories and making calls.

Lira Unit of currency used in Italy and Malta.

Lis. Lisbon.

List Inclination of a ship to one side due to uneven lading, cargo shifting or flooding of a compartment(s) on one side.

Listed Company A Public Company, *q.v.*, listed on the Stock Exchange, *q.v.* Also called a Quoted Company.

Listed Stocks Stock Exchange, *q.v.*, term for Bonds, *q.v. See* **Stocks and Shares**.

Lit. Literature.

LIT *or* **L.I.T.** Longitude in Time *nautical*.

Lith. Lithuania.

Litre Metric unit of liquid capacity equivalent to 1.760 pints. *Abbrev.* L.

Litt.B. Literarum Baccalaureus *Latin*—Bachelor of Letters.

Litt.D. Literarum Doctor *Latin*—Doctor of Letters.

Littera scripta manet *Latin*—The written letter remains.

Littoral The area of land situated along the coastline.

Liv. Liverpool, *UK*.

Live Oil Drilled out oil that contains gas.

Livestock Horses, cattle, sheep, hogs, goats. The carrier has the option to carry these animals on deck 'at shippers' risk'. However, in livestock ships there is appropriate accommodation and fittings in order to allow the livestock to be protected from the weather.

Livestock Carrier Ship fitted with pens on several decks with feeding, watering and ventilation arrangements for the transport of live animals, usually intended for slaughter at destination. Known as Trot-on Trot-off. *Abbrev.* To/To *or* TO/TO.

Live Wool Wool taken from live sheep.

L.J. Lord Justice.

l.j. Lifejacket. *See* **Lifebuoy**.

L.JJ. Lords Justices.

Lkg. Leakage.

Lkg. & Bkg. Leakage and Breakage *insurance term*.

Lkr. Laker *q.v.*; Locker.

LL *or* **L.L.** Live Load; Lower Limb *navigation*; Late Latin; Low Latin; Leave Loose; Load Line, *q.v.*

L/L Lutlag, *q.v. Norwegian*—Limited Company; Load Line, *q.v.*

LLAT *or* **L.L.A.T.** Latitude Lowest Astronomical Time.

LL.B. Legum Baccalaureus *Latin*—Bachelor of Laws.

LLC *or* **L.L.C.** Load Line Certificate; Load Line Convention.

l.l.c. Low Left Centre.

Ll & Cos. Lloyd's and Companies.

LL.D. Legum Doctor *Latin*—Doctor of Laws.

Llds. Lloyd's.

LLEC *or* **L.L.E.C.** Load Line Exemption Certificate.

LLL *or* **L.L.L.** Light Load Line, *q.v.*

Ll.L.Rep. Lloyd's List Law Reports.

LL.M. Legum Magister *Latin*—Master of Laws.

LLMC 1976 International Conference on Limitation of Liability for Maritime Claims 1976. *IMO*.

Lloyd (Norddeutscher) The North-German Lloyd Steamship Company. *Abbrev.* NDL.

Lloyd's Lloyd's of London is the world's leading insurance market. It is one of those singular British institutions that has grown by slow degrees from very humble origins. It is not known exactly when Edward

Lloyd established his coffee house in the City of London or even why his modest emporium, out of hundreds of similar businesses popular in London at that time, became the acknowledged centre of marine insurance underwriting. We do know, however, that it existed in 1688 and by the early years of the 20th century had evolved into a unique international insurance market renowned for flair and innovation and offering a secure policy second to none.

Few contemplating Lloyd's building in Lime Street can begin to imagine what the dimly-lit and smoke filled establishment where it all began must have been like. No contemporary description of the coffee house exists and little is known about Edward Lloyd himself.

From the outset it seems likely that Edward Lloyd encouraged a clientele from among men engaged in foreign trade, mostly merchants, shipowners, and ships' masters. His house was close enough to the Thames Street wharves to be a convenient rendezvous for seafarers who no doubt brought him the latest news.

By the time of Edward Lloyd, marine underwriting had long been established in London. The Lombard merchants first introduced the practice of insuring ships and cargoes in the 15th century. In the days before the existence of insurance companies, marine insurance was done entirely by private individuals on an informal basis and as a profitable sideline to their usual business.

The normal method of insuring a ship in the seventeenth century was for the broker—often referred to as an 'insurance office keeper'—to take a policy round the City for subscription by wealthy merchants, men of substance who would be able to meet their share of a claim to the full extent of their personal fortunes if need be. Those taking a share of the risk signed their names with the amount, one beneath the other, at the foot of the policy wording, and in consequence were known as 'underwriters'.

As enterprising as he was, Lloyd took no part in underwriting. He contented himself with providing his customers with congenial premises, reliable shipping news and the simple facilities they needed to do business with each other. In his day he was described as a 'coffee-man' and so he remained until his death in 1713. His principal legacy to the future was his name, and the coffee house which bore it.

During the ensuing centuries, Lloyd's gradually assumed its present day appearance. Membership was regulated and the elected management committee took on increasing responsibility for the security of the Lloyd's market. The Society was incorporated by the 1871 Lloyd's Act of Parliament which provided it with a sound legal basis for the next hundred years. From the 1880s, under the influence of the legendary Cuthbert Heath, Lloyd's was transformed from a somewhat moribund club of marine underwriters to a thrusting innovative market for a bewildering variety of international risks. By the turn of the century Lloyd's pre-eminence as a world centre for insurance of almost every type had been established—especially in the United States, which remains a major source of its business.

The rapid and accelerating expansion of the Lloyd's market since 1900 is vividly demonstrated by the three buildings it outgrew within 50 years. The fourth, which Lloyd's moved into in 1986, was designed to allow for any further growth well into the next century. Richard Rogers' startling solution to the housing problem has already become a focal point of interest and not only in architectural circles.

Lloyd's then is not a company. It has no shareholders and accepts no corporate liability for risks insured there. Lloyd's is a society of underwriters, all of whom accept insurance risks for the personal profit or loss.

A policy is subscribed at Lloyd's today by private individuals with unlimited liability and corporate members with limited liability.

Individual underwriting membership is open to men and women of any nationality provided that they meet the stringent financial requirements of the Council of Lloyd's. Lloyd's membership today is drawn from many sources in more than 70 countries. Industry, commerce and the professions are strongly represented, while many members work at Lloyd's, either with brokerage firms or underwriting agencies.

Currently more than 17,000 individual Lloyd's members and 95 corporate members are grouped into about 179 syndicates varying in size from a few score to more than a thousand Names. The affairs of each syndicate are managed by an underwriting agent who is responsible for appointing a professional underwriter for each main class of business.

Expansion, which saw Lloyd's underwriting membership increase from 6,000 to over 30,000 in 20 years, brought serious problems during the 1970s. The Fisher report in 1980 recommended sweeping constitutional reforms, chiefly the setting up of an elected council to take over the membership's legislative powers. The Lloyd's Act of Parliament which enabled these reforms to be carried through was passed in July 1982 and for the first time in its history, gave the Society a sound working constitution.

The 28-strong Council of Lloyd's has overall responsibility for, and control of, the affairs of the Society, including rule-making and market discipline. It consists of 12 working and eight external members of Lloyd's together with eight nominated persons with no business connections with the market.

The constitutional reforms carried out by the Council in its first eight years were phenomenal. This needed to be done without stifling the market's enterprise and innovative tradition.

Lloyd's

Today, Lloyd's is an insurance market unique in the world. Almost anything can be insured there: fleets of ships and aircraft, civil engineering projects, factories, oil rigs and refineries to name but a few of the thousand-and-one risks which are placed at Lloyd's each year. This business flows from all parts of the world and represents an income of about £30 million in premiums each working day.

This is not to say that Lloyd's is all about the big and the exotic. It is, for example, collectively the largest insurer of private cars in Britain with 17 per cent of the market. It also provides cover for homes and contents, sailing and motor craft, livestock and race horses, even the event of a village fete being ruined by rain.

The Underwriting Room at Lloyd's—known to everyone simply as 'the Room'—is essentially a market for insurance where the placing of risks is a matter of negotiation between broker and underwriter.

The insurance broker is a key figure in the Lloyd's market. Lloyd's underwriters have no other contact with the insuring public and their premium income is entirely dependent on the initiative and enterprise of Lloyd's brokers in obtaining business throughout the world. The Council of Lloyd's demands the highest professional standards from the 220 accredited brokerage firms permitted to place risks in the Room.

The Lloyd's broker's prime duty is to negotiate the best terms for his clients. To this end he is free to place risks wherever he thinks fit whether at Lloyd's, with the insurance companies, or both.

On receiving a request for insurance cover, a Lloyd's broker first makes out the 'slip'—a sheet of folded paper with details of the risk. Many risks today are broked using electronic placing in support. The next step is to negotiate a rate of premium with underwriters expert in that particular type of business. Lloyd's thrives on competition and the broker may obtain several quotes before deciding on the best one—bearing in mind what his client will be prepared to pay and what level of premium is required to get the risk adequately covered in the market. The leading underwriter, having set the rate, takes a proportion of the risk on behalf of his syndicate.

Armed with this 'lead' the broker approaches as many other syndicates as are needed to get the slip fully subscribed. Large risks are usually spread over the whole London market, cover being shared by Lloyd's underwriters and the insurance companies.

Spreading a risk as widely as possible is one of the cardinal principles of insurance which enabled Lloyd's and the London market to withstand the pressure of heavy claims which might otherwise be ruinous. The famous preamble to an insurance Act of Parliament passed during the reign of Elizabeth I puts it succinctly '. . . it cometh to pass that on loss or perishing of any ship there followeth not the undoing of any man, but the loss lighteth rather easily upon many than heavilie upon fewe . . .'.

The security of a Lloyd's policy, universally regarded as second to none, is built on solid foundations. Every member of Lloyd's is subject to a searching statutory test of solvency designed to detect any weakness at the earliest moment and to ensure that provision is made to protect the assured. All members are required to show that the value of their underwriting assets is sufficient to meet their liabilities.

The cornerstone of the security backing Lloyd's policies is the Central Fund. Members are required to pay a proportion of their premium income into the Fund which, in the event of any members ultimately being unable to meet their underwriting commitments, is used to protect the policyholder. The Fund is worth around £1 billion.

Probably nowhere in the world is there so much collective underwriting expertise under one roof as at Lloyd's. Although syndicates can compete with each other there is a wealth of shared experience within the market. This subtle blend of competition and co-operation combines with an unshakeable belief in the old insurance dictum of 'utmost good faith' to give Lloyd's its unique quality.

Throughout its history, Lloyd's has survived and prospered through the innovation shown by its underwriters and brokers in pioneering new insurance products to meet the changing demands of society as a whole. It was the first to cover satellites and space launches, it was the first to cover computers and even computer fraud.

To this end, during the years immediately following the passing of Lloyd's Act in 1982, the Council of Lloyd's devoted its efforts to creating and implementing the regulatory framework within which the market could operate at its most effective. Having completed this long and complex task, the Council focused its attention on supporting the market in developing desirable new business by introducing a series of radical measures aimed at making the market more competitive and more attractive to Lloyd's brokers who are not restricted to dealing with Lloyd's underwriters.

First, steps were taken to enable individual syndicates to accept more than one class of business. Traditionally they operated as marine, non-marine, aviation, motor or short-term life syndicates but to enable brokers to place composite risks involving more than one class of business without the necessity of speaking to more than one underwriter, these so-called 'market barriers' have been removed. Lloyd's underwriters may now therefore compete on more equal terms with their counterparts in the company market.

260

Second, the Council put in place legislation to facilitate new routes to Lloyd's underwriters for personal lines business. Householders and personal accident cover, for example, may now be placed by intermediaries other than Lloyd's brokers subject to certain safeguards. Syndicates may also accept this kind of business through their own service companies although commercial motor insurance continues to involve the services of a Lloyd's broker.

Lloyd's A & CP Lloyd's Classification notation for Anchors and Cables tested.

Lloyd's Agency System The Lloyd's Agency system, as we know it today, was organised in 1811 when the subscribers to Lloyd's agreed to formalise the rather haphazard and cumbersome means previously employed to appoint representatives to gather information and care for their interests as insurers of ships and cargo in casualty.

The Committee of Lloyd's was invested with the responsibility of establishing the system and today Lloyd's Agents are asked to carry out specific duties for the Corporation of Lloyd's in addition to their normal business activities. When acting as agents of the Corporation, their duties are to collect and transmit to the Corporation information of likely interest to the Lloyd's Market and insurers worldwide.

The basic qualifications for a Lloyd's Agent have remained unchanged since 1811, namely that they should be resident and well established at the place concerned and be of high commercial status and integrity.

It became common practice for insurers, both Lloyd's and companies worldwide, to include in their policies and certificates a requirement that surveys to establish the extent and cause of loss or damage to ships, their cargo and goods in transit by land, sea and air be held by Lloyd's Agents or by surveyors appointed by them. It should be noted however that in carrying out surveys or other tasks, whether for Insurers or other Principals, the company would be employed directly by these parties and would not be acting as an agent of Lloyd's itself.

Where a Lloyd's Agent is asked to carry out a survey, the surveyor may be responsible not only for reporting the condition of the subject matter, the extent and cause of damage, but if requested, also recommend steps to minimise loss, alternative uses for damaged goods and even may suggest means to prevent recurrence of such losses. In agreeing with an applicant for survey a figure for depreciation endeavours would be made to see that justice is done to the insurers, the claimant, carriers and/or all other interested parties.

Knowledge of a wide range of commodities and long experience of various modes of transport greatly aids the surveyor who may be the Agent himself, an employee, or a non-staff expert commissioned to hold a specific survey. Agents may employ the services of such experts if they consider the nature of the survey requires it. From time to time they may also utilise the services of analysts and other specialists to investigate and establish the cause of damage.

The introduction of containers in recent decades has resulted not only in changes in the transport field but required Agents and surveyors to learn a completely new set of environmental conditions affecting the transport of goods and the losses and damage caused thereby.

In 1866 certain selected Lloyd's Agents were authorised by the Committee of Lloyd's to adjust, settle and purchase claims on Lloyd's policies and certificates which made special provisions for the settlement of claims abroad. This does not mean paying the claim on behalf of Underwriters but only that the Agent concerned may, at his discretion, purchase title to the claim on behalf of the Corporation of Lloyd's who thereafter become the claimants on Underwriters.

The 20th century has seen an increase in the scope and complexity of insurance and has involved Lloyd's Agents in an ever widening field of activity including both Aviation and Non-Marine surveys and investigations.

Inevitably the Agency system has varied in size both to meet the needs of the time and in view of communications systems available. Originally the horse carriage and sailing ship limited the speed of communications and the ability of surveyors to respond rapidly to requests for their services. Lloyd's and Lloyd's Agents were early users of telegraphic facilities and strong backers of Marconi's early radio stations before a Post Office monopoly was established in the United Kingdom. Today the latest developments in public document transmission, computers and printing will be found employed in the gathering and distribution of information.

Lloyd's American Trust Fund An office at Lloyd's where all American funds, whether premiums or payments of claims are transacted. *Abbrev.* LATF *or* L.A.T.F.

Lloyd's Average Bond *See* **Average Bond**.

Lloyd's Broker An intermediary who negotiates insurance contracts with Lloyd's underwriters on behalf of his clients, the assured. For a broker to be admitted as a Lloyd's broker he must satisfy the Committee of Lloyd's that he is a suitable person to become a Lloyd's broker. Only Lloyd's brokers are permitted to enter the Underwriting Room at Lloyd's to transact business with underwriters.

Lloyd's C.A. Account Lloyd's Central Accounting Account.

Lloyd's Central Accounting Office Where premiums and claims are dealt with.

Lloyd's Intelligence Department From its worldwide sources Lloyd's Intelligence Department gathers information of interest to the shipping and insurance communities. The service dates back to the 17th century coffee house of Edward Lloyd where details of vessels arriving and sailing, and those reported in casualty, would be made available to his underwriting clients.

Three hundred years later, a much expanded service is still being provided to the Lloyd's underwriters and brokers, and to an even wider audience through the specialist services and publications produced by LLP Limited, of which the Intelligence Department is now a part.

With such a wealth of current information, together with extensive historical records, the Department has become known as the world centre for shipping information. Hardly a merchant ship moves anywhere in the world without Lloyd's knowing about it. Much of the information comes from Lloyd's Agents, who were first appointed in 1811 and today cover over 2,000 ports around the world. Each year more than two million shipping movements are received in the Department for processing directly into the LLP computer for publication in the daily *Lloyd's List* newspaper and the weekly Lloyd's Shipping Index and Voyage Record.

Lloyd's Agents also provide much of the casualty information which is received in the Department. The Agents have close links with the local maritime authorities and radio stations, while distress messages received by United Kingdom marine radio stations and by the Coastguard Service are immediately communicated to Lloyd's. Over recent years this co-operation has been extended to include the United States Coast Guard, European and Australian search and rescue authorities and the major maritime organisations around the world.

Loss Book If an incident is serious enough that it may result in either the loss or extensive damage to the vessel which may then become a potential total loss, an entry is made in the Casualty Book which stands in the centre of the underwriting room. The purpose of the entry is to alert the Market to the seriousness of the case. The Casualty Book, also known as the Loss Book, is one of the traditional features of Lloyd's and is often shown to visitors. The old Loss Books, dating back to 1774, are in the safe keeping of the City of London library at Guildhall. The Intelligence Department staff who decide to enter a casualty into the Book still use a quill pen. They also act as a link and liaison between LLP Limited and the whole of the insurance/shipping markets.

Casualty Reporting Service As a direct result of maintaining a 24-hour watch, the Intelligence Department is able to offer a round-the-clock casualty reporting service to tug and salvage operators, shipowners, dock/repair yards, newsagencies and a variety of other marine companies by a computerised telex facility.

Shipowners, brokers or ship masters with news of a marine casualty are invited to contact the Department, day or night—telephone Colchester (01206) 772277 or telex 987321 LLOYDS G or fax 46273. If the vessel requires tug or salvage assistance, details will be sent to the appropriate companies.

Lutine Bell In the underwriting room hangs the Lutine Bell. It is rung on authorisation of the Manager of the Intelligence Department to announce definite news about either the loss or safety of an overdue vessel. As soon as the bell is rung all insurance trade in the Lloyd's overdue market is effectively stopped. Due mainly to the modern-day speed and reliability of maritime communications it is now seldom rung.

Lloyd's Newswire Service This service provides the latest news of marine casualties, non-marine incidents such as strikes and fires, together with aviation casualties and general news items affecting insurance. The Newswire Service, which operates during normal business hours, is transmitted directly from Colchester to a mini-teleprinter installed in the subscriber's premises.

Non-Marine and Aviation Information An Intelligence Department service to the Market continues to be maintained in liaison with the Aviation Department of Lloyd's. Aviation Casualty reports received are posted as 'Blue' notices in the underwriting room. In addition a wide range of non-marine information is disseminated in the form of 'Pink' notices.

Overdue Vessels Owners and other interested parties who are concerned about the safety of a vessel because of a non-arrival in port or absence of radio communication are advised to contact the Intelligence Department.

The Department has facilities for broadcasts to be made to shipping requesting later news. Lloyd's Agents and search and rescue organisations can also be alerted.

Yacht Reporting Yachtsmen and women undertaking overseas voyages and wishing to keep their relatives informed of their whereabouts should signal to passing merchant vessels with the International Code Signal ZD2 ('Please report me to Lloyd's London.')

Prior to the voyage, the Intelligence Department should be informed to enable the necessary reporting arrangements to be made.

Lloyd's List See **Lloyd's** *and* **Lloyd's Intelligence Department**.

Lloyd's Maritime Information Services Lloyd's Maritime Information Services Limited, (LMIS), is a private limited company owned by Lloyd's Register of Shipping and LLP Limited. The Company was formed to enable the maritime community to take advantage of the wealth of unique maritime information and specialist consultancy skills available from LMIS for both operational, marketing, research and management information purposes and is probably the world's largest source of commercially available maritime information. It is also able to offer its clients high level consultancy services.

Databases can be accessed in various ways to provide output on printout, diskette, magnetic tape, as well as on-line to meet individual client specifications and requirements. New construction vessel characteristics and ownership data are also available in PC packages with analytical software.

The main computer databases are:

New Construction File, containing accurate and comprehensive details of all known propelled sea-going merchant ships of 100 GRT and over, on order or under construction worldwide. In addition to daily updates, every three months an order book check is undertaken utilising Lloyd's Register offices in over one hundred countries worldwide.

Ship Particulars File, updated daily, contains accurate and comprehensive (technical and other) details of over 76,000 known propelled sea-going merchant ships of 100 GRT and above worldwide.

Shipping Movements File, containing details of over 40,000 merchant ships currently engaged in sea-going world trade. On average the file is updated nearly 6,000 times daily seven days a week of which over 4,000 of the daily entries relate to new movement information. Voyage histories are also available from the computer files from 1st January 1976.

Shipowner and Parent Company File, updated daily, contains the names and addresses of Shipowners and Managers as well as their respective parent companies with addresses. Interrogation of this file can complement information abstracted from the other databases but also facilitates analysis or listing of owners and fleets by company groups as well as by nationality or country of residence.

Casualty Information System containing comprehensive details of reported serious casualties (including total losses) to all known propelled sea-going merchant ships in the world of 100 GRT and over since January 1978 and all reported incidents, serious and non-serious, to tanker types of ships broken up or otherwise disposed of not consequent upon casualty. The file is updated daily.

LMIS On-Line Services include Viewdata and Seadata. LMIS also offers a variety of manual services which relate to verification of cargo loading dates, ownerships, movements, casualties (either current or researched from our extensive manual casualty records dating back to 1921). It is also possible to have the movements of a single vessel or group of vessels monitored for a given period.

SEA Group, the consultancy division of LMIS, provides high level, confidential consultancy services internationally. Typical of the range of consultancy services undertaken are:

- short and long term shipping forecasts, including trade, market and rate analysis;
- strategic and new venture studies including investment and disinvestment feasibilities;
- logistics and distribution planning

for a wide range of interests in the shipping industry

- shipowners
- shipping companies
- storage and distribution companies
- marine equipment manufacturers
- government departments
- financial institutions

SEA Group also offer multi-subscriber surveys, including regular monitoring and assessment of worldwide oil and tanker movements, liner trade analysis and forecast, World Fleet Forecast, and an 'early warning service' for dry bulk carrier and tanker markets.

Lloyd's Mardata, with offices in Stamford, CT. USA, offer a range of on-line services to clients based upon the following Libraries:

- Ships on Order
- Ship Casualty
- Ships Characteristics
- Ship Movements
- Charter Fixtures

Lloyd's Maritime Inc. is also LMIS representative in the United States.

Lloyd's Open Form (Standard Form of Salvage Agreement) A printed form issued by Lloyd's and used as a contract for any salvage, *q.v.*, operation. The agreement is signed by the master or owner of a salving ship and the master or owner of a ship in need of salvage assistance, containing their names, the amount of salvage payable and a provision that the service is provided on a no cure no pay basis, that is, that salvage is not payable unless the property is salved in accordance with the agreement. *Abbrev.* LOF 95.

Lloyd's Register of Shipping The origins of ship classification go back to 1760 when a group of underwriters and others in the City of London formed a society to produce a register of shipping as a guide to the assessment of maritime risks. This small register, which 'classified' ships according to their general condition, was invaluable to the marine underwriter. Shipowners and merchants also realised the advantages of classification for their business. The society was reconstituted in 1834 as Lloyd's Register of British and Foreign Shipping, with all sectors of the marine industry represented on its General Committee, and the first rules were published. The society quickly gained prestige and authority and its influence spread worldwide. New rules were introduced and existing ones amended to reflect the knowledge and experience gained over the years, a process that continues to the present day. Although its activities now extend to many other areas, LR's objectives within the shipping world have changed little over the centuries. They are

Lloyd's Register of Shipping

simply the formulation and application of proper standards of safety in ships. LR's marine work accounts for two-thirds of its overall business. LR is the largest ship classification society in the world, currently with a 21% share (by tonnage) of the world fleet.

Classification can be defined as setting and maintaining standards of safety and reliability by establishing rules for the design, construction and maintenance of merchant ships. It provides an assurance that the vessel has been surveyed impartially to the standards set out in its rules. Recent developments in this field include LR's ShipRight, a set of procedures addressing ship safety from the design and construction stage and throughout the ship's life. In addition, LR has been authorised to carry out statutory surveys and certification covering IMO conventions, codes and protocols on behalf of 130 national administrations, and it provides advisory and specification services in the marine field. Recognising the growing importance of the human factor in marine safety, LR has introduced a new service aimed at helping owners and operators develop their safety management systems to comply with the new IMO Safety Management (ISM) Code in advance of its mandatory implementation in 1998 for most ship types.

LR's work also covers a wide range of land-based industries. It first set up its industrial services in the early 1930s to satisfy a demand for professional, independent inspection and assessment. Today, the Industrial Division has its headquarters in Croydon and handles a variety of projects in the chemical, petrochemical, power, civil and general engineering industries. Its services range from technical appraisal at design stage, survey during construction and commissioning to in-service inspections throughout the life of the installation.

LR's involvement in the offshore industry began in the 1950s when it surveyed the first drilling platform built in Europe. Since the LR's Offshore Division has acted as a certification or verification agency for all types of offshore structures and has certified over 700 fixed platforms worldwide, including over 90% of the fixed installations in the North Sea. While the North Sea continues to be important, a growing proportion of LR's offshore business today is in areas such as Canada, Mexico, Vietnam, Australia and West Africa, as the focus of oil exploration gradually moves to new areas.

As well as geographical changes in its markets, LR has responded to the changing needs of its clients and in 1994 it established an Advisory Services Group. This has brought together the quality, safety and environmental services to provide a 'one-stop solution' to businesses wishing to improve their technical and management systems. In addition to advisory services in the quality, safety and environmental fields, Advisory Services provides a wide range of technical services ranging from safety case preparation, risk assessment, hazard identification and analysis and safety-critical systems assessment to environmental reviews, audits and verification. The work of LR's Engineering Services Group is equally wide-ranging. Its activities include machinery plan approval and other engineering aspects of marine, offshore and industrial projects dealt with by LR's three operating divisions, as well as technical investigation and advisory projects undertaken on behalf of clients. Specialist departments deal with materials, fuel and lube oil analysis, non-destructive testing, electrical and control engineering, refrigeration, lifting appliances and materials handling.

In the field of shipping information, LR's Register of Ships has grown into a three-volume reference work covering over 78,000 ships. Other volumes published by LR contain details of shipowners, shipbuilders, docks and offshore units.

There are ten other major classification societies. In common with Lloyd's Register, some of the older societies owed their inception to the initiative of underwriters, but brought in shipowners soon afterwards. In some cases the foundation of a society coincided with a new and developing phase in the history of a particular country.

In 1821 and 1822, exceptionally bad weather had led to the loss of some 2,000 European ships of various nationalities at the cost of 20,000 lives.These losses led to financial difficulties for certain underwriters in Paris and a few years later, in 1828, two Antwerp underwriters and a broker opened a 'Bureau de Resignements pour les Assurances Maritimes'. A year later, with financial help from King William I of Holland, this Bureau published the first 'Registre de Rensignements' under the name of Bureau Veritas. In 1831, it transferred its head office to Paris. Formal recognition by the French Government was accorded in 1908 for purposes connected with the enforcement of a Safety of Navigation Law passed the previous year. The decade between 1860 and 1870 saw the foundation of four more classification societies. The Registro Italiano Navale was founded in Genoa by a group of marine insurance companies headed by the Mutua Assicurazione. Nowadays the RI is a 'para national' body, for which the Government of Italy is responsible, which runs its own affairs autonomously.

In 1862, a year after the foundation of RI, the American Ship Masters Association was founded in New York, once more by a group of marine insurance companies. In 1889 the name was changed to the American Bureau of Shipping, and the US Merchant Marine Act of 1920 gave a comprehensive directive to all Government departments and boards to recognise the ABS for all relevant purposes. The ABS is still an independent body incorporated under the laws of New York state, without power to distribute any profits. Since the 1920 Act the Government has appointed two members to the society's governing committee.

Det Norske Veritas was founded in 1864, with the initiative again coming from underwriters. Previously, Norwegian ships had been surveyed by several diverse mutual interest societies, and the need for a common standard became evident. Further, the Bureau Veritas was beginning to secure a strong foothold in Norway, and the national feelings characteristic of the age stimulated a desire to be free from dependence on a foreign organisation. As with Lloyd's Register, ship-owners, and subsequently shipbuilders and engine builders, were later brought into the controlling body.

The fourth society to be founded in these years was Germanischer Lloyd. Here too, Government recognition was eventually given, and at the present time, although GL is formally autonomous, many of its activities are under authority delegated by the Seeberufsgenossenschaft, to which body the German Government in turn entrusts much of its control of national shipping.

The last of the older classification societies to be established was the Japanese Nippon Kaiji Kyokai, founded in 1899. The promoters were primarily shipping lines—the NYK, the OSK and others—and the purpose was stated as the development of general maritime affairs and promotion of the design of ships. Despite the stated purpose of its foundation, the NKK did not in fact start classification until 1915, and in 1920 was accorded recognition by the Japanese for purposes of the Japanese Inspection Law. *See* **International Association of Classification Societies**.

Lloyd's Surveyor A technical person nominated by Lloyd's Register of Shipping to act as a surveyor on their behalf.

LLR *or* **L.L.R.** Lloyd's Law Reports.

LLT *or* **L.L.T.** London Landed Terms.

LLTM Long Lead Time Material.

LLTV *or* **L.L.T.V.** Low-Light-Level Television, *q.v.*

LLW *or* **L.L.W.** Low Level Waste; Lower Low water, *q.v.*

LLWL *or* **L.L.W.L.** Light Load Water Line. *Similar to* Light Load Line.

l.m. Livello del mare *Italian*—Sea Level.

Lm *or* **lm** Middle Latitude; Lumen *luminous flux.*

L.M. Licenciate in Midwifery; Licentiate in Medicine; Long Major.

LMA London Maritime Association.

LMAA *or* **L.M.A.A.** London Maritime Arbitrators' Association.

LMC Lloyd's Register Machinery Certificate.

l.m.c. Low Middling Clause *cotton charterparty.*

LMCLQ *or* **L.M.C.L.Q.** Lloyd's Maritime and Commercial Law Quarterly.

LMC UMS *Lloyd's Register notation* for Lloyd's Machinery Certificate and Unmanned Machinery Space.

LMDRC London Market Drilling Rig Contract.

LME *or* **L.M.E.** London Market Exchange; London Metal Exchange.

LMF Last Memory Function *computer.*

L.Med. Licentiate in Medicine.

LMMC Labor Management Maritime Committee *USA.*

LMO *or* **L.M.O.** Light Machine Oil; Light Marine Oil.

LMP Lunar Module Pilot.

L.Mq. Lourenço Marques.

LMR *or* **L.M.R.** London Money Rate.

LMS Least Mean Square; Licentiate in Medicine and Surgery; Loading/Mooring/Storage, *q.v.*; Load Monitoring System; London Mathematical Society; London Medical Schools.

LMSR Large Medium Speed Roll on/Roll off ship maintained by US Department of Defense for strategic sealift of the Army.

LMT *or* **L.M.T.** Local Mean Time *navigation* or Ship Mean Time, *Abbrev.* SMT *or* S.M.T.; London Market Tanker.

LMTS London Market Treaty Standard.

LMUA *or* **L.M.U.A.** Lloyd's Motor Underwriters' Association.

LNAWP *or* **L.N.A.W.P.** Lloyd's Names Association Working Party.

Lndg. Landing.

Lndg. a/c Landing account.

Lndg & dly Landing and delivery.

LNG *or* **L.N.G.** Liquefied Natural Gas.

LNGC *or* **L.N.G.C.** Liquefied Natural Gas Carrier.

LNG Carrier Liquefied Natural Gas Carrier. A ship designed for this purpose.

L.N.G.I. Liquefied Natural Gas Insurance.

LNG/LPG *or* **L.N.G./L.P.G.** Liquefied Natural Gas/Liquefied Propane Gas.

LNRS Liverpool Nautical Research Society.

LNS *or* **L.N.S.** Land Navigation System.

L.N.Y.D. *or* **l.n.y.d.** Liability Not Yet Determined *insurance.*

L/O Loss Overboard.

L.O. Liaison Officer; Length Overall; Lubricating Oil.

l.o. Lubricating Oil.

LOA *or* **L.O.A.** *or* **l.o.a.** Life Offices Association; Length Overall.

Load Displacement *or* **Loaded Displacement** *See* **Displacement**.

Load Draft *or* **Load Draught** The distance or height from the keel of the ship up to the load mark.

Loaded Displacement *See* **Displacement**.

Loading Coefficient The ratio of the stowage factor to the weight ton needed to fill up the ship's cargo holds to her marks. By dividing the cubic bale space of the cargo holds by the deadweight capacity of a ship the loading coefficient is obtained. *Ex.* If the deadweight is 4000 tons and the bale space is 176,000 cu.ft. the loading coefficient is 176,000 divided by 4000 = 44 cu.ft.

Loading Dates Dates given by the Loading Broker for the vessel to start loading.

Loading Disc. *Similar to* **Loadline**.

Loading of Rate Enhancing or increasing the standard insurance premium.

Loading on the Berth A shipping term meaning that the ship or liner is ready to load cargo at her present berth at any time.

Loading Spot When the loading quay is named or specified in a contract of affreightment, *q.v.*, or charterparty, *q.v.*

Loading Turn The vessel is to wait for her turn to start loading cargo.

Loadline Marks Official marks cut in or painted on a ship's hull indicating the depth to which the ship can be loaded.

Loadline Zones The navigable waters of the world divided into climatic zones—summer, winter and tropical—indicating the marks to which a ship can be loaded.

Load on Top Related to tank cleaning on petroleum tankers. The mixture of oil residues and water resulting from cleaning with water, steam and/or chemicals is pumped into the Slop Tanks where separation takes place. The separated water can then be pumped out while the oil remains at the bottom. Subsequent loading of new petroleum cargo is on top of the remaining oil. *Abbrev.* LOT.

Load Port *or* **Loading Port** The port at which loading is to take place. Relevant in the Charterparty, *q.v.*, Tramping and Liner Services.

Load Sheet A document used in aviation for clearance by aircraft. It contains particulars of registration, nationality, details of the flight, total weight of the aircraft including crew, passengers and baggage, etc.

Load Waterline Former name for Load Line.

Loan The act of borrowing money or articles by a person or company or any other organisation from others with or without interest being payable.

Loan Capital Loan Stock or Debentures or Debenture Stock, *q.v.*

Loan Stock *Similar to* **Debenture** *or* **Debenture Stock**, *q.v.*

Loanable Funds Disposable money which is meant for lending to individuals or companies.

L.O.B. Location of Offices Bureau; Loss Overboard.

LOB Line of Balance; Loss Overboard.

LOBAL Long Base Line Buoy.

LOBAR Long Base Line Radar.

LOC *or* **L.O.C.** Launch Operation Centre; Letter of Compliance *USCG*; Letter of Credit.

l.o.c. Letter of Credit, *q.v.*

Local Haulage The charge for transporting containers from the shippers' premises to the loading quay or from the discharging quay to the consignees' premises.

Localiser Steering guidance device in the landing system of an aircraft.

Location Book A record book often used in USA ports and kept by a Shipping Clerk showing the location of landed cargoes of each consignment on the pier. *See* **Location Clerk**.

Location Clause A clause in a cargo insurance contract limiting insurance cover at any place prior to loading on to the vessel.

Location Clerk A dock clerk in US ports very similar to a Tally Clerk whose duty is to find out full details and positions of the discharged cargo and to assist delivery. *See* **Location Book**.

Loc. cit. Loco citato. In the place cited.

Lockage A charge or toll for the use of a lock.

Locker A small cupboard; in a nautical sense it is a chest or compartment for clothes, stores, valuables, ammunition, etc. *See* **Lock-up Space**.

Lock In (1) Action by strikers to shut themselves into work premises and admit no-one else. (2) To move a ship into a wet dock or navigation system through a Lock.

Lock Keeper A person in charge of Locks.

Lock-out An employer's action in closing a factory or premises to prevent employees entering while on strike. This is done as a counter action in order to try to force the strikers to agree to the employers' terms.

Lock-up A stock exchange word meaning securities which are worth buying.

Lock-up Cargo Cargo which, due to its high value, is safeguarded by being locked up in special spaces or compartments that are under lock and key. *Ex.* Bullion, jewellery, money, etc. *See* **Locker** and **Lock-up space**.

Lock-up Space Rooms or lockers in the hold of a ship where cargoes of an expensive or precious nature are stored and secured. *See* **Lock-up Cargo** and **Locker**.

Loc. laud. Loco laudato *Latin*—In the place cited with approval.

Loco *or* **Loco Price** Quotation for merchandise excluding the on-carrying expenses from the warehouse up to the place of destination. In other words the quotation is 'net' and all expenses are for the account of the consignees or receivers.

Loco Citato *Abbrev.* Loc. Cit., *q.v.*

Loco Laudato *Abbrev.* Loc. Laud. *q.v.*

Locomotive An engine for hauling a railway train.

Loc. ten. Locum tenens *Latin*—Deputy, a substitute; deputy for another.

Locum tenens *Abbrev.* Loc. ten., *q.v.*

Locus in quo *Latin*—The place in which.

Locus poenitentiae *Latin*—A place of repentance.

Locus sigilli *Latin*—The place of the seal.

Locus standi *Latin*—The right to appear.

L.o.d. *or* **LOD** *or* **L.O.D.** Loss Occurring During *insurance.*

LOE Level of Effort *in USA*—award of contracts.

LOF *or* **L.O.F.** Line of Fire.

LOF 90 Lloyd's Open Form (Standard Form of Salvage Agreement).

L of C Lines of Communication.

LOFI *or* **L.O.F.I.** *or* **lofi** Liner Out/Free In.

Log (1) Instrument for measuring ships speed and distance run. This may be a Towed or Patent Log with a rotator at the end of a long line streamed astern, or a Bottom Log, using an impeller, pressure or electromagnetic system. There is also a Dutchman's Log which is simply a chip of wood dropped over the side to gauge the ship's movement. (2) The official record of a ship's voyage and events on board—e.g. measures taken to care for the cargo; illness of crew.

Log. Logarithm; Logistic.

Log 10 Common Logarithm to the base 10.

Logger A heavily constructed bulk carrier designed for the carriage of logs.

Logs *or* **LOGS** Lloyd's Ocean Going Spares. *Lloyd's Register notation* concerning engine spares that are carried on board a ship.

LOH *or* **L.O.H.** Light Observation Helicopter; Loss of Hire *in charterparty. See* **Breakdown Clause**.

LOI *or* **loi** Letter of Intent; Limit of Indemnity *insurance*; Lunar Orbit Insertion.

L.O.L. *or* **LO life** Loss of Life.

L.O.L. Limitation of Owner's Liability, *q.v.*

LOLA Lunar Orbit Landing Approach.

L.O.L. & P.I. Loss of Life and Personal Injury *insurance.*

LOMA Life Office Management Association.

London Airport Cargo Electronic Data Processing Scheme IATA (International Air Transport Association) automated customs clearance. *Abbrev.* LACES.

London Commodity Exchange Formed in 1954 to merge the major soft commodity futures trade associations in the UK, it is now recognised as Europe's primary centre for the trading of soft commodities futures and options contracts, including cocoa, coffee, sugar, wheat, barley, potatoes and BIFFEX (dry cargo freight futures).

London Inter-Bank Offering Rate The rate of interest London banks offer to each other on occasions. *Abbrev.* LIBO *or* LIBOR.

London Maritime Arbitrators Association —LMAA.

London Market Referring to Lloyd's and the London insurance companies.

London Metal Exchange A market in London dealing in various metals such as copper, zinc, lead, silver, etc.

London Stock Exchange A society for the conduct of the sale or purchase, on behalf of non-members, of Government securities and stocks or shares in public companies. The members of the 'House' (as it is called) are re-elected annually and pay a substantial annual subscription.

The London Stock Exchange is an unincorporated private company. It has many rules which, unless they are recognised by the law, are binding upon its members only to such extent as they can be enforced by disciplinary action, such as expulsion, on the part of the committee. Upon persons who are not members the rules are not binding, unless in the eye of the law they are reasonable, and the law will not regard as reasonable any custom which is in violation of law. Thus all contracts for the sale of bank shares must set out the distinguishing number of each share sold. Members of the Stock Exchange habitually, as between themselves, disregard this enactment; but the custom under which they do this is not binding on persons who are not members of the Stock Exchange and who have no knowledge of the custom. It is indeed a question whether any such contract is not void to all intents and for all purposes.

The Stock Exchange does not recognise in its dealings any other parties than its own members, every bargain, therefore, whether for account of the member effecting it, or for account of a principal, must be fulfilled according to the rules, regulations and usages of the Stock Exchange.

Long. Longitude, *q.v.*; also US stock market version of Bull, *q.v.*

Long Blast *Similar to* **Prolonged Blast**, *q.v.*

Long-Dated Paper *Similar to* **Long-Dated Bill**, *q.v.*

Long-Dated Bill A Bill of Exchange term where payment is to be settled on long term. *Ex.* A bill to be paid three months or six months after sight.

Long Draft Foreign Bill of Exchange. An alternative phrase for a Bill of Exchange, *q.v.*

Long Duration Exposure Facility An orbital self-contained flight capsule for astronautical experiment. *Abbrev.* LDEF.

Long Ends A stevedoring expression with various meanings. (1) Detailed cargo information as regards to quantity unloaded from every Hatch, *q.v.*, Hold, *q.v.*, and time allotted to complete the work. (2) The calculation of the underdeck heights and dimensions of the Hatches, *q.v.*, and Holds, *q.v.* (3) To discharge or unload long pieces of cargo.

Long Hundred Fishing expression for 132 fresh herrings or 120 mackerel.

Longitude The angular distance of any place on the globe eastward or westward from a standard meridian, as in Great Britain that of Greenwich. Each degree of longitude represents 4 minutes of time so that 15 degrees of longitude represent an hour. *Abbrev.* Long.

Longitudinal Framing Hull framing of a ship running from forward to aft.

Longitudinal Position of the Centre of Buoyancy The point in the fore-and-aft line of a ship where buoyancy may be assumed to act. It is the centre of gravity of her underwater form. If the ship's centre of gravity (CG) is vertically above this point, the ship will be on an even keel. If weights are shifted so that the CG moves forward or aft there will be a moment causing the ship to tip and trim by the head or the stern. This has a bearing on the planning of cargo stowage and taking ballast. *Abbrev.* LCB.

Long Rate Long term policy interest on Bonds.

Long Room A spacious room in a customs house where clearance of customs papers and customs recording takes place.

Longs Long-dated government stock, i.e., with redemption dates of over 15 years.

Long Sea Regular long swell.

Longshoreman The American equivalent of Stevedore, *q.v.*

Long Term Policy A policy aiming to have a long duration in business or other specific work as opposed to a Short Term Policy, *q.v.*

Long Time Charter A time charterparty contracted for a long period.

Long Ton 2,240 lbs. *Abbrev.* LT or L.T. *or* l.t. *or* l.t.u.

Loofa *or* **Loofah** The dried fibrous part of the fruit of the tropical luffa gourd. Used in filters and for body washing and massage.

Look-out *or* **Look-outs** Visual watch while at sea. The look-out is bound to report any objects and/or lights during the course of navigation to his officer of the watch and this is recorded in the Log Book, *q.v.*

Loom (1) That part of the oar inside the rowlock. (2) Vague first appearance of land or a light at sea.

Loop *or* **LOOP** Louisiana Offshore Oil Port. Offshore oil complex in Louisiana, *USA*.

Loose Fiscal Policy *See* **Fiscal Policy**.

Loose Monetary Policy *See* **Monetary Policy**.

Loose Tools Light hand or machine tools that can be carried about.

LOP *or* **L.O.P.** Line of Position *nautical*; Loss of Profit; Local Operations Plot *naval*.

LOPAR Low-Power Acquisition Radar.

Loq. Loquitur *Latin*—He or she speaks.

LOR Letter of Readiness, *q.v.*; Light Output Ratio; Lunar Orbit Rendezvous.

LORAC Long Range Accuracy.

LORAD Long Range Detection.

LORAN Long Range Navigation.

Loro Accounts Third Party Accounts.

LOS Law-of-the Sea.

LOS *or* **L.O.S.** Line of Sight; Loss of Signal.

Losing Rate The daily error recorded of a chronometer when running slow. Opposite to Gaining Rate, *q.v.*

Loss In the law of marine insurance, the losses which arise from the various perils insured against may be either total or partial; they are total when the subject-matter of the insurance is wholly destroyed or injured to such an extent as to justify the owner in abandoning it to the insurer, and partial when the thing insured is only partially damaged or where, in the case of an insurance on goods, the owner of them is called upon to contribute to a general average. Total losses may again be divided into actual and constructive total losses. Actual total losses arise where the ship or cargo is totally destroyed or annihilated, or where they are placed by any of the perils insured against in such a position that it is wholly out of the power of the assured to procure their arrival. Thus, where by means of a peril insured against, a ship founders or is actually destroyed, or even where she is so much injured that she becomes a wreck, the loss is total and actual, although the form of the ship may still remain and in these cases the assured may recover for a total loss without abandonment. Losses are constructively total when the subject-matter of the insurance, although still in existence, is either actually lost to the owners or beneficially lost to them, and notice of abandonment has been given to the underwriters. Thus, where the ship, although existing as a ship, is captured or laid under an embargo, and has not been recaptured or restored before action brought, so that she is lost to the owners, or where she is so damaged by a peril insured against as to be unnavigable, and is so situated that either she cannot be repaired at the place in which she is, or cannot be repaired without incurring an expense greater than her value when repaired, the assured may abandon and treat the loss as total. *See* **Marine Insurance**.

LOSS *or* **L.O.S.S.** Large Object Salvage System.

Loss Adjusters Specialists with expertise in investigating and arranging settlement of claims on behalf of insurers and underwriters.

Loss Leader When a certain commodity is offered at a loss to attract customers to other products.

Loss of Hire Referring to **Breakdown Clause**, *q.v.*

Loss of Hire Insurance Insurance taken by a shipowner on a new vessel's completion date. This allows a sum of money per day for every lost day beyond the completion time.

Loss of Licence Insurance Insurance covering air pilots who may temporarily have lost their employment due to incidents such as their aircraft having been grounded.

Loss of Specie An actual total loss where insured property is so damaged as to lose its identity, or specie, and be unfit for its purpose.

Loss of Use Insurance Hull Insurance clause covering loss of freight income after an accident which is the subject of a claim. It is not normally relevant in the case of Total Loss, *q.v.*

Loss Overboard Deck cargo washed overboard. *Abbrev.* LOB.

Loss Payable Clause A Marine Insurance, *q.v.*, clause where a ship on Mortgage, *q.v.*, is involved in a major accident. The clause is a security for the bank that advanced the money for a mortgaged ship. In the event of a Total or Constructive Total Loss, *q.v.*, or in excess of a stated amount of claim the insurance settlement will go to the Mortgagor.

LOT Large Orbit Telescope; Load on Top, *q.v.*; Polskie Linie Lotnicze *Polish State Airlines*; Loss of Time.

Lot Referring to parcels of cargoes. *Ex.* The whole consignment to be shipped in two or three lots.

Lot Label Various small packages air freighted under one Air Waybill, *q.v. Similar to* **Collective Bill of Lading**, *q.v.*, when referred to sea freight.

LOTELA London Overseas Trades Employers Association.

Lot Money The auctioneer's charge for each item sold under his hammer.

Lou. Louisiana, *USA*.

Low *or* **LOW** Low pressure area *weather observation*.

Low Clouds Clouds averaging a height of 6,500 feet.

Low Cost Non-Returnable Referring to the cheap wood material used for pallets which render them non-returnable to clients. *Abbrev.* LCNR.

Low Density Cargo An aviation term for cargo that is light in relation to its volume. Opposite to Low Stowage Factor, *q.v.*

Lowland Land which is low with respect to higher land adjacent to it.

Low–Light Level Television Since the performance of under-water television cameras is substantially reduced, a special camera has been produced for adequate films to be shot in much deeper waters.

Low Profile Vessels Ships built with a low superstructure and/or hydraulic bridge/accommodation which can be lowered to permit navigation beneath bridges and overhead obstructions.

Low Speed Diesel A diesel engine ranging from 100 to 250 revolutions per minute. *Abbrev.* LSD *or* L.S.D.

Low-Stowage Factor Heavy deadweight cargo. Opposite to Low Density Cargo, *q.v.*, or High Stowage Factor.

Low Water Low tide.

Lower High Water The lower of two high tides occurring in a period of 24 hours. *Abbrev.* LHW *or* L.H.W.

Lower Low Water The lower of two low tides occurring in a period of 24 hours. *Abbrev.* LLW *or* L.L.W.

Lox *or* **L.o.x.** Liquefied oxygen.

L.P. *or* **l.p.** Legal Procurator; Life Policy. *See* **Life Insurance**; Liquid Petroleum; Lord Provost; Low Pressure.

L/P *or* **L.P.** Life Policy. *See* **Life Insurance**.

LPAN London Premium Advice Note.

l.p.c. Low Pressure Chamber.

LPEA London Port Employer's Association.

LPG *or* **L.P.G.** Liquefied Petroleum Gas; Liquid Propane Gas.

LPGC *or* **L.P.G.C.** Liquefied Petroleum Gas Carrier; Liquid Propane Gas Carrier.

l.p.i. Lines per inch.

L. Plms Las Palmas.

LPM *or* **L.P.M.** Lines per Millimetre; Lines per Minute; Long per Metre.

LPN Licensed Practical Nurse; National Padi and Rice Authority *Malaysia*.

LPN *or* **L.P.N.** Licensed Practical Nurse *USA*.

L'Pool Liverpool, *UK*.

L.P.S. Lord Privy Seal.

L.P.S.O. *or* **LPSO** Lloyd's Policy Signing Office. In the past all insurance policies at Lloyd's were signed by the individual underwriters, who accepted a share of the risks. Today this function is performed by Lloyd's Policy Signing Office—LPSO—a group within the Corporation, under the control of a Board of Management responsible to the Committee of Lloyd's. LPSO provides a range of services to underwriters and brokers including the checking of policies, endorsements, etc., with brokers' slips, signing on behalf of all subscribing syndicates and embossing with the LPSO seal. In 1961 a system of central accounting for the Lloyd's market was introduced, based on details provided by LPSO. Settlement is made centrally each month on the agreed figures issued by the Office. A new system of direct data entry involving the use of visual display units in LPSO departments was introduced during 1978. The computers of Lloyd's Data Processing Services are used in connection with the central accounting and LPSO services. LPSO has been moved out of London to Gun Wharf at Chatham, where a new Lloyd's building now stands close to the historic dockyard.

LPT Low Pressure Turbine.

LPTB London Passenger Transport Board, now the London Transport Executive, *Abbrev.* LTE.

l.q. Lege quaeso *Latin*—Please read.

269

LQP Living Quarters Platform.

LQS Lubricant Quality Scan, *q.v.*

LQT *or* **L.Q.T.** Liverpool Quay Terms.

LR *or* **L.R.** Lloyd's Register (Lloyd's Register of Shipping).

LR Lloyd's Register. Letter markings on each side of a ship above the horizontal line and passing through the disc or Plimsoll Line. This indicates that the freeboard was assigned by Lloyd's Register. *See* **Lloyd's Register of Shipping**; Law Report; Long Range; Loading Rate.

L/R Last Refusal; Left to Right; Lloyd's Register, *q.v.*

Lr *or* **lr** Lugger, *q.v.*

Lr Lawrencium *radioactive element*.

l.r. Landing Report; Long Range; Laufen Rechnung *German*—Current Account.

L.R.A.D. Licentiate of the Royal Academy of Dancing.

L.R.A.M. Licentiate of the Royal Academy of Music.

LRBC *or* **L.R.B.C.** Lloyd's Register Building Certificate.

L.R.C.P. Licentiate of the Royal College of Physicians.

L.R.C.S. Licentiate of the Royal College of Surgeons.

L.R.C.V.S. Licentiate of the Royal College of Veterinary Surgeons.

LRHC *or* **L.R.H.C.** Lloyd's Register Hull Construction Certificate.

LRIS *or* **L.R.I.S.** Lloyd's Register Industrial Services.

LRMC *or* **L.R.M.C.** Lloyd's Refrigerating Machinery Certificate.

LRMR Long Range Marine Reconnaissance.

LRN Lloyd's Register Number. *See* **LR Number** and **IMO Number**.

LR Number Recording number used in the Lloyd's Register Book of Ships.

LRO Long Range Objectives.

LROSG Lloyd's Register Offshore Services Group.

LR PASS *or* **LR Pass** Lloyd's Register's Plan Approval System of Ships.

LRQA *or* **L.R.Q.A.** Lloyd's Register Quality Assurance Ltd.

LRR *or* **L.R.R.** Long-Range Requirements; Lower Reduced Rate *income tax*.

LRS *or* **L.R.S.** Lloyd's Register of Shipping.

L.R.S.C. Licentiate of the Royal Society of Chemistry.

LR Safe Lloyd's Register's system for the structural analysis of ships.

LRT London Regional Transport.

L.S. Lesotho; Locus sigilli *Latin*—The place of the seal *documents*; Leading Seaman; Letter Service; Summer Deck Cargo load line markings on each side of the ship for timber; Lump Sum; Landing Ship, *q.v.*

l.s. Landing Ship, *q.v.*; Litres per Second; Local Sunset; Lump Sum.

LS Livestock.

L/S Lump Sum.

LSA *or* **L.S.A.** Licentiate of the Society of Apothecaries; Life Saving Appliances; Lloyd's Standard Form of Salvage Agreement, *see* **Lloyd's Open Form**; Liverpool Shipowners' Association; Low Specific Activity.

LSAT Law School Admission Test *USA*.

LSC *or* **L.S.C.** Liberian Shipowners' Council; Liberian Shipping Council.

l.s.c. Loco supra citato *Latin*—In the place cited above.

L.S. Cls. Livestock Clauses.

lsd *or* **l.s.d.** Last Safe Date *insurance*.

L.S.D. *or* **LSD** Librae, solidi, denarii *Latin*—Pounds, shillings and pence, English money until 1972 before decimalisation; Lightermen, Stevedores and Dockers; Last Safe Date *insurance*; the hallucinatory drug (lysergic acid diethylamide).

LS&D Charges *or* **L.S.&D. Charges** Landing Storage and Delivery Charges.

L.S.E. London School of Economics and Political Science; London Stock Exchange.

Lse. Loose.

L.s.e. Limited Signed Edition.

LSHWCA Longshoremen's and Harbor Workers' Compensation Act *USA*.

LSHW. liab *or* **L.S.H.W. Liab.** Longshoremen's and Harbor Workers' Liability *USA*.

LSI Large Scale Integration *computer control system*.

LSM *or* **L.S.M.** Linear Synchronous Motor; Litera scripta manet *Latin*—The written word stays.

LSO Leading Signal Officer; Landing Signal Officer; London Symphony Orchestra.

LSS *or* **L.S.S.** Life Saving Station *USA*; Life Saving Service *USA*.

LSSBS London Shipowners' and Shipbrokers' Benevolent Fund.

L.S.T. *or* **l.s.t.** Local Standard Time; Landing Ship Tank.

LSV Logistics Support Vessel *USA*.

LSWC Liverpool Seaman's Welfare Centre.

LT *or* **L.T.** Law Times Reports; Less Than; tropical load line markings which are shown on each side of the ship for deck timber cargo assignment; Local Time; Long Ton, *q.v.*; Letter Telegram; Turkish Pound (the lira of 100 kuru).

Lt. Left; Lieutenant.

l.t. Landed Terms; Large Tug; Local Time; Low Tension; Locum tenens *Latin*—Substitute for another.

l.t. *or* **l.tn** Long ton, *q.v.*

LTA or **L.T.A.** Lawn Tennis Association; Line Throwing Appliance or Line Throwing Apparatus. *See* **Marine Pyrotechnics**; Long Term Agreement; Lighter Than Air; Lost Time Accident *oil drilling*.

LTB or **L.T.B.** or **l.t.b.** Low Tension Battery.

Lt Bends or **LTBENDS** Liner Terms Both Ends.

LTC or **L.T.C.** Lloyd's Training Centre.

LTCF Less than Carload Freight, *q.v.*

Ltd Limited, *q.v.*

Lt.D. or **LTD** Light Displacement.

LTD Long Term Discount.

LTDW or **Ltdw.** Long Ton Deadweight.

LTE London Transport Executive.

Ltée *French*—Limited.

LTF or **L.T.F.** Tropical fresh water marking on each side of the ship for timber. The 'L' stands for Lloyd's Register assignment.

LTG Low Temperature Gas.

Ltg. or **ltge** Lighterage.

Lth. Length.

L.Th. Licentiate in Theology, Durham.

LTH or **L.T.H.** Light Training Helicopter.

LTI Lost Time Injury. An injury resulting in a person being unable to work for the day; or for the remainder of the shift.

LTIF Lost Time Injury Frequency. The number of LTI's, *q.v.*, recorded for a group of workers per million hours.

LTK or **RTK** Load Tonne Kilometre Revenue Tonne. *See* **Revenue Ton**.

LTM Load Ton Mile *aviation*; Long Term Mooring *anchors and chains*.

Ltr. Letter; Lighter, *q.v.*; Litre, *q.v.*

LTT or **L.T.T.** Less Than Truckload; Liquid Toner Transfer.

Lt.V. Light-vessel.

Lu. Lucerne *Switzerland*.

L/U or **l/u** Laid Up, *q.v.*, or Laying Up; Leading Underwriter, *q.v.*

LUA or **L.U.A.** Liverpool Underwriters' Association; Lloyd's Underwriters' Association; Leading Underwriters' Agreement.

L.U. Agreement Leading Underwriters' Agreement.

LUAA or **L.U.A.A.** Lloyd's Underwriting Agents Association.

LUALLS Leading Underwriter Agreement Liability Line Slips.

LUAMC or **L.U.A.M.C.** Leading Underwriter Agreement for Marine Cargo.

LUAMH or **L.U.A.M.H.** Leading Underwriter Agreement for Marine Hull.

Lub. or **Lubr.** Lubricant; Lubricate; Lubrication; Lubricator.

Lubber's Line Indication inside a compass in a form of a line to mark the ship's bow or head.

Lubricant Quality Scan A Lloyd's Register service established 1990 for analysing the lubricating oils

in marine and industrial engines and consultancy on the condition of machinery. *Abbrev.* LQS.

LUCO Lloyd's Underwriters' Claims Officer.

LUCRO Lloyd's Underwriters' Claims and Recovery Office.

Lucrum cessans *Latin*—Ceasing profit, i.e. Potential loss of earnings, for example as a result of an accident.

LUF Lift Unit Frame, *q.v.*

Luff (1) Windward side of a ship, opposite to Lee Side, *q.v.* (2) To head a ship into or closer to the wind. (3) The widest part of a ship's bow. (4) The leading edge of a fore-and-aft sail. (5) To lift up a derrick or jib of a crane. (5) To manoeuvre so as to force another boat to sail up into the wind.

Lufthansa German Airlines. *Designator LH.*

Luftpost *German*—Airmail.

Lug Literally ear. Projection to assist lifting or slotting into position; handle; loop.

Luggage Baggage containing personal effects of a passenger.

Lugger Small sailing vessel or boat with one or more masts carrying a quadrilateral fore-and-aft sail, a lugsail, on each. *Abbrev.* Lr *or* lr.

LUHF or **L.U.H.F.** Lowest Useful High Frequency.

LUM Lunar Excursion Module.

LUMAS Lunar Mapping System.

Lumber Sawn or split logs, planks, props, beams and poles. Used in America to mean timber, tree trunks and logs generally.

Lumberers Men engaged in felling trees and transporting them to the lumber mills.

Lumbering Felling trees.

Lumber Loadline A loadline on certain bulk carriers to which a vessel can load because the effect of a deck cargo of lumber is to increase the Freeboard, *q.v.*

Lumber Port An opening generally positioned in the fore part of a ship to ease loading and unloading operations.

Lump Sum An agreed sum of money for freight or the like irrespective of the amount of cargo carried.

lun Lundi *French*—Monday; Lunedi *Italian*—Monday; Lunes *Spanish*—Monday.

Lunar Day A day of 24 hours 48 minutes.

Lunar Month About 28 days.

Lunation The time from one new moon to the next, about $29\frac{1}{2}$ days.

Lun. Int. Lunitidal Interval; Longitudinal Interest.

LUR Lying Up Returns, *q.v.*

LUSI Lunar Surface Inspection.

LUT Landing Vehicle Truck *USA*.

Lutine Bell Originally belonging to the frigate *Lutine*, which was lost in 1799 with all hands and a valuable cargo, the bell hangs over the Caller's rostrum in the Underwriting Room at Lloyd's. Traditionally the ringing of the bell meant that an announcement was to follow about the loss or safe arrival of a vessel known to be overdue—the aim being to inform all underwriters simultaneously of important news, so that none had the advantage of special knowledge. Since the development of modern communications, however, the bell is now seldom rung except on ceremonial occasions.

Lutlag *Norwegian*—Limited Company. *Abbrev.* L/L.

Lux. Luxembourg.

Lux mundi *Latin*—The light of the world.

LV *or* **L.V.** Largest Vessel; Light-vessel; Luncheon Voucher; Liverpool.

l.v. Low Voltage.

LVA Landing Vessel Assault *US Navy*.

LVI Low Viscosity Index.

LVN *or* **L.V.N** Licensed Vocational Nurse *USA*.

LVNH Licensed Victuallers' National Homes.

LW *or* **L.W.** Lichte Weite *German*—Internal Diameter; Long Water; Low Water *chart symbol*; Lloyd's Winter load line marking on each side of the ship for deck timber cargo.

LW *or* **lw** Low Water *chart symbol*; Lumens per watt.

LWD Logging While Drilling *oil drilling*; Low Water Datum *chart symbol*.

L.W.H. Length, Width and Height.

LWL *or* **L.W.L.** Length on Water Line; Load Water Line, *q.v.*

LWM *or* **L.W.M.** Low Water Mark.

LWONT *or* **L.W.O.N.T.** Low Water Ordinary Neap Tides.

LWOST *or* **L.W.O.S.T.** Low Water Ordinary Spring Tides.

LWP Load Waterplane. Also called Design Waterplane.

LWR Light Water Reactor.

LWT *or* **Lwt.** Lightweight; Light Weight Ton.

lx Lux *physics*, unit of illuminance.

LX Roman numeral for 60.

LXX Roman numeral for 70.

Ly Lyons, *France*.

Lying Off Said of a vessel or boat waiting to take up a berth alongside and either under way while keeping clear or anchored temporarily in reasonable shelter.

Lying Up Returns *or* **Laid-Up Returns** The return of Premiums, *q.v.*, given by Underwriters, *q.v.*, to the Insured, *q.v.*, for a reasonable length of time the ship has been laid up. Generally the insured takes out Port Risk, *q.v.*, insurance to cover the ship for any accidents which may happen while she is laid up. *Abbrev.* LUR *See* **Cancelling Returns**.

LYNX Liquid Yield Exchangeable Notes *USA finance*.

LYONS Liquid Yield Option Notes *USA finance*.

Lzb Landeszentralbank, *q.v.*

M

m German Mark; Magnetic; Meridian *lower branch*; Meridional Difference; mile(s); Minute, *q.v.*; Statute Mile; Metre.

M Mach number, *q.v.*; Magnetic direction; Male; Married; Masculine; Member; Meridian, *q.v.*; Miles; Million; Monsieur; Nautical Mile; Roman numeral for 1000; Transverse metacentre of a ship; Mega.

mA Milliampere *electricity.*

MA Management Appraisal; Maritime Administration; Massachusetts, *USA.*

M.A. Magister Artium *Latin*—Master of Arts; Memorandum of Agreement, *q.v.*

M/A Mediterranean/Adriatic; Memorandum of Agreement, *q.v.*; My account.

MAA Maritime Appropriations Authorisation *USA.*

MAA *or* **M.A.A.** Manchester Airport Authority; Master at Arms, *q.v.*; Mutual Assurance Association *USA.*

MAB Marine Amphibious Brigade *USA*; Maritime Advisory Board *US Navy.*

MABS Maritime Applications Bridge System, *q.v.*

MAC Maritime Arbitration Commission, *USSR*; Maximum Acceptable Concentration, *q.v.*; Medical Advisory Committee; Merchant Aircraft Carrier; Multiple Access Computer; Municipal Assistance Corporation *USA.*

M.ACC. Master of Accountancy.

M.A.C.D. Member of the Australian College of Dentistry.

MACEE Malaysian-American Commission on Educational Exchange *USA.*

Mach *or* **Machy** Machinery.

MACH Modular Automated Container Handling.

Machinery This word covers the main engine(s) and auxiliaries in a ship including deck machinery such as capstons, winches, pumps, lifts and door and ramp operating systems.

Machinery Damage Co-Insurance A clause in a hull policy deducting 10% of net claims for damage to ship's machinery which is attributable to negligence on the part of the master, officers or crew.

Machmeter An instrument to measure the speed of an aircraft in relation to the speed of sound.

Mach Number An aviation speed which does not represent a fixed one. It is the air speed in relation to the speed of sound, the latter of which changes according to the altitude, *q.v.*, and air pressure. *Ex.* Mach 1 refers to an approximate speed of 670 miles per hour, Mach 2 is 1,350 m.p.h., Mach 3 is 2,000 m.p.h. and so on.

Mackerel Sky *See* **Cirrocumulus Clouds.**

MACS Multi-activity Cruise Ship.

MAD Magnetic Airborne Detector *aviation*; Magnetic Anomaly Detection *aviation*; Maintenance Assembly and Disassembly; Mean Absolute Deviation *aviation.*

Madag Madagascar.

MADAP Maastricht Automatic Data Processing *electronics.*

Made Good *See* **Amount made Good.**

Made Merchantable Reconditioned damaged merchandise which is fit to be sold. *Abbrev.* m/m *or* m.m.

MADGE Microwave Helicopter Guidance Equipment.

Mad. Is. Madeira Islands.

MADRE Magnetic Drum Receiving Equipment.

M.A.E. Master of Aeronautical Engineers; Master of Arts in Education.

Ma.E. Master in Engineering.

m.a.e. Mean Absolute Error.

M.A.(Econ.) Master of Arts in Economics.

M.A.(Ed.) Master of Arts in Education.

Mae West An inflatable life-jacket named after film actress.

m.a.f. Major Academic Field.

MAFF Ministry of Agriculture, Fisheries and Food.

Mafi Trailer A flat trailer with small wheels which can carry heavy loads of cargoes on board Roll-on/Roll-off, *q.v.* ships.

Mag. Magazine; Maggio *Italian*—May; Magnesium; Magnet; Magnetic; Magnitude *nautical.*

Magazine A store for arms, ammunition and provisions in time of war; a store for gunpowder or other explosives.

magg. Maggio *Italian*—May.

MAGLEV Magnetic Levitation.

Magnalium Alloy of magnesium and aluminium.

Magnetic Meridian Magnetic direction of a compass needle affected by the attraction of the pole. *See* **Compass.**

Magnetic Pole The North Pole of the earth's magnetism which wanders in the Canadian Arctic and to which the needle of a compass will point.

Magnetic Storm A period of intense solar flare or eruption when a flash of ultra-violet light will cause abnormal ionisation in the atmosphere disrupting radio communications while the magnetic field of powerful currents may deflect a compass needle. These effects may last several days.

Magnetometer An instrument for measuring magnetic forces.

Magni nominis umbra *Latin*—The shadow of a great name.

Magnum A wine bottle of 50 liquid ounces capacity.

Magnum bonum *Latin*—A great good.

Magnum opus *Latin*—An author's chief work; a writer's or other artist's main production.

M. Agr. Master of Agriculture.

Maghreb Collective word for Morocco, Algeria and Tunisia.

MAH *or* **M.A.H.** Marine and Hull Insurance.

MAIB Marine Accident Investigation Board *UK Department of Transport*.

MAIB *or* **M.A.I.B.** Member of the Association of Insurance Brokers; Marine Accident Investigation Branch.

Maiden Voyage The first voyage of a newly built ship after she has undergone trials successfully.

MAIIF Marine Accident Investigators International Forum.

Mail Correspondence collected and delivered by the Post Office.

Mail Carrier *or* **Mail Ship** A ship carrying mail, usually in the liner trade.

Mail Order A system of ordering goods by post.

Mail Ship *See* **Mail Carrier**.

Mail Transfer Facility whereby someone can request a bank to transfer a sum of money to another country giving the name and address of the payee and stating how the payment is to be made. Postal, cable or telex services may be used.

Main Mast The principal mast of a ship.

Main Stay The stay secured from the top of the Main Mast, *q.v.*, down to the deck.

Mainstream The principal channel of a river having many tributaries.

Maître d'hôtel *French*—Hotel Manager or Head Waiter.

Major Currency One of the stronger currencies traded in international exchange.

Majority Owners *Similar to* **Part Owners**, *q.v.*

Make Port To arrive at a port generally; to reach a port, possibly diverting from the planned route, for shelter, repairs, to land a sick sailor or some other emergency.

Making Full and Sound Making up deficiencies in the weight or contents of packages to bring them to their original state.

Making Water Leaking *nautical*.

Makuta Currency unit of the Republic of Zaire equal to one hundredth of a Zaire.

Mal. Malaysia; Malaysian; Malta; Maltese.

Mala fide *Latin*—In bad faith.

Mala in se *Latin*—Acts which are intrinsically wrong.

Mal à propos *French*—Inopportunely.

Malaysian Dollar Currency unit of Malaysia, also called a Ringgit.

Mal.d. *or* **M.D.** Malicious Damage *insurance*.

M.A.L.D. Master of Arts in Law and Diplomacy.

Mal de mer *French*—Sea sickness.

Malentendu *French*—A misunderstanding.

MALEV Hungarian State Airlines.

Malfeasance Evil-doing, especially official misconduct.

Malheur *French*—Bad luck; disaster.

MALTA An island in the Mediterranean midway between Gibraltar and Alexandria; Microprocessor Aircraft Landing Training Aid.

Malta Group A group of government legal representatives from western countries led by UK which meets to discuss air transport matters. It was first convened in 1974 with the intent of reaching an agreement which would require air license holders to cover every air passenger, for a sum greater than the minimum established by the Warsaw Convention and Hague Protocol, and equivalent to £25,000, as compensation for bodily injury or death in an accident.

Maltese Cross A symbol of the Knights of Malta.

MAMBO *or* **M.A.M.B.O.** Mediterranean Association for Marine Biology and Oceanography.

Man. Management; Manager; Manchester, *UK*; Manitoba, *Canada*.

MAN *or* **M.A.N.** Maschinenfabrik Augsburg-Nurnberg. German diesel engine manufacturers.

Management Accountant *Similar to* **Cost Accountant**, *q.v.*

Management by Objectives Technique whereby the manager and his subordinates agree on and quantify the goals to be attained. It aims to improve performance and motivation since the subordinates 'own' the goals set. *Abbrev.* MBO.

Management of Affairs The management of a company when in the hands of a person from outside.

Management Performance Appraisal Systematic assessment by a manager of the performance of subordinates in the pursuit of established goals which assists in the review of work patterns and results, training needs, the control of promotions and changes in remuneration.

Manager Person who controls or directs a business or a section of a large enterprise.

Managership The office of a manager.

Managing Director A director appointed to be in charge of the management of a company. *Abbrev.* Man. Dir. *See* **Company Director.**

Managing Owner One of several co-owners, to whom the others, or those of them who join in the adventure, have delegated the management of the ship. He has authority to do all things usual and necessary in the management of the ship and the delivery of the cargo, to enable her to prosecute her voyage and earn freight, with the right to appoint an agent for the purpose. He is not entitled to make an extra profit for himself by accepting secret commission.

Mandamus *Latin*—We command. A prerogative order issued in certain cases to compel the perform-ance of a duty.

Mandarin Chinese official in any of nine grades (*historical*); standard spoken Chinese.

Mandate A direction or command. A right given to a person to act in the name of another for a specific purpose or for administrative convenience. Thus a cheque is a mandate from the drawer to his banker to pay the stated amount to the payee named, or to the bearer of the cheque.

Mandatory Injunction A court's order requesting a person to do a special act.

Man. Dir. Managing Director, *q.v.*

Manet Currency unit of Turkestan.

Mang. B. Manganese Bronze.

Manhole An opening around the crankcase of an engine or boiler room or on top of a Double Bottom, *q.v.*, or Tank Top, *q.v.*, of a ship. The opening will allow a normal sized person to pass through in order to carry out an inspection and do any necessary work.

Manned Orbiting Laboratory A vehicle used for the study of outer space *USA. Abbrev.* MOL.

Manning The act of engaging seamen for service on a ship; a vessel's complement of officers and ratings.

Manning Certificate A certificate issued by the maritime department of a place or the Shipping Master, *q.v.*, where a ship is registered, under a guideline issued by the International Maritime Orga-nisation, *q.v.*, laying down the minimum number of qualified officers and crew to be engaged on a ship to make her officially termed 'Safely Manned'.

Manning Scales The minimum number of officers and crew members that can be engaged on a ship to be considered as sufficient hands with practical ability to meet every possible eventuality at sea.

Man-of-war An armed ship; a warship.

Manometer A pressure gauge for gases and vapours.

Manpower (1) The total amount of male and female labour of a country or enterprise. (2) An assumed unit equal to the rate of mechanical work and which is taken to be 1/10th horsepower.

Manropes The ropes used as hand rails on a Gangway, *q.v.*, or Hatchway.

MANS *or* **M.A.N.S.** Member of the Academy of Nautical Science *USA.*

MANTIS Manchester Technical and Commercial Information Services.

Mantissa Decimal part of a logarithm.

Manual Training Training emphasising the skill of the hands. *Ex.* engineering, carpentry, sculpturing, etc.

Manufacture The making of articles by physical labour or machinery, especially on a large scale.

Manufacturer's Agent An agent who is appointed by a manufacturer to act for him in obtaining orders from clients on a commission basis.

Manuscript *Latin*—Handwritten; an author's work written by hand; an author's copy for the printer. Usually referred to as typescript when a typewriter is used.

Manxman An inhabitant of the Isle of Man.

M.A.O. Magister Artis Obstetricae *Latin*—Master in Obstetrics *mid-wifery.*

Map Derived from the *Latin* Mappa, meaning a cloth when in olden days maps were drawn on cloths, skins, and parchment. It is a graphic representation of part of the earth's surface showing the names of countries, rivers, ports, cities, state boundaries, etc. Also seas and oceans are illustrated, depending on the kind of map one is using.

M.A.P. Maximum Average Price; Medical Aid Post.

Maple Leaf The emblem of Canada, having a maple leaf against a white background.

MAPONY *or* **Mapony** Maritime Association of the Port of New York.

MAR *or* **M.A.R.** Marginal Age Relief *tax*; Master of Arts in Religion, Mid-Apprenticeship Release.

Mar. March; Marine; Maritime; Martedi *Italian*—Tuesday; Marzo *Italian*—March.

M.Ar. *or* **M.Arch.** Master of Architecture.

MARAD *or* **Mar.Ad.** Maritime Administration *USA.*

March. Marchioness.

M.Arch.E Master of Architectural Engineering.

MARDEC Malaysian Rubber Development Cor-poration.

MARDI Malaysian Agricultural Research and Development Institute.

Mare clausum *Latin*—Closed sea; A sea which is within the Territorial Waters; *q.v. Opposite to* Mare liberum, *q.v.*

Mare liberum *Latin*—A sea open to all. This term refers to the liberty enjoyed by whoever is on the High Seas, *q.v. Opposite to* Mare clausum. *See* **Freedom of the Seas.**

Marena *Italian*—Marshy coastland mostly seen in Italy.

Mare nostrum *Latin*—Our sea. The Roman saying for the Mediterranean Sea.

Mares' Tails *See* **Cirrus Clouds**.

MAR form MAR Policy Form, *q.v.*

Margaritae ante porcoi *Latin*—Pearls before swine.

MARICHEM Marine transportation, handling and storage of bulk chemicals.

MARIDAS Maritime Data System.

Marginal Cost The aggregate of direct and variable cost involved in changing the normal volume of a product or service by one unit.

Marginalia *Latin*—Marginal notes.

Marginal Income Additional income received apart from standard or basic wages, such as bonuses and overtime.

Margin Plate Flanged longitudinal plate at the boundary of a double bottom tank.

MARIN Maritime Research Institute *Holland*.

Marina A small creek, port or basin where boats or yachts are accommodated during the winter season or to carry out repairs.

Marina *or* **MARINA** Maritime Industry Association, *Philippines*.

Marine Adventure Exposure of ship and cargo to maritime perils for profit and with liability to a third party.

Marine Audio-Visual Instruction Systems A navigational system for navigators, masters, officers, fishermen and yachtsmen of deep sea and coastal vessels. *Abbrev.* MAVIS.

Marine Business Marine, aviation and transit insurance.

Marine Engine An engine installed on a sea-going vessel for its propulsion.

Marine Environment Protection Committee A section of the International Maritime Organization (I.M.O.) whose primary objective is the elimination of wilful pollution of the seas by oil and the minimising of accidental spills. *Abbrev.* MEPC.

Marine Insurance Marine insurance is a contract of indemnity against certain perils or sea risks to which the ship, freight or cargo, and the interests connected therewith, may be exposed during a particular voyage, or a fixed period of time. The practice of marine insurance is older than insurance against fire and upon lives, and the law is codified in the Marine Insurance Act, 1906. The Act renders void any policy of marine insurance in which the insured has not an insurable interest, or expectation of such interest, as being in the nature of gaming or wagering. The Marine Insurance Act, 1745, which it repeals, made provision much to the same effect, and the Life Assurance Act, 1774, has similar provisions as to assurances on lives or on any other event. An attempt is also made still further to restrain gambling by the Marine Insurance (Gambling Policies) Act, 1908, under which 'p.p.i.' (policy proof of interest) policies are prohibited. There is an *ad valorem* duty on policies of sea insurance.

While fire and life insurances are made at the risk of companies which include within themselves the requisites of security, wealth and numbers, a large proportion of marine insurances is made at the risk of individuals called underwriters. The underwriters meet in a room at Lloyd's. Agents (who are commonly styled Lloyd's agents) are appointed in all the principal ports of the world, who forward regularly to Lloyd's accounts of the departures of ships from, and arrivals at, such ports, as well as of losses and other casualties; and, in general, all such information as may be supposed of importance towards guiding the judgments of the underwriters.

Merchants and shipowners who manage their own insurance business procure blank policies, which they fill up to meet the case, and submit them to underwriters, by whom they are subscribed or rejected. Each policy is handed about in this way until the amount required is complete. Merchants and shipowners also give orders to insurance brokers, who undertake and are responsible for the business of insuring; and to them likewise are transmitted the orders for insurance from the outports and manufacturing towns.

Besides individual underwriters and companies, there are associations formed by shipowners, who agree, each entering his ships for a certain amount, to divide the losses sustained by any of them. These are institutions of long standing, but since the alteration of the law in 1824, appear to be on the decline.

The losses against which a merchant or shipowner is not protected by insurance in this country in usual form are the following:—

(1) Acts of our own Government. (2) Breaches of the revenue laws. (3) Breaches of the law of nations. (4) Consequences of deviation. (5) All loss arising from unseaworthiness. Unseaworthiness may be caused in various ways—such as want of repair, want of stores, want of provisions, want of nautical instruments, insufficiency of hands to navigate the vessel, or incompetency of the master. (6) All loss arising from unusual protraction of the voyage. (7) All loss to which the shipowner is liable when his vessel does damage to others. (8) Average clause.

'Average' is a name applied to a certain description of loss to which the merchant and shipowner are liable. There are two kinds of average—general and particular.

General average comprehends all loss arising out of a voluntary sacrifice of a part of either vessel or cargo, made by the captain for the benefit of the whole. If a captain throws part of his cargo overboard, cuts loose an anchor and cable, or cuts away his masts, the loss is distributed over the value of the ship and cargo as general average.

Particular average comprehends all loss occasioned to ship, freight, and cargo, which has not been wholly or partly sacrificed for the common safety or which does not otherwise come under the heading of general average or total loss.

Losses where the goods are saved, but in such a state as to be unfit to forward to their destination, and where the ship is rendered unfit to repair, are called partial or salvage loss. The leading distinction between particular average and salvage loss is that in the first, the property insured remains the property of the assured, the damage sustained being made good by the insurer; in the second, the property is abandoned to the insurer, and the value insured claimed from him, he retaining the property so abandoned.

All the elements of general average may be classed under four heads: (1) Sacrifice of part of the ship and stores. (2) Sacrifice of part of the cargo and freight. (3) Remuneration of service required for general preservation to be distinguished from the 'sue and labour clause' in marine insurance policies, which does not cover general average contributions or salvage. (4) Expense of raising money to replace what has been sacrificed, and to remunerate services.

Marine and fire insurances being contracts of indemnity, the assured is only entitled to be paid once, and if he obtains payment of his loss from any other source, he is not entitled to claim against his insurer. Moreover, the insurer, upon paying the loss, is entitled to be put in the place of the assured; therefore if the assured had a claim which he might have enforced in respect of the property (*e.g.*, against a person whose negligence or wrongful act caused the loss), the insurer will thenceforth be entitled to enforce that claim for his own benefit; or if after the insurer paid the assured's claim, the latter receives compensation from other sources, the insurer is entitled to recover from the assured any sum which he may have received in excess of his actual loss.

Double insurance takes place when the same interest and the same risk are insured twice; a second insurance is often necessary where the precise value of the interest is not at first known. But if it appears, when the value of the interest becomes known, that there has been an over-insurance, that is to say, that the sum of the two or more insurances exceeds the interest of the assured, the excess cannot be recovered. And if the same person insures the same property with two insurers, and it turns out that there has been an over-insurance, then the insurers are entitled to make one another contribute rateably, so that they may all bear the loss in proportion to the amounts for which they have respectively insured the property, instead of the loss being thrown on one. *See* **Average**; **Constructive Total Loss**; **General Average**; **Particular Average**; **Loss**; **Policy Proof of Interest**; **Perils of the Sea**; **Sue and Labour Clause**.

Marine Law *Similar to* **Maritime Law**, *q.v.*

Marine Lien *or* **Maritime Lien** *See* **Lien**.

Marine Market That section of Underwriters, *q.v.*, and insurance brokers specialising in marine risks.

Marine Perils Perils pertaining to the sea.

Marine Policy *See* **Maritime Policy**.

Marine Pollution Control Unit *Abbrev.* MPCU. This has the responsibility within COASTGUARD, *q.v.*, for planning and operating counter-pollution measures at sea, when spilled oil (or other dangerous substances like chemicals) from ships threaten the UK coast, important fisheries or concentrations of marine wildlife. It also provides advice and assistance to local authorities who are primarily responsible for dealing with pollution which comes ashore, as well as to port and harbour authorities who deal with spills within ports and harbours.

Stockpiles of specialist beach cleaning equipment are maintained for use as agreed with local authorities. The Unit funds a research programme into both sea and shore cleaning operations. It advises local authorities, port and harbour authorities and offshore operators on their contingency plans; and assists in training local authority and port and harbour authority staff in shoreline clean-up management and techniques.

In a major coastal pollution incident, overall command of counter pollution operations is exercised from the Marine Emergency Operations Room in COASTGUARD's Southampton headquarters. MPCU staff are also sent to a convenient location near the coast, where local command of at-sea operations would be established, and to a local authority response centre, where beach clean-up operations are coordinated.

The MPCU also has aircraft fitted with specialised remote sensing electronic equipment; this detects the position, nature and extent of oil slicks on the surface. These aircraft provide accurate information about oil slicks and can be used to control MPCU's resources for cleaning up the oil. The aircraft fly frequent patrols around the coast of the UK to deter ships' masters and platform operators from discharging oil illegally and to catch those who do.

When a vessel threatens to cause oil pollution at sea following an accident, the MPCU is initially concerned with trying to stop the oil spilling. The MPCU maintains a stockpile of cargo transfer equipment which can be lifted by helicopter to a stricken vessel if necessary. In a major oil spill incident, MPCU may deploy its fleet of seven DC3 aircraft equipped with the necessary dispersant spraying equipment.

The MPCU has equipped a number of commercial tugs around the country with modern dispersant spraying equipment for use in the event of a spill near their home port, or to deal with a continuing spill at sea. It also maintains four oil recovery systems for combating spills at sea.

Marine Pyrotechnics Although this word refers to fireworks, it plays an important part in saving life at sea. There are various types of pyrotechnics but the most commonly used are the following:–

Hand Flare—It shows a red coloured flare when ignited and can be seen during the day and night from a reasonable distance.

Line Throwing Apparatus—A small rocket which is attached to a cord of about 600 feet. Generally used in aiming at a particular place which cannot be reached in rough seas during an emergency. It is fired from the bridge of a ship and also from lifeboats or liferafts.

Smoke Flare or Smoke Float—This is only used during the day as it emits a considerable amount of orange coloured smoke that is not visible during darkness. It can be left floating in the sea near to the distressed person(s) or object(s).

Parachute Flare or Rocket—A rocket held by the hand and when ignited a flare shoots into the air and descends slowly by means of a parachute. The colour of the flare is red. The parachute flare can be seen at a greater distance than the hand flare.

Self-Igniting Light—This is currently used with the lifebuoy, *q.v.*, and is manufactured to have a long life as opposed to the other methods mentioned above which have an approximate lifetime of three years. The lights are operated by means of batteries which are easily obtainable and replaced.

Marine Riser A large diameter pipe used in oil drilling operations at sea.

Marine Safety Information Broadcast messages to ships giving navigational and meteorological warnings, weather forecasts and urgent safety-related advice. *Abbrev.* MSI. *See* **Weather Services for Ships**.

Marine Underwriter A person who joins with others in entering into a policy of insurance as insurer. Except where an insurance is effected with a company, a policy of marine insurance is generally entered into by a number of persons, each of whom makes himself liable for a certain sum, so as to divide the risk; they subscribe or underwrite the policy in lines one under the other, and hence to subscribe a policy is sometimes called 'taking a line'. Policies are usually effected through brokers. Lloyd's syndicates—they include marine, non-marine, aviation, motor and short-term life syndicates—have over 19,000 underwriting members between them. The marine market at Lloyd's remains the world's pre-eminent insurance market. Huge sums of cover are sought not only for conventional fleets but the oil rigs, VLCCs and container ships which are the modern equivalent of the barques and schooners insured in the coffee house. *See* **Maritime Policy, Marine Insurance** and **Lloyd's**.

Mariner An alternative word for a sailor or seaman.

Marine Safety Agency *Abbrev.* MSA. Formerly the Surveyor General's Organisation, this was established as an executive agency within the Department of Transport on 1 April 1994. It is responsible for implementation of the Government's strategy for marine safety and prevention of pollution from ships. The Agency's overall aim is to develop, promote and enforce high standards of marine safety and to minimise the risk of pollution of the marine environment from ships. The MSA is headed by a Chief Executive who is supported within the MSA by four Divisional Directors who comprise the Management Board. A brief description of the duties of each division follows:

Ship Construction and Navigation Division (MSCN): navigation and communications, vessels traffic systems, ship routeing; novel craft; navigational and radio equipment; hydrographic advice; passenger and cargo ship construction; load line policy; intact and damaged stability; crew accommodation; tonnage; bulk cargoes; fishing vessel safety; recreational craft safety.

Marine Engineering, Equipment and Pollution Prevention Division (MSEP): dangerous goods; tanker safety; marine engineering pollution prevention policy and operational equipment; fire protection; submersibles, diving; life saving appliances; audits of classification societies; Quality Assurance Scheme.

Operations and Seafarers' Standards Division (MSOS): survey and inspection policy (including port state enforcement); surveyor training; central survey and inspection records; occupational safety for seafarers; Inspectorate of ships' provisions; general non-legislative matters; marine office management; training standards and examinations for certificates of competency; safe manning.

Finance, Personnel and Corporate Services Division (MSCS): finance, contracts, personnel management, manpower planning, marine office administration and property management, research, IT purchasing and support.

The Agency's headquarters are in Southampton with Marine District Offices at Aberdeen, Glasgow, Liverpool, London, Newcastle and Southampton and a number of sub-district offices. The MSA took over the responsibility for the Register of Shipping and Seamen, (RSS), *q.v.*, Cardiff in March 1994 and this operates under a service level agreement. The RSS maintains a central register of UK merchant ships, fishing vessels and yachts; issues and revalidates seafarers' certificates; recruits and maintains a record of the Merchant Navy reserve; and registers births and deaths at sea.

Marine Society, The This is the oldest public maritime charity in the world. Founded on 25 June 1756 by Jonas Hanway to encourage men and boys of good character to join the Royal Navy at the start of the Seven Years War, the Society had, by the end of that war, recruited 5,451 men and 5,174 boys. Incorporated by Act of Parliament in 1772 to apprentice poor boys to the Royal Navy and the Merchant Service, it gave them not only clothing but also pre-sea education. In 1783, Jonas Hanway presented his detailed proposal for County Naval Free Schools in every seaport to provide a regular supply of suitably trained boys. The Society, unable to afford so large a concept, commissioned a ship which, on 13 September 1786, took onboard 30 boys in the charge of a superintendent assisted by a mate, schoolmaster, boatswain and cook. The number of boys under training remained at some 200 a year until 1940. Thus started the era of pre-sea training ships. The Society's example was followed in 1810 by the West India Dock Company (until 1834), in 1826 by Admiral Sir Isaac Coffin in Connecticut, USA, and in the United Kingdom from 1856 by a number of organisations, each specifically concerned either with officer cadets, poor boys of good character, boys in need of care or reform school boys.

By 1940, the Society had provided 39,910 men and 36,047 boys for the Royal Navy and 34,776 boys for the Merchant Service. In that year, the ship was decommissioned, as it was a potential target for enemy bombers. The Society continued its support for young people with its pre-sea training camp fund for sea cadets wishing to join the Royal Navy (until 1945), bursaries at nautical colleges, uniform loans, books and sextants, as well as financial support for maritime youth organisations. The Society was influential in the formation of King George's Fund for Sailors, Sea Cadet Corps, Outward Bound Sea Schools, Sail Training Association, Nautical Institute and the Annual National Service for Seafarers.

In 1976 The Marine Society merged with a number of other charities, with all of whom it had close connections. The Sailors' Home and Red Ensign Club (1830) in Dock Street provided accommodation for seafarers from 1830 to 1974, and its London School of Nautical Cookery (1893) trained over 104,000 ships' cooks. The Incorporated Thames Nautical Training College (*HMS Worcester*) (1862) provided over 26,000 officer cadets. The Seafarers Education Service (1919) and College of the Sea (1938) provided libraries for ships and further education facilities for seafarers. The Merchant Navy Comforts Service Trust (1940) helped with Arctic clothing during the war, and many ships took part in the British Ship Adoption Society's (1936) scheme linking schools with ships.

The Society's main operations today are:

Seafarers Libraries: Library service for ships, oil rigs and nautical establishments. Books on loan and for purchase.

The College of the Sea: Further education and advice service (including GCSE, A-levels, Open University).

Sea Lines: Marine education service for schools through seafarers linking with schools, and the provision of teaching resources.

Training Ships Jonas Hanway and *Earl of Romney*: Basic and advanced sea training.

Jonas Hanway Scholarships: for professional/degree courses for Merchant Navy new entrants.

Worcester Scholarships: for those intending to go to sea and the career development of seafarers.

Scholarship Schemes: for ratings.

Financial Assistance: for professional seafarers, retired seafarers and their dependants in need, and for other charities whose objects support those of the Society.

MARISAT Maritime Satellite System, *q.v.*

MARISEC Maritime International Secretarial Services. An organisation serving the International Chamber of Shipping, the International Shipowners' Federation, the International Maritime Employers' Committee and the International Ship Managers' Association.

Marit. Maritime.

Maritime Applications Bridge Systems Generally refers to a ship's handling and accurate course-keeping radar system in *USA. Abbrev.* MABS.

Maritime Courts Courts of law dealing with maritime cases. *See* **Maritime Law.**

Maritime Industries Forum *Abbrev.* MIF. An organisation founded in 1992 by EC and EFTA shipbuilding associations to improve competitiveness within Europe and the world through a common policy.

Maritime Law The law relating to harbours, ships, and seamen. No system or code of maritime law has ever been issued by authority in Great Britain. The laws and practices that obtain are founded on the practice of merchants, the principles of the civil law, the laws of Oleron and Wisby, the works of jurisconsults, the judicial decisions of our own and foreign countries, etc. The decisions of Lord Mansfield did much to fix the principles of the maritime law of England as also did those of Lord Stowell. The decisions of the latter chiefly have reference to questions of neutrality, and of the conflicting pretensions of belligerents and neutrals. It has been alleged that he favoured the claims of belligerents.

Maritime Lien A claim which attaches to the *res, i.e.,* the ship, freight, or cargo. A maritime lien consists in the substantive right of putting into operation the Admiralty court's executive function of arresting and selling the ship, so as to give a clear title to the purchaser and thereby enforcing distribution of the proceeds among the lien creditors in accordance with their several priorities and subject thereto rateably. It may arise *ex delicto, e.g.,* compensation for damage by collision, or *ex contractu,* for services rendered to the *res*; but it is strictly confined to services such as salvage, supply of necessaries to the ship, and seamen's wages. Thus for ordinary work done upon a ship, such as repairs, there will be no maritime lien, but there may be a possessory lien so long as possession is retained. The privilege when once it attaches will not be affected by any change in the possession of the *res. See* **Lien.**

Maritime Perils Any perils at sea such as fires on board ship, storms, war action, etc.

Maritime Policy Maritime policies are either unvalued (or open) or valued. An unvalued policy is where the value of the thing insured is not stated in the policy, and must therefore be proved if a loss happens. A valued policy is where the value of the thing is settled by agreement between the parties and inserted in the policy. An insurance may be effected either for a voyage or for a number of voyages, in either of which cases the policy is called a voyage policy; or the insurance may be for a particular period, not exceeding twelve months, irrespective of the voyage or voyages upon which the vessel may be engaged during that period and the policy is then called a time policy; in addition to the two last-mentioned kinds of policy, there is a third, which is usually called a mixed policy, as, for instance, where a ship is insured from one port to another for a year, this being in effect a time policy with the voyage specified.

Before the Acts forbidding insurances by persons having no interest in the subject-matter of insurance, it was sometimes provided in the policy that it should be valid whether the insurer had any interest or not, in order to dispense with proof of interest in case of loss; these were called 'interest or no interest policies' or 'wager policies'. See **Marine Insurance**.

Maritime Research Information Service Established in 1970 and sponsored by the US Department of Commerce through the Transportation Research Council and National Research Council. It provides various services to American shipping and industry. *Abbrev.* MRIS.

Maritime Satellite System The satellite communications element of the International Maritime Satellite Organisation (INMARSAT), *q.v. Abbrev.* MARISAT.

Maritime Subsidy Board One of the branches of the US Maritime Administration. *Abbrev.* MSB. See **CDS** *or* **C.D.S.**

Maritime Welfare Organisations (U.K.)

The Royal National Mission to Deep Sea Fishermen: For over 100 years the R.N.M.D.S.F. has provided a complete welfare, spiritual, moral and material service to fishermen and their families. Its work began by sending Mission vessels to sea with the fishing fleets to combat the indifference and lack of concern for the scores of hundreds of men who would be confined to the area 'Nor'ard of the Dogger' for up to three months at a time.

Today, long family separations and difficult conditions still bring many problems. It is the work of the Mission Superintendents to endeavour to meet every need and try to find an answer to every problem. All the Mission's service is inspired by its Christian basis and conviction which is expressed in its motto 'Preach the word: heal the sick'.

Shipwrecked Fishermen and Mariners' Royal Benevolent Society: Instituted 1839, incorporated and registered under the Charities Act 1960. The Society operates through some 600 Honorary Agents in the British Isles. Its object is to relieve distress among the seafaring community, including: immediate financial aid to the dependants of seamen lost at sea; assistance to all seamen shipwrecked on the coasts of the British Isles; financial aid in time of need to seamen and their widows and orphans; any other objects for the benefit and welfare of seafarers. The Society also makes Awards for skill and gallantry in preventing loss of life at sea.

Seamen's Hospital Society: The Seamen's Hospital Society was founded in 1821 and incorporated by Act of Parliament in 1833 to provide for the charitable and, in particular, hospital needs of merchant seamen. It was responsible for the financing and control of several seamen's hospitals until 1948, when the responsibility for these was taken over by the National Health Service.

Since 1948 the Society has continued to function for the charitable relief of merchant seamen of all nations and their immediate dependants, including widows, who from sickness, misfortune or age are in need of assistance. The management of the Society is by Committee and Secretary. Income is derived by subscriptions and donations.

Mark Unit of currency used in Germany. One hundred pfennig equal one mark.

Market A public time and place of buying and selling; also purchase and sale. It differs from the *forum*, or market of antiquity, which was a public market-place on one side only, the other sides being occupied by temples, theatres, etc. A market can only be set up by virtue of a royal grant, or by long and immemorial usage, which presupposes a grant. As to disturbance of a market, it was held that any member of the public has a right of access to a franchise market on payment of tolls and observance of by-laws for the purpose of conducting sales, and that a sale by auction in the vicinity was not a disturbance. See **Markets and Fairs**.

Market Capacity The maximum amount an insurance market can absorb as liability to its policy holders while maintaining a proper solvency margin.

Market Forces The pressure of supply and demand which ultimately decide the value of an article and hence its price.

Market Maker A trading firm of the London Stock Exchange dealing in shares for its clients or itself.

Market Overt Open market. Market overt in ordinary market towns is only held on the special days provided for the particular town, by charter or prescription; but in the City of London every day, except Sunday, is market day. The market place, or spot of ground set apart by custom for the sale of particular goods, is also in ordinary towns the only market overt, but in the City of London every shop in which goods are exposed publicly for sale is market overt, though only for such things as the owner professes to trade in. That part of London not within the City does not appear to have the privilege of market overt. The doctrine of market overt is that all sales of goods made therein are binding not only on the parties but also on all other persons: so that if stolen goods are sold in market overt, the purchaser, if acting in good faith, acquires a valid title to them against the true owner, unless the latter has prosecuted the thief to conviction, in which case the goods revest in the true owner, and the buyer is left to obtain compensation out of any money which may have been taken from the thief on his apprehension. The doctrine of market overt does not apply to goods belonging to the Crown, and in the case of horses it is subject to statutory restrictions.

Market Price The current price generally quoted.

Market Research The careful study made by a company or any other business concern in relation to the products and various services in connection with their respective markets.

Marketable Security For the purposes of the Stamp Act, 1891, the expression 'marketable security' is defined by s. 82 of that Act as a security of such description as to be capable of being sold in any stock market in the United Kingdom.

Markets and Fairs At common law a market or a fair is a franchise or privilege to establish meetings of persons to buy and sell. The privilege is in each case derived either from royal grant or from prescription implying such grant. It may be possessed by a lord of a manor or by a municipal corporation or other body corporate. The franchise carries with it the right to restrain, by *scire facias* or by injunction, any person from setting up another market or fair so near as to interfere with the exercise of the franchise; and it also implied the right to hold a court of pie powder; but the right to levy tolls is not implied and exists only where given by the grant, though it may be established by prescription.

The difference between a fair and a market is that the fair is the larger gathering, and is held, say, only once or twice a year, while a market is held once or twice a week.

In course of time many fairs had ceased to fulfil any useful purpose, and had become mere gatherings of undesirable persons. The Fairs Act empowers the Home Secretary to abolish any fair upon the representation of the justices of the petty sessional district in which it was held, and with the consent of the owner of the franchise. Under the Fairs Act, the Home Secretary may alter or lessen the days on which a fair is held upon the like representation of the justices, or upon the representation of the franchise owner. The holding of unlawful fairs within the metropolitan police district can be summarily prevented under the Metropolitan Fairs Act.

In the great modern centres of population, markets are held either under special local Acts, generally embodying the Markets and Fairs Clauses Act, or under the Public Health Act. Powers are given for the provision and regulation of markets by the Food and Drugs Act.

The owner of the market is under no obligation to provide stalls or pens. His duty is limited to providing a space in which buyers and sellers can meet and conduct their trading.

Markka Unit of currency used in Finland. One hundred pennia equals one markka.

Mark-up *or* **Mark-up Price** What is added to the Wholesale, *q.v.*, Price or Production Cost to arrive at the Retail Price, and is thus the Profit. *Example* A product is sold for US $20 when the cost from the Wholesaler was US $15. Then the Mark-up will be US $5.

Marline Two-stranded left-hand laid cordage. It may be tarred or not. Used for light lashings and for parcelling the ends of ropes, especially wire ropes.

Marlinspike A pointed metal or wooden tool which is inserted between the strands of nylon, fibre and steel ropes to separate them for Splicing, *q.v.*

MARLO Maritime Liaison Officer *US Navy.*

Mar. Mere. Marina Mercantile *Italian*—Merchant Marine.

MARPOL 73/78 The International Convention for the Prevention of Pollution from Ships 1973, as modified by the Protocol of 1978 relating thereto (MARPOL 73/78). This IMO, *q.v.*, instrument is thus a combination of two international treaties. Its history goes back to 1954 when, in response to growing concern about the danger of marine pollution from tanker accidents and other causes at a time when the sea transport of oil was growing rapidly, the UK organised a conference which resulted in a Convention for the Prevention of Pollution of the Sea by Oil. Responsibility for this 1954 OILPOL Convention, which was principally concerned with the discharge and disposal of oily residues, passed to IMCO, *q.v.*, (later IMO) on its establishment in 1959. Amendments were adopted in 1962, 1969, and 1971. Over this period, other related conventions were also adopted: Intervention on the High Seas in Cases of Oil Pollution Casualties 1969, Civil Liability for Oil Pollution Damage 1969, Establishment of an International Fund for Compensation for Oil Pollution Damage 1971, and Prevention of Marine Pollution by Dumping of Wastes and Other Matter 1972 with various amendments.

Because of industrial and maritime developments, the IMO Assembly decided in 1969 that a completely new instrument was required to control pollution of the land, sea and air by ships. This resulted in the adoption in 1973 of the first ever comprehensive anti-pollution convention, the International Convention for the Prevention of Pollution by Ships (MARPOL) covering all aspects of pollution from ships and from cargoes, except the disposal of waste into the sea by dumping, and applies to ships of all types with respect to noxious and harmful substances, garbage and sewage as well as oil. New restrictions on the discharge of oily ballast water were introduced, with a total prohibition in specified seas. Tankships were required to be able to retain oily wastes by operating a Load on Top *q.v.*, system or for discharge to shore reception facilities. New tankers over 70,000 dwt ordered from 1976 were to have Segregated Ballast Tanks (SBT), *q.v.*, and meet certain subdivision and damage stability standards for survival after collision or stranding.

Certain technical problems made it difficult for many states to ratify MARPOL 1973. Meanwhile, a series of tanker accidents in the winter of 1976–77 led to demands for further action. IMO convened the International Conference on Tanker Safety and Pollution Prevention in 1978, which adopted Protocols to SOLAS, *q.v.*, as well as a Protocol to MARPOL. The latter extended the SBT requirement to vessels over 20,000 dwt and modified certain construction standards. Crude Oil Washing (COW) was prescribed for certain classes of existing, as well as new, vessels. A number of amendments were adopted in following years, including those for the introduction of the Harmonised System of Survey and Certification (HSSC) into MARPOL 73/78, and requiring ships to carry an oil pollution emergency plan. The 1992 amendments, following the stranding of the *Exxon Valdez* and other developments, which came into force in July 1993, significantly affected the design and construction of tankers. In general terms, tankers laid down from 1994 were to have double hulls. As an alternative, their design could incorporate the 'mid-deck' concept, under which the pressure within cargo tanks may not exceed the external water pressure. This means that the vessels have double sides but not a double bottom. Instead, another deck is installed with the venting arranged so that there is an upward pressure on the hull. Other designs ensuring the same level of protection in the case of an accident may be permitted with the approval of IMO's Marine Environment Protection Committee (MEPC). Existing tonnage would be subject to enhanced inspections at periodical surveys while, from 1995, 25-year-old tankers of a certain size would have to be converted to double bottoms and double sides.

MAR Policy Form A simplified marine insurance policy introduced by Lloyd's in 1982 for cargo and in 1983 for hull insurance, replacing the SG Form of Policy, *q.v.*. *Abbrev.* MAR form.

Marq Marquis.

Marry To bring two objects together. To lay two ropes in parallel so that they can be taken up together for hauling and veering, as in handling a boat's falls to keep her level when hoisting or lowering. To join the ends of two ropes to pass through a block. To bring strands together in Splicing, *q.v.*

MARSIM International Conference on Marine Simulation and Ship Manoeuvrability.

Mart Short for Market; Martinique.

MART Maintenance Analysis and Review Technique.

Maru *Japanese*—Merchant Ship.

MARV Manoeuvrable Re-entry Vessel.

MARVS Maximum Available Relief Valve Setting *of a liquid tanker tank*.

Mas *Latin*—a male.

MAS *or* **M.A.S.** Master of Applied Science; Malaysian Air System, Malaysia State Airlines; Maritime Advisory Service *USA*; Maximum Amount Subject *re-insurance*; Military Agency for Standardisation *NATO*.

MASCE *or* **M.A.S.C.E.** Member of the American/Australian Society of Civil Engineers.

MASER *or* **Maser** Microwave Amplification by Stimulated Emission of Radiation, *q.v.*

Maskapai *Indonesian*—Shipping Company.

MASME *or* **M.A.S.M.E.** Member of the American/Australian Society of Mechanical Engineers.

Mass. Massachusetts, *USA*.

MASS Maintenance Activities Sub-Sea Surface; Malta Agency for Scientific Services.

Mass Production The production of large quantities of a standardised article by standardised mechanical processes.

MAST Missile Automatic Supply Technique.

Mastable *or* **Mast-table** A support attached around the mast of a ship or sailing boat to hold the pivot of derricks or boom ends respectively.

Master Air Waybill When consolidated air cargo is despatched the Consolidator, *q.v.*, is issued with a Master Air Waybill. *Abbrev.* MAWB.

Master-at-Arms An officer whose duties are those of a security or police officer on board. *Abbrev.* MAA.

Master Cover *Similar to* **Broker's Cover**, *q.v.*

Master of the Rolls Originally the chief of a body of officers called the Masters in Chancery, of whom there were 11 others, including the Accountant-General. The Master of the Rolls was originally keeper of the records and acted as assistant to the Lord Chancellor, like the other Masters in Chancery. Subsequently, in the reign of Edward I, he acquired judicial authority in matters within the jurisdiction of the Court of Chancery, and in the reign of Henry VI bills for relief were addressed to him as well as to the Lord Chancellor. In more modern times, any suit, petition, etc., could be heard in the first instance by the Master of the Rolls, as well as by the Vice-Chancellors. But down to 1827 the Master of the Rolls sat in the evening from six to ten o'clock, the theory being that he, as being only the deputy of the

Lord Chancellor, ought not to sit at the same time as the Lord Chancellor, who up to then had dealt with such matters as now come before a judge of the Chancery Division. Until 1873 he was qualified to sit in the House of Commons. By the Judicature Act, 1873, the Master of the Rolls was made a member of the High Court of Justice, and by s. 6 an *ex officio* member of the Court of Appeal, while under the Judicature Act, 1881, he became a judge of the Court of Appeal only, but he still retains his non-judicial duties as custodian of the records. He admits solicitors of the Supreme Court. His powers as to matters formerly appertaining to the Petty Bag Office were preserved by the Solicitors Act. He is Chairman of the Advisory Council on Public Records and is responsible for the records of the Chancery of England.

Master Owner The owner of a vessel who is also its Master. This is frequently the case with coasters.

Master Pallet Heavy duty platform with hooks used for loading and unloading Pallets, *q.v.*, and Unit Load Devices.

Master's Declaration Outwards Also known as 'Master's Declaration & Stores Content for Vessels Outwards with Cargo'. It is a customs document which has to be signed by the master of a cargo ship certifying that all the particulars and requirements for the outgoing vessel have been observed. Ships sailing in ballast have to make a declaration known as 'Master's Declaration Outward or in Ballast'.

Masthead The highest part of the mast of a ship.

Masthead Light A white light positioned over the fore and aft centreline of the vessel showing an unbroken light over an arc of the horizon of 225 degrees and so fixed as to show from right ahead to 22.5 degrees abaft the beam on either side of the vessel.

Mast Riser A mast on a tanker vessel used as a ventilator in order to keep vapour coming from liquid petroleum clear of the decks.

Mat Material.

MAT *or* **M.A.T.** Marine, Aviation and Transport Insurance *hovercraft*; Master of Arts in Teaching; Moving Annual Total.

MATAC Marathon Air Terminals & Air Canada.

Mate *or* **Chief Mate** *or* **Chief Officer** *or* **First Officer** An officer on a merchant ship who sees to the execution of the master's commands and takes command in his absence. *Abbrev.* C/O for Chief officer. *See* **Chief Officer**.

Material Circumstance *See* **Material Fact**.

Material Fact The disclosure of any material circumstance by the assured to the Underwriter, *q.v.*, before the insurance risk is accepted. The information produced will affect the quote of the Insurance Premium, *q.v.* Thus the insurance is accordingly accepted, altered or rejected on this basis. Non-disclosure of any kind makes the insurance automatically null and void.

It is also called Material Circumstance, *q.v. See* **Utmost Good Faith**.

MATI Moskovskiy Aviatsiomyg Tecknologicheskiy Institut. Moscow Institute of Aviation Technology.

MATS Military Air Transport Service.

Matured A bill is said to be matured when payment is due. *See* **Maturity**.

Maturity *Also called* Maturity Date.

Maturity Date *Same as* **Maturity**, *q.v.*

MATV Master Antenna Television.

MATZ Military Aerodrome Traffic Zone *aviation*.

Maund A herring trade measure of $\frac{1}{4}$ cran (1 cran $= 3\frac{3}{4}$ cwt.) or 1,100 sprats.

Maur. Mauritania.

Mauritore Code name for Charterparty, *q.v.*, for the carriage of ore from the island of Mauritius.

MAWP Maximum Allowable Working Pressure.

Max Maximum.

Maximum Acceptable Concentrations Definitions as expressed in parts per million (ppm) for gases, vapours, fumes, dust, etc., where limits are not to be exceeded.

Maximum Distance Range *Similar to* **Distance Endurance**, *q.v. Abbrev.* MDR or M.D.R.

Maximum Permitted Mileage The total of the distances flown between two places with stop-overs elsewhere enroute which can be included for the air fare between those two places direct. Intermediate stops involving further deviation from the route and additional mileage would incur an additional fare. *IATA. Abbrev.* MPM.

Maximum Probable Loss An insurance expression generally used in reinsurance policies.

Mayday A spoken distress signal sent by radio telephony. Not to be confused with **SOS**, *q.v.*

mb Millibar(s).

m.b. Magnetic Bearing; Motor Boat.

Mb Megabyte.

M.B. Magnetic Bearing *nautical*; Maritime Board; Medicinae Baccalaureus *Latin*—Bachelor of Medicine; Motor Boat; Moulded Breadth of a ship.

Mba. Mombasa.

M.B.A. Marine Biological Association; Master of Business Administration.

M.B.A.C. Member of the British Association of Chemists.

mbc Maximum Breathing Capacity.

M.B.C.S. Member of the British Computer Society.

M.B.D. Machinery Break Down.

MBF Maximum Breaking Force.

m.b.h. mit beschrankter haftung *German*—Limited Company.

MBL Maximum Breaking Load.

M.B.L. Marine Biological Laboratory.

M.B.M. Master of Business Management.

MBO Management by Objectives, *q.v.*, Management Buy-Out.

MBS Main Boiler Survey, *Lloyd's Register notation* for British Machinery Corporation Classification.

M.B. Sc. Master of Business Science.

MBTA Massachusetts Bay Transportation Authority.

Mc Megacycle.

MC *or* **M.C.** Machinery Certificate; Magnetic Course *nautical*; Magistrates' Court; Maritime Commission *USA*; Master Commandant; Master of Ceremonies; Medical Certificate; Member of Congress; Metallic Clause; Metallic Currency; Military Cross; Morse Code.

mc *or* **m.c.** Mois courant *French*—This month.

M/C Machinery Certificate; Magnetic Course *nautical*; Manchester, *UK* Marginal Credit; Metallic Clause.

MCA Manufacturing Chemists Association *USA*; Maritime Conventions Act 1911 *USA*; Maximum Credit Account *finance*; Monetary Compensatory Amounts *finance*.

M.C.A. Management Consultants Association; Master of Commerce and Administration; Ministry of Civil Aviation.

MCA Cargo Similar to Cargo Information Card, *q.v.*

MCAT Medical College Admissions Test *USA*.

MCB Mobile Construction Battalion *USA*.

MCB *or* **M.C.B.** Maritime Commission Board *Canada*.

m.c.b. Miniature Circuit Breaker.

MCC Marylebone Cricket Club; Mid-Course Correction *nautical*; Mission Control Centre *satellite operation*.

MCCS Multi-Use Container Control System.

MCDFI Marine Catering and Duty Free *international magazine*.

MCDS Management Control Data System.

M.C.E. Master of Civil Engineering.

Mcf 1,000 Cubic Feet.

Mcfd 1,000 Cubic Feet per day.

Mcfh 1,000 Cubic Feet per hour.

MCFH Micro Cubic Feet per hour, *q.v.*

Mcfm 1000 Cubic Feet per month.

M.Ch. Magister Chirurgiae *Latin*—Master of Surgery.

M.Ch.D. Magister Chirurgiae Dentalis *Latin*—Master in Dental Surgery.

M.Ch.E. Master of Mechanical Engineering.

M.C.I.T. Member of the Chartered Institute of Transport.

M.Ch.Orth. Master in Orthopaedic Surgery.

Mcht Merchant.

Mchy Machinery.

Mchy. Aft. Machinery Aft, referring to the position of the engines of a ship.

Mchy. dge. Machinery damage *insurance*.

Mchy. Fwd. Machinery Forward, referring to the position of the engines of a ship.

m.c.i. Malleable Cast Iron.

MCJ *or* **M.C.J.** Master of Comparative Jurisdiction *USA*; Master of Comparative Jurisprudence.

MCL *or* **M.C.L.** Master of Comparative Law *USA*.

M.C.L. Master of Civil Law.

MCM Mine Counter Measures *US Navy*.

MCO Miscellaneous Charges Order. *IATA. Refers to air passage refund.*

Mco. Morocco.

M.Com. Master of Commerce.

M.Comm. Master of Commerce; Minister of Commerce.

MCP *or* **M.C.P.** Master of City Planning.

M.C.P.S. Member of the College of Physicians and Surgeons.

m.c.p.s. *or* **mc/s** Megacycles per second.

MCPU Marine Control Pollution Unit.

MCR *or* **M.C.R.** Master of Comparative Religion; Maximum Continuous Revolutions; Mid-Cadetship Release.

MCS Maritime Communications System *USA*.

M.C.S. Master of Commercial Science; Military College of Science.

MCSC Movement Control Sub-Committee. *IATA.*

M.C.S.P. Member of the Chartered Society of Physiotherapy.

MCT Minimum Connecting Time *travel*.

MCT *or* **M.C.T.** Moment to change trim *nautical*.

MCT 1 cm *or* **M.C.T. 1 cm** Moment to Change Trim 1 centimetre.

m.c.w. Modulated Continuous Wave.

M/cy Machinery.

M.d. Months after date.

MD Maryland, *USA*; Measured Depth *oil drilling*.

MD *or* **M.D.** Malicious Damage *insurance*; Managing Director; Market Day; Medicinae Doctor *Latin*—Doctor of Medicine; Memorandum of Deposit; Moulded Depth, *q.v.*

M.D. *or* **Mal D.** Malicious Damage *insurance*.

M/D Malicious Damage *insurance*; Managing Director; Memorandum of Deposit; Moulded Depth.

MDAPAD Machinery Design Approval Department of Lloyd's Register.

MDC Management Development Course.

Mddx. Middlesex, *UK*.

M.D.E. Master of Domestic Economy.

M.Dent. Sc. Master of Dental Science.

M.D.G. Medical Director General.

MDH *or* **M.D.H.** Maritime Declarations of Health; Medical Department of Health.

MDHC *or* **M.D.H.C.** Mersey Docks and Harbour Company.

MDI Management Development International.

M. Dip. Master of Diplomacy.

M'dise Merchandise.

M.Div. Master of Divinity.

M.DK. Main Deck.

Mdlle *or* **Mlle** Mademoiselle *French*—Miss.

Mdme *or* **Mme** Madame *French*—Mrs.

Mdnt Midnight.

MDO *or* **M.D.O.** Medical District Officer: Marine Diesel Oil.

MDR *or* **M.D.R.** Maximum Distance Range.

MDS Motion Damping System; Multipoint Distribution Service: the technology of using microwaves to beam television programmes, etc.

M.D.S. Master of Dental Surgery.

M.D.Sc. Master of Dental Science.

Mdse Merchandise.

MDST *or* **M.D.S.T.** Mountain Daylight Saving Time.

MDU Medical Defence Union; Memory Decoder Unit: Mobile Diving Unit.

MDV Mats, Dunnage and Ventilators *hold accessories*.

MDWT *or* **mdwt** *or* **M.D.W.T.** *or* **m.d.w.t.** Metric Deadweight Ton.

Mdx Middlesex, *UK*.

ME *or* **M.E.** Managing Editor; Main Engine; Marine Engineer, Master of Education; Master of Engineering; Mechanical Engineer; Middle East; Mining Engineer; Most Excellent; Moulded Edge, *of a ship*.

Me *or* **ME** Maine, *USA*.

Mᵉ Maître *French*—Notary Public; Solicitor; Advocate.

m.e. Maximum Effort.

MEA *or* **M.E.A.** Master of Engineering Administration; Middle East Airlines.

Mea culpa *Latin*—My fault.

Mean Draft Average draft of a vessel.

Mean Effective Pressure The regular or constant pressure acting upon the pistons of an engine. *Abbrev.* M.e.p.

Mean Nautical Mile Sea distance of 6,076.12 feet or 1.8532 kilometres.

Mean Sun Fictitious sun moving in celestial equator at mean rate of real sun. *See* **Mean Time**.

Mean Time The calculated time of the rotation of the earth in relation to the Mean Sun, *q.v.*

Mean Time Between Failures General technical term to show the reliability of engine operations. This is calculated by taking the total operation time divided by the number of failures and/or defects *Abbrev.* MTBE.

Meas. Measurement.

Measured Mile The normal distance of a mile which is used to determine the speed of a vehicle on land or a ship at sea. The nautical mile is of 1.8532 kilometres and the statute mile is 1.6093 kilometres.

Measure of Indemnity The amount of coverage or guarantee for which the insurer or Underwriter, *q.v.*, assumes responsibility in accordance with the clauses of an Insurance Policy.

Meat Clauses *or* **Meat Cls.** Clauses appearing on Bills of Lading, *q.v.*, when meat is carried.

Meat Ship A ship designed for the carriage of frozen meat.

MEBA *or* **Meba** Marine Engineers Beneficial Association *USA*.

M.E.C. Master of Engineering Chemistry; Member of Executive Council; Mercato Europeo Comune *Italian*—European Common Market; Marine Electronic Certificate.

M.Ec. Master of Economics.

MECAS *or* **M.E.C.A.S.** Middle East Centre for Arab Studies.

MECO *or* **M.E.C.O.** Main Engine Cut Off.

M.Econ. Master of Economics.

M.E.D. Marine Electronic Diploma.

Med. Medium.

M.Ed. Master of Education.

Med. *or* **Medit.** Mediterranean.

Med. Exp. Medical Expenses.

Median Line A straight line drawn from the angular point of a triangle to the middle of the opposite side.

Mediation Intervention in a dipute or strike with the object of bringing about a compromise between the opposing parties.

Meditore *or* **MEDITORE** Code name for Chamber of Shipping Mediterranean Ore Charterparty.

Medium Frequency *See* **MF** *or* **M.F.**

Medium Speed Diesel Engine A diesel engine of from 250 to 1,000 r.p.m. which can run on heavy fuel.

Mediums Medium-dated gilt edged stock, i.e., with a life of between five and 15 years.

Med luftpost *Swedish*—By airmail.

285

MEDMECON Mediterranean Middle East Conference.

MEDPOL Mediterranean Pollution Monitoring.

M.E.E. Master of Electrical Engineering.

MEES Middle East Economic Survey.

meg. Megohm. Unit of resistance (one million ohms *electricity*).

Mega One million times.

Megacycle One million cycles. *Abbrev.* Mc.

Megger Apparatus for measuring insulation resistance *electricity*.

MEIC *or* **M.E.I.C.** Member of Engineering Institute of Canada.

Me judice *Latin*—In my opinion.

MEL Marine Equipment Leasing.

Melb. Melbourne *Australia*.

Mel udara *Malay*—By airmail.

Mem. Memento *Latin*—Remember; Memorandum, *q.v.*

M.E.M.A. Marine Engine Manufacturers' Association.

Member Banks American expression for Commercial Banks, *q.v.*, and Clearing House, *q.v.*, associates.

MEMIS Maintenance and Engineering Management Information System.

Memo Memorandum, *q.v.*

Memorandum A note to help the memory; A record of events for future use; A condition expressed at the foot of the S.G. policy form applying and F.P.A. warranty to certain specified cargoes and a franchise to particular average claims for other interests. In modern practice it applies only to cargo policies.

Memorandum of Association When a company is formed. *See* **Articles of Association.**

Memorandum of Satisfaction A mortgage document confirming that the outstanding debt had been paid or cleared. *Abbrev.* MOS.

Memorandum of Understanding An official statement made out by the competent authorities in respect of actions to be fulfilled in due course. *Abbrev.* MOU.

Memorandum of Understanding on Port State Control Concluded in Paris 1982. An agreement between the maritime authorities of 14 European nations, with which the US, Canada and Russia are also associated, on the exercise of powers by port authorities to inspect ships of all flags for compliance with international conventions (IMO and ILO) and appropriate certification with regard to the state of ships and equipment, operational standards and procedures, competence of crews, safety and living conditions; and to detain ships found to be substandard in such respects. This also provides a check on controls by Flag States and classification societies.

Memorial of a Deed Brief details of a deed for registration purposes.

MENA *or* **M.E.N.A.** Marine Engine Manufacturers' Association; Middle East News Agency.

MENAS *or* **M.E.N.A.S.** Middle East Navigation Aids Service.

M. Eng. Master of Engineering.

Men of War *or* **Men O'War** Warships.

Mens sana in corpore sano *Latin*—A sound mind in a sound body.

Meo periculo *Latin*—At my own risk.

M.e.p. Mean Effective Pressure, *q.v.*

MEP *or* **M.E.P.** Master of Engineering Physics; Member of the European Parliament.

MEPA (UK) Marine Environment Protection Association of the UK.

MEPC Marine Environment Protection Committee *IMO.*

ME(PH) *or* **M.E.(PH)** Master of Engineering (Public Health).

mer Mercantile; Meridian; Mercoledi *Italian*—Wednesday; Mercredi *French*—Wednesday.

MER *or* **M.E.R.** Marine Engineer's Review *magazine.*

MERB *or* **M.E.R.B.** Mechanical Engineering Research Board.

Merblanch Chamber of Shipping & White Sea Conference Charterparty for the carriage of wood to the United Kingdom.

Mercantile Anything that is connected with commerce.

Mercantile Law Law which deals exclusively in commercial matters.

MERCAST Merchant Ship Broadcast System.

Mercator's Projection A method of map-making, from Latinized form of surname of Gerhard Kremer, 1512–94, Flemish-born German cartographer.

Mercer Textile, silk or woollen merchant.

Merchandise Commodities of commerce, goods for sale.

Merchandise Marks Whenever goods are shipped overseas they are invariably marked to enable them to be checked and distinguished from other items stowed with them.

Merchantable Where goods are bought by description from a seller who deals in goods of that description (whether he is the manufacturer or not), there is an implied condition that the goods shall be of merchantable quality, but if the buyer has examined the goods, there is no implied condition as regards defects which such examination ought to have revealed.

Merchantable Quality *See* **Merchantable**.

Merchant Banks *Similar to* **Investment Banks**, *q.v.*

Merchantman A name commonly used to define a vessel engaged in the carriage of merchandise. Alternatively called Merchant Vessel, *q.v.*, or Merchant Ship.

Merchant Navy Welfare Board, The The Second World War had an enormous impact on seafarers, especially during the first three years. Ships and their crews were subject to air, surface, submarine attack and mining in many parts of the world. These increased in severity as they approached Britain, where the approaches and the ports were particularly subject to both air attack and mines. Not only did this entail heavy casualties at sea in men and ships both for the Royal Navy and the Merchant Fleet, but also in harbours and in the ports where sailors' dependants lived. It also required the use of ports and harbours with minimal facilities for seafarers. Public and voluntary support for seafarers was both enthusiastic and generous, but relatively ill-organised.

Shipping was directed and controlled by Government and ship-owners were somewhat divorced from their ships and seafarers. This situation was changed when Mr Ernest Bevin set up the Seamen's Welfare Board in 1940, and subsequently provided funds from the Exchequer as the problems involved were beyond the resources of the voluntary organisations. In 1943 Mr Bevin appointed a Committee to study seamen's welfare in ports to recommend how this subject should be tackled after the war. This led to the establishment of the Merchant Navy Welfare Board in 1947. Its principal objective is the co-ordination of the work of the many societies and charitable organisations concerned with the welfare of serving merchant seafarers of all nationalities, retired British merchant seafarers and seafarers' dependants. The Board maintains Port Welfare Committees in major port areas throughout the UK and its work recognises the Country's obligations under ILO recommendations.

Although permitted under its Constitution to make appeals for funds, the Board is almost entirely self-financing and is empowered to make financial grants for projects of a capital nature undertaken by other British-based marine charities, both in this country and overseas.

Merchant Officers Protective Syndicate American insurance protection for masters, deck and engineer officers, pilots, etc. *Abbrev.* MOPS.

Merchant's Risk *See* At Merchant's Risk.

Merchant Ship *or* **Merchant Vessel** A vessel owned by a private company or a person or a country which carries goods against payment of freight. *Abbrev.* MS *or* M.S. *or* M.V. *or* M/V.

Merchant Shipper One who buys from manufacturers, usually FOB, to sell overseas, and ships the goods.

Mercury Planet nearest to the sun; Another name for quicksilver, a metallic liquid element mostly used for thermometers and barometers because of its high density and uniform rate of volume expansion and retraction in hot and cold conditions.

Merger An amalgamation of two or more companies to form one big organisation. Also called

Horizontal Integration, *q.v., or* Lateral Integration.

Meridian A semi-great circle joining the terrestrial poles (or celestial poles) and passing through any position on the globe (or its zenith). It marks the Longitude, *q.v.*, common to all places along it. The meridian at 0 degrees longitude is known as the Greenwich Meridian. *See* **Greenwich Mean Time**.

Meridian Line *Similar to* **Meridian**, *q.v.*

Merit Payment Remuneration of some employees for their special skill or progress, in addition to normal wages.

Merit Rating An employee rating method which is an assessment in regard to the quality of the employees in the organisation.

Mer. Parts Meridional parts.

Mer. pass. Meridian passage.

MERSAR Merchant Ship Search and Rescue Manual *IMO. Conduct of search and rescue operations.*

M.E.S. Master of Engineering Science.

Mesic (1) Related, or adapted, to life with little moisture. (2) A nucleus with an orbital meson. *Atomic.*

Mesne Profits Profits derived from land whilst the possession of it has been improperly withheld: that is, the yearly value of the premises. Mesne profits are the rents and profits which a trespasser has, or might have, received or made during his occupation of the premises, and which therefore he must pay over to the true owner as compensation for the tort which he has committed. A claim for rent is therefore liquidated, while a claim for mesne profits is unliquidated. The jury are not bound by the amount of the rent, but may give extra damages. But ground-rent paid by the defendant should be deducted from the damages. A plaintiff may recover in this action the costs of the action of ejectment. A claim for mesne profits may be joined with an action for the recovery of land.

Messagerie *French*—Transport by sea or rail.

Messenger An endless rope passing from a capstan to a cable in order to haul it in.

Messcook Messman; seaman's duty in the galley or kitchen.

Messieurs *French*—Plural of monsieur; used especially as a prefix to the name of a firm, or introducing a list of gentlemen.

Messuage A dwelling house with out buildings and land assigned to its use.

Met. Meteorological; Meteorology; Metropolitan.

Metallic Currency Money composed of metal alloys, such as gold, silver, nickel, bronze, etc.

Metallurgy The art of working metals, especially of extracting metals from their ores.

METAR Meteorological Aerodrome Report *aviation.*

Met Area Metropolitan Area *USA.*

METAREA One of 16 geographical sea areas, identified by Roman numerals, for the purpose of

Global Maritime Distress and Safety System (GMDSS) meteorological warnings and forecasts.

METC *or* **M.E.T.C.** Marine Engineering Technical Certificate.

Met. E. Metallurgical Engineer.

Meteor An atmospheric phenomenon, especially a 'shooting star', a small mass of matter from outer space rendered luminous by collision with earth's atmosphere.

Meteorological Office *See* **Weather Services for Shipping.**

Meteorology The study of motions and phenomena of the atmosphere, especially for weather forecasting.

Methyl Bromide Fumigant used in infested areas. Symbol CH_3Br.

Met. O. Meteorological Office; Meteorological Officer.

Metre Unit of length in the metric system equivalent to 3.2808 feet.

Met. Rep. Meteorological Report.

Metric System The use of metric units in contracts was made legal by the Metric Act of 1864, and the Weights and Measures (Metric System) Act of 1897 legalised the use of the metric system for most purposes. The transition to metric in the UK has been going on since 1965 when the Government's decision to support the change was announced in Parliament. The Government set up a Metrication Board in 1969. Also in 1969 the Government accepted, as a basis for the change-over to the metric system, the International System of Units (SI), which had been recommended in 1960 by the General Conference of Weights and Measures in which Britain took part.

Metric Ton Unit of weight in the metric system equivalent to 1,000 kilogrammes.

Metrology The science of weights and measures.

Metsat Meteorological Satellite.

MeV Mega-electron-Volt.

meV milli-electron-Volt.

M.E.W. Microwave Early Warning.

Mex. Mexico.

M.Ex.C. Marine Extension Clause *insurance.*

Mexicana de Aviacion Mexican National Airline.

MEZ *or* **M.E.Z.** Middle European Zone.

MF *or* **M.F.** Machine finished *paper;* Master of Forestry; Medium Frequency *radio.* From 300 up to 3000 kHz per second.

MFA Major Floating Aid *navigation;* Malta Football Association; Master of Fine Arts; Metal Finishing Association; Motor Factors' Association; Multi-Fibre Arrangement.

MFAG Medical First Aid Guide. Supplement to the IMDG Code, *q.v.,* IMO.

MFB *or* **M.F.B.** Metropolitan Fire Brigade.

MFC Multi-Function Console.

MFC *or* **M.F.C.** Maximum Foreseeable Cost *marine reinsurance;* Motorised Film Control *photography.*

mfd Manufactured.

mfg Manufacturing.

MFH Master of Foxhounds.

MFL *or* **M.F.L.** Maximum Foreseeable Loss. *Similar to* **Maximum Probable Loss,** *q.v.*

MFN *or* **M.F.N.** Most Favoured Nation. *See* **Most Favoured Nation Clause.**

MFO *or* **M.F.O.** Marine Fuel Oil; Medium Fuel Oil.

MFPE Maritime Fraud Prevention Exchange for developing countries. An organisation under the auspices of UNCTAD involving BIMCO, IMB, ICS and Lloyd's Register.

mfr/s Manufacturer(s).

mfst Manifest.

MFT *or* **M.F.T.** Motor Freight Tariff.

MFW *or* **M.F.W.** Maritime Federation of the World usually *Abbrev.* MWF *or* M.W.F, Maritime World Federation.

mg Milligramme.

Mg Symbol for Magnesium.

mg. cu.m Milligrammes per Cubic Metre.

mgd *or* **MGD** Million Gallons per Day.

M.G.I. Member of the Institute of Certificated Grocers.

MGPD Million Gallons of Distilled Water Per Day.

Mgr Monseigneur *French*—My Lord; Monsignor *title of RC bishop;* Manager.

m.g.s. Metre Gram Second.

mgt Management.

M.H. Master of Hygiene; Medal of Honour; Master of Horticulture.

MH Malaysian Airline System *airline designator;* Megaherz.

mh Millihenry.

MH *or* **M.H.** Main Hatch; Magnetic Heading *nautical;* Ministry of Health; Manhole, *q.v.*

m/h Main Hatch.

MHA *or* **M.H.A.** Master in Hospital Administration.

MHAWB Master House Air Waybill.

MHB Materials Hazardous Bulk; *US Coast Guard regulations.*

MHCI *or* **M.H.C.I.** Member of the Hotel and Catering Institute.

m.h.c.p. Mean Horizontal Candle Power.

MHD *or* **M.H.D.** Medical Health Department.

MHE *or* **M.H.E.** Master of Home Economics; Materials Handling Equipment; Mechanical Handling Equipment.

m.h.f. Medium High Frequency.

MHL *or* **M.H.L.** Master of Hebrew Literature.

MHM Manifold Heater Module *electronics.*

mho Unit of conductivity (electricity).

M.Hon. Most Honourable.

MHR Member of the House of Representatives.

MHT Mild Heat Treatment.

MHW *or* **M.H.W.** Mean High Water.

MHWH *or* **M.H.W.H.** Mean Higher High Water *of double higher tides.*

MHWI *or* **M.H.W.I.** Mean High Water Lunitidal Interval.

MHWN *or* **M.H.W.N.** Mean High Water Neaps.

MHWNT *or* **M.H.W.N.T.** Mean High Water Neap Tide.

MHWS *or* **M.H.W.S.** Mean High Water Springs.

MHWST *or* **M.H.W.S.T.** Mean High Water Spring Tide.

MHz MegaHertz.

MI Michigan, *USA.*

MI *or* **M.I.** Malleable Iron; Marine Inspector *USA*; Marine Insurance, *q.v.*; Military Intelligence.

M/I Migration Inversion *geophysical.*

Mi. Mississippi, *USA.*

MIA Maritime Industry Authority *Philippines*; Marine Insurance Association.

M.I.A. Marine Insurance Act. *See* **Marine Insurance**; Missing in Action.

M.I.A.A. Manchester International Airport Authority.

M.I. Act Marine Insurance Act 1906. *See* **Marine Insurance**.

MIAS Marine Information and Advisory Service.

MIB *or* **M.I.B.** Marine Index Bureau; Motor Insurers' Bureau.

MICA Mutual Insurance Companies' Association.

MICC Malaysian International Chamber of Commerce.

M.I.C.E. Member of the Institute of Civil Engineers.

Mich. Michigan, *USA.*

M.I.C.M. Member of the Institute of Credit Management.

MICR Magnetic Ink Character Recognition *electronics.*

Micro One-millionth part.

Micron Measure of thickness or length of one-millionth of a metre equivalent to .00003937 of an inch; film thickness of paints is measured in microns or mills.

Micro Cubic Feet per hour The standard measure of liquid in relation to testing for leaks. *Abbrev.* MCFH.

Microfiche Pocket-size film negative containing the much reduced images of a very large number of pages of original documents. These images can be scanned and read with the aid of a magnifier and can be re-printed with the appropriate equipment. Microfiches are a useful method of storing the contents of bulky documents in a very small space while permitting instant retrieval.

Micro-processor Aircraft Landing Training Aid A low-cost aircraft landing system based on a micro-processor which provides realistic simulator training for student pilots. *Abbrev.* MALTA.

Microsecond One millionth of a second.

Microwave Amplification by Stimulated Emission of Radiation. An instrument to amplify electromagnetic waves used to pick up signals from satellites. *Abbrev.* MASER *or* Maser.

Microseismograph A device to detect storms from a distance.

M.I.C.S. Member of the Institute of Chartered Shipbrokers formerly A.I.C.S.

Mid Middle; Midlands, *UK*; Midnight; Midshipman.

MIDAS Maritime Industrial Development Area Schemes; Measurement Information and Data Analysis System.

MIDC Maritime Industry Development Commission. *Australia.* Committee of shipowners', unions' and government representatives.

Mid–Channel Similar to Fairway, *q.v.*

Middle Clouds Clouds from 6,500 up to 20,000 feet in height.

Middleman Agent or broker, *q.v.*, who acts as an intermediary between the seller and the buyer.

Middle Market Price The mean or average price between the buying and selling prices from the official price list of securities. It can also mean the market price in general business.

Middle Market Value On a stock exchange the prices of Stocks, *q.v.*, and Shares, *q.v.*, are pronounced during the finishing time of the day to be the mean value or Middle Price, *q.v.*

Middle Price The average price between the quotation of the seller and the buyer. The middle price is arrived at by adding both prices and dividing the result by two. *See* **Middle Market Value**.

Middle Rate The average of buying and selling rates in currency exchange.

Middle Watch *or* **Mid Watch** Seaman's watch while at sea from midnight to 4 a.m.

Middle Way Mixed economy of a state.

Middx *or* **Middlx.** Middlesex, *UK.*

MIDF Malaysian Industrial Development Finance.

Mi. Dk. Middle Deck.

Mid. lat. Middle Latitude *nautical.*

Mid. lat. corr. Middle Latitude Correction *nautical.*

Midshipman Junior naval officer under the rank of Sub-Lieutenant; a naval cadet before he becomes a sub-lieutenant.

Midships Midway between fore and aft or stem and stern. Also expressed as amidships.

Mid Watch *Similar to* **Middle Watch**, *q.v.*

MIE *or* **M.I.E.** Master of Industrial Engineering; Master of Irrigation Engineering.

M.I.E.E. Member of the Institution of Electrical Engineers.

M.I.Ex. Member of the Institute of Export.

MIF Maritime Industries Forum. An organisation founded in 1992 by EC and EFTA shipbuilding associations to improve competitiveness within Europe and the world through a common policy.

MIFASS Marine Integrated Fire and Air Support System *USA*.

M.I.F.T. Manchester International Freight Terminal.

MIG Metal Inert Gas *welding*.

M.I.G. Magnesium Inert Gas.

M.I. Gas E. Member of the Institution of Gas Engineers.

MII Maritime Institute of Ireland.

M.I.I.E. Member of the Institution of Industrial Engineers.

MIILA *or* **M.I.I.L.A.** Member of the International Institute of Loss Adjusters *USA*.

Mij Maatschappij *Dutch*—Joint Stock Company.

MIJ *or* **M.I.J.** Member of the Institute of Journalists.

Mike International Radio Telephone phonetic alphabet for 'M'.

Mil. Milan, *Italy.*

Mile An English statute mile equals 5,280 feet.

Mille A board measurement of 1,000 feet of lumber from the USA; a mille measures 83.333 cubic feet.

Millimicron One millionth of a metre or one thousandth of a millimetre.

Millisecond One-thousandth of a second.

Mill/s One-thousandth of an inch. Paint film thickness is measured in mills or microns, *q.v.*

Millscale A thin film of oxide formed when steel is hot-rolled which causes corrosion when in contact with seawater.

MILS Missile Impact Location System.

MILSPECS Military Specifications *USA*.

MILSTRIP Military Standard Requisitioning and Issuing Procedure *USA*.

MILU *or* **M.I.L.U.** Member of the Institute of London Underwriters.

Milw Milwaukee, *USA*.

MIM Machinery Interface Module *computer control system.*

M.I.M.A. Member of the International Marine Association.

M.I. Mar. E. Member of the Institute of Marine Engineers.

MIMCC Marconi International Marine Communications Company.

MIMgt Member of the Institute of Management. *See* **IM**.

mi/min Miles per minute.

MIMS Monthly Index of Medical Specialities *medical index book.*

min. Minimum; Minute.

MIN *or* **M.I.N.** Member of the Institute of Navigation.

Min. B/L Minimum Bill of Lading.

MINDAC Marine Inertial Navigation Data Assimilation Computer.

MIN/DEP *or* **Min/Dep** Minimum Deposit Premium *insurance.*

Mineral Oil This is classed under petroleum, being a mixture of various hydrocarbon compounds obtained from the bottom layers of rocks both under the sea and under the ground.

Min. Plenip. Minister Plenipotentiary.

M.Inst. C.E. *Similar to* **M.I.C.E**, *q.v.*

M.Inst.E.E. *Similar to* **M.I.E.E**, *q.v.*

M.Inst. Mar.E. *Similar to* **M.I. Mar.E.**, *q.v.*

M.Inst. M.S.M. Member of the Institute of Marketing & Sales Management.

M.Inst. T. Member of the Institute of Transport.

M in T *or* **Mint** Money in Transit.

Mint A place where money is coined, usually under state authority.

Mintage The charge levied by a mint in the manufacture of the grade of coins for a country.

M. Int. Met. Member of the Institute of Metals.

Mint par of Exchange The rate of exchange between two countries on a Gold Standard.

Minute/s (1) A brief summary of proceedings of an assembly, committee, etc.; official memorandum authorising or recommending a course of action. (2) The units of latitude and longitude. *q.v.*, each equivalent to one-sixtieth part of a degree. *Abbrev.* m. or min.

min. wt. Minimum weight.

MIOM *or* **M.I.O.M.** Member of the Institute of Office Management.

MIP *or* **M.I.P.** *or* **m.i.p.** Magnetic Induced Polarity *geophysical*; Malleable iron pipe; Manual Injection Panel *computer*; Manila International Port; Marine Insurance Policy. *See* **Marine Insurance** *also* **Maritime Policy**; Mean Indicated Pressure; Mis-erasure Prevention *computer*; monthly investment plan.

MIPE *or* **M.I.P.E.** Member of the Institute of Production Engineers.

MIPS Millions of Instructions per second *standard measure of computer power.*

MIRA *or* **M.I.R.A.** Motor Industry Research Association.

Mirabile dictu *Latin*—Wonderful to relate.

Mirage An optical illusion or distortion of the appearance of objects near or beyond the horizon. It is caused by abnormal refraction of light rays near the sea or land surface when a difference of temperature affects the density of air close to it. One effect of this phenomenon is to make objects, such as ships and lights, beyond the natural horizon visible and closer than expected. Occasionally, an inverted image may be seen above the real object. Another effect is to reduce the apparent distance of the horizon which appears to be shimmering with objects floating above it. On land, especially in a desert, this gives the illusion of water. Navigators have to be particularly careful in taking sights with the sextant when mirage occurs.

MIRAS Mortgage Interest Relief at Source, *q.v.*

MIRID Maritime Institute for Research and Industrial Development *USA*.

MIRV Multiple Independent Re-Entry Vehicle.

MIS *or* **M.I.S.** Management Information System *USA*; Master of International Service; Mining Institute of Scotland.

MISC *or* **M.I.S.C.** Malaysian International Shipping Corporation, Malaysian Shipping State Line.

Misc Miscellaneous; Miscellany.

Misdemeanour Misdeed; a lesser offence than a felony.

Misfeasance The improper performance of a lawful act, as where a person is guilty of negligence in performing a contract.

Misprision Neglect, negligence, oversight. In its larger sense, it is used to signify every considerable misdemeanour which has not a certain name given to it by the law; and it is said that a misprision is contained in every treason or felony whatsoever, and that one who is guilty of felony or treason may be proceeded against for a misprision only. Upon the same principle, while the Court of Star Chamber existed, it was held that the sovereign might remit a prosecution for treason, and cause the delinquent to be censured in that court, merely for a high misdemeanour, as in the case of the Earl of Rutland, 1601, concerned in Essex's rebellion. Every great misdemeanour, according to Coke, which has no certain term appointed by the law, is sometimes called a misprision. The term is, however, rarely if ever used in this sense, it being now practically confined to the two phrases 'misprision of treason' and 'misprision of felony.'

Misprision of treason is where a person who knows that some other person has committed high treason does not within a reasonable time give information thereof to a judge of assize or justice of the peace. At common law the punishment is imprisonment for life and forfeiture of the offender's goods, etc.; but prosecutions for the offence are now unknown.

Misprision of felony is where a person who knows that some other person has committed felony conceals or procures the concealment thereof. In ordinary cases the offence is a common law misdemeanour; if the offender is a sheriff or coroner, or the bailiff of either officer, a special punishment is inflicted. But prosecutions for this offence also are obsolete.

Misrepresentation Misrepresentation may be *suggestio falsi* in a matter of substance essentially material to the subject, whether by acts or by words or by manoeuvres or by positive assertions, or material concealment (*suppressio veri*) whereby a person is misled and damnified.

Fraudulent misrepresentation is a representation contrary to the fact, made by a person with knowledge of its falsehood, or without belief in its truth, or recklessly, not caring whether it is true or false, and being the cause of the other party's entering into the contract. A misrepresentation by an agent from whom the facts have been withheld by his principal does not amount to fraudulent misrepresentation by the principal.

Innocent misrepresentation is where the person making the representation believed it to be true.

Miss. Mississippi, *USA*.

Missing Flight Term used when an aircraft is overdue.

Missing Ship A ship is deemed to be 'missing' when, following extensive inquiries, she is officially posted as 'missing' at Lloyd's. She is then considered to be an 'actual total loss' and policy claims for both hull and cargo are settled on that basis. *See* **Posted for Information**.

Missions to Seamen An Anglican charity which has been serving the seafarers of the world since 1856. Both missionary society and welfare agency, and a founder member of the International Christian Maritime Association, it ministers to seamen of all races and creeds through 83 clubs, home missions and visits to ships in another 200 ports worldwide. Its work includes helping distressed seamen, family welfare and supporting the cause of those subjected to substandard conditions onboard ship. Its symbol is the Flying Angel.

MIT *or* **M.I.T.** Massachusetts Institute of Technology *USA*.

MITA Multilateral Interline Traffic Agreements *aviation*.

MITI Ministry of International Trade and Industry *Japan*.

MITMA *or* **M.I.T.M.A.** Member of the Institute of Trade Mark Agents.

MITTS Market Index Target Terms Securities *USA Finance*.

MIWM *or* **M.I.W.M.** Member of the Institution of Works Managers.

Mixed Economy The economy of a country involving a mixture of state and private enterprise.

Mixed Policy An insurance policy combining the coverage of both the voyage and the time of the adventure.

Mizzen Fore and aft sail set on a Mizzen Mast, *q.v.*, or right aft in a ship or boat.

Mizzen Mast The aftermost mast of a three-masted ship.

MK *or* **M.K.** Mohammedan Killed, *q.v.*

mk *or* **M.K.** Mark. Unit of currency used in Germany. One hundred pfennige equals one mark.

MK Metre Kilogramme; Air Mauritius *airline designator.*

MKS Metre Kilogramme Second.

MKSA Metre Kilogramme Second Ampere.

mkt Market.

ml Millilitre.

ML *or* **M.L.** Magister Legum *Latin*—Master of Laws; Mean Level; Middle Latin; Motor Launch; Moulded Line of a ship.

Ml Longitudinal Metacentre of a ship.

m.l. Mean Level.

MLA *or* **M.L.A.** Malta Legislative Assembly; Marine Librarians' Association; Maritime Law Association; Master of Landscape Architecture; Mean Line of Advance; Modern Language Association.

M.Lat. Mean Latitude *nautical.*

MLAW *or* **M.L.A.W.** Members of the London Association of Wharfingers.

MLB Maritime Labor Board *USA;* Motor Lifeboat.

MLC *or* **M.L.C.** Member of the Legislative Council.

mld Moulded.

Mlf Multilateral Force.

MLF Member of the Legal Fraternity *UK.*

Mlle Abbreviation for Mademoiselle, Miss.

MLM Multi-level Marketing.

MLR *or* **M.L.R.** Minimum Landing Rate *discharging cargo;* Minimum Lending Rate *banking.*

MLRS Multi-launch Rocket System, *q.v.*

MLS *or* **M.L.S.** Master of Library Science.

MLS Microwave Landing System *aviation.*

MLW *or* **M.L.W.** Mean Low Water *nautical.*

MLWI *or* **M.L.W.I.** Mean Low Water Interval *nautical.*

MLWN *or* **M.L.W.N.** Mean Low Water Neaps *nautical.*

MLWNT *or* **M.L.W.N.T.** Mean Low Water Neap Tide *nautical.*

MLWS *or* **M.L.W.S.** Mean Low Water Springs usually *abbrev.* (Low Water Springs) LWS *or* L.W.S.

MLWST *or* **M.L.W.S.T.** Mean Low Water Spring Tide *nautical.*

mm Millimetre(s).

m/m *or* **m.m** *or* **M/M** Made Merchantable, *q.v.;* Mutatis Mutandis *Latin*—With the necessary changes in points of detail.

MM Messieurs *French*—Messrs.

MM *or* **M.M.** Master Mariner; Mercantile Marine; military medal.

MMA *or* **M.M.A.** Manual Metal Arc *welding;* Merchandise Marks Act; Merchant Marine Act 1970 *USA.*

MMbpd Million Metric Barrels per day.

MMC Maximum Metal Condition; Merchant Marine Committee, a branch of the US House of Representatives; Metro Manila Commission *Philippines;* Monopolies and Mergers Commission *UK.*

MME *or* **M.M.E.** Master of Mechanical Engineering; Master of Mining Engineering; Master of Music Education.

Mme Abbreviation for Madame.

MMF Merchant Marine and Fisheries *USA.*

m.m.f. Magnetomotive Force.

mm. Hg. Millimetre(s) of Mercury.

MMI Man Machinery Interface, *computer control system.*

MMIAA Member of the Manchester International Airport Authority.

mmm Millimicron.

MMM Merseyside Maritime Museum.

MMO Marine Manager's Office; Medium Machine Oil.

MM&P Masters, Mates and Pilots *USA union.*

MMR Monthly Maintenance Rate *USA: Estimate of cost of living in various institutions.*

MMRBM Mobile Medium Range Ballistic Missile.

MMRS *or* **M.M.R.S.** Member of the Market Research Society.

MMS Marine Surveys and Services Bureau; Merchant Marine Safety *USA;* Minerals Management Service *USA;* Motor Minesweeper.

MMSA Mercantile Marine Service Association.

MMSCF Million Standard Cubic Feet.

MMSI Maritime Mobile Satellite Identification.

MMT Multiple Mirror Telescope.

MMTC Metro Manila Transit Corporation.

MMtoe Million Metric Tons of Oil Equivalent *oil drilling.*

MMU Million Monetary Units; Manned Manoeuvring Unit.

M.Mus. Master of Music.

Mn Symbol for manganese.

MN *or* **M.N.** Magnetic North, Merchant Navy, Minnesota, *USA.*

MNAOA *or* **M.N.&A.O.A.** Merchant Navy and Airline Officers' Association.

MNB Moscow Narodny Bank Limited.

MNC Major Nato Commander; Multi-national Company.

Mn. Drn. Main Drain.

MNE *or* **M.N.E.** Merchant Navy Establishment.

MNEA *or* **M.N.E.A.** Merchant Navy Establishment Administration.

MNEO *or* **M.N.E.O.** Merchant Navy Establishment Office.

MNETA *or* **M.N.E.T.A.** Merchant Navy Establishment Training Allowance.

MNI *or* **M.N.I.** Member of the Nautical Institute.

MNOF *or* **M.N.O.F.** Merchant Navy Officers' Federation.

MNOPF or M.N.O.P.F. Merchant Navy Officers' Pension Fund.

'M' Notice or MSN Merchant Shipping Notice.

MNR or M.N.R. Mean Neap Rise nautical.

MNSA or M.N.S.A. Mercantile Marine Service Association.

MNSC Managed Network Steering Committee.

MNT or M.N.T. Mean Neap Tide nautical.

MNTB or M.N.T.B. Merchant Navy Training Board.

MNWB or M.N.W.B. Merchant Navy Welfare Board.

MO or Mo Missouri USA.

MO or M.O Marine Office insurance; Medical Officer; Money Order, q.v.

Mo Monday.

Mo/s Month(s).

MOA or M.O.A. Memorandum of Agreement. Ministry of Aviation; Ministry of Agriculture.

Mobile Shop A vehicle or van containing groceries and other items for sale while moving from one place to another.

MOD or M.O.D. Ministry of Overseas Development; Mail Order Department; Ministry of Defence.

Model Most Developed Liner.

Modo et forma Latin—In the manner and form.

MODU Mobile Offshore Drilling Units.

MODU Code Code for the Construction and Equipment of Mobile Offshore Drilling Units 1989 IMO.

Modulation Variation of radio frequency in the transmission of Speech.

Modus Latin—Method; Way.

MODUS Mobile Offshore Drilling Unit.

Modus operandi Latin—Working plan; the way of proceeding.

Modus vivendi Latin—A way of living; a temporary compromise.

MOF or M.O.F. Ministry of Food.

MOH or M.O.H. Master of Otter Hounds; Medical Officer of Health; Ministry of Health.

Mohammedan Killed An expression referring to mutton, favourite meat of the Mohammedans. The animal is killed by Mohammedans in the presence of their priest in accordance with their religious rites. Certificates of MK mutton are issued after this procedure has been followed. Abbrev. MK.

MOI or M.O.I. Ministry of Information.

Moiety Half, especially in legal use; one of two parts into which a thing is divided.

MOIV Mechanically Operated Inlet Valves.

MOL Manned Orbiting Laboratory, q.v.; More or Less.

mol Molecule(s); Molecular.

M.O.L. Ministry of Labour.

MOLCHOP or MOLCO More or Less in Charterer's Option, q.v.

Mollusc An animal belonging to the Mollusca, sub-kingdom of soft-bodied and usually hard-shelled animals, including limpets, snails, oysters, etc.

MOLOO More or Less in Owner's Option, q.v.

mol.wt. Molecular weight.

MOMA Museum of Modern Art, New York.

Mon Monday.

Monel A nickel-copper alloy mostly used for strengthening in connection with rust or corrosion resistance. It is commonly used in water meters, propeller shafts, etc.

Monetary Policy This term refers to the interest and exchange rates applied by the countries. The Monetary Policy is Tight when interest or exchange rates are high, while Monetary Policy is Loosened when interest or exchange rates fall.

Monetary Unit The standard unit of currency of a country.

Money at Call Loans by commercial banks which can be withdrawn immediately or at short notice. Sometimes called overnight loans. They form part of the banks' liquid assets.

Money At Call At Short Notice Similar to At Call, q.v.

Money Order A Post Office document issued in the form of a cheque to enable a person to send money abroad up to a certain restricted amount. Abbrev. MO.

Money Up Front Money available and required on the spot for a business transaction.

Monkey Block Single Block, generally on a small scale with a swivel.

Monkey Business Deceitful business or behaviour slang. Also termed Monkey Shine.

Monkey Face Similar to Monkey Plate, q.v.

Monkey's Fist A sizeable knot at the end of a Heaving Line, q.v.

Monkey Island An open deck above the Wheelhouse, q.v., of a ship where another compass is positioned.

Monkey Jacket A short coat or jacket worn by sailors.

Monkey Plate A triangular metal plate having a hole at each corner linking three ropes or chains together. It is also termed Monkey Face.

Monkey Rail Railings constructed around the poop and sides of the main deck of a ship.

Monkey Shine Monkey Business, q.v.

Monoplane An airplane with only one set of wings.

Monopoly Exclusive possession of the trade in some commodity; this can be conferred as a privilege by the state.

Monorail A railway with one rail upon which carriages ride or from which they are suspended.

Monsoon Seasonal winds in Southern Asia, especially in the Indian Ocean. There are the wet monsoons from the Southwest during the months of May up to September and the dry monsoons from the Northeast during the months of October up to December.

Montreux Agreement An international treaty of 1936 governing transits of the Turkish Straits.

MOO or **M.O.O.** Money Order Officer.

Moonlighting The act of performing a second part-time job when the person is already employed elsewhere on a full-time basis.

Moonpool Protected area on a drilling rig through which equipment is lowered and drilling conducted. Also called Moonwell.

Moon's Age The number of days since the previous new moon.

Mooring Line A cable or line to tie up a ship. Known by various other names according to position or function as, for example, Headrope, Headfast, Bowfast, Sternrope, Sternfast, Breast, Spring, Fore Spring, After Spring.

MOP Mobile Offshore Production.

MOP or **M.O.P.** Mother of Pearl; Muriate of Potash.

MOPGC Malaysian Oil Palm Growers' Council.

MOPS Merchant Officers Protective Syndicate *USA.*

MOPED or **Moped** Motorised Pedal Cycle.

Mor. Moroccan; Morocco.

Moratorium Postponement of payment by a court decree in favour of a debtor.

Mor. Dict. More dicto *Latin*—In the manner directed.

More *Latin*—In the manner of; According to the custom; *Russian*—Sea.

Mori or **MORI** Market and Opinion Research International.

MORIS or **M.O.R.I.S.** Monitoring of Flexible Risers in Situ *oil drilling.*

Morning Star Venus or other planet or bright star seen in the East before sunrise.

Morning Watch Seaman's watch at sea from 4 a.m. to 8 a.m.

Morse Code A telegraphic code in which letters are represented by variations on two signs, e.g. dot and dash, long and short flash, etc. The signals are transmitted either by sound or light. Invented by S.F.B. Morse (d. 1872).

Morse Lamp An electrical or dry battery signalling lamp used for the transmission of signals or messages by Morse Code, *q.v.*

Morska *Polish*—Maritime.

Mort. Mortality.

Mortgage A mortgage is the creation of an interest in property, defeasible (*i.e.*, annullable) upon performing the condition of paying a given sum of money, with interest thereon, at a certain time. This conditional assurance is resorted to when a debt has been incurred, or a loan of money or credit effected, in order to secure either the repayment of the one or the liquidation of the other. The debtor, or borrower, is then the mortgagor, who has charged or transferred his property in favour of or to the creditor or lender, who thus becomes the mortgagee. If the mortgagor pays the debt or loan and interest within the time mentioned in a clause technically called the proviso for redemption, he will be entitled to have his property again free from the mortgagee's claim; but if he did not comply with such proviso, at common law he lost the property to the mortgage equity. However, until the mortgage has been foreclosed, or unless the property has been sold under powers to satisfy the debt, it has always been redeemable upon payment of the debt or loan, with interest and expenses, at any period within twelve (formerly twenty) years after the last recognition of the mortgage security by the mortgagee; and this because equity deems the non-compliance with the proviso for redemption a penalty, against which it always relieves when practicable.

Mortgagee The creditor or lender. See **Mortgage.**

Mortgage Interest Relief at Source Interest on mortgages which within a certain limit is tax deductible. *Abbrev.* MIRAS.

Mortgagor The debtor or borrower. See **Mortgage.**

MOS Military Occupational Speciality *USA.*

M.O.S. Ministry of Supply.

MOST Metal oxide silicon transistors; Ministry of Surface Transport *India.*

Most Favoured Nation Clause A stipulation in a treaty according to which one state grants to the other state the same treatment as it has granted, or may grant, to a third state.

m.o.t. Monthly Overtime, Charterparty, *q.v.*, term.

M.O.T. Ministry of Transport.

M.O.T.H.Q. Ministry of Transport Headquarters.

Mot juste, le *French*—The precise word; The word that fits.

MOTNE Meteorological Operational Network, Europe.

Motor Insurance Policy A Policy, *q.v.*, taken out to cover all consequent damage to a motor vehicle or other vehicles and to the driver, passengers and third parties. There are various ways as to how the indemnity can be effected. *Abbrev.* MIP or M.I.P.

Motor Insurers' Bureau An incorporated body to satisfy any unsatisfied judgment arising from any liability required to be covered by insurance under the UK Road Traffic Acts. It is a condition of recovery from the Bureau that notice of proceedings should be given to the Bureau either before action or within 21 days of the commencement of proceedings.

Motor Spirit Same as gasoline or petrol, a fraction from the distillation of petroleum.

Motorway An express highway.

Motu proprio *Latin*—Of his own accord; A document issued by the Pope without consultation with others.

MOU Memorandum of Understanding *q.v.*

MOUPS Memorandum of Understanding on Port State Control.

Movables Any property regarding goods, furniture, personal property.

Movement Summary Daily confidential movements of all merchant vessels in US ports. Also expressed as Daily Movement Summary. *Abbrev.* MOVSUM.

MOVSUM Movement Summary, *q.v.*

Moz. Mozambique.

MP *or* **M.P.** Manipulation-proof *security locks*; Market Price, *q.v.*; Medium Pressure; Meridian Passage *nautical*; Member of Parliament; Mile Post; Months After Payment; Multi-purpose.

m.p. Melting Point.

MPA Marine Preservation Associates *Washington*; Marine Preservation Association, based in Scottsdale, Arizona, 54 shipowners were in membership in 1993; Maritime Patrol Air/Aircraft.

MPA *or* **M.P.A.** Master Printers Association; Master of Professional Accountancy; Master of Public Accounting; Master of Public Affairs; Master of Public Administration; Medway Ports Authority.

M.P.A. Monomeles Protodikion Athinon *Greek* —One-Member First Instance Court of Athens.

m.pack Missing Package.

MPC *or* **M.P.C.** Marginal Propensity to Consumer; Member of Parliament, Canada; Metropolitan Police College; Multi-Purpose Converter.

MPCA Marine Pollution Control Administration.

MPCF Maritime Pollution Claim Fund.

MPCU Marine Pollution Control Unit, *q.v.*

MPD *or* **M.P.D.** Multi-Purpose Dredger; Maximum Permissible Draft (Draught).

MPDU Multiple Product Distribution Unit *oil drilling*.

MPE Master of Physical Education; Maximum permissible exposure *radiation*.

MPG *or* **mpg** *or* **M.P.G.** Miles per gallon.

MPH *or* **mph** *or* **M.P.H.** Miles per hour; Master of Public Health.

M. Phil. Master of Philosophy.

MPI Magnetic Particle Inspection.

M.P.I. Mean Point of Impact.

M.P.L. Maximum Probable Loss, *q.v.*; Maximum Possible Loss; Maximum Permissible Level.

MPM Maximum Permitted Mileage, *q.v.*

M.P.M *or* **m.p.m.** Metres per minute.

MPP Master of Physical Planning; Member of the Provisional Parliament; Most Probable Position *nautical*.

M.P.P. Monomeles Protodikion Pireos *Greek* —One-Member First Instance Court of Piraeus.

MPS Minimum Performance Standard; Maritime Prepositioning Ship.

M.P.S. Member of the Pharmaceutical Society of Great Britain; Member of the Philological Society; Member of the Physical Society.

m.p.s. Megacycles per Second.

MPSS Multi-purpose Semi-Submersible.

m.pt. Melting Point.

Mque Martinique.

MR Money Rate; Marginal Revenue *economics*.

m.r. Moment of Resistance.

Mr Master, Mister.

M.R. Master of the Rolls; Mate's Receipt, *q.v.*

M/R Mate's Receipt; Master of the Rolls.

M & R Maintenance and Repair.

M.R.A.S. Member of the Royal Asiatic Society.

M.R. Ae. S. Member of the Royal Aeronautical Society.

MRC *or* **M.R.C.** Medical Research Council; Movement Report Center *USA*.

MRCC *or* **M.R.C.C.** Maritime Rescue Co-Ordination Centre.

M.R.C.O.G. Member of the Royal College of Obstetricians and Gynaecologists.

M.R.C.P. Member of the Royal College of Physicians.

M.R.C.P.(E). Member of the Royal College of Physicians, Edinburgh.

M.R.C.S. Member of the Royal College of Surgeons.

M.R.C.V.S. Member of the Royal College of Veterinary Surgeons.

MRE *or* **M.R.E.** Master of Religious Education; Microbiological Research Establishment; Mining Research Establishment.

MRELB Malaysian Rubber Exchange and Licensing Board.

mrem/h *or* **MREM/H** Millirem per hour.

M.R.G.S. Member of the Royal Geographical Society.

M.R.I. Member of the Royal Institution.

M.R.I.N.A. Member of the Royal Institution of Naval Architects.

MRIS *or* **M.R.I.S.** Maritime Research Information Service, *q.v.*

MRIT Marine Radar Interrogation Transponders, *navigational aids. See* **CAS**.

MRNC *or* **M.R.N.C.** Marine Radio and Radar Technicians Certificate.

MRNT *or* **M.R.N.T.** Maritime Radar and Navigation Trainer.

MRO Maintenance, Repair and Overhaul; Movement Report Office *USA*.

MRP Machine Readable Passport; Marginal Revenue Product; Master of Regional Planning.

MRRDB Malaysian Rubber Research and Development Board.

Mrs Missis, Missus (corruptions of Mistress).

MRS *or* **M.R.S.** Market Research Society.

MRSA Multi Role Strategic Aircraft.

M.R.S.C. Member of the Royal Society of Chemistry; Maritime Rescue Sub-Centres.

M.R.S.H. Member of the Royal Society of Health.

M.R.S.L. Member of the Royal Society of Literature.

MRU Much Regret Unable *naval*.

MS Egyptair *airline designator*; Manuscript; Mississippi, *USA*.

m.s. Millisecond, one-thousandth of a second; Mail Steamer; Motor Ship.

Ms. Microsecond.

Ms Title of woman whether or not married.

M.S. Master of Science; Master of Surgery.

M/S Machinery Survey, *Lloyd's Register notation*. Months After Sight.

M/S *or* **M.S.** Member State.

MS *or* **M.S.** Multiple Sclerosis

m/s Metres per second, SI unit of speed.

M&S Maintenance and Supply.

MS/S Manuscript(s).

MSA Maritime Safety Agency; Merchant Shipping Act, originally enacted in UK in 1894; Merchant Shipping Association; Mutual Security Agency *USA*.

MSAE *or* **M.S.A.E.** Master of Science in Aeronautical Engineering.

MSAM *or* **M.S.A.M.** Master of Science in Applied Mathematics.

M.S. Arch Master of Science in Architecture.

MSB Maritime Subsidy Board *USA*; Minesweeping Boat *USA*.

M.S.B.A. Master of Science in Business Administration.

MSBC *or* **M.S.B.C.** Master of Science in Building Construction.

M.S. Bus Master of Science in Business.

MSC Marine Safety Committee *IMO.*; Major Subordinate Commander *NATO*.

MSC *or* **M.S.C.** Manchester Ship Canal; Manpower Services Commission; Marine Safety Council *US Coast Guard*; Maritime Safety Committee, *IMO*; Maritime Safety Committee *USA*; Maritime Shipping Committee *USA*; Merchant Shipping Code; Military Sealift Command *USA*.

M.Sc. Master of Science.

Msc. Miscellaneous.

m.s.c. Mandatum sine clausula *Latin*—Authority given without restriction.

M. Sc.D. Master of Dental Science.

MSCE *or* **M.S.C.E.** Master of Science in Civil Engineering.

M. Sc.CR.E. Master of Science in Chemical Engineering.

MSCP *or* **M.S.C.P.** Master of Science (Community Planning).

M.S.C.P. Mean Spherical Candle Power.

MSCUSCG Marine Safety Council of the United States Coast Guard.

MSD *or* **M.S.D.** Doctor of Medical Science; Marine Sanitary Devices; Master of Scientific Didactics; Master Surgeon Dentist; Medium Speed Diesel (Engine), *q.v.*

M.S. Dent Master of Science in Dentistry.

M.S. (Dent) Master of Surgery (Dental Surgery).

MSDS Material Safety Data Sheet. Standardised document giving details of hazardous or toxic goods showing characteristics and precautions required in handling.

MSE Malaysia Shipyard and Engineering.

M.S.E. Master of Science in Engineering; Member of the Society of Engineers.

m.s.e. Mean Square Error.

M. S. Ed. Master of Science in Education.

MSEE *or* **M.S.E.E.** Master of Science in Electrical Engineering.

MSEM *or* **M.S.E.M.** Master of Science in Engineering Mechanics.

MSF Module Support Frame *oil drilling*. Multi-Stage Flash, *reverse osmosis distillation*.

MSF *or* **M.S.F.** Main Support Frame; Master of Science in Forestry; Médecins Sans Frontières *French*; Multiple Shops Federation.

MSGM *or* **M.S.G.M.** Master of Science in Government Management.

MSH Minesweeper Hunter *US Navy*; Master of Staghounds.

MSHA Mine Safety and Health Administration *USA*.

MSHEc *or* **M.S.H.Ec.** Master of Science in Home Economics.

M.S. Hort. Master of Science in Horticulture.

M.S. Hyg. Master of Science in Hygiene.

MSI Medium Scale Integration; Marine Safety International.

MSI *or* **M.S.I.** Marine Safety Information, *q.v.*; Member of the Chartered Surveyors' Institution.

M.S. Ind. E. Master of Science in Industrial Engineering.

MSIS Marine Safety Information System. US Coast Guard data bank of ships and owners and flag states to aid the targeting of examination under Port State Control, *q.v.*

M.S.J. Master of Science in Journalism.

MSL or **M.S.L.** or **m.s.l.** Mean Sea Level; Mean Sphere Level.

MSL or **M.S.L.** Master of Science in Linguistics.

MSM or **M.S.M.** Master of Sacred Music.

MSMus or **M.S.Mus.** Master of Science in Music.

MSN or **M.S.N.** Master of Science in Nursing.

MSN or **'M' Notice** Merchant Shipping Notice. One of a numbered series of notices issued by the UK Marine Safety Agency for the guidance and regulation of shipowners and masters.

MSNAME or **M.S.N.A.M.E.** Member of the Society of Naval Architects and Marine Engineers.

MSOA Manchester Steamship Owners' Association.

MSP Maritime Security Program, *USA*.

MSPE or **M.S.P.E.** Master of Science in Physical Education.

MSPH or **M.S.P.H.** Master of Science in Public Health.

M.S.Pharm. Master of Science in Pharmacy.

MSR or **M.S.R.** or **m.s.r.** Mean Spring Rise *nautical*; Mean squared residual; Mechanical Strain Recorder; Missile Site Radar.

m.s.r. Mean Supply Route; Main Supply Route.

M.S. & R. Merchant Shipping and Repairs.

MSRA Mersey Ship Repairers' Association.

MSRP Marine Spill Response Corporation *Washington*.

MSS Manuscripts; Marine Specification Services.

MSSCS Manned Space Station Communications System.

MSSVFI Manufacturers Standardisation Society of Valves and Fitting Industry *USA*.

MST or **M.S.T.** or **m.s.t.** Mean Spring Tide; Mountain Standard Time; Mineral Slurry Transportation *oil drilling*.

mst. Measurement.

MSTS Military Sea Transportation Service *USA*.

MSV Motor Sailing Vessel; Multifunctional Support Vessel, *q.v.*

mt Empty; Motor Transport; Motor Tanker.

Mt. Moment; Mount, Mountain.

MT or **m.t.** Metric ton; Montana *USA*.

MT or **M.T.** Empty; Mean Tide; Motor Tanker.

M/T Mail Transfer *banking*.

MTA Manage Transfer Agent (EDI).

MTA or **M.T.A.** Marine Trades Association; Marine Transit Association *USA*.

MTB Material Transportation Bureau *USA*; Motor Torpedo Boat.

MTBD or **M.T.B.D.** Mean Time Between Defects.

MTBF or **M.T.B.F.** Mean Time Between Failures, *q.v.*

MTBM or **M.T.B.M.** Mean Time Between Maintenance.

MTBR or **M.T.B.R.** Mean Time Between Rejections.

MTC Marine Traffic Control; Maritime Transport Committee, in Paris *OECD.*

M.T.C.A. Ministry of Transport and Civil Aviation.

MT Container Empty Container.

MTDAT or **M.T.D.A.T.** Metric Tons Deadweight All Told.

MTDW or **M.T.D.W.** or **mtdw** Metric Tons Deadweight.

MTE or **M.T.E.** Monthly Total of Entries.

MTELP Michigan Test of English Language Proficiency *USA*.

MTFS Medium-Term Financial Strategy *finance*.

mtge Mortgage, *q.v.*

mtgee Mortgagee. *See* **Mortgage**.

mtgor Mortgagor. *See* **Mortgage**.

M.T.H. Master of Trinity House.

Mth. Month.

MTI or **M.T.I.** Moment to Alter Trim One Inch; Moving Target Indicator.

MTIB Malaysian Timber Industry Board.

MTL or **M.T.L.** Maritime Transport Law; Mean Tidal Level *nautical*; Medium Term Loan.

MTM or **M.T.M.** Methods Time Measurement.

MTN Multilateral Trade Negotiation.

Mtn Mountain.

MTO Multimodal Transport Operators, *UN Convention*; Motor Transport Office(r).

MTTA Machine Tools Trade Association.

m/Ton Measurement Ton or T/M or t/m Ton measurement.

MTRB Maritime Transportation Research Board.

MTR Mass Transit Railway

MTRC Maritime Training and Research Center *Toledo USA*; Mass Transit Railway Corporation *Hong Kong.*

MTS Marine Technology Society *USA*.

MTTF Mean Time to Failure.

MTFF Mean Time to First Failure.

MTTR Maximum Time to Repair; Mean Time to Restore.

MTWA Maximum Take-Off Weight Authorised *aviation*.

MU or **M.U.** Maintenance Unit; Monetary Unit.

Mucking Out The process of removing sediments following the discharge of dirty oils from a tanker.

Mucking Winch A winch on a tanker used to haul a small bucket of sediments remaining in a tank after the discharge of dirty oil.

Mud A substance carried in a dry form by oil well operators to circulate around a well during drilling operations. The mud cleans and cools the drilling bits. Another word used is Circulating Fluid. *See* **Blow Out Preventer.**

Mud Berth Mooring berth of a ship where she settles on soft ground made up of mud when the tide is low. *See* **Always Afloat** *or* **Always Safely Afloat.**

Mud Box A steel square-shaped box situated alongside the suction pipes of the bilges adjacent to the valve chest to stop any solid matter entering and interrupting the functions of the valves.

MUF *or* **M.U.F.** Maximum Usable Frequency.

Muffler Exhaust silencer *USA.*

MUL Manufacture(d) Under Licence.

Mule Electrically driven cotton machine.

Multi–Currency Describes accounting in several currencies at the same time.

Multifunctional Support Vessel A multipurpose supply vessel attached to oil rigs at sea. *Abbrev.* MSV.

Multilateralism Dealings between several nations on a common basis.

Multi–Launch Rocket System Nicknamed the Deadly Dozen. A powerful weapon which can devastate an area equal to 12 football pitches. *Abbrev.* MLRS.

Multi–Pack Package containing several small ones.

Multiple Store A large shop offering a wide variety of items. Large bazaars or supermarkets.

Multi–Product Firm A firm or factory producing various products and by-products. By so doing it may avoid business fluctuations. *See* **Cyclical Fluctuation.**

Multum *Latin*—Much; very; far.

Multum in parvo *Latin*—Much in little.

Mumental Temporary magnet with an alloy of iron and nickel.

Muntz–metal An alloy of 60% copper and 40% zinc used for sheathing ships, etc.

Murdulugu *Turkish*—Administration.

Muriatic Hydrochloric Acid. *Abbrev.* HCl.

MUS *or* **M.U.S.** Manned Underwater Station.

Muster Nautical term for the assembling of ship's officers, crew members and passengers, if any, for inspection and drill purposes, etc.

Muster Roll A record book wherein all details of the officers, crew members and passengers of a ship are kept. An alternative word for Crew List, *q.v.*, and Passenger List.

MUT Mutilated.

Mutatis mutandis *Latin*—With the necessary changes in points of detail.

Mutiny Revolt or collective disobedience.

Mutual Funds This can mean mutual insurance, such as that provided by Protection and Indemnity Clubs. A P. & I. Club is a mutual association formed by shipowners to provide protection from large financial loss to one member by contribution towards that loss by all members. The P. & I. Club covers liabilities not insurable by the shipowner in the running of his ship, such as cost of defending claims made by cargo owners. *See* **Protection & Indemnity** *or* **P. & I. Club**; The American term for Unit Trusts. *See* **Unit Trust Company.**

Muzzler A strong headwind.

MV Master Vehicle *oil drilling.* Motor Vehicle; Motor Vessel; Muzzle Velocity.

M.V. *or* **m.v.** Motor Vessel; Million Volts; Market Value; Mean Variation; Medium Voltage; Merchant Vessel.

M&V Meat and Vegetables.

mV Millivolt *electricity.*

MVA *or* **M.V.A.** Million Volt Amperes *electricity.*

M. Val. Market Value.

MVD *or* **M.V.D.** Maximum Variable Deadload.

M.V.Ed. Master of Vocational Education.

MVTR Moisture Vapour Transmission Rates.

MW Malawi; Million Watts; Megawatt; Mine Warfare.

mW Milliwatt.

M.W. Medium Wave.

M&W Marine & War Risks *insurance.*

M–Way Motorway.

M.W.B. Metropolitan Water Board.

MWD Measurement while Drilling *oil drilling.*

MWe Megawatts *electrical.*

mWh Megawatts Hour.

MWNT *or* **M.W.N.T.** Mean Water Neap Tide.

MWO Maritime Welfare Organisation. Meteorological Watch Office. *See* **Mariners**, **Weather News for.**

MWS Maritime Work Station *USA.*

MWT Metric Weight Ton. *See* **Metric Ton.**

MWt Megawatts Thermal.

MX Mexicana de Aviacion *airline designator.*

MY Air Mali *airline designator.*

MY *or* **M.Y.** Motor Yacht.

Myanmar Formerly Burma.

Myriad Ten thousand; of indefinitely great number.

Myxomatosis A contagious and destructive infection of rabbits.

M.Z. Mangels Zahlung *German*—for non-payment.

MZD *or* **M.Z.D.** Meridian Zenith Distance *nautical.*

MZn Magnetic Azimuth.

N

n nano *prefix mathematics*; an indefinite integer, whence (generally) *to the nth degree*; Neutral; Neutron; Newton *SI unit of force*; Symbol for amount of substance *chemistry*.

N Symbol for nitrogen; Naira; Near; North; *Lloyd's Register notation* for New; Navigation; Norway; November; Nullity.

N₂ Nitrogen.

Na Nadir, *q.v.*; Non Acceptable; Sodium.

NA *or* **N.A.** National Airlines; Nautical Almanac; Nautical Institute; National Academy; Naval Architect; Nebraska, *USA*; Nearest Approach; Net Absolutely; North Africa; North America; Not Above.

N/A *or* **N/a** New Account; No Advance; No Account; No Advice; Not Acceptable; Not Applicable; Not Available.

n/a No Account; No Advice; Not Applicable; Not Available.

NAA National Aeronautic Association *USA*.

NAA *or* **N.A.A.** *or* **n.a.a.** Not Always Afloat.

NAABC National Association of American Business Clubs.

NAABSA *or* **N.A.A.B.S.A.** *or* **n.a.a.b.s.a.** Not always Afloat but Safely Aground.

NAACP *or* **N.A.A.C.P.** National Association for the Advancement of Coloured People.

NAAF *or* **N.A.A.F.** North African Air Force.

NAAFI *or* **N.A.A.F.I.** Navy, Army and Air Force Institutes.

NAAS *or* **N.A.A.S.** Naval Auxiliary Air Station.

Naamlooze Vennootschap *Dutch*—Limited Company. *Abbrev.* NV.

NAB National Association of Broadcasters; Naval Air Base.

NABM *or* **N.A.B.M.** National Association of British Manufacturers.

NAC Names Advisory Committee. *Lloyd's*; North Atlantic Conference *shipping*; North Atlantic Council *NATO*.

NACA National Air Carriers Association.

NACCB National Accreditation Council for Certification Bodies *UK*.

NACE National Association of Corrosion Engineers *USA*.

Nacelle Car of a balloon or airship accommodating the crew and cargo and/or passengers; Housing external to the main fuselage which encloses an aeroplane engine.

N.A.D. National Academy of Design.

NADGE Nato Air Defence Ground Environment.

Nadir Point of heavens directly under observer, opposite to zenith; lowest point, place or time of greatest depression. *Abbrev.* Na.

NAEGE North American Export Grain.

NAF Naval Air Facility *USA*.

NAFCO Northwest Atlantic Fisheries Consultative Organisation.

NAFEC National Aviation Facilities Experimental Center *USA*.

NAFED National Agency for Export Development, *Indonesia*.

NAFF Nordic Association of Freight Forwarders.

NAFEM National Association of Food Equipment Manufacturers.

N. Afr. North Africa.

NAFSA National Association for Foreign Student Affairs *USA*.

NAFTA New Zealand–Australia Free Trade Agreement; North American Free Trade Agreement, between Canada, USA and Mexico; North Atlantic Free Trade Area.

Nag. Nagasaki, *Japan*.

n.a.g. Net Annual Gain.

NAGA North American Grain Association.

NAHR No Annual Hull Rate.

NAIC National Association of Insurance Commissioners *USA*.

Naiko *Japanese*—shipping line.

Nail to Nail Insurance covering the move of wall-hung art and paintings from one place to another as specified in the Policy, *q.v.*

NAIR National Arrangements for Incidents Involving Radioactivity.

Naira Unit of currency used in Nigeria. One hundred kobo equals one naira. *Abbrev.* N.

NAIWC *or* **N.A.I.W.C.** National Association of Inland Waterways Carriers.

Nakagai *Japanese*—Broker.

Naked Debenture Unserved or unsecured debenture.

NALCOMIS Naval Air Command Management Information System *USA*.

NALGO *or* **N.A.L.G.O.** National and Local Government Officers' Association *UK*.

NAM *or* **N.A.M.** National Association of Manufacturers. *USA*.

Nam *or* **Naam** The taking of another's movable goods either by lawful distress or otherwise.

N. Am. North America.

NAMAS National Measurement Accreditation Service.

Namation Taking or impounding.

Name An underwriting member at Lloyd's who pledges his wealth with unlimited liability on insured risks and perils. He becomes a member of a Syndicate, *q.v.*

NAME National Association of Marine Engineers *Canada and USA*; National Association of Marine Exchanges *USA*.

Name Board A wooden board with carved or painted name of the ship.

Named Policy A marine insurance policy where the name of the vessel carrying the cargo insured is inserted in the policy.

Name of Ship An important word for clearing customs documents and other official matters. It is chosen by the owner on approval by the Ministry of Shipping or/and Transport or other state shipping authorities. The name is subsequently recorded by Lloyd's Register of Shipping with its Official Number, *q.v.*, and Call Sign, *q.v.* Every vessel is to have her name clearly shown on both bows and also at the stern. Under the name of the ship at the stern there must also be the name of the Port of Registry.

NAMS National Association of Marine Services Inc. *USA*.

nano A prefix used in mathematics meaning one thousand millionth. *Abbrev.* n.

NAOS *or* **N.A.O.S.** North Atlantic Ocean Station.

NAPE National Assocation of Port Employers *UK*.

Naphtha Colourless, volatile and inflammable liquid distilled from petroleum, coal and shale. An intermediate product between benzine and gasoline used as a fuel and a solvent.

Naphthalene White crystalline volatile strongly odorous substance obtained from coal-tar or crude oil. $C_{10}H_8$. Used in making dyes, resins and moth repellent.

NAPLC North Atlantic Passenger Line Conference.

NAPs Noise Abatement Procedures *aviation*.

NAPTS North Atlantic Passenger Traffic Study.

NAPVO National Association of Passenger Vessel Owners.

NARL National Aero Research Laboratory *Canada*.

Narrow Market A stock exchange expression used when only a limited number of securities are available in the market.

Narrows A Strait, *q.v.*, or part of a waterway having restricted navigable width.

NARTU Naval Air Rescue Training Unit.

NAS Naval Air Station; Noise Abatement Society.

N.A.S. National Academy of Sciences *USA*; Naval Air Station; Netherlands Association of Shipsuppliers. *See* **N.V.S.**; National Association of Sciences.

NASA *or* **N.A.S.A.** National Aeronautics and Space Administration *USA*; North Atlantic Shippers' Association; North Atlantic Space Agency.

NASBA *or* **N.A.S.B.A.** National Association of Ship Brokers and Agents.

NASCO National Academy of Sciences Committee of Oceanography *USA*; National Steel and Shipbuilding Company *USA*.

NASD National Amalgamated Stevedores and Dockers.

Nash. Nashville, *USA*.

NASS Naval Academy Sailing Squadron *USA*.

N.A.S.S. Naval Air Signals Schools.

Nass. Nassau, *Bahamas*.

NASFT National Association for the Speciality Food Trade.

NASN National Administration of Shipping and Navigation *Sweden*.

Nat. Natal, *South Africa*.

NAT *or* **N.A.T.** National Arbitration Tribunal; Normal Allowed Time.

N. At. North Atlantic.

NATCS *or* **N.A.T.C.S.** National Air Traffic Control Service.

National Cargo Bureau A United States Maritime Department having representative offices throughout the main harbours in America. It is empowered to inspect cargoes of a hazardous nature and issue certificates. These certificates are automatically approved as far as the United States Coast Guard is concerned. *Abbrev.* NCB *or* N.C.B.

National Currency Currency belonging to the local country.

National Defense Transportation Association A US organisation for the organisation of emergency transportation between the Department of Defense and the US sea, air and railway industries. *Abbrev.* NDTA.

National Dividend National Income, *q.v.*

National Engineering Laboratory One of the Industrial Research establishments of the Department of Trade and Industry in the UK. *Abbrev.* NEL.

National Flag *or* **Ensign** The flag carried by a ship to show her nationality.

National Flag Ship *or* **National Flag Vessel** Vessel trading under the flag of the country in question.

National Income The total value of goods and services produced by a country over a period of a fixed time which in normal cases will be for one year. Also said to be National Dividend.

National Shipbuilding Research Program Instituted in the USA in the year 1970 for government and industry collaboration in research for shipbuilding under the direction of SNAME, *q.v.*

N. Atl. North Atlantic.

NATLAS National Testing Laboratory Accreditation Scheme.

NATO North Atlantic Treaty Organisation.

NATS Naval Air Transport Service *USA*.

N. Att. Naval Attaché.

NATTS National Association of Trade and Technical Schools *USA*.

Natural Gas Unprocessed gas produced from a gas well. Also known as Dissolved Gas.

NAUA *or* **N.A.U.A.** National Automobile Underwriters' Association *USA*.

naut. Nautical, *q.v.*

Nautical Pertaining to sea, seamen, navigation or ships. *Abbrev.* naut.

Nautical Assessors are active Elder Brethren of Trinity House. If the decision of a question arising in an Admiralty action requires technical knowledge and experience in navigation, the judge or court is usually assisted at the hearing by two Trinity House Elder Brethren who sit as assessors and advise the court on questions of a nautical character.

Nautical Institute, The A society constituted in 1972 with the principal aim of 'promoting a high standard of knowledge and competence amongst those in control of seagoing craft'. The Institute is an international professional body for qualified mariners and its membership standards are based upon a foreign-going master mariners' certificate from a recognised maritime administration or a naval command qualification for a major warship. It publishes the monthly journal, 'Seaways', with an international circulation to over 6,000 members and subscribers in more than 70 countries. Included are reports from the Institute's Confidential Marine Accident Reporting Scheme (MARS) covering near-miss incidents and potentially dangerous occurrences at sea worldwide.

Although international, the Institute has consultative status with the UK Department of Transport and is recognised by the Admiralty Board. It is often represented in national delegations to the IMO.

The Institute runs career development programmes and courses in Command; The Work of the Nautical Surveyor; The Work of the Harbour Master; Pilotage; and Personal effectiveness in management. It is also a major publisher of nautical text books and monographs.

The Institute validates an international programme of training and certification for officers in charge of a watch on board Dynamic Positioning Vessels and runs a Square Rig Sailing Ship Examination Scheme.

The Institute is directed by a Council, of whom two-thirds must be actively seagoing, supported by technical committees made up of members with special interests.

The Institute has 14 branches in the UK and 13 branches overseas, with more expected. Their committees organise programmes of technical meetings and consult locally on maritime affairs.

Nautical Mile The nautical mile of 6,080 ft. (1853.18 m), which has been used by the U.K. for many years, is roughly 0.06 per cent greater than the international nautical mile of 1852 m (6,076.12 ft.). In September, 1970, the Hydrographic Department of the Ministry of Defence issued a Notice to Mariners (1518/70) stating that the Department had adopted the international nautical mile, in place of the UK one, to conform with other Government Departments and with the recommendation of the International Hydrographic Bureau. The international nautical mile is equivalent to the average length of a minute of latitude, and corresponds to a latitude of 45°. The knot is based upon this mile, whereas the knot previously used by the UK was equal to 0.514773 m/s.

NAUTIS Naval Autonomous Tactical Information System, computerised aid to minehunting.

N. Aux. B. *Lloyd's Register notation* for New Auxiliary Boiler. *Accompanied by date of fitting.*

nav. Naval; Navigable; Navigate; Navigation; Navigator.

n.a.v. Net Asset Value.

Nav. *or* **Navig.** Navigating; Navigation; Navigational.

NAV *or* **n.a.v.** Net Asset Value.

Navaids Navigational Aids.

NAVAIR Naval Air Systems Command *USA*.

Navais *Portuguese*—Naval.

Naval Courts Courts held abroad in certain cases to inquire into complaints by the master or officers or seamen of a British ship, or as to the wreck or abandonment of a British ship, or when it is required in the interests of the owners of the ship or cargo. A naval court consists of three, four or five members, being naval officers, consular officers, masters of British merchant ships or British merchants. It has power to supersede the master of the ship with reference to which the inquiry is held, to discharge any of the seamen, to decide questions as to wages and send home offenders for trial or try certain offences in a summary manner. A naval court is quite distinct from a naval court-martial.

Navale *French*—Naval.

Navali *Italian*—Naval.

Naval Ports

Naval Ports Ports of a maritime nation possessing a navy whereby warships and/or submarines are stationed for possible action in time of war and for participating in naval exercises.

NAVAREA One of 16 geographical sea areas, identified by Roman numerals, for the purpose of the Worldwide Navigational Warning Service.

Nav.Cl. Navigation Clause. *Hull insurance.*

Navegaçao *Portuguese*—Navigation.

Navegaçion *Spanish*—Navigation.

NAVFAC Naval Facilities Engineering Command *USA.*

Navicert A certificate to the effect that a ship's cargo does not contravene war contraband regulations; a commercial passport for a particular consignment.

Navicular Boat-shaped.

Naviera *Spanish*—Shipping.

Navig. Navigable; Navigate; Navigation, *q.v.*

Navigable Affording passage for ships; seaworthy.

Navigate To direct the course of a ship or aircraft; to sail over or up or down a sea or river.

Navigation The art and study of setting a safe course for a ship or aircraft or vehicle from one place to another by means of astronomy and/or instruments. In the legal sense, the right of navigation is the right of the public to use an arm of the sea, a river, or other piece of water, as a highway for shipping, boating, etc., including the right to anchor in it. It is a right of way and not a right of property, and therefore the owner of the bed of a river over which the public have by user acquired a right of navigation may erect any structure on it which does not interfere with the navigation warranted by that user. In the case, however, of estuaries and navigable tidal rivers, the beds of which are prima facie vested in the Crown, the ownership of the soil is wholly subject to the public right of navigation, and no part of it can be used so as to derogate from or interfere with that right. A river which is subject to a right of navigation is said to be 'navigable.'

The question whether a river is navigable or not seems to depend partly on its size and the formation of its bed, and partly on the use to which it has been put; if a river will admit ships and it has been used for shipping purposes by the public, it is a navigable river, whether it is tidal or non-tidal, and whether it flows through or over private land or land belonging to the Crown. As a rule, however, an arm of the sea or a tidal river with a broad and deep channel is deemed navigable. Where the public have acquired the right of navigation on a private or non-tidal river, the original exclusive right of the riparian owners to fish in it is not thereby affected. Obstructing the navigation of a navigable water is a public nuisance. As to navigation on the sea, the high seas are common to all nations, and no part of them can become the property of any one state. They are, therefore, open to the navigation, fishery and commerce of all the world, and no nation

has the right to exercise civil or criminal jurisdiction over the ships of other nations while passing over the high seas. The courts of a country may however exercise jurisdiction over foreign ships in respect of injuries committed by them on the high seas, if they subsequently come within the territorial waters of that country: and the English courts have in fact jurisdiction in such cases under the Merchant Shipping Acts. As regards the right of fishing on the high seas, where a particular locality is habitually frequented by fishermen of different countries, and a custom is established regulating the mode and time of fishing, and such matters, the custom is binding on all those who frequent the locality.

The bed of the high seas appears to follow the same general rule, that it is common to all nations. There is, however, this difference, that it is capable of permanent occupation, and it would consequently appear that if any state keeps uninterrupted and exclusive possession of a portion of the bed of the high seas, for a certain length of time, the state thereby acquires a right to it as against all others, by analogy to the doctrine of possession.

Navigation Risks Risks to a vessel while at sea. Such a clause is not applicable to accidents in port.

Navigazione *Italian*—Navigation.

NAVITEX Navigational and Meteorological Warning Broadcast Service *US Coast Guard.*

NAVOCEANO Navy Oceanographic Office *USA.*

NAVSAT Navigational Satellite.

NAVSEA *or* **NAV-SEA** Naval Sea Systems Command *Contracting Agency USA.*

NAVSEC Naval Ship Engineering Center *USA.*

NAVASSES Navy Ship Systems Engineering Stations *USA.*

NAVTEX A narrow-band telex direct-printing service providing automatic reception of maritime safety messages up to a range of about 100 nautical miles from a transmitting station.

Navvy A labourer employed in excavating, etc., for canals, railways, roads, etc.

Navy An assemblage of ships, commonly ships of war; a fleet. The whole of a state's ships of war with their crews and all the organisation for their maintenance.

NAWFA North Atlantic Westbound Freight Association.

NAWFC North Atlantic Westbound Freight Conference.

NAWK National Association of Warehouse Keepers *UK.*

NAWSA North Atlantic Winter Seasonal Area.

NAWSZ North Atlantic Winter Seasonal Zones.

naz. Nazionale *Italian*—National.

NB *or* **N.B.** *Lloyd's Register notation* for New Boilers; Narrow Bore; Naval Base; New Brunswick,

Canada; New Business; North Borneo; Nota Bene *Italian*—Note Well.

NBBPVI National Board of Boiler and Pressure Vessel Inspectors *USA*.

NBC National Broadcasting Company *USA*.

NBCC Nigerian British Chamber of Commerce.

N.b.E. *or* **N.by.e.** North by East *compass bearing*.

NBL *or* **N.B.L.** No Berth List.

NBPI National Board for Prices and Incomes. *UK*. Also *PBI*.

n.br. Naval Brass.

nbre. Noviembre *Spanish*—November.

NBS *or* **N.B.S.** National Bureau of Standards *USA*.

nb.st. Nimbostratus Clouds, *q.v.*

NC *or* **N.C.** National Certificate; New Caledonia; Non-continuous Liner; North Carolina, *USA*; Numerically Controlled.

N/C New Charter; New Crop; Net Charter, *q.v.*

N.C. *or* **n.c.** New Charter; New Crop; No Change; Numerically Controlled.

N.C.A. No Copies Available.

NCAD Notice of Cancellation at Anniversary Date *Insurance*.

NCACC *or* **N.C.A.C.C.** National Civil Aviation Consultative Committee.

NCAPC National Center for Air Pollution Control *USA*.

NCAR National Center for Atmospheric Administration *USA*.

NCAR *or* **N.C.A.R.** No Claim for Accident Reported, *q.v.*

NCB *or* **N.C.B.** National Cargo Bureau *USA*; National Coal Board; No Cargo On Board; No Claims Bonus, *q.v.*

NCBAE *or* **N.C.B.A.E.** No Claims Bonus As Earned.

NCBFAA National Customs Brokers' and Forwarders' Association of America.

NCBS National Cargo Bureau Surveyor *USA*.

NCC *or* **N.C.C.** Nature Conservancy Council *UK*; Nautical Catering College; No Collection Commission when freights are collected at destination.

NCCL National Council for Civil Liberties.

NCD *or* **N.C.D.** Naval Construction Department.

NCF National Freight Consortium *USA*.

NCGA National Computer Graphics Association.

NCH *or* **N.C.H.** National Clearing House(s).

n.c.i. No Common Interest.

NCI *or* **N.C.I.** National Cancer Institute *USA*; No Common Interest.

NCIS Naval Criminal Investigation Service *US Navy*; National Criminal Investigation Service *UK*.

NCITD National Council for International Trade Documentation *New York*.

NCLL *or* **N.C.L.L.** National Council of Labour Colleges.

NCMA National Contractors Management Association *USA*.

NCNA New China News Agency.

NCO *or* **N.C.O.** Non-Commissioned Officer.

NCOB No cargo on board.

NCP *or* **N.C.P.** Normal Circular Pitch.

NCPS *or* **N.C.P.S.** Non-Contributory Pension Scheme.

NCQA National Committee for Quality Assurance.

NCR *or* **N.C.R.** National Cash Register; No Carbon (paper) Required.

NCRE *or* **N.C.R.E.** Naval Construction Research Establishment.

NCRP National Council on Radiation Protection and Measurements *USA*.

NCSCI National Center for Standards and Certification Information. *USA*.

N.C.T. National Chamber of Trade.

N.C.U. Nitrogen Control Unit; National Cyclists' Union.

N.C.U.P. No Commission Until Paid.

NCV Non-Carrying Vessel. *See* **Non-Carrying**.

NCV *or* **N.C.V.** No Commercial Value.

NCW National Council of Women.

ND Not Dated; No Date; North Dakota, *USA*.

ND *or* **N.D.** National Debt; National Diploma; New Discount; No Discount; *Lloyd's Register notation* for New Deck; No Date; Not Dated.

N.D. *or* **N/D** *or* **n.d.** No Date; No Discount; Non-Delivery; Not Dated.

nd. Niederdruck *German*—Low Pressure.

n.d. Next Day; No Date; No Decision; No Delay; Not Dated; Not Decided; Non-Delivery.

N.d.a. Nota dell'Autore *Italian*—Author's Note.

n.d.a.a. Not Dated At All.

N. Dak. North Dakota, *USA*.

NDB *or* **N.D.B.** Non-Directional Beacon *aviation*; *Lloyd's Register notation for* New Donkey Boiler(s).

NDC *or* **N.D.C.** National Debt Commissioners, National Defence Contribution.

NDD *or* **N.D.D.** National Diploma and Design; Navigation and Direction Division.

NDF National Defense Feature *USA*.

NDL Norddeutscher Lloyd.

NDLB National Dock Labour Board.

NDO *or* **N.D.O.** Number and Date Only.

n.d.p. Normal Diametric Pitch.

NDRF National Defense Reserve Fleet *USA*.

NDT *or* **N.D.T.** Non Distributive Trade; Non Destructive Testing, *q.v.*

NDTA National Defense Transportation Association, *q.v.*

ndw *or* **NDW** *or* **n.d.w.** *or* **N.D.W.** Net Deadweight.

NE *or* **N.E.** Naval Engineer; New Edition; No Effects *banking*; *Lloyd's Register notation* for New Engine(s); North East or North Easterly *compass reading*; Not Entered; Not to exceed.

Ne Neon.

N/E New Effect; *Lloyd's Register notation* for New Engines; Not Exceeding.

n/e No Effect; Not Essential; Not Excluding.

n.e. Not exceeding; Not excluding.

NEAC North Europe US Atlantic Conference.

NEAFC *or* **N.E.A.F.C.** North East Atlantic Fisheries Commission.

Near Cash Similar to Near Money, *q.v.*

Near Clause *Similar to* **As Near As She Can Safely Get,** *q.v.*

Near Continental Trade Any ports or places on the coast of Europe and Scandinavia, including the Baltic and North West European coasts up to the Spanish mainland.

Near Money Some documents which, although they are not actually money in themselves, are factors of money. *Ex.* A Bill of Exchange, *q.v.*, Irredeemable Debentures, etc. *See* **At Call**.

Neaped A vessel trapped at the beginning or during a period of neap tides—too deep to enter or to leave a port.

Neap Tide A tide happening near to the time of the first and last quarter of the moon at which time the tide generally tends to go down.

'neath Beneath.

NEB *or* **N.E.B.** National Enterprise Board.

Neb. Nebraska, *USA*.

NEBA National Marine Engineers' Beneficial Association.

NEBSS National Examination Board for Supervisory Studies, *for Catering Officers Certificate*.

N.E by E. North East By East *compass bearing*.

N.E by W. North East By West *compass bearing*.

NEC National Electric Code; National Exhibition Centre *Birmingham UK*; Navy Enlisted Classification *USA*; North East Coast; North European Command *NATO*; Not Elsewhere Classified.

n.e.c. Not Elsewhere Classified.

Necessitas non habet legem *Latin*—Necessity has or obeys no law.

NECIES *or* **N.E.C.I.E.S.** North East Coast Institution of Engineers and Shipbuilders.

Necore Code Name for Chamber of Shipping Mediterranean Ore Charterparty.

NECS National Electric Code Standard *USA*. *See* **NEC**.

necy. Necessary; Necessity.

NEDC *or* **N.E.D.C.** National Economic Development Council. *UK. Nick-named Neddy.*

Nederlandse Vereniging Van Scheepsleveranciers Netherlands Association of Shipsuppliers *Members of the International Shipsuppliers Association*.

Neddy National Economic Development Council.

Née Born.

Ne exeat regno *Latin*—Let him not go out of the realm. A writ which issues from the Chancery Division to restrain a person from going out of the kingdom without licence of the Crown, or leave of the court. It is a high prerogative writ, which was originally applicable to purposes of state only, but was afterwards extended and confined to private transactions. It is or may be employed when a person, against whom another has an equitable claim for a sum of money actually due, is about to leave the country for the purpose of evading payment, and the absence of the defendant would materially prejudice the plaintiff in the prosecution of his action.

neg. Negative; Negotiate; Negligence; Negotiable.

Negara *Malay/Indonesian*—The State.

Negative In signals and messages is used for saying no, denial, refusal of assent, dissent, disapproval. Opposite to **Affirmative**, *q.v. Abbrev.* Neg.

Negative Income Tax Similar to Reverse Income Tax. Concept of paying allowances using the taxation system. *Abbrev.* NIT.

Negotiable Documents Documents which legally transfer the right of property from one or more persons to others. Some negotiable documents are Bills of Exchange, *q.v.*, Promissory Notes, *q.v.*, and Cheques.

Negotiorum gestior *Latin*—Manager of Business. In law, this means someone who acts to another's advantage without the latter's authority but may be entitled to compensation, as in the case of salvage and others. *See* **Management of Affairs**.

negt. Negociant *French*—Merchant; Negligent.

Negus A drink consisting of sweet spiced wine and boiling water.

n.e.i. Not Elsewhere Indicated/Included.

NEL National Engineering Laboratory, *q.v.*

NELC Naval Electronic Laboratory Center *USA*.

NEMA National Electrical Manufacturers Association *USA*.

Nem. con. Nemine contradicente *Latin*—No one contradicting; Unanimously.

Nem. diss. Nemine dissentiente *Latin*—No one disagreeing or dissenting.

Nemo dat quod non habet *Latin*—No one can give another any better title than he himself has.

NE/ND New Edition/No Date.

N.Eng. New England, *USA*; Northern England.

Neodymium Metallic element used to colour glass purple. *Symbol* Nd.

N.E.OF *or* **N.E.of** *or* **n.e.of** North East Of.

Neolite Darkish green material silicate of aluminium and magnesium which is found in basalt.

n.e.p. New Edition Pending; Not Elsewhere Provided.

Nephoscope Instrument to indicate and measure the direction, velocity and altitude of cloud.

Ne plus ultra *Latin*—No more beyond; The ultimate attainable; unsurpassable.

n.e.p/s. Not Elsewhere Provided/Stated.

Neptune Mythical god of the sea.

Ne quid nimis *Latin*—Nothing to excess.

NERC National Environment Research Council.

NEREC Nuclear Energy Risks Exclusion Clause *insurances of fuel in transit.*

n.e.s. Not Elsewhere Specified/Shown; Not Elsewhere Stated.

NESC National Electric Safety Code *USA.*

Ne cit vox missa reverti *Latin*—The word spoken cannot be taken back.

NESS National Energy Supply Strategy *USA.*

NEST *or* **N.E.S.T.** Naval Experimental Satellite Terminal.

Nest Egg A sum of money in reserve.

Nesting (1) Paper or other similar material used in packing earthenware such as bowls or cups or any other similar goods. The paper or straw is inserted in the hollow parts while nesting the items on top of each other to protect them from damage. (2) Stowage of items of cargo within the hollow spaces of others to economise on freight charges.

Net *or* **Netto** *Italian*—The bare or the lowest. Free from deduction, remaining after necessary deductions. Net profit is true profit, i.e. the actual gain after working expenses have been paid.

NETC No Explosives in Total Contents.

Net Capacity The actual deadweight cargo capacity a ship can carry. *See* **Deadweight Cargo**.

Net Charter (1) *Similar to* **Bareboat Charter** and **Demise Charter**, *q.v.*, (2) A Charterparty, *q.v.*, where the charterers pay for all port expenses leaving the owners to receive the freights without any deductions. (3) A Charterparty, *q.v.*, with Free of Address Commission, *q.v.*

Neth. Netherlands.

Neth. Ant. Netherlands Antilles (Curaçao).

Net Liquid Funds The net amount of capital held as cash in hand or at the bank.

Net Loss on Time Clause A charterparty clause whereby no charter hire or charter money is to be paid for loss of time. This is quite similar to the Breakdown Clause, *q.v.*

Net Present Value The method of finding out the anticipated investment in a project is to obtain the net expenditure of the expected adventure. *Abbrev.* N.P.V. *See* **Discount Cash Flow**.

Net Price The exact price.

Net Terms Cargo-handling is the responsibility and for the account of the charterers, shippers or receivers.

Net Value Clause A marine insurance clause covering Particular Average, *q.v.*, on damaged cargo, i.e. the net arrived value of the damaged cargo before withdrawal from the ship.

Net Weight The exact weight of the contents and excluding the weight of the container.

Nev. Nevada, *USA.*

NEWCI *or* **N.E.W.C.I.** Not East or West Coast of Italy.

Newf. Newfoundland, *Canada.*

New for Old When new material or parts replace damaged material or parts during repairs to a ship. Underwriters are entitled to make deduction from the claim as a result of betterment but they waive this right in practice. Average adjusters may apply the principle in general average for vessels of 15 years of age.

New Hebr. New Hebrides.

New M. New Mexico.

New Peso Currency unit of Chile.

New Taiwan Dollar Currency Unit of Taiwan.

New Uruguayan Peso Currency Unit of Uruguay.

New World The USA and Americas as opposed to Old World, i.e., Europe, Asia and Africa.

New York Prime Loan Rate *Abbrev.* NYPLR. *See* **Base Rate**.

New York Stock Exchange One of the main world organisations for market securities since 1792. The sale or purchase of Government securities and stocks or shares in public companies takes place on a worldwide scale. *See* **London Stock Exchange**.

New Zealand Dollar Unit of currency of New Zealand.

NF *or* **N.F.** Newfoundland, *Canada.*

n.f. No funds, *q.v.*

N.F. *or* **N/F** No Funds, *q.v.*

n.f.a. No further action.

NFAS National Federation of American Shipping, *in Washington DC USA.*

NFC *or* **N.F.C.** National Freight Corporation.

n.f.c. Not favourably considered.

Nfd Newfoundland, *Canada.*

NFEC Naval Facilities Engineering Command *USA.*

NFO *or* **N.F.O.** National Freight Organisation.

NFPA National Fire Protection Association *USA.*

n.f.r. No further requirements.

NFS *or* **N.F.S.** National Fire Service; Non-Federated Ships; Not for Sale.

NFTA *or* **N.F.T.A.** Niagara Frontier Transportation Authority.

NFTC National Foreign Trade Council *USA*.

NFTZ Non-Free Trade Zone.

NFU *or* **N.F.U.** National Farmers' Union.

N.G. National Giro; National Guard.

n.g. Narrow Gauge; No Good.

NGI Norwegian Geotechnical Institute.

NGL Natural Gas Liquids.

NGO Non-governmental Organisation *USA*.

NGS *or* **NGV** Natural Gas Ship or Natural Gas Vessel.

NGSA Narrow Gap Submerged Arc *welding*.

NH *or* **N.H.** New Hampshire, *USA*.

N.H.B.C. National House Builders Council.

NHI *or* **N.H.I.** National Health Insurance.

NHO Navy Hydrographic Office.

NHP *or* **N.H.P.** *or* **n.h.p.** Nominal Horse Power.

NHS *or* **N.H.S.** National Health Service.

Ni Symbol for Nickel.

NI *or* **N.I.** National Insurance; Northern Ireland; Nautical Institute; Naval Instructor; Naval Intelligence.

NIAID National Institute of Allergies and Infectious Diseases.

NIC Network Interface Card, used in connecting computers.

NIC *or* **N.I.C.** Names' Interests Committee *Lloyd's*; National Incomes Commission; Newly Industralised Country.

Nicar. Nicaragua.

NiCd Nickel Cadmium Cells used in electric batteries for lighthouses and other purposes.

Nickel US expression for a 5 cent coin; Metallic element very resistant to corrosion. *Abbrev.* Ni.

Nickel-Silver Alloy of nickel, copper and zinc.

NIEO New International Economic Order.

Nieuw *Dutch*—New.

Nig. Nigeria; Nigerian.

NIG National Industry Group *UK*.

Night Shift Work period at night-time or in the Silent Hours, *q.v.*, gang or team of workpeople so employed.

Nihil *Latin*—Nothing.

Nihil ad rem *Latin*—Nothing to the point.

NII Nuclear Installations Inspectorate *USA*.

Nikkei The Tokyo Stockmarket Index.

Nil admirari *Latin*—To wonder at nothing; to be superior and self satisfied; proud.

Nil desperandum *Latin*—Despair of nothing.

NILOS Netherlands Institute for the Law of the Sea.

Nil sine numine *Latin*—Nothing without divine will. Motto of Colorado, USA.

Nimbo-stratus clouds Form of clouds in solid and dense grey appearance which may produce rain and occasionally snow. *Abbrev.* Ns *or* nb.

Nimbus A low heavy cloud clearly showing that rain is imminent.

NIN National Information Network.

NIO National Institute of Oceanography.

NIOC National Iran Oil Company.

NIOSH National Institute for Occupational Safety and Health *USA*.

Nipped A ship is nipped when she is Beset, *q.v.*, and the ice is actually pressing against the sides with the possibility of damaging the hull, rudder and propellers.

NIPPONCOAL A voyage charterparty published by the Japan Shipping Exchange and used for shipments of coal.

NIPPONSALE Code name for Japanese sale form issued in 1965 and revised in 1977 by the Japan Shipping Exchange.

NIRA Northern Ireland Shipowners' Association.

NIRNS National Institute of Research in Nuclear Science.

NIS National Insurance Scheme; National Insurance Surcharge.

n.i.s. Not in stock.

NISCON National Industrial Safety Conference.

Ni. stl. *Lloyd's Register notation* for Nickel Steel.

Nisi Dominus frustra *Latin*—Unless God is with us, all is in vain. Motto of Edinburgh.

Nisi prius *Latin*—Unless before. In the practice of the High Court, a trial at *nisi prius* is where an action is tried by a jury before a single judge, either at the sittings held for that purpose in London and Middlesex, or at the assizes.

NIT Negative Income Tax, *q.v.*

NITL National Industrial Transportation League, *USA*.

NITPRO Nigerian Committee on Trade Procedures and Facilitation.

Nitrogen Colourless, tasteless, scentless gas forming four-fifths of the earth's atmosphere; one of the Inert Gases, *q.v.*, used in Crude Oil Washing, *q.v. See also* **Carbon Dioxide** *and* **Inert Gas System**.

N.J. New Jason Clause; New Jersey, *USA*.

NJC *or* **N.J.C.** National Joint Council.

NK *or* **N.K.** Not Known.

NKK Nippon Kaiji Kyokai. *See* **Classification Societies**.

NKORL *or* **Nkorl** Not Known or Reported Loss *insurance*.

N.Kr. Norwegian Krone, unit of currency used in Norway—one hundred ore equals one krone.

NL Air Liberia *airline designator*; Netherlands; No liner.

nl non licet *Latin*—it is not permitted; non liquet *Latin*—it is not clear; non longe *Latin*—not far.

N.Lat. North Latitude.

NLB Northern Lighthouse Board.

NLGI National Lubricating Grease Institute.

NLI *or* **N.L.I.** National Lifeboat Institution.

NLL National Lending Library for Science and Technology.

N.L.O. Naval Liaison Officer.

NLOW No Limit on Warranty *insurance*.

NLRB National Labor Relations Board *USA*.

NLRSS Naval Long Range Strategic Study *USA*.

n.l.t. Not less than, Not later than.

NLUR *or* **N.L.U.R.** No Laying Up Returns *insurance*.

NM *or* **N.M.** New Mexico *USA*; Normal Moveout *geophysical*.

n.m. Nautical Mile, *q.v.*; New Moon.

n/m *or* **N.m.** *or* **NM** No Mark; Non-Marine *insurance*.

NMA *or* **N.M.A.** Non-Marine Association.

NMB *or* **N.M.B.** National Maritime Board.

n.m.c. No more credit.

NMD Norwegian Maritime Directorate.

NMEA National Marine Electronics Association *USA*.

NMFS National Marine Fisheries Service *USA*.

N.Mex. New Mexico.

NMI National Maritime Institute; Netherlands Marine Institute *or* Netherlands Maritime Institute.

n. mile Nautical Mile, *q.v.*

NMM National Maritime Museum.

NMRC National Maritime Research Center *USA*.

NMRO Navy Mid Range Objectives.

NMS National Maritime Safety *USA*; National Market System *finance*; Navy Mid-Range Study *USA*.

n.m.t. Not more than.

NMU National Maritime Union *USA*.

NMU *or* **NMUA** National Maritime Union of America *USA*.

N/N *or* **n.n.** No name/s; Not to be noted; Not North of, *q.v.* No noting.

N/N *or* **n/n** Not North of, *q.v.*

nn Names.

n.n. No Name.

NNE *or* **N.N.E.** North North East *compass bearing*.

n.n of *or* **N.N. of** Not North of, *q.v.*

NNP *or* **N.N.P.** Net National Products.

NNRF Non-negotiable Report of Findings.

NNSS Navy Navigation Satellite System *USA*.

NNW *or* **N.N.W.** North North West.

No. Numero *Latin*—In number; *Italian*—Number.

N/o *or* **N.o.** No Orders; Not Out *cricket*.

NO New Orleans, *USA*.

N.O. Naval Officer; Navigation Officer.

NOA National Oil Account; Not Otherwise Authorised.

NOAA National Oceanographic and Atmospheric Administration *USA*.

nob. Nobis *Latin*—On our part; For our part.

No Bottom Sounding When the Sounding Tape, *q.v.*, does not record the depth of water needed.

NOBU New Orleans Board of Underwriters.

NOC Not Otherwise Classified; Notice of Cancellation.

No Claims Bonus An incentive by insurance companies for the insured to refrain from claiming for any accidents during the period of the policy risk. It allows an annual rebate on the premiums charged. If no claim is made in the first year a rebate of 10% is allowed and then up to a maximum of 60% after four or five years without a claim being made. *Abbrev.* NCB *or* N.C.B.

No Claim for Accident Reported A condition that underwriters may reject a claim on a 'lost or not lost' basis if the ship is not reported lost by Lloyd's before or during the attachment of the policy. *Abbrev.* NCAR *or* N.C.A.R.

No Commercial Value Marked or stamped in full or abbreviated NCV *or* N.C.V. on packages, generally via parcel post, as a declaration by the sender that the contents are of insignificant value and therefore not taxable.

No Cure No Pay *or* **No Cure/No pay** The principle of pure salvage whereby the salvor who fails in his task receives no reward for his efforts. *See* **Lloyd's Open Form (Lloyd's Standard Form of Salvage Agreement).**

NOD *or* **N.O.D.** Night Observation Device; Naval Ordnance Department.

NOE *or* **N.O.E.** Nap-of-the Earth, Low Flying *aviation*.

n.o.e. Notice of Exception; Not Otherwise Enumerated.

NOFI National Oil Fuel Institute *USA*.

No Funds Expression used by banks indicating that a cheque which has been issued is not cashed or honoured due to the lack of money in the drawer's account. *Abbrev.* N.f *or* N/F *or* n.f.

No Higher Order not to steer any closer to the wind *sailing*.

n.o.h.p. Not otherwise herein provided. *Insurance*.

NOIA Newfoundland Ocean Industries Association.

n.o.i.b.n. Not otherwise indexed by name. *Insurance*.

NOIC National Oceanographic Instrumentation Center *USA*; Naval Officer in Charge.

Noise Induced Hearing Loss Excessive noise which over a period of time can cause deafness.

NOK *or* **N.O.K.** Not otherwise known.

n.o.k. Next of Kin.

NOL Naval Ordnance Laboratory *USA*.

Nol. con. Nolo contendere *Latin*—I plead guilty.

Nolens volens *Latin*—Willy nilly; perforce.

Nolle prosequi *Latin*—*Abbrev.* nol. pros., *q.v.*

Nol. pros. Nolle prosequi *Latin*—To be unwilling to prosecute. A proceeding in the nature of an undertaking by the plaintiff when he had misconceived the nature of the action, or the party to be sued, to forbear to proceed in a suit altogether, or as to some part of it, or as to some of the defendants. It differed from a *non pros.*, which put a plaintiff out of court with respect to all the defendants. In criminal prosecutions by indictment or information, a *nolle prosequi* to stay proceedings may be entered by leave of the Attorney-General at any time before judgment: it is not equivalent to an acquittal and is no bar to a new indictment for the same offence.

Nom. Normal; Nominal.

NOMA National Office Management Association.

NOMAD Navy Oceanographic Meteorological Device *USA*.

Non. cap. Nominal Capital, *q.v.*

Nom collectif *French*—A partnership operating under a name. Its obligations are guaranteed by the unlimited joint and several liability of the partners.

Nom de guerre *French*—Assumed name.

Nom de plume *French*—A pseudonym, fictitious name.

Nomen novum *Latin*—New name.

Nominal In name only, not real or actual such as Nominal Capital, *q.v.*, Nominal Horse Power, *q.v.* etc.

Nominal Capital The authorised or registered capital of a company under the Memorandum of Association.

Nominal Horse Power Unit of power to lift 33,0001lbs one foot in one minute. *Abbrev.* NHP *or* N.H.P *or* n.h.p

Nominal Partner A person authorising his name to be inserted in the name or title of a business concern and who does not have an actual interest in the trade or business, or its profits.

Nomination Generally refers to the Nominated Vessel.

Nominee Name Name under which a security, *q.v.*, is registered which does not reveal the name of the Beneficial Owner, *q.v.*

NOMSS National Operation Meteorological Satellite System *USA*.

nom. std. Nominal Standard.

Non-Carrying In cargo claims the vessel which carries the cargo if a collision occurs is called the 'carrying vessel'. The ship that does not carry the cargo is called the 'Non-Carrying Vessel'. *Abbrev.* NCV.

Non-Com. Non-Commissioned Officer.

Non-Compatibility Opposite to Compatibility, *q.v.*

Non compos mentis *Latin*—Not of sound mind.

Non-Contrib. Cl. Non-Contribution Clause.

Non culpabilis *Latin*—Not guilty. *Abbrev.* Non cul.

non-cum. Non-cumulative.

Non-Dangerous Oils Oils which have their flash points at 151° and over.

Non-Delivery Short Shipped, *Abbrev.* non.d.

Non-Destructive Testing Ultrasonic testing of metals. *Abbrev.* NDT.

Non-Drying Oils Natural oils and fats consisting of almond, coconut, groundnut, lard, olive, palm, rape, rice and tallow.

Non-Durable Goods Goods which are consumed in the daily way of life of a person such as food, beverages, etc.

Non est *Latin*—It is not.

Non est inventus *Latin*—He is not found. The return which was made by the sheriff upon a writ commanding him to arrest a person who was not within his bailiwick.

Non-Ferrous Goods Materials containing lead, copper, aluminium, tin, zinc as opposed to iron and steel, which are ferrous metals. The latter are metals which are attracted by magnets.

Non licet *Latin*—Prohibited.

Non lignet *Latin*—It is not clear, not proven.

Non obstante *Latin*—Notwithstanding.

Non obstante veredicto *Latin*—Notwithstanding the verdict.

Non placet *Latin*—It does not please. Expresses a negative vote.

Non plus ultra *Same as* **Ne plus ultra**, *q.v.*

Non possumus *Latin*—We cannot. Expresses inability to act.

Non pros. Non prosequitur *Latin*—He does not prosecute.

Non reperat. Non reperatur *Latin*—Do not repeat.

Non-Separation Agreement Document usually attached to an Average Bond, *q.v.*, or Underwriters' General Average Guarantee whereby ship and cargo interests agree that their rights and liabilities in general average shall not be affected by the forwarding of cargo from a port of refuge.

Non seq. Non sequitur *Latin*—It does not follow logically.

Non-suit The judge orders a non-suit when the plaintiff fails to make out a legal cause of action or fails to support his pleadings by any evidence; whether the evidence which he gives can be considered any evidence at all of a cause of action is a question of law for the judge. Before 1875 the advantage of this practice (which was peculiar to the common law courts) was that the plaintiff could, on paying all costs, bring another action against the defendant for the same cause of action; but under Ord. 41, r. 6, of the R.S.C., 1875, any judgment of non-suit, unless the court otherwise directed, had the same effect as a judgment on the merits, that is, it barred the plaintiff from bringing another action for the same cause; but in case of mistake, surprise or accident, a judgment of non-suit might be set aside by the court. Order 41, r. 6, was cancelled in 1883 and there is now in strictness no such thing as a non-suit in the High Court. A plaintiff cannot now elect to be non-suited, and if he offers no evidence it is the duty of the court to direct the jury to find a verdict for the defendant, and the usual consequences of such verdict will follow; but a judge cannot order a non-suit on plaintiff's opening without the consent of his counsel. The term is still loosely used instead of the term 'judgment for the defendant,' where the judge withdraws the case from the jury and directs a verdict for the defendant. Non-suit still applies in the county court.

Non-Working Holiday A Charterparty, *q.v.*, phrase meaning a public holiday on which work is charged against overtime.

NOO National Oceanographic Office *USA*.

Noon Watch *or* **Afternoon Watch** Seaman's watch from midday (noon) up to 4 p.m.

NOP Navy Objective Plan; Net Original Premium *insurance*; Non-Oxidizing Paint.

n.o.p. Not otherwise provided; Non-Oxidizing Paint.

No-Par-Value Shares Shares which are currently in the stock exchange market.

NOR *or* **N.O.R.** Net Original Rate; Notice of Readiness, *q.v.*; Not Otherwise Rated.

Nor *or* **Norw.** Norway; Norwegian.

NORA *or* **N.O.R.A.** Notice of Readiness Accepted. A charterparty phrase.

NORAD North American Air Defense Command; Norwegian Agency for International Development.

NORDCO Newfoundland Oceans Research and Development Corporation.

Norges Rederforbund Norwegian Shipowners Association, Oslo.

No Risk After Discharge An insurance term which indicates that once the merchandise is free of the ship's tackle underwriters are no longer liable. *Abbrev.* N.R.A.D. *or* n.r.a.d.

No Risk After Shipment In marine insurance this clause relieves the underwriters from liability as soon as the cargo is shipped on board. *Abbrev.* N.R.A.S. *or* n.r.a.s.

No Risk Till Waterborne An insurance clause limiting liability until the vessel is floating in the sea. This refers to the ship when she is in dry dock. *Abbrev.* N.R.T.W.B *or* n.r.t.w.b

Norske Veritas *or* **Det Norske Veritas** The Norwegian ship classification society. *See* **Classification Societies** *and* **Lloyd's Register of Shipping**.

North Cardinal point on the compass, directly opposite to South.

Northern Lighthouse Board The Board was created in 1786 for the construction and administration of four lighthouses in Scotland. It now consists of the Local Advocate, Solicitor General, the Sheriffs-

Principal of Scotland, the Lords Provost of Edinburgh, Glasgow and Aberdeen, the Provost of Inverness, the Chairman of Argyll and Bute District Council, a nominee from the Isle of Man and five co-opted Commissioners. Its responsibilities cover 84 major lighthouses of which 13 are manned (1994), 111 minor lights, 43 beacons, 29 racons, 9 radio beacons and 122 buoys around the Scottish and Manx coasts, with a further 59 buoys serviced under contract. It inspects navigation aids on oil rigs in the Scottish sector. It owns two ships which have flight decks for helicopter operations for the support or remote lighthouses. The headquarters are in Edinburgh and there are depots at Granton, Oban and Stromness.

Manned lighthouses are divided into two main categories; mainland stations, and rock or relieving stations. At the former, a principal lightkeeper and up to three assistant lightkeepers live in the immediate vicinity of the light, accompanied by wives and families. At rock stations, six lightkeepers are assigned. A principal and two assistants serve a month on duty and a month ashore with their families at the shore station, usually in the nearest major town. A relieving station is in effect a rock station on the mainland, but so remote that it is unsuitable for families. The Board has a programme to automate all manned lighthouses and convert lightbuoys to solar power. The Northern Lighthouse Board, together with Trinity House (England, Wales and the Channel Isles) and the Commissioners for Irish Lights (Eire and Northern Ireland) are the General Lighthouse Authorities for the United Kingdom and Ireland. Running costs are met from a 'General Lighthouse Fund' administered by the Department of Transport and financed by the collection of Light Dues, paid by ships loading or discharging cargoes at British and Irish ports, and by fishing vessels over 10 metres in length.

Northern Range A range of eastern seaboard ports in the USA comprising Norfolk, Newport News, Baltimore, Philadelphia, New York, Boston and Portland. *Abbrev.* NR *or* N.R *or* N/R.

Norw. Norway; Norwegian.

Norwegian Saleform A standard printed memorandum of agreement widely used for the sale and purchase of merchant ships.

Norwegian Shipping Academy An international training school for personnel from all nations. Correspondence courses are also undertaken. *Abbrev.* NSA.

Norwegian Underwater Institute An organisation in Bergen for research and development in the underwater sciences and technology. It is jointly owned by Det Norske Veritas and the Royal Norwegian Council for Scientific and Industrial Research. *Abbrev.* NUI.

No/s Numbers.

NOS National Ocean Survey *USA*; Not Otherwise Specified, *see* **Not Otherwise Enumerated**.

NOSAC National Offshore Safety Advisory Committee *USA*.

NOSC Naval Ocean Systems Center *USA*.

Nosce te ipsum *Latin*—Know thyself.

Noscitur a sociis *Latin*—A man is known from his associates.

No. Si. *or* **No. S.I.** No Short Interest.

No Sparring A *Lloyd's Register notation* signifying that the ship has no Cargo Battens fitted.

NOTAM Notice to Airmen.

NOTAR No Tail Rotor *helicopter*.

Not Before After this phrase a date follows specifying the date for loading. *Ex.* Time of loading to start not before 6 December. If the ship is ready before the appointed day and hour and the cargo is also ready, this will be advantageous for the owners of the cargo.

Not East Of An expression used in the trading limits of a ship when she is chartered or insured. *Abbrev.* N.E. of *or* N.E.OF.

Not Negotiable *or* **Non-Negotiable** Either one of these phrases are to be seen stamped or printed on documents like cheques, Bills of Lading, *q.v.*, etc., to denote that the rights or property shown on the face of the bills cannot be transferred or is restricted.

Not North Of An expression used in the trading limits of a ship when she is chartered or insured. *Abbrev.* N.N of *or* N.N.OF.

NOTO Non-official Trade Organisation.

Not On Risk *or* **Not At Risk** Underwriters are not liable for any claims.

Not South Of An expression used in the trading limits of a ship when she is chartered or insured. *Abbrev.* N.S. of *or* N.S.OF.

Not to Inure Clause A clause in a cargo policy stating that the policy shall not inure to the benefit of a carrier or other bailee. The intention is to deny the right of carriers to benefit from the insurance when they claim such right in the contracts of carriage.

Not Under Command A disabled vessel. She has to show that she is not under command in accordance with the rules laid down in the International Regulations for Preventing Collisions at Sea, *q.v.*

Not Under Repair An insurance expression usually seen in the Hull Policies' Return Clause. This phrase frees the underwriters from any claims by the owners in case of repairs under the Lying Up Returns, *q.v.*

Not West Of An expression used in the trading limits of a ship when she is chartered or insured. *Abbrev.* N.W. of *or* N.W.OF.

Notary *or* **Notary Public** An official certified to take affidavits and depositions from members of the public.

Nothing Off Order to steer as close to the wind as possible *sailing*.

Notice of Abandonment The initial action to be taken by an assured who wishes to claim a constructive total loss. Notice to underwriters must be given with reasonable diligence as soon as the assured is aware of the circumstance. Its purpose is to give the underwriter the opportunity to take action to prevent or minimise the loss. *See* **Abandonment**.

Notice of Readiness A notice indicating a vessel's readiness to load or to discharge.

Nottm. Nottingham, *UK*.

Notts. Nottinghamshire, *UK*.

Nov. November; Novembre *Italian/French*—November.

Novation Substituted contract to the one in existence.

November International Radio Telephone phonetic alphabet for 'N'.

Novus actus interveniens *Latin*—A new act intervening. An intervention by a third party which may affect a person's liability for damage *law*.

Novus homo *Latin*—A new man; Newcomer.

NOx Abbreviation for Nitrous Oxide. *Symbol* N_2O.

n.o.y. Not Out Yet.

n/p *or* **N.P.** Net Proceeds.

N.P. No Protest; Notary Public, *q.v.*, Net Proceeds.

NPA *or* **N.P.A.** National Packing Association of Australia; National Petroleum Association *USA*; National Production Authority; Nigerian Ports Authority; Newspaper Publishers' Association.

N.Pac. North Pacific.

N.Pac. Cur. North Pacific Current.

NPC *or* **N.P.C.** National Ports Council; National Petro-Chemical Company, *Iranian Government*; National Petroleum Council *USA*; National Parks Commission.

NPCAA National Passenger and Cargo Agents Associations.

NPD Norwegian Petroleum Directorate.

N.P.D. North Polar Distance.

NPDC Naval Project Development Company *Holland*.

n.p.f. Not provided for.

N.Ph. Nuclear Physics.

N.P.I. Net Premium Income.

NPL *or* **N.P.L.** National Physical Laboratory.

NPN Negative-Positive-Negative.

n.p.n.a. No protest for No Acceptance.

NP/ND Not Published, No Date Given.

n.p *or* **d.** No Place or Date.

n.p.p. No Passed Proof.

n.p.r. Noise Power Ratio.

NPRA National Petroleum Refineries Association.

N.P.S. Non-Professorial Staff, *referring to university employees*.

NPS Noise Protection Service.

n.p.s. Nominal Pipe size.

n.p.s.h. Net Positive Suction Head.

n.p.t. Normal Pressure and Temperature.

n.p.u. Ne plus ultra *Latin*—Perfection.

N.P.V. *or* **n.p.v.** Net Present Value, *q.v.*; No Par Value.

N.Q.A. Net Quick Assets.

NR Non returnable; Northern Range, *q.v.*, Nuclear Reactor; Norges Rederforbund *Norwegian Shipowners' Association*.

nr Near, Number.

N/R Not Reported; Northern Range, *q.v.*

NRA National Restaurant Association; National Rivers Authority *UK*.

N.R.A.D. *or* **n.r.a.d.** No Risk After Discharge, *q.v.*

N.R.A.L. *or* **n.r.a.l.** No Risk After Landing.

N.R.A.S. *or* **n.r.a.s.** No Risk After Shipment, *q.v.*

NRC *or* **N.R.C.** Netherlands Red Cross; Nuclear Research Council *USA*.

N.R.D.C. National Research Development Corporation *UK*.

NRDF National Reserve Defense Fleet *USA*.

NRDO National Research Development Organisation *USA*.

NRL Naval Research Laboratory *USA*.

NRO *or* **N.R.O.** Non Returnable Outer.

NRR Noise Reduction Rating.

NRT *or* **N.R.T.** *or* **n.r.t.** Net Registered Ton or Net Registered Tonnage.

NRTA National Response Team Agency *USA*.

N.R.T.O.B. *or* **n.r.t.o.b.** No Risk Till on Board. *Similar to* **No Risk After Shipment**, *q.v.*

N.R.T.O.R. *or* **n.r.t.o.r.** **No Risk Till On Rail.**

N.R.T.W.B. *or* **n.r.t.w.b.** No Risk Till Waterborne, *q.v.*

N.R.V. Non Return Value.

NRVOCLONL Non Returnable Vessel and/or Cargo Lost or not Lost. Referring to 'Freight Payable in Advance' in a Charterparty, *q.v.*

Nrw. Norway; Norwegian.

NS *or* **N.S.** Nachschrift *German*—Postscript; New Series; Newspaper Society; Nova Scotia, *Canada*; No Sparring, *q.v.*; Nuclear Ship; Nickel Sheet; Nimbostratus Clouds, *q.v.*

n/s *or* **n.s.** Not Sufficient; New Style; New Service.

NSA National Shipping Adjusters *US*; National Shipping Authority. Non-Separation Agreement, *q.v.*; Norwegian Shipowners Association. Norwegian Shipping Academy, *q.v.*

NSB National Savings Bank; National Savings Bond; National Science Board *USA*.

NSC National Savings Committee; National Security or Safety Council *USA*; Newfoundland Science Council.

NSEA National Sales Executive Association *USA*.

NSEZ North Sea Economic Zone.

NSF *or* **N.S.F.** Norges Standardiserings-Forbund, *Norwegian Standards Institute*; Norwegian Sale Form; National Sanitation Foundation *USA*.

NSGT *or* **N.G.S.T.** Non-Self-Governing Territories.

NSMB Netherlands Ship Model Basin, now part of the Maritime Research Institute, Holland.

N.S.O. Naval Staff Officer; Naval Stores Officer.

N.S.OF *or* **N.S.Of** *or* **n.s.of.** Not South of, *q.v.*

n.sp. New Species.

N.S.P.C.C. National Society for the Prevention of Cruelty to Children.

n.s.p.f. Not specially provided for.

NSR Northern Sea Route *for Arctic passage*.

NSRDC Naval Research and Development Center *USA*.

NSRDF Naval Supply Research and Development Facility. US Navy department concerned with logistics.

NSRP National Shipbuilding Research Program *USA*, *q.v.*

NSS *or* **N.S.S.** National Savings Stock.

NSSC Navy Sea Systems Command *US Navy*.

NSSN National Standard Shipping Note.

NST Newfoundland Standard Time; Nomenclature Standard Transport. The standard nomenclature for transport statistics, being EUROSTAT classifications.

NSTC *or* **N.S.T.C.** National Sea Training College.

NSTT *or* **N.S.T.T.** National Sea Training Trust.

NSU Netherlands Shipping Union.

NSW *or* **N.S.W.** New South Wales, *Australia*.

NT *or* **N.T.** National Theatre; National Trust; Neap Tide; New Testament; Not Titled; Net Terms, *q.v.*; Net Tonnage, *or* Net Ton.

nt *or* **n.t.** *or* **N.t.** *or* **N/T** Normal Temperature; Net Terms, *q.v.*; Net Tonnage *or* Net Ton.

NTA National Transport Act.

n.t.c. Negative Temperature Coefficient.

NTD *or* **N.T.D.** Non-Tight Door.

NTDO Navy Technical Data Office *USA*.

NTDS Naval Tactical Data System *USA*.

NTEC Naval Training Establishment Center *USA*.

ntfy. Notify.

Nthb. Northumberland, *UK*.

Nthn. Northern.

NTIS National Technical Information Service, Springfield, *USA*.

n.t.l. No time lost.

N.t.m. Net Time Mile.

NTM Notice to Mariners.

n.t.o. Not Taken Out.

NTO National Tourist Office.

N.T.O. Naval Transport Officer.

n.t.p. No Title Page.

NTP *or* **N.T.P.** Normal Temperature and Pressure.

n.t.s. Not to scale.

NTSB National Transportation Safety Board *USA*.

NTSBAL National Transportation Safety Board's Audio Laboratory *USA*.

NTSC National Television System Committee.

n.t.u. Not taken up *insurance*.

NTU Network Terminating Unit *computer*.

Nt.Wt. *or* **nt.wt.** Net Weight.

N.U. Nations Unies *French*—United Nations.

n.u. Name Unknown; Number Unobtainable.

NUA Network User Address *computer*.

NUBE *or* **N.U.B.E.** National Union of Bank Employees.

NUBOPS Nippling Up Blow-out Preventers *oil drilling*.

NUC Naval Undersea Center *USA*; Not Under Command, *q.v.*

Nuclide An atom having a specified type of nucleus *physics*.

Nuda pacta obligationem non parit *Latin*—A naked promise does not beget an obligation. A nudum pactum, or naked contract is a bare agreement without consideration and not formally sealed.

NUDETS Nuclear Detectional System.

NUGMW *or* **N.U.G.M.W.** National Union of General and Municipal Workers.

NUI National University of Ireland; Network User Identification *computer*. Norwegian Underwater Institute, *q.v.*

NUIW National Union of Insurance Workers.

N.U.J. National Union of Journalists.

Nulla nuova, buona nuova *Italian*—No news is good news.

Nulli secundis *Latin*—Second to none.

NUM *or* **N.U.M.** National Union of Mineworkers.

Numismatics Of coins and coinage.

Numdah Pad under a saddle.

NUMAST National Union of Marine Aviation and Shipping Transport Officers.

Nun Buoy A buoy circular in the middle and tapering to each end.

Nunc aut nunquam *Latin*—Now or never.

Nunc pro tunc *Latin*—Now for then. The court sometimes directs a proceeding to be dated as of an earlier date than that on which it was actually taken, or directs that the same effect shall be produced as if it had been taken at an earlier date. Thus, it will direct a judgment to be ante-dated if the entry of it has been delayed by the act of the court, or if the plaintiff has died between the hearing and the date when the judgment was delivered. This is called entering a judgment *nunc pro tunc*.

Nunquam non paratus *Latin*—Always ready; never unprepared.

NUPE *or* **Nupe** National Union of Public Employees.

N.U.R *or* **n.u.r.** National Union of Railwaymen; Not Under Repair, *q.v.*

NUS *or* **N.U.S.** National Union of Students; National University of Singapore.

NUSC Naval Underwater System Center US Oceanographic Observation Centre.

NUT *or* **N.U.T.** National Union of Teachers.

N–u–T Newcastle-upon-Tyne.

NV Nevada, *USA*; New Version.

NV *or* **N.V.** Naamlooze Vennootschap, Dutch Limited Company; New Version; Norske Veritas, *q.v.*

n.v. Needle Valve.

NVD *or* **N.V.D.** No Value Declared.

n.v.m. Non-Volatile Matter.

NVO Non Vessel Operating.

NVOC Non Vessel Owning Carrier.

NVOCC Non-Vessel Owning Common Carrier.

NVQ National Vocational Qualification *UK education*.

n.v.r. No Voltage Release.

NVS Nederlandse Vereniging Van Scheepsleveranciers, Netherlands Association of Shipsuppliers—Members of the International Shipsuppliers Association (ISSA).

N.W. North Wales; North West; North Western.

n.w. Net Weight; No Wind *weather observation*.

NWA North West Africa.

NWB Natural Wet Bulb *temperature*.

N.W by N. North West by North *compass bearing*.

N.W by W. North West by West *compass bearing*.

NWC National Water Council; National Waterways Conference *founded 1960 in USA*; Naval Weapons Center *USA*.

NWECS North West Euorpean Continental Shelf.

NWF National Wildlife Federation *USA*.

Nwfld. Newfoundland, *Canada*.

N.W.G. National Wire Gauge.

NWH Normal Working Hours.

N.W.I. Netherlands West Indies.

n.w.l. Natural Wave Length.

N.W. OF *or* **N.W. of** *or* **n.w. of.** North West of.

N.W.R. No War Risk *marine insurance*.

NWS *or* **N.W.S.** Normal Water Surface; National Weather Service.

N.W.T. North West Territories, *Canada*.

n.w.t. Non-watertight.

n.wt. Net Weight.

NWTA National Waterways Transport Association.

n.w.t.d. Non-Watertight Door.

N.X. NAFTICA CHRONICA, marine annuals in Greece.

NY New York.

N.Y. New Year; New York.

NYC New York City.

NYCCI New York Chamber of Commerce and Industry.

N.Y. DOT. New York Department of Transportation.

NYH New York Harbor.

NYIE New York Insurance Exchange.

NYK *or* **N.Y.K.** Nippon Yusen Kaisha, Japan Mail Steamer Company.

nyl Nylon.

Nylon Contraction for New York/London.

NYLS New York Law School.

NYMA New York Maritime Association.

NYME New York Mercantile Exchange.

NY/NJ New York/New Jersey.

NYO Not Yet Out.

NYPA New York Port Authority.

NYPE New York Produce Exchange; New York Produce Exchange Time Charter Form.

NYPLR New York Prime Loan Rate. *See* **Base Rate**.

n.y.r. Not yet returned.

NYS *or* **N.Y.S.** New York State.

NYSA New York Shipping Association; New York State Assembly.

NYSB New York's Security Bureau Inc.

NYSBA New York State Bar Association.

NYSBB *or* **N.Y.S.B.B.** New York State Banking Board.

NYSE New York Stock Exchange; New York Selling.

NYT *or* **N.Y.T.** New York Standard Time.

NYU *or* **N.Y.U.** New York University.

NYWC New York Wings Club.

NZ New Zealand.

NZEI *or* **N.Z.E.I.** New Zealand Electronic Institute.

NZIE *or* **N.Z.I.E.** New Zealand Institution of Engineers.

NZIM *or* **N.Z.I.M.** New Zealand Institute of Management.

NZMS *or* **N.Z.M.S.** New Zealand Meteorological Service.

NZUKCC New Zealand UK Chamber of Commerce.

O

O of; Ohio, *USA*; *Lloyd's Register notation* for Outer; Oscillator; Oxygen.

O_2 Symbol for Dioxygen.

O_3 Symbol for Ozone.

o.a. Onder andere *Dutch*—Among others.

O.a. On Account Of; Over All, i.e. measuring from the extremes.

O.A. Office Address; Old Account; Over All; Observed Altitude *nautical*.

O/A On Account; On or About; Overage, *q.v.*; Our Account.

OAA Orient Airlines Association.

o.a.a.o.o.p On arrival at or off the port.

O/A AP Overage Additional Premium *when cargo is to be carried on an overaged vessel*.

OABA *or* O.A.B.A. *or* oaba *or* o.a.b.a. Owners' Agent Both Ends, *in Charterparty, q.v.*

OACI *or* O.A.C.I Organisation d'Aviation Civile Internationale *French*—International Civil Aviation Organisation, *Abbrev.* ICAO *or* I.C.A.O.

o.a.d. Over All Depth.

OAF Office of Alcohol Fuels.

OAG *or* O.A.G. Official Airline Guide *USA*.

o.a.h. Over All Height.

Oakum Loose fibre obtained by picking old rope to pieces and used especially in caulking. *See* **Caulk**.

o.a.l. Over All Length.

O&M Organisation and Methods.

OAO Orbiting Astronomical Observatory.

o.a.o. Off and On.

O.A.P. Old Age Pension(er).

OAPEC *or* O.A.P.E.C. Organisation of Arab Petroleum Exporting Countries.

Oarlock Same as Rowlock, *q.v.*, but rarely used.

OAS Ordinary Ammunition Storage.

OAS *or* O.A.S. Old Age Security; On Active Service; Organisation of American States.

OASI *or* O.A.S.I. Old Age and Survivors' Insurance *social security in USA*.

Oasis A fertile patch or area in a desert where water exists.

o.a.t. Outside Air Temperature.

OATC *or* O.A.T.C. Oceanic Air Traffic Control.

OAU *or* O.A.U. Organisation of African Unity.

OAW *or* O.A.W. Overall Width.

OAWB Ocean Airway Bill.

OB *or* O.B. Observed Bearing *navigation*; Official Board; Oil-bearing; Old Bailey, *q.v.*; On Board; Ordinary Business; Outboard; Outside Broadcast.

o/b On or Before.

O/B On Board; Ordinary Business; Outboard; Outside Broadcast.

OBA Optical Bleaching Agent; Oxygen Breathing Apparatus.

OBC Out Bill for Collection *banking*.

ob.dk. Observation Deck.

obdt. Obedient.

OBITB Offshore Petroleum Industry Training Board *USA*.

Obiter *Latin*—By the way; In passing.

Obiter dictum *Latin*—An opinion not necessary to a judgment and therefore not binding as a precedent. *See* **Dictum**.

Object Clause The clause or clauses in the Statute or Memorandum of Association which state(s) the purpose for which the company, firm or association has been set up.

Oblig Obligatory.

OBO *or* O.B.O. Ore/Bulk/Oil Carrier or Oil/Bulk/Ore Carrier, *q.v.*

o.b.o. On behalf of.

Obre. Octobre *French*—October.

obs. Obligations; Observed; Obsolete.

OBS Omnibearing Selector *radio bearing*.

Obs. Alt. Observed Altitude, *q.v.*

Observation Manipulator Bell A battery-powered submersible bell-shaped container used for many sub-aqua purposes. It can operate in depths of about 3,000 feet. *Abbrev.* OMB.

Observed Altitude An altitude by sextant. *Abbrev.* Obs. Alt.

Obs. Posn. Observed Position *nautical*.

OBU Offshore Banking Unit *finance*.

OBV *or* O.B.V. Ocean Boarding Vessel.

O.C. Opere citato *Latin*—in the work cited; Off Cover *insurance*; Open Cover *insurance*; Open Charter, *q.v.* Open Cheque, *q.v.*

O.C. *or* O/C. Off Cover *insurance*; Officer Commanding; On Completion; Open Cover *insurance*; Old Crop; Open Charter, *q.v.*

O/c Open Cheque, *q.v.* Overcharge(d).

OCA Offshore Contractors Association, based in Aberdeen; Operational Control Authority *naval*; Outstanding Claims Advance *insurance*.

OCAM Organisation Commune Africaine et Malgache *French*—African and Madagascar Common Organisation.

OCAS *or* **O.C.A.S.** Organisation of Central American States. *See* **ODECA**.

Oc. B/L Ocean Bill of Lading, *q.v.*

Occ. Occupation; Occulting Light, *navigation*; Occurrence.

occas. Occasion; Occasional.

Occupational Disease *or* **Occupational Hazard** The hazards or dangers encountered in certain jobs.

Occur. Occurrence.

Occurrence Chain of events leading to an Insurance Claim. *Example* An oil tanker struck a hard object whilst at sea; fire broke out; and she subsequently sank. There were three events to this occurrence—collision, fire and foundering.

O/c.d. Other Cargo Damaged *insurance*.

OCDE Organisation de Coopération et de Développement Économiques. *French*—Organisation for Economic Co-operation and Development *OECD*.

O/Cdt Officer Cadet.

OCE Officer Conducting the Exercise.

Ocean Bill of Lading *Similar to* **Order Bill of Lading**, *q.v.* The American corresponding term is Port Bill of Lading.

Ocean Freight Freight income obtained from cargoes loaded on ocean-going vessels.

Ocean Going Losses Losses by evaporation of liquid cargoes in the period between loading and discharge.

Ocean Going Pusher Barge *or* **Tug Barge Combination.** *See* **CATUG** *and* **Artubar**.

Ocean Marine Insurance Actual Marine Insurance as opposed to Inland Marine Insurance *USA term*.

Oceanog. Oceanography.

Ocean Routeing Charts Charts which are part of a ship's equipment showing the prevailing currents, and seasonal weather conditions.

Ocean Station Vessel A ship stationed within a specified area to take weather observations. *Abbrev.* OSV *or* O.S.V.

Ocean Surface Current Radar Device for measuring surface currents using radiation reflected from the waves. *Abbrev.* OSCAR.

OCGO *or* **O.cgo.** Other cargo. Damages sustained by other cargo *insurance*.

OCL *or* **O.C.L.** Obstacle Clearance Limit *aviation*. Off Centre Line.

OCIMF *or* **O.C.I.M.F.** Oil Companies International Marine Forum.

OCNUS Outside Continental United States.

OCR *or* **O.C.R.** Optical Character Recognition or Reading *electronics*; Ordinary Capital Resources *finance*.

OCS *or* **O.C.S** On-Board Courier Service, *q.v.*; Outer Continental Shelf; Outside Controlled Space *aviation*.

OCSLA Outer Continental Shelf Lands Act *USA*.

OCT Octuple Screw *of ship*; Onward Clearance Time *aviation*; Associated Overseas Countries and Territories *EEC*.

oct. Octavo; octo *Latin*—Eight; October.

Octant An instrument in the form of a graduated eighth of a circle used in astronomy and navigation. Similar to a Sextant, *q.v.*, but with a smaller elevation.

O.C.T.U. Officer Cadets Training Unit.

Oculis subjecta fidelibus *Latin*—Under one's very eyes.

OD *or* **O.D.** Organisation Development; Orlop Deck; Outside Diameter; Over Deck; Overdraft, *q.v.*

O/d *or* **o/d** On Deck; On Demand, *q.v.*; Out Dated; Outside Diameter; Over Deck; Overdraft, *q.v.*; Oil Damage *insurance*.

ODA Overseas Development Administration; Overseas Development Aid *EEC*.

ODAS *or* **O.D.A.S.** Ocean Data Acquisition System. *UNESCO/ICO*.

ODB Omnidirectional Beacons, *q.v.*

ODECA Organization de Estados Centro-Americanos *Spanish*—Organisation of Central American States. *Abbrev.* OCAS.

ODETTE Organisation for Data Exchange and Teletransmission in Europe.

ODF Ozone Depletion Factor.

O.d.g. Ordine de giorno *Italian*—Order of the day.

ODI Overseas Development Institute *UK*.

O.Dk. Observation Deck.

ODM Office of Defense Mobilization *USA*; Oil Discharge Monitoring.

O.D.O. Out Door Officer *customs officer*.

Odograph An instrument to register the course taken by a ship on a chart.

Odometer An instrument used to obtain the distance travelled by a wheeled vehicle. It is generally found alongside or works in conjunction with a speedometer.

Odorous Cargoes Cargoes which emit repugnant smells and which are therefore liable to contaminate others in the vicinity.

ODP Open Door Policy; Ozone Depletion Potential; Ocean Drilling Programme.

ODS Operational Downtime Simulator *oil drilling*.

ODS *or* **O.D.S.** Ocean Data Station; Operating Differential Subsidy, *Maritime Subsidy Board. See* **CDS** *or* **C.D.S.**

O.Ds. Other Denominations.

O.D.V. Eau de Vie *French*—Cognac.

OE *or* **O.E.** Omissions Excepted. *See* **Errors Excepted**; Original Error.

O&E Operations and Engineering.

OEC *or* **O.E.C.** Overpaid Entry Certificate, *customs form*.

OECD Organisation for Economic Co-operation and Development.

OECF Overseas Economic Co-operation Fund.

o.e.c.o. Outboard Engine Cut Off.

OECON Offshore Exploration Conference *USA*.

OED Ocean Engineering Department *USA*.

OEDO Ordnance Engineering Duty Officer *USA*.

OEEC Organisation for European Economic Co-operation. Obsolete on becoming OECD, *q.v.*, in 1960.

OEIC Open-ended Investment Company.

O.E.I.U. *or* **OEIU** Office Employees International Union.

OEM Original Equipment Markers *computer*.

Oersted Unit of magnetic field strength *physics*.

OESH Ocean Engineering Ship Handling *USA*.

OETB Offshore Energy Technical Board.

OF *Lloyd's Register notation* for fitted for Oil Fuel with date; Oceanographic Facility; Oil Flat; *Old French*.

O.F. Oceanographic Facility; Officers' Federation.

o.f. Outside Face; Oxidizing flame, *Lloyd's Register notation* for Oil Fired.

Of a Like Kind *Similar to* **Ejusdem generis**, *q.v.*

OFC Overseas Food Corporation.

Off. Office; Officer.

Offals Meats which are carried in a hard frozen condition from 10 to 18°F. They include oxtails, hearts, casings or intestines, livers, kidneys, etc. They are packed in bags, cartons, cases and kegs.

Off Charter When the time of a Charterparty, *q.v.*, expires and the owner takes over the vessel from the charterers.

Off.C.L. Off Centre Line.

Offer The initial introduction before business can be transacted or a contract negotiated.

Offer Price The buying price of stocks and shares as opposed to Bid Price, *q.v.*

Off hire Clause *or* **Off-Hire Clause** In a Time Charter, the owner is entitled to a limited time for his vessel to be off hire until such time as she may be repaired or drydocked. The permissible time is prearranged by the contracting parties. Normally 24/48

hours are allowed, after which time the hire money ceases until the vessel is seaworthy again in all respects. This clause is also known as Breakdown Clause, *q.v.*

Office Bearer Chairman, Director, Secretary or Treasurer of a company or society as specified in its rules.

Official Book *or* **Official Log Book** *See* **Log Book.**

Official List A price list of stocks, shares, securities, etc. The official list is issued by the Stock Exchange, *q.v.*

Official Number The number allotted to all registered ships and yachts. *Abbrev.* o.n. or O.N *or* ON. *See* **Name of Ship**.

Official Quotations Part of the contents of the Official List, *q.v.*

Offing Towards the horizon. The most distant part of the sea that is visible by an observer.

Off-Licence A licence to sell alcoholic liquors for consumption off the premises only; a shop with such a licence.

Off-Load Another word for the discharge of cargo from a ship, or of air parcels to reduce the overall load on an aircraft.

Off-Peak Not at the time of highest demand. *See also* **Rush Hour**.

Off-Risk When the liability of the insurance risk ceases to exist.

Offshore (1) At a distance from the coast. (2) Relates to financial and investment operations by corporations established in countries or territories which offer special terms, concessions including tax exemptions, and facilities with the aim of strengthening their financial positions.

Offshore Activity Operations of various kinds conducted offshore, particularly in oil and gas exploration and extraction.

Offshore Installation Any type of Offshore, *q.v.*, structure including drilling rigs, sea platforms, etc.

Offshore Oil Drilling Drilling within territorial seas. *See* **Offshore Installation**.

Off Soundings Sea areas where no sounding can be taken with the deep sea lead and which are considered to be over 100 fathoms (600 feet).

Off-Trade Selling alcoholic products to supermarkets and Off-licence, *q.v.*, shops.

OFM *or* **O.F.M.** Ordino Fratrum Minorum *Latin*—Order of Friars Minor. *Franciscan Society.*

O.F.S. Orange Free State, *South Africa*.

OFT *or* **O.F.T.** Office of Fair Trading.

Oftel *or* **OFTEL** Office of Telecommunications.

O.G. *or* **o.g.** Original Gravity; Operating Gear; On Gross *insurance gross premium*; *Lloyd's Register notation* to indicate that the propeller shaft is fitted with Oil Glands. *See* **Gland Packing.**

O.G.Dk.P. Operating Gear Deck Place.

o.g.e. Operational Ground Equipment.

OGIL Open General Import Licence.

O.G.L. Open General Licence.

O.G.M. Ordinary General Meeting.

OGO Orbiting Geophysical Observatory.

OGP *or* **O.G.P.** Original Gross Premium, *q.v.*

OGPI *or* **O.G.P.I.** Original Gross Premium Income, *q.v.*

OGR *or* **O.G.R.** Original Gross Rate, *q.v.*

O.G.S Osservatorio Geofisico Sperimentale. One of the most important Italian organisations working in the field of applied geophysics and oceanography.

O/H Overzuche Handelsmaatschappij *Dutch*— Overseas Trading Company.

OH Ohio, *USA*.

o.h. Observation Helicopter; Office Hours; On Hand.

OHC Overhead Cam or Camshaft.

OHD Offene Handelsgesellschaft *German*—Normal Partnerships.

OHDETS Over Horizon Detection System.

Ohm. Ohmmeter, *q.v.*

Ohmmeter Instrument to measure electrical current and resistance. *Abbrev.* Ohm.

O.H.M.S. On Her/His Majesty's Service.

OHP Overhead Projector.

OHRG Official Hotel and Resort Guide.

OHV *or* **O.H.V.** Overhead Valve or Vent.

OI Office Instructions: Operating Instructions; Oceanology International *USA*.

OIC *or* **O. in C.** Officer in Charge.

OIA *or* **O.I.A.** Oil Import Association *USA*.

OIE International Office of Epizootics *animal diseases*.

OIL Orbiting International Laboratory.

Oil Bag An oval or triangular shaped canvas bag about 15 inches long which may be used to quell rough seas when oil is poured from it around a lifeboat or other small craft.

Oil/Bulk/Ore Carrier *See* **Ore/Bulk/Oil Carrier.**

Oil Engine *Lloyd's Register notation* indicating that the vessel is powered by a diesel engine.

Oiler Sometimes refers to an oil tanker.

Oilfield An area where the exploration for oils and/or gases has been successful.

OILPOL 1954 International Convention for the Prevention of Pollution of the Sea by Oil 1954, amended in 1962 and 1969 *IMO*. Now effectively superseded by MARPOL 73/78, *q.v.*

Oil Pollution The contamination resulting from the escape or discharge of oil from a vessel or any other property.

Oil Record Book A book or log kept by the master of an oil tanker wherin every discharge or escape of oil is recorded.

Oil Slick Film of oil on the surface of the water caused by accidental or illegal discharge and giving rise to pollution.

Oil Well A hole appropriately fitted for the extraction of oil.

Oils Liquids which may be classified in five different ways. (1) Petroleum base, extracted by means of heating distillation of crude petroleum producing petrol/gasolene, paraffin/kerosene, light fuel oils, lubricating oils, and thick asphalt. (2) Animal oils, e.g. Lard Oil. (3) Vegetable Oils, e.g. Linseed, tung, cotton seed, palm and castor oils. (4) Coal base, extracted by means of heating distillation of coal and shale producing benzol, etc. (5) Marine animal and fish oils, extracted from cod, sharks, seals, whales and other fish.

Oilskins Waterproof clothing, widely used by seamen as a protection from rain and seawater.

OIML International Organisation for Legal Metrology.

OIPH *or* **O.I.P.H.** Office of International Public Health.

oiro Offers in the region of . . .

o.j. Open Joint; Open Joist.

O.J.R. Old Jamaica Rum.

o.j.t. On Job Training.

OK *or* **O.K.** All Correct; Outer Keel; Oklahoma, *USA*.

o.K. Ohne Kosten *German*—Without Cost.

o.k.a. Otherwise Known As.

Okla Oklahoma, *USA*.

o.l. Overflow Level; Overhead Line.

OLB *or* **O.L.B.** Official Log Book. *See* **Log Book.**

OLBM Orbital Launched Ballistic Missile.

OLC Oil Liability Convention.

Old Bailey The Central Criminal Court in London, *Abbrev.* OB or O.B.

Old Glory The US stars and stripes flag.

Old Man The common name used by officers and crew of a ship when referring to their Master, *q.v.*

Old Salt Award An award given in Portland, Oregon, USA for outstanding service in the shipping industry. *Abbrev.* OSA.

Old World *See* **New World.**

Oleaginous Having properties of or producing oil; oily, fatty, greasy.

'O' Licence Operator's Licence.

Oligarchy A government or state governed by the few.

OM Air Malawi *airline designator*.

O.M. Of Merit; Old Measurement; Order of Merit.

O&M Organisation and Method.

o.m. Old Measurement.

omarb. Omarbetad *Swedish*—Revised.

OMB Observation Manipulator Bell, *q.v.*

O&MB Operation and Maintenance Budget *USA*.

OMBO One-Man Bridge Operation *navigation*.

OMCAS Outstanding Marine Claims Advice Scheme.

Omega Very Low-Frequency Beacon Navigation Aid.

Omnibus A passenger vehicle used for public services.

Omnibus Bill of Lading The American expression for Collective Bill of Lading, *q.v.*

Omnibus Clause A clause in a hull insurance policy extending liability cover to embrace, in addition to the assured's legal liability, the liability of other organisations who are connected with the ship. It usually excludes liability of shipyards, repair yards and others to whom underwriters do not wish to extend cover.

Omni Carrier Multi-purpose vessel capable of operating as a L1/L0, *q.v.*, RO/RO, *q.v.*, vehicle carrier, container ship, general cargo ship or bulk cargo carrier; and which can trade world-wide carrying any type of cargo.

Omnidirectional Beacons Aids to aircraft navigation. *Abbrev.* ODB.

o.m.s. Output per man shift.

OMSA Offshore Marine Service Association *USA.*

ON *or* **O.N.** *or* **o.n.** Official Number, *q.v.* On Net, *q.v.*

ONA Optical Navigation Attachment.

on a/c On Account.

On appro On Approval.

On Ballast *See* **In Ballast.**

On Board Courier Service An urgent consignment by air accompanied by a courier throughout the flight and which is delivered immediately during business hours at destination. Speed and security are the objects of this service. *Abbrev.* OCS.

On Consignment *See* **Consignment.**

On Cost *or* **On Cost Charges** *or* **Oncost Expenses** Fixed and variable overheads.

On Demand (1) A Bill of Exchange, *q.v.*, term demanding payment on presentation. *Abbrev.* O/D or o/d. (2) Any uncrossed cheque. (3) Articles which are popular in the market.

One Occurrence *See* **Occurrence.**

On Even Keel When the fore and aft draughts of a vessel are equal.

One Way Lease Arrangement where a Container is leased for the voyage but after arrival at its destination it must be returned to its original base or to such location as is agreed.

On Gross When the insurance premium is subject to a discount from the gross amount *Abbrev.* O.G. *or* o.g.

ONHRO Original Net Rate to Head Office. A term used in reinsurance, referring to the net rate of premium received by the reinsurer from the original assured.

ONI Office of Naval Intelligence.

On Line Rate An air cargo freight rate applicable only to one Specific Carrier as opposed to an Interline Rate.

ONM Office of Naval Material Command *USA.*

On Net When the insurance premium is subject to a discount from the net figure arrived at after other deductions. *Abbrev.* O.N. *or* o.n.

O.N.O. *or* **o.n.o.** Or Near Offer.

ONP *or* **O.N.P.** Original Net Premium, *q.v.*

On Passage In transit or on the way to.

ONPI *or* **O.N.P.I.** Original Net Premium Income, *q.v.*

ONR *or* **O.N.R.** Original Net Rate; Office of Naval Research *USA.*

On Right and True Delivery In some voyage charterparties the payment of freight is settled on completion of discharge. This procedure is a handicap on the owner or his agents since they cannot exercise any lien on the cargo if needs be. In some cases payment of freight is agreed to be effected at a percentage on the commencement of discharge and the remainder on completion or even concurrently with the discharge of the cargo.

On Risk When the insurance starts to take effect. *Opposite to* **Off Risk**, *q.v.*

Onshore Wind or sea which is coming towards the shore.

On Stern To be on the after part of a vessel. Opposite to the forepart (stem).

Ont. Ontario, *Canada.*

ONT *or* **O.N.T.** Ordinary Neap tide.

On the Berth (1) When the ship is berthed and ready to receive or discharge cargoes. (2) When a ship is advertised to receive cargoes at a scheduled time and she continues to receive cargo even when alongside the quay until her holds are full. *See* **Berth** and **Dock.**

On the Bow Nautical term. Direction between right ahead and abeam.

On the Quarter Nautical term. In a direction between abeam and right aft; 45 degrees abaft the beam.

One Bottom The whole lot. *Ex.* To ship the whole cargo in one bottom, i.e. all in one vessel.

One Man Company *See* **Private Company.**

One Minute of Latitude The angular distance on a meridian of latitude which is one sixtieth part of a degree. *See* **Latitude.**

One Safe Port *See* **Safe Port.**

Onus probandi *Latin*—The burden of proof. The most prominent canon of evidence is that the point in issue is to be proved by the party who asserts the affirmative, according to the civil law maxims, *ei incumbit probatio qui dicit, non qui negat; actori incumbit onus probandi;* and *affirmanti non neganti incumbit probatio.* The burden of proof is shifted when he has adduced sufficient evidence to raise the inference that what he alleges is true. The burden of proof lies on the person who has to support his case by proof of a fact which is peculiarly within his own knowledge, or of which he is supposed to be cognisant.

O/o On order; Order of; Oil/Ore Carrier.

O/O Oil/Ore Carrier; Order Of; Percentum *Latin*—Per cent or Per centage.

OO *or* **O.O.** Observation Officer; Office of Oceanography; Orderly Officer; Owner's Option *charterparty term*.

o/o/o *or* **O.O.O.** Out of Order.

OOC Ore/Oil Carrier.

OOD *or* **O.O.D.** Officer of the Deck; Offshore Oil Drilling.

OOG Out of Gauge; Officer of the Guard *naval*.

Oomiak Large open Eskimo boat.

OOW *or* **O.O.W.** Officer of the Watch.

o.p. Over Proof.

OP *or* **O.P.** *or* **o.p.** Open Policy. *See* **Floating Policy**; Opus *Latin*—Work; Out of Print; Over Proof; Osmosis Pressure.

OPA 90 Oil Pollution Act of 1990 *USA*; Office of Price Administration *USA*.

OPAL Optical Platform Alignment Linkage.

OPAS Occupational Pensions Advisory Service; Programme for Provision of Operational Assistance *UN*.

Op. cit. Opere citato *Latin*—In the work quoted.

OPCON Operational Control *USA*.

OPD *or* **O.P.D.** Out Patient's Department.

OPDAR Optical Detection and Ranging.

OPEC *or* **O.P.E.C.** Organisation of Petroleum Exporting Countries,

Open Account One which has not been stated or settled between the parties.

Open Berth A berth located in a roadstead or open to the elements.

Open Charter Where the charterparty specifies neither the kind of cargo nor the ports of destination.

Open Cheque One not crossed, payable either to bearer or to order. *Abbrev.* O.C. *or* O/C.

Open Coast A coastline considered to be safe for navigation.

Open Contract A complete contract of which the meaning admits the implications of law without special conditions, or except so far as such conditions may modify those implications, as a contract to sell land without mentioning the day for completion of the purchase, or without stipulations as to title or otherwise.

Open Court (UK) Every court of justice is open to every subject of the Crown. By statute the place where justices try offences summarily is an open court, but not so the place where they determine whether to commit a prisoner for trial at assizes or sessions. Whether a coroner's court is an open court is a matter of doubt. The general rule is that all courts of justice are open to all so long as there is room.

Open Cover A form of long term cargo insurance contract. It has no aggregate limit but, subject to a limit to the amount at risk in any one vessel, and often a limit to the amount at risk in any one location, the contract covers all shipments forwarded by the assured during the currency of the open cover. Underwriters have the right to cancel at any time by giving the requisite notice of their intention to cancel, but shipments that have commenced transit before the notice period expires continue to be covered until final delivery within the terms of the transit clause. *Abbrev.* O.C. or O/C

Open Credit A Letter of Credit, *q.v.*, requesting categorical payment. *See* **Clean Credit**.

Open Deck Ship's deck exposed to the weather when at sea.

Open-door Policy Opportunity for free trade.

Open Hatch Where a hatch cover opens wide both to port and starboard side and fore and aft leaving most of the hold easily accessible.

Open-hatch Bulker A ship with continuous hatches the length of the cargo space.

Open Hold Container Ship Container ship with a relatively high freeboard and no hatch covers, which can transport containers in continuous stacks from the tank tops to above the upper deck. Special draining and pumping arrangements cope with rainwater and seawater which may enter the holds. Also known in the USA as an Open Top Container Ship.

Open Indent *See* **Indent**.

Opening(s) *See* **Cargo Door** *or* **Cargo Port**.

Open Jaw Term used when the return destination of an aircraft is different from the place where the outward flight originated.

Open Market A free market for both buyers and sellers.

Open Outcry Bidding at auction in the markets for certain commodities. Admission to these public proceedings is sometimes limited to members of the same society as in the case of the Baltic Exchange, *q.v.*

Open Policy A cargo insurance policy designed to cover all consignments forwarded by the assured subject to a limit in any one vessel and, usually, a time limit during which declarations must attach. Unlike the floating policy it does not have an aggregate limit, but the underwriter can invoke a cancellation clause if he wishes to withdraw cover. *See*. **Floating Policy**.

Open Port A port without protection from the sea and unsafe in bad weather for lack of shelter.

Open Registry Shipping *Abbrev.* ORS. *Similar to* **Flags of Convenience**, *q.v.*

Open Sea Generally the sea or ocean clear of the land and navigational dangers.

Open Seas *Same as* **High Seas**, *q.v.*

Open Slip An Insurance Slip, *q.v.*, used to record cargo and parcels to be shipped up to the amount originally covered by the total premium.

Open Storage Pound or storage area without a roof to protect it from the weather. Cargo landing quays are normally open storage with merchandise unprotected until withdrawn by the Consignees, *q.v.* Cargo kept in open storages is usually such as will not be harmed by exposure to the elements, for example, vehicles, steel products (if oiled or coated), pipes, plates.

Open Water An area of navigable water where concentrations of sea ice cover less than a tenth.

Operating Costs Cost of production in running a business concern.

Operating Profit The total income of a business concern less the operating expenses.

Operational Research The study of various problems encountered in a business operation in order to find the best possible way to make profits.

Operation Job Card *or* **Job Card** A card with the number and the name allotted to each employee of a factory or large company which records the actual working hours covered per day/week/month and year.

OPEVAL Operational Evaluation *USA*.

OPF Overseas Project Fund *UK*.

OPH Ocean Patrol Hydrofoil *USA*.

OPIC Overseas Private Investment Corporation *USA*.

OPITB Offshore Petroleum Industry Training Board *USA*.

OPITO Offshore Petroleum Industry Training Organisation.

OPM Other People's Money; Office of the Prime Minister.

OPN Other Procurement *USA*.

opn. Option, *q.v.*

OPNAV Office of the Chief of Naval Operations *USA*.

Opp. Opposite.

OPP *or* **O.P.P.** Off-Lease Processing Platform *oil drilling*; Out of Print at Present.

Opportunity Lost The value of an alternative course of action foregone. Similar to Alternative Lost, *q.v.*

OPPR Oil Pollution Preparedness and Response.

OPRC The International Convention on Oil Pollution Preparedness, Response and Co-operation adopted by the IMO in 1990 and to take effect in 1995.

ops. Operations.

OPS Oceaneering Production Systems; Office of Pipeline Safety *USA*.

OPSP Office of Products Standards Policy *USA*.

OPT Other Productive Time *insurance*.

opt. Option, *q.v.*

Optimum The best or most favourable.

Option The privilege (acquired for a consideration) of calling for delivery or of making delivery of, or both, within a specified time, of some particular stock or article at a specified price to a specified amount. An option is valid if—which is not always the case—it is intended that the stock or article shall actually change hands.

Optional Insurance *See* **Standard Cover**.

Optional Stowage Arrangement for the stowage of an Optional Cargo, *q.v.*, so that it is readily accessible at the selected port without charges for shifting other cargo.

Oqud *Arabic*—Legal framework of a contract.

OR Oregon, *USA*; Operational Research.

O.R. *or* **O/R** Official Reference; Original Rate; Orderly Room; Other Rank *military*; Overriding Commission, *q.v.*; Operational Risk; Owners' Risk, *q.v.*

Orange Smoke Flare *See* **Marine Pyrotechnics**.

ORB *or* **O.R.B.** Oceanographic Research Buoy; Omnirange Beacons *aviation*.

ORBIS Orbiting Radio Beacon Ionospheric Satellite.

Orbit The route taken by a heavenly body around another body. *Ex.* The orbit of earth around the sun or the orbit of the moon around the earth.

ORC *or* **O.R.C.** Overseas Research Council; Owner's Risk of Chafing *insurance*.

ORD *or* **O.R.D.** Owner's Risk of Damage.

ord. bkge. Ordinary Breakage, *q.v.*

Order Mandate, precept; command; also a class or rank. In its simplest sense, an order is a mandate or direction. Thus, bills of exchange, cheques, etc., are said to be drawn to order when the payee is entitled to transfer the right to claim payment to any person whom he may direct. More commonly, however, order signifies a direction or command by a court of judicature, and directions or commands termed 'rules' are included in 'orders.'

Order Bill of Lading A Bill of Lading, *q.v.*, made 'To Order' which can be negotiated by any Consignee, *q.v.*, presenting it.

Ordinary Breakage Breakage of fragile cargo by its regularity has become accepted as inevitable loss during transit. *Abbrev.* ord. bkge.

Ordinary Leakage The natural loss occurring in liquid cargoes like petroleum, water from grains, logs, etc.

Ordinary Loss The normal and usual loss any businessman expects to encounter.

Ordinary Oils Oils having flashpoints ranging from 73 degrees up to 150 deg. F.

Ordinary Partner A person involved in business who is entirely responsible for all the debts incurred by the company as opposed to a Limited Partner, *q.v.*

Ordinary Seaman One of the members of a crew who, although he may have sailed for some time, has not had sufficient continuous service at sea to qualify as an Able Bodied Seaman, *q.v. Abbrev.* OS *or* O.S.

Ordinary Ticket A normal air, sea or rail ticket conferring no special advantages.

ord. lkge. Ordinary Leakage.

ord. w.t. Ordinary Wear and Tear.

Ore Mineral extracted from rocks containing a worthwhile amount of metals.

Ore. or **Oreg.** Oregon, *USA*.

Ore/Bulk/Oil Carrier A combination carrier, which has cargo compartments which may be used for the carriage of oil on one voyage and dry bulk cargo—(e.g. ore) on another.

Ore Carrier A dry cargo vessel constructed for the carriage of ore. Usually of heavy construction and with a small grain cubic capacity, *q.v.*, in the cargo compartments.

Ore/Oiler A combination carrier which has separate cargo compartments for oil and for ore.

OREDA Offshore Reliability Database.

Oreg. Oregon, *USA*.

Ores Solid native mineral aggregate from which valuable constituents, not necessarily metal, may be usefully extracted.

ORESCO Overseas Research Council.

OREVOY The BIMCO, *q.v.* Standard Ore Charterparty.

OREVOYBILL Bill of Lading for Shipments on the OREVOY Charterparty.

O.R.F. Owner's Risk of Fire *insurance*.

Or. F.S. Orange Free State, *South Africa*.

ORG or **O.R.G.** Operational Research Group.

ORI Operational Readiness Inspection *USA*.

Orient Oriental.

Original Gross Premium The premium charged by the reinsurer to the insured without reductions. *Abbrev.* OGP or O.G.P.

Original Gross Premium Income A reinsurance phrase regarding the gross and net income arrived at by the reinsurer, before quoting for the premium to the insured, *Abbrev.* OGPI or O.G.P.I.

Original Gross Rate Reinsurance phrase referring to the gross premium charged to the insured, *Abbrev.* OGR or O.G.R. Sometimes also called Gross Original Rate and *abbrev.* G.O.R or GoR.

Original Net Premium Income The net premium income due to the reinsurer. *Abbrev.* ONPI or O.N.P.I.

Original Slip An original short and abbreviated note presented by insurance brokers to Underwriters, *q.v.*, for acceptance or otherwise. *Similar to* **Slip**, *q.v.*

Originating Bank A bank from which documents have originated or have been despatched to the corresponding bank.

Origo mali *Latin*—The origin of evil.

ORIL Open Reversible Inflatable Liferaft.

ORIMS Ontario Risk and Insurance Management Society. *See* **RIMS.**

Ork or **Orkn.** Orkney Islands.

ORL or **O.R.L.** Owner's Risk of Leakage *insurance*.

Orlop The lowest deck of a ship.

ORR or **O.R.R.** Owner's Risk Rate *insurance*.

ors Others; The other members of a Lloyd's Syndicate represented by an Underwriter.

ORS Open Registry Shipping. *Similar to* **Flags of Convenience**, *q.v.*

ORSA Operations Research Society of America.

Orsat Apparatus Instrument used to measure the gas emitted by the exhausts of engines.

Ortaklari *Turkish*—Partners.

OS Austrian Airlines *airline designator*.

OS or **O.S.** Off Slip; Open Slip, *q.v.*; Opposite Side; Ordinary Seaman, *q.v.*; Old Style, Outstanding.

O/S or **O/s** Ocean Station. *See* **Ocean Station Vessel**; Off Slip; Open Slip; On Sale or return; On Sample; On the Spot; Out of Stock; On Sale.

OSA Ocean Shipping Act 1980 *USA*: Official Secrets Act; Old Salt Award, *q.v.*

Osakeyhtio *Finnish*—Limited Company. *Abbrev.* O/Y or OY or Oy.

OSAS Overseas Service Aid Scheme.

OSCAR Ocean Surface Current Radar, *q.v.*; Orbiting Satellite Carrying Amateur Radio *USA*.

Oscar International Radio Telephone phonetic alphabet for 'O'.

OSCE Organisation for Security and Cooperation in Europe, formerly the CSCE, *q.v.*, with 53 participating states (1996).

OSD or **O.S.D.** Open Shelter Decker.

OSD/CSD or **O.S.D.C.S.D.** Open Shelter Deck/Closed Shelter Deck. Open Shelter Decker/Closed Shelter Decker.

O-seas Overseas.

OSG *Lloyd's Register notation* for Offshore Services Group.

OSI Open Systems Interconnection. Model of data Communications established by ISO, International Standards Organisation of the UN.

o.s.l. On signed lines *insurance*.

Osl. Oslo, Norway.

O/S L.Res. Outstanding Loss Reserve *re-insurance*.

Osmosis Tendency to percolation and intermixture of fluids separated by porous septa *physics*. *See* **Reverse Osmosis.**

OSO Orbiting Solar Observatory.

OSP or **O.S.P** One Safe Port *charterparty*.

OSPAR Oil Spill Preparedness and Response. Profect established in 1993 to give assistance in setting up bases in the ASEAN countries for oil spill equipment.

OSR Optical Still Recorder; On Signing and Releasing *referring to the payment of freight to be effected when the bills of lading are signed and released*.

OSR B/L *Similar to* **OSR**, *q.v.*

OSR B/L IN FTC On Signing and Releasing Bills of Lading in Free Transferrable Currency. Similar to **OSR**, *q.v.*, and the money paid for freight must be free currency with no restrictions.

OSRD Office of Scientific Research and Development.

OSRO Oil Spill Removal Organisation.

OSRV Oil Spill Recovery Vessel.

OST *or* **O.S.T.** Offshore Storage and Treatment *oil drilling*; Ordinary Spring Tide(s).

OSV Offshore Support Vessel.

OSV *or* **O.S.V.** Ocean Station Vessel, *q.v.*

OT *or* **O.T.** Oil Tight; Old Testament; Overtime; Ordinary Tide(s) *nautical*.

O/t Oil Tight; Old Term *in grain trade*; On Truck; Overtime.

OTA Office of Technology Assessment *USA*.

OTAR *or* **O.T.A.R.** Overseas Tariffs and Regulations.

OTC Officer in Tactical Command *naval*; Organisation for Trade Co-operation.

OTC *or* **O.T.C.** *or* **o.t.c.** Offshore Technology Conference; Over the Counter; Officers' Training Corps; Overseas Trading Corporation.

OTCO Over the Counter Options *payment contract*. See **Futures Contract.**

OTD *or* **O.T.D.** Ocean Travel Development; Overseas Travel Development *passenger-sea travel organisation*.

OTE Opportunity to Earn *measure of salesman's potential remuneration*.

Other Ship's Cargo An Insurance, *q.v.*, collision clause indemnifying a third party.

Other Term The insured is covered for any damages sustained to other cargoes.

OTS Orbital Test Satellite.

Ott. Ottawa, *Canada*.

Otto Cycle Engine Engine performing four stroke actions, namely, suction, compression, explosion and exhaust, per cycle of two revolutions.

OTU Offshore Technical Unit.

OTV Orbital Transfer Vessel.

O.U. Offshore Unit; Offshore Use; Oxford University.

Ouguiya Currency unit of Mauretania.

OUK *or* **O.U.K.** Out of United Kingdom, *q.v.*

Ounce *Abbrev.* oz. A unit weight. 16 ozs make 1lb.

Out Voice message on radio telephone indicating that transmission is finished and no reply is expected.

Outbd. Outboard, *q.v.*

Outboard External side of the ship's hull as opposed to inboard, the inner side. This can also refer to the engine or engines of a launch which are attached to the stern.

Outboard Profile Ship's outline plan, detailing masts, rigging, funnels, deck erections, etc.

Outbound Freight Freight collected or to be collected on outward bound cargo. Also expressed as **Outward Freight.**

Outbound Cargo *See* **Outward Cargo.**

Outer Bottom The outer shell of a double-bottom ship.

Out of Commission Opposite to In Commission, *q.v.*

Out of Date Cheque Stale cheque. When a certain number of months elapse from the date of the cheque it cannot be cashed. Some bankers accept a limit of three months and others up to six months.

Out of Pocket Expenses Casual and minor expenses met from cash in hand.

Out of Trim Ship not adequately trimmed or ballasted.

Out of United Kingdom Code name for foreign seamen engaged on British ships. *Abbrev.* OUK or O.U.K.

Output The result of production by labour or other means.

Out Shipment Cargo or passengers not accepted for loading or embarkation before sailing due to shortage of space. The term Shut Out is also used. *See* **Short Shipment.**

Outturn The quantity of a cargo discharged from a ship and its condition as recorded by tallying.

Outward Bill of Lading This is simply a Bill of Lading *q.v.*, where goods are actually being exported to another country and not to a port of the same country. *Abbrev.* Outward B/L.

Outward B/L Outward Bill of Lading, *q.v.*

Outward Cargo *or* **Outbound Cargo** Cargo loaded for export.

Outward Freight Similar to Outbound Freight, *q.v.*

Outward Manifest A form of a customs manifest declaring the cargo shipped outwards. *See* **Manifest.**

OV *or* **O.V.** Observed Vessel; Orbiting Vehicle.

O/V On Voyage.

Ov. Bd. Overboard.

OVC *or* **O.V.C.** Overcast *weather observation*.

Overall Length Length between two extreme ends. *Abbrev.* OAL.

Over and Out The very last words used and pronounced in radio-telephony to signify that the message is concluded and no more conversation is intended.

Over-Deck The stowage of cargo on deck, as opposed to under deck, i.e. in a hold or tweendeck. For insurance and contract of carriage purposes, goods are assumed to be stowed under deck except where custom or regulations, as with some hazardous products, dictates otherwise.

Overdraft The amount a customer has been permitted by his bank to draw in excess of the money paid in. Security for repayment is usually demanded. *Abbrev.* OD or O.D.

Overdue Past the time of payment.

Overdue Ship A ship of which news has not been received for such a time as to give rise to the presumption or probability that she has been lost.

Overflight An aircraft flying over a specific land or sea area.

Over Freight *Opposite to* **Dead Freight**, *q.v.* Cargo which is loaded over and above the agreed amount shown in the Charterparty, or cargo bookings. Naturally the charterer or the shipper pays the Over Freight on a pro rata basis as per charter rate.

Overhead Charges *or* **Overhead Expenses** Expenses over and above direct expenses, such as taxes, rent, salaries, rates, water/electricity bills, etc.

Overheight Cargo Container cargo whose dimension exceeds the height of the Container, *q.v.* It may be accepted under certain conditions.

Over-insurance When a risk is insured for more than its actual value. In the event of a loss the insurers will refuse to pay more than the actual value of the subject-matter.

Overland Describes a travel or transport service or what is carried across a land mass by road or rail.

Overlap When the chartered ship is redelivered to her owners after the stipulated date.

Overlay A descriptive term for **Demurrage**.

Overnight (1) In the course of the night; Travel at night; A stop for the night. (2) Very short-term deposit in a bank.

Overprice To put the price higher for larger than normal profit.

Over-Price Arrangement A mutual arrangement arrived at between the principal or exporter and his representative agent whereby the difference of the higher price of the merchandise secured by the latter is to be shared with the principal or as prearranged.

Overproof Containing more alcohol than does proof spirit.

Overrider Overriding Commission, *q.v.*

Overriding Commission Commission additional to the ordinary one to compensate for extraordinary expenses involved in the nature of the work. *Example.* A Commission Agent, *q.v.*, gets 10% commission on the invoice value of goods he orders. But because of extra work performed and labour costs the Principal,

q.v., allows him a further $2\frac{1}{2}$% overriding commission.

Over-Subscribed Shares When more shares are applied for than actually needed. The normal method applied in this case is either (a) The applicants are arranged on a pro rata basis or (b) on a first-come first-served basis. *See* **Fully Subscribed.**

Overt Open to the public. See **Market Overt.**

Overture The start or opening of negotiations.

Ov. Fl. Overflow.

O.W. Ohne Werte *German*—Without Value.

O/W *or* **o/w** Oil in Water; One Way.

OWE *or* O.W.E. Operating Weight Empty *aviation.*

OWL Original Written Line *insurance*; On Written Line.

Owner's Declaration A shipowner's declaration to confirm his right of shares in the property of the ship. The ship is thus registered on the strength of this declaration.

Owner's Risk The Carriage of Goods Acts 1924 and 1936, *q.v.*, imposes certain responsibilities on the owner of the vessel. However in some cases the cargo owners or Consignees, *q.v.*, become liable for damage in transit between ship and warehouse and they may alternatively be insured against such claims.

OWS *or* **O.W.S.** Ocean Weather Ship; Ocean Weather Station.

OWV *or* **O.W.V.** Ocean Weather Vessel. *See* **Ocean Station Vessel.**

OXFAM Oxford Committee for Famine Relief.

Oxon. Oxford; Oxfordshire; Oxonia *Latin*—Oxford; Oxonienis *Latin*—of Oxford.

Oxter Plate In a hull's construction, a specially shaped plate above the Stern Post, *q.v.*, where the side plating ends in way of the after Deadwood.

O/Y Osakeyhtio *Finnish*—Limited Company.

oz. Ounce, not now in scientific use.

oz. ap. Apothecaries' Ounce.

oz. av. *or* **oz.** Avoirdupois Ounce.

Ozone Allotropic form of oxygen which is naturally formed between 15 and 30 kilometres above the earth's surface. It protects the earth from strong solar radiation through the formation of layers that regulate the climate. Symbol O_3.

P

p pico: 10^{12} *physics*.

p. Polar Distance.

P *Lloyd's Register notation* indicating that the ship can carry more than 12 passengers. *See* **Passenger Ship**; Symbol representing phosphorus; Departure flag letter P also called Blue Peter; Page; Paid; Parking; Parallax; Parallel; Pastor; Peso, *q.v.*; Peseta, *q.v.*; Planet; Pole; Polar Distance; Atmospheric Pressure; Wave Period *water*; President; Prince.

Pa Pascal *unit of pressure equal to one newton per square metre*.

Pa. *or* **PA** Pennsylvania, *USA*.

PA *or* **P.A.** Paddle; Particular Average, *q.v.*; Per Annum; Performance Appraisal; Personal Accident; Personal Assistant; Position Approximate; Poop; Pour ampliation *French*—True Copy; Press Association; Public Address, *q.v.*; Publishers Association.

P/A Particular Average, *q.v.*; Personal Account; Private Account; Power of Attorney, *q.v.*

p.a. *or* **per an.** Per annum *Latin*—yearly.

PAA Pan American Airlines.

PABX *or* **P.A.B.X.** Private Automatic Branch Exchange.

PAC Parents Advisory Committee; Public Accounts Committee.

Pac. Pacific; Pacific Coast Ports.

PACE Port Automated Cargo Environment.

Pace Performance and Cost Evaluation.

PACIS Port and Cargo Information Service.

Pac.Is. Pacific Island.

Pack *Lloyd's Register notation* for Package.

Pack Ice Large areas of pieces of floating ice of a considerable size.

Package/s A general shipping term for cases, crates, bundles, bales, bags, kegs, drums, barrels, etc.

Package Cargo Cargo consisting of bags, bales, cartons, crates, cases, drums, kegs, wrapped packages, etc. Also called Package Freight.

Package Deal A special deal or sale offer for a group of items, passengers, products or labourers contracted as a whole.

Packaged Timber Timber packed in bundles. Bundles of equal length pieces of timber are known as length packaged. Bundles of unequal lengths of timber pieces are known as truck packaged.

Package Freight *Similar to* **Package Cargo**, *q.v.*

Package Insurance Combination insurance against fire, marine and other risks very much in use in the USA. *See* **Package Policy**.

Package Policy Insurance policy covering Package Insurance, *q.v.*

Packed Ice *Similar to* **Hummocked Ice**, *q.v.*

Packetfahrt–gesellschaft *German*—Packet Company.

Packet Port *or* **Ferry Port** A port invariably made use of by ferry boats regularly plying between ports and embarking and disembarking passengers as well as loading and unloading passenger/commercial vehicles by means of drive on/drive off methods.

Pact Contract of Agreement.

PAD Packet Assembly/Disassembly *computer*.

PAD *or* **P.A.D.** Payment Against Documents. *See* **Documents Against Payment**; Payable After Death; Predicted Areas of Danger *nautical*.

Padar Passive Detection Ranging.

PADC Philippine Aerospace Development Corporation.

Paddle Implement, shorter than an oar, with a flat blade, used without a rowlock to propel a boat; the act of using a paddle.

Paddy Rice with husks. When the husks are removed it is then called clean rice.

P.Adr. Per Adresse *German*—Care of.

Padrone *Italian*—A master of a vessel whose certificate enables him to command a ship in coastal waters and within restricted open seas; A business operator.

PAEC Philippines Atomic Energy Commission.

P.af. Puissance au frein *French*—Brake Horse Power.

PA & GAL *or* **P.A. & G.A.L.** Particular Average & General Average Loss *marine insurance*.

PA &/or GAL & DEP Particular Average and/or General Average Loss and Deposit.

PAGC Port Area Grain Committee.

PAHO Pan American Health Organisation.

PAI Personal Accident Insurance.

Paid Freight *Similar to* **Advanced Freight**, *q.v.*

Paid Up Capital The actual money paid by the shareholders of a company or firm to form the capital. *See* **Paid Up Shares**.

Paid Up Shares The actual money paid by the shareholders. *See* **Paid Up Capital**.

Painter Length of small rope made fast at the stem of a boat, used to secure the boat to a ship or alongside a berth, wall, jetty or landing stage.

Paint Locker A locked compartment quite often in the fore part of a ship where paints and similar items liable to pilferage are stored for protection.

Pak. Pakistan; Pakistani.

Paketvaart Maatschappij *Dutch*—Packet Company. *Abbrev.* Paket. Maats.

PAL *or* **P.A.L.** Phase Alternating Line *colour T/V system*; Particular Average Loss. *See* **Particular Average**; Philippine Air Lines; Prisoner At Large.

Palletisation The loading of goods onto pallets. The process is used for a great variety of durable and perishable cargoes and there are ships specially adapted as Pallet Carriers.

Pallets A flat wooden frame on which goods can be stored and secured—e.g. bagged goods.

PAMI Personnel Accounting Machine Installations *USA*.

PAN Premium Advice Note.

Pan. Panama.

PANAM Pan American World Airlines.

Panamerica Northern, Central and Southern States of America.

Pan Can Panama Canal.

Pandi Club *or* **P. & I. Club** *See* **Protection and Indemnity Club**.

P and L *or* **P&L** Profit and Loss.

P and L A/C *or* **P&L A/C** Profit and Loss Account.

P.A. *and/or* **GAL & DEP** Particular Average and/or General Average Loss and Deposit. *See* **Particular Average** *and also* **General Average Deposit**.

PANS *or* **P.A.N.S.** Procedures for Air Navigation Services.

Panamax A term used to describe vessels designed to the maximum beam (a shade over 32.2 metres) to transit the Panama Canal, e.g. Panamax bulk carriers.

Panting Flexing, working in and out or oscillations of the hull plating, particularly at the bow, as a ship pitches when driven into a heavy sea. *See* **Panting Beams**.

Panting Beams Strengthening built in near the bow of a ship against Panting, *q.v.*

Pantograph An instrument for copying a plan, etc., on any scale.

PANTRAC Pan American Tracing and Reservations System.

P.A.P. Port Autonome de Paris *French*—Port of Paris Authority.

Papa International Radio Telephone phonetic alphabet for 'P'.

Paper Bid An offer for Shares, *q.v.*, in the process of take-over bidding.

Paper Money Bank notes, bills of exchange, and promissory notes. A system used the world over

involving the issue of paper money in place of gold and silver coins as in the past. The paper has no value in itself but there is a declaration that the value will be honoured.

Papers, Ship's *See* **Ship's Papers**.

PAPI Precision Approach Path Indicator, *q.v.*

PAR *or* **P.A.R.** Port Autonome de Rouen *French*—Port of Rouen Authority; Project Area Ratio; Precision Approach Radar.

Par Equality or equal footing especially par rate of exchange—the recognised value of one country's currency in terms of another's; Paraguay; Precision Approach Radar.

Para Paragraph; Paraguay.

Parachute Flares and Rockets *See* **Marine Pyrotechnics**.

Paragrapher A vessel constructed with a gross tonnage marginally below a significant tonnage size at which more stringent regulations come into force.

Parallax Apparent displacement of an observed heavenly body due to the position of the navigator between two different points of observation. *Abbrev.* Plx.

Parallel Importation Importation of the same branded product from more than one origin or source.

Parallels of Latitude The angular distance of places North or South of the Equator, *q.v.* It is measured in degrees, each degree representing 60 geographical or nautical miles.

Paravane A torpedo-shaped device towed at a depth regulated by its vanes or planes to cut the moorings of submerged mines.

Par Avion *French*—By airmail; By aircraft.

Parbuckle A rope for raising or lowering casks and cylindrical objects, the middle being secured at the upper level, and both ends passed under and around the object and then hauled or let slowly out.

Parcel Tankers Tankers equipped with a variety of cargo tanks—either coated and/or stainless steel—and able to carry a variety of chemical and natural liquids at the same time.

Parceria *Portuguese*—Partnership.

Parent Company A company in control of other subsidiaries. Alternatively known as a Holding Company, *q.v.*

Par excellence *French*—Pre-eminently.

Par exemple *French*—For Example.

Pari passu *Latin*—At the same rate; on an equal footing.

PARS Programmed Airline Reservation System *electronics*.

Pars *or* **Para/s** Paragraph(s).

Pars pro toto *Latin*—A part for the whole.

Partafelag *Faröese*—Limited Company.

Partenreederei *German*—Part Shipowners.

Part Owners Persons having claim to the possession or ownership of a vessel and each having a stated share in the company.

Partial Loss In the law of Marine Insurance, *q.v.*, the losses which arise from the various perils insured against may be either total or partial; they are total when the subject-matter of the insurance is wholly destroyed or injured to such an extent as to justify the owner in abandoning it to the insurer, and partial when the thing insured is only partially damaged or where, in the case of an insurance on goods, the owner of them is called upon to contribute to a General Average, *q.v. See* **Abandonment** *and* **Loss**.

Particular Average A fortuitous partial loss to the subject-matter insured, proximately caused by an insured peril but which is not a general average loss.

Particular Charges An expense incurred by an assured in relation to an insured loss. This can be a means of preventing further loss in transit (e.g. sue and labour charges), of assessing loss (e.g. survey fee) or making good a loss at destination (e.g. repacking); sometimes referred to as an 'extra charge' when it is the subject of a claim on the policy.

Partim *Latin*—Partly; in part.

Partner, Active *See* **Active Partner**.

Partner, Nominal *See* **Nominal Partner**.

Partnership The relationship which exists between persons carrying on a business with a view to profit.

Partrederi *Danish, Faröese, Norwegian, Swedish* —Part Shipowners.

Parts per million The measure of a low concentration of a substance when mixed in another as millionth parts of the whole. *Abbrev.* ppm.

PASI Personal Accident and Sickness Insurance.

pas Power-assisted steering.

PAS Port Auxiliary Service.

Passage Passing, transit; transition from one state to another; liberty, right to pass through; voyage, crossing, from port to port; journey by air, sea or land.

Passage Freight The bare minimum charged on cargo for the sake of earning money to offset expenses on a ballast voyage.

Passage Money The fare paid by the propective passenger.

Passage Risk *See* **Delivery Trip**.

Passage Ticket The ticket or coupon issued by a shipping company, airline, railway company, etc., which entitles the holder to travel. It can show various details, depending on the type of travel to which it applies.

Passageway A corridor on board a ship.

Passenger A traveller in a public conveyance by land or sea or air.

Passenger-mile *or* **Passenger-kilometre** A method of calculating the revenue earned by the carrier of passengers over particular distances, mostly used in aviation.

Passenger Return *or* **Passenger List** A list showing the names of travellers on a ship which is required for various official purposes. *See* **Breakdown**.

Passenger Ship A ship that is authorised to carry over twelve passengers.

Passenger Throughput Recorded number of passengers passing through an air, rail or sea terminal in a given period.

Passenger Toll Port authority charges imposed on every passenger embarking and/or disembarking.

Passim *Latin*—Here and there throughout.

Passing Off The wrong committed by a person who sells goods or carries on business, etc., under such a name, mark, description or otherwise in such a manner as to mislead the public into believing that the goods or business, etc., is that of another person. The latter person has a right of action in damages or for an account, and for an injunction to restrain the defendant for the future.

Passing Over Quay A saying in shipping where arrangements are made by the port authorities or companies to give direct cargo deliveries from the ship's hold on the the quays. Cargo is normally stored in transit sheds. *Abbrev.* POQ *or* P.O.Q. *or* poq *or* p.o.q.

Passing-ticket A kind of permit, being a note or check which the toll-clerks on some canals give to the boatmen, specifying the lading for which they have paid Toll, *q.v.*

Passive Balance A balance of payments that is in deficit.

Passive Debt A debt upon which, by or without agreement between the debtor and creditor, no interest is payable, as distinguished from an Active Debt, i.e., a debt upon which interest is payable. In this sense the terms 'active' and 'passive' are sometimes applied to debts due from defaulting foreign governments. This use of the terms is to be distinguished from the French use of *actif* (assets) and *passif* (liabilities).

Passport Control A customs and police control section at international air or sea terminals.

Past Debt A debt which is in existence before the doing of some subsequent act in the law relating to it, e.g., a debt which is secured by a promissory note given after the advance had been made and not when it originally was made.

PA System Public Address System.

PAT Port Authority of Taiwan; Professional Association of Teachers.

Pat. Patent.

Pata. Patagonia.

Pataca Currency unit of Macao.

Patent-right The exclusive privilege granted by the Crown to the first inventor of a new manufacture of making articles according to his invention.

Patents (UK) A patent means letters patent for an invention. A patent is obtained by making to the

Patent Office an application accompanied by a specification. If the specification is provisional, the complete specification must be lodged within 12 months of the date of the application, or such further time not exceeding three months as the Comptroller may direct. The provisional specification secures priority as from the date of the application: for, on acceptance, every patent is sealed as from the date of the application.

A patent, when granted, is sealed with the seal of the Patent Office. Its normal duration is 16 years but this may be extended by five, or in exceptional cases ten, years on the ground of inadequate remuneration, or war loss. Patents are assignable, and assignments, as well as licences, are to be registered in the register of patents. In addition to licences which the patentee may grant, there are, since 1919, licences which any patentee must grant, upon certain terms, if his patent is endorsed by the Comptroller with the words 'licence of right'. Such endorsement is one of the means of preventing abuse of the monopoly given by the patent.

The remedy for infringement of a patent is an action for damages or an injunction, or both.

A patent will be granted only for something which is both novel and of utility; but utility means merely utility for the purpose indicated by the patentee, although that purpose may itself be useless.

Patent Slip A cradle on inclined rails from the level of low spring tides to above that of high spring tides on which a small vessel can be supported and hauled out of the water for bottom cleaning and repairs.

PATO Pacific Area Travel Association.

Patt. Pattern.

P.Av. Particular Average, *q.v.*

PAW Powered All The Way.

Pawnbroker A person who accepts articles in a form of a guarantee for money he lends. *See* **Pledge**.

PAX Private Automatic Exchange.

Pax. Parallax *nautical*; Passenger.

Payable A sum of money is said to be payable when a person is under an obligation to pay it. 'Payable' may therefore signify an obligation to pay at a future time, but when used without qualification 'payable' means that the debt is payable at once, as opposed to 'owing'.

Payable Abroad Agents Authorised representatives of the Underwriters, *q.v.*, who are involved in settling claims abroad.

Payable to Bearer A phrase put on Bills of Exchange, *q.v.*, and bank cheques which have to be endorsed by the bearer himself to be cashed or transferred.

Payable to Order A phrase typically used on Bills of Exchange, *q.v.*, and bank cheques on which the payee's name is inserted. The difference between Payable to Order and Payable to Bearer is that the former can be cashed by anyone endorsing it whereas the latter has to be endorsed by the person to whom it is addressed.

Pay as Paid Settlement of the buyer's account with his supplier as and when he is paid by his clients.

Pay As You Earn The name given to the income tax system whereby the particular amount of tax levied on the taxpayer's income is automatically deducted at source. His or her employer accounts to the Revenue for the amount so deducted.

Payback Method A way of comparing the profitability of alternative projects. The object is to determine over what period the net cash generated or saved by an investment will repay the cost of the project. *See* **Payback Period**.

Payback Period The number of years required to restore the financial balance with regard to a capital outlay. *See* **Payback Method**.

Payee One to whom a bill of exchange or promissory note or cheque is made payable; he or she must be named or otherwise indicated therein, with reasonable certainty.

Payer The person who pays; the person who draws a cheque or bill of exchange.

Paying Bank The bank which is requested to make a payment by cheque, remittance or transfer. This might be on the instructions of a client to his bank to disburse a sum of money to another person or bank abroad through another local bank.

Paying-in-Book A bank register book for customers to record all cash and cheques deposited in their account with the bank.

Paying-in-Slip A banking system whereby a slip of paper is filled in and signed for by the depositor. Similar to Paying-in-Book, *q.v.*

Payload An aviation term. The maximum revenue that can be obtained in the carrying capacity of passengers and air freight on an aircraft.

Payment A transfer of money from one person (the payer) to another (the payee). When made in pursuance of a debt or obligation it is sometimes called payment in satisfaction.

Payment in fact is an actual payment from the payer to the payee; payment in law is a transaction equivalent to actual payment; thus, payment in fact by a debtor to one of two or more joint creditors is payment in law to all; and retainer, set-off, allowance in account, acceptance of security, goods or other means of obtaining actual payment, and payment into court, are said to be equivalent to payment, because they produce a satisfaction of the debt.

Payment in satisfaction is said to be absolute when the debt is completely discharged, as by payment of cash without stipulation; conditional payment is where the debt may afterwards revive if the mode of payment does not result in actual payment, as where a creditor is paid by a cheque or bill which is afterwards dishonoured; whether the acceptance by the creditor

Payment Against Documents

of a negotiable security operates as conditional or absolute payment is a question of fact in each case.

The payment of money before the day appointed is in law payment at the day; for it cannot, in presumption of law, be any prejudice to him to whom the payment is made to have his money before the time; and it appears by the party's receipt of it that it is for his own advantage to receive it then, otherwise he would not take it.

In the law of bills of exchange, payment for honour is where a person pays a dishonoured bill for the honour of some one of the parties. The payer has the rights of a holder against the person for whose honour he has paid, and against all antecedent parties, but the subsequent parties are discharged.

Payment Against Documents *See* **Documents Against Payment**.

Payment in Kind Settlement of payment effected in goods or other services instead of actual money. *Abbrev.* PIK.

Payment into Court The deposit of money with the official of or banker to the court for the purpose of proceedings commenced in that court. Payment into court is not strictly a defence; it is rather an attempt at a compromise. No such plea was known to the common law; it is entirely the creature of statute.

Payment on Account Part payment of a full account.

Pay Off To discharge and pay all wages due to the crew by the master after the ship's articles are terminated. This procedure is generally undertaken in the presence of the Shipping Master, *q.v.*, or Consul, *q.v.*, of the registered flag of the ship.

Pay Out To slacken or ease out a cable, wire, rope or chain *nautical*.

Payroll Detailed records of the conditions regarding wages, salaries, names and rank of each employee in a company. *Abbrev.* PR *or* P.R.

Pays du droit coutumier *French*—Country of the law established by custom and usage.

Pays du droit écrit *French*—Country of the written law.

Pay Slip A slip of paper enclosed in an envelope and handed to employees on Pay Days, giving a detailed account of the respective wage or salary.

Payt. Payment.

Pb Symbol representing the element of lead.

P.B. *or* **PB** Permanent Bunkers; Pharmacopoeia Britannica.

PBA Paid by Agent(s).

p.b.f. Poop, Bridge and Forecastle. *See* **Three Island Ship**.

PBAS Personal Buoyancy Aids Standard.

PBH *Lloyd's Register notation* for Partial Bulkhead.

PBPC Passenger and Baggage Processing Committee.

PBR *or* **P.B.R.** Payment by Results.

PBS *or* **P.B.S.** Panama Bureau of Shipping.

PBT *or* **P.B.T.** President of the Board of Trade; Profit Before Tax.

P.b.Wt. Parts by weight.

P.B.X. *or* **PBX** Private Branch Exchange.

p.c. Per centum *Latin*—Per hundred or every hundred; Prime Cost, *q.v.*; Part Cargo; Postcard; % Per cent.

PC Politically Correct.

PC *or* **P.C.** Part Cargo; Passenger Certificate; Per Capita; Per Cent; Per Compass; Per Contra; Personal Connection; Personal Computer; Platform Crane; Police Constable; Postcard; Prime Constructor, *q.v.*; Privy Council; Profit Commission; Propulsive Coefficient *shaft horse power*.

P/C Particular Charge, *q.v.*; Per Cent; Petty Cash; Price Current, *q.v.*, Profit Commission.

PCA *or* **P.C.A.** Permanent Court of Arbitration; Portland Cement Association; Precondition Air System which involves connection between a GPU, *q.v.*, and a parked aircraft.

PCB *or* **P.C.B.** Petty Cash Book; Private Car Benefits; Printed Circuit Board *electronics*; Polychlorinated Biphenyl.

PCC *or* **P.C.C.** Panama Canal Company; Pour copie conformée *French*—Certified True Copy; Pure Car Carrier, *q.v.*; Panama Canal Commission.

PCD *or* **P.C.D.** Polar Cap Disturbances.

PCEC Pacific Coast European Conference.

P.C.F. *or* **p.c.f.** Pounds per Cubic Foot; Power per Cubic Foot.

PCG Patrol Chaser Missile *USA*.

P.Chgs Particular Charges.

Pcl Parcel.

PCM Production Choke Module *electronics*; Pulse Code Modulation *electronics*.

P.Cmdr. *or* **P.Cr.** Paymaster Commander.

PCMSG Pacific Coast Marine Safety Code.

PCNT Panama Canal Net Tons.

PCO Pest Control Operator, *q.v.*; Prospective Commanding Officer *USA*; Principal Control Officer *naval*; Professional Committee/Council Organiser.

PCP Potential Collision Points *electronics*; Production Control Pod *electronics*.

PCRV Prestressed concrete reactor vessel, *USA*.

Pc/s Price(s).

PCT Pacific Coast Terminals; Polychlorinated triphenyl.

PCTC Panama Canal Tonnage Certificate.

PC/TC Pure Car/Truck Carrier.

PCU Programmable Control Unit *electronics*.

P.C.U. *or* **p.c.u.** Power Control Unit.

PCV *or* **P.C.V.** Power Control Vehicle.

PCVT Prestressed Concrete and Steel Pressure Construction.

PCZ Panama Canal Zone.

PCZST Panama Canal Zone Standard Time.

PD *or* **P.D.** Polar Distance *nautical*; Port Dues, *q.v.* Power Density; Position Doubtful *nautical*; Property Damage.

Pd *or* **pd.** Paid; Past; Per Day.

P/D Port of Disembarkation.

PDA Personal Digital Assistance *computer*.

PDB Pipe Drive Bushing *oil drilling*.

PDC Polycrystalline Diamond Cutter, *q.v.*

PD *or* **p.d.** Per Day; Per diem *Latin*—per day; Pharmacopoeia Dublinensis; Poop Deck; Port Dues; Post Dated, *See* **Postdated**; Position Doubtful *nautical*; Postal Dues; Potential Difference.

P. de C. Pas de Calais, *France*.

PDG *or* **P.D.G.** Paymaster Director General; President-Directeur General *French*—President Managing Director.

PDI Personal Disposable Income.

P.D.I. *or* **p.d.i.** Pre-Delivery Inspection.

P.Dk. Poop Deck.

pdl Poundal, *q.v.*

PDM Passive Distribution Manifold *oil drilling*; Physical Distribution Management; Positive Displacement Motor *oil drilling*; Procurement Division Memoranda *USA*.

PDMS Point Defence Missile System *military*.

PDO Plan Development Operation *oil drilling*.

p.doz. Per Dozen.

PDR Yemen *or* **P.D.R. Yemen** People's Democratic Republic of Yemen.

PDRK People's Democratic Republic of Korea *North Korea*.

PDSR *or* **P.D.S.R.** Principal Director of Scientific Research.

PDST *or* **P.D.S.T.** Pacific Daylight Saving Time.

PDT Pacific Daylight Time *nautical*.

PDU Polycrystalline Diamond.

PE *or* **P.E.** Personal Effects; Private Effects; Physical Exercise; Permissible Error.

P/E Port of Embarkation; Price Earnings Ratio.

p/e *or* **PE** Price Earnings Ratio.

p.e. Par exemple *French*—For example; Per esempio *Italian*—For example.

Peak Rate Highest obtainable exchange rate, or Freight, *q.v.*

Peculator An embezzler or fraudster.

Peculatus Embezzling public money.

Pedlar A travelling vendor of small wares.

PEEK *or* **Peek** Polyetheretherketone, *q.v.*

PEEP *or* **P.E.E.P.** Pilots Electronic Eye Level Presentation.

PEG Polyethylene-glycol. A timber preserving fluid.

Peggy An ordinary seaman or deckboy acting as a steward in a mess room.

P.E.I. Prince Edward Island, Gulf of St. Lawrence, *Canada*.

Pekoe Superior kind of black tea.

PEL Persistently Enlarged Lymph glands *Aids symptom*.

Pelican Hook A slip hook which can be released by a tripping line or unfastening a latch.

Pelorus Originally a dumb compass, aligned by hand to the ship's heading with a moveable ring for observing bearings. Now, the ring fitted to the rim of the compass bowl is for this purpose, and pelorus has come to mean the gyro compass and its mounting as a whole. The name is from that of Hannibal's pilot.

PEM Production Engineering Measures *USA*.

PEMEN Panhellenic Union of Merchant Marine Mechanics.

Pemex Petroleos Mexicanos. State-owned Mexican oilwells.

Penalty Clause A common clause in a contract providing access to damages in case of default in the agreement by either party. In a Charterparty, *q.v.*, the damages or claims are calculated not to exceed the gross amount of freight. *See* **Breach of Contract** *and* **Misrepresentation**.

Pendant *See* **Pennant**.

Pendente lite *Latin*—During the trial.

P.Eng. Member of the Society of Professional Engineers.

Penetration Price Price set for something with a view to introducing it into the market and capturing a large share.

Peninsula A piece of land almost surrounded by water.

Penn. *or* **Penna.** Pennsylvania, *USA*.

Pennant *or* **Pendant** A long tapering flag occasionally flown at the masthead of a ship.

Pennsylvania Rule US law established in 1873 in a collision case involving the *Pennsylvania* which presumes fault when a ship is shown to have violated the rules of the road or COLREGS.

Penumbra The less dark border of a sun-spot; Partial shadow outside the shadow of an eclipse. *See* **Umbra**.

PEP Political & Economic Planning; Personal Equity Plan *finance*.

Peppercorn Rent When it is desired to reserve only a nominal rent for any period.

Per By means of; Via; Through; By; Period.

Per ann. Per annum *Latin*—Yearly.

PERC Powered Emergency Release Coupling, *q.v.*

Per capita *Latin*—Per head; Individually.

Per capita income The total income of a group divided by the number of people in the group. It can be used in comparing standards of living between different economies.

Perceived Noise Decibel A unit of noise representing the annoyance level of different frequencies.

Percentage of Depreciation The proportion of the total value of cargo that is the subject of loss from an insured peril. The percentage is applied to the sum insured by the policy to determine the amount of claim payable.

Per contra *Latin*—On the other hand.

Per curiam *Latin*—By the Court.

Per diem *Latin*—Daily; Daily allowance.

Perestroika *Russian*—Restructuring; Reconstruction of the state and of society.

Perfecting the Sight Inserting necessary details on a Bill of Lading, *q.v.*, when such had been previously omitted.

Performance Bond A guarantee to supply, demanded of the Seller by the Buyer and arranged through a third party, usually a bank, frequently set as a percentage of the value of the goods or services. The American term is Standby Letter of Credit. *See* **Guarantee**; Loading Guarantee.

Performance Certificate *or* **Casualty Certificate** A certificate enforced by US maritime law which is required by the Federal Maritime Commission for all passenger vessels operating in US commercial areas. This certificate ensures that the cruise liner or operator has covered the regulations as to passengers' insurance against any subsequent claims regarding casualties or inability to undertake the scheduled voyages.

Performance Guarantee *See* **Performance Bond**.

Per Hatch per Day *or* **Per Workable Hatch Per Day** *or* **Per Working Hatch per Day.** A Charter Party, *q.v.*, term quantifying the cargo a ship is to load or discharge.

Periculum in mora *Latin*—Danger in delay.

Period Policy Similar to **Time Policy**, *q.v.*

Perigee That point in a planet's (especially the moon) orbit at which it is nearest to the earth. *Opposite to* **Apogee**, *q.v.*

Peril A term used in the Marine Insurance Act to denote a hazard. The principle of proximate clause is applied to an insured peril to determine whether or not a loss is recoverable. In modern practice the term 'risk' replaces 'peril'.

Peril Insured Against The actual risk covered by the insurance.

Perils of the Sea Maritime dangers such as collision, stranding, sinking, hitting submerged objects, etc.

Per infortunium *Latin*—By mischance.

Per. Inj. Personal Injury.

Period Bill A Bill of Exchange, *q.v.*, or note due for payment on a particular day and not at sight.

Period Charter A Time Charter for a stated period.

Period of a Ship The interval of time of a ship rolling from one side to the other and back to the first position.

Period of Insurance The time during which the assured is covered or the Underwriter, *q.v.*, is at risk in relation to the terms of the Insurance Policy, *q.v.*

Period of Waves The interval of time between two consecutive wave crests.

Period Policy *Similar to* **Time Policy**, *q.v.*

Per mensem *Latin*—Per month.

Per mille *Latin*—Per thousand, symbol ‰.

Perp. Perpendicular.

Perpetual Day *or* **Perpetual Night** The day or night in the far North or Arctic zones where the sun never goes down during the summer season and never rises during the winter season.

Perpetual Debentures Debentures, *q.v.*, which cannot be redeemed.

Perpetual Night *See* **Perpetual Day**.

Per pro *or* **p.p.** Per procurationem *Latin*—A person who signs for another person or firm; On Behalf Of; By Proxy.

Per procurationem *Abbrev.* Per pro, *q.v.*, or p.p.

Pers. Acc. Personal Account, *q.v.*; Personal Accident.

Persaluan *Indonesian*—Association.

Per se *Latin*—By himself, herself, itself *or* themselves.

Persona *Latin*—A person according to law.

Personal Account/s Money transactions between a person or company with another person or companies. *Abbrev.* Pers. Acc.

Personal Flotation Device Approved floats meant as life preservers and carried on board American ships. *Abbrev.* PFD *or* P.F.D.

Personal Loan A bank loan advanced to an individual.

Persona non grata *Latin*—Unacceptable person. *Abbrev.* P.N.G. *or* p.n.g.

Personnel Officer Large commercial and government concerns often employ a personnel officer responsible for the welfare of the employees. His or her duty is also to interview prospective applicants for work.

Perspective A science which can be defined as a geometrical method of producing three dimensional objects on a plane surface.

Per Standard Compass This refers to the standard magnetic compass currently used by navigators. *Abbrev.* P.s.c.

PERT *or* **P.E.R.T.** Programme Evaluation and Review Technique.

Perusahaan *Indonesian*—Company.

PES *or* **P.E.S.** Petroleum Engineering Society; Programmable Electronic System.

PESA Petroleum Equipment Suppliers' Association; Philippines Electrical Suppliers' Association.

Peseta Unit of currency used in Spain. One hundred centimos equals one peseta.

Peso Unit of currency used in Argentina, Bolivia, Colombia, Cuba, Dominican Republic, Mexico, Uruguay, and the Philippines.

Pest Control Operator A trained and state-certified person specialising in the fight against vermin. *Abbrev.* pco.

Peta- Prefix meaning 10^{15}. *Abbrev.* P.

PETCOM Petroleum Coke-oil mixture.

Petite vitesse *French*—Low speed. An International Freight Train at the slower of two transit speeds covering 200 kilometres per day. Now obsolete in France but sometimes referred to in documents. *Abbrev.* PV or P.V.

Petition A request in the legal sense; Written form presented to the courts to grant rights or redress.

Petrel Kinds of oceanic bird spending nearly all their lives at sea.

Petrography *or* **Petrology** The science of the description and composition of rocks.

Petroleum A mineral oil found in the upper strata of the earth, used as a fuel for heating and in internal combustion engines. *See* **Mineral Oil.**

PETROMIN *or* **Petromin** Petroleum and Mineral Organisation *Saudi Arabia.*

Petties *or* **Petty Expenses** Expenses of a small nature in business. *Ex.* Transport, tips, etc.

Petty Cash A company's money kept for on-the-spot expenses.

Pewter A grey alloy of lead and tin with an occasional addition of copper and antimony. Mostly used to make utensils.

p. ex. Par exemple *French*—For example.

PEXA Practice and Exercise Area *nautical.*

PF *or* **P.F.** Poop and forecastle; Power Factor; Procurator Fiscal.

pf. Pfennig, a German unit of currency one-hundredth of a mark; *Lloyd's Register notation* for Plain furnace-boiler.

PFA Private Fliers Association *USA*; Pulverised Fuel Ash.

PFC *or* **Pfc.** Passed Flying College; Passed Flying Colours; Passed First Class; Parallel Flange Channels *ship's construction.*

PFD *or* **Pfd** *or* **pfd** Preferred *financial.*

P.f.d. Position Fixing Device *ship's navigational aid.*

PFD *or* **P.F.D.** Position Fixing Device *ship's navigational aid*; Personal Flotation Device, *q.v.*

PFF *or* **P.F.F.** Path Finder Force.

P.f.m. Path Finder Meter.

pfo Portfolio.

pfo.tr. Portfolio Transfer.

PFPC Passenger Forms and Procedures Committee.

PFR Prototype Fast Reactor.

PFRT Preliminary Flight Rating Test.

PG *or* **P.G.** Paying Guest.

p.g. Paying Guest; Persona Grata *Latin*—Accredited or Acceptable Person; Proof Gallons.

pgc *or* **p.g.c.** Per Gyro Compass *nautical.*

PGCE *or* **P.G.C.E.** Post Graduate Certificate of Education.

PGFO *or* **P.G.F.O.** Persian Gulf for Orders.

PGR *or* **P.G.R.** Population Growth Rate.

p.g.t. Per Gross Ton.

P/H Policy Holder.

Ph. Phase, *q.v.*

PHA Port Health Authority; Port of Houston Authority; Process Hazard Analysis *oil industry.*

Phase Condition of a heavenly body including the moon. *Abbrev.* Ph.

Phat. Photostat.

Ph. B. Philosophiae Baccalaureus *Latin*—Bachelor of Philosophy.

ph. brz. Phosphor Bronze.

Ph. D. Philosophiae Doctor *Latin*—Doctor of Philosophy.

PHD Doctor of Public Health.

PHE Plate Heat Exchanger.

PHG *or* **P.H.G.** Postman Higher Grade.

Phil. Philippines.

Phil. I. *or* **Phil. Is.** Philippines Islands.

phos. Phosphate.

Phosphate Rock Phosphate deposits mostly found in the USSR.

Phot Unit of illuminance. *Abbrev.* ph.

Photo/s Photograph(s).

Photocomposition *or* **Photosetting** Setting copy on to film, etc., instead of in metal type.

Photocopier A machine for reproducing documents, etc.

Photochromy Colour photography.

Photoelectric Marked by or utilising emission of electrons from solid, liquid, or gaseous bodies when exposed to light of suitable wavelengths.

Photoelectric cell A cell or vacuum tube that uses the photoelectric effect to produce an electric current.

Photometer An instrument for measuring the intensity of light.

Photosphere A luminous envelope of the sun or a star from which its light and heat radiate.

Photovoltaic Energy Electromotive force produced across the function of dissimilar materials when exposed to light or ultra violet radiation. Used in satellites. *Abbrev.* PV.

p.h.p. Pounds per horsepower.

PHPA Partially Hydrolysed Polyacrylamide.

PHV Propeller Hull Vortex.

PH Value The measure of alkalinity or acidity of a substance. pH7 is neutral or nil. pH14 is the maximum.

P.I. Pasteur Institute; Personal Injury *insurance*; Petrol Injected; Philippine Islands.

PIA *or* **P.I.A.** Peril Insured Against *insurance*; Pakistan International Airline, Pakistan State Air Lines.

Piastre Currency unit of Egypt, Lebanon, Syria and Sudan.

PIB *or* **P.I.B.** Petroleum Information Bureau.

PIC *or* **P.I.C.** Pilot in Command *aviation*.

Picked Ports In a Charterparty, *q.v.*, loading and/or discharging ports are specifically named because of quick despatch. If other ports are used other than those picked and which may be considered as slow working ports, the charter or hire rate will be increased accordingly. *Abbrev.* P.P. *or* p.p.

Picketing In a strike the union engages some of its own members to act as pickets who endeavour to stop men from returning to work.

Pickle A colloquial word used in shipping for a chemical solution mostly used as a preservative or oil dispersant on board a ship.

Pick Up Act of the Carrier or his agent who makes arrangements for merchandise to be collected from the warehouse of the Shipper or Consignor, *q.v.*, in preparation for shipment.

PID Photo Ionization Detector.

P & ID Piping and Instrument Diagram *oil drilling*.

PIEA Petroleum Industry Electrical Association *USA*.

Pièce de résistance *French*—The best item.

Piece Goods Items sold out piecemeal generally referring to fabric or wool, cotton, canvas and the like.

Piece Rate A payment given in relation to the output of work. *See* **Piece Work** *and also* **Time Rate**.

Piece Work Work paid according to the quantity of work done per hour, per day or per week. Generally it is calculated per hour. *See* **Piece Rate, Time Rate** *and* **Time Wages**.

Pied à terre *French*—An occasional residence.

Pig Device used to scrub the internal surfaces and to check for malformations in pipelines.

Pig-iron An oblong mass of metal, usually iron, which comes from a smelting furnace.

Pigment Colouring-matter used as paint or dye; natural colouring-matter of a tissue.

Pigment Volume Concentration A term commonly used in the manufacture of paints. *Abbrev.* p.v.c. *See* **Pigments**.

PIK Payment-in-Kind, *q.v.*

PIL Payment in Lieu; Premium Income Limit; Pest Infestation Laboratory.

pil Pilula *Latin*—Pellet, Pill.

Pilf. Pilferage, *q.v.*

Pilfer To steal in small quantities. Commonly done on board conventional cargo ships during loading or discharge. *See* **Pilferage**.

Pilferage The act of stealing from cargo while in transit from the port of loading until customs clearance at the port of delivery; loss from the same cause.

Pilgrim Nut A patented nut used to secure a Propeller, *q.v.*

Pillar Box Letter Box, Post Box. Steel container with aperture to receive letters for collection by postmen.

Pillars Deck supports in a ship.

Pilot Flag Every pilot boat, when afloat, must fly in some conspicuous position a pilot flag, i.e. a large flag (as compared with the pilot boat) divided horizontally into two colours, the upper half being white and the lower half being red (Pilotage Act 1983, s. 46(1)). It is the duty of the master of a pilot boat to see that the pilot flag is kept clean and distinct so as to be easily discerned at a reasonable distance. If he fails to do so he is liable in respect of each offence to a maximum penalty of level 2 on the standard scale in England, Wales and in Scotland or in Northern Ireland, £50.

Lights to be carried by pilot boats The lights to be carried by pilot boats are specified in r. 29 of the International Regulations for Preventing Collisions at Sea which states:

'(a) A vessel engaged on pilotage duty shall exhibit:
    (i) at or near the masthead, two all round lights in a vertical line, the upper being white and the lower red;
    (ii) when underway, in addition, sidelights and a sternlight;
    (iii) when at anchor, in addition to the lights prescribed in sub-paragraph (i), the anchor light, lights or shape.
(b) A pilot vessel when not engaged on pilotage duty shall exhibit the lights or shapes prescribed for a similar vessel of her length.'

Pilot in Command The First Pilot or one who is in charge of an aircraft in flight.

Pilot Ladder *See* **Jacob's Ladder**.

Pilot Station A place on shore, or the position at sea in which a pilot vessel is placed, whence pilots are sent to ships using a pilot boat.

Pilot Waters Areas where the use of pilots is compulsory.

PIN Personal Indentity Number *Credit card identity number*.

Pin A small cask of $4\frac{1}{2}$ gallons.

p. in² Parts per square inch.

p. in³ Parts per cubic inch.

P. in A. Parallax in Altitude *astro-navigation*.

Pinnace A small sailing vessel; Boat or Launch propelled by oars, sail, motor or steam engine used for harbour work. Formerly carried aboard large warships.

PINRO The Russian Polar Scientific Research Institute of Marine Fisheries and Oceanography.

Pintles Vertical bolts or pins that serve as an axis for the rudder to swing on the sternpost or rudder post of a ship.

PIO *or* **P.I.O.** Public Information Office; Public Information Officers.

PIOPIC Protection and Indemnity Oil Pollution Indemnity Clause.

PIOSA *or* **P.I.O.S.A.** Pan-Indian Ocean Science Association.

PIP *or* **P.I.P.** Picture in Pictures *television*.

Pipe *Similar to* **Butt**, *q.v.*; a wine cask usually having a capacity of 105 gallons.

Pipe Layer *or* **Pipe Laying Ship** Vessel designed for laying pipes on the Seabed, *q.v.*

Pipeline End Manifold The termination of a submarine pipeline for connection into another facility like SBM, *q.v. Abbrev.* PLEN.

Pipeline Under the Ocean A petroleum pipeline laid under the English Channel during the 1939–45 world war to provide fuel for the Allied Forces' invasion of France. *Abbrev.* PLUTO.

PIR Property Irregularity Report, *q.v.*

Piracy An assault on a vessel, cargo, crew or passengers at sea by persons owing no allegiance to a recognised flag and acting for personal gain. It also includes acts of rioters who attack a ship from the shore and of passengers who mutiny. This peril, along with war perils, was excluded from the SG Policy, *q.v.*, by the F.C. & S. clause; to be reinstated if the cover conditions covered war risks. Thus, for many years piracy was related to war risks. The 1982 cargo war clauses do not cover piracy, and it is not embraced within the risks covered by the B or C cargo clauses. The 1983 hull clauses incorporate piracy among the risks covered by the standard marine clauses; piracy not being included in the war risks cover (1983).

PIRO Petroleum Industry Response Organization *USA*.

PIS *or* **P.I.S.** Port Information Service.

Pitch A distilled residue from crude petroleum or coal tar largely used for road surfacing, paint manufacturing, sealing cracks, caulking, etc.; the twist of a propeller blade.

Pitch *or* **Pitching** The downward and upward movement of a ship at sea forward and aft.

Pitting Small indentation or uneven surface on metal plates due to corrosion.

Pit Trading The act of bargaining in the Futures Market, *q.v.*, in face-to-face contacts.

PITWOODCON Chamber of Shipping code name for a Charter Party, *q.v.*, for the carriage of pitwood timber, pit props.

Pixpinus Chamber of Shipping approved Charter Party, *q.v.*, for the carriage of pitch pine timber.

P.J. Presiding Judge; Probate Judge.

PK Pakistan International Airlines *airline designator*.

Pk. Park; Peak.

PKD *or* **P.K.D.** Partially Knocked Down. *See* KDC.

Pkg. Package; Packing.

PKN Polski Komitet Normalizacyjny, Polish Standards Commission.

PKP Polski Kolejo Panstwowo *Polish State Railways*.

Pkt. Pocket.

PL Position Line *navigation*.

P.L. *or* **P/L** Partial Loss, *q.v.*; Public Liability; Plimsoll Line, *q.v.*; Poet Laureate; Public Loss; Pharmacopoeia Londiniensis *Latin*—Pharmacopoeia of London.

PLA *or* **P.L.A.** Port of London Authority.

Plaintiff A person or firm bringing an action in law against another.

Planet A heavenly body revolving in approximately circular orbit around the sun. The major planets are Mercury, Venus, Earth, Mars, Jupiter, Saturn, Uranus, Neptune and Pluto.

Planimeter An instrument for measuring the area of plane surfaces.

Plank Long flat piece of timber; One of a number of such pieces in the hull construction of a boat. Also a political viewpoint or point of argument.

Plan Position Indicator Visual radar indication of the neighbouring area of the vessel. *Abbrev.* PPI or P.P.I.

Plant Business tools, machinery, fixtures and all other related items.

Plat. Plateau.

Plate Sheet of metal of uniform thickness, flat when manufactured.

Plate Bill of Lading Bill of Lading *q.v.*, generally used for any merchandise from the River Plate.

Platform Jackets Legs of an offshore oil drilling platform.

Plating (1) The sheets of metal, plates, used in building the external form of a ship, and its decks and other parts; also in boilers and a variety of structures. (2) The process of coating a metal with a layer of another.

Platinum A white heavy ductile malleable metallic element unaffected by simple acids and fusible only at very high temperature.

PLB *or* **P.L.B.** Personal Locator Beacon.

PLC *or* **P.L.C.** *or* **plc** *or* **p.l.c.** *or* **Plc** Public Liability Company; Public Limited Company.

PLC Programmable Logic Control(ler) *oil drilling*.

Plea The defendant's answer to the plaintiff's declaration; also a suit or action.

Pleadings Written statements delivered alternately by the parties to an action to one another, until the questions of facts and law to be decided in the action have been ascertained.

Plebe The lowest grade at a US naval or military academy.

Pledge Anthing put to pawn or given by way of warrant or security; also a surety, bail, or hostage. *See* **Pawnbroker**.

PLEN Pipeline End Manifold, *q.v.*

PL (Fire) Passenger Liability *insurance*; Public Liability (Fire) *insurance*.

PLHA Port of London Health Authority.

PLI Periodical Loadline Inspection.

Plimsoll(s) Cheap rubber-soled canvas shoes.

PLMB Production Linear Block Manifold *oil drilling*.

PLOFR Please Offer.

Ploughed Back Profit A part of a business profit set aside for future development.

PLR Public Lending Right *publishing*.

PLSS Personal or Portable Life Support System.

Plt. Plate.

PLUNA Primeras Lineas Uruguayas de Navigacion Aerea, *Uruguay State Airline*.

PLUTO *or* **Pluto** Pipeline Under the Ocean, *q.v.*

Plutocracy When the wealthy side of the population has the monopoly in state affairs.

Plx. Parallax, *q.v.*

Ply. Plymouth.

p.m. Post Meridiem *Latin*—Afternoon; Per mensen *Latin*—Monthly.

P/m Put of More, *q.v.*

P.m *or* **pm.** Premium, *q.v.*

PM Phase Modulation *electronics*.

PM *or* **P.M.** Paymaster; Performance Monitoring; Post Master; Post Mortem *Latin*—Autopsy; Post Meridiem *Latin*—Afternoon; Powder Metallurgy; Provost Marshal; Prime Minister; Pulse Modulation *nautical*.

P/M *or* **p.m.** Pounds per minute.

PMA Pacific Maritime Association; Panama Maritime Administration.

Pmaapb *or* **PMAAPB** Pacific Maritime Association's Accident Prevention Bureau.

PMAR Private Medical Attendant's Report, *q.v.*

PMB *or* **P.M.B.** Parallel Middle Body.

PMBX Private Manual Branch Exchange.

PMG *or* **P.M.G.** Postmaster General; Paymaster General.

PMH *or* **P.M.H.** Per Man Hour, production per man hour.

pmk Postmark.

PML *or* **P.M.L.** Possible Maximum Load; Possible Maximum Loss; Probable Maximum Loss.

PMO Port Meteorological Officer.

PMO *or* **P.M.O.** Port Medical Officer; Principal Medical Officer.

pms Post Marketing Surveillance.

PMS Planned Maintenance System.

pmt *or* **PMT** *or* **p.m.t.** Per Metric Ton; Payment; Prompt.

PMTS Predetermined Motion Time Signal.

PMX *or* **P.M.X.** Private Mutual Exchange *telephone*.

PN *or* **P.N.** *or* **Pn** *or* **p.n.** Part Number; Promissory Note, *q.v.*; North Pole *nautical*; North Celestial Pole.

p.n. Please Note; Position Negative.

P/N *or* **P.N.** Part Number; Pakistan Navy; Promissory Note, *q.v.*

Pna Panama.

PNA Panhellenic Seamens Federation.

PNC Power Development Corporation.

PNdB Perceived Noise Decibel, *q.v.*

PNEU *or* **P.N.E.U.** Parents' National Education Union.

Pneumafathometer Diving instrument indicating the depth and rate of ascent of a diver.

Pneumatic Electrical Chipping Hammer *See* **Chipping Hammer**.

PNG Papua New Guinea.

P.N.G. *or* **p.n.g.** Persona non grata *Latin*—Undesirable person in a country or state.

P.N.L. Prodotto Nazionale Lordo Grosso. *Italian*—Gross National Product.

PNPF Pilots National Pension Fund.

PNR Passenger Number Request *Confirmation of air ticket number in air passage reservation system*.

p.n.r. Prior Notice Required.

PNSA Polynuclear Aromatic Hydrocarbon.

Pnt *or* **PNT** Pentagon, *USA*.

PNYA *or* **P.N.Y.A.** Port of New York Authority.

PO *or* **P.O.** *or* **p.o.** Parcel Office; Part of; Postal Order; Post Office; Power Operated; Province of Ontario, *Canada*; Previous Orders; Public Office; Public Officer.

P.O. Posta Ordinaria. *Italian*—Regular *or* Normal Post.

P/O Part of.

POA *or* **P.O.A.** Price on Application.

POAC Port and Ocean Engineering Under Arctic Conditions.

POAS Poly oxy-aluminium stearate, waterpoofing compound.

POB *or* **P.O.B.** Pilot On Board; Post Office Box, *q.v.*

POC *or* **P.O.C.** Port of Call.

Pocket A bag or sack, especially as a measure of hops (168 lbs.) or wool.

POCRS Post Office Coast Radio Station.

POD Probability of Detection.

POD *or* **P.O.D.** Pay on Delivery or Payment on Delivery, similar to C.O.D., i.e. Cash on Delivery;

Payment on Demand; Port of Distress; Port of Discharge; Post Office Department; Port of Destination; Port of Disembarkation.

POE Plan of Exploration *oil drilling*.

POE *or* **P.O.E.** Port of Embarkation; Port of Entry.

POEA Philippines Overseas Employment Administration.

POFI *or* **P.O.F.I.** Pacific Ocean Fisheries Investigations.

POGO Polar Orbiting Geophysical Observatory.

Point-line A hemp or nylon rope knotted to one end of a mooring rope and having a padded weight to enable it to be thrown from the shore to the ship or vice versa. It is pulled up until the mooring rope is also picked up and fastened to the Bollards.

Point of No Return The point at which an aircraft on a long-distance flight has to proceed because of insufficient fuel to enable it to return to its starting place.

Point of Origin Country from which merchandise originated or where it was made.

Point of Rest A terminal or area reserved for the reception of containers, trailers, etc., which are then forwarded to destination by Roll-on Roll-off, *q.v.*, vessels.

POL *or* **P.O.L.** Patent Office Library; Petrol, Oil and Lubricants; Port of Loading; Property Owners' Liability.

Pol Poland; Policy; Polish.

Polar Distance The angular distance of a heavenly body from the nearest Celestial Pole.

Polarity Condition of magnetised material causing its extremities to be attracted to one or other magnetic pole; Electrical condition as positive or negative.

Polar Star *or* **Pole Star** *See* **Ursa Major**.

Pole Each of two terminal points (positive, negative) of electric cell, battery, etc.; the North and South extremities of the earth's axis.

Pol.ex. Policy Exclusion.

Polgen A Polish charterparty principally for the carriage of Polish-held cargoes and used both by Polish and foreign-flag vessels.

Policy (1) A course of action adopted by a government, political party, business organisation, etc. (2) A document containing a contract of assurance or insurance.

Policyholder The person or firm in possession of the policy of assurance/insurance.

Policy Proof of Interest A policy wherein the underwriter agrees to waive proof that an insurable interest is enjoyed by the assured, at the time of loss, as a condition of claim payment. In other policies the Underwriter, *q.v.*, is not liable for any claim where the assured is unable to prove that his interest in the subject-matter of the insurance exists at the time of loss. P.P.I. policies are invalid in a court of law, but are not illegal except where no interest exists or where

there was no reasonable expectation that it would exist at the time the policy was effected. *See* **Interest** *or* **No Interest** *also* **Maritime Policy**.

Policy Year A Marine Insurance, *q.v.*, phrase meaning a year from noon on any 20 February up to noon of the next 20 February.

Polish Coal Warranty One of the marine insurance exception clauses not to carry Polish coal as cargo except for bunkers on voyages to Europe and north of Cape Finisterre.

poll. Pollution, *q.v.*

Pollards Animal foodstuffs consisting of meal and bran which are liable to heat and sweat.

Pollution Waste material in the form of liquids, dust, solids, odours, etc., which create inconvenience to the general public as a whole. This can cause air, sea and soil pollution to the detriment of health of all human beings. Shipowners, most especially tanker owners, are bound to insure their ships against polluting the seas and shores. *See* **CRISTAL** and **TOVALOP**.

Pollution Hazard Clause A clause in a hull policy, whereby underwriters cover deliberate damage or loss to the ship caused by Governmental authority in attempts to mitigate a threat of pollution hazard; where such relates to damage to the ship and where it has not resulted from a want of due diligence on the part of the assured, owners or managers of the ship.

Polski Rejestr Statkow The Polish ship classification society. *Abbrev.* PRS or PR.

Polycrystalline Diamond Cutter A particular type of drilling bit giving increased penetration. *Abbrev.* PDC.

Polyetheretherketone A fibre-reinforced material used in oil well environments capable of withstanding temperatures up to 250 degrees Centigrade. *Abbrev.* PEEK.

Polynia *or* **Polynya** Open water among Arctic sea ice.

Polypropylene A thermoplastic material made by polymerisation of propene. *Also known as* polypropene.

Polytetrafluoroethylene *Abbrev.* PTFE. A synthetic material that can withstand up to 327 degrees Celsius. It is used in cooking utensils, engine gaskets, electrical insulation and non-lubricated bearings.

Pontoon (1) A low flat-bottomed boat used as a float to paint or do work on the hulls of ships. (2) An inflatable device commonly used in salvage operations to raise submerged vessels. (3) A float on a seaplane. (4) A temporary floating support bridge across a small or medium sized river.

Pontoon Dock *Similar to* **Floating Dock**, *q.v.*

Pontoon Hatch A hatchway such as a slab of steel which fits over an opening in a tweendeck into a lower hold.

POO *or* **P.O.O.** Post Office Order.

Pool (1) An arrangement between competing parties by which prices are fixed and business divided to do away with competition. (2) A common fund, e.g. of the profits of separate firms. (3) A common supply of persons, commodities, etc. (4) To share in common. (5) Collective stakes in a joint gambling venture. (6) An underground cavity of oil or gas found in certain regions having porous rocks.

Poop The aftermost and highest deck of a ship.

Poop Deck The enclosed part under the poop deck.

Poor Market *See* **Weak Market**.

PoP Point of Presence The link between a subscriber and an Internet system.

POP Point of Purchase; Point of Preference; Colloquialism for popular music.

P.O.P. Perpendicular Ocean Platform *USA*; Plaster of Paris; Print Out Paper.

POPA Property Owners Protection Association.

POPI *or* **P.O.P.I.** Post Office Positioning Indicator.

Po-Po *or* **PO/PO** Push On/Push Off Vessel.

Poppet (1) A shutter used to close a rowlock in an oared boat to restore the integrity of the gunwhale when the boat is not being rowed or pulled. (2) Shores or timbers under the bilge of a ship being launched and forming part of the cradle.

Population Bulge When the birth-rate of a country rises above the normal.

POQ *or* **P.O.Q.** *or* **poq** *or* **p.o.q.** Passing Over Quay, *q.v.*

POR *or* **P.O.R.** *or* **p.o.r.** *or* **P.o.r.** Port of Refuge; Port of Registry; Payable on Receipt; Pay on Report.

PORIS *or* **P.O.R.I.S.** Post Office Radio Interference Station.

Port (1) A sheltered place from the open sea where a ship can enter to load or unload its merchandise. The major or principal ports are the registered places of ships. (2) Round, oval, square, or rectangular shaped opening in the side of a ship to serve as a means of looking through while receiving light and fresh air. This hole is protected by thick iron-framed glass which opens as a window. This window can be protected from the sea by a steel cover. *See* **Portlight** *or* **Porthole**. (3) The left-hand side of a ship when facing the bows. During darkness it is indicated by a red light.

Port. Portugal; Portuguese.

Portage The cost of carrying goods; the transportation of goods between two navigable waters; the services executed by crew members while the vessel is under repair, such as watch keeping, carpentry, etc.; the carrying or moving of luggage by a porter.

Portage Bill A bill giving the statement of wages of each member of the crew at the end of a voyage.

Portainer A particular type of crane used on quays to lift containers.

Port and Ocean Engineering Under Arctic Conditions Developments in Arctic water operations including oceanography, exploration, marine transportation and logistics, harbours and terminals, offshore structures, fisheries. *Abbrev.* POAC.

Port and Tanker Safety Act A law enacted in the USA relating to tanker design and operation. *Abbrev.* PTSA.

Port Area The land and facilities administered by the legal port authority.

Port Captain *See* **Superintendent**.

Port Clearance Customs permission for the master to take his vessel to sea.

Porthole *Similar to* **Portlight**, *q.v.*

Porthole Container A refrigerated container with holes in its sides for connecting trunking to a chilled air blower, or 'blow-freeze'.

PORTIA *or* **Portia** Port Operations Research Transport and Internal/Accounting.

Portlight *or* **Porthole** An opening on the sides of a ship generally in a circular form sealed with hinged reinforced heavy glass and protected by a metal cover. The cover can be opened to allow light to enter the cabins and compartments and the glass can also be opened to allow fresh air. Sometimes it is called Port, *q.v.*

Port of Call The port nearest or commercially viable for a ship passing en route from port to port. Normally this port caters for bunkering, provisions and stores. *Abbrev.* POC *or* P.O.C.

Port of Distress Port of Refuge, *q.v.*

Port of Entry That port where the ship and/or merchandise is cleared inwards or outwards. *Abbrev.* POE *or* P.O.E.

Port Operations Research Transport and Internal Accounting A computer cargo-control system developed by the port of Liverpool and used in Houston in the new container terminals. *Abbrev.* PORTIA *or* Portia.

Port R. Port Risks, *q.v.*

Port Risks A marine insurance policy covering the vessel when in port. *Abbrev.* PR *or* P.R. *or* P/R *or* p.r. *or* Port R.

Port State Control The examination of ships for compliance with IMO Conventions and resolutions. *Abbrev.* PSC. *See* **Memorandum of Understanding on Port State Control**.

Ports and Waterways Safety Act A United States Coast Guard comprehensive scheme to regulate traffic routes, operations, pilotage and safety device specifications on tankers. *Abbrev.* PWSA *or* P & WSA.

Port Tack When the vessel sails with the wind on the port side. *Opposite to* Starboard Tack, *q.v.*

POS Point of Sale.

POS *or* **P.O.S.** Point of Sale; Port of Spain, *Trinidad*.

POSB *or* **P.O.S.B.** Post Office Savings Bank *UK*. *Now the* National Savings Bank *UK*.

POS-DP Additional notation by the Norwegian Norske Veritas classification society concerning the construction of dynamic positioning systems for ships and mobile offshore units.

Posn. Position.

Postal Cheque A cheque issued by the Postal Giro, *q.v.*

Postal Giro A system of transferring money through the Post Office organisation.

Postal Order A form of a cheque issued by the Postmaster General through a Post Office clerk to the public. Postal Orders are for a limited sum of money in excess of which Money Orders are used.

Post Date To put a belated date. Generally refers to a post-dated cheque. If such cheque is given it cannot be cashed until the date inserted on the cheque.

Post Dated *See* **Post Date**.

Poste Restante *French*—Postal deposit. A department of the Post Office where letters and other correspondence are deposited, generally in numbered boxes, for collection by individual subscribers. *See* **Post Office Box**.

Posted Reported Missing. *See* **Posted for Information**.

Posted for Information Lloyd's of London preliminary notice for a vessel presumed lost after a period of time overdue. If no information of any sort is at hand after a reasonable time the vessel will be listed as 'missing'. *See* **Missing Ship**.

Posted Missing *Similar to* **Posted**, *q.v.*

Post hoc ergo propter hoc *Latin*—Therefore on account of this.

Post Meridiem *Latin*—Afternoon. *Abbrev.* P.M. *or* p.m.

Post mortem *Latin*—Medical examination of a dead body; After Death.

Post Office Box A numbered box kept at the Poste Restante, *q.v.*, where all letters and correspondence are kept for collection during office hours by the holder who has access to the box by means of a key. *Abbrev.* POB *or* P.O.B. *or* P.O. Box.

Post Town Town or city having a Post Office.

Post Weld Heat Temperature A successful development in underwater welding techniques which eliminates compulsory drydocking with considerable saving in time and money.

Potable Suitable for drinking.

Potable Water Drinkable water.

Potale Waste grain residue left after distillation of whisky.

Potentiometer Instrument to adjust and/or measure the strength or potentiality of electricity.

Potlife This word is commonly used as the maximum time allowed after mixing two paints before they are to be applied and after which time the mixture will not be fluid enough to use.

POUCE or P.O.U.C.E. Post Office Users Co-Ordination Committee.

Pound In the Imperial system of weights and measures the pound equals 16 ounces. In the Metric system it is 0.454 of a kilogramme. The pound is also a unit of currency used in the United Kingdom, Cyprus, Egypt, Gibraltar, Lebanon, Sudan, Syria.

Poundage When a Postal Order, *q.v.*, or Money Order, *q.v.*, is issued by the Post Office a cover charge or poundage is added.

Poundal Foot pound per second, unit of force. *Abbrev.* pdl.

Pour Point The lowest temperature at which oil can be stored and still be capable of flowing. The pour point gives a rough indication of the oil's wax content.

POV *or* **P.O.V.** *or* **p.o.v.** Privately Owned Vessel or Vehicle.

POW Prisoner of War.

P.O. Waggon Privately Owned Waggon.

Powder Flag International Code flag B which when hoisted on a vessel warns that highly inflammable fuel or powder is being loaded or discharged.

Powder-room *or* **Powderroom** A compartment on a ship for the transportation of explosive cargoes; a ladies' cloakroom. *Abbrev.* PR *or* P.R.

Powered Emergency Release Coupling A device in the loading arms for bunkering purposes used only in emergencies. Its function is to release rapidly the loading arm while bunkering and avoid damage to it or the ship. *Abbrev.* PERC.

Power of Attorney An official document empowering a person to act on behalf of another or others. *Abbrev.* P/A; Proxy.

Power Station An electricity generating plant forming part of the electrical supply system.

PP *or* **P.P** *or* **p.p.** Per pro, *q.v.*; Picked Port(s) *q.v.*; Port payé *French*—Carriage Paid; Profets et pertes *French*—Profits and Losses; Per procurationem *Latin*—On behalf of, by Proxy; Parcel Post.

pp. Pacco Postale *Italian*—Parcel Post; Picked Ports; pages.

PP Patent Packing.

P & P *or* **p & p** Packing and Postage.

PPA Philippines Ports Authority.

P.P.A. Polimeles Protodikion Athinon *Greek*—Multi-Member First Instance Court of Athens.

PPB *or* **P.P.B.** Parts Per Billion; Private Posting Box.

PPBAS Planning, Programming, Budgeting, Accounting System.

PPBS Planning, Programming, Budgeting System.

P.P.C. Pour prendre congé *French*—To take permission to leave.

Ppd *or* **ppd** Prepaid.

PPI *or* **P.P.I.** *or* **p.p.i.** Plan Positioning Indicator, *q.v.*; Policy Proof of Interest, *q.v.*; Parcel Post Insured.

PPL *or* **P.P.L.** Private Pilot's Licence.

ppl Proposal *insurance.*

PPM *or* **P.P.M.** *or* **ppm** Parts per million.

ppn Proposition.

P.P.P. Polimeles Protodikion Pireos *Greek*—Multi-Member First Instance Court of Piraeus.

P.P.R. Printed Paper Rate.

p.p.s. Post postscriptum *Latin*—Additional post-script.

P.P.S. Parliamentary Private Secretary; Post post-scriptum *Latin*—Additional postscript.

PPSO Prevention of the Pollution of the Sea by Oil.

PPT *or* **Ppt.** Prompt, *q.v.*

PPT.L. *or* **ppt.l.** Prompt Loading. *See* **Prompt.**

P.Q. Parliamentary Question; Province of Quebec, *Canada.*

PR Philippines Airlines *airline designator*; Puerto Rico.

PR *or* **p.r.** Port Risks, *q.v.*; Postal Receipt; Pro Rata, *q.v.*

P.R. *or* **P/R** Parcel Receipt; Payroll, *q.v.*; Polski Rejestr Polish ship classification society; Port of Registry, *q.v.*; Proportional Representation; Provisional Release; Polish Register;

Pr. Pair; Per; Price; Provisional Release.

P.R.A. President of the Royal Academy; Puerto Rico Association.

Practice Shipping terminology for the customs and usage of the place.

Pram Perambulator; Type of small boat or dinghy, often with a squared-off bow and serving as a tender to a yacht, *q.v.*

PRC People's Republic of China.

PRCA Pro Rata Conditions of Average *insurance.*

Preamble Introductory part of a contract or a charterparty. The preamble in a charterparty consists of the names of the contracting parties, details of the vessel including her classification, the date and place of delivery and redelivery.

Precept A magisterial warrant; an order for payment by the local authority.

Precious Cargo A shipping term for articles of special value such as gold, silver, jewels, etc. The Freight Charged, *q.v.*, for this type of cargo is ad valorem. *See* **Ad Valorem Freight.**

Precision Approach Path Indicator A system of lights on a runway to increase accuracy while approaching and landing an aircraft. *Abbrev.* PAPI.

Pre-emption The right given to a person to purchase before another, or to buy compulsorily purchased lands which were formerly owned by him. *See* **First Refusal.**

Preference Shares Shares on which a dividend is paid before any is paid on ordinary stock.

Preferential Duty The favouring of a country by admitting its products at a lower import duty. *See* **Discriminating Duty.**

Prem. *or* **PREM** Premium, *q.v.*

Prem. I.F. Premium in Full, *q.v.*

Premium in Full The situation where the maximum total of Risks, *q.v.*, have been placed in the market and Underwriters, *q.v.*, are not interested in further cover. *Abbrev.* Prem. I.F.

Prem. Red. Premium Reducing *insurance.*

Prem. Res. *or* **PREM. RES.** Premium Reserve *reinsurance.*

Premium (1) The sum which exceeds the par value of stocks and shares. (2) At a premium—referring to above par. (3) Reward. (4) A fee payable by the Insured, *q.v.*, to the Insurer, *q.v.*, in consideration of a contract of insurance.

Premium Advice Note An accounting procedure by Lloyd's and the insurance companies. A note is issued by the insurer to the insured advising him when the Insurance Policy, *q.v.*, is due for renewal. *Abbrev.* PAN *or* P.A.N.

Premium Hunter *See.* **Stag.**

Prepaid *See* **Advance on Freight.**

Pre-paid Ticket Advice Passenger's ticket for an air passage which has been paid for in advance in a different country or place and is to be collected from a specified office. *Abbrev.* PTA.

Prepared Timber Cut timber used for joinery and building. *See also* **Tongued and Grooved.**

Pressure Vessels Quality Assurance Industrial activity recognised by Lloyd's Register of Shipping and identical to ASME, *q.v.* Qualified surveyors under the auspices of the Institute of Mechanical Engineers provide an assurance that materials used in pressure systems, pressure vessel components, oil refineries and petro-chemical plants, etc., are manufactured to British Standards. *Abbrev.* PVQA.

Presumed Total Loss When the insured subject-matter is considered to be a total loss for the purpose of settling the insurance claim.

Prf. Proof.

PRF *or* **P.R.F.** Pulse Repetition or Recurrence Frequency *radio bearing.*

PRI Price Commission Index.

Price Current Price list of commodities produced by brokers, dealers and similar businessmen. This can also mean the up-to-date rates of bullion, securities, etc.

Price Cutting Cut-throat competition, *q.v. See* **Price War.**

Price Maintenance Protection against price cutting in connection with goods sold to consumers.

Price out of Market When the wages and cost of living of an exporting country rise to an unrealistic level to the extent of exceeding those of other competitor countries. This can easily cause the loss of the export market.

Price Rates A method of payment by piece work, namely the worker is paid in accordance with the amount of work he does per day. This system is advantageous for both the 'fast' worker and his employer.

Price Ring *or* **Price Association** Group of companies or firms who combine to stabilise their prices and not compete with one another.

Price War Cut throat competition, *q.v.*

Prima facie *Latin*—On the first impression; At first sight; On the face of; On the first appearance.

Prima facie case A case which initially seems to be true. *See* **Prima facie**.

Prima facie Evidence A clear indication of truth unless it can be proved to be otherwise. *See* **Prima facie**.

Primage An additional percentage charge on freight. Originally primage used to be reserved for the master to attend and care for the cargo on board. Subsequently this additional percentage was retained by those owners who formed a pool under the title of Conference Lines, *q.v.*, and eventually refunded to their respective regular clients or shippers. The slang term is Hat Money.

Primary Products Products appertaining to agriculture, fisheries, mining, forestry and other raw materials.

Prime Constructor *or* **Prime Constructors** Privately owned shipyards that are awarded contracts by the federal states of Amercia. *Abbrev.* PC.

Prime Cost (1) A sum of money paid to the merchant excluding all trade discounts. This normally refers to the price given by a commission agent, who imports the goods for his various clients en block on which he is allowed a Trade Discount, *q.v.*, for his efforts. (2) Overall direct expenditure on materials, labour and overheads. Also termed Direct Expenditure.

Prime Entry (1) Before a ship is allowed to discharge her merchandise the cargo manifest is to be handed over to the customs authorities. (2) An entry in book-keeping.

Prime Rate The interest rate most devoid of financial risk *USA*. The Federal Reserve Bank's rate on advances to the banks.

Primers The first treatment to be applied when painting on bare metal or wood, etc., before the finishing paints are applied. *See* **Anti-Corrosive Paint**.

Prime Meridian *See* **Zero Meridian**.

Primus inter pares *Latin*—First among equals.

Principal (1) First in rank or importance. (2) The head of certain institutes or colleges. (3) Constituting the original sum of money invested or lent. (4) One to whom another, particularly an agent, is subordinate. (5) Capital sum as distinguished from interest or from income.

Printed Matter A sender's declaration on packages or parcels containing printed papers or advertisements or books in order to obtain reduced rates. One of the conditions is to leave the envelope unsealed.

Print Flow A Royal Mail service handling printed material in bulk for delivery abroad.

Private Bond A store or warehouse belonging to an individual or a company used to deposit or store imported dutiable goods pending their ultimate destination or consumption. A private bond is found very useful for companies who are in the business of re-exportation and/or for the supply of ships in harbour as duty is saved when the stores are re-exported or supplied as free of duty items on ships. The private bond is locked by the keys of the proprietor and also by those of the customs authorities.

Private Carrier A carrier involved in carrying a specific kind of merchandise as opposed to a common carrier who undertakes to accept any type. It could be a contractor whose main business is to deal and offer his services to a named or specific client and is not committed to the general public.

Private Company A company formed by a number of individuals where the shares are restricted to certain named persons and are not issued to the public.

Private Medical Attendant's Report A practitioner's medical report on the assured proposing Life Assurance. *Abbrev.* PMAR.

Private Sector That part of a country's economy engaged in private enterprise.

Private Ship A warship which is not a flagship.

Private Wording Scheme *See* **Special Wording Scheme**.

Privatise To remove a business or service from State or Government operation and control and put it in the hands of a private enterprise or public company.

Privileged Ship *or* **Privileged Vessel** A vessel having the right of way in navigation when nearing others. *See* **Burdened Vessel** *or* **Burdened Ship**.

Privileges American commercial term for Options.

Privity A common word used in shipping and insurance referring to the actual knowledge and consent in an action.

PRMSA Puerto Rico Maritime Shipping Authority.

PRO *or* **P.R.O.** Public Record Office; Public Relations Officer whose duty is to boost the company or the state, whichever the case may be.

Pro Procurationem *Latin*—When signing for a firm or a company the word 'pro' or 'for' is inserted before the name denoting that the person signing is not one of the principals or proprietors but simply an employee or who does not bind himself or commit himself on final decisions. *See* **Per Pro**.

Probate Official proving of a will.

PROBO Product/Ore/Bulk/Oil Carrier.

Proceeds Net value obtained in the sale of a property or similar negotiations; the yield or produce.

Proceeds t.b.a.f. Proceeds to be accounted for *insurance claims.*

Proceed With All Due Despatch A charterparty clause. Under English common law it is implied in contracts of carriage that the shipowner must perform the voyage with reasonable speed.

Produce The amount produced, especially agricultural and natural products collectively.

Produce Exchange An exhibition of the products in a specified building or market to be sold to interested buyers. *See also* **Commodity Exchange**.

Productivity Output. In most cases this refers to labour productivity.

Productivity Bargain A wage agreement designed to improve output and labour productivity.

Productivity Bonus Bonus given over and above wages and related to the height of the output achieved over a period.

Product Tanker A tanker used for the distribution of refined oil products.

Prof. Professor.

Prof. Com. Profit Commission *reinsurance.*

Profit & Loss Account In book-keeping accounts a summary of the revenue and the general expenses for the business are set out in addition to the Balance Sheet. In the Profit and Loss Account there are two columns showing the Assets and Liabilities.

Profits à prendre *French*—Advantages for the taking. A legal right of a person to make use of another's property by, for example, crossing over a field or grazing animals on it.

Profit Sharing Distribution of a percentage of the profit earned by a company to its employees in addition to their normal wages. It may be in the form of shares in the company. The aim is to strengthen ties or loyalties with the company. However, the practice could turn out to be unfavourable for the employees in the sense that if the company makes a loss there will be nothing to share out.

Pro Forma As matter of form; an account showing the market price of specified goods.

Programmed Traverse Light A narrow beam composed of three different lights, red, white and green, with the white light in the middle and the others on either side. They are operated individually or collectively to direct and instruct pilots and masters on vessels approaching terminal ports as to the required speed and positioning. This electronic device is a good step forward for safety in the inland navigation of large tankers and bulk carriers in US waters. *Abbrev.* PTL.

Progressive Tax A tax imposed in proportion to the increase of income.

Pro hac vice *Latin*—On this occasion.

Prohibitory Injunction *or* **Restrictive Injunction** *See* **Injunction**.

Projected Cash Flow This is perhaps one of the most important items a shipowner has to consider in his ownership of a vessel, particulary when on Mortgage, *q.v.* He has to investigate in detail the overall expenses vis-à-vis the total income for say a period of one year. By so doing he will be in a position to estimate the profit and loss situation. *See* **Cash Flow**.

Prolonged Blast *or* **Long Blast** The sound or blast of a ship's siren for a time of about five seconds *nautical.*

PROM Programmable Read-Only Memory *electronics.*

Prom.dk. *Lloyd's Register notation* for Promenade Deck.

Promissory Note A note promising to pay a certain person a stated sum on a specified date. *Abbrev.* PN *or* P.N.

Promoter The originator in the formation of a company.

Prompt Immediate availability of the cargo or vessel offered in the charterparty or any other contract. *Similar to* **Spot**.

Prompt Cash Payment to be settled within a certain time limit, say ten days.

Prompt Delivery *See* **Delivery Terms**.

Prompt Ship A ship which is available on the spot and which can proceed to the port of loading almost immediately.

Proof Load Maximum test load. A lift and tackle calculation generally taken as twice the safe working load.

Proof of Interest *See* **Interest or no Interest** and **Policy Proof of Interest**.

Prop. Propeller, *q.v.*; Proportion.

Propane A colourless liquefied gas being a by-product of petroleum gas widely used in petrochemical industries. It is symbolised C_3H_8.

Propellant Liquid and solid materials commonly used in rocket engines as fuel.

Propeller The screw of a ship. *Abbrev.* prop.

Proper Return Port A seaman's engagement port. This could also mean the national place of the seaman. When a seaman is repatriated the proper return port becomes that port where he originally signed on. *See* **DBS** *or* **Distressed British Seaman**. *Abbrev.* PRP *or* P.R.P.

Proper Route In the Contract of Affreightment, *q.v.*, the contracted ship is bound to pass through the proper route and deviation is not allowable unless to save life at sea or in case of other emergencies.

Property Accounts A book-keeping term dealing in various items.

Property Irregularity Report A form issued by airlines when a complaint is lodged by a passenger for any baggage found missing, contents missing or damaged. *Abbrev.* PIR.

Propocon The code name for a BIMCO charterparty for shipments of pit props from the Baltic to the European Continent.

Proposal Form A printed form used for insurance purposes. The would-be insured will be asked to answer the questions on the form before the insurance is accepted and confirmed.

Proprietary Of a proprietor; held in private ownership as proprietary medicines, the sale of which is restricted by patent, etc.

Proprietary Rights The right of ownership. This generally refers to the remaining damaged property transferred to the Underwriters, *q.v.*, after the settlement of a Total Loss, *q.v. See* **Letter of Subrogation**.

Propter hoc *Latin*—Because of this.

Pro Rata In proportion.

Pro Rata Freight *See* **Pro Rata Rate**.

Pro Rata Rate The calculated rate in proportion to the whole; according to the proper share or scale.

Pro rege, lege et grege *Latin*—For the king, the law and the people. For ruler, rule and ruled.

Prospectus A call, invitation or offer to the public for the purchase of Shares, *q.v.*, or Debentures, *q.v.*, of a company; proposed commercial operation outlined in the form of a pamphlet.

Protection and Indemnity Club *or* **Protection and Indemnity Association (P & I Club)** A mutual association formed by shipowners to provide protection from large financial loss to one member by contribution towards that loss by all members. The P & I Club covers liabilities not insurable by the shipowner in the running of his ship, such as cost of defending claims made by cargo owners.

Pro.t.b.a.f Proceeded to be accounted for *insurance*.

Pro tem Pro tempore *Latin*—For the time being.

Protest Flag Manifestation of a protest during a yacht race which must be decided upon before the winner is officially declared. This is shown by hoisting the international B flag.

Provisional Certificate In general terms it is a temporary certificate until a complete or full one is obtained. A classification society issues a provisional certificate for a ship to sail from one port to another and on arrival the certificate is no longer valid.

Provisional Premium Premium Deposit.

Proviso A stipulation; a clause or condition in a deed or contract.

Proximate Cause The most effective cause of a loss in a chain of events leading to the loss. A basic principle of insurance in that, unless the policy provides otherwise, the underwriter is not liable for any loss that is not proximately caused by an insured peril.

Proxy *Similar to* **Power of Attorney**, *q.v.*

PRP *or* **P.R.P.** Petrol Refilling Point; Proper Return Port, *q.v.*

p.r.p. Proper Return Port, *q.v.*

PRR *or* **P.R.R.** Pulse Repetition or Recurrence Rate *radio bearings*.

P.R.S. President of the Royal Society (London).

PRS Polski Rejestr Statskow *Polish classification society.*

P.R.S.A. President of the Royal Scottish Academy.

P.R.S.E. President of the Royal Society of Edinburgh.

PRT Petroleum Revenue Tax.

PRT Pressure Relieve Valve.

P.s *or* **P.S.** *or* **P/S** Paddle Steamer; Parliamentary Secretary; Permanent Secretary; Part Shipment; Passenger Steamer; Passenger Service; Private Secretary; Privy Seal; Postscriptum *Latin*—Postscript; Public Seal; Public Sale.

PS *or* **P.S.** Paddle Steamer.

P&S Port & Starboard; Purchase & Sale; Purchasing and Supply Department.

PSA Particularly Sensitive Area *connected with the risk of oil pollution by tankers*; Property Services Agency, *government agency within the UK Department of the Environment having responsibility for design, building and construction support*; Prostate Specific Antigen *medical*.

PSA *or* **P.S.A.** Passenger Sales Agency; Port of Singapore Authority; Passenger Shipping Association *UK*; Pressure Swing Absorption; Post Shakedown Availability *USA*.

PSAR Passenger Sales Agency Rules.

PSAT Preliminary Scholastic Aptitude Test.

PSBR Public Sector Borrowing Requirement, *q.v.*

PSC *or* **P.S.C.** Pakistan Shippers Council; Public Service Commission.

Psc *or* **psc** Per Standard Compass, *q.v.*

PSCD Patrol Service Central Depot.

P.sett. Previous Settlement.

P.S.F. *or* **p.s.f.** Pounds per Square Foot.

PSG *or* **Psg.** Passing.

PSHFA Public Servants Housing and Finance Association.

PSI *or* **p.s.i.** Pounds per Square Inch; Pressurised Sphere Injector.

PSIA *or* **p.s.i.a.** Pounds per Square Inch Absolute.

P/side Port Side.

PSIG *or* **p.s.i.g.** Pounds per Square Inch Gauge.

PSK *or* **P.S.K.** Pulse Shift Keying.

psn Pension.

PSO Pollution of the Sea by Oil.

PSQC Philippines Society for Quality Control.

PSS Port Safety and Security; Principal Support Unit or Ship; Platform Supply Ship.

PSSC Passenger Ship Safety Certificate.

PST *or* **P.S.T.** Pacific Standard Time, *US Pacific coast.*

p.stg.c Per Steering Compass *nautical.*

pstl. Postal.

PSTN Public Switched Telephone Network *computer.*

PSV Platform Supply Vessel; Public Service Vehicle.

PSW *or* **P.S.W.** Psychiatric Social Worker.

Psychrometer Wet and dry bulb thermometer for measuring the amount of moisture in the atmosphere.

PT *or* **P.T.** Perte totale *French*—Total Loss *insurance*; Parcel Ticket; Physical Training; Premium Transfer; Pacific Time; Preferential Treatment; Public Trustee; Purchase Tax; Private Terms; Post & Telegraphic Service.

pt. Pint; Part; Point.

p.t. Private Terms; Pro tempore *Latin*—For the time being.

PTA *or* **P.T.A.** Port of Tyne Authority; Passenger Terminal Amsterdam; Prepaid Ticket Advice *aviation*; Public Transport Aircraft *aviation.*

PTA *or* **P.T.A.** Passenger Transport Authority.

Pt.B. *or* **pt.B.** Part Bunkers.

Pt.C. *or* **pt.C.** Part Cargo Tanks.

PTC *or* **P.T.C.** Personnel Transfer Capsule; Produce Time Charter.

Pte Private Company *Singapore*; Private soldier.

PTFE Polytetrafluorethylene *synthetic material, q.v.*

Ptg.Std. Petrograde Standard *timber trade.*

PTI *or* **P.T.I.** Pre Trip Inspection; Preventive Technical Inspection.

PTL *or* **P.T.L.** Partial Total Loss. *Similar to* **Partial Loss**, *q.v.*; Presumed Total Loss, *q.v.*; Programmed Traverse Light, *q.v.*

PTM Phase Time Modulation.

PTO *or* **P.T.O.** Please Turn Over; Power Take Off *aviation*; Public Trustee Office; Public Trustee Officer.

PTP Paper Tape Printer.

PTPFT *or* **P.T.P.F.T.** Policy Third Party Fire and Theft *vehicle insurance.*

pt/pt Point to Point.

PTS Production Testing Ship *oil drilling.*

PTSA Port and Tanker Safety Act, *q.v.*

pts/hr Parts or pieces per hour.

Ptsmh Portsmouth, *UK.*

PTT *or* **P.T.T.** Postes, Télégraphes, Téléphones *French*—Mail, Telegraph and Telephones.

PTTI *or* **P.T.T.I.** Postal Telegraph and Telephone International.

Pt/tm *or* **pt.tm** Part Time.

PTV *or* **P.T.V.** Propulsion Test Vehicle.

ptw Per thousand words.

Pty Proprietary.

PU *or* **P.U.** Public Utilities.

p.u. Paid Up.

Pub. Public.

Public Address Addressing or speaking to the passengers on an aircraft by the pilots or stewards; a public address system. *Abbrev.* PA *or* P.A.

Public Company A company, *q.v.*, or incorporated body which may have a minimum of shareholders and the bulk of the shares are offered to the general public to buy.

Public Enemies Persons considered to be enemies of a country.

Public Limited Company A firm or company formed by public shareholding, *Abbrev.* PLC *or* P.L.C. *or* plc *or* Plc.

Public Ownership Nationalised industries.

Public Policy The principles under which freedom of contract or private dealings is restricted by law for the good of the community.

Public Revenue Tax and other similar state revenue.

Public Sector Borrowing Requirement The excess of the government's spending over its receipts. The UK Government adopted the term in 1980. It can be financed either by increasing the money supply or by borrowing, depending on fiscal policy. *Abbrev.* PSBR.

Public Utility Undertaking A company or other body or person authorised by or under any Act, or under any order having the force of an Act, to construct, work, or carry on a railway, canal, inland navigation, dock, harbour, tramway, gas, electricity, water or other public undertaking. *Abbrev.* PUU *or* P.U.U.

Pudoris causa *Latin*—for reasons of shame.

Pull To row with an oar; to so propel a boat.

Pulpit Safety rail in the bow of a yacht.

Pumping Clause A Charterparty, *q.v.*, in oil tanker hire in which the owner guarantees the capacity of discharging the total cargo within 24 hours or maintaining 100 PSI, *q.v.*, at Ship's Rail, *q.v.*, always provided the receiver can cope with this out turn.

Pump Room A compartment on a tanker from which pumps are controlled.

Puncheon A large cask to contain wine, beer, rum and other liquids having an average capacity of 72 to 120 gallons.

Punica fides *Latin*—Punic faith; Treachery.

Punt (1) Irish pound, currency of the Irish Republic. (2) Flat-bottomed boat for river work or for maintenance work near the waterline on ships.

Purchase Also refers to getting a secure hold or leverage in order to heave. *See also* **Chain Purchase**.

Purchase Option *See* **Option**.

Purchaser Buyer.

Purchasing Officer An executive in charge of buying articles for his company's ships.

Pure Car/Carrier A high-sided Ro/Ro, *q.v.*, vessel of distinctive somewhat box-like appearance with many decks to accommodate a cargo of motor vehicles which are driven on and off over its ramps. *Abbrev.* PCC.

Pure Car and Truck Carrier As a PCC, *q.v.*, but with the added facility of being able to carry high sided vehicles—such as trucks—in addition to cars. *Abbrev.* PCTC.

Purser A ship's officer who is in charge of accounts, etc., especially on a passenger ship; a paymaster on a naval ship. *See* **Chief Steward**.

PURV *or* **P.U.R.V.** Powered Underwater Research Vehicle.

PUS *or* **P.U.S.** Parliamentary Under Secretary; Permanent Under Secretary; Pharmacopoeia of United States; Permanently Unfit for Service.

Pushpit Safety rail at the stern of a yacht.

Put *See* **Call** *or* **Calls**.

Put About To go on to the other tack. To change a sailing ship's or boat's heading to bring the wind on the other bow.

Put and Call *See* **Call** *or* **Calls**.

Put Back Return, referring to a ship returning to her original port of departure for some unforeseen reason.

Put Forth Leave port, set sail, go out to sea.

Put In To enter a port of refuge for repairs.

Put of More The right to sell more on a stock exchange. *Abbrev.* P/m.

Putrefaction The deterioration of perishable merchandise.

Putting Back Clause *or* **Put Back Clause** A standard Time Charter clause. Should the vessel deviate from her course or put back during a voyage the hire rate shall cease until the resumption of the voyage. If the cause is due to some fault of the charterer then the expenses are to be borne by the latter.

Put to Sea To sail.

PV *or* **P.V.** Petite Vitesse *French*—Freight Train; Prime Vertical *nautical*; Petrol Vessel; Piccola Velocita *Italian*—Slow Train.

PV Photovoltaic Energy, *q.v.*

PIV Valve Pressure/vacuum valve designed to relieve overpressurising the structure of a cargo tank.

PVA Passenger Vessel Association; Polyvinyl Acetate.

PVC Polyvinyl Chloride *plastic material*.

p.v.c. Pigment Volume Concentration, *q.v.*

PVD Pumping Vapour Deposition.

PVDC Polyvinylidene Chloride.

PVF Polyvinyl Fluoride.

PVG Polyvinyl Glycol.

PVQA Pressure Vessels Quality Assurance, *q.v.*

p.v.t. Par Voie Télégraphique *French*—Telegraphically.

p.w. Per Week.

PW Pulse width; Prisoner of War *American*.

PWA Public Works Administration *USA*.

pwd. Powered.

PWD *or* **P.W.D.** Public Works Department.

PWHT Post Weld Heat Temperature, *q.v.*

PWR Pressurised Water Reactor *nuclear power plant system*.

PWSA *or* **P&W.S.A.** Ports and Waterways Safety Act, *q.v.*

pwt. Pennyweight.

PX Passenger.

Px. Physical Examination; Post Exchange; Private Exchange; Please Exchange.

P.X. Please Exchange.

PXM Pigging Crossover Module *oil drilling*.

PX Ship Passenger Ship.

Pylon A gateway, especially of an Egyptian temple; A tall compound structure erected as a support (especially for power cables) or as a boundary or decoration.

Pyrometer An instrument for measuring high temperatures. Alternatively for measuring the intensity of the sun's rays or any solar power of identical intensity.

Pyrotechnics The art of making and displaying fireworks as well as rockets and flares commonly used to show distress signals during emergencies. *See* **Marine Pyrotechnics**.

Q

Q *Lloyd's Register notation* for a ship with quadruple screws, also for quadruple expansion engines; Polaris Correction *air navigation*; Q Flag, *q.v*; Quartermaster; Quebec, *Canada*; Queensland, *Australia*; Query; Question; Quarto.

Q 10¹⁸ British Thermal Units (1.05 × 10²¹ joules).

QA Quality Assurance; *Lloyd's Register notation* for Quality Assurance, *q.v.*

q.a. Quick Assembly.

Q & A Question & Answer.

QAAS Quality Assurance Advisory Scheme.

QAD *or* **Q.A.D.** Quality Assurance Data System.

QAFCO CHARTER Fertiliser Voyage Charterparty.

Qantas *or* **QANTAS** *or* **Q.A.N.T.A.S.** Queensland & Northern Territory Aerial Service, National Australian Airline.

QAP Quality Assurance Programme.

Q.A. Scheme *or* **Q.A.S.** Quality Approved Scheme, *q.v.*

Q.A. Services Quality Assurance Services.

Q.B. Queen's Bench.

Qbc Quebec, *Canada.*

QBL Qualified Bids List *USA.*

QC *or* **Q.C.** Quick Change, *q.v.*; Queen's Counsel; Quality Control, *q.v.*; Quantity at Captain's Option, *q.v.*

QCC Quality Control Circle.

Q.C.Isl. Queen Charlotte Islands.

QCO *or* **Qco** Quantity at Captain's Option.

Q.C.R. Quality Control Reliability.

Q.C.T. Quality Control Technology.

q.d. Quasi dicat *Latin*—As if one should say; Quasi dictum *Latin*—As if said.

q.e. Quod est *Latin*—That is to say; which is.

QEA *or* **Q.E.A.** Quantas Empire Airways.

QED *or* **Q.E.D.** Quod erat demonstrandum *Latin* —Which was to be demonstrated.

QEF *or* **Q.E.F.** Quod erat faciendum *Latin* —Which was to be done.

QEI *or* **Q.E.I.** Quod erat inveniendum *Latin* —Which was to be found out.

Q.F. Quality Factor; Quick Firing.

QF Quantas Airways *airline designator.*

QFE Quartz-Fibre Electroscope *used to detect radio-active sources.*

Q Flag A small square yellow flag denoting that the master of a ship is requesting customs authority presence as he is not yet cleared and is under quarantine. Also called Quarantine Flag and Yellow Duster. *See* **Quarantine Restrictions**.

QGA Code for 'May I Transmit Radio Signal?'

Q.H.M. Queen's Harbour Master or K.H.M. for King's Harbour Master.

Qindarka Currency unit of Albania.

qk Quick.

q.l. Quantum libet *Latin*—As much as you please.

Qld Queensland, *Australia.*

Qlty Quality; Quarterly.

Q.M. Quartermaster, *q.v.*; Queen's Messenger.

q.m. Quo modo *Latin*—By what means.

QMED Qualified Members Engine Department, *q.v.*

Q.Mess. Queen's Messenger or K. Mess. for King's Messenger.

QMG *or* **Q.M.G.** Quartermaster-General.

QMS *or* **Q.M.S.** Quartermaster-Sergeant.

qn Question; Quotation.

QNS *or* **Q.N.S.** Quantity not Sufficient.

qnty Quantity.

q.pl. Quantum placet *Latin*—As much as seems good; at your discretion.

QPC *or* **Q.P.C.** Quasi Propulsion Coefficient.

QPL Qualified Products List.

QPM Queen's Police Medal.

QPPA Qatar Petroleum Producing Authority.

QQ Celestial Equator *nautical.*

QQC Quantity and Quality Certificate, *q.v.*

qq.v. Quae vide *Latin*—Which see (refers to plural).

QRA Quantified Risk Assessment.

Q.R. & A.I. Queen's Regulations and Admiralty Instructions.

QRC Quality Radio Certificate.

Q/S Q.S. Quarter Sessions; Quota Share *re-insurance*; Queen's Scholar.

q.s. Quantum sufficit *Latin*—As much as suffices; sufficient quantity.

QSCS Quality System Certification Scheme. *See* **180, 9000**.

Q.S.M. Quadruple Screw Motor/Steamer.

QSO Quasi-Stellar Object *astronomy.*

QSRS Quasi Stellar Radio Source *astronomy.*
QSL Qualified Services List.
Q.S.S. Quadruple Screw Steamer.
Q.S.T.S. Quadruple Screw Turbine Ship.
Q.T.E.V. Quadruple Turbo-Electric Vessel.
QTOL Quiet Take Off and Landing.
Qt Quart.
Qtr Quarter.
Qtly Quarterly.
Qtrs Quarters.
Qty Quantity.
qu. Question.
qua *Latin*—In the character of; as; as being.
Quad Quadrangle; Quadruped; Quadruple; Quadruplet.
Quadrant (1) An instrumeent used for the purpose of navigation by measuring altitudes. Similar to the marine sextant; (2) A flat steel plate in a form of a quadrant fixed at the top of the rudder to which the steering chains are connected.
Quadrennial Every four years.
Quadrivial Pertaining to where four roads meet.
Quads *or* **QUADS** Quality Assurance Data System.
Qualified Acceptance A Bill of Exchange, *q.v.*, term referring to various conditions which affect its payment, as opposed to General Acceptance.
Qualified Members Engine Department Qualified engineers in the USA who attend to a fully automated engine room. *Abbrev.* QMED.
Quality Approved Scheme *Lloyd's Register* sign of approval of a strict quality control method in accordance with international standards of manufacture. *Abbrev.* Q.A Scheme *or* Q.A.S.
Quality Assurance Procedure whereby every stage of, and method used in, the project management, design, calculation and manufacture, construction or provision of a product or structure is subject to examination and certification by an independent authority such as a Classification Society, *q.v.*, to ensure the achievement of the standards required. *Abbrev.* QA.
Quality Control A system undertaken by the manufacturers of various products to ensure that their products are being produced to certain standards of quality for the public good.
Quality & Quantity Unknown A Bill of Lading, *q.v.*, clause commonly inserted to exonerate the owners from liability in so far as the quantity and quality of the contents of packages is shown. If the freight is paid on weight or Ad Valorem, *q.v.*, the clause does not apply.
Quality of Bills of Exchange The grading of Bills of Exchange, *q.v.*, which are taken into account according to their popularity and apparent solvency.
Quam proxime *Latin*—As nearly as possible.
Quango Quasi-autonomous non-governmental organisation.

Quantity and Quality Certificate A document or certificate occasionally used for the purpose of reporting on the quantity and quality of every shore tank during the process of loading a tanker. *Abbrev.* QQC.
Quantity at Captain's Option The maximum amount of merchandise to be loaded on board a ship or an aircraft is at the option of the captain who is the person responsible for the safety of the vessel or plane. *Abbrev.* QC *or* Q.C.
Quantity Rebate A rebate allowed to a buyer who is engaged in purchasing a large quantity of merchandise.
Quantum *Latin*—A concrete quantity; a natural minimum quantity of an entity.
Quantum meruit *Latin*—As much as he has earned. If a person enters into a contract to perform services for another, and either the contract is put an end to before they are completed, or they are not rendered in the manner provided by the contract, the contractor is obviously not entitled to be paid his contract price, but in some cases he is entitled to be paid the actual value of his services; and if he brings an action to recover it, he is said to sue on a *quantum meruit*. Thus, where the failure to complete performance of the contract is due to the fault of the other party, the party not in default has the right to sue on a *quantum meruit* for the services which he has done under it.
Quantum of Interest Invisible partial interest. This refers to the insurance covering both the mortgagee and the mortgagor who are distinct and separate as far as insurance purposes are concerned. Consequently the contribution between their respective underwriters, as regards over insurance or double insurance, does not apply.
Quantum valebant *Latin*—As much as the goods are worth.
Quantum vis *Latin*—As much as you will.
Quarantine Isolation of persons or animals with infectious disease or potentially carrying an infection. Isolation at an anchorage of a ship which does not have Pratique. The term is from the Italian quaranta meaning forty, the period in days for confinement to a Lazaretto, *q.v.*
Quarantine Restrictions Restrictions imposed by the health authorities of a state as a precaution against the spreading of infectious diseases. If a ship arrives at a port with a suspected infectious disease on board, all officers, crew and the ship itself are kept under quarantine unless it can be clearly shown that they are actually recovered from the suspected disease. When the ship is kept under quarantine the Yellow Flag (Q Flag) is hoisted. During the dark hours quarantine is shown by red and white lights clearly seen against the main mast. As long as the ship is on quarantine no

person is allowed to go on board except the appropriate authorities and no one from the ship is authorised to disembark. In the United Kingdom cattle, etc., imported and not slaughtered at the port of landing are required to be kept in quarantine. Dogs and cats imported from abroad must be kept in quarantine for six months.

Quart A liquid measure in the Imperial system equal to two pints. Four quarts are equal to one Imperial gallon.

Quartation The mixing of three quarters of silver to one portion of gold to purify gold.

Quarter (1) A wine cask with a capacity of one-quarter of a Pipe, *q.v.* (2) Eight bushels of grain and equivalent to 400 lbs. weight and 320 lbs. for oats; also 28 lbs., the fourth part of a hundredweight. (3) Every three months, being one-quarter of 12 months or a year. (4) At an angle to the stern of a craft.

Quarterage Quarterly payment or allowance. Wages paid every three months of a year.

Quarter-Bill (1) A list showing the stations or duties of every officer and man to take when summoned to their 'quarters' for action or exercises on a ship for emergency purposes. (2) A bill produced or furnished every three months of the year.

Quarter-Deck Better known as a raised quarter-deck which is an erection at the stern end of a ship to allow more cargo to be carried.

Quarter Girt A timber measurement term. The measurement of logs is generally calculated by a string around the log and after it is measured or folded into a quarter or four parts the cubic measurement can then be taken.

Quarterly Trade Accounts Accounts falling due during the end of March, June, September and December of every year. Quarter-days are 25 March, 24 June, 29 September and 25 December.

Quartermaster *or* **Helmsman** A seaman on an ocean-going vessel who steers on orders from the captain or other officers. Also, in a warship, a seaman keeping the watch on deck in harbour.

Quarter Mile A distance of 440 yards, taken as a standard measure for a foot race.

Quartern Quarter of a pint or one gill. Also quarter of a peck, which is a measure of capacity for dry goods equal to two gallons.

Quarter Side *Similar to* **Quarter Wind**, *q.v.*

Quarter Wind A wind blowing on the Quarter, *q.v.*, of a ship. Also termed 'Wind blowing on the Quarter Side'.

Quartzite Metamorphosed sandstone.

Quartz Crystal form of silica found in granite, rocks, sandstone, etc., occasionally containing gold. *Abbrev.* Qz.

Quasi Contract A legal obligation arising in the absence of a contract, where for the purpose of substantial justice there is an implied contract. For instance, a salvage claim could be made in the absence of mutual agreement.

Quasi dicat *Abbrev.* q.d., *q.v.*

Quasi dictum *Abbrev.* q.d., *q.v.*

Quasi Partner A person pretending to be a partner of a company but who in fact is not.

Quayage Charges for the use of a quay.

Quay Dues Dues assessed against a ship for use of a quay. *Also* **Droit de Quai**.

Quay to Quay An insurance phrase. The merchandise is held covered as from the port of loading up to the port of discharging.

Que. Quebec, *Canada*.

Quebec International Radio Telephone phonetic alphabet for 'Q'.

Quebec Standard Measure of timber equal to 299 cubic feet.

Queen's Bench The Court of Queen's Bench or King's Bench was one of the superior courts of common law, having in ordinary civil actions concurrent jurisdiction with the Courts of Common Pleas and Exchequer; it was, however, considered superior to them in dignity and power, its principal judge being styled the Lord Chief Justice of England, and taking precedence over the other common law judges, and there being formerly an appeal to it from the Exchequer and the Common Pleas. It also had special jurisdiction over inferior courts, magistrates and civil corporations by the prerogative writ of mandamus, and (concurrently with the two other courts) by prohibition and certiorari and in proceedings by *quo warranto* and habeas corpus. It was also the principal court of criminal jurisdiction: informations might be filed and indictments preferred in it in the first instance, and indictments from inferior courts might be removed into it by certiorari, subject to certain limitations. By the Judicature Acts, 1873–75, the jurisdiction of the court was transferred to the High Court. Its judges (namely, the Lord Chief Justice and the five puisne justices) formed a separate division of the High Court, called the Queen's Bench Division, to which was assigned all business (civil and criminal) which was formerly within the exclusive cognisance of the Court of Queen's Bench. In 1881, however, the Common Pleas and Exchequer Divisions were consolidated with it into one division called the Queen's Bench Division.

Queen's Bench Division The jurisdiction of the Court of Queen's Bench was assigned by the Judicature Act, 1873, s.34, to the Queen's Bench Division of the High Court, and by Order in Council under s.32 of the Act, the Common Pleas and Exchequer Divisions were, in February, 1881, merged in the Queen's Bench Division, which was styled from the death of Queen Victoria in January, 1901, until the death of George VI in February, 1952, the King's Bench Division.

Queen's Chambers Those portions of the seas, adjacent to the coasts of Great Britain, which are enclosed within headlands so as to be cut off from the open sea by imaginary straight lines drawn from one promontory to another. They appear to have always formed part of the territorial waters of the Crown

Queen's Counsel Barristers who have obtained the appointment of counsel to the Crown by reason, as old writers put it, of their learning and talent. They wear silk gowns, sit within the bar, and take precedence in court over utter barristers, that is to say, the ordinary barristers, who sit outside the bar. They have no active duties to the Crown to perform, but they could not formerly be employed in any cause against the Crown (e.g., in defending a prisoner) without special licence. Now, however, by a general dispensation granted in 1920, they can appear against the Crown without such licence. There used to be also Queen's Counsel in the County Palatine of Lancaster (Judicature Act, 1873, s.78), who took precedence of other barristers in the Palatine Courts. All Queen's Counsel at the death of Queen Victoria became King's Counsel, and all King's Counsel at the death of George VI became Queen's Counsel without any new appointment.

Queen's Evidence A prisoner who, instead of being put upon his trial, is permitted by the Crown to give evidence against those associated with him in crime, is said to turn Queen's evidence.

Queen's Regulations The Queen's Regulations and Orders for the Army are issued by the Crown, through the Army Council, for the government of the army. Their legal force depends upon the constitutional principle that the Crown commands the army subject to statutory regulation.

The Queen's Regulations and Admiralty Instructions are made by Order in Council countersigned by the Lords of the Admiralty, and they, along with the Articles of War, constitute the code under which the navy is governed.

Queen's Warehouses Places provided by the Crown and approved by the Commissioners of Customs and Excise for the deposit of goods for the security thereof and of the duties thereon, including unclaimed goods or seized goods, e.g., contraband.

Quetzal Unit of currency used in Guatemala. One hundred centavos equals one quetzal.

Quick Assets Liquid Assets, *q.v.*, which are easy and ready to cash.

Quick Change A prompt action in offering additional airfreight space to another operator having insufficient space for all the cargo in hand.

Quick Despatch *See* **Quick Turn Round**.

Quick Ratio. *See* Acid Test Ratio.

Quicksilver Mercury, *q.v.*

Quick Turn Round *or* **Quick Despatch** A commercial term for a speedy loading and/or discharging operation.

Quid pro quo *Latin*—Something in return; an equivalent.

Quieta non movere *Latin*—Not to move quiet things; Let sleeping dogs lie.

Qui facit per alium facit per se *Latin*—He who acts through another is deemed to act in person. A principal is liable for the acts of his agent within the scope of his authority.

Quintal A weight of 100 kilograms, equal to $220\frac{1}{2}$ lbs.

Quintant A sextant, *q.v.*, with a range angle of 114 degrees.

Quire Twenty-four sheets of writing paper or 26 copies of newspapers.

Qui tacet consentit *Latin*—He who remains silent consents.

Quittance An acquittance;. a release (from debt or other similar responsibilities).

Quod Slang for prison.

Quod hoc *Latin*—As far as this is concerned.

Quod erat demonstrandum *Abbrev.* QED, *q.v.*

Quod erat inveniendum *Abbrev.* QEI, *q.v.*

Quod vide *Latin*—Which see (refers to singular). *Abbrev. q.v.*

Quoin *See* **Sconce**.

Quoit Ring of steel, wire or rope used at the end of a rope, which can be thrown to encircle a post or similar projection, or a hook for mooring; a rope grommet or rubber ring used in the game of Quoits and other shipboard pastimes.

Quondam *Latin*—Former; Formerly.

Quor. Quorum, *q.v.*

Quorum The minimum number of persons required to be present in order to make proceedings valid. *Abbrev.* quor.

Quot Quotation.

Quota A share that an individual person or company is bound to contribute to or entitled to receive from a total; a quantity of goods which under government controls must be manufactured, exported, imported, etc.; the number of yearly immigrants allowed to enter the United States from any one country.

Quota Share Reinsurance Coding a specified portion of insurance business a reinsurer. Premiums are paid and claims reimbursed in a like proportion.

Quotation Statement of the premium rates and conditions for insurance given by an insurance broker to a shipowner or Merchant or an individual to enable

Quoted Company

the insured to calculate his costs before giving a firm order for placement. Rates are valid for a limited time. The seller of goods or services may give a quotation, or quote, to a buyer and this is sometimes in the form of a Pro-forma Invoice.

Quoted Company Similar to Listed Company, *q.v.*

Quotient The result given by dividing one quantity by another.

Quo vadis *Latin*—Whither goest thou?

QPC Quasi-propulsive Coefficient.

q.v. Quod vide *Latin*—Which see.

Qy Quay; Query.

qz. Quartz, *q.v.*

R

r Rare; recto; residence; resides; right; rises; rouble; rupee.

R Air fare symbol for 'Concord' class; Radius; Rain *weather observation*; Rand *South African currency*; Railway; Regiment; Rankine *temperature*; Refraction; Réaumur *temperature*; Retarder *on time-piece regulator*; River; Relative Direction; Roentgen, former unit of ionizing radiation dose, now expressed in coulomb per kilogram; Rupee, *q.v.*

R *or* **Rcvd** Received.

(R) Refrigeration.

Ra Radium. Radioactive metallic element occurring in uranium ores, e.g., pitchblende.

R/A Rederiaktiselskab, *q.v.*; Refer to Acceptor.

RA *or* **R.A.** Right Ascension *astronomy*; Rear-Admiral; Royal Academy; Royal Artillery.

RAA Royal Academy of Arts.

RAAA *or* **ROAAA** Reachable on Arrival or Always Accessible. *Charter Party, q.v., reference to Port of Loading or Discharge.*

RAC *or* **R.A.C.** Royal Agricultural College; Royal Armoured Corps; Royal Automobile Club.

RACE Radiation Adaption Compression Equipment; Rapid Automatic Checkout Equipment; Railways of Australia Container Express.

Racer Rankin Closed Energy System.

RACON Radio Beacon, *Abbrev.* R. Bn.

RAD *or* **R.A.D.** Radar, *q.v.*; Radiation; Raised After Deck, Lloyd's Register *notation*; Rear Admiral Destroyers *naval*; Royal Academy of Dancing.

rad. Radian, *q.v.*; Radiation Absorbed Dose; Radiator; Radius.

r.a.d. Rapid Automatic Drill; Radiation Absorbed Dose.

R/D Refer to Drawer (of a cheque).

R & D Research and Development.

RADA *or* **R.A.D.A.** Royal Academy of Dramatic Art.

RADAR Radio Detection and Ranging, *q.v.*

RADAS Random Access Discrete Address System.

RADCM Radar Counter Measure.

Radeau *French*—A form of float or raft employed to transport men and goods.

RA-DE SALM Rapid Deployment Single Anchor Leg Mooring.

RADIAC Radio Activity Detection Indication and Computing *computer.*

Radian SI unit of plane angle (approx. 57.296 degrees) *Abbrev.* rad.

Radiation Fog Fog mostly occurring over the land and formed through the heat emitted. *See* **Advection Fog**.

Radioactive Minerals Radioactive metallic elements such as uranium, thorium, etc.

Radio Aid to Navigation Information transmitted through radio as an aid to navigation. *Abbrev.* RAN *or* R.A.N.

Radio Beacon A fixed transmitter radiating regular signals which ships and aircraft can read to obtain bearings for their navigation; or a mobile one radiating distress signals to assist search and rescue.

Radio Detection and Ranging (RADAR) Apparatus used regularly on land, sea and in the air to ascertain the direction and range of aircraft, ships, coasts and other objects by means of the electromagnetic waves which they reflect. *Abbrev.* RADAR or RAD.

Radio Fix Obtaining a position or a bearing of a vessel by means of a Direction Finder, *q.v.*

Radio Log Book officially recording all radio traffic transmitted and received and kept by the ship's Radio Officer, *q.v.*

Radio Magnetic Indicator An aircraft instrument showing the magnetic heading. *Abbrev.* RMI.

Radio Officer A Certificated, *q.v.*, officer of a ship who keeps watch at the radio transmitter/receiver whilst at sea.

Radiosonde A miniature radio transmitter carried aloft in a balloon and descending by parachute for broadcasting pressure, temperature and humidity at various levels.

Radio Time Signal Time signals regularly transmitted by various stations by radio for use by navigators and others. *Abbrev.* RTS *or* R.T.S.

Radio Urgency Signal An emergency radio message transmitted by the master or other responsible officer of a vessel in a situation of distress. *See* **Distress Signals**.

Radius Vector An imaginary variable line drawn to curve from a fixed point, especially in astronomy from the sun or a planet to the path of a satellite. *Abbrev.* RV *or* R.V.

RADM Rear Admiral *US Navy.*

RADOP Radio Operator.

rad/s Radians per second.

RAE *or* **R.A.E.** Royal Aircraft Establishment(s); Radio Astronomy Explorer.

R.A.E.C. Royal Army Education Corps.

R.Aero.C. Royal Aero Club of the United Kingdom.

R.Ae.S. Royal Aeronautical Society *UK*.

RAF *or* **R.A.F.** Royal Air Force.

RAFA *or* **R.A.F.A.** Royal Air Forces Association.

R.A.I. Royal Anthropological Institute.

R.A.I.A. Royal Australian Institute of Architects.

R.A.I.C. Royal Architectural Institute of Canada.

Rail Air Link A scheduled train service between a city's airport and a central station.

Railex A Post Office express mail delivery service.

RAILS Remote Area Instrument Landing Sensor.

Rake The inclination from the vertical of masts, funnels and other poles on a ship; protruding parts of the stem and stern of a ship.

R.A. Keel Raised Keel.

RAM *or* **R.A.M.** Random Access Memory *electronics*; Right of Ascension of the Meridian; Royal Academy of Music (London); Royal Air Morocco, Morocco State Airlines.

RAMAC Radio Marine Associated Companies.

R.A.M.C. Royal Army Medical Corps.

RAMNAC Radio Aids to Marine Navigation Application Committee.

Ramps The means of access by wheeled vehicles to Ro/Ro vessels, *q.v.*, being capable in modern designs of sustaining heavy loads and sometimes of being rotated in an arc (e.g. a slewing ramp) to accommodate whatever berth is available.

RAMS *or* **R.A.M.S.** Right Ascension of the Mean Sun *astronomy*.

RAN *or* **R.A.N.** Radio Aid to Navigation; Request for Authority to Negotiate; Royal Australian Navy.

RANCON Random Communication Satellite.

Rand Unit of currency used in South Africa. One hundred cents equals one rand.

Random Access Memory A computer microchip assembly containing programmes and data which can be changed by the user. The information generated is stored on tape or disc and can be printed out. *Abbrev.* RAM or R.A.M.

Random Stow The resulting stow of bagged cargo dropped roughly into a cargo hold to economise on labour costs in places where stevedore services are expensive.

Range The distance a vessel can steam or sail with her full bunkers in relation to speed.

Range Finder A device to find out the distance of the target or object.

Range of Visibility The range of sight of the naked eye.

Ranging Clause A clause in a hull policy exempting underwriters from liability under the policy (for damage received as well as for damage done) in the event that the insured ship collides with another ship while the vessels are approaching or ranging alongside each other at sea for the purpose of transferring cargo. Cover may be obtained by arrangement with the underwriters, in practice provided they are notified of the intention to transfer cargo in this manner in advance. The exemption does not apply to customary transhipment in port areas involving inshore harbour craft.

RANN Research Applied to National Needs *USA*.

RAO Response Aptitude Operators.

R.A.O.C. Royal Army Ordnance Corps.

R.A.P.C. Royal Army Pay Corps.

Rap Full Applies to a ship on a wind 'Keep her rap full' means do not sail too close to the wind or let the sails shiver.

Rapid Intervention Vehicle A heavy duty vehicle able to cross rough ground within a short time. It is used during an emergency landing by an aircraft. *Abbrev.* RIH.

Rapid Transport System A transportation concept within an airport area using computerised controlled vehicles without drivers running smoothly on parallel guided elevated tracks. This system originated in the USA. *Abbrev.* RTS.

RAPL *or* **R.A.P.L.** Rotterdam/Antwerp Pipeline.

RAPPI Random Access Plan Position Indicator.

RAR *or* **R.A.R.** Radio Acoustic Ranging; Restricted Articles Regulation.

Ra. Ref. Radar Reflector.

RAS *or* **R.A.S.** Royal Agricultural Society; Royal Asiatic Society; Royal Astronomical Society; Rectified Air Speed; Replenishment at Sea; Rubber Association of Singapore.

RASRO Rescue at Sea Resettlement Officer/Offer.

RAT *or* **R.A.T.** Rapeseed Association Terms.

RATAN Radar and Television Aid to Navigation.

RATAS Research and Technical Advisory Services. *See* **Lloyd's Maritime Information Services**.

RATCC Radar Air Traffic Control Center *USA*.

Rate Class The class of fare or freight rate applied to a specific type of passenger or cargo.

Rate Construction Calculating or building up the Tariff, to be charged by an organisation or association.

Rated Load Value The maximum load which is technically and officially applied to an object when on test. *Abbrev.* RLV. Other terms used are Safe Working Load, Working Load, Working Load Limit and Resultant Safe Working Load. *See* **Proof Load**.

Rates of Exchange The reciprocal rates by which moneys of countries are officially exchanged. These give some idea of the day-to-day financial state of the countries concerned. *Abbrev.* R/E.

Rate War Cut-throat competition pursued by the sellers of certain products and services.

Rat Guard A sizeable round-shaped metal sheet, normally of heavy duty tin, with a hole in the centre and having a radius slit (from the hole up to the end of the arc) to allow the Mooring Line, *q.v.*, to pass through while the ship is moored to the quay. The rats will find it difficult to jump over the guard disc and the ship is prevented from becoming infested.

Ratification Confirmation. Agency may be created by ratification. Where A purports to act as agent for B, either having no authority at all or having no authority to do that particular act, the subsequent adoption by B of A's act has the same legal consequences as if B had originally authorised the act. But there can be no ratification unless A purported to act as agent, and to act for B; and in such a case B alone can ratify. Nor can there be any binding ratification of any agreement which was void or where the principal was not in existence at the time of the act, either in fact or in the contemplation of law as in the case of persons such as trustees in bankruptcy or personal representatives who acquire title by relation. The doctrine of ratification is also applied to torts.

Rating (1) A seaman below the status of officer. In the navy his precise grade is his Rate. (2) Measure of the operating power of machinery as in Horse-Power or kilowatts. (3) Calculation of a Yacht's, *q.v.*, comparative performance for the racing rules.

Ratio decidendi The reason or ground of a decision, as opposed to an *obiter dictum*, which is an opinion not necessary to a judgment and therefore not binding as a precedent.

Ratione tenurae By reason, or in respect of, his tenure. If a person is liable for the repair of a highway by reason of his having the tenure or occupation of the adjoining lands, his obligation is said to be *ratione tenurae*.

Ratline or Rattling One of the horizontal steps forming a rope ladder and commonly used on ships for climbing the rigging or for the sailors to go aloft.

RATO Rocket Assisted Take Off.

RATS Restricted Articles Terminal System *IATA*; Right Ascension of the True Sun.

Rattling *Similar to* **Ratline**, *q.v.*

RAV. Restricted Availability.

RAVC *or* **R.A.V.C.** Royal Army Veterinary Corps.

RAWIN Radar Wind Sounding.

RAWINSONE Radar Wind Sounding and Radiosonde.

Raw Materials That out of which any process of manufacture makes the articles it produces.

RB Relative Bearing *nautical*; Republic of Bolivia; Syrian Arab Airlines *airline designator*.

R/B Rederibolag, *q.v.*

RBA Royal Brunei Airlines, designator B1.

RBA *or* **R.B.A.** Reinsurance Brokers' Association; Royal Society of British Artists.

RBC Ready Berth Clause, *similar to* **Whether in Berth or Not.**

RBD Rapid Beam Reflector; Return of Births and Deaths *occurring on ships*.

r.b.i. Require Better Information.

Rbl. Rouble, Russian unit of currency. One hundred copecks equals one rouble.

RBL Raised Base Line.

RBN Riser Base Manifold *oil drilling*.

RBN *or* **R.B.N.** Registry of Business Names.

R.Bn Radio Beacon, *also abbrev.* RACON.

RBS Royal Society of British Sculptors.

Rbt *or* **RBT** Re-built; Roundabout.

RBV Ram-based Video *electronics*.

RC *or* **R.C.** Red Cross; Reinforced Concrete; Reinstatement Clause; Republic of China, Taiwan; Research Centre; Roman Catholic; Royal College.

R/C *or* **r/c** Return(ed) Cargo; *Portuguese*—ground floor.

r.c. Radio Coding; Reverse Course; Rotary Combustion.

RCB Registro Cubana de Buques *Cuban Ship Register.*

RCC *or* **R.C.C.** Rescue Co-ordination Centre; Riots, Civil Commotions; Roman Catholic Church.

RCCA Route Capacity Control Airline.

RCC & S *or* **R.C.C. & S.** Riots, Civil Commotions and Strikes.

R & CC Riots and Civil Commotions, *q.v.*

RCD *or* **R.C.D.** Regional Co-operation for Development.

RCDC Rangoon City Development Committee.

RCDS Royal College of Defence Studies.

Rcd Received.

r.c.f. Relative Centrifugal Force.

RCFCA *or* **R.C.F.C.A.** Royal Canadian Flying Club Association.

RCGP *or* **R.C.G.P.** Royal College of General Practitioners.

RCH Railway Clearing House.

RCI *or* **R.C.I.** Radar Coverage Indicator.

RCL *or* **R.C.L.** Ruling Case Law.

RCLIBA *or* **R.C.L.I.B.A.** Reinsurance Committee of Lloyd's Insurance Brokers' Association.

RCM *or* **R.C.M.** Radar Counter Measures *electronics*; Royal College of Music (London).

RCN *or* **R.C.N.** Royal Canadian Navy; Royal College of Nursing.

RCO Royal College of Organists.

RCP *or* **R.C.P.** Royal College of Physicians.

RCS Radar Cross Section. Quantification in square metres of the effective radar reflective area of a target, e.g. a ship or radar reflector. Used in comparing the visibility on radar of objects encountered at sea.

RCS or R.C.S. Reactor Control System; Royal College of Science; Royal College of Surgeons; Royal Commonweath Society; Royal Corps of Signals.

RCSC or R.C.S.C. Radio Components Standardisation Committee.

RCSS Random Communication Satellite System.

RCST or R.C.S.T. Royal College of Science and Technology.

Rct. Receipt.

RCT or R.C.T. Royal Corps of Transport.

RCU Rate Construction Unit; Remote Control Unit; Road Construction Unit; Rocket Counter Measure Unit.

RCV or R.C.V. or Rcv Receive; Receiver; Remote Control Vessel.

Rcvd Received.

r.d. Relative density.

RD or R/D Refer to Drawer, *q.v.*; Relative Density, *q.v.*

RD or R.D. or rd. Running Days.

R & D Research and Development.

Rd Btm or RDBTM Round Bottom *Ship.*

RDC or R.D.C. or r.d.c. Running Down Clause, *q.v.*

RDCJ Research Development Corporation of Japan.

RDD or R.D.D. Required Delivery Date.

RDE or R.D. & E. Research, Development and Establishment.

RDF Radio Direction Finder; Rapid Deployment Force *USA.*

rd. hd. Round Head.

R dk. Raised Deck *Lloyd's Register notation.*

rdo. Radio.

Rdr or RDR or R.D.R. Radar, *q.v.*

Rds. Roads *q.v.*

RDT & E Research, Development, Test and Evaluation.

RDZ Radiation Danger Zone.

re Reference; Referring; With or in reference to; Relative to.

R/E Rates of Exchange, *q.v.*

RE Symbol for Fidenavis S.A. *Spanish classification society.*

REA or R.E.A. Radar and Electronics Association; Request for Engineer's Authorisation.

Reach (1) Same as Outreach of a crane. (2) More or less straight section of a river or inlet. (3) Sailing with the wind on the beam.

Reachable From a Marine Insurance, *q.v.*, viewpoint this word refers to the sufficient depth of water available at the vessel's destination.

Reachable on Arrival or Reachable on Arrival Always Afloat A Charter Party, *q.v.*, term meaning that the Charterer, *q.v.*, undertakes to provide a loading and/or discharging berth immediately on the ship's arrival. *Abbrev.* RAAA or ROAAA.

Reach and Burden The full cargo capacity including the lawful deck space of the ship.

Reachstacker Vehicle with a hydraulically operated arm for lifting and stacking containers.

Read Back A radio telephonic instruction to the other party to repeat the whole of a message just transmitted to him.

Read Only Memory A computer microchip assembly containing a data pattern that cannot be altered by the user, unlike the Random Access Memory, *q.v. Abbrev.* Rom.

Ready Berth Clause *Similar to* **Whether in Berth or Not**.

Ready Delivery *See* **Delivery Terms**.

Ready Reserve Force Force of merchant ships maintained by MARAD, *q.v.*, as a component of the U.S. government's strategic sealift fleet. *Abbrev.* RRF.

Ready Ship When a ship is ready to receive or discharge her cargo.

Ready Willing and Able A banking term in buying transactions. The buyer is 'Ready Willing and Able' in all respects to guarantee the amount of money agreed prior to the opening of a Letter of Credit, *q.v. Abbrev.* RWA or R.W.A.

Reagent A substance used to detect the presence of another by reaction; a reactive substance or force.

Real (1) Actually existing as a thing or occurring in fact, objective, genuine. (2) In law consisting of immovable propery such as lands or houses, especially real estate. (3) Former silver coin and money of account used in Spain and Spanish-speaking countries.

Real Accounts A book-keeping term commonly used in conjunction with nominal and personal accounts. The nominal concerns the income and revenue, the personal deals with debtors and creditors while the real accounts relate to capital and assets.

Real Estate Stationary property: land, houses, villas, etc.

Realised Accounts The actual total proceeds or amounts.

Ream A paper measure consisting of 480 sheets (often 500 to allow for waste) making 20 quires. A printer's ream is 516 sheets. *Abbrev.* rm.

REAP Regional Export Assistance Programme. Offered by the Port Authority of New York and located at the World Trade Center.

Reasonable Despatch One of the warranties or implied undertakings in the Contract of Affreightment, *q.v.*, referring to the normal port clearance and speed of the vessel during the execution of the work as per contract.

Reassd or REASSD Reassured.

Réau or Réaum Réaumur. The name of the French physicist appended (*abbrev.* R) to readings of the thermometer introduced by him with freezing point O deg. and boiling point 80 deg.

REB Radar Evaluation Branch.

Rebate A refund or discount allowed on the payment of the services rendered. There are some shipping lines that give a rebate on freight earnings after a certain number of shipments. This no doubt encourages the shipper to load regularly with the same carrier.

recd. Received.

Receiver (1) The person appointed to 'receive' and administer the rents and profits, or other moneys accruing to an estate or business undertaking which is administered or wound up under the supervision of the court. Official Receivers are officials permanently employed to act in that capacity in bankruptcy proceedings or the winding-up of joint-stock companies. (2) The person or firm named in the contract of carriage to accept the delivery of goods at their destination. (3) A person who knowingly receives stolen goods.

Receiver of Wreck An official in the UK responsible to the Department of Trade for wreckage or salved objects found on the coastline or salved at sea and brought to a British port. His duty includes avoiding pilferage and preparing for the disposal of a wreck subject to claims from owners and finders. He may demand any help required to assist a stranded vessel and preserve human life. The wreck may consist of Flotsam, *q.v.*, Jettison, *q.v.*, Lagan, *q.v.*, Derelict, *q.v.*

Receiving Carrier The Carrier who is responsible for the transhipment of the merchandise.

Receiving Note A note made out by the agent of the ship giving details of the cargo meant to be loaded. This is given to the shipper who presents it to the receiving officer or the chief officer when the cargo is ready to be loaded. This note is signed by the chief officer or other responsible person on board and receipted. On presentation of this receipt note the shipper will be given a Bill of Lading, *q.v.*, by the agent. This is also termed Shipping Note and Shipping Order and Mate's Receipt, *q.v.*

Reception Depot A warehouse where merchandise for export is collected for eventual shipment.

Reception Facilities for Oily Residues and Oily Wastes Fixed installations and mobile or floating processing plants exist at ports around the world for receiving and treating oily waste and residues or Slops, *q.v.*, from ships. These are listed in IMO publications.

Recession A slump in trade leading to unemployment. *Also called* **Depression**, *q.v.*

Reciprocating In mechanics this means an engine with an alternate backward and forward motion.

Reciprocity A reciprocal condition, mutual action; the principle or practice of give and take, especially the interchange of privileges between states as a basis of commercial relations.

Reclamation Barren land brought under cultivation or made ready for the building of houses, factories, etc. Reclamation can also mean land which has been reclaimed from the sea or marshes.

Recmd. Recommissioned *Lloyd's Register notation.*

Record Book *See* **Log** and **Log Book**.

Recorded Delivery A Post Office delivery service against the payment of a small fee on production of a receipt by the addressee when the letter or package is delivered.

Recourse The right to payment requested by the Endorser, *q.v.*, of a negotiable document such as a Bill of Exchange, *q.v.*, or cheque, *q.v.*, against the Guarantor, *q.v.*, or the Endorsee, *q.v.*

Recovery (1) Acquisition of something through a court judgment. (2) Process of claiming compensation for loss covered by a paid-up insurance policy. (3) Salvage of something that was lost, for example, overboard. (4) Separation and return to good use of elements of waste products, for example serviceable oil from slops, *q.v.*

Rec. TBAF Recovery to be accounted for.

Recto The right hand page of an open book.

RED *or* **R.E.D.** Rate, Extras and Demurrage. Worldscale terms of a charter.

red. Reduced; Redwood.

Re-del *or* **Re/dly.** Redelivery.

Red Duster The British Merchant Navy flag—the Red Ensign.

Redemption The re-purchasing for cash of Securities, *q.v.*, at the end of the term, or on the date, fixed at the time of their issue. Commonly applies to Government stocks.

Redemptoris Missio *Latin*—The Mission of Christ the Redeemer.

Redeployment Transfer of labour or machines from one task or one place to another to make better use of skills and resources.

Redeployment of Labour The transferring of a number of workers from a hard hit redundancy area to another where they are in demand.

Rederi Danish, Norwegian, Swedish—Shipowning Company.

Rederiaksjeselskap Norwegian—Shipowning Company Limited. *Abbrev.* R/A.

Rederiaktiselskab Danish—Shipowning Company Limited. *Abbrev.* Rederi A/S *or* R/A.

Rederiaktiebolag Finnish, Swedish—Shipowning Company Limited. *Abbrev.* Rederi A/B.

Rederibolag Finnish, Swedish—Shipping Company. *Abbrev.* R/B.

Rederierne Danish, Norwegian—Shipping Company.

Rederiselskab Danish—Shipowning Company.

Redicon Refrigerated Digital Control.

353

Red or Green Light System A method used in customs halls at certain ports or airports where passengers have a choice as to which exit to use. Red lights or signs indicate that passengers should use those areas if they have items to declare before clearance whereas passengers passing through the green light exits are presumed to have nothing to declare for customs inspection. However, spot checks are frequently made in the green areas and if anyone is caught smuggling the penalty can be severe.

Redress Clause A charterparty expression which would limit the owner's liability as per Carriage of Goods by Sea Act, *q.v.*, limits; the charterers are responsible for excess cargo claims.

Redundancy The state of being superfluous to the manpower requirement of an employer, which usually means that arrangements are made for the redundant person to leave the company or organisation with certain rights in compensation. *See* **Redundancy Payment**.

Redundancy Payment When the services of an employee are no longer required through shortage of orders for the product or because of a recession the employer will compensate the person by a Redundancy Payment or Severance Pay. This payment can also by called a Golden Handshake, *q.v, or* Terminal Benefit.

Reed (1) Chart notation for Reed Foghorn. (2) The thin metal reed which under the pressure of compressed air produces the sound in a Reed Foghorn.

Reederei *German*—Shipowning company.

Reef (1) A patch, ridge or chain of rocks, or coral, with little depth over them, sometimes just above the sea surface. (2) To reduce the area of a sail(s).

Reefer Refrigerated; Refrigerated Space; Refrigerating Space; Refrigerated Ship or Refrigerating Ship.

Reefer Carrier A vessel constructed to accommodate frozen products. Also called a Reefer Ship or Reefer, *q.v.*

Reefer Container A container constructed to accommodate frozen goods fitted with its own motor and/or capable of being connected to an outside supply source.

Reefer Plug A facility on a ship to connect to and to sustain a reefer container.

Reeve To pass a rope through a hole, ring, ringbolt, cringle, fairlead or sheave of a block.

Ref. Reference; Refrigerating Machinery.

Referee A person to whose judgement a matter may be referred for arbitration.

Reference The submission of a matter in dispute to an arbitrator for his award.

Refer to Drawer A remark put on an uncashed or dishonoured cheque by the bank signifying (a) there are no funds available in favour of the drawer to the amount specified on the cheque or (b) it could be insufficiently or incorrectly filled out. *Abbrev.* R/D or RD.

Refined Oils *See* **Clean Oils** and **Dirty Oils**.

Ref. Machy. Refrigerated Machinery, *Lloyd's Register notation.*

Refrig. Refrigerator *also abbrev.* Frig.

Refrigerated Box Van A vehicle equipped with refrigerating machinery that makes it capable of carrying refrigerated cargoes while constantly maintaining them at the correct temperature.

Refrigerated Cargo Cargo which, due to its inherent nature, requires cold storage to preserve its natural condition while being transported. Refrigerated cargo can be categorised as (a) Air cooled, for fresh fruits, vegetables, etc. (b) Chilled, for tinned meats, vegetables, etc., and (c) Frozen, for meats, fish, vegetables, etc.

Refrigerated Container *See* **Reefer Container**.

Refrigerated Ton A unit of refrigeration being the heat absorbed by 2,000 lbs of pure ice at 32 degrees F, in 24 hours corresponding to 12,000 BTU per hour.

Refund Rebate, *q.v.*

Refurbishment Modernisation, renewal, repair and redecoration of premises and plant to improve efficiency, the working environment or the public perception of the business. It stops assets depreciating and enhances their value in the event of disposal.

Reg or Regd Registered.

REGAL *or* **Regal** Range and Evaluation Guidance for Approach and Landing *aviation*.

Registered Capital Authorised Capital.

Registered Office The office of a company for legal purposes, which receives correspondence of that nature. It is not always located with the place of business.

Register of Shipping and Seamen At Cardiff, UK. From 1994, this organisation became part of the Marine Safety Agency, *q.v.*, under the UK Department of Transport. It maintains a central register of ships; issues seafarers' certificates; recruits the Merchant Navy Reserve; and registers births and deaths at sea. *Abbrev.* RSS.

Register of Shipping Book A book or books published every year by a ship classification society giving details regarding ships registered with the society.

Registered Stock Inscribed Stock, *q.v.* Stock having a certificate but registered in a record book of the company.

Registration of British Ships In 1994, under the Merchant Shipping (Registration) Act 1993, a new central register of British ships and fishing vessels kept by the Registrar General of the Register of Shipping and Seamen, *q.v.*, at Cardiff, UK came into being, and all existing UK registrations were transferred to it on 21 March. British citizens and nationals, and those from states in the European Economic Area who are established in the UK, are qualified to register owned or bare-boat chartered ships and have to specify one of 140 ports of choice (120 for fishing vessels) whose name will be painted on the vessel's stern. The registration of a ship is valid for a period of five years and must be renewed after that if required. Small vessels, not being fishing vessels and under 24 metres l.o.a., are on a register managed by the Driver and Vehicle Licensing Agency at Swansea which establishes proof of nationality but not title.

Regnat populi *Latin*—The people rule. Motto of Arkansas, USA.

Reg P. Registered Post.

Reg/s Regulation(s).

Regt. Regiment.

Reg. T.M. Registered Trade Mark.

Regular Turn. A charterparty expression to the effect that time for loading/discharging is to start counting only when the vessel is berthed and is then to wait for her turn for loading or discharging. *Abbrev.* R/T.

Reid Vapour Pressure The measure of substances emitting inflammable vapour. *Abbrev.* RVP.

Reimburse To repay equivalent money disbursed.

Reinsd *or* **REINSD** Reinsured.

Reinst *or* **REINST** Reinstatement.

Reinsurance A system of insurance whereby the insurer procures the whole or part of the sum which he has insured (i.e. contracted to pay in case of loss, death, etc.) to be insured again to him by another person.

Relative Density Similar to Specific Gravity.

Relative Humidity The percentage of moisture in the atmosphere. *Abbrev.* r.h.

Relative Mass The mass of a body moving at a velocity compared with the speed of light.

REM Rapid Eye Movement *in sleep*; Roentgen Equivalent in Man.

rem Remainder; remaining; remarks; remittance.

REMAD Remote Magnetic Anomaly Detection.

REMC *or* **R.E.M.C.** Radio and Electronics Measurements Committee.

Remittance Money despatched through the post or via a bank to a person or a firm.

Remotely Operated Vehicle *or* **Remotely Operated Vessel** Generally refers to a submersible vessel which can be operated electronically by remote control. *Abbrev.* ROV.

Remotely Piloted Vehicles Unmanned aircraft with electronics systems used for target practice. *Abbrev.* RPV.

Remoteness of Damages *See* **Damages.**

REMPEC Regional Pollution Emergency Response Centre, *Malta, established 27 April 1991.*

Rendu *Similar to* **Franco,** *q.v.*

Rengokai *Japanese*—Association.

Rentcon Rental Container.

Rentier *French*—A person of independent means; one who lives on income by providing capital for the sake of interest and dividend.

Renunciation The act of giving up a right. A renunciation is a document by which a person appointed by a testator as his executor, or a person who is entitled to take out letters of administration of the estate of an intestate in priority to other persons, renounces or gives up his right to take out probate or letters of administration; the document is filed in the Probate Registry. The word renunciation is also commonly used as synonymous with disclaimer where the latter word is used generally in connection with a right, interest, or office.

ReO Reinsurance Office.

R & EOU Radio & Electronic Officers Union.

Rep. Repair; Repeat; Report(er); Representation; Representative; Reprint; Republic.

REP *or* **R.E.P.** Regional Employment Premium.

Repatriate To send a seaman back to his country. Repatriation expenses are for the account of the shipowner or the employer unless it is the fault of the seaman or the employee.

Repeal To revoke, rescind or annul.

Repetitive Strain Injury Pain suffered by manual operators where machines require the same rapid and continuous movement without a break, notably in the case of some keyboard users.

Replacement Clause A marine insurance clause which specifies that Underwriters, *q.v.*, shall make good for only the damaged parts of machinery under a claim.

Reply Coupon International postal voucher which covers in advance the cost of franking a letter in reply to the one enclosing it.

Reporting a Vessel On arrival the master of a ship is to produce to the customs authorities all the particulars of the ship by showing certificates concerning cargo gears, load line, ship's registry, deratting, crew and passenger lists, quantities of items in bond and other dutiable objects used as personal effects by the officers and crew, the cargo manifest and corresponding bills of lading, etc. When the ship is carrying no cargo she is said to be 'On Ballast'.

Reporting Day The day when the Notice of Readiness, *q.v.*, is presented by the master of a ship to the agents or the charterers.

Report of Cargo Cargo which is to be reported by the ship or the agent to the customs authorities on arrival at a port. *Also called* Cargo Report.

Representation *Similar to* **Agency,** *q.v.*

Representation Accounts *or* **Representation Expenses** Accounts or expenses incurred by the master of a ship while entertaining businessmen and officials in connection with the publicity of the ship and company.

Reprieve The commuting or suspension of a court sentence.

REPS *or* **R.E.P.S.** Rail Express Parcel Service.

Repudiation The renunciation of a contract, which renders the repudiator liable to be sued for breach of contract, and entitles the other party, on accepting the repudiation, to treat the contract as at an end.

Repugnant Contrary to or inconsistent with.

Requisition Clause A standard time charter clause. If, during the charter of a vessel, the ship's government requisitions her, then she automatically becomes off charter but the hire rate paid by the government will be retained by her owners.

Res. Reservoir; Residue.

Res *Latin*—A thing. Any physical or metaphysical existence in which a person may claim a right. In an Admiralty action *in rem*, the *res* is the ship, cargo or other property proceeded against.

Rescission Rescission, or the act of rescinding, is where a contract is put an end to by the parties, or one of them. Thus, a contract is said to be rescinded where the parties agree that it is to be at an end, or where one of the parties is entitled to avoid it by reason of the act or default of the other party, and elects to do so, by giving notice of his election to the other party, by setting up the invalidity of the contract as a defence to proceedings taken by the other party, or by instituting proceedings to have the contract judicially set aside (judicial rescission). The most frequent instances of rescission by one party occur where there is fraud or misrepresentation and in certain cases where there is a failure of performance by one of the parties in an essential part of the contract. If a party to a contract fails to comply with a condition precedent, or by his own act makes the performance of the contract impossible, the other party may rescind the contract.

Reserve Buoyancy The watertight volume of the portion of a floating object, or a vessel, that is above the waterline. This reserve counteracts the effect of sinkage due to flooding or movement in heavy weather.

Reserve Currency Foreign currency held by a state to finance its international trade.

Reserve Funds Business money kept aside in case of need.

Reserve Price The rock bottom price the seller is willing to accept for his articles during negotiations. *Opposite to* without Reserve.

Res gestae *Latin*—The things done. The facts surrounding or accompanying a transaction which is the subject of legal proceedings.

Resiant A resident. The term was chiefly used in speaking of manors.

Residual Fuel Oil Heavy Fuel Oil usually containing much sulphur, sodium, vanadium, water and fines. Termed residual because it is what remains when lighter products have been refined from crude oil.

Residual Value The subsequent or remaining value of assets at the termination of an agreement.

Residue (1) The surplus of a testator's or intestate's estate after discharging all his liabilities. (2) That which remains after oil is pumped out of the tanks of a tanker, or that which is left after unloading bulk cargoes.

Resin An adhesive substance insoluble in water secreted by most plants and exuding naturally or upon incision. Synthetic resin is one of the ingredients used in sophisticated paint compositions and also in plastic materials.

Res integra *Latin*—A point governed neither by any decision nor by any rule of law and which must therefore be decided upon principle.

Res inter alios acta alteri nocere non debet *Latin*—A transaction between strangers ought not to injure another party.

Res ipsa loquitur *Latin*—The thing speaks for itself.

Res nullius *Latin*—A thing which has no owner.

Res perit domino *Latin*—The loss falls on the owner.

Respice finem *Latin*—Consider the result.

Res. Prem. Reserve Premium.

Respondeat superior *Latin*—Let the principal answer for the act of his servant.

Respondent A person or firm against whom a petition is presented, a summons issued, or an appeal brought.

Rest. *Lloyd's Register notation* for 'Classification Restored'.

Restitutio in integrum *Latin*—The rescinding of a contract or transaction, so as to place the parties to it in the same position, with respect to one another, as they occupied before the contract was made, or the transaction took place.

Restraints of Princes, Rulers or Peoples This expression occurs in Marine Insurance, *q.v.*, policies, bills of lading, etc., being one of the group of contingencies against which provision is made. It covers any forcible interference with the voyage at the hands of the constituted government or ruling power of any country.

Restricted Letter of Credit A Letter of Credit, *q.v.*, which can be amended without prior notice to the receiver. There is no doubt that this can be to the disadvantage of the receiver since definite preparations to meet the conditions cannot be made before it is confirmed.

Restricted Visibility Weather conditions such as mist, fog, rainstorm, sandstorm, etc., which severely reduce visibility.

Restrictions Prohibitions; Embargoes.

Restrictive Injunction *See* **Injunction**.

Resultant Safe Working Load *Abbrev.* RSWL. *See* **Rated Load Value**.

Ret. Retired; Return of Premium *insurance q.v.*

Retailer A seller of goods after buying them in gross from the Wholesaler, *q.v.*

Retail Price The price of goods charged by the Retailer, *q.v.*, and applicable to the general public.

Retard of Tide The interval between a spring tide and the conjunction, or opposition of, the sun and the moon which causes it.

Retd. Retired; Returned.

Retention Money A percentage of money retained by the paying party in a contract involving work being done. When the work is satisfactorily completed and in accordance with the contract the money retained is released to the contractor. Also termed as Deposit, *q.v.*

Retired Bill *or* **Retiring a Bill** A paid Bill of Exchange, *q.v.*, before or on the due date at a discount.

Retiring from a Line An American insurance term. When the Underwriters, *q.v.*, refuse to extend the policy after expiration or if they want to discontinue covering the existing policy before expiration.

Ret. P. Retired Pay.

Return A tally of goods or cargo.

Return of Premium Refund of Premium by the Underwriter, *q.v.*, to the insured. This is effected whenever the service of the risk has not actually taken place. *Ex.* The laying up of a vessel during a Time Policy, *q.v.*, over one month either because of no work or for repairs. *Abbrev.* Ret.

Returns The coming in of proceeds or profit of an undertaking.

Returns Clause *See* A.C.V. Form

Rev. Revenue, *q.v.*; Reversed; Revolutions of engine.

Rev. ed. Revised Edition.

Revenons à nos moutons *French*—Let us return to our subject.

Revenue Income, especially of a large amount, from any source; taxes and duties collected from the general public by the state.

Revenue Account Income and Expenditure Account, alternatively Profit and Loss Account.

Revenue Passenger Kilometre The charges or income derived from passage money per person per kilometre. *Abbrev.* RPK.

Revenue Ton The type of ton which yields more revenue when freight is paid on cargo. Whichever is the greater of the weight or measurement is the Revenue Ton. *See* **Freight Per Revenue Ton.**

Reverse Osmosis Purifying brackish water or sea water through a desalination procedure. *Abbrev.* RO *or* R.O.

Reverse Pitch The putting into reverse motion the propeller blades of an aircraft in order to assist braking while landing.

Reversible Propeller *See* **Controllable Pitch Propeller.**

Revictualling To take in a fresh supply of provision. *See* **Victualling.**

Rev/min Revolution per Minute.

Revs. Revolutions of an engine or propeller or propeller shaft.

RF *or* **R.F.** Radio Frequency; Range Finder; Replacement Factor; Representative Fraction; Relative Azimuth.

r.f. Radio Frequency; Range Finder; Rheumatic Fever.

R-factor Resistance Factor.

RFC *or* **R.F.C.** Reconstruction Finance Corporation *USA.*

RFD *or* **R.F.D.** Radio Frequency Devices; Raised Fore Deck; Reporting for Duty.

RFDS *or* **R.F.D.S.** Royal Flying Doctor Service *Australia.*

RFI Radio Frequency Interference; Request for Information; Request for Interest.

R. Fix. Running Fix *nautical.*

RFP Request for Proposal.

r.f.p. Retired on Full Pay.

RFQ Request for Quotation.

R.F.R. Royal Fleet Reserve; Radio Frequency Radiation.

RFR *or* **Rfr** Required Freight Rate.

RFW Reserved Feed Water; Radial Friction Welding.

RGA Returned Goods Authorization *USA.*

Rga. Riga.

Rgd. Registered.

rge. Range.

RGN *or* **R.G.N.** Registered General Nurse.

RGS *or* **R.G.S.** Registrar General for Shipping; Royal Geographical Society.

RGV Remote Guidance Vehicle.

RH Air Zimbabwe *airline designator.*

R.H. Right Hand; Royal Highness.

r.h. Relative Humidity, *q.v.*

RHA *or* **R.H.A.** Road Haulage Association.

r.h.c. Rubber Hydrocarbon Content.

RHC *or* **R.H.C.** Road Haulage Cases.

RHE *or* **R.H.E.** Road Haulage Executive.

RHI *or* **R.H.I.** Range Height Indicator.

RHKJC Royal Hong Kong Jockey Club.

R.H.N. Royal Hellenic Navy, *obsolete.*

RHP *or* **R.H.P.** *or* **r.h.p.** Rated Horse Power.

R.H.S. Royal Humane Society; Royal Horticultural Society.

R. ht. *Lloyd's Register notation* for engine fitted with reheaters.

Rhumb A line cutting all meridians at the same angle; the angular distance between two successive points of a compass.

RI Rhode Island, *USA.*

RI *or* **R.I.** Refraction Index; Reinsurance; Registro Italiano Navale—Italian ship classification society;

Rubber Insulated; Rubber Insulation; Royal Institution; Rhode Island.

R/I *or* **RI.** Reinsurance.

RIA Radio Immune Assay *electronics*; Royal Irish Academy.

Rial A unit of currency used in Iran and Oman. One hundred Iranian dinars equals one rial; one hundred Omani baiza equals one rial.

RIAS *or* **R.I.A.S.** Royal Incorporation of Architects in Scotland.

RIB Rigid Inflatable Boat.

RIBA *or* **R.I.B.A.** Royal Institute of British Architects.

Riba *Arabic*—Interest.

RIC Royal Institute of Chemistry, now part of Royal Society of Chemistry.

RICA *or* **R.I.C.A.** Reinsurance of Common Account *reinsurance*; Research Institute for Consumer Affairs.

Rickers Light poles of 20 to 60 feet used for building purposes such as scaffolding.

RICS *or* **R.I.C.S.** Royal Institution of Chartered Surveyors.

RID/ADR *or* **ADR/RID** European Agreements for Rail and Road Transport of Hazardous Cargoes.

Ride at Anchor To be anchored, to lie to an anchor or anchors.

Ride Easy *or* **Riding Easy** A nautical phrase describing a ship at anchor in relatively easy and calm seas and exerting no strain on the chain cables. *Opposite to* **Ride Hard** *or* **Riding Hard,** *q.v.*

Ride Hard *or* **Riding Hard** *Opposite to* **Ride Easy** *or* **Riding Easy**, *q.v.* This refers to a ship at anchor pitching heavily during a rough swell and putting continuous strain on her cables.

Ride Out *or* **Riding Out** When a ship is at anchor and though pitching heavily is still considered safe and seaworthy despite the strain on the chain cables.

Rider Frame A welded frame that serves as a stiffener or strengthener. *See* **Riders**.

Riders A word used to signify 'Additional Clauses' in a contract, etc.; ropes or part of ropes lying on top of others; extra layer of plates or timbers used as strengthening frames of a ship. *See* **Rider Frame**.

Riding Anchor *See* **Ride at Anchor**.

Riding Easy *See* **Ride Easy**.

Riding Hard *See* **Ride Hard**.

Riding Lights Lights which are compulsory during darkness when a ship is at anchor or moored.

Riding Out *See* **Ride Out**.

RIDM Revue Internationale des Droits Maritime *French*—International Review of Maritime Rights.

Riel Currency unit of Campuchea, formerly Cambodia.

RIF *or* **R.I.F.** Resistance Inducing Factor; Royal Irish Fusiliers.

RIFT Reactor in Flight Test.

Rig The way a ship's masts, sails, etc., are arranged; to equip with rigging; Oil Rig.

Rigging Ropes and chains used to hold the masts, booms, etc., of a vessel.

Rigging the Market Artificial propaganda advertised by a newly-formed company with a view to influence interested people to buy shares.

Right-hand Lay Anti-clockwise direction turn. This generally refers to the twist direction of the Strands, *q.v.*, of ropes which are also right-handed.

Right of Action Advantageous right to start civil proceedings.

Right of Hot Pursuit (The) *See* **Hot Pursuit**.

Rights Issue The right for the holder of shares to have additional shares from the company as a priority over outsiders.

Rigid Rotor A helicopter rotor with pitch variations.

Rig Ten. Rig Tender.

Rig Up To start preparing the rig for the oil/gas drilling operation.

R.I.I.A. Royal Institute of International Affairs.

Rime Hoar-frost, which is frozen water vapour deposited in clear still weather on lawns, etc.

RIMS Risk and Insurance Management Society, Canadian insurance management society. *See* **ORIMS**.

RINA Registro Italiano Navale—Italian ship classification society. *See* **Classification Societies**.

R.I.N.A. Royal Institution of Naval Architects.

Ring A combination of businessmen or companies to control a market.

Ringgit Currency unit of Malaysia. The Malaysian dollar.

Ring Money Precious articles worn in the form of rings, bracelets, armlets and chokers that are convertible to money.

RINSMAT Resident Inspector of Naval Materials *USA*.

RIO Reporting In and Out.

Rio Rio de Janeiro, *Brazil*.

Riots and Civil Commotions A marine insurance warranty exempting Underwriters, *q.v.*, from the consequences of any riots or civil commotions. *Abbrev.* R & CC *or* R. & C.C.

Riots, Civil Commotions and Strikes A marine insurance warranty similar to Riots & Civil Commotions, *q.v.*, but in addition underwriters are also exempted from liability for the effect of strikes. *Abbrev.* RCC & S *or* R.C.C. & S.

RIP Requiescat in pace *Latin*—May he or she rest in peace.

RIPH & H *or* **R.I.P.H. & H.** Royal Institute of Public Health and Hygiene.

Rip. viet. Riproduzione vietata *Italian*—Copyright Reserved.

RIS *or* **R.I.S.** Research Information Service.

RISE Research in Supersonic Environment.

Rising Tide Similar to Flood Tide, *q.v.*

Risk of Craft A marine insurance policy covering risks against the merchandise carried on lighters and barges from ship to shore or vice versa. *Also called* Risk of Boat Clause.

Risk of Boat Clause *Similar to* **Risk of Craft**, *q.v.*

Risks A word widely used in the Insurance, *q.v.*, business meaning the hazards or losses to which a venture may be unexpectedly exposed. Formerly known as Perils, *q.v.*

Risques des routes *French*—Transportation risks.

RITE Rapid Information Technique for Evaluation.

RIV Rapid Intervention Vehicle, *q.v.*

Riv. River.

R.I.V. Repayment Issue Voucher.

Rival Supply Competitive Supply.

River Dues Charges for the use of a river.

River Port A port located at the entrance or on the sides of a navigable river.

Rivs. Rivets *also abbrev.* Rts.

RJE Remote Job Entry, Lloyd's Register's Computer Services Department.

RK Rubbing Keel *ship.*

RL Radio Location; Research Laboratory; Rhumb Line.

RLCE Request Level Change *aviation.*

Rl/s. Roll(s).

r.l. *Lloyd's Register notation* for Rubber Lined.

R.L.O. Returned Letter Office.

RLS Regjistri Laknori Shqiptar *Albanian Classification Society.*

RLSS *or* **R.L.S.S.** Royal Life Saving Society.

RLV Rated Load Value, *q.v.*

RM Relative Motion *radar.*

Rm *or* **rm** Room.

rm *See* **Ream**.

R.M Resident Magistrate; Registered Midwife; Royal Mail; Royal Marines.

r.m. Radio Monitoring.

r/m Revolutions per minute.

R&M Reliability and Marketing; Repair and Maintenance.

RMA Royal Military Academy.

r. mast Radio Mast.

RMC *or* **R.M.C.** Radio Memory Computer; *Lloyd's Register notation* for Refrigeration Machinery Certificate.

(RMC) *or* **(R.M.C.)** *Lloyd's Register notation* for Refrigeration Machinery Suspended by the Society.

RMCS Royal Military College of Science.

r.m.d. Ready Money Down.

R. Met. S. Royal Meteorological Society.

RMI Radio Magnetic Indicator, *q.v.*

r/min Revolutions per minute.

RMNS *or* **R.M.N.S.** Royal Merchant Navy School.

RMP *or* **R.M.P.** Royal Military Police.

RMS *or* **R.M.S.** Root Mean Square; Royal Mail Service; Royal Mail Steamer.

RMSDS Reserve Merchant Ship Defence System *USA.*

RMSS *or* **R.M.S.S.** Royal Mail Steamship.

RMT Union of Rail, Maritime and Transport Workers *UK.*

RMUV Remotely Manned Underwater Vessel. *Similar to* **Remotely Operated Vehicle**, *q.v.*

RMV Remote Maintenance Vehicle *oil drilling.*

RMVA Remote Mechanical Valve Actuator.

r.n. Reception Nil; Research Note.

RN *or* **R.N.** Release Note; Royal Navy.

R$_N$ Reynolds' number *naval architecture.*

RNA Ribonucleic Acid *chemistry.*

RNAC Royal Nepal Airlines Corporation.

RNF Receiver Noise Figure.

RNFP Radar Not Functioning Properly.

RNLI *or* **R.N.L.I.** Royal National Lifeboat Institution. *See* **Royal National Lifeboat Institution**.

rnl(s) Renewal(s).

R.N.M.D. Registered Nurse for Mental Defectives.

R.N.M.S. Registered Nurse for the Mentally Subnormal; Royal Navy Medical School.

RNR *or* **R.N.R.** Rate Not Reported; Registru Naval Roman *Rumanian Classification Society*; Renewal Not Required; Royal Naval Reserve.

RNSA Royal Netherland Shipowner's Association; Royal Naval Sailing Association.

R.N.T. Registered Nurse Tutor.

RNTH Raised Non Tight Hatch.

rnwy. Runway.

R.N.Y.C. Royal Northern Yacht Club.

RNZAC *or* **R.N.Z.A.C.** Royal New Zealand Aero Club.

ro. Recto, *q.v.*; Rood, *q.v.*; Slight Rain *weather observation.*

RO Retired Officer, retained in service.

RO *or* **R.O.** Radar Observer; Radar Operator; Registered Office; Receiving Officer; Routeing (Routing) Order; Returning Officer; Reverse Osmosis, *q.v.*, Royal Observatory.

R/O Routeing Order.

R.Ö. Republik Österreich. *Austria.*

ROA *or* **R.O.A.** Reinsurance Offices Association; Return on Assets.

Road Haulage The act of carrying goods by road transport.

Roads *or* **Roadstead** An area in the sea, away from the shore and out of harbour where ships can safely lie at anchor.

Road Transport The means of transportation of goods on trucks, lorries, carriers, trailers, etc.

ROAMA Rome Air Material Area.

R.O.A.R. Right of Admission Reserved.

Roaring Forties The belt of ocean between 40 and 50 degrees south latitude where the unobstructed westerly winds blow right round the earth.

ROB *or* **rob** *or* **R.O.B.** *or* **r.o.b.** Remaining on Board; Referring to Bunkers; Round of Beam, *q.v.*

ROBIN Rocket Balloon Instrument.

ROC Reference our Cable; Reserve Officer Candidate *USA*;

ROCE *or* **R.O.C.E.** Return on Capital Employed.

Rock Hound An oil industry term for Geologist.

Rock Pressure The fluid pressure existing in an oil/gas reservoir during oil drilling.

ROCP Radar out of Commission for Ports.

ROD Roughneck or Rotating Driller. *Similar to* **Floorman**, *q.v.* Rust Oxidation and Discoloration.

Rode A fisherman's technical word for Anchor Rope, commonly used by fishermen from Canada and New England.

ROE Reflector Orbital Equipment.

R of T Range of Tide.

ROG Receipt of Goods.

Rogatory Letter *See* **Letter Rogatory.**

Roger Word of acknowledgement for 'Message well received' or 'All in Order' or 'OK'.

ROI *or* **R.O.I.** Return on Investment.

ROJA *or* **R.O.J.A.** Reinsurance of Joint Account.

ROK Received OK; Acknowledgement of receipt of a message; Republic of Korea *South Korea.*

ROL Reference Our Letter.

Rolling Settlement London Stock Exchange procedure, introduced in 1994 to replace Account Days and Settlement Day, which makes transactions due for settlement ten business days after dealing.

Rolling Stock An expression usually applied to railways, meaning the goods wagons and/or passenger carriages employed in the industry.

Roll on/Roll off Vessels, such as ferries, where the wheeled cargo (e.g. trailers) is loaded by means of stern ramps, etc. and discharged by the same means.

Roll-over The deferral of capital gains tax where the proceeds from the sale of assets are used to purchase new and similar assets.

Roll-over Loan A loan with a fixed repayment period and provision for review of interest at regular intervals.

ROM *or* **R.O.M.** Read Only Memory *electronics*; Record Office Memorandum; Red Oxide of Mercury.

rom. Roman type.

Romeo International Radio Telephone phonetic alphabet for 'R'.

ROMV Remotely Operated Maintenance Vessel *oil drilling*.

RON Remaining Overnight.

R of N Range of Neap Tides.

R of T Range of Tide.

Rood Quarter of an acre. *Abbrev.* ro.

Roof of the World A mountain range in Tajikistan in the region of Central Asia called 'The Pamirs'.

Room This refers to 'The Room' at Lloyd's.

ROP *or* **R.O.Pr.** Rate of Penetration—referring to drilling for oil.

r.o.p. Run on Paper.

RORC *or* **R.O.R.C.** Royal Ocean Racing Club.

RORCE Rate of Return on Capital Employed.

Ro/Ro *or* **RO-RO** *See* **Roll-on/Roll-off.**

Rose Box That perforated section of a bilge suction in a ship which is so made to keep the pipe free of obstructions.

ROSIE Reconnaissance by Orbiting Ship Identification Equipment.

ROSLA Raising of School Leaving Age.

RoSPA Royal Society for the Prevention of Accidents.

Roster A list detailing the order or sequence of duties and duty periods to be performed by personnel, individually or by sections.

Rostrum (1) The Beak, *q.v.*, of an ancient galley which was used in battle to ram enemy ships. (2) Elevated platform for public speaking deriving the name from that in the Roman forum which was decorated with the Beaks of captured vessels.

ROT Reference our Telex or Telegram; Remedial Occupational Therapy; Remotely Operated Tool.

Rot. Rotterdam, *Holland.*

Rotary Helper Similar to Floorman or Roughneck, *q.v.*

Rotation Number *See* **Range (of ports).** *Abbrev.* rotn. no.

ROTC Reserve Officers Training Corps *USA*.

rotn. no. Rotation Number. *See* **Range (of ports).**

ROTS Robotic Auto Tooling System *computer.*

Rouble Unit of currency used in Russia. One hundred copecks equals one rouble.

Roughneck *or* **Rough Neck** *See* **Floor Man** *or* **Rotary Helper.** *Also* **Drilling Crew.**

Round Charter A charter contract for a round trip voyage, starting from one port, calling at others and ending at the original port of departure.

Round C/P Round Charterparty. *See* **Round Charter.**

Round of Beam The transverse curvature on the deck of a ship. Also termed Camber. *Abbrev.* ROB *or* R.O.B.

Round Robin A petition, originally written with signatures in a circle to conceal the order in which they were written.

Round Trip A voyage out and back to the original port of departure. *Abbrev.* R/V.

Roup A Scottish word for 'auction.'

Roustabout A dock worker or wharf labourer; a deckhand on an oil rig or in an oilfield. *See* **Drilling Crew.**

Routeing Alternative measures taken so as to avoid the risk of collision in traffic separation schemes; the direction taken for travelling or sailing.

ROV Remotely Operated Vehicle or Remotely Operated Vessel.

Rovers Sea robbers; Pirates.

ROW *or* **R.O.W.** Removal of Wreck; Right of Way.

Rowlock A U-shaped metal contrivance with a pivot to be inserted on the sides of a boat to hold an oar during rowing. Also, rarely, called Oarlock.

Royal Alfred Seafarers' Society The original Royal Alfred was founded in 1865 to provide residential homes for veteran seafarers of the Merchant Navy and fishing fleets of British nationality, irrespective of rank, class, creed or colour. The first Royal Charter was granted in 1950, the second in 1977. As a result of the latter Royal Charter the title was up-dated to the Royal Alfred Seafarers' Society, and the Society was also allowed to admit to its Homes seafarers of foreign nationality who have served in British ships and, for the first time, officers and ratings of the Royal Navy. This Charter was amended in 1892 to enable those with long service in shore-based industries connected with the sea to be regarded as seafarers for accommodation purposes.

Various benevolent funds were established to give financial assistance to retired seafarers, their widows and children, and sick and disabled serving seamen who were in need. The Society administers the National War Fund for Merchant Seamen. Grants from this source are mainly for war widows, war disabled and those who were discharged physically unfit either during or after their war service, although not necessarily as war casualties.

Royal Assent The act by which the Crown agrees to a bill which has already passed both Houses of Parliament (the House of Commons and the House of Lords).

Royal Courts of Justice The statutory name, by the Judicature Act, 1925, s. 222, replacing the Judicature (Officers) Act, 1879, s. 28, of the Law Courts, on the north side of the Strand, London, between St. Clement Danes Church and Chancery Lane, in which the business of the Supreme Court is transacted. The erection of buildings for bringing together into one place all the superior courts of law and equity, the Probate and Divorce Courts and the Court of Admiralty, recommended by a Royal Commission in 1858, was authorised by the Courts of Justice Building Act, 1865, and the Courts of Justice Concentration (Site) Act, 1865. The Royal Courts were formally opened by Queen Victoria on 4 December 1882, and opened for business on 11 January 1883, the judges' chambers and other offices having been opened for business in

January, 1880. Prior to the opening, the Chancery Division occupied courts at Lincoln's Inn, and the Queen's Bench and Probate, Divorce, and Admiralty Division occupied courts adjoining Westminster Hall.

Royal Draught A right to take a limited kind of fish throughout the whole of a river by the officer of the Crown as conservator, or the officer of the grantee of such conservancy, who in some rivers appears to have had a right to take a net down the river through all the private fisheries in it at stated periods.

Royal Fish Whale and sturgeon. These, when either thrown ashore, or caught near the coast, are the property of the Crown, on account of their superior excellence. Porpoises are also said to be royal fish. The right to royal fish may be vested in a subject by grant from the Crown or prescription.

Royal Merchant Navy School Foundation, The For over 150 years, the Royal Merchant Navy School Foundation has been caring for the education and general welfare of those orphan boys and girls whose parents have been lost at sea. The School is entirely independent of state aid.

Royal National Lifeboat Institution *Abbrev.* **RNLI** Founded in 1824 by Sir William Hillary to co-ordinate nationally the several lifeboat societies around the coasts of the British Isles, some dating to the 18th century, it took its present name in 1854. It is a charity supported entirely by voluntary contributions and is the oldest lifeboat service in the world. There are over 200 lifeboat stations in Great Britain, Ireland and the Channel Islands providing a 24-hour sea rescue service out to 50 miles from shore. There are 278 boats ranging from inshore inflatables of 16 feet to all-weather craft up to 54 ft 6 ins in length with an additional relief fleet of 100 boats. The boats are crewed by some 3,500 volunteer men and women from all walks of life, while all-weather boats are cared for by full-time mechanics. Only one lifeboat, at the Humber, has a full-time crew.

In Britain and Northern Ireland, lifeboats are usually launched when alerted by the Coastguard, *q.v.* In the Republic of Ireland, the request is usually from the Marine Emergency Service, the police or the public. Lifeboats are launched either from a slipway, or from a tractor-towed carriage, or from moorings.

In 1993, there were 5633 lifeboat launches, average 15 each day, and 1314 lives were saved, average 3 each day.

The RNLI acts as the secretariat for the International Lifeboat Federation.

Royalty This can mean the payment to a patentee for the use of a patent or to an author, for each copy of his or her book which has been sold.

RP Reporting Point; Received Pronunciation; Rinave Portuguesa *Portuguese Classification Society.*

RP *or* **R.P.** Reception Poor; Regimental Police; Reinforced Plastic; Reply Paid; Resale Price; Return of Premium, *q.v.*, Rules of Procedure.

R/p *or* **R/P** Return to Port, *for orders*; Return of Post; Reprinting; Reply Paid.

RPA Record of Personal Achievement; Rotterdam Port Authority.

RPC *or* **R.P.C.** Request the Pleasure of your Company *naval message*; Royal Pioneer Corps.

RPD *or* **R.P.D.** Regional Port Director; Rerum Politicarum Doctor *Latin*—Doctor of Political Science.

RPE *or* **R.P.E.** Radio Production Engineer; Rocket Propulsion Engineer.

r.p.g. Radiation Protection Guide.

RPH *or* **R.P.H.** *or* **r.p.h.** Revolutions per Hour.

RPI Radar Precipitation Integrator; Retail Price Index.

RPK Revenue Passenger Kilometre, *q.v.*

RPM *or* **R.P.M.** *or* **r.p.m.** Revolutions per Minute: Reliability Performance Measure; Resale Price Maintenance.

RP/ND Reprinting No Date.

R.P.O. Railway Post Office.

RPQ Request Price Quotation.

R.P.R. Romanian People's Republic.

RPS *or* **R.P.S.** Radiation Protection Supervisor; Radiological Protection Service; Rapid Processing System; Regeltechnischen Plannung und Steuerung *German*—Network Analysis System; Revolutions per Second.

rpt. Repeat; Report.

RP/UC Reprint Under Consideration.

RPV Remotely Piloted Vehicle, *q.v.* Reactor Pressure Vessel.

R/Q Request for Quotation.

RQD *or* **R.Q.D.** Raised Quarter Deck.

RQL Reference Quality Level.

Rqst. Request.

rr *or* **RR** Continuous Heavy Rains *weather observation*; Radio Ranging.

RR *or* **R.R.** Railroad *USA*; Recommended Repair/Repairer *insurance*; Return Rate; Right Reverend; Road Risk *insurance*.

R & R Rest and Recreation. Period in port for the recuperation of the crew of a military vessel.

RRA *or* **R.R.A.** Radiation Research Association; Rubber Reclaiming Agent.

RRB *or* **R.R.B.** Radio Research Board.

RRC Ratcheting Reclosable Circulating *valve*.

RRE Relative Rotative Efficiency *ship's propeller*; Royal Radar Establishment.

RRF Ready Reserve Force, *q.v.*

RRG Real Rate of Growth *finance*; Reverse Reduction Gear.

RRMIP Rapid Reconnaissance Magnetic Induced Polarisation *geophysical*.

RRP Recommended Reproduction; Recommended Retail Price.

RRR Required Rate of Return *or* YRR—Yield Rate of Return *ship design*.

R.r.r. Raccomandata con Ricevuta di Ritorno *Italian*—Registered receipt requested letter.

RRS *or* **R.R.S.** Radiation Research Society; Radio Research Station; Royal Research Ship.

RRV *or* **R.R.V.** Raise or Rise of Voltage.

Rs. Rupees, *q.v.*

RS Registry of Shipping *Russian classification society.*

RS *or* **R.S.** Revised Statutes *USA*; Radio Survey, referring to the radio telegraphy survey of a ship; Running Survey *Lloyd's Register notation*; Right Side; Residual Resistance.

RSA *or* **R.S.A.** Royal Society of Arts; Royal Scottish Academy.

RSAC *or* **R.S.A.C.** Reactor Safety Advisory Committee *Canada.*

RSB *or* **R.S.B.** Range Safety Beacon.

RSC Royal Shakespeare Company; Royal Society of Chemistry.

R.S. & C.C. Riots, Strikes and Civil Commotions.

RSCN *or* **R.S.C.N.** Registered Sick Children's Nurse.

RSD *or* **R.S.D.** Raised Shelter Deck; Recovery Salvage and Disposal; Rolling Steel Door; Royal Society of Dublin.

RSE Royal Society of Edinburgh.

RSFSR River Register; Russian *classification society*; Russian Soviet Federated Socialist Republic.

RSGB Radio Society of Great Britain.

RSH Royal Society of Health.

RSI Repetitive Strain Injury, *q.v.*

RSJ Rolled Steel Joint.

RSL Returned Services League of Australia; Review of Social Law; Royal Society of Literature.

RSLA *or* **R.S.L.A.** Raising School Leaving Age.

RSM Regimental Sergeant-Major; Royal Society of Medicine.

RSNA *or* **R.S.N.A.** Radiological Society of North America.

RSO *or* **R.S.O.** Railway Sorting Office.

RSPB Royal Society for the Protection of Birds.

RSPCA Royal Society for the Prevention of Cruelty to Animals.

RSS Register of Shipping and Seamen, *q.v.*

RSS *or* **R.S.S.** Royal Society of Scotland.

RSU Road Safety Unit.

RSV *or* **R.S.V.** Rhine-Schelde Verolme; Rough Sea Barge.

RSVP *or* **R.S.V.P.** Random Signal Vibration Projector; Répondez, s'il vous plaît *French*—Please reply.

RSW Refrigerated Sea Water.

RSWC *or* **R.S.W.C.** Right Side Up With Care.

RSWL *or* **R.S.W.L.** Resultant Safe Working Load. *See* Rated Load Value.

RT *or* **R.T.** Radio Telegraph; Radio Telegraphy; Room Temperature; Radio Telephone; Radio Telephony; Return Ticket; Rye Terms, *q.v.*

R/T Radio Telegraph; Radio Telegraphy; Regular Turn, *q.v.*; Radio Telegraphist; Radio Telephone; Radio Telephony.

rt. Right.

RTA *or* **R.T.A.** Rate to be Arranged; Rate Type Agreed; Reciprocal Trade Agreements; Right Time of Arrival; Road Traffic Act; Rubber Trade Association.

RTAC *or* **R.T.A.C.** Rubber Trade Association Clauses.

r.t.b. Return to base.

RTBA *or* **R.T.B.A.** Rate to be arranged; Rate to be agreed.

RTCA Radio Technical Commission for Aeronautics *USA*.

RTCM Radio Technical Commission for Maritime Services *USA*.

RTD Right Time of Departure.

rtd. Retired; Returned.

RTDS Real Time Data System.

rte. Route.

RTEB *or* **R.T.E.B.** Radio Trades Examination Board.

RTF Resistance Transfer Factor.

RTFO *or* **R.T.F.O.** Rye Terms Full Outturn, *q.v.*

Rtg. *or* **rtg.** Rating.

RTh Radio-Telephone (High Frequency) *Lloyd's notation*.

RTHF *or* **R.T.H.F.** Radio Telephone High Frequency.

Rt. Hon. Right Honourable.

RTI *or* **R.T.I.** Round Table International.

RTIC Radio Telephone Interim Certificate.

RTM Radio Telephone Medium *frequency*; Rapid Tuning Magnetron; Receiver Transmitter Modulator; Riser Turret Mooring *oil drilling*.

RTm Radio-Telephone (Medium Frequency) *Lloyd's notation*.

rtm *or* **RTM** Registered Trade Mark.

RTN *or* **R.T.N.** Registered Trade Name.

Rtng. Returning.

RTO *or* **R.T.O.** Railway Transport Officer; Request to Off-Load.

RTOW *or* **R.T.O.W.** Regulated Take-Off Weight *aviation*.

RTP Reduction to the Pole *geophysical*.

RTPA Restrictive Trade Practices Act 1976.

RTPI *or* **R.T.P.I.** Royal Town Planning Institute.

RTR Reinforced Thermosetting Resin.

RTS *or* **R.T.S.** Radio Time Signal *q.v.*; Rapid Transport System, *q.v.*

RTSA *or* **R.T.S.A.** Retail Trading Standards Association.

RTSD *or* **R.T.S.D.** Rye Terms Sound Delivered, *q.v.*

RTTY Radioteletype.

RTV Radio Television.

RTV *or* **R.T.V.** Radio Telephone Very High Frequency.

RTW *or* **R.T.W.** Ready to Wear; Road Tank Wagon; Round the World *air ticket*.

RTYC *or* **R.T.Y.C.** Royal Thames Yacht Club.

RU Rumania; Rumanian.

rud. Rudder.

Rudder A vertical control structure hinged to the Rudder Post, *q.v.*, or Sternpost of a vessel or craft, or hung by a Rudder Stock which is the steering apparatus governing the manoeuvring of the vessel and its course when underway. *See* **Rudder Post**.

Rudder Post A vertical or nearly vertical member of the ship's structure upon which the steering rudder is hung or supported. Synonymous with Sternpost where the latter is fitted with gudgeons for the hanging of the rudder by its Pintles, *q.v.*

R.U.E. *or* **r.u.e.** Right Upper Entrance.

Rules of Court Orders regulating the practice of the courts; orders made between parties to an action or suit.

Rules of Practice Rules relating to the Association of Average Adjusters which are to be adhered to in the assessment of Particular and General Average, *q.v.*

RUM *or* **Rum** Remote Underwater Manipulator.

Rummage A search by customs officers to ascertain that no dutiable articles are hidden. Sometimes called Jerque. *See* Jerque Note.

Rummaged A ship which has been searched by customs officers. Also called Jerqued. *See* **Rummage**.

Run (1) The total distance covered during 24 hours by a ship *nautical*. This is generally termed a 'day's run'; (2) The narrow stern part of the aft hold of a ship. (3) To continue without falling due in a Promissory Note, *q.v.*, or other bill. (4) To engage any service. (5) To desert.

Run Aground A vessel in a stranded position. *See* **Stranding**.

Run Into To collide with another vessel.

Run in With To steam close to another vessel.

Runnel A rivulet or a small brook.

Runner (1) Any rope passing through a Block, *q.v.* (2) A wire for mooring ships; (3) A messenger or representative of a shipping company calling on board a vessel to obtain orders for supplies. *See* **Ship Suppliers**.

Running Agreement Articles of Agreement or ship's articles used by ships engaged in foreign-going trade. Such voyages may take months that would render time a factor for the officers and crew members. A running agreement enables them to continue on the basis of the contract for over a period of one year and until the vessel has finished her consecutive voyages.

Running a Reciprocal Examining a ship's magnetic compass to check its accuracy.

Running Broker A Bill Broker, *q.v.*, who is always roaming about, acting as a go-between.

Running Costs *or* **Running Expenses** The running expenses of a merchant ship, normally calculated under separate headings such as crewing, insurance, stores and provisions, drydocking and repairs, and converted to an overall daily rate.

Running Down Clause A clause incorporated in a Hull Insurance Policy to cover legal liability amounts paid by the insured consequent upon collision of the insured ship with another vessel.

Running Expenses *or* **Running Cost** *See* **Voyage Estimating.**

Running Lights The navigation lights of a vessel displayed when steaming.

Run Off In long term Insurance, *q.v.*, some risks still remain in effect until the adjustment of claims is finalised. Thus insurance on time risks will have to stay in force until it has 'run off'.

Run on Bank An unusual request for the withdrawal of deposits and repayment of moneys from a bank or banks; a sudden demand from many customers for immediate repayment.

Runway A specially prepared surface in an airfield or airport for the taking off and landing of aircraft. *Abbrev.* Rw *or* R.W.

Rupee A unit of currency used in India, Maldive Islands, Mauritius, Nepal, Pakistan, Seychelles, and Sri Lanka. The Indian rupee comprises 100 paise.

Rupiah A unit of currency used in Indonesia. One hundred sen equals one rupiah.

Rus *or* **Russ** Russia; Russian.

Rush Hour The peak hour for traffic when offices, factories, schools and other organisations start and finish work.

Russcon Code name for a Chamber of Shipping Black Sea Berth Contract, 1912. Amended 1931.

Russwood Code name for Chamber of Shipping White Sea Wood Charter 1933. From White Sea to UK and Continent.

Rutelag *Norwegian*—Transport Company.

RUTEX Rural Transport Experiments.

Rv. Rendezvous.

RV *or* **R.V.** Radius Vector, *q.v.*; Rateable Value; Research Vessel; Revised Version.

R/V Round Voyage; Rendez-vous.

R & V.A. Rating and Valuation Association.

RVP Reid Vapour Pressure, *q.v.*; Regional Vice-President.

RVR *or* **R.V.R.** Runway Visual Range *aviation*.

RVU Research Vessel Unit.

RW *or* **R.W.** Rainwater *weather observation*; Riveted and Welded, *Lloyd's Register notation*; Royal Warrant; Runway.

Rw Rwanda.

RWA *or* **R.W.A.** Ready Willing and Able, *q.v.*

RWF *or* **R.W.F.** Radio Weather Forecasts.

RWS Round the World Service.

RWTH *or* **R.W.T.H.** Raised Water Tight Hatch.

Rx Radio Receiver.

RYA *or* **R.Y.A.** Royal Yachting Association.

RYC Re Your Cable.

Rye Terms In the negotiations for the selling of rye, the seller guarantees to the buyer the condition of the merchandise up to final destination even if it gets damaged by sea water on the way. *Abbrev.* RT *or* R.T. *or* R/T.

Rye Terms Full Outturn Rye is bought and accepted against that quality delivered. *Abbrev.* RTFO *or* R.T.F.O. *See* **Rye Terms.**

Rye Terms Sound Delivered Rye is bought in accordance with the state and sound condition until the time of delivery by the receiver. *Abbrev.* RTSD *or* R.T.S.D. *See* **Rye Terms.**

RYL Re Your Letter.

RYS *or* **R.Y.S.** Royal Yacht Squadron.

RYT *or* **R.Y.T.** Re Your telegram/telephone/telex.

RYT *or* **RYTX** Re Your telex.

RZn. Relative Zenith *nautical*.

S

s Snow *weather observation*; Second *time.*

S Symbol for Sulphur; Load Line, *q.v.*; a letter seen each side of a cargo ship amidships indicating the level she is to be when fully loaded during the summer season; Salvage; Shipping; Starboard; *Lloyd's Register notation* for Simple Expansion Engine and Steamship; Subject to Approval; Subject to Acceptance *insurance*; Slow; South; Special, Loran PRR; Speed *nautical*; Summertime.

$ Dollar.

SA Safety Approved, *q.v.*; Salvage Association, *q.v.*

S/A Shipsaksjeselskäp, Norwegian Liability Insurance; Salvage Association, *q.v.* Safe Arrival, Subject to Acceptance; Subject to Approval, *See* **Subject To Approval No Risk**; Societa Anonima, Italian Limited Company.

S.A. Shaft Alley; Salvage Agreement, *q.v.*

S.A. *or* **s.a.** Safe arrival *marine insurance*; Salvage Association, *q.v.*; Saudi Arabia; Sociedad Anonima, Spanish Limited Company; Societé Anonyme, Belgian or French Limited Company; Sociedade Anonima, Portuguese Liability Company; South Africa; South America; South Australia/Australian; Secretary of the Army *USA*; Salvation Army.

Sa Speed of apparent wind in units of ship's speed.

s.a. Sine anno *Latin*—Without date.

S&A Signing and Accounting as *Lloyd's underwriting accounting procedure.*

SAA *or* **S.A.A.** South African Airways; Standards Association of Australia; Surface Active Agent.

S.A. a/c. Salvage Association Account.

SAAF *or* **S.A.A.F.** South African Air Force.

SAAP Super-Adaptive Autopilot.

SAB *or* **S.A.B.** Shipping Advisory Board.

Sabah Part of Malaysia.

SABAS *or* **S.A.B.A.S.** Shipping Agents and Brokers Association of Singapore.

SABENA Société Anonyme Belge d'Éxploitation de la Navigation Aérienne, Belgian State Airline.

Sabotage To inflict damage clandestinely on persons or property.

SABA *or* **S.A.B.A.** South African Bureau of Association.

SABIC Saudi Basic Industries Corporation.

S.A.B.S. South African Bureau of Standards.

SACD South African Container Depot.

SACEUR *or* **Saceur** Supreme Allied Command Europe.

SACLANT *or* **Saclant** Supreme Allied Commander Atlantic/*NATO.*

Sacr. Sacramento, *USA.*

Sacrifice To forsake something valuable for a higher consideration, higher good or advantage. For a specific meaning *see* **General Average Sacrifice**.

Sacrificial Anode An anode or zinc plate attached either to the interior cargo tanks or to the exterior underwater parts of a ship to prevent deterioration of the metal through electrochemical reaction. Anodes are generally composed of zinc and aluminium though there are also pure zinc ones.

S.A.C.S.I.R. South African Council for Scientific and Industrial Research.

S.A.C.T.U. South African Congress of Trade Unions.

SACVT Society of Air Cushion Vehicle Technicians *Canada.*

SAD Single Administrative Document, *q.v.*

SADCC South Africa Development Co-ordinating Conference.

Saddle Tanks Side tanks used for ballast on a ship. These are usually constructed in bulk carriers to adjust the stability when light cargo is carried. Also called Side Water Tanks.

SAE *or* **S.A.E.** *or* **s.a.e.** Society of Automotive Engineers *USA*; Stamped Addressed Envelope.

SAFA South African Shipowners' Association.

SAFCON *or* **Safcon** Safety Construction Certificate, *q.v.*

SAFCORD Safety Co-ordination Unit.

S.A.F.E. *or* **s.a.f.e.** Stamped Addressed Foolscap Envelope.

Safe Custody The strong room of a bank used as a security to accommodate valuable articles and documents belonging to commercial companies and clients.

Safe-Life The period applicable to various components regarding their usefulness and efficiency, after which time they either should be or must be replaced.

Safe Port A port which a ship can reach, use and depart from, without being exposed to danger—perils of the sea and nature, a political situation or risks to health—which cannot be avoided by good navigation and seamanship.

Safety Construction Certificate

Safety Construction Certificate A certificate issued by Ship Classification Societies. The certificate confirms that the ship conforms to certain minimum and specified standards of construction, fire protection, equipment, machinery, etc. Certain items are subject to regular surveys and in some instances continuous surveys as far as machinery and the hull are concerned. *Abbrev.* SAFCON *or* Safcon.

Safety Equipment Short Term A safety equipment certificate provided by the ship classification societies on a temporary or short term basis.

SafetyNET Code name for INMARSAT, *q.v.* Enhanced Group Calling System. An international service for broadcast via satellite and automatic reception of Maritime Safety Information.

Safety of Life at Sea *See* **SOLAS**.

Safe Working Load The maximum approved lift on a derrick or crane. *Abbrev.* SWL.

Sag, Sagging, Sagged These terms refer to the longitudinal bending of a ship as the bow and stern are raised while the centre is lowered. This can be a severe strain on the hull and arises (1) at sea when the length and angle of incidence of waves is such that the crests support the ends of the ship while the centre is in a trough, or (2) when heavy cargo is concentrated in the centre of the ship's length. The opposite is Hogging, *q.v.*

S.Ahd. An order for 'Slow Ahead' *nautical.*

SAI *or* **S.A.I.** Single Accident Indemnity *insurance.*

S.A.I.C.F. Sociedad Anonima Industrial y Fiduciaria, *q.v.* Societa Anonima Italiana, Italian Incorporated Company.

Sailed as per List A remark on the insurance slip. *Abbrev.* Sld.a.p.l. or S.a.p.l. The Insurance, *q.v.*, cover becomes effective as soon as the vessel officially sails and is so recorded in Lloyd's List.

Sailing Ice *Similar to* **Drift Ice**, *q.v.*

Sailing List *Similar to* **Sailing Card**.

Sailing Ship Vessel under sail and with no engine propulsion.

Sailors Families Society Founded in 1821 as a seamen's missionary society, it fitted out a floating chapel at Hull known as the *Valiant*. Work then extended to the welfare of seafarers and their families and then concentrated on orphans as the Sailors' Children's Society. In 1863 the Society came under Royal Patronage. In 1895 it established the Newland Homes for the children at Hull, now caring to 30 to 40 children. The Society supports over 400 children of seafarers, who are in the care of widows or relatives in over 200 families, in their upbringing and education. Families are also assisted with clothing and holidays. The Society also cares for aged seafarers, wives and widows in homes at Hull and South Shields. Working with other charities, it is concerned with seafarers docking in the Rivers Trent and Humber.

SAIMA *or* **S.A.I.M.A.** Shipbuilding and Allied Industries Management Association.

S.A.I.Y.C. Sociedad Anonima Industrial y Commercial, Spanish Industrial and Commercial Association.

SAJ Shipbuilders Association of Japan.

S.A.K. Société Anonyme Kuwaiti.

Salary A fixed payment usually on a monthly basis as a remuneration for work done.

SALC Ship Acquisition Licensing Committee *Indian Government.*

Sale or Return A system whereby the wholesaler or principal supplies the retailer or representative agent with goods on the basis that they may be returned if not successfully disposed of by a certain date or within a reasonable time. *Abbrev.* SOR.

Sale Plate *Similar to* **Bed Plate**, *q.v.*

SALESCRAP 87 BIMCO, *q.v.* Standard Contract for the Sale of Vessels for Demolition.

Sales Journal A register book for the daily recording of sale entries.

Sales Manager A senior officer of a company engaged in the sales department.

Salesmanship The special art of convincing an audience or the would-be buyer to buy.

Sales Tax *Similar to* **Value Added Tax** *or* **Value After Tax**, *q.v.*

Salinity The amount of salt content expressed by weight in parts of thousands in a liquid. *See* **Salinometer**.

Salinometer An instrument to find out the salt percentage of seawater. *See* **Salinity**.

SALM Single Anchor Leg Mooring, offshore mooring buoy.

Saloon Deck The deck reserved for saloon passengers.

SALRAM Single Anchor Leg Rigid Arm Mooring, *q.v.*

SALT Strategic Arms Limitation Talks.

Saltcon A BIMCO charterparty for salt from the Mediterranean to Norway. Also for other salt cargoes from elsewhere.

Salt/Dep. Salt Deposit.

S/A L/U Subject to Acceptance (by the assured) to be advised to the Leading Underwriters Only.

Salus pupuli suprema lex esto *Latin*—Let the welfare of the people be the supreme law. Motto of Mississippi, *USA.*

Salvage This may mean (1) money paid to those who assist in saving a ship or goods from the dangers of the sea; (2) the goods so saved; (3) property saved from a fire on land.

Salvage Agreement An agreement entered into by the master of a distressed vessel and the Salvor. This is generally agreed upon the conditions and remunerations laid down under the Lloyd's Open Form and must be signed by both parties. The agreement is usually on a No Cure No Pay, *q.v.*, basis.

Salvage Association An association, closely connected with the London insurance market, whose function is to take instructions from interested parties (e.g. underwriters) to investigate casualties and to make recommendations for the preservation and protection of property; also to determine the extent and proximate cause of loss when required. *Abbrev.* SA *or* S.A.

Salvage Award An amount awarded to a salvor for services rendered in the salvage of property in peril at sea. The award may be made by a court or by arbitration, depending on the terms of the salvage contract. Underwriters, *q.v.*, contribute towards a salvage award insofar as the award is in respect of insured property in peril from an insured risk, subject to any restrictions imposed by the Policy, *q.v.*

Salvage Loss A compromise settlement on a cargo Policy; *q.v.*, usually when the adventure has been terminated short of destination and damaged goods are sold at the intermediate port. The underwriter pays the difference between the sum insured by the policy and the proceeds of the sale.

Salvage Tug A powerful vessel specifically designed to tow a distressed vessel needing assistance in any type of weather. Its manoeuvrability enables it to control the tow without difficulty. The American equivalent is Wrecking Tug.

Salvage Value The assessed value of a salvaged vessel and/or the cargo after arrival in a Port of Refuge, *q.v.*

Salvage Working Group Established in 1990, an international body comprising shipowners, insurers, salvors, IMO and others.

Salvi juri *Latin*—The law being safeguarded; without prejudice.

SAM Surface to Air Missile.

SAM *or* **S.A.M.** Service Assistance Management.

S. Am *or* **S. Amer.** South America; South American.

SAMA Saudi Arabian Monetary Authority.

Samband *Icelandic*—Association; Federation.

SAMI Speed of Approach Measuring Instrument, *q.v.*

Samlogufelagid *Icelandic*—Non-limited Company.

SAMOS Satellite and Missile Observation System.

Sampan A light Chinese boat.

Sampling A clause in a tanker Charterparty, *q.v.* requiring the obtaining of samples of oil cargoes for analysis purposes. This protects the owners and charterers with regard to the qualities of oils carried. Normally the samples are kept for a reasonable time, say for 14 months, which is beyond the 12-month time bar, enabling the claim, if any, to be produced.

Sampling Order A small order given by a merchant to a supplier of commodities on the understanding that if the quality or grade meets with the approval of the customers the order will eventually increase.

Sampling Orders Authority given by merchants to the warehousemen to give out samples from stored goods belonging to them.

Samson Post A short strongly built tubular or square mast on which a Derrick, *q.v.* is rigged. It is also called **Derrick Post** *or* **King Post**.

San. Sanitary.

Sanction A punishment or penalty imposed by law to enforce an order.

S. and/or N.D. Shortage and/or Non-Delivery *insurance.*

Sandwich Course Course of instruction consisting of alternating periods of study and relevant practical work enabling students to acquire the fullest knowledge of theory and practice alike. They often attain higher qualifications even up to the highest degrees.

S and P Broker Sale and Purchase Broker.

San Fran. San Francisco.

Sangyo Kaiun *Japanese*—Industrial Transport.

S.A.N.R. *or* **s.a.n.r.** Subject to Approval No Risk, *q.v.*

Sans appel *French*—Without appeal.

Sans frais *French*—Free of Expense or Charge.

Sans Recourse *French*—Without recourse or without prejudice. These words are sometimes inserted near to, or at the end of, the signature in a contract, Bill of Exchange, *q.v.*, etc., for the endorser to limit liability.

Sans souci *French*—Without Care.

SANZ *or* **S.A.N.Z.** Standards Association of New Zealand.

SAO *or* **S.A.O.** Senior Accountant Officer.

SAP *or* **S.A.P.** *or* **s.a.p.** Soon as Possible.

SapA Societa Anonima per Azioni, *q.v.*

S.a.p.l. Sailed as per List, *q.v.*

Saponification The conversion of vegetable oils and fats into soap.

SAR *or* **S.A.R.** Search and Rescue Center *USA*; Service and Repair; Sons of the American Revolution; South African Railways; Special Absorption Rate; Synthetic Aperture Radio.

SARBE *or* **S.A.R.B.E.** Search and Rescue Beacon Equipment.

SARC Subject Area Review Committee.

Sargasso Kinds of seaweed. *See* Sargasso Sea.

Sargasso Sea An area in the south-west quarter of the north Atlantic Ocean between 20° and 40° North latitude where a strong circulatory current causes the Sargasso, *q.v.*, weed to collect and float on the surface in island-like masses.

S.A.R. & H. South African Railways and Harbours.

S.A.R.L. Sociedade Anonima de Responsabilidade Limitada, Portuguese Limited Company; Société a

Responsabilité Limitée, Belgian or French Limited Company.

SARP Search and Rescue Planning.

SARPS Standards and Recommended Practices.

SARSAT Search and Rescue Satellites, *q.v.*

SART Search and Rescue Radar Transponder.

SAS *or* **S.A.S.** Safety at Sea; Scandinavian Airline System; Special Air Service.

S.a.s. Società in Accomandata Semplice, Italian Limited Company.

SASA South African Shipowners' Association.

SASAR *or* **SAS&R** *or* **SASR** Singapore Association of Shipbuilders and Repairers.

SASMEX Safety at Sea and Marine Electronics Exhibition.

SASO Saudi Arabian Standards Association.

SASS Seafloor Anchor and Support Ship *oil drilling tender vessel*; Statistics Analysis of Steel Structures, *Lloyd's Register notation.*

S.Astn. Slow Astern *nautical.*

Sastrugi Ridges on a snow surface caused by the wind. May occur where snow lies on floating ice.

Sat. Saturday.

SAT Scholastic Aptitude Test; Ship's Apparent Time; Sulpheric Acid Tanker.

Satis verborum *Latin*—Enough of words.

SATCO Signal Automatic Air Traffic Control System.

SATCOM Satellite Communication.

SATCOMA Satellite Communication Agency.

SATCP Surface-Air à très Courte Portée *French* —Surface to Air Short Range Missile.

Satellite A body in orbit around a larger body, such as the moon revolves around the earth and the earth is a satellite of the sun. It can also be an object launched into space to orbit above the earth for a specific purpose, such as Satellite SOS System Aids, *q.v.*

Satellite SOS System Aids An air-sea rescue aid system which has proved to be effective. It was developed by the electronics division of the Aerospace Research Institute in Oberfaffenhofen in conjunction with Messerschmitt-Bolkow-Blohm of Munich. It is in the form of a waterproof buoy which automatically transmits signals to the nearest satellite on touching the water. Its compactness makes it easily accommodated in a ship's lifeboat and it is capable of transmitting the location and nature of assistance required.

SATEX Semi-Automatic Telegraph Exchange.

Satis verborum *Latin*—Enough of words; enough said.

SATNAV Satellite Navigation.

SATO *or* **S.A.T.O.** South American Travel Organisation.

Saturated Steam *See* **Steam.**

Saturation Vapour Pressure The exerted vapour pressure produced by heated liquid in an enclosed area or space. *Abbrev.* SVP *or* S.V.P. Also known as Vapour pressure.

SAU Search and Attack Unit *USA.*

SAUDIA Saudi Arabian Airlines.

Saudi Ar. Saudi Arabia.

SAV *or* **S.A.V.** Sale at Valuation.

Save-All (1) Container to catch drips or leaks and avoid spillage of water or oil, or save wastage. (2) Additional small sail laced to the foot of a studding sail in a square-rigged ship to take full advantage of a light following breeze. Also called a Bonnet or Water Sail. (3) Sheet of canvas spread when cargo is being lifted in and out of a ship's hold.

Save As You Earn A state backed scheme where a wage earner is encouraged to deposit part of his wages in a bank. In return a concession from income tax can be claimed. *Abbrev.* SAYE *or* S.A.Y.E.

Save Our Port A port authority in New York/ New Jersey in regard to dredging and ocean dumping. *Abbrev.* SOP.

Savings Ratio The ratio of total savings or deposits to total income.

S.A.W. Sale at Will.

SAYE *or* **S.A.Y.E.** Save As you Earn, *q.v.*

Sazcnan Raden Bandi Iranian *classification society.*

SB Safe Berth; Single-Ended Boiler(s) *Lloyd's Register notation*; Short Bill; *q.v.*

SB *or* **S.B.** Safe Berth; Sales Book; Saving Bank; Scientiae Baccalaureus *Latin*—Bachelor of Science; Selection Board; Short Bill; *q.v.*; Signal Book; South Britain; Special Branch *police*; Statute Book; Stretcher Bearer.

S/B Savings Bank Account.

SBA *or* **S.B.A.** School of Business Administration; Small Business Administration *USA.*

SBAC *or* **S.B.A.C.** Society of British Aerospace Companies; Society of British Aircraft Constructors.

SB.B *or* **sb.b.** Submersible Barge.

SBB *or* **S.B.B.** Schweizerische Bundesbahnen, Swiss Federal Railways; Ship Borne Barge.

SBBNF *or* **S.B.B.N.F.** Ship and Boat Builders' National Federation.

s.b.d. Scratching, Bruising and Denting *insurance.*

SBL Short Base Line, *acoustic systems.*

SBM *or* **S.B.M.** Single Buoy Mooring, *q.v.*

SBN *or* **S.B.N.** Standard Book Number now ISBN (International Standard Book Number).

SB.RG. *or* **sb.rg.** Submersible Rig.

SBS *or* **S.B.S.** Small Business System; Surveyed Before Shipment *insurance*; Ship fitted with submarine signalling equipment, *Lloyd's Register notation.*

s.b.s Surveyed Before Shipment *insurance.*

SBSV Standby/Safety Vessel.

SBT Segregated Ballast Tank.

SBTC Sino-British Trade Council.

SBT/PL SBT-PL Protectively Located Segregated Ballast Tank.

SBV *or* **S.B.V.** Sea Bed Vehicle.

S. by W. South by West *compass bearing.*

SC *or* **S.C.** *or* **s.c.** Sailing Club; Salvage Charges *insurance*; School Certificate; Self Contained; Service Certificate; Slip Coupling; Staff Captain; Staff College; South Carolina, *USA*; Steel Cover(s).

Sc. Science; Screw; Solicit; South Carolina, *USA*; Strato Cumulus Clouds, *q.v.*

Sc *or* **Scan** Scandinavia; Scandinavian.

S&C Shipper and Carrier.

SCA Shipbuilders' Council of America.

SCA *or* **S.C.A.** Settlement of Claims Abroad *insurance*.

SCA Software Conformity Assessment, a quality assurance service of Lloyd's Register.

Scab A person who disregards union instructions to strike. *See also* **Blackleg**.

SCAD Self Contained Automatic Fire Detection System; Subsonic Cruise Armed Decoy, *US Air Force*.

SCADA Supervisory Control and Data Acquisition *electronic engine control*.

SCADAR Scatter Detection and Ranging.

SCAF Supreme Commander Allied Forces *1939–45 war*.

SCALDS Shipboard Containerised Air Force System.

Scale Discharge In some Charterparties, *q.v.*, a scale of unloading fees is charged to the Carrier, *q.v.*, at the port of discharge.

Scale of Provisions Most states of registered ships issue regulations regarding the entitlement of stores and provisions in connection with the catering of the master, officers and crew. This is a scale of detailed items the owner is to provide daily for the vessel's account.

Scale of Rates An insurance schedule referring to an open or cover policy against each consecutive voyage. This information is generally attached to the policy. *See* **Floating Policy**.

Scaling Hammer *See* **Chipping Hammer**.

Scan. Scandinavia; Scandinavian.

Scantlings (1) A set of dimensions for all components of a ship's structure, including the thickness of steel plate, timber, etc. (2) A small beam under five inches in breadth and depth. (3) Ships which are said to be with full scantlings have been built in accordance with the highest requirements of the classification society concerned. Consequently they are assigned to the maximum Draught (draft), *q.v.*, possible with the result of a minimum Freeboard, *q.v.* On the other hand, ships without full scantlings are not permitted to load the maximum draught (draft) and so they have a higher freeboard and less cargo capacity.

SCAP Supreme Command Allied Powers, *1939–45 war*.

SCAR Scientific Committee on Antarctic Research.

SCARAB Submersible Craft Assisting Repair and Burial.

Sc.B Scientiae Baccalaureus *Latin*—Bachelor of Science.

SCBA Self-Contained Breathing Apparatus.

SCC *or* **S.C.C.** Sea Cadet Corps; Senate Commerce Committee *USA*; Single Cotton Covered; Sociedad Clasificadora de Colombia. *Classification Society of Colombia*; Stockholm Chamber of Commerce.

SCCUK Swedish Chamber of Commerce in UK.

Sc. D. Scientae Doctor *Latin*—Doctor of Science.

SCECSAL Standing Conference of East, Central and Southern African Librarians.

SCH Austrian Schilling; Hotel Combined *insurance*.

Scend (1) Of a ship, to rise after pitching heavily and suddenly in the trough between two waves; to pitch violently by the bow or the stern in the trough of a heavy sea. (2) Vertical movement of waves or swell against rocks and cliffs, or alongside quays and jetties where it can endanger berthed vessels and their moorings. *Also called* Send.

Sch. Schedule; Schilling; Schooner.

Scharl Boards The outer part of wood logs that are so cheap in cost and quality, they are sold for packaging.

SCHDV Surface Controlled Hydraulic Downhole Safety Valve *oil drilling*.

Sched. Schedule, *q.v.*

Schedule Itinerary; Inventory; Catalogue; Printed or written list. *Abbrev.* Sched.

Scheduled Territories The countries which were formerly part of the sterling area.

Schiffahrts-gesellschaft *German*—Navigation Company.

Schiffahrtskontor *German*—Shipping Company.

SCI Supervisory Customs Inspector *USA*; Society of Chemical Industry.

SCIA Supervisory Customs Inspectors' Association *USA*.

SCICI Shipping Credit and Investment Company of India.

Scienter *Latin*—Knowingly.

Scil. Scilicet *Latin*—Namely; Being understood. *Abbrev.* also sc. or s.c.

SCIP Solid Cast Iron Propeller.

SCIT *or* **S.C.I.T.** Special Commissioners of Income Tax.

SCM *or* **S.C.M.** Satellite Control Module *electronics*; State Certified Midwife.

Sc.M. Scientiae Magister *Latin*—Master of Science

SCMES Society of Consulting Marine Engineers and Surveyors.

SCN Navy Shipbuilding and Conversion *US Navy*.

SCNE Select Committee on Natural Expenditure.

SCNRT Suez Canal Net Registered Ton.

SCNT Suez Canal Net Tonnage.

SCOD *or* **S.C.O.D.** South Coast One Design *UK Yachting*.

Sconce A wedge or expanding device for locking type in a forme; a wedge used to stop a barrel from rolling. It is also called **Quoin**.

Scoop *Similar to* **Bailer**, *q.v.*

SCOP *or* **S.C.O.P.** Steering Committee of Pilotage.

SCOR *or* **S.C.O.R.** Scientific Committee on Oceanic Research.

Scot. Scotland.

SCOTEC *or* **S.C.O.T.E.C.** Scottish Technical Education Council.

SCOTUS Supreme Court of the United States.

Scow (1) A flat-bottomed boat square shaped at both sides mostly used as a barge or a lighter. (2) Derogatory term for an unhandy or ill-found craft.

SCP Simplified Clearance Procedure *customs*.

SCPC Single Channel per Carrier *telecommunications*.

SCR *or* **S.C.R.** Specific Commodity Rates, *q.v.*; Silicon Controlled Rectifier

Scraping the bottom Cleaning the underwater part of a ship by taking off incrustation, weeds, etc. Not covered by a hull policy. *See* **Bottom Scraping.**

Scratch To start negotiating an agreement.

Scr. Bh. Screen Bulkhead.

SCRC *or* **S.C.R.C.** Specific Commodity Rates Committee. *See* **Specific Commodity Rates.**

SCREAM Sufferers' Campaign to Resolve the European Aviation Mess *air transport*.

Screw Propeller of a ship.

Scrip A provisional certificate of money subscribed to a bank or company entitling the holder to a formal certificate in due time and to dividends, etc.

Scrip Issue A bonus share. Extra shares given over the original ones as a means of strengthening or backing the capital of a company which may be in need of additional finance.

Scrivener A copyist, drafter of documents, notary, broker, formerly a money-lender.

Scruple A weight unit of 20 grains (apothecaries' weight).

SCS Sea Control Ship; Seawater Circulation System.

SCSCT *or* **SCSC&T** Said to contain Shippers Count and Tally.

SCSP Solid Cast Steel Propeller.

SCSSV Surface Controlled Subsea Safety Valve *oil drilling*.

SCT Scattered Clouds *weather observation*; Suez Canal Tonnage.

SCTC Suez Canal Tonnage Certificate.

SCTW Standards of Certification, Training and Watchkeeping Convention.

SCUA Suez Canal Users' Association.

SCUBA Self-Contained Underwater Breathing Apparatus.

Scuppers Holes throughout the sides of a ship in line with the deck to allow seawater and rainwater to be washed overboard and leave the deck clear. Also termed **Deck Drains**.

Scuttle To make holes in a ship or to open the seacocks in order to sink her. Such an action is common in time of war so as to prevent the ship falling into enemy hands. It is also an action undertaken by maritime fraudsters.

Scuttles (1) Large metal pails to carry coals. (2) Openings in the topside and superstructure of a ship allowing light to pass through. They are closed as a protection against rough weather but can be opened in calm seas for fresh air.

SCWS Scottish Co-operative Wholesale Society.

SD *or* **S.D.** Scientiae Doctor *Latin*—Doctor of Science; Semi-diameter; Send Direct; Short Delivery, *q.v.*; Standard Deviation; Standard Design, *q.v.*; Steam Drifter; Steel Diesel; South Dakota, *USA*.

s.d. Sea Damage, *see* **Sea Damage Terms**; Sine die *Latin*—Adjourned indefinitely, without a day being named.

S/D *or* **s/d** Sea Damage, *see* **Sea Damage Terms**; Short Delivery, *q.v.*; Short Delivered, *q.v.*

S.d. Short Delivery, *q.v.*; Short Delivered, *q.v.*

S.D. *or* **s.d.** Safe Deposit; Same Date; Sight Draft, *q.v.*; Special Delivery; State Department *USA*; Submarine Detector; Suction Dredger; Semi-Diameter *nautical*; Standard Design, *q.v.*

SD *or* **S/D** Sailing Date; Standard Design.

SD 14 A standard design, *q.v.*, general cargo ship of about 14,000 tons d.w. popular for its running economy and adaptable for canals and port restrictions.

SDA Scottish Development Agency.

S.Dak. South Dakota, *USA*.

S.D. Aug. *or* **s.d. Aug.** Semi-Diameter Augmented *nautical*.

SDBL *or* **S.D.B.L.** Sight Draft Bill of Lading.

SDC Submersible Decompression Chamber, *q.v.*; Stevedore Damage Clause, *q.v.*

SDC/DDC Submersible Decompression Chamber/Deck Decompression Chamber.

SDD Sbornie Deistvuiushehik Dogovorov Soglashenii i Konventsii *Russian*—Agreements and Conventions Concluded with Foreign States.

SDFC *or* **S.D.F.C.** Shipping Development Fund Committee, *q.v.*

SDHF *or* **S.D.H.F.** Standard Dutch Hull Form or Dutch Hull Form, *q.v.*

Sdg. Sounding.

SDHE Spacecraft Data Handling Equipment.

S.dk Shelter Deck; Shelter Decker.

SDL Special Duties List.

SDLC Synchronous Data Link Communication *electronic data handling*.

SDNR Valve Screw Down, non-Return Valve.

SDP Social Domestic and Pleasure *insurance* (motor).

SDPM Simultaneous Drilling Production and Maintenance *oil drilling*.

SD & PP Social, Domestic and Pleasure Purposes *insurance*.

SDR Special Despatch Rider; Special Drawing Rights, *q.v.*; Silicon Controlled Rectifier, Shaft Driven Generators.

SDS Shallow Draft (Draught) Survey, for offshore drilling units, *Lloyd's Register notation*.

SDT Saddle Tank(s), *q.v.*; Side Tank(s).

SDW *or* **S.D.W.** Side Wheel; Summer Dead-weight.

SDWT *or* **S.D.W.T.** Summer Deadweight Tons.

Sdz Ship of the future *USA*.

se Self-Elevating *Platform*.

SE *or* **S.E.** Safety Equipment; Senior Engineer; Special Equipment; South East; South Eastern; Staff Engineer; Static Equipment.

S.E. *or* **S/E** Society of Engineers; Stock Exchange, *q.v.*

s.e. Single Engine; Single Ended; Single Entry; Straight Edge.

SEA *or* **S.E.A.** Subterranean Exploration Agency; South East Asia; Strategic Environmental Assessment.

Sea-Air Air adjacent to the sea.

Sea Anchor A towed object in the form of a canvas cone which serves as a floating anchor at sea to prevent the vessel from drifting or to keep the bows to the wind. It is used on small craft like lifeboats in emergencies. Termed also as **Drag Anchor** *or* **Drogue**.

Sea-ape Shark or sea otter.

Sea Bag A large cylindrical canvas bag into which sailors put their clothing and personal effects. It can be closed at one end by a thick cord.

Seabank A mole or bank or any other material to prevent the sea from washing over the land.

Sea Beach A beach covered with shingle or sand.

Seabed The bottom of the sea or ocean.

Seabee (1) A person serving in the US Navy engaged in harbour work and on air bases. (2) Similar to the term LASH, *q.v.*, or Lash ship meaning Lighter Aboard Ship. The only difference is that the self-propelled loaded barges are themselves loaded on board as cargo and are considerably larger than those loaded on Lash ships. Consequently the volume and capacity of the Seabees are larger. Also called **Barge Carrying Vessel**, *q.v.*

Sea Biscuit A hard dry biscuit commonly used on board ships, especially as emergency biscuits, namely as lifeboat rations if an emergency occurs. Also called **Sea Bread** or **Ship's Biscuits**, *q.v.*

Seaboard Coastline of a country.

Seaboat (1) Used in describing a vessel's behaviour in a seaway, e.g. 'a good seaboat'. (2) A boat kept ready at sea for immediate launching in an emergency.

Seaborne Anything carried by a ship at sea.

Seaborne Trade The carriage of passengers and general merchandise on merchant vessels.

Sea Bread *Similar to* **Sea Biscuit**.

Sea Breeze A breeze blowing from the sea to the land due to convection currents.

SEAC South East Asia Command, *1939–45 war*; Standard Electronic Automatic Computer.

Sea Calf A seal.

Sea Canary Sailors' expression for a whale.

Sea Captain A naval master or captain as distinguished from a military one.

Sea Carriage Act 1924 Commonwealth of Australia The Transportation of Merchandise by Sea Act enacted on 17 September 1924 in Australia. *See* **Hague Rules**.

Sea Chests (1) Cupboards or chests in sailors' cabins for their personal effects. (2) Water intake compartments for the purpose of feeding filtered seawater into the engines of a ship for cooling purposes. They are situated on either side under the waterline near to the engine room and are protected by metal grills being level with the hull. *Also called* **Suction Boxes**. *See* **STRUM**.

Sea Chestnut A sea urchin.

Seacoast Coastline. Land touching the sea.

Sea Cock A valve by which water can be let into a ship's interior.

SEACON *or* **Seacon** Sea Construction, *q.v.*

Sea Construction The first ever offshore platform constructed for the US Naval Facilities Engineering Command. *Abbrev.* CHESNAVFACENG COM. *Abbrev.* SEACON *or* Seacon.

Seacor Systems Engineering Associates Corporation.

Sea Damage Terms *or* **Sea Damaged Terms** A grain contract term covering the buyers or receivers in case of damage sustained by the grain by either seawater or condensation. Normally the sellers are to make good for subsequent damage.

SEADD South-East Asia Development Division.

Sea Dog An old experienced sailor.

Seafarer Sea traveller or sailor.

Seafaring One's occupation at sea; seagoing.

Sea Fight When ships are in combat at sea.

SEAFIRST *or* **Seafirst** Seattle First National Bank.

Seafolk Sailors: seamen; fishermen.

Sea Front Land or buildings developed along the shore and facing the sea.

Seagate A waterway leading to the sea.

Seagoing Occupied at sea, as opposed to Shore-bound, Shore Service or Harbour Service.

Sea-Going Ships Ships which are well fitted to encounter the normal strains of a voyage in open waters.

Sea Horses Visible white crested waves.

Sea Keeping General term covering the behaviour and performance of a ship in a seaway.

Sea Keeping Ability A ship's ability to maintain normal functions at sea.

Sea Kindliness Property of a ship in behaving comfortably with an easy motion in a seaway.

Sea Ladder A rope ladder which is lowered down to sea level to pick up the pilot. This ladder is also used sometimes by seamen when the Accommodation Ladder, *q.v*, is not available. Also called **Pilot Ladder** *or* **Jacob's Ladder**.

Sea Lawyer A seaman who is always arguing and pretending that he is always right in his way of reasoning. In any trouble he may be the ring-leader.

Sea Legs The act of withstanding seasickness; Walking well on a ship while pitching and rolling at sea without becoming seasick.

Sealer (1) An official engaged in confirming by a stamp the weights and measures in accordance with official standards. (2) A painting term meaning a surface preparation to prevent any solvent reaction on to the previous paint applications. (3) A seal hunter.

Sea Letter (1) An American Licence which is issued to commercial vessels of 20 tons and over. (2) Document issued to a neutral vessel in wartime, with details of her voyage and cargo, which can be inspected by a boarding officer.

Sea Level Surface of the sea.

Sealink *or* **Sea Link** A sophisticated method of ocean tug pushing or tug barge connection *Ex.* **Artubar**, *q.v.*

Seaman A sailor employed on a ship.

Seaman's Card An identification card issued to a seaman by the shipping authorities of the state of his nationality as a means of identification in lieu of a passport. All the particulars of the seaman are shown therein.

Seaman Watches *See* **Bell** *and* **Watch Duties**.

Seamark An elevated object on land or floating on the sea serving as a guide to navigators. It could be a buoy at sea which flashes or a flag or beacon on land or any other conspicuous reference mark or signal.

Seamen's Hospital Society The society was founded in 1821 to provide for sick and suffering mariners with a hospital ship in the Thames until replaced by a hospital ashore, the *Dreadnought*, at Greenwich in 1870. It was incorporated by Act of Parliament in 1833. In following years, other hospitals and convalescent homes were opened. Control passed to the Minister of Health in 1948 but in 1982 the facilities for seafarers were transferred from Greenwich to a Dreadnought Unit at St Thomas's Hospital. The

Society continues to administer the unit, which provides free medical treatment for all seafarers and their immediate families, and a hostel nearby for the dependants of the sick. The Samaritan Fund makes grants to men and women of the Merchant Navy regardless of nationality, and their dependants, for their welfare in cases of need arising from sickness, misfortune or old age.

Seamen's Mail This generally alludes to a sticker or a stamp on the envelope to indicate that the contents are letters for seamen, who naturally look forward to receiving them from families and friends after a long spell abroad. Such mail is supposed to be given special attention for quick delivery.

Sea Mile A nautical mile of 6,080 feet or 1,853.18 metres. *See* **Geographical Mile** *and* **Nautical Mile**.

SEAN *or* **S.E.A.N.** State Enrolled Assistant Nurse.

Sea Perils Marine Perils, *q.v.*

Seaplane An aircraft having floats in place of the Undercarriage, *q.v.*, to enable it to land and take off from the sea. *Also called* **Float Plane**.

Seaport An area by the sea, river or lake having berthing or anchorage facilities as well as cargo gears for loading and discharging.

Sea Power A show of naval strength.

Sea Protest *See* **Note Protest**.

Sea-Purse (1) Mermaid's Purse. The egg-case of the skate and species of rays and sharks. (2) Small whirlpool on the surface of the water occurring on the US Atlantic Coast and dangerous to bathers.

SEAQ The Stock Exchange Automated Quotations System, for continuous computer updating of bids and offers on securities.

Seaquake An earthquake from under the sea.

Search and Rescue Satellite A satellite station originally developed by the USA, Canada and France in 1970. It is capable of locating both marine and aviation distress signals practically anywhere in the world. *Abbrev.* SARSAT.

Searcher A customs officer who visits ships in harbour prior to their sailing and is authorised to give the ship clearance when he is satisfied from his search that no contraband is present. *See* **Searching Note**.

Searching Note A note or advice issued by the customs authority of a port after the ship has been searched. *See* **Searcher**.

Search Warrant A legal authority to search for specific items or other objectives as decided by the state.

Sea Return Reflected echoes from the sea. Generally referring to radar and similar aids.

Sea Risks Marine Perils, *q.v.*; Perils of the sea.

Sea Room The sufficient or adequate space that allows a vessel or group of vessels to manoeuvre clear of navigational hazards and other shipping.

Sea Rover A ship moving about at sea with the master's and crew's intention to act as pirates.

Sea-Shed An extra large multi-purpose container base capable of handling all sizes of containers.

Seasickness Vomiting or the inclination to vomit caused by the motion of a ship. This can be counteracted by medication in some cases.

Sea Side (1) That side or end of a valve or pump connected to, or admitting, sea water. (2) Beach and sea coast.

Sea Slug A small slimy fish.

Seasonal Ports Ports which are only open at certain periods during the year, such as ports which are icebound in the winter season.

Seasonal Unemployment Unemployment caused by seasonal weather conditions. *Ex.*: Oil drilling in certain zones where seas become too strong for work to be done.

Seasonal Zones Each registered ship has to have a load line sign on each side amidships limiting her immersion when loading to the corresponding mark. The lines are white coloured 25 mm. thick and 230 mm. long. The load line markings are assigned as follows: Line with letter S for summer; Line with letter W for winter; Line with letters WNA for winter North Atlantic; Line with letter T for Tropical fresh water. The navigation map is divided into permanent and seasonal zones. The former is sub-divided into seasonal winter and seasonal tropical areas. *See* **Load-line Marks**, **Loadline Zone** *and* **Annual Load Line Certificate**.

Season Ticket A ticket which is issued at reduced rates for any number of journeys taken, performances attended, etc., within a year, six months, or other period.

Sea Smoke *or* **Frost Smoke** Caused by very cold air passing over warm seawater areas. Water vapour rises from the sea forming this fog.

Seastainers Containers, *q.v.*, carried on specially designed ships for containerisation. These containers are also carried on road trailer vehicles. Seastainers is an American expression for Sea Containers.

Sea Stocks *or* **Sea Stores** Collective essential stores needed for the maintenance and efficient running of a ship. These consist of various items and are categorised under different departments.

Seastrand Seashore.

SEATAC South East Asian Agency for Regional Transport and Communications.

Sea Tangle Seaweed.

Seat Belt *or* **Safety Belt** A strong strap which is fastened around a person while seated to prevent him or her from being thrown out of the seat. On aircraft the fastening of seat belts is absolutely compulsory during take-off, landing and in turbulent weather.

Seat Kilometre The revenue derived from the sale of seats, principally on aircraft. Also called Seat Mile. *See* **Freight Tonne Kilometre**.

Seat Mate A passenger sitting next to another in one of a pair of coupled seats in an aeroplane or vehicle.

SEATO South East Asia Treaty Organisation.

Seat Pitch Measure of the interval between rows of seats in an aircraft when these are in the upright position.

Seatrain A ferryboat or a specially constructed ship equipped to carry railway engines and/or carriages.

Sea Transport Goods and/or passengers carried on a ship.

SEATS The Stock Exchange Alternative Trading Service. Handling shares very infrequently traded.

Sea Wall A strong wall near to the shore built to protect the adjacent land from the sea.

Seaway (1) Sea Traffic lane. (2) Navigable channel or inland waterway connected to the sea suitable for ocean-going vessels, e.g., the St. Lawrence Seaway, a joint Canadian–US project connecting the Great Lakes and the North Atlantic Ocean. (3) A heavy sea.

Seaway and Great Lakes Trading Clause A standard time charter clause published by BIMCO.

Seawaymax Term applied to ships within the limiting dimensions for passage of the Welland locks on the St Lawrence Seaway.

Seaweed Plants that grow on the seabed, submerged rocks, and ships' bottoms. *See* **Sea-Whack**.

Seaworthiness The fitness of a ship to encounter the hazards of the sea with reasonable safety. In addition to having a sound hull the ship must be fully and competently crewed and sufficiently fuelled and provisioned for the contemplated voyage. All her equipment must be in proper working order and, if she carries cargo, she must be cargoworthy. The right to claim under a hull policy may be prejudiced if the ship puts to sea in an unseaworthy condition.

Sea-Wrack Any material cast ashore, especially coarse sea-grass or Seaweed, *q.v.*

SEB Single-Ended Boiler.

S.E.b.E. *or* **S.E. by E.** South-East by East *compass bearing.*

S.E.b.S. *or* **S.E. by S.** South-East by South *compass bearing.*

SEBS Single-Ended Boiler Survey; Submarine Emergency Buoyancy System *USA*.

Sec. Secant; Second; Secondary; Secretary; Section; Secundum *Latin*—According to.

SEC *or* **S.E.C.** Simple Electronic Computer; Securities and Exchange Commission *USA*.

Sec. art. Secundum artem *Latin*—According to art.

Sec. L. Sectional Line.

Sec. Leg. Secretary to Legation; Secundum Legum *Latin*—According to law.

Sec. nat. Secundum naturam *Latin*—Naturally.

Second Bottom Working (1) Loading/unloading cargo to or from a lighter when the latter is between the ship and quay. (2) Loading/unloading cargo from a ship to a lighter, then on to another ship alongside the lighter, namely the lighter will be between two ships.

Second Carrier The Carrier who undertakes to pick up a transhipment cargo at an airport, seaport or elsewhere and convey it to its final destination. Carrier may also mean a ship, aircraft or other means of transport.

Second Class Mail Mail or letters posted as second class at a lower postage.

Second Dog Watch A seaman's watch from 6 p.m. to 8 p.m. *See* **Bell** and **Watch Duties**.

Second Hand Replacement Clause An insurance term covering the shipments of second hand machines, which if lost or damaged, are to be replaced by old or second hand ones.

Second Liner/s Medium-sized companies quoted on the Stock Exchange. *See* **FT-SE**.

Seconds Imperfect manufactured goods.

Secretary (1) A person employed by another to assist him or her in correspondence, literary work, getting information, and other confidential matters. (2) An official appointed by a society or company or corporation to conduct its correspondence, keep its records, and deal in the first instance with its business. (3) A Minister in charge of a Government Office, such as the Secretary of State for Air, etc.

Sec. reg. Secundum regulam *Latin*—According to rule.

Sect. Section; Sector.

Secular (1) Occurring once in or lasting for an age or a century. (2) Lasting or going on for ages or an indefinitely long time. (3) Concerned with the affairs of this world, worldly, not sacred, not monastic, etc.

Secured Creditor A creditor whose debt is backed by immovable and other properties.

Securities Documents or stocks and shares representing title of investment. This can also relate to bills of exchange, etc.

Securities and Exchange Commission Similar to Stock Exchange, *q.v. Abbrev.* SEC.

Securities Market Stock Exchange, *q.v.*

Security In marine insurance, this defines the insurers with whom insurance has been effected.

Secus *Latin*—Otherwise; Not so; On the contrary.

SEDA Safety Equipment Distributors Association

SEDAM Société d'Etude et de Developpement des Aeroglisseurs Marine *French*—Society for the Marine Study and Development of Hovercraft.

SEDAR Submerged Electrode Detection and Ranging.

SEDB Singapore Economic Development Board.

Se defendendo *Latin*—In defending himself (herself).

SEE *or* **S.E.E.** Senior Electronic Engineer.

Seel Sudden heeling over of a vessel of momentary duration.

SEF *or* **S.E.F.** Shipbuilding Employers' Federation.

SEG. Society of Exploration Geologists *USA*.

Segregate To set apart or isolate one heterogeneous body or commodity from others. It could concern cargo, or personnel, or passengers. *Examples.* Dangerous cargo segregated from General Cargo, *q.v.*; Oranges segregated from bagged coffee; passengers segregated from the officers and crew. The term is commonly used when petroleum tankers carry different grades of oils. So the vessel must have Segregated Tanks and Segregated Water Tanks.

SEIA Solar Energy Industries Association *USA*.

Seiche The oscillation of lake waters due to changes in barometric pressure.

SEILOD Special Examination in Lieu of Dry Dock *US Coast Guard Inspection*.

Seisin *or* **Seizin** The possession of land by freehold; the act of taking such possession; what is so held.

Seismograph An instrument showing the force, place, etc., of an earthquake.

Seismology The scientific study of earth tremors and earthquakes.

Seized of the Case Legal term meaning 'Had been Informed'.

Seizure In Marine Insurance, *q.v.*, this includes capture or seizure by revenue customs officers of a foreign state and covers every act of taking possession of a ship either by lawful means or by any overpowering force.

SEK Swedish kroner.

Self Discharger A term used to describe a vessel capable of discharging itself—e.g. by means of grab-fitted cranes.

Self-Employed Describes a Freelance, Sole Proprietor, *q.v.*, or Partner.

Self Insurance Relates to a risk which is not covered by Underwriters, *q.v.*, but for which the assured himself assumes responsibility in case of accident.

Self-service Store A retail store where the customer has the advantage of looking at the many different items with their individual prices. After selecting and collecting the items needed the client pays at the cash desk.

Self-trimming Hatch A specially designed hatch which allows bulk cargo to be directed and to fall freely into the hold without the necessity for manual trimming. *See* **Self-trimming Ship**.

Self-trimming Ship A bulk carrier with sloping upper wing tanks to port and starboard under the weather deck, which helps trim bulk cargo.

Sell-By Date The date by which perishable consumable products must be sold or withdrawn from sale to the general public. This date should be displayed on the goods.

Seller's Market When the seller is at an advantage. A market in which goods are scarce and high prices favour the sellers.

Sellers' Over A business expression meaning that the number of sellers exceed those of the buyers in the market or where there are only seller dealers. This is opposite to **Buyers' Over**, *q.v.*

Selling Costs Costs incurred in creating and maintaining the market for a product including marketing, advertising, sales staff and their expenses. Distribution costs are not normally included but sometimes are.

Selling Out A stock exchange term. The selling of Shares, *q.v.*, at the best available price.

Selvage *or* **Salvedge** The woven edge of canvas and bunting.

Selvagee Parallel rope yarns laid in a circle and marled together, which makes a strop for handling a spar, lifting a bale, tailing a jigger or tackle to a larger rope, temporarily holding the weight of a hawser, etc.

SEM Scanning Electronic Microscope *electronics*; Standard Electronic Modules *US Navy.*

Semaphore A signalling apparatus of a post with oscillating arms, an arrangement of lanterns, etc., for use on railways by day and night; military or naval signalling by operator's two arms or two flags.

Semble *French*—It appears. Used in judgments and textbooks to introduce a proposition of law which is not intended to be stated definitely.

Semester A period of six months; Half an academic year.

Semi–containership A vessel which carries containers as well as break bulk general cargo and may be equipped for Roll-on/Roll-off, *q.v.*, as well.

Semi–Drying Oils Natural oils and fats consisting of:– Cod liver, cotton seed, herring, maize, sardine, sesame, shark, sunflower, whale, wheat, etc.

Semidur Semi-duration.

Seminar From the *Latin*—Seminarium, a seed plot. A class or group of students at a college or university attending a course of study or research under the guidance of a director or professor. Alternatively a conference of specialists, a short intensive course of study, a discussion group.

Semi–Submerged Catamaran A passenger vessel originally developed by Japanese engineers and shipbuilders suitable for all weathers. The two torpedo-shaped airtight hulls float on the sea and support the structure above the waterline, thus eliminating sea friction or resistance. *Abbrev.* S.S.C.

Semi–Submersible Rig A type of offshore drilling rig capable of submerging to a suitable depth and providing a stable base for the necessary drilling work. *Abbrev.* SSR *or* S.S.R.

SEMP Superconductive Electromagnetic Propulsion.

Sempaku *Japanese*—Ships.

Semper fidelis *Latin*—Always faithful.

Semper idem *Latin*—Always the same.

Semper paratus *Latin*—Always ready.

SEN *or* **S.E.N.** State Enrolled Nurse.

Send (1) Same as **Scend**, *q.v.* (2) To transmit a message.

S. en C. Société en Commandite *French*—Company Limited.

S. en N.C. Société en Nom Collectif *French*—Joint Stock Company.

Send *or* **Scend** The driving impulse of a wave against a ship; The lift after plunging of a ship in the trough of a wave; surging of the sea-swell within a harbour which can be dangerous for a ship moored alongside.

Sendirian Berhad *Malay*—Limited Company.

Senegal West African republic.

Senior Ordinary Seaman A seaman over the age of $17\frac{1}{2}$ years having served 18 months at sea or three years as an apprentice. *Abbrev.* SOS.

Seniores priores *Latin*—Elders first.

SENIT Système d'Exploitation Navale des Informations Tactiques *French*—Naval Tactical Information Evaluation System.

Sennit Flat cordage, for various purposes, made by plaiting rope yarns.

Sentence A law court's decision. Judgment.

Sentimental Loss A market loss to goods brought about by fear that goods may have suffered as casualty whereas no such loss exists in fact.

Senyosen *Japenese*—Motor Transport.

S.E. & O. Salvis erroribus et omissis *Latin*—Errors and Omissions Excepted, *q.v.*

SEOCT Safe and Efficient Operation of Chemical Tankers, at the ship Research Institute in Norway.

SEP *or* **S.E.P.** Self-Elevating Platform; Subject to Endorsement on Policy.

Sep. *or* **Sep.proc.** Separation procedure.

Separation Cloths Tarpaulins or similar used in Separation of parcels of bulk cargo.

Separation Error Error in the collimation or optical axis of a sextant or telescope.

Separator A device used for the separation of mixed oil and water (and gas in oil well production).

Separ. t. Separating Tank.

seq. Sequens *Latin*—The following; Sequitur, *q.v.*

Sequens *Abbrev.* seq., *q.v.*

Sequestration The legal act of holding the property from its owner for the purpose of covering oneself for the recovery of debts, profits, satisfying claims, etc. This is similar to **Maritime Lien** and **Lien**, *q.v.*

Sequestrator A person appointed by the courts to hold a property or properties until claims are met.

Sequitur *Latin*—It follows. *Abbrev.* seq.

Serang A Lascar, *q.v.*, Boatswain, *q.v.*

SERI Solar Energy Research Institute *USA*.

Seriatim *Latin*—In strict order; serially; point by point.

Se.rig Self Elevating Rig.

Seriki *Japanese*—Associate.

SERFOR Service Force *USA*.

Seroon *or* **Seron** A bale wrapped in hide. Originally from Spanish America with the hair left on the outside.

SERS Ship Emergency Response Service. Lloyd's Register 24-hour advisory service for dealing with and minimising damage to a vessel casualty.

SERT *or* **S.E.R.T.** Society of Electronics and Radio Technicians.

Serum The fluid that separates from clotted blood, an antitoxin.

Servants *or* **Master & Servants** Persons employed in the management of a ship. They are the agents and other persons engaged in the administration, clearance or any other work for a ship.

Service Charge A gratuity for services rendered. Hotels, restaurants, bars and others add a service charge on all bills.

Service of Suit Clause A clause binding insurers and underwriters to accept the time allowed for legal acts to be instituted on cargo claims against the carriers or shipowners, generally within the prescribed time of one year.

Service Pick-Up Location at an airport where cargo is to be picked up or lifted for despatch.

Service Speed *See* **Commerical Speed** and **Economic Speed**.

SES *or* **S.E.S.** Satellite Earth Station; Seafarers' Education Service; Société Européenne des Satellites *French*—European Satellite Society; Standards Engineers Society *USA*; Surface Effect Ship.

Sess. Session.

SEST *or* **S.E.S.T.** Safety Equipment Short Terms, *q.v.*

SET *or* **S.E.T.** Selective Employment Tax; Settlement.

Set (1) Velocity and direction of sea or tidal current. (2) Attitude of a sail to the wind. (3) Influence of the elements on a ship's direction, as in being set towards the rocks.

Set. Settlement.

Set of Bills of Lading The number of Bills of Lading, *q.v.*, needed by the shippers to prepare for the shipment and delivery of the merchandise by the consignees at the port of destination. The number may vary in accordance to the procedure of the shipping companies and the consignees. Normally one original bill of lading and four copies of non-negotiables are needed but then if requested two or three originals and more than four copies may be required to form a full set.

Set Sail To commence a voyage with a sailing boat, sailing vessel or motor/steam vessel.

Sett. Agt. Settling Agent, *q.v.*

Settee Mediterranean decked boat with two lateen sails.

Settlement Property settled by a contract; Payment of Accounts.

Settlement Calendar Fixed monthly dates taken by Lloyd's Central Accounting Office regarding the settlement of underwriters' and brokers' accounts in various Canadian, UK sterling and US dollar denominations.

Settling Agent An agent appointed by the insurance Underwriters, *q.v.*, who is authorised to settle claims. *Abbrev.* Sett. Agt.

Settling Tanks Ship's fuel tanks constructed to separate contaminated oil from water. This is done by letting the water settle at the bottom of the tanks whence it is pumped out. *Abbrev.* Sett.T.

Sett.T. Settling Tanks, *q.v.*

SEV *or* **S.E.V.** Soviet Ekonomicheskoy Vzaimopomoshehi *Russian*—Council for Mutual Economic Aid, COMECON; Surface Effect Vessel *air cushion vessel*.

Seven Seas Consist of seven oceans, namely: Arctic, Antarctic, North Pacific, South Pacific, North Atlantic, South Atlantic and Indian.

Severance Pay *Similar to* **Redundancy Payment**, *q.v.*

SEW *or* **S.E.W.** Safe Equipment Worker.

SEXT. *or* **Sext.** Sextant, *q.v.*, also *abbrev.* SXT.

Sext. Alt. Sextant Altitude *nautical*.

Sextant A navigational instrument by which the postion of the ship is fixed. *Abbrev.* SEXT *or* Sext *or* SXT.

Seych Seychelles.

SEZ Special Economic Zone.

s.f. Signal Frequency; Sinking Fund, *q.v.*, Surface Feet; Surface Foot; Sub finem *Latin*—Towards the end.

SF *or* **S.F.** French Thread Standards; Shear Force; Statement of Facts, *q.v.*, Sparring Fitted, re ship's cargo battens *Lloyd's Register notation*; Stowage Factor, *similar to* **Storage Factor**, *q.v.*; Summer Freeboard.

S/F *or* **S/Fee** Survey Fee.

S&F Spot and Forward *finance*.

S.&F.A. Shipping and Forwarding Agent(s).

SFA Securities and Futures Association, to which all Licensed Traders of the LCE, *q.v.*, belong; Société Française d'Astronautique, French Astronautical Society.

SFC *or* **S.F.C.** Specific Fuel Consumption.

SFCN Société Française de Constructions *French*—Society of French Naval Construction.

SFE Sydney Futures Exchange.

SFEA Safe Arrival.

SFECH Sydney Futures Exchange Clearing House.

Sfee *or* **S/F** Survey Fee.

S.ff Se faz favor *Portuguese*—If you please.

SFFA Singapore Freight Forwarders' Association.

SFI Société Financière Internationale *French*—International Finance Corporation.

SFR Sinking Fund Rate of return; Swiss franc.

SFRP Structural Fibreglass Reinforced Plastic.

SFSR *or* **S.F.S.R.** Soviet Federated Socialist Republic.

SFV 1977 Torremolinos International Convention for the Safety of Fishing Vessels 1977. *IMO. Relating to construction, stability and equipment.*

SFWTH *or* **S.F.W.T.H.** Semi Flush Water Tight Hatch.

Sg. Surgeon.

S.g. *or* **s.g.** Specific Gravity.

S.G. S.G. Form of Policy, *q.v.*; Salutis gratia *Latin*—For Safety's Sake; Scots Guards; Secretary-General; Solicitor General; Standing Group *NATO.*

SG1A Seaman Grade 1A.

Sgd. Signed.

S.G. Form of Policy A policy form adopted by the Lloyd's market in 1779 as a basic contract for Marine Insurance, *q.v.* The form was designed to cover both ship and goods on a specified voyage, but has been adapted as required to cover either type of interest; and has been extended in regard to hull insurance to provide the basic contract for time policies. The form has been falling into disuse since the introduction of the MAR type of policy form, and will be eventually superseded by the latter, in practice.

S.G.H.W.R. *or* **SGHWR** Steam Generation Heavy Water Reactor.

Sgl Single.

SGML Standard Generalised Markup Language *ISO 8879.*

SGS Steam Heated Generator Survey *Lloyd's Register notation.*

SH Ships Head *nautical*; Super-heated.

Sh. Shore.

S/H Second Hand.

S/H *or* **SHE** Sundays and Holidays Excepted, *q.v.*

SHA *or* **S.H.A.** Sidereal Hour Angle *nautical.*

Shacho *Japanese*—Executive of a company.

Shackle (1) A heart-shaped or U-shaped piece of metal having at the ends a pin attached which passes from a hole from one end and is screwed or bolted from the other end. (2) Another word is 'Shot' or 'Length of Cable'. It is the standard length of a chain cable measuring $12\frac{1}{2}$ fathoms and 15 fathoms in the UK and US navy respectively. In the UK merchant navy a shackle measures 15 fathoms. A fathom is 6 feet.

Shaft Horse Power The actual power shown on the shaft connected to the engine after passing from the gear box of the ship. It is calculated to be at 90% of I.H.P. *See* **Torsion Meter**.

Shaft Tunnel The part of the ship which accommodates the propeller shaft.

SHAM *or* **S.H.A.M.** Sidereal Hour Angle of the Mean Sun *nautical.*

SHAPE *or* **S.H.A.P.E.** Supreme Headquarters Allied Powers Europe *NATO.* Located in Belgium.

SHARE Society to Help Avoid Repetitive Effort *USA.*

Share Broker Stockbroker, *q.v.*

Share Capital The authorised or total capital of a company.

Share Certificate A document of title of the shares.

Shareholder A person in possession of shares.

Shares Equal portions of a capital in business generally supported by certificates.

Share Warrant A share certificate issued by the company which can be transferred from hand to hand.

Shariah The Islamic Law.

SHAS *or* **S.H.A.S.** Sidereal Hour Angle of a star *nautical.*

SHATS *or* **shats** Sidereal Hour Angle of the True Sun *nautical.*

Sh. del. Short Delivery/Delivered.

Sh.Dl. Short Delivery/Delivered.

S/HE *or* **S/H** Sundays and Holidays Excepted, *q.v.*

Sheave A running wheel in a Block, *q.v.*, which the rope passes over.

Shedding Cargo The act of receiving and storing merchandise in the transit sheds or verandahs.

Sheer The upward curve of the deck and structural lines of a ship from amidships to bow and stern and to the sides. The sheer of a merchant ship is more or less calculated as 1% of the overall length. *See* **Sheerstrake**.

Sheerstrake The way or course of all deck shell plating. *See also* **Sheer**.

Sheet Ice A layer of ice which lies on the smooth and level-looking sea.

Shekel Currency unit of Israel.

Shelf A ridge of rock near to the surface of the sea.

Shelf Life The period for which a product may be stored before natural deterioration renders it unfit for its intended use.

Shellac A hard resin extracted from the bark of a tree. It is soluble in alcohol and is used for insulation purposes in electricity.

Shell Plating The actual steel plates forming the hull and decks of a ship.

Shellvoy A standard form of charterparty to carry oil.

SHEX *or* **S.H.E.X.** *or* **S.Hex.** A charterparty phrase; Sundays and Holidays Excepted, *q.v.*

S&H exct. Sundays and Holiday Excepted, *q.v.*

SHF *or* **S.H.F.** Super High Frequency, *q.v.*

Shift (1) To manoeuvre or move from one berth to another or to shift from one port to another adjacent port. (2) When the wind changes direction. (3) A regular period of work.

Shifting In shipping this refers to movements or changing positions of cargo from one place to another. This can easily endanger the seaworthiness or cargo-worthiness of the ship. Shifting generally occurs in very rough seas and may, though not necessarily, be caused by bad stowage. Bulk cargoes especially have a tendency to shift unless good trimming is done or proper care is taken in the use of Shifting Boards, *q.v.*

Shifting Between Berths A common practice while a ship is loading or discharging at one berth and is shifted so as to continue the operation on another or other berths. In normal cases time lost for such shifting becomes the owner's account unless the number of movements does not exceed that which was agreed by the parties to the contract or charterparty.

Shifting Boards Wooden frames erected to reduce the size of a cargo compartment and to restrict the movement of bulk cargo at sea.

SHINC *or* **S.H.I.N.C.** *or* **S.H. Inc.** Sundays and Holidays Included, *q.v.*

Shintaku *Japanese*—Trust.

SHIP Small Hauliers International Partnership; Seamen's Health Improvement Program *USA*.

Ship Agent *or* **Shipping Agent** A person who acts as an agent of a ship, representing the owner and/or the charterer in all the formalities needed while entering port, during customs clearance up to the time of final sailing. He is to arrange with the authorities for the allocation of berthing place to load and/or unload, advise import/export cargo owners, organise loading/unloading of merchandise. There are many other services rendered while the ship is in port. With a passenger ship the agent handles all that is needed for the disembarking and/or embarking of passengers. *See* **Ship's Husband** *or* **Ship's Husbandry**.

Ship Breaker A person or firm buying old ships for breaking up purposes. Parts are resold to metal manufacturers or to shipbuilding yards or as scrap.

Shipbuilders and Shiprepairers Association, The Formed in 1989, and based on earlier associations, it represents the interests of the UK shipbuilding and shiprepairing industries. *Abbrev.* SSA. The SSA provides a forum for members to discuss topics of interest to them. It also acts as a focal point for industry representation and as facilitator of collaborative action in the industry, for example as a lead body in research projects, sponsor of trade missions and trade exhibitions. There is an active and influential European industry body, the Association of West European Shipbuilders (AWES), to which SSA is the recognised UK representative. Membership of AWES brings with it access to valuable information about European and international building and repair capacity, demand and industry-generated production statistics. AWES membership is wider than that of European Community countries alone, so a special group within AWES, known as CESA, handles representation with the Commission and as necessary with the Council and European Parliament. Within AWES there are also special groups on shiprepair and on small ships, both of which are intended to provide member yards with an opportunity to meet European colleagues and to promote their common interest. Through AWES, the Association has links with shipbuilder associations in Japan, South Korea and the USA.

Shipfitter A person working in a dockyard or shipbuilding yard whose job is to cut and form the steel sheets and after which they are welded or joined together on ships.

Ship Improvement Guide A ship's record book demonstrating a clear picture of all the items of class maintained and those recommended by the Classification Society. *Abbrev.* SIG.

Shipload A cargo constituting a full load for the ship.

Ship Lost or Not Lost A phrase inserted on Bills of Lading, *q.v.*, especially when freight is paid in advance or prepaid. By this no refund of any freight is given in case the vessel does not fulfil the whole voyage due to some accident provided she was seaworthy during the initial journey and it was not the fault of the owner.

Shipman *or* **SHIPMAN** BIMCO, *q.v.* Standard Ship Management Contract form.

Ship Managers *See* **Ship Management**.

Shipmaster General term with the same meaning as Master, *q.v.*, of a merchant ship.

Ship Mortgage *See* **Mortgage**.

Ship Reporting System An organised system of shore stations positioned so as to cover various sea areas in case of emergency messages being received from ships. *Abbrev.* SRS.

Shipowner A person or firm that owns one or more ships.

Shipowner's Claims Bureau One of the oldest mutual clubs and insurance underwriters in the USA. The American Steamship Owners Mutual Protection and Indemnity Association. *Abbrev.* ASOMPIA and also American Steamship Owners Mutual Protection and Indemnity Association Inc (The), *q.v.*

Shipowner's Lien A legal right of a shipowner to hold in possession cargo and freight pending the full settlement of chartered freight, demurrage, and all other expenses due.

Shippable Suitable to the shipped or carried.

Shipped Bill *Similar to* Shipped (on board) Bill of Lading, *q.v.*

Shipping The merchant shipping fleet of a country; the transportation of merchandise.

Shipping Agent *See* **Ship Agent**.

Shipping Bill A customs form for the exportation and re-exportation of stores which are to be consumed or used on board a ship.

Shipping Cards *Similar to* **Sailing Card**, *q.v.*

Shipping Clerk A tally clerk, Checker, *q.v.*

Shipping Development Fund Committee A branch in the Ministry of Transport and Shipping of India which deals with loan applications to Indian shipping companies for the purchase of ships from overseas yards. *Abbrev* SDFC *or* S.D.F.C.

Shipping Forwarding Agents *Similar to* **Freight Forwarders**, *q.v.*

Shipping Information Services A computerised general service set up by Lloyd's Register of Shipping and Lloyd's of London Press offering complete ship data on the marine industry. *See* **Lloyd's** *and* **Lloyd's Register of Shipping**.

Shipping Lane *or* **Shipping Traffic Lane** A normal route taken by commercial vessels when trading. In some areas its limits may be determined by international convention.

Shipping Master A port superintendent whose duties are to carry out the registration and supervision of ships. The shipping master must have a good knowledge of shipping laws and be conversant with the conditions and other pertinent details of engaged and discharged seamen.

Shipping Officer A customs officer in charge of imported and exported merchandise.

Shipping Order *Similar to* Shipping Note.

Shipping Permit An American shipping term for Shipping Note.

Shipping Specifications Customs formalities regarding the merchandise shipped. Also detailed information given by the manufacturers or shippers regarding each individual item shipped to the consignees.

Shipping Ton *See* **Freight Ton**.

Ship's Bag *or* **Steamer's Bag** A bag or large envelope where letters or parcels belonging to the consignees of the cargo are kept. It is collectively addressed to the ship's agent for necessary distribution to the consignees. *See* **Flight Bag**.

Ship's Biscuits Hard biscuits or, as they are usually called sea biscuits or sea bread, for consumption by ships' passengers, officers and crew in case of emergency. They form part of the lifeboat emergency stores. See **Sea Biscuit**.

Ship's Certificate of Registry A certificate given by the Registrar of a port where the ship is registered. It gives all the particulars of the ship, including the name of her owners, master, classification and other pertinent details which the customs authorities of every port request to know for their records.

Ship's Clearance Inwards *or* **Ship's Clearance Outwards** *See* **Clearance Inwards** and **Clearance Outwards**.

Ship's Log *See* **Log (book)**.

Ship's Manifest One of the essentials for a commercial ship to have Clearance Inwards and Outwards, *q.v.* The manifest gives a clear picture of the various cargoes loaded for every port. For a passenger ship the manifest or passenger list shows the names and passport details of the passengers. If it is a passenger-cargo ship then both types of manifests are to be produced to the customs authorities.

Ship's O.N. O.N. stands for Official Number, *q.v.*, in the registry of the ship with the Registrar of Shipping.

Ship's Papers These are a Ship's Certificate Registry, *q.v.*, Bill of Lading, *q.v.*, Bill of Health, *q.v.*, Charterparty, *q.v.*, log, and other documents which show the character of the ship and cargo. They are the documents required for the manifestation of the property of the ship and cargo. They are of two sorts: those required by the law of a particular country, as the certificate of registry, licence, charterparty, bills of lading and of health, required by the law of England to be on board all British ships; and those required by the law of nations to be on board neutral ships, to vindicate their title to that character, as the passport, sea-brief or sea-letter, proofs of property, muster-roll or *rôle d'équipage*, charterparty, bills of lading and invoices, logbook or ship's journal, and bill of health.

Ship's Rail Applies where cargo loading is at the risk and expense of charterers or shippers up to the point where the goods cross the ship's rail, whereupon the shipowner becomes responsible for the risk and expense of the goods. The reverse applies at the discharge port.

Ship's Register The documentary particulars of a ship under registry. *See* **Certificate of Registry**.

Ship's Registry The port at which the vessel is registered; its name is painted on the stern.

Ship's Report A report declared by the master of a ship to the customs on a standard form giving information of the cargo or if in Ballast, *q.v.*

Ship's Stability The property of a ship in maintaining its equilibrium under the influence of gravity, buoyancy and external forces.

Ship's Store Bond A sealed magazine in a ship where duty free bonded stores like liquors, spirits, cigars, tobacco, cigarettes, perfumes, etc., are sealed by the customs officers in a port to ensure that no such duty free items are smuggled ashore.

Ship's Stores *or* **Ship Stores** Stores and provisions, etc., which may be liable to duty if landed ashore.

Shipsuppliers *or* **Ship-Suppliers** *or* **Ship Chandlers**.

Ship-worm A marine worm that lives in submerged wood. For protection from such organisms the underwater parts of boats and ships are sheathed with copper.

Shipwreck A ship lost completely at sea because of a storm or accident in collision, sinking, running ashore or on rocks, hitting hard or pointed submerged objects, etc. *See* **Wreck.**

Shipwrecked Fishermen and Mariners' Royal Benevolent Society Known as the Shipwrecked Mariners' Society for short. In 1838, the concern of Mr John Rye, a medical man, and Mr C.G. Jones, an ex-pilot, for the wrecks of fishing boats on the north coast of Devon and the families of drowned seamen led to the foundation of the society in 1839 under Admiral Sir George Cockburn as President. It was incorporated by Act of Parliament in 1850 and the reigning monarch is patron. Although originally founded to provide practical and financial assistance to the survivors of shipwrecks, the Society's main function today is to make financial grants to former mariners and fishermen, their widows and families in cases of need. Priority is given to the elderly, chronically sick or disabled, and to widows with young children. The Society, which operates through some 450 honorary agents, still provides assistance to seamen shipwrecked on the coast of the British Isles, and also makes awards for skill and gallantry in preventing loss of life at sea.

Shipwright (1) A seaman engaged as a carpenter who is also called a 'chippy'. (2) A skilled metal worker in a repair yard in shipbuilding.

Shl.Dk Shelter Deck.

Shl. & Wdks Shell and Weather Deck Survey, *Lloyd's Register notation.*

SHM Simple Harmonic Motion.

Shoji *Japanese*—Commercial.

Shokai *Japanese*—Company.

Shoko *Japanese*—Commerce and Industry.

Shooting Star *Similar to* **Meteor**, *q.v.*

Shoot Loose To unload grains or other bulk cargo direct from a ship's hold into barges or lighters.

Shop Assistant A person employed to sell goods in a shop or store.

Shop Floor (1) The space occupied by machines in a factory, or by all the workpeople. (2) The manual labour force as a whole. (3) Relating to the consensus of the members of a union.

Shop Hours The legal period allowed for shops to be kept open for business and for employees to work.

Shopkeeper A person who owns a shop and sells by retail.

Shoplifting Stealing from a shop.

Shopman A man who serves in a shop.

Shop Steward A workers' trade union representative generally in a factory who is in direct contact with the management.

Shopwalker A supervisor and customers' attendant in a supermarket; A supervisor engaged in a large store.

SHORAN Short Range Navigational Aid.

Shore/s (1) The limit(s) of the land towards the sea. (2) A wooden prop used when a ship is in dry dock to prevent her from moving and to keep her upright. These props are inserted at the bottom and sides of the ship.

Short Shortage.

Short Bills Bills, *q.v.*, drawn for a short time, normally within a period of ten days from their maturity. *Abbrev.* SB *or* S.B. *See* **Short Dated.**

Short Blast A whistle or fog horn signal lasting less than three seconds.

Short Bridge When the length of the bridge accommodation is less than about 15% of the ship's length.

Short-Dated Negotiable bills having an early date of maturity. *Similar to* **Short Bills**, *q.v.*

Short Delivery The quantity of cargo delivered is less than the bill of lading quantity.

Shortfall Freight When freight is rated under a tariff but the shipment is below the minimum to which it applies, this is the increase in charge to cover the balance.

Short-Handed An inadequately or under-manned ship resulting in unseaworthiness.

Short Interest When the amount insured in the Floating Policy, *q.v.*, exceeds the actual value finally declared, the premium difference is refunded to the assured. *Abbrev* S.I. *or* s.i. *or* S/I.

Short Rate A premium on an insurance Policy, *q.v.*, paid at intervals over a year. Normally the premium is paid annually.

Short Sea Refers to shipping on routes which involve short crossings between ports.

Short Sea Tonnage Cargo ships employed on short sea crossings and in coastal trade.

Short Term Certificate *See* **Provisional Certificate.**

Short Term Insurance Insurance which expires before the usual 12-month period.

Short Term Policy (1) Acting on a principle of the present time and neglecting what may happen in the future. *Opposite to* **Long Term Policy**, *q.v.* (2) An insurance cover for a short period of less than the usual 12 months.

Short Term Rate of Interest When money is deposited in a bank for a short time the interest allowed will be lower than if deposited for a long time.

Short Time Charter A time charter of a vessel for a short period, say for 15/20 weeks.

Shosen *Japanese*—Merchant Ship.

Shot Blasting The blasting of grit by a high pressure machine against a ship's steel plates to eliminate rust.

Shoten *Japanese*—Trading Company.

Show of Hands A system of voting at a meeting by counting the number of raised hands for or against a motion.

SHP *or* **Shp.** *or* **S.H.P.** Shaft Horse Power, *q.v.*

Shr. Share; Sheer.

Shrops Shropshire, *UK*.

Shrouds Wire ropes extending from the masthead to either side of the ship to stay the mast upright, as well as to take the extra strain when Cargo Derrick(s), *q.v.*, are in use.

Shroud Laid A wire rope having four strands right-hand laid, generally with a centre core.

Shs. Shares.

SHSV Super Heater Safety Valve.

Sht. Pc/s Short Piece(s).

Shut-Down The period during which a factory stops production, allowing employees to take the paid holidays to which they are entitled, as well as the overhaul and maintenance of machinery.

Shuttle Tanker A tanker, usually with special fittings for mooring, which lifts oil from offshore fields and transports it to a shore storage or refinery terminal on repeated trips.

Si Symbol for Silicon.

SI Système International d'Unités, *q.v.*

S/i *or* **S.I.** *or* **s.i.** Short Interest; Sum Insured.

SIA *or* **S.I.A.** Singapore Air Lines.

SIAP Shipbuilding Industry Assistance Program *USA*.

SIB Securities and Investments Board *regulatory authority accountable to the UK Treasury.*

SIB *or* **S.I.B.** Special Investigation Branch.

sic *Latin*—Thus, so. Printed or written in brackets as a parenthetical comment on quoted words.

SIC *or* **S.I.C.** Specific Inductive Capacity; Standards Industrial Classification, British timber standards; Soluble Inorganic Compounds.

S.I.C.A. Society of Industrial and Cost Accountants *Canada*.

SICC Singapore Indian Chamber of Commerce.

Sic in originali *Abbrev.* sic, *q.v.*

Sick Bay Any cabin or compartment on a ship allocated for the treatment of casualties, sick passengers, officers and crew members.

SID Standard Instrument Departure *aviation*.

SID *or* **Sid** Single Decker *vessel*.

sid. Sidereal, *q.v.*

SIDA Swedish International Development Authority.

Side-Lift Carrier A fork-lift truck or vehicle used to load Containers, *q.v.*, from the side of a Container Ship, *q.v.*

Side Lights 'Side lights' means a green light on the starboard side and a red light on the port side each

showing an unbroken light over an arc of the horizon of 112.5 degrees and so fixed as to show the light from right ahead to 22.5 degrees abaft the beam on its respective side. In a vessel of less than 20 metres in length the sidelights may be combined in one lantern carried on the fore and aft centre line of the vessel.

Side Ports Openings built into the side of ships to facilitate cargo-handling—e.g. in Reefers, *q.v.*, for the passage of Pallets, *q.v.*

Sidereal Pertaining to the stars. A sidereal year is the time in which the earth makes one complete revolution around the sun.

Sidereal Day The time the earth takes to make a complete revolution on its axis. Approximately 23 hours 56 minutes 4 seconds.

Side-rolling Hatch Covers Modern hatch covers on large bulk carriers usually open to either port or starboard (usually alternately) rather than fore and aft as on smaller vessels.

Side Scuttles Holes situated in the sides of a ship admitting light and fresh air, otherwise portholes or square ports, fitted with hinged glass ports and/or blank covers or deadlights for watertightness when required at sea.

Side Water Tanks Similar to Saddle Tanks, *q.v.*

SIDF Saudi Industrial Development Fund.

Sierra (1) Spanish for mountain range. (2) International Radio Telephone phonetic alphabet for 'S'.

Sift Cl. Sift Clause, *q.v.*

Sift Clause A marine insurance clause covering the assured in the case of a cargo of grains or flours infected during the voyage. It covers the loss of weight due to sifting and related expenses so incurred. Delays, deterioration and market loss are excluded. *Abbrev.* Sift Cl.

sig. Signature; Signor *Italian*—Mister.

Sigd. Signed.

Sight Bill A bill of exchange to be paid at sight. *Similar to* **Sight Draft**, *q.v.*

Sight Draft A Bill of Exchange, *q.v.*, which is to be paid when presented. *Similar to* **Sight Bill.** Abbrev. S.Dr.*or* sd.

Sight Entry Sight Bill, *q.v.*

Sighting Bills of Lading The scrutinising and checking of the contents in the Bills of Lading, *q.v.*, against the actual quantities discharged.

Sighting the Bottom *or* **Bottom Sighting** Testing or examining the underwater part of a ship to trace any damage which might have occurred after an accident.

Sigill. Sigillum *Latin*—A seal.

Signalling Lamp *See* **Morse Lamp**.

Signal Letters Official letters assigned to a registered ship and endorsed in the Certificate of Registry. The four letters are used in hoisting flags which show a ship's identity and as radio call-signs.

Signal of Distress An upsidedown flag displayed on a ship as a sign that she is in distress apart from the use of pyrotechnics. *See* **Marine Pyrotechnics** *and* **Pyrotechnics.**

Signatory A person representing a company who signs documents or agreements on behalf of the company.

Signature Bonus Name for money paid to a principal by a contractor on the award of a substantial contract, as a sign of good will.

Signed Referring to the insurance policy signed by Underwriters, *q.v.*

Signed Under Protest A person signing a contract or similar document under protest will be covered legally to pursue the dispute further at a later date.

Signing of a Charterparty *or* **Bill of Lading** The broker or agent has to be careful as to the capacity in which he is signing the contract otherwise he will be legally answerable for any inaccuracies or mispresentations. There are various titles under which a broker or agent may sign, These are: Agent; As Agent Only; As Agent for Charterer; As Agent for Shipowner; with additions such as: By telegraphic or telex authority; Subject to Owners' Approval.

Signing Off *or* **Sign Off** The closing of articles. Every seaman or officer of a ship whose term of contract with the master or owner ends is obliged to sign off.

Signing On *or* **Sign On** The signing of ship's articles. Every person initially employed on a ship is obliged to sign on.

Sigtto *or* **SIGTTO** Society of International Gas Tanker and Terminal Operators.

Silent Hours *or* **Dead Hours** Night Time; Non-Working Hours.

Silepcilik *Turkish*—Tramping Company.

Silica Silicon dioxide. Represented by SiO_2. A colourless or white vitreous solid found in various quartz crystal forms including rock, sand and gemstones. Used in furnace linings and the manufacture of glass, glazes and enamels. *See Silica Gel.*

Silica Gel Amorphous form of hydrated Silica, *q.v.*, used as an absorbent drying agent, as an inhibitor against the humidity of merchandise, and to protect articles susceptible to moisture. *See* **Silica.**

SIMA Scientific Instrument Manufacturers' Association. *UK.*

SIMI Sea Ice Mechanics Initiative, an activity of the US Navy's ONR, *q.v.*

Similia similibus curantur *Latin*—Like cures like.

Sim. Pay. Cl Simultaneous Pay Clause *insurance.*

Simplification of International Trade Procedures An international organisation set up to guide and assist in the unification and simplification of international trade, documentation and procedure. *Abbrev* SITPRO *or* S.I.T.P.R.O.

SIN Subject Indicator Number.

Sin Sine *Latin*—Without; Sine *mathematics.*

Sine cura *Latin*—Without care; without obligation or responsibility.

Sine die *Abbrev.* s.d., *q.v.*

Sine dubio *Latin*—Without doubt.

Sine mora *Latin*—Without delay.

Sine qua non *Latin*—An indispensable condition.

Sing. Singapore.

Single Administrative Document Form adopted by the EEC in 1988 for the improvement of international trade procedures, embracing the function of some seventy customs documents formerly used. *Abbrev.* SAD.

Single Anchor Leg Rigid Arm Mooring Generally used in conjunction with a Floating Production Storage and Offtake, *q.v.*, vessel for mooring and oil transfer purposes. *Abbrev.* SALRAM.

Single Buoy Mooring A large buoy constructed in a way as to be permanently anchored away from the coast where a ship can safely lie at anchor for loading or discharging petroleum liquid products. *Abbrev.* SBM *or* S.B.M.

Single Level Gauging A suitable method to monitor all types of liquid cargoes ballast or bunker levels. *Abbrev.* SLG.

Single Point Mooring A fixed structure or floating mooring positioned outside a shallow draught (draft) harbour but in deep waters to enable large Tankers to moor by the bows and enable them to load or unload their cargoes of oil. *Abbrev.* SPM *or* S.P.M.

Single Screw A ship having one propeller.

Single Trip Container A drum of about 55 US gallon size conforming to the US National Cargo Bureau Regulations containing flammables and having the initial STC or S.T.C. on top of the drum which can only be used for one voyage.

Single-Use Goods *Similar to* **Non-Durable Goods,** *q.v.*

Single Well Deck Ship *or* **Single Well Decker** A ship with the bridge and poop joined by a continuous superstructure. The extra space thus gained compensates for that which is lost in the Shaft Tunnel, *q.v.*

Single Well Oil Production System A method of collecting oil from the well into a tanker stationed above the production well and plugged on to the seabed tap through a length of pipe. *Abbrev.* SWOPS.

sinh Hyperbolic sine.

Sinking Fund Moneys set aside for the purpose of sinking or wiping out a state's or corporation's debt by degrees. The surplus of revenue over expenditure devoted to reduction of the national debt.

Sinking Platform Lighter Aboard Ship An identical vessel to Lash (Lighter Aboard Ship) but having a different system in loading the barges or lighters with cargo contents on to the ship. *Abbrev.* SPLAH.

SINOAGENT China Marine Shipping Agency.

SINOCHART China National Chartering Corporation.

Sinotime 1980 A time charterparty, issued by the China Chartering Corporation.

SINOTRTRANS China National Foreign Trade & Transportation.

SINS or **Sins** Ships Inertial Navigation System.

SINTEF Foundation for Scientific and Industrial Research at the Norwegian Institute of Technology.

SIO Senior Information/Intelligence Officer.

Si quaeris peninsulam amoenan circumspice *Latin*—If you seek a pleasant peninsula, look about. Motto of Michigan, USA.

SIRE Ship Inspection Report: a database system of the OCIMF, *q.v.*, for recording details of tankers achieving its standards.

SIRIM Standards and Industrial Research Institute in Malaysia.

Sirketi *Turkish*—Company.

SIRS Support Information Retrieval System.

SIS Security Investigation Services; Shipping Information Services, *q.v.*; Shipping Intelligence Service, *Lloyd's Intelligence Department*.

SISIR Singapore Institute of Standards and Industrial Research.

Sister Ship Clause A clause in a policy covering collision liability and salvage contributions, usually amongst other things, whereby the underwriters agree to treat sisterships as if they were separately owned (and capable of legal liability, one to the other) in regard to collision liability claims and claims for salvage charges.

SIT Silicon Intensified Target; Spontaneous Ignition Temperature; Stopping in Transit; Storing in Transit.

SIT or **S.I.T.** Stopping in Transit; Storing in Transit.

SITA or **S.I.T.A.** Société Internationale de Télécommunications Aeronautiques *French*—International Society of Aeronautic Telecommunications; System International Tinplate Area; Students' International Travel Association.

SITC or **S.I.T.C.** Standard International Trade Classification *United Nations*.

Sit-Down Strike A strike by workers who obstruct the premises by sitting down and refusing to move.

SITREP Situation Report *USA*.

SITPRO or **S.I.T.P.R.O.** Simplification of International Trade Procedures.

Sitting Aground Another phrase for Touching Soft Ground, *q.v.* In a charterparty some shipowners agree to the vessel loading or discharging on soft ground during tidal hours. See **Always Afloat**.

Sittings The sittings of the Supreme Court of Judicature in the United Kingdom are four in number: Hilary, Easter, Trinity and Michaelmas. Sittings at the Royal Courts of Justice include sittings in chambers as well as in Court. *See* **Vacation**. The dates of the sittings vary slightly from year to year.

SITU Flexible connections in oil drilling.

Si vis pacem, para bellum *Latin*—If you want peace, prepare for war.

Sixth Freedom The Carriage of passengers between two airports of different countries by an aircraft of a third, which lands at some time during the flight at one of its own national airports in transit.

s.j. Sub judice *Latin*—Under legal consideration.

SJAB or **S.J.A.B.** St. John Ambulance Brigade.

SJAC or **S.J.A.C.** Society of Japanese Aircraft Constructors.

SJC or **S.J.C.** Supreme Judicial Court *USA*.

S.J.D. Scientiae Juridicae Doctor *Latin*—Doctor of Juristic Science; Doctor of Juridical Science.

Sjofartsverket Swedish Maritime Bureau.

Sk. Sack.

SK SAS Scandinavian Airlines System *airline designator*; Storekeeper *USA*; Streptokinase, *q.v.*

SKD Semi-Knocked Down, Partly dismantled cargo. See **Knocked Down Condition**.

Skeg The aft projecting part of a ship's keel supporting the rudder.

Skein Coil or reel or twist of uniform size of thread or yarn.

Skeppningsaktiebolag *Swedish*—Shipping Company Limited.

Skiff A small light boat which can comfortably be rowed and manoeuvred single-handed.

Skimming Price An unusually high price which would normally be considered only by persons who can really afford to pay.

Skip (1) A cage, bucket, etc., in which men or materials are lowered and raised in mines and quarries. (2) A large steel container used for the collection of waste material.

Skipper A sea captain, especially master of a small trading vessel.

Skipping A customs house word to temporarily transfer the contents of packages into others so as to determine the weight.

Skipsaksjeselskap *Norwegian*—Shipping Company Limited.

SKP Senbak Kamdok Phenian *North Korean Ship Register*.

S.Kr. Swedish Krona unit of currency; one hundred ore equals one Krona.

Skyborne Airborne.

Skylight A window on decks or roofs to allow light through to the internal cabins, accommodations, engine rooms and rooms.

Skyline or **Sky Line** Visible horizon.

Skysail A sail of a square rigged vessel.

Sky Wave A radio wave passing outwards from the earth and then reflected back by the Ionosphere, *q.v.*

Skyway An elevated highway constructed to eliminate traffic jams in a city especially during rush hours; an air lane or passage way.

Skywriting Legible smoke-trails made as an advertising method by an aircraft.

s.l. Sine loco *Latin*—Without place; Secundum legem *Latin*—According to the law.

SL *or* **S.L.** Salvage Loss, *q.v.*, Sea Level: Solicitor-at-law; Safe Locker.

S/L Salvage Loss, *q.v.*

S & L Sue and Labour Charges, *q.v.*; Sue and Labour Clause, *q.v.*

Slack Water The interval at the turn of the tide when there is little or no current or tidal effect.

SLAET *or* **S.L.A.E.T.** Society of Licensed Aircraft Engineers and Technologists.

SLAM Stand-Off Land Attack Missile. *Military*; Supersonic Low Altitude Missile.

Slamming The impact of water on the bows of a ship.

s.l.a.n. Sine loco anno vel nomine *Latin*—Without place, year or name.

Slander Oral defamatory statement.

Slant of Wind Nautical term for Favourable Wind.

SLAR *or* **S.L.A.R.** Sideways looking Airborne Radar *aviation*.

SLATE Small Lightweight Altitude Transmission Equipment.

Slav Slavic, Slavonic.

SLBM Submarine Launched Ballistic Missile.

S/LC *or* **S&L Clause** Sue & Labour Clause, *q.v.*

S.L. & C. Shipper's Load and Count.

S/L. Ch. Sue and Labour Charges. *See* **Sue & Labour Clause**.

SLCM Sea-Launch Cruise Missile.

Sld.a.p.l. Sailed as per List, *q.v.*

SLEEP Ship Life Extension Program *US Navy*.

Sleeping Car Railway carriage with sleeping berths or compartments.

Sleeping Partner *Similar to* **Dormant Partner**, *q.v.*

Sleet A mixture of melting snow with cold rains.

SLEP Service Life Extension Program *US Navy*.

Slewing Ramp *See* **Ramps**.

s.l.f Straight Line Frequency.

Slick Smoothness on the surface of the sea produced by a thin coating of oil.

Sliding Scale Ad Valorem, *q.v.*; Device for adjusting such things as wages, allowances, taxes, and commissions on such criteria as value, cost of living index, economic need and fiscal policy.

Sling L Sling Loss, *q.v.*

Sling Loss Cargo lost by falling from a ship's lifting tackle during loading or unloading. *Abbrev.* Sling L.

Slings The means whereby packages, bales, pallets, trays, containers, vehicles, animals, hatches, etc., are supported and attached to the hook of a crane or derrick for lifting. They can be of rope, wire, chain or canvas. They may be passed around the load or shackled or hooked to its four corners. Some Packages have integral slings attached by the exporter before despatch for shipment.

Slip (1) An artificial slope of stone used as a landing-stage; (2) an inclined plane on which ships are built or repaired; (3) a note presented to underwriters by brokers for acceptance or otherwise of part of an insurance risk; (4) in certain conditions of trim a propeller may be slightly exposed above the waterline, or close to the surface or be adversely affected by opposing currents or bad weather, causing the propeller to slip—i.e. not perform to its full potential.

Slipstream The backward pressure of air formed by an aircraft's propeller.

Slipway A shipbuilding or landing slip.

SLM Sound Level Meter.

Slop Book A record book of items like cigarettes, spirits, clothing, shaving kits, accessories, etc. This book is kept by the master, purser or chief steward of a ship who also is in charge of issuing these commodities against payment to the passengers, officers and crew.

Slop Pail Receptacle for garbage and dirty water.

Slops Tank cleaning residues retained on board a tanker in the slop tank for eventual disposal ashore or mixture with incoming cargo.

Slopwork Cheap inferior work; The making of cheap garments.

SLOTHIRE Code name for Standard Slot Charterparty.

Slowdown Deliberate reduction in the pace of normal work by labourers to obtain their demands from the employer. Similar to Go-Slow, *q.v. See also* **Work to Rule**.

SLR Single-Lens Reflex *photography*.

SLSA St. Lawrence Seaway Authority, Ottawa, Canada.

Sluice (1) A sliding watertight door in a passenger ship's bulkhead. (2) An artificial passageway to regulate the flow of water in a river or canal through a sliding gate or other contrivance.

Slump A period of sudden decline in business and trade which may ultimately cause depression in a state.

SLURP Self Levelling Unit to Remove Pollution.

Slurry A sticky muddy residue separated from coal at the pithead washing plants; a special cement and water mixture to plug drilled holes in oil or gas wells.

Slush Ice *or* **Young Ice** When sea water starts freezing even though the water is still navigable. *Also called* **Sludge Ice**.

SLV Standard Launch Vehicle.

SM *or* **S.M.** Sales Manager; Senior Magistrate; Shipment Memorandum; Shipping Master; Short

Metre *music*; Screw Motor Ship; Station Master; Sergeant Major.

SMA *or* **S.M.A.** Society of Maritime Arbitrators, established in the USA in 1936.

Small Battens *See* **Battens**.

Small Circles Similar to Parallels of Latitude, *q.v.*

Small Hours Early hours of the day after midnight.

Small Parcel Express Small air parcel express service performed by airlines. Abbrev. SPEX.

Smalls Consignments of merchandise which are less than a ton in weight meant for road or sea transport. A minimum freight rate is charged.

S. Marino San Marino.

Smart Money Money invested on a risk or in betting.

SMATV Satellite Master Antenna Television.

SMC Safety Management Certificate, required of ship operators by the ISM Code.

SMD Short Metre Double *music*.

SMDG Ship Planning Message Development Group; a body working on the application of electronic data interchange (EDI) to shipping.

Smelling the Bottom A nautical expression indicating the vessel is manoeuvring in shallow waters.

S. Met.O. Senior Meteorological Officer.

SMFC Ship Mortgage Finance Corporation.

SMG Speed Made Good *nautical*.

SMM Ship, Machinery & Marine Technology International.

SMMT Society of Motor Manufacturers and Traders.

SMO *or* **S.M.O.** Senior Medical Officer.

Smog A combination of smoke and fog, applied to air pollution.

Smoke Chaser Lightly equipped firefighter.

Smoke Eater Slang for fireman.

SMOM Sovereign Military Order of Malta.

SMPIA Shipowners' Mutual Protection & Indemnity Association; Suspended Manoeuvring System *aviation*; Sea floor Mapping System.

SMS *or* **S.M.S.** Safety of Merchant Seamen; Special Marine Structure.

S/MSD Special Marine Sanitation Device.

SMT *or* **S.M.T.** Ship Mean Time or Local Mean Time *nautical*.

SMTRB *or* **S.M.T.R.B.** Ship and Marine Technology Requirements Board.

Smuggling *See* **Contraband**.

S-MUX Subscriber's Multiplexer *computer*.

SMW *or* **S.M.W.** Standard Metal Window.

S.N. *or* **S/N** Shipping Note; Signal to Noise Ratio; Short Notice.

Sn Stannum (tin); Snow *met*.

s.n. Secundum naturam *Latin*—Naturally; Sine nomine *Latin*—Without name; Sub nomine *Latin* —Under a specified name; serial number.

SNAFU Situation normal—all fouled up! *military slang*.

SNAJ Society of Naval Architects of Japan.

Snake *See* **Unit of Account**.

SNAME Society of Naval Architects and Marine Engineers. *See* **National Shipbuilding Research Program**.

Snatch Block A single wood or metal block which can be opened and locked from one side.

S.N.C.B. Société Nationale des Chemins de Fer Belges. Belgian National Railways.

S.N.C.F. Société Nationale des Chemins de Fer Français. French National Railways.

SNF System Noise Figure.

SNFCC Shippers National Freight Claim Council *USA*.

SNFL Standing Naval Force Atlantic. *NATO*.

SNG Synthesised Natural Gas; Synthetically Produced Substitute Gases. These are derived from liquefied natural gases.

Sng. Singapore.

s.n.g. Sans notre guarantie *French*—Without our guarantee.

SNLR *or* **S.N.L.R.** Services No Longer Required.

SNO *or* **S.N.O.** Senior Navigation Officer; Senior Naval Officer.

Snowdrift Snow piled up by the wind.

Snow Train Train carrying winter sports passengers.

SNR Signal to Noise Ratio *radio*; Society for Nautical Research.

SNSA Singapore National Shipowners' Association.

SNSP Soonest Possible.

Snug To secure a vessel in readiness for a storm by lashing deck cargo, make fast movable gear, etc. *See* **Lashings**.

Snug Harbour Sheltered Harbour.

S.O. *or* **s.o.** Seller's Option; Ship Owner; Shipping Order, *q.v.*

S/O Ship Owner.

SO *or* **S/O** Senior Officer; Section Officer.

SOA Speed of Advance *nautical*.

SOB Shipped on Board; Society of Bookmen.

Soc. Society, *q.v.*

S.O.C.G.P.A. Seed, Oil Cake and General Produce Association.

SO₃ Acid Vapour. Sulphur trioxide.

Social Insurance National Insurance.

Social Overhead Capital Infrastructure, *q.v.*

Social Wealth Properties and assets owned by the state comprising schools, buildings, libraries, ships, aircraft, industries, etc. The whole capital of the country.

Sociedad Anonima Industrial y Commercial *Spanish*—Industrial and Commercial Company Limited. *Abbrev.* S.A.I.C.

Sociedad Anonima *Spanish*—Limited Company. *Abbrev.* SA.

Sociedad Anonima Industrial y Fiduciaria *Spanish*—Industrial & Commercial Society Limited. *Abbrev.* S.A.L.C.F.

Sociedade *Portuguese*—Company; Society.

Sociedade Anonyma *Portuguese*—Limited Company. *Abbrev.* SA.

Sociedade Anonyma de Responsibilidade Ltda *Portuguese*—Limited Liability Company.

Sociedad en Comandita *Spanish*—General Partnership.

Società *Italian*—Society; Company.

Società Anonima *Italian*—Joint Stock Company. *Abbrev.* SA.

Società Anonima per Azioni *Italian*—Limited Company. *Abbrev.* SapA.

Società con Responsibilita Limitata *Italian*—Limited Liability Company. *Abbrev.* Srl.

Société *French*—Society; Company.

Société Anonyme *French*—Limited Company. *Abbrev.* SA or S.A.

Société en Commandite *French*—Joint Stock Company.

Society (1) The social mode of life, the customs and organisation of a civilised nation. (2) Any social community. (3) The upper classes of a community, the socially distinguished, fashionable and well-to-do. (4) An association of persons united by a common aim or interest or principle.

Society of World Wide Interbank Financial Telecommunication An association of bankers with a view to hastening information and payment requests, offering standardisation of transactions among various banks in different countries and promoting many other advantages for smooth and fast running services. *Abbrev.* SWIFT *or* S.W.I.F.T.

Socio *Italian*—Business Partner.

Sodium Hydroxide Caustic Soda.

SOFAR Sound Fixing and Ranging.

SOFCS Self Organising Flight Control System.

S of S Secretary of State.

Soft Coal Bituminous Coal.

Soft Conversion *or* **Software Conversion to Metric System** The use of the dual scale units in English and metric. *Ex.* a speedometer in a car showing both miles and kilometres per hour.

Soft Currency Money which is easily exchanged into other foreign currency in the sense that no restrictions exist in the country involved. *Opposite to* **Hard Currency**, *q.v.*

Soft Goods Materials pertaining to cloth such as wool and cotton.

Soft Loan A loan without any rate of interest or a rate below that officially quoted.

Soft Metal *Similar to* **Babbitt-metal**, *q.v.*

Soft Sell Diplomatic or suggestive method of selling or advertising.

Soft Water Pure water; Rain Water; Water containing no calcium or magnesium salts.

SOG Speed on Ground *nautical*.

SOIP *or* **S.O.I.P.** State Organisation of Iran Ports.

Sol Unit of currency used in Peru. One hundred centavos equals one sol.

S.O.L. *or* **s.o.l.** Shipowners' Liability *to cargo*.

SOLA Solitary, *q.v.* In drawing Bills of Exchange, *q.v.*, or Bills of Lading, *q.v.*, only one original copy is issued.

SOLACS Solicitors' Accounting and Costing System.

SOLAIR Solomon Airways. Airline of The Solomon Islands.

Solar Pertaining to the sun.

Solar Day The time taken for a complete rotation and back to the original position of a place directly under the sun. A day of 24 hours.

Solar System Heavenly bodies, including the planets, which revolve around the sun.

Solar Time Time calculated by taking the angle hour measurements of the sun.

SOLAS Safety of Life at Sea. The International Convention for the Safety of Life at Sea (SOLAS) 1960 and 1974, adopted by the International Maritime Organisation (IMO). The most important of all international treaties concerning the safety of merchant ships and one of the oldest. The first version was adopted by a conference of 13 nations in 1914 following the loss of the *Titanic* in 1912, but the First World War prevented its full entry into force. In 1929, a conference of 18 nations adopted a new SOLAS Convention updating the previous regulations, as well as revising those for the prevention of collisions at sea. Technical developments led to a third conference under UN auspices in 1948 which adopted a new SOLAS Convention, coincident with that for the establishment of IMCO, *q.v.*, (later IMO). Ratification took many years but IMCO came into being in 1959 and its first major task was revision of the 1948 SOLAS Convention which was adopted by a conference of 55 nations in 1960. The 1960 Convention came into force in 1965 covering a wide range of measures including watertight subdivision and stability; fire protection; life-saving appliances; radio; safe navigation; carriage of grain; carriage of dangerous goods; and nuclear ships. Several amendments to this basic international instrument followed, but there were procedural difficulties in making them effective. This led to the Convention of 1974, which included all the measures agreed up to that date and procedures which would bring changes into force more speedily. The Convention entered into force in 1980.

The Convention's objective is to specify minimum standards for the construction, equipment and operation of ships, compatible with their safety. Flag States are responsible for ensuring that ships under their flags comply with its requirements, and a number of certificates are prescribed as proof that this has been done.

There have been a number of important amendments since 1980 dealing with: construction and fire safety (1981); life-saving appliances and other matters (1983); improvements to the safety of ro-ro passenger ships (following the capsize of the *Herald of Free Enterprise*) (1988); stability of passenger ships after damage and other improvements (1988); introduction of GMDSS, *q.v.* (1988); watertight doors and fire safety (1989); damage stability of cargo ships (1990); fire safety in passenger ships, carriage of grain and other dry cargoes and pilot safety (1991); and further measures for the safety of passenger ships (1992). Additionally, two Protocols have been adopted: modified inspection and survey procedures and annual surveys of tank ships' bottoms (1978); the harmonised system of survey and certification and other matters (1988).

Sole Proprietor A person who entirely owns a firm. He has in other words a 'one man' business. He takes the whole risk upon himself in that he has to manage the financial position of the business. On the other hand he has the advantage of being able to make decisions without internal or external interference and can keep in close personal contact with his employees. Also termed Self-employed.

Sole Trader A person running a company or business on his or her own, including capital, profits, losses, etc.

Solh *Arabic*—Agreement.

Solid Propeller When the boss and the propeller are constructed as one.

Solicitor A person employed to conduct the prosecution or defence of an action or other legal proceeding on behalf of another, or to advise him on legal questions, or to frame documents intended to have a legal operation, or generally to assist him in matters affecting his legal position. To enable a person to practise as a solicitor, he must serve a term as an articled clerk, pass certain examinations, be admitted and enrolled as a solicitor of the Supreme Court, and take out a yearly certificate authorising him to practise.

A solicitor is an officer of the Supreme Court, who, and who only, is entitled to sue out any writ or process, or commence, carry on, or defend any action, suit or other proceeding on behalf of another in any court.

Solitary Referring to one set of Bills of Lading, *q.v.*, i.e. when only one original bill of lading is issued. Also refers to one copy of a Bill of Exchange, *q.v. Abbrev.* SOLA or Sola.

Solv. Solvent, *q.v.*

Solvent Having the power of dissolving or forming solution with something; having money enough to meet all pecuniary liabilities; a chemical used to disperse oil pollution, also for cleaning the tanks of oil carriers.

Somali shilling Unit of currency used in Somalia. One hundred centesimi equals one Somali shilling. *Abbrev.* SomSh.

S.O.M.S School of Maritime Studies.

SOMSS Submarine Off-board Mine Search System.

Sonar *or* **SONAR** Sound Navigation and Ranging. This is an apparatus using ultra high frequency sound for detecting submarines and locating other underwater objects such as mines, wrecks, rocks and shoals of fish. The same principle is used in Echo Sounding, *q.v.*, to measure the depth of water.

Sonarman Sailor operating and maintaining Sonar, *q.v.*, equipment *naval*.

Sonic Of or relating to sound or sound-waves.

Sonic Barrier *or* **Sound Barrier** Excessive resistance offered by air to objects moving at a speed near that of sound.

Sonic Boom A loud noise made when an aircraft crashes through the sound barrier.

SONOAN Sonic Noise Analyser.

So og Handelskompagni *Danish*—Marine and Trading Company.

SOP Standard Operating Procedure;

SOPA Senior Officer Present Afloat *USA*.

Sophisticated Paints *or* **Sophisticated Coatings** Paints or coatings which are much superior in quality and results than conventional ones. They are chlorinated rubber, synthetic resin paints, etc.

S.O.R. *or* **s.o.r.** Sale or Return.

SORC *or* **S.O.R.C.** Southern Ocean Racing Conference *yachting*.

SORD *or* **S.O.R.D.** Submerged Object Recovery Device.

Sorter A postman whose job is to sort out incoming and outgoing mail. Also an American expression for a person responsible for the sorting of unloaded cargo into the assigned warehouses so as to be readily identified when the consignees demand delivery.

SOS Radio code signal (Morse) of extreme distress; Scheduled Oil Sampling, *q.v.*; Senior Ordinary Seaman, *q.v.*; Ship's Operational Safety; Service of Suit, *see* **Service of Suit Clause**; Singapore Organisation of Seamen.

SOSS *or* **S.O.S.S.** Shipyard Oceanographic Survey System.

SOT Slop Oil Retention.

Sotto Paranco *Italian*—Similar to Alongside, *q.v.*

Sotto Voce *Italian*—Said in an undertone.

Sou. South; Southern.

Sound This can mean an area of water connecting two seas or a sea with a lake or strait; to test the depth of the sea, channel, pond, etc., and the quality of the bottom by using a sounding line or apparatus. *See* **Sounding Rod** and **Sounding Line**.

Sound Buoy A buoy attracting the attention of navigators in the area by the sound of a horn, trumpet, whistle, gong or bell.

Sounding Line Coiled steel or hard cloth material used to check the depth or level of oil carried in oil or other liquid tankers. It is weighted at one end which allows the line to become straight before taking the measure. Also termed **Ullage Tape**, *q.v.*, or **Sounding Tape**.

Sounding Pipe *or* **Sounding Tube** A pipe running vertically down from the top of a tank to serve as a guide to the Sounding Rod, *q.v.*, or Sounding Line, *q.v.*, or Sounding Tape, *q.v.*, used to measure the depth of the liquid contents. Its existence makes it possible to dip the tank without the necessity of lifting the hatch or lid.

Sounding Rod A graduated wooden or metal rod which is used to ascertain the depth of water or fuel. Sounding rods are commonly used to check the depth of fuel, ballast, bilge tanks, water tanks, etc.

Sounding Tape *Similar to* **Sounding Line**, *q.v.*

Sound Value The value of an article as it would have been before it was damaged, used in calculating an insurance claim.

Source Language The computer language in which a program is first written.

Sour Gas Odorous gas by the name of hydrogen sulphide obtained from gas wells.

Sous palan *French*—Under the hook.

South Cardinal point of the compass, directly opposite to North.

Southbound Describes a vessel or aircraft proceeding in a southerly direction towards its destination.

Southern Cross Four principal stars in the form of a cross constituting a bright constellation in the southern hemisphere.

Southwester A storm coming from the southwest; a waterproof oiled linen hat shaped in a way to protect a seaman from rain and sea spray. With the hat there are also the raincoat, rainproof jacket and trousers.

Sou'wester *See* **Southwester**.

s.o.v. Shut off Valve.

Sov/s Sovereign(s).

SOx Sulphur Oxides.

sp. Specie, *q.v.*; Specimen; Spirit.

s.p. Self-propelled; Single Phase; Sine prole *Latin*—Without issue.

Sp. Spanish.

SP Shore Patrol; Special Police; Starting Price, *in purchase and sale negotiations*; Safe Port, *q.v.*

SP *or* **S.P.** Safe Port, *q.v.*; Supra Protest, *q.v.*; Stern Post.

S & P Sale and Purchase.

SPA Saudi Ports Authority; Subject to Particular Average, *q.v.*

S.P.A. Subject to Particular Average; Sales Promotion Association.

S.p.a. *or* **S.p.A.** Società per Azioni *Italian*—Joint Stock Company.

Spacecraft *or* **Spaceship** A manned or unmanned vehicle launched into space above the earth's atmosphere.

Space Platform *or* **Space Station** A manned vehicle launched into space as an observation post or as a base for maintaining or launching Spacecraft, *q.v.*

Space Port Storage, testing and launching base for Spacecraft, *q.v.*

Space Probe An unmanned vehicle in orbit outside the earth for exploration purposes.

SPADES Ship Protection and Design Engineering System.

Span (1) Length of rope or wire stretched between two points; rope secured at each end to a spar or yard to take a halyard or halyard block in the bight. (2) Overall measurement between extremes of an aircraft's wings.

Spar (1) Stout pole. (2) General term for Derricks, *q.v.*, masts and yards of ships. (3) Principal member of an aeroplane's wing.

Spar Buoy A buoy showing a mast above the waterline.

Sparred A system of packing used to protect breakable containers from rough handling.

S.P.A.S. Societatis Philosophicae Americanae Socius *Latin*—Member of the American Philosophical Society.

SPASUS Space Surveillance System *USA*.

Sp.bt.cl. Speedboat Clause *insurance*.

SPC *or* **S.P.C.** Self-Polishing Copolymer *antifouling*; Ships' Production Committee; Silicon Photo Cell; Society of Philippines Electrical Contractors; South Pacific Commission; Statistical Process Control; Stored Program Control *electronics*; Stowage Planning Centre *containerisation*; Solent Points Race *yachting*.

Sp.Chgs. Special Charges.

S.P.C.K. Society for Promoting Christian Knowledge.

spd. Sparred, *q.v.*; Speed.

SPD *or* **S.P.D.** Steamer Pays Dues, *q.v.*; South Polar Distance; Subject to Permission to Deal; Small Package Dispatch.

SPD/CONS Speed and Consumption.

SPE Senior Project Engineer *USA*; Society of Petroleum Engineers.

SPEC Society for Pollution and Environmental Control of Canada.

Spec Special; Specification.

Special Valuable cargo such as money, precious metal, jewellery, etc.

Special Buyer An authorised agent in the discount market of the Bank of England.

Special Cargo *or* **Specials** A shipping expression which refers to cargo needing special attention or supervision. This may be due to either its value or fragility. Freight on valuable cargo is paid in proportion to the value known as 'Ad Valorem Freight', *q.v.* *See* **Lock-up Cargo**.

Special Drawing Rights A term, introduced in 1984, to facilitate exchange rating regarding limitation of liability under international law. The SDR replaces the 'gold franc' in such calculations, from November 1984. The value of the SDR varies daily, and is obtainable from the customary financial sources of information. *Abbrev.* SDR *or* S.D.R.

Specialist Broker A Jobber, *q.v.*, in the New York stock market.

Special Wording Scheme Coded wordings used at Lloyd's Leading Underwriters, *q.v.*, for ready reference and uniformity. *Also called* Private Wording Scheme.

Specie Coin as opposed to paper money.

Specific Commodity Rates Special air cargo freights controlled by IATA Special commodity Rates Committee. *Abbrev.* SCR *or* S.C.R.

Specific Density *See* **Specific Gravity.**

Specific Heat The heat required to raise the temperature of a given substance to a given extent (usually one degree).

Specific Impulse Efficiency of rocket engine.

Specific Performance The remedy sought by a plaintiff who, instead of damages for a breach of contract, seeks the enforcement of the terms of the contract.

Specifications Specified detail, especially a detailed description of construction, workmanship, materials, etc., of work undertaken by architects, engineers, etc.

Spectrograph An apparatus for photographing or otherwise reproducing the spectrum.

Spectroheliograph An instrument for taking photographs of the sun from light of one wavelength only.

Spectrometer An instrument for measuring the refraction of light rays in passing through a prism.

Spectroscope An instrument for forming and analysing the spectra of rays.

Speed of Approach Measuring Instrument A Marconi marine device developed to assist pilots to estimate to an accurate degree the speed of large ships approaching mooring places. *Abbrev.* SAMI.

Speronara Type of sailing boat once found in the central Mediterranean.

SPEX Small Parcel Express, *q.v.*

SPG Speed Made Good *nautical.*

SP/FR Semi-Pressurised/Fully Refrigerated.

SPGCR Simplified Procedure for Goods Carried by Rail.

sp.gr. Specific Gravity, *q.v.*

Sph. Spherical.

Sphere Globe; Round Solid Body; Orbit; Ball.

Spherical Buoy A buoy showing a domed top above the water.

Spherograph Stereographic projection of the earth on disc, with meridians and parallels of latitude marked in single degrees.

Spherometer An instrument for finding the radius of a sphere and for exact measurement of the thickness of small objects.

SPHL Self-Propelled Hyperbaric Lifeboat.

Sp.ht. Specific Heat, *q.v.*

SPI Selected Period Investment.

SPIA Shipowners' Protection and Indemnity Association.

Spider Band *or* **Spider Hoop** Iron band around a mast to hold belaying pins.

Spider's Web Crosswires in a sight or telescope.

Spirit Room Compartment where rum used to be stowed in British warships.

Spiv A petty swindler; or black marketeer often a disreputable and dandified character person living by his wits.

SPL Sound Pressure Level; Supplementary Flight Plan Message *aviation.*

SPLASH Sinking Platform Lighter Aboard Ship.

Splash Zone An area at sea level where tidal and wave actions result in continuous wetting and drying and where corrosion on metal surfaces occurs.

Splice *or* **Splicing** The joining of two ropes or wires by unlaying their Strands, *q.v.*, and tucking or relaying them together. A Marlinspike, *q.v.*, or a Fid, *q.v.*, may be used as tools.

SPM *or* **S.P.M.** Single Point Mooring, *q.v.*

SP Model Special Performance Model.

Sponson Structure projecting from ship's side.

Sponte sua *Latin*—Of one's own accord.

Spoorweg *Dutch*—Railway.

S.P.Q.R. Small Profits and Quick Returns. Senatus Populusque Romanus *Latin*—The Senate and People of Rome.

SPR Strategic Petroleum Reserve *USA.*

Spread The difference between buying and selling prices.

Spreadeagle American term for Straddle. *See* **Call** *or* **Calls.**

Spreading Rate An area covered by a given paint film thickness, per litre or gallon.

Spring A mooring rope or Hawser, *q.v.*

Spring Tide The highest range of tide during a new or full moon.

Spring Tide Port A port particularly subject to tidal ranges with most of its trade being conducted during periods of spring tides.

SPS Solar Power Satellite; Steampipe Survey *Lloyd's Register notation*; System Performed Specification *computer control system.*

SPSNA Steampipe Survey Not Applicable *Lloyd's Register notation.*

SPSSC Special Purpose Ship's Safety Certificate.

Spt. Spot; *Lloyd's Register notation* for Superheaters.

SPT Signal Pressure Tranducer *oil drilling*.

SPU Stationary Production Unit *for deep water drilling*.

Spudding Setting the legs of an oil drilling rig in the seabed. *Opposite to* **Breaking Out**, *q.v.*

Spud Pole A telescopic arm which rests on the seabed and secures the position of a ro/ro ship or semi-submersible ship.

Spunyarn *or* **Spun Yarn** A rope of two or four untwisted yarns of tarred hemp for Lashing, *q.v.*

Sq. *or* **sq.** Square; Squadron.

SQ *or* **S.Q.** Special Quality *referring to steel cable*.

Sq.ft Square feet; Square foot.

Sq.in. Square Inch.

Sq.m. Square metre.

Squall A short period of sudden strong wind followed by heavy rain or a hail storm.

Square Mile Said of the City of London, where Merchant Banks, Insurance Companies, Lloyd's, the Stock Exchange, etc., are located.

sq.yd. Square yard.

Squawk An aviation term, being a four-figure code set by the crew members of an aircraft on a Transponder, *q.v.*, which serves as an identification number on the radar screen of the air traffic controller.

sr Stere, a cubic metre of timber; Steradian, unit of solid angle.

s.r. Shipping Receipt; Short Rate; Standard Rate.

Sr Senior; Sister; Señor; Strontium.

S.R. Store Room; Storage Room; Shipping Receipt.

SRA Special Rules Area *aviation*; Selected Restricted Availability *USA*; Surveillance Radar Approach *aviation*.

SRAM Short-Range Attack Missile.

srb sulphate reducing bacteria. They attack exposed metal surfaces and cause corrosion in tanks and bilges.

SRBM Short Range Ballistic Missile.

SRC *or* **S.R.C.** Safety Radio/Radiotelegraphy Certificate.

S.R.C. Science Research Council; Swiss Red Cross.

SR & CC *or* **s.r.c.c.** Strikes, Riots and Civil Commotions. *Similar to* **Riots, Civil Commotions and Strikes**, *q.v.*

SRCC & MD *or* **S.R.C.C. & M.D.** Strikes, Riots, Civil Commotions and Malicious Damage.

SRCP Special Recovery Capital Project *USA*.

S.R.C.R.A. Shipowners' Refrigerated Cargo Research Association,

SRE Surveillance Radar Element; Steam Reciprocating Engine.

SRI Stafford Research Institute *USA*; Standards Research Institute.

S.R.L *or* **S.R.Liab.** Ship Repair's Liability *marine insurance*.

S.r.l. *or* **S.R.L.** Società a Responsabilità Limitata, Italian Limited Company.

SRM Short Range Missile; Speed of Relative Movement *nautical*.

SRMC Special Refrigeration Machinery Certificate, *Lloyd's Register notation*.

SRN *or* **S.R.N.** State Registered Nurse.

SRNA *or* **S.R.N.A.** Shipbuilders and Repairers National Association.

SRO Self Regulating Organisation *financial*; Standing Room Only; Statutory Rules and Orders.

SRP Supply Refuelling Point.

SRPC Search and Rescue Point of Contact.

SRS Ship Reporting System, *q.v.*

S.R.S. Societatis Regiae Sodalis *Latin*—Fellow of the Royal Society.

SRSA *or* **S.R.S.A.** Scientific Research Society of America.

SRZ Special Rules Zone *aviation*.

SS *or* **S.S.** Special Survey *q.v*; Secretary of State; Standard Size; Steam Ship; Supra Scriptum *Latin* —Above Written; Signal Station; Saints.

SS *or* **S/S** Signal Station; Screw Streamer; Stainless Steel; Special Survey; Soeriges Skeppshandlareförbund. *See* **S.S.A.**

ss. *or* **Scil** *or* **Sciz** Scilicet *Latin*—Namely.

S.S.A. Swedish Ship Suppliers Association.

SSADM Structured Systems Analysis and Design Methodology *Computer*.

SS adv. Special Survey Advanced.

S.S. and C. Salvage and Salvage Charges *marine insurance*.

SSA Shipbuilders and Shiprepairers Association UK.

SSAP Statement of Standard Accounting Practice.

SSB Stranded, Sunk or Burned *insurance*.

s.s.b. Single Side Band; Stranded, Sunk or Burned *marine insurance*.

SSBL Super Short Base Line *accoustic systems*. *Similar to* **USBL**, *q.v.*

SSC *or* **S.S.C.** Semi-Submerged Catamaran; Simultaneous Settlements Clause; Solicitor to the Supreme Court.

SSCC Sulphide Stress Corrosion Cracking.

S.S. & C. Same Sea and Country or Coast *marine insurance*.

SSCV Semisubmersible Crane Vessel; Semi-Submersible Construction Vessel.

SSD Safety Shutdown System *electronics*; Slow Speed Diesel.

S.S.D. Sacrae Scripturae Doctor *Latin*—Doctor of Sacred Scripture.

SS.D. Sanctissimus Dominus *Latin*—Most Holy Lord *The Pope*.

SSE *or* **S.S.E.** South-South-East.

SSG Schweizerische Speisewagon-Gesellschaft, dining car service of Swiss national railway.

SSG (UK) Sea Safety Group *UK*, established 1992.

SSI Small Scale Integration *electronics.*

SSIA Shiprepairers and Shipbuilders Independent Association *USA.*

S/side Shipside.

SSIT *or* **S.S.I.T** Submersible Icebreaker Tanker.

S.S.L. Sacrae Scripturae Licentiatus *Latin*—Licentiate of Sacred Scripture.

SSM *or* **S.S.M.** Single Sideband Modulation; Surface to Surface Missile.

SS(M) Special Survey for Hull Modified *Lloyd's Register notation.*

SSMUA *or* **S.S.M.U.A.** Steamship Mutual Underwriting Association. Mutual or P. & I. Club.

SSN Standard Shipping Note; Nuclear-powered submarine.

S.S.N. Servizio Sanitario Nazionale *Italian*—National Health Service.

S.S.O. *or* **S.S.O.** Struck Submerged Object.

SSOA Standby Ship Operators Association.

s.s.o.b. *or* **S.S.O.B.** Stranded, sunk or burnt *marine insurance.*

SSP *or* **S.S.P.** Semi-Submerged Platform *oil drilling craft*; Stainless Steel Propeller.

SSPC Steel Structures Painting Council *USA.*

SSPU *or* **S.S.P.U.** Ship's Service Power Unit.

SSR *or* **S.S.R.** Semi-Submersible Rig, *q.v.*; Soviet Socialist Republic; Secondary Surveillance Radar *aviation*; Special Survey of Refrigerating Machinery.

SSRC Shipping and Shipbuilding Rationalization Council *USA.*

SSS Strategic Sealift Ship(s) *USA.*

SSS *or* **S.S.S.** Selective Service System; Single Screw Ship; System Safety Society *USA.*

SSSO Specialised Satellite Service Operators, *UK licensees.*

SSSU Sub-Sea Separator Unit *oil drilling.*

SST *or* **S.S.T.** Supersonic Transport *aviation*; Supersonic Travel.

SS(TD) *Lloyd's Register notation* for Special Survey for Hull Thickness Determination.

SSTG Short Sea and Service Transport Group *UK.*

SSU Swedish Seafarers' Union.

S.Sup. *or* **S.sup.** Submersible Support.

SSW *or* **S.S.W.** South-South-West.

st. Stet *Latin*—Let it remain; Starboard; Strainer.

St Saint; Stratus Clouds; Stokes or Stoke *unit of kinematic viscosity, named after the physicist Sir George Stokes.*

ST *or* **S.T.** Short ton of 2,000 lbs or 907.19 kilos; Steam Trawler; Shaft Tunnel; Shipping Ticket; Side Tank *Lloyd's Register notation*; Spring Tide; Summer Time; Surtax; Sounding Tube; Steam Turbine; Saddle Tank or Side Tank, *q.v.*; Stern Thruster.

S&T Supply and Transport.

STA *or* **S.T.A.** Sail Training Association; Shaft Tunnel Alley; Short Tunnel Alley; Solar Trade Association; Subject to Average *insurance.*

Sta Santa *Spanish, Italian, Portuguese*—Saint.

Stabilisers Steel fins which can be extended at right angles to the hull to reduce rolling in heavy seas.

Stability The state or quality of a vessel remaining afloat and able to recover equilibrium at sea while pitching and rolling.

Stacking The positioning of packages in a uniform and secured way, in the holds of a ship, aircraft, store, etc.

Staff Captain *or* **Master** On large passenger vessels the Master is assisted by this senior officer who has particular responsibility for the welfare of the passengers.

Stag An entrepreneur in stock exchange business who buys shares in anticipation of an increase in price. He then expects to sell them within a short period at a good profit. Also known as Premium Hunter.

Staggers *or* **The Bends** *or* **Decompression Sickness** If a deep-sea diver is brought to the surface too quickly bubbles of nitrogen form in his bloodstream, leading to an agonising condition known as 'the bends' or 'staggers'. It is essential that he is placed in a decompression chamber for a number of hours.

Staithe *or* **Staith** Berth, *q.v.*, for ships alongside with a projecting wall or rails that allow cargo to be tipped directly from railway trucks into ships' holds. Also a mooring or landing place in some inland waters.

Stale Bill of Lading A bill of lading which is received after the vessel has arrived at her discharging port. The Consignees, *q.v.*, generally produce a bank guarantee to the agents to withdraw the merchandise if same is urgently needed.

Stale Cheque A cheque which appears on its face to have been in circulation for an unreasonable time. A person who takes such a cheque takes it subject to equities and at his own risk. A banker may refuse payment of a stale cheque if the drawer has not directed him to pay it. What is an unreasonable time depends on the circumstances. By the custom of bankers, confirmation from the drawer is usually required when the cheque has been in circulation for six months before presentation.

Stale Demand A claim which has not been made for so long that it must be taken to have been waived.

Stall An aircraft can become unstable and may crash if it 'stalls', i.e. if it loses its airspeed.

Stallage The liberty or right of pitching or erecting stalls in a fair or market or the money paid for the same; a payment due to the owner of a market in respect of the exclusive occupation of a portion of the soil within the market.

The right of stallage is a right for a payment to be made, to the owner of the market, in respect of the exclusive occupation of a portion of the soil, for the purpose of selling goods in the market. A voluntary bargain cannot be reopened on the ground that the stallage charge is unreasonable.

STANAVFORLANT Standing Naval Force Atlantic. A squadron formed of warships from several nations. *Abbrev.* SNFL.

Stanchion A fixed or collapsible post such as is fitted to some timber carrying vessels along the bulwarks to secure a deck cargo of timber or packaged lumber.

Stand (1) Section of floor area at a trade exhibition or fair allocated to an exhibiting organisation. (2) Parking position for an aircraft. (3) Witness-box *USA*. (4) Support for an engine under overhaul.

Standard A timber trade measurement. *See* **Standard Measure.**

Standard Form of Charterparty A charterparty formulated by the Chamber of Shipping or the Baltic White Sea Conference or similar societies and institutions.

Standard Cover Minimum cover under the basic form of an Insurance Policy, *q.v.* Anything additional will need to be covered under Optional Insurances.

Standard Design The construction of a ship or aircraft or any other object to a general standard pattern. *Abbrev.* SD *or* S.D. *or* s.d.

Standard Gross 10,000 matches.

Standard Measure A timber measure of 165 cubic feet known as Petrograd Standard. The weight of one timber standard ranges from two to five tons depending on the type of wood.

Stand By (1) To await orders. (2) During a Salvage, *q.v.*, operation or in a distress situation the vessel rendering assistance 'Stands By'. (3) Time during which a person is unable to work due to unforeseen circumstances and is standing by until the order to work is given.

Standby Letter of Credit A Guarantee, *q.v.*, raised by the bank on behalf of the buyer as a line of recourse available to the seller should the buyer default on a direct payment.

Stand By Money Charges paid to workmen waiting for work.

Stand Clear A nautical word meaning to stay out of the way.

Stand In To head for an anchorage or for a harbour *nautical.*

Standing Deposit A deposit of money left on credit against a regular importer of merchandise with the customs authorities to cover any future duty to be imposed.

Standing Lights Lamps providing permanent subdued illumination in alleyways and passages within a ship's accomodation for safety. Sometimes called Police Lights. May be red where it is necessary to preserve the night vision of people going on deck.

Standing Order An official order to be adhered to. *Abbrev.* STO *or* St.O.

Stand On To keep on course in navigation.

STAR Ship Technology and Research; Satellites for Telecommunications Applications and Research, European Consortium; Scientific and Technical Aerospace Reports *USA*.

Starboard Side *Similar to* **Starboard**, *q.v.*

Starboard Tack When a vessel sails with the wind on the starboard side.

Star Finder A graphic representation of the heavens to enable a navigator to locate and identify celestial bodies.

Stat. Statistics; Statute.

Statement of Accounts Itemised accounts. A list of debits and credits for services rendered. *Ex.* As soon as a ship leaves port her agents prepare a Statement of Accounts with supporting vouchers to the owners. This covers all the expenses and charges during her stay balanced by any money advanced by her owners or freights collected.

Statement of Affairs Balance Sheet, *q.v.*

Statement of Claim The first step in the pleadings to an action in which the plaintiff particularises his claim and the legal grounds on which it is based.

Statement of Facts A statement of events relevant to the charterparty and to the calculation of laytime—usually prepared by the port agent and signed by the master.

Stateroom A large cabin or suite of rooms for officers or for first-class passengers on liners.

Static Load Acting as a weight but not moving.

Stationary Engine An engine installed in a fixed place, i.e. not mobile.

STATOIL Norwegian state oil company.

Statsbaner *Danish*—State Railways.

Status Social position or rank in relation to others; relative importance; position of business affairs, etc.

Status Enquiry An enquiry regarding the financial soundness or otherwise of a person or firm. Many such enquiries are produced confidentially on behalf of clients, who need the relevant information for guidance in their business contacts.

Status quo *Latin*—The same state as now.

Statute An Act of Parliament; a permanent rule enacted by the governing body of a company, institution or corporation, etc.

Statute Barred Causes of action in respect of which proceedings cannot be brought because the periods laid down by the Statutes of Limitation have elapsed. *See* **Time Bar** *or* **Time Barred**.

Statute Law Law enacted by an Act of Parliament, which cannot be altered or cancelled unless by new government legislation.

Statutory Books Books kept by a company that are compulsory by law. Generally they are the Memorandum of Agreement, Register of Directors, Register of Members and the Minute Book.

Statutory Companies Companies formed and covered by the laws of the state.

Statutory Obligation Any obligation, liability or provision imposed by any legislative or similar enactment, decree or regulation, having the force of law in any country.

Stays Ropes of hemp, synthetic or wire material to support masts, topmasts, etc., of a ship.

STB *or* **S.T.B.** Sacrae Theologiae Baccalaureus *Latin*—Bachelor of Sacred Theology; Said to be; Scientiae Theologiae Baccalaureus *Latin*—Bachelor of Theology; Sun True bearing *nautical*.

S.T.B. Form of Tanker Time Charter A code word for a tanker time charterparty.

STBC Self-trimming Bulk Carrier. *See* **Self-trimming Ship.**

Stb. cl. Stability Clause *insurance*.

Stbd Starboard.

STB Voyage An improved version of the Exonvoy 1969 charterparty employed by international companies for the chartering of tankers.

STC *or* **S.T.C.** Short Term Certificate; Sensitivity Time Control *electronics*; Said to Contain; Single Trip Container, *q.v.*

s.t.c. Said to Contain.

STCW International Convention on Standards of Training, Certification and Watchkeeping of Seafarers 1978 *IMO*.

STD *or* **S.T.D.** Subscriber Trunk Dialling *telephone*; Sea Transport Department.

S.T.D. Sacrae Theologiae Doctor *Latin*—Doctor of Sacred Theology.

Std/s Standard(s) *timber trade*.

Std. Time Standard Time.

Ste Sainte *French*—Saint.

Steady Wind A constant and regular wind blowing in the same direction.

Steam Vapour of water, especially the gas into which water is changed by boiling, largely used as motive power owing to its elasticity. Saturated steam is that which is in contact with and at the same temperature as boiling water. Superheated steam is that which has a higher temperature at a given pressure and greater volume for a given weight than saturated steam.

Steamer Generic word for a Cargo Vessel whether she is a steamship or a motorship. *Abbrev.* Stmr *or* Str.

Steamer Pays Dues Under this term in a charterparty the shipowner pays all the expenses for the use of a port, quays, canals, jetties, etc. *Abbrev.* SPD *or* S.P.D.

Steaming Distance The actual distance to be run by a ship from one point to another.

Steaming Range *Similar to* Cruising Radius, *q.v.*

Steaming Time The time, in days, hours and minutes taken to sail from one port to another.

Steam Ship *or* **Steam Vessel** Any ship or vessel propelled by machinery.

Steel Kinds of malleable alloy or iron and carbon largely used as material for tools, weapons, etc., and capable of being tempered to many different degrees of hardness.

Steel Wire Rope Twisted steel strands in left or right spiral directions on a central core in natural, synthetic, metallic or combined materials, forming one whole steel wire rope. *Abbrev.* SWR *or* S.W.R.

Steelyard A kind of balance with a short arm to take the thing weighed and a long graduated arm along which a weight is moved until it balances this.

Steenkolen-Handelsvereeniging Dutch Shipowning and Trading Group. *Abbrev.* SHV.

Steerage Passengers All passengers except those that are cabin passengers.

Steerman He is theoretically a pilot on a Mississippi River towboat who stands watch alongside the captain.

STEL Short Term Exposure Limit.

STEM Scanning Transmission Electron Microscope.

Stem The bows or forward part of a vessel. *Opposite to* **Stern**.

Stemdate The date when Laydays, *q.v.*, start to count in the loading operation. *See* **Subject to Stem**.

Stemming The ship agent's application to the customs authorities requesting the allocation of a berth for the anticipated arrival of the vessel. Making preparations to supply bunkers to the ship. Also used with the tide or current as object to mean holding the ship steady without making headway over the ground —stemming the tide.

Stemming List A roster of ships awaiting berths at a port. Also the list of ships booked to be drydocked or to be repaired in a repair yard.

Steppe Large expanses of grassland as in European Russia and South-west Siberia.

Stere A cubic metre or 35.314 cubic feet.

Sterling (1) Describes silver of 92.5% purity. (2) British currency, referred to as 'pounds sterling'.

Stern The rear, back or aft part of a ship. *Opposite to* the bows or stem.

Stern Anchor An anchor used from the stern of a ship.

Stern Gear The general meaning for the propeller and shaft.

Sternlight A white light placed as nearly as practicable at the stern of a vessel showing an unbroken light over an arc of the horizon of 135 degrees and so fixed as to show the light 67.5 degrees from right aft on each side of the vessel.

Sternpost The main wooden or iron post in the stern of a vessel which is fastened to the keel and which serves as a rudder support. Also called Rudder Post, *q.v.*

Stern Thruster A propeller at the after end of a ship to give a transverse movement which assists in berthing or unberthing and in turning in a restricted space. It can save the expense of using tugs. *See also* **Bow Thruster**.

Sternway The reverse movement of a vessel.

Stern-Wheeler A boat or ship propelled by a paddle wheel at the stern.

Stet *Latin*—Let it stand; Do not delete.

Stevedore Damage Clause A standard time charter clause approved by the Documentary Council of BIMCO. Responsibility for damage caused by stevedores and any time lost in consequence is for the account of charterers and is to be recorded in the Time Charter Rate.

Stevedores *or* **Stevedoremen** Port workers engaged in the stowage of cargo in the holds of a ship. This may also mean contractors in general loading and discharging cargoes from ships.

Stev. Liab. Stevedores Liability.

Steward (1) A person entrusted with the management of another's property, especially a paid manager of a great house or estate. (2) A purveyor of provisions, etc., for a college, guild, club, ship, etc. (3) A passengers' attendant and waiter on a ship or aircraft. (4) Any of the officials managing a race meeting, ball, show, etc.

St. Ex *or* **St. Exch.** Stock Exchange.

Stg Sterling.

Stiffening The ballasting of a vessel for Stability, *q.v.*

Stipend A fixed periodical money allowance for work done; a salary, especially a clergyman's official income.

Stk. Stock, *q.v.*

Stkhld Stockholder.

STL Submerged Turret Loading, *q.v.*

Stm. Hpr. Bge. Steam Hopper Barge.

Stn. Station.

STO *or* **S.T.O.** Sea Transport Officer; Senior Technical Officer; Standing Order, *q.v.*

St. O. Standing Order, *q.v.*

STOA *or* **S.T.O.A.** *or* **stoa** *or* **s.t.o.a.** Subject to Owners' Approval.

Stock Capital of a corporation or company.

Stockbroker Member of the Stock Exchange, *q.v.*, who buys and sells Stocks and Shares, *q.v.*, either on his own account or on behalf of others. He works on a commission basis.

Stock Book Stock control. A register book kept to record new stocks and goods that are sold.

Stock Company The capital of the company or corporation is contributed by individuals for the prosecution of some undertaking and is divided into Shares, *q.v.*, entitling the holders to a proportion of the profits.

Stock Exchange A market operating on an international basis where stocks and shares are publicly bought and sold. The world's major stock exchanges are to be found in London, New York, Tokyo, Paris, Amsterdam, Zurich, Milan and Hong Kong. *Abbrev.* SE.

Stockholder The recorded owner of stocks and shares; a livestock dealer or owner in Australia.

Stock-in-Trade Goods to be sold in a store; goods and appliances, tools, equipment, etc., necessary for a particular trade or occupation.

Stockmarket Another word for Stock Exchange, *q.v.*

Stockpiling The accumulation of commodities, raw materials, etc., to be held in reserve.

Stocktaking A review of one's stock for an accurate knowledge of what one has in stock.

Stocks Stock Exchange securities; Commodities.

Stocks and Shares Money documents invested in commercial or industrial companies or corporations. In turn they are loaned to the state. They are sold and bought at a Stock Exchange, *q.v.* While private shares are negotiated by individual agreement in accordance with the company's Articles of Association, *q.v.*, UK stocks and shares are loaned on a fixed interest to the government.

Stockyard An enclosure purposely fitted to accommodate cattle.

STOL *or* **S.T.O.L.** Short Take Off and Landing *aviation*.

STOLVCD *or* **S.T.O.L.V.C.D.** Short Take Off and Landing Vertical Climb and Descent *aviation*.

S'ton Southampton, *UK*.

Stone A weight of 14 lbs avoirdupois.

Stoomboot *Dutch*—Steamboat.

Stoomvaart *Dutch*—Steam Shipping.

Stop for Freight A Possessory Lien which can be exercised either by the shipowner or agent of a ship before freight is settled and while the merchandise is being unloaded and stored in a government warehouse.

Storage The act of storing.

Storekeeper A person who keeps a range of commodities in his custody to have them ready when they are requested or needed for consumption. A storekeeper can be engaged ashore or at sea.

Stores (1) Places or shops where goods are kept for eventual sale. (2) Supplies of fresh and dry provisions; deck, engine and other stores needed or belonging to ships. (3) Warehouses privately owned or belonging to a government to hold Bonded Goods, *q.v.*

Storm A violent disturbance of the atmosphere with thunder, strong wind, or heavy rain or snow or hail; a tempest, usually accompanied by lightning.

Stormbound Prevention of sailing owing to severity of weather. *Also* weatherbound.

Storm Centre The point to which wind blows spirally inward in a cyclonic storm; the centre of minimum barometric pressure.

Storm Signal A cone, drum or other device for warning of an approaching storm.

Storm Tide An increased water level in consequence of a storm.

sto-ro A vessel with capacity for break-bulk cargo as well as vehicles or trailer borne cargo, as in the forest products trade.

S to S Station to Station *railway charges.*

Stowage Plan A plan of a laden ship showing the location of cargo parcels in various compartments.

Stowage Factor The average cubic space occupied by one tonne weight of cargo as stowed aboard a ship. Normally described in terms of cubic feet per tonne.

Stowaway An unauthorised person boarding a ship and keeping out of sight for the purpose of sailing without paying the passage, naturally without the knowledge of the officers on board and other authorities from ashore.

STP Submerged or Sub-sea Turret Production, *q.v.*

STP *or* S.T.P. Scandinavian Tanker Pool; Standard Temperature and Pressure.

STPD Ship Type Procedural Documents, *Lloyd's Register* Documentary System.

Stp. L. Stop Loss *insurance.*

Str. Steamer; Strait, *q.v.*; Strasse *German*—Street.

S.T.R. *or* s.t.r. Surplus to Requirements.

STRAD *or* S.T.R.A.D. Signal Transmitting Receiving and Distribution.

STRADAP Storm Radar Data Processor.

Straddle *See* **Call** *or* **Calls**.

Straight B/L *See* **Straight Bill of Lading**.

Straight Bill of Lading A Bill of Lading, *q.v.*, which stipulates that the goods are to be delivered only to the named Consignee, *q.v. Opposite to* Order Bill of Lading, *q.v.*

Strainer Plate *See* **Strum Plate**.

Strait A narrow passage of water connecting two seas or large bodies of water. *Abbrev.* Str.

Strake A continuous line of planking or steel plates from stem to stern on a ship.

Stranding, Stranded In the case of a rope, the words apply to the unlaying of its strands or the breaking of one due to wear or rough treatment.

Stratocumulus Clouds Low clouds below 8,000 feet or 2,400 metres in a form of globular masses or rolls. *Abbrev.* Sc.

Stratosphere The layer of atmospheric air lying above the troposphere in which the temperature ceases to fall with height, remaining constant. Throughout the ten-mile-thick layer of stratosphere the temperature is usually about minus 60 degrees Centigrade. *See* **Tropopause** and **Troposphere**.

Stratus Cloud in the form of a level or horizontal sheet.

Str. Cl. Strike Clause.

Strd. Stranded *marine insurance.*

Street Market *See* **Curb Market**.

Streptokinase An enzyme used to dissolve blood clots in the arteries and restore blood circulation to the heart for cardiac victims.

Strike The cessation of work by a body of persons employed in any trade or industry acting in combination, or a concerted refusal under a common understanding of persons so employed to continue to work or to accept employment; 'lock-out' is the closing of a place of employment or the suspension of work, or the refusal by an employer to continue to employ any number of persons employed by him in consequence of a dispute, done with a view to compelling persons employed by him to accept terms or conditions or conditions of or affecting employment.

There is, of course, no question as to the right of an individual workman to cease to work, provided that he does so without Breach of Contract, *q.v.*

Strike a Bargain Conclude an agreement between buyer and seller on a price convenient to both parties.

Strike Expenses Expenses incurred as a result of a strike, such as forwarding costs for goods that cannot be discharged at the scheduled destination port or extra freight charged for overcarriage to another port when the scheduled discharge port is strike-bound. These expenses are not covered by the marine policy with the standard strikes clauses attached hereto.

Stringer One of the longitudinal members in the structure of ships and boats connected to the frames or ribs respectively on either side which provides stiffening.

Stripped Gas Natural gas which has been processed by removing condensation.

Stripper Paint remover. Also a non-commercial oil or gas well.

Stripping a Container The emptying of a cargo container by the consignee or his agent.

Stripping Pump A small capacity pump used to strip remaining small quantities of liquid from a cargo tank following discharge of the majority of the tank's contents by cargo pump.

Strong Market When prices of commodities rise and are in demand. *Opposite to* **Weak *or* Poor Market**, *q.v.*

Strop Cargo handling gear; Cargo Sling. *See* **Sling**.

Str/s Strait(s), *q.v.*

Strum *or* Rose *or* Strainer Plate A filter fitted in the suction section of the Bilges, *q.v.*, of a ship to prevent dirt from entering and clogging the water circulating systems for the engine room. *See* **Sea Chests**.

STS *or* **S.T.S** Spare Time at Sea; Special Treated Steel; Spare Tail Shaft.

Stuffing The act of filling up a container with merchandise.

Stuffing Box A box device on a ship generally behind the tail end shaft of a ship which is fitted with water-resistant grease to stop water entering the rotating shaft and at the same time functioning as a lubricant for the shaft; a chamber in machinery through which a rod can work without allowing the passage of air, etc., all vacant spaces being filled with stuffing or packing.

S. Turb. Steam Turbine.

Stuven *Swedish*—Shipping Company.

STV Subscription Television.

Stv *or* **Stvdr.** Stevedore.

STVP Short Term Vehicle Park.

Stw. Stowage.

Stylography Writing or engraving with a special stylo or pen.

SUAA *or* **S.U.A.A.** Standard Underwriting Agency Agreement.

Sub-agent When an agent employs a person as his agent, to assist him in transacting the affairs of his principal, the person so employed is called a sub-agent. In the absence of an agreement to the contrary, there is no privity between the principal and the sub-agent; therefore, the principal is not liable to the sub-agent for his remuneration, and he cannot sue the sub-agent for negligence or misconduct; only the agent can be sued. But if the agent has an express or implied authority to employ a sub-agent, privity of contract arises between the principal and the sub-agent, and the principal may sue or be sued by the sub-agent. Similarly, when a contractor makes a contract with a sub-contractor to carry out his contract, or part of it, there is no privity between the principal and the sub-contractor.

SUBC Self-Unloading Bulk Carrier.

SUBCAP Submarine Combat Air Patrol *USA*.

Sub. chgs. Substituted Charges *insurance*.

Sub. exps. Substituted expenses.

Subject Receivers A Charterparty, *q.v.*, term giving the charterers the right to check with the cargo receivers that the vessel can be moored at the discharging quay or terminal.

Sub judice In course of trial.

Submerged Turret Loading A system for loading a Shuttle Tanker, *q.v.*, from seabed offshore oil storage. A moored submerged buoy of conical shape carries the riser. The ship has a loading compartment in her bow section below which is a mating cone or aperture in the keel. When the ship is manoeuvred using her thrusters, and by DGPS and underwater camera, over the buoy it is winched up to marry with the hull and locked so that the ship rides to the buoy's moorings.

The ship's manifold is then connected to the buoy. *Abbrev.* STL.

Submerged *or* **Sub-Sea Turret Production** A system similar to Submerged Turret Loading, *q.v.*, which allows a Floating Production Storage and Off-Loading Vessel to be positioned over, and receive, a submerged buoy with risers carrying oil, gas and water. Swivel connections are so designed that the ship, while locked to the buoy, can swing freely to wind and current. *Abbrev.* STP.

Submersible A container constructed of a strong material which can be filled with various liquid cargoes. When full it will float partly submerged and can be towed to its destination. When emptied it can be folded or rolled.

Submersible Decompression Chamber A strong capsule supported by a ship's lifting gear complete with life support systems and electrically compressed air to accommodate a limited number of underwater researchers and explorers. It is used for the exploration of continental shelves, underwater construction, laboratory, salvage and oceanographic work.

Submission A submission to arbitration is an instrument by which a dispute or question is referred to arbitration. When the reference is made by the order of a court or judge, the order is called a reference; submission denotes an agreement between the parties. Such an agreement may be either general (that is, an agreement to refer to arbitration all future disputes arising out of a specifed matter), or particular (that is, an agreement to refer to arbitration a dispute which has already arisen). A general submission is commonly contained in articles of partnership.

Subpoena A writ or summons issued in an action or suit requiring the person to whom it is directed to be present at a specified place and time, and for a specified purpose, under a penalty (*sub poena*) for non-attendance. The varieties of subpoena now in common use are the subpoena *ad testificandum*, called a subpoena *ad test.*, used for the purpose of compelling a witness to attend and give evidence, either in court or before an examiner or referee, and the subpoena *duces tecum*, used to compel a witness to attend in court or before an examiner or referee, to give evidence and also bring with him certain documents in his possession specified in the subpoena. The subpoena bears a close analogy to the citation, or *vocatio in jus* of the civil and canon laws. A person attending under a subpoena *duces tecum* to produce a document need not be sworn, and in that case he cannot be cross-examined.

Subrogation The substitution of one person or thing for another, so that the same rights and duties as attached to the original person or thing attach to the substituted one. Thus, if a person insures a ship which is lost by a collision caused by the negligence of another ship, he may recover the value of the ship from the underwriters, and then they are subrogated to his rights so as to be able to bring an action against the person who caused the collision. A similar rule applies in fire insurances and employer's liability insurances.

As between insurer and insured, the insurer is entitled to the advantage of every right of the insured connected with the insurance which was effected between them. The doctrine is not confined to insurance, but extends in equity to many other cases, e.g., where money has been borrowed without authority but applied in payment of existing debts; in such cases the quasi-lender is entitled to stand in the shoes of the creditors thus paid. *See* **Letter of Subrogation**.

Sub rosa *Latin*—In strict confidence; Between ourselves; Secretly; Confidentially.

Sub. rts. Subrogation rights. *See* **Subrogation**.

Subscribed Capital Paid up or issued capital of a company.

Subsidiary Company A company which is managed by a parent or another company. *See* **Holding Company**.

Subsidy A parliamentary grant of money to the sovereign for state needs; a tax levied on a particular occasion; a money grant from one state to another in return for military or naval aid or other equivalent; money contributed by a state to keep down the price of commodities (food, housing) or to expenses of commercial undertaking, charitable institution, etc., held to be of public benefit.

Subsonic Slower than the speed of sound *Opposite to* Supersonic.

Sub specie *Latin*—Under the appearance of.

Substituted Expenses Expenses incurred in place of loss or expense which would be allowed as general average (e.g. cost of removal of a ship, with general average damage, to a place where repairs would cost less).

Sub.U/rs.App. Subject to Underwriters' Approval.

Sub voce *Abbrev.* s.v., *q.v.*

Sucre Unit of currency used in Peru and Equador.

Suct. Drg. Suction Dredger.

Suction Boxes *Similar to* **Sea Chests**, *q.v.*

Sue and Labour Charges Expenses incurred short of destination by a marine assured to prevent or minimise loss for which underwriters would have been liable.

Sue and Labour Clause A clause in the SG Policy, *q.v.*, which, as a supplementary contract, entitles the assured to claim expenses incurred in preserving, or attempting to preserve, the insured property from loss which would be recoverable under the policy (the Sue and Labour Clause was abolished in cargo policies, with the passing of the SG policy form in 1982).

Suf *or* **suff** Sufficient; Suffix.

Sufferance Wharf A place for landing cargo, apart from a congested port's legal quays, which was licensed for the clearance of imports.

Suggestio falsi *Latin*—An indirect lie.

Sui generis *Latin*—Peculiar; Of its kind.

Sui juris *Latin*—Of full age and capacity.

Suitcasing Anchors The lifting of the Anchors, *q.v.*, of a pipe-laying barge using the power of the anchor handling tug in attendance.

Sullage Garbage landed from ships.

Summer Solstice The longest day of the year, when the sun reaches its maximum altitude from the equator.

Summons A call of authority, an admonition to appear in court, a citation. A summons is a document issued from the office of a court of justice, calling upon the person to whom it is directed to attend before a judge or officer of the court for a certain purpose.

Suomen *Finnish*—Finnish.

Suomi *Finnish*—Finland.

Sunk Costs Past costs which are beyond recovery and therefore not relevant to current decisions.

SUNS Sonic Underwater Navigation System.

SUNY *or* **S.U.N.Y.** State University of New York.

Sup. Supra *Latin*—Above; Superior.

Super *or* **Supt.** Superintendent, *q.v.*

Sup. Ct Superior Court; Supreme Court.

Superannuation Payment A reduction from salary or wages in order to provide a pension in retirement.

Superficial Foot A timber measurement calculated on a piece of wood measuring 12 ins. × 12 ins. × 1 in. thick.

Superheat Steam at a temperature above that of Saturated Steam. *See* **Steam**.

Super-het Super-heterodyne *radio receiver electronics.*

Super High Frequency Communications radio frequency in the range 3 to 30 Ghz.

Superintendent An expert generally engaged in the supervision of work on a ship concerning (1) General Maintenance—General Superintendent. (2) Stores and Provisions—Victualling or Catering Superintendent. (3) Engine maintenance—Engineer Superintendent. (4) Hull and Deck Superintendent or Port Captain. There are some companies that use the word Inspector. *Abbrev.* Supt. or Super.

Superior Planets Mercury, Venus, Earth, Mars, Jupiter, Saturn, Uranus, Neptune, Pluto.

Supersonic Speed faster than that of sound.

Supersonic Flight Flight with a speed of the aircraft greater than the speed of sound—approx. 670 miles per hour. *See* **Mach Number**.

Superstructure That part of a ship which is built on top of the Upper Deck, *q.v. Abbrev.* Super.

Supp. Supplement.

Suppressio veri *Latin*—Suppression of the truth. Misrepresentation by concealing part of the facts.

Supra *Latin*—Above. This word occurring by itself in a book refers the reader to a previous part of the book, like *ante*.

Supra Protest On protest. There may be either acceptance or payment of a bill of exchange by a person other than the drawee or acceptor or other person liable, after it has been protested for non-acceptance or non-payment. The full term is 'acceptance (or payment) supra protest for honour', i.e. for the honour or in relief of the person liable. The rights and liabilities of the parties are regulated by the Bills of Exchange Act 1882.

Supreme Court of Judicature The court formed by the Judicature Acts, 1873–75, in substitution for the various superior courts of law, equity, admiralty, probate and divorce then existing, including the Court of Appeal in Chancery and Bankruptcy, and the Exchequer Chamber. It consists of two principal divisions, *viz.*, a court of original jurisdiction, called the High Court of Justice, and a court of appellate jurisdiction, called the Court of Appeal. Its title 'supreme' is a misnomer, as the superior appellate jurisdiction of the House of Lords and Privy Council, which was originally intended to be transferred to it, has been allowed to remain.

Supt. Superintendent, *q.v.*

SUR Surgery *insurance.*

SURANO Surface Radar and Navigation Operation.

Surcharge An extra charge imposed to cover insufficient or extraordinary expenses. *Ex.* A liner company makes a surcharge on the normal freight from London to a Mediterranean port due to congestion and heavy port expenses.

Surety Guarantee, *q.v.*

Surety Company *Similar to* **Bonding Company**, *q.v.*

Surface Active Agent Generally a Detergent, *q.v.*, which is added to water for cleaning purposes.

Surface Carrier A road transport service connected with the transportation of air cargo to and from airports.

Surface Freight Sea freight.

Surface Mail Mail sent by land or sea as opposed to airmail.

Surf Days *or* **Surfdays** In a charterparty where loading/discharging is prevented by heavy swell or surfs the Charterer, *q.v.*, inserts this expression to cover himself when these days occur in order to be excluded from Laytime or Laydays, *q.v.*

Surf Port *Similar to* **Craft Port**, *q.v.*

Surg. Surgeon.

Surge (1) To ease or slacken a wire or rope round a Capstan, *q.v.*, bollard or bitts. (2) Impulse or movement of the swell or a large wave. (3) The fore-and-aft movement of a vessel under the pressure of heavy swell

particularly when moored in a harbour so exposed. *See* **Send**. (4) Momentary acceleration of the engines as wave motion eases the water resistance to the hull or propeller(s). (5) The difference in height between predicted and observed tides due to abnormal weather conditions.

SURIC Surface Ship Integrated Control System.

Surrender Value The cash price which an insurance company will pay for the surrender by the Policy Holder, *q.v.*, to the company of the policy and all claims thereunder. *See* **Cash Surrender**.

SURTASS Surveillance Towed Array Sensor *USA.*

SURV Standard Underwater Research Vessel.

Survey To view and examine the condition of an object and determine the recommendations needed, if any, to update its standard according to the classification needed by the authorities or any authorised corporation. *See* **Surveyor** *also* **Surveys**.

Surveyor A person employed or engaged by a shipping association or private person to inspect cargoes, ships, etc. He may be a Lloyd's Register surveyor who undertakes to inspect ships and issue the appropriate classification certificates. He may also be asked to inspect and report on a damaged ship or a ship for delivery and redelivery in a Time Charterparty, *q.v.*

Surveys The act of examination by an independent or impartial group of persons on behalf of others. Surveys may be effected on request by insurance companies when insurance claims are submitted for damaged cargoes or ships. Surveys are also held by the Classification Societies, *q.v.*, on ships. Vessels are to be surveyed every year and a more detailed inspection and report is done every four years and called a Special Survey, *q.v. See* **Survey** and **Surveyor**.

Survey, Veto and Tender Clause A Marine Insurance, *q.v.*, clause found in hull policies extending the conditions of the Tender Clause. Underwriters have the full right of deciding if, when and where to proceed to dry dock and repair the damage claimed by the assured. They have also the option to invite any number of tenders or deal directly and immediately with the repair work.

Susp. Suspended.

Suspected Bill of Health When a contagious disease exists or is suspected among the passengers, officers or crew of a ship, the port authorities will produce a Suspected Bill of Health instead of a clean one before departure. It can also be termed **Foul Bill of Health**, *q.v.*

SUT Society of Underwater Technology.

Suum cuique *Latin*—To each his own.

s.v. Sub voce *Latin*—Under a word or heading, as in a dictionary; Sailing Vessel.

SV Safety Valve; Saudia, Saudi Arabian Airlines *airline designator*; Stop Valve.

S.V *or* **s.v.** *or* **S/V** Sailing Vessel.

SVA Shareholder Value Analysis.

Sveriges Allmänna Sjofartsforening Swedish General Shipping Association.

Sveriges Redareförening Swedish Shipowners' Association, Stockholm.

Sveriges Skeppshandlareförbund Swedish Ship Suppliers Association. Members of the International Ship Suppliers Association.

Sveriges Skeppsklarerare-Och Skeppsmäklareförening Swedish Shipbrokers' Association.

SVP or **S.V.P.** Saturation Vapour Pressure; S'il Vous Plaît *French*—If you please.

SVPT Sound Velocity Pressure and Temperature.

SW or **S.W.** Salt Water; Shippers' Weight; South West; South Westerly; Steel Wire; Snow Showers *weather observation*; South Wales; Sweden.

S.W. or **S/W** Sea Water; Seaworthy; Seaworthiness.

Sw or **Swed** Sweden; Swedish.

SWA or **S.W.A.** South West Africa.

SWACS Space Warning and Control System.

SWAD or **S.W.A.D.** Salt Water Arrival Draft (Draught).

SWAG Standard Written Agreement *USA*.

Swaged Plate Plate which is corrugated to make it stronger than flat plate.

Swallow The cavity between the cheeks of a block which takes the sheave.

Swap Facilities The Bank of International Settlement based in Basle, Switzerland, acts as an intermediary for members of its association in the event of a bank seeking aid from another bank. The swap may concern bullion, securities, francs, etc.

SWAPO South West Africa People's Organisation.

SWASH Small Waterplane Area Single Hull *ship*.

SWATH Small Waterplane Area Twin Hull *ship*.

SWB or **S.W.B.** Short Wheel Base.

Swbd. Switch Board.

SWBW or **S.W. by W.** South West by West.

SWBM or **S.W.B.M.** Still Water Bending Moment.

SWD or **S.W.D.** or **s.w.d.** Sea Water Damage; Sea Water Draft (Draught).

Sweep (1) A long oar used for steering a craft in the absence of a rudder, for manoeuvring barges and propelling becalmed sailing vessels. (2) A curve in a vessel's hull shape or lines. (3) The conduct of a search for something, afloat or submerged, e.g. survivors, wrecks, mines, submarines, enemy forces.

SWG or **S.W.G.** Standard Wire Gauge.

SWID or **S.W.I.D.** South West India Dock.

SWIFT or **S.W.I.F.T.** Society for Worldwide Inter-Bank Telecommunications, *q.v.*

Swinging a Ship Turning a ship at rest and holding her on several headings while correcting her magnetic compass for Deviation, *q.v.*

Swinging Derrick A system of lifting cargo from or into the Hold, *q.v.*, of a ship with the use of one derrick. The Derrick, *q.v.*, is made to swing by pulling the Guys, *q.v.*, attached to it to the required direction and position.

Swissair National Swiss Airlines.

Switching Shifting or changing over from one investment to another because it is convenient to do so. Such as from stocks to shares and vice versa or from equities to gilt-edged securities.

Switz. Switzerland.

SWL or **S.W.L.** Safe Working Load, *q.v.*, referring to load on the lifting gear.

SWOB Sewage Waste Offloading Barge; Ships' Waste Offloading Barge US barge used by US Navy.

SWOPS Single Well Oil Production System, *q.v.*

SWORD Shallow Water Oceanographic Research Data System *USA*.

SWP Safe Warning Pressure; Sea Water Pump.

SWPS Submerged Wellhead Platform System *oil drilling*.

SWR or **S.W.R.** Steel Wire Rope.

SWS or **S.W.S.** Static Water Supply; Special Wording Scheme, Lloyd's Policy Signing Office system, *q.v.*

Swt. dge. Sweat Damage.

Swtz. Switzerland.

Sx. Sussex, *UK*

SXT or **SEXT** or **Sext** Sextant, *q.v.*

S.Y. or **s/y** Sailing Yacht; Steam Yacht.

S.Yd. Scotland Yard.

Syli Currency unit of Guinea.

Syll. Syllable; Syllabus.

Sym. Symbol.

Symp. Symposium.

Sympathetic Damage Loss suffered by cargo following damage to other goods in the same ship. An example would be taint arising from odour given off by another cargo which has been damaged by seawater.

Sympiesometer 19th century oil barometer used at sea.

Syn. Synonym; Synonymous; Syndicate.

Syndic An appointed person acting on behalf of a corporation or a group in certain matters.

Syndicate An association of two or more persons, constituted for carrying out a projected commercial or public undertaking.

Syndicate (Lloyd's) A group of underwriting members at Lloyd's whose acceptances and liabilities are handled jointly by an underwriting agency acting on their behalf, while each member remains legally liable solely for his/her own share of the syndicate's liability.

Synjet Fuel made from coal or oil shale which is expensive to manufacture.

Synop Synopsis.

Synoptic Charts Weather Maps. See **Weather Stations**.

Syn sht Synopsis Sheet.

Synth. Synthetic.

Synthetic Fibres A generic word for nylon, terylene, etc., materials produced from chemicals, mineral sources and general petroleum products. They are used for durability and lightness along with relative economy.

Syst. System.

Systems Engineering Associates Corporation US navy engineering services corporation which employs over 800 people in 13 offices nationwide. *Abbrev.* SEACOR.

Système International d'Unités A coherent system of scientific units based on the metre, kilogram, second, ampere, Kelvin, mole, and candela, with supplementary units radian and steradian; adopted by the General Conference of Weights and Measures (CGPM) and endorsed by the International Organisation for Standardization (ISO). *Abbrev.* SI Units.

Syzygy Conjunction or opposition *astronomy*.

Sz. Size.

T

t Dry Bulb Temperature *nautical*; Thunderstorm *weather observation*; Tonneau *French*—Ton, *q.v.*; Tun, *q.v.*; Thickness; Troy.

t¹ Wet Bulb Temperature.

T Air temperature correction *altitude*; Table; Temperature; Time; *Lloyd's Register notation* for Triple Expansion Engine; Meridian Angle; Tonneau *French* —Ton, *q.v.*; Thickness; Towards *altitude difference*; Top; Tanker; Tropical Load Line; True direction *navigation*, Tesla, mag.flux density; Tera (10^{12}); Prefix.

T1 *and* **T2** Community Transit Declaration *documents*.

T2 A type of US tanker built during the 1939–45 war with speed and efficiency to help the Allied war effort. Most of the vessels, if not all, have now gone for demolition.

Ta. Tank(s), *Lloyd's Register notation*.

TA *or* **T.A.** Table of Allowance; Telegraph Address; Telegraphic Address; Territorial Army.

T/A *or* **t/a** Trading as; Thence Annually *insurance*.

t.a. Take Action.

TAA *or* **T.A.A.** Test of Academic Aptitude; Trans-Australian Airlines; Transportation Association of America.

TAAG Tropical Africa Advisory Group.

TAB Technical Assistance Board *United Nations*.

TABA *or* **T.A.B.A.** Timber Agents and Brokers Association.

Tabernacle Structure on the deck of a ship which supports the heel of a mast in such a way that the mast can be hinged up and down thus permitting the ship unhindered passage under bridges.

Tabulate To arrange or display or compute data in the form of tables for ease of comprehension.

TAC Tanker Advisory Center, New York, *USA*; Total Allowable Catches, relative to the catch in deep sea fishing.

TACA Trans-Atlantic Conference Agreement.

TACAN *or* **Tacan** Tactical Air Navigation, *q.v.*

Tach. Tachometer, *q.v.*

Tachograph A device for recording the speed and travel time of a vehicle.

Tachometer A mechanical apparatus to find out the shaft revolutions per minute of a ship. *Abbrev.* Tach.

Tackle Mechanism, especially of ropes, pulley-blocks, hooks, etc., for lifting weights, managing sails

or spars, etc.; a windlass with its ropes and hooks; requisites for a task or sport, such as fishing.

TACMAR Tactical Multi-Function Array Radar *USA*.

TACS Tactical Air Control System.

TACTAS Tactical Towed Array Sonar Systems *electronics*.

Tactical Air Navigation A device used in Distance Measuring Equipment (DME) to find out the time pulse travelling from radio beacon to the aircraft apart from the bearing information it provides. *Abbrev.* TACAN *or* Tacan.

TACV Tracked Air Cushion Vehicle *hovercraft*.

Tael Chinese unit of weight for precious metals, equal to 37.429g.

TAF *or* **T.A.F.** Terminal Area Forecast *aviation*; Time and Frequency.

Taffrail A rail around the stern of a ship. The uppermost deck area at the stern of a ship.

Tag A label of suitable material with the name and address of a passenger attached to his luggage as a means of identification in case it is lost or left unattended.

TAI Taxpayers' Association International.

Tai. Taiwan.

Tail-lift Rear platform mechanism of heavy goods vehicles to assist lifting or unloading heavy items.

Tail Light A rear light on a vehicle or train.

Tail Series An insurance term referring to the last packages to be discharged which total an insufficient number to reach the agreed 'series' for application of a franchise or deductible.

Tail Shaft The extreme section at the aft end of a ship's propeller shaft. *Abbrev.* TS.

Taint Cargo damage sustained by strong or pungent smells emitted by other adjacent cargo stowed in the Holds, *q.v.*, of a ship or warehouse. *See* **Cargo Worthiness**.

Taka Unit of currency used in Bangladesh. One hundred paisa equals one taka.

Take-Off-Weight The weight of an aircraft inclusive of the fuel. *Abbrev.* TOW *or* T.O.W.

Take-Off The lifting into the air by an aircraft after accelerating along a runway.

Take-Over Bid When a company or individual offers to buy shares or part of the shares of another company to enlarge production and to gain the controlling interest, thus being in a position to monopolise the entire business organisation.

Taker-In Opposite to **Giver-on**, q.v.

Take Ullage or **Taking Ullage** Obtaining the measure of tanks in oil tankers by means of Ullage, q.v.

T.A. Round Transatlantic round voyage.

Takings at Sea The act of commandeering a neutral ship at sea during a war, it being suspected of having objects detrimental to the nation at war.

TAL or **T.A.L.** Traffic or Accident Loss.

Tale quale This signifies that goods are as sample when shipped, the liability for loss or deterioration in transit being the responsibility of the buyer.

Talis qualis Latin—Such as it is.

Tallage Taxes generally. Tall(i)age was a form of taxation abolished in the 14th century.

Tall-oil Product extracted from pine woodpulp.

Tallyman (1) Same as Tally Clerk, q.v. (2) One who sells on credit by instalments, or keeps a tally-shop, i.e. both seller and buyer keep accounts which agree, or tally.

Tally Trade Items given on credit or by instalments.

Talon A slip of paper attached to share certificates which can be used at a later date, or when necessary, to apply for further dividend coupons or shares from the company.

TALUS Transportation and Land Use Study.

Tambala Currency unit of Malawi equal to one hundredth of a Kwacha.

Tan or **tan.** Tangent.

T and/or C and/or TLOA Total and/or Constructive and/or Total Loss Only Arranged insurance.

T and/or CTL Total and/or Constructive Total Loss insurance.

T and/or C and/or ATLO Total and/or Constructive and/or Arranged Total Loss Only insurance clause.

Tango International Radio Telephone phonetic alphabet for 'T'.

Tanker Owners' Voluntary Agreement Concerning Liability for Oil Pollution An agreement subscribed by most tanker owners worldwide, whereby owners agree to pay for the clean-up costs incurred by governments in respect of oil discharge from tankers belonging to such owners. Abbrev. TOVALOP.

Tanker Structure Cooperative Forum A group of oil companies, tanker owners and classification societies involved in aspects of tanker design.

TANKERVOY 87 INTERTANKO Voyage Charterparty.

Tank Farm An area enclosing a number of tanks for the storage of petroleum products or chemicals.

Tank Tops The bottom of a bulk carrier's or tweendecker's hold above the double bottom tanks. See **Ceiling**.

Tank Top Dimensions The inner length and breadth of the tank tops used for planning a stow of non-bulk cargo—e.g. steel sheets.

TANS Terminal Area Navigation System.

Tant pis French—So much the worse.

Tanz. Tanzania.

TAO or **T.A.O.** Technical Assistance Operations.

TAP Total Action Against Poverty; Transportes Aereos Portugueses Portuguese—Portuguese Air Lines; Technical Assistance Program USA.

Tapline Trans-Arabian Pipeline, in Saudi Arabia.

T&P or **t and p.** Theft and Pilferage insurance.

TAPCODE Turbocharging and Process Calculation of Diesel Engine.

TAPS Trans-Alaska Pipeline System; Turret Anchored Production System, q.v.

Tar A very heavy viscous black material extracted from wood, coal, etc., by means of distillation. Used for deck caulking, sealing, a preservative of timber and iron, etc. Tar-macadam is a mixture of tar and stone or slag used for road surfacing. See **Caulk**.

Tare and Tret The weight of a container (Tare) and the weight of wastage of the contents (Tret).

Target Risk An insurance policy on a large risk accepted and shared by several Underwriters, q.v.

Tarpaulin Waterproof material, generally of tarred canvas, to cover ships' hatches, cargo, decks and other open spaces to prevent them from getting wet from sea and rain spray or to protect them from the hot rays of the sun.

TAR/V or **T.A.R./V** Trans-Atlantic Round Voyage charterparty.

Tas Tasmania.

TAS Telecommunications Authority of Singapore; Torpedo Anti-Submarine naval branch; Towed Array Sonar naval; True Air Speed.

T&S or **TANDS** or **t and s** Touch and stay, q.v.

TASC Technical Advisory Sub-Committee.

TASS or **T.A.S.S.** The official news agency of the USSR.

TAT Thematic Apperception Test; Thrust Augmented Thor NASA; Transitional Automated Ticket;

T/Atl. Trans-Atlantic.

TAV or **T.A.V.** Tax on Added Value.

TAWB Through Air Waybill.

Tax Credit Tax deducted from Dividends before payment which is credited against income tax due for the year UK.

Tax Reliefs Income Tax Allowances; Non-taxable Income.

Tax Reserves Amounts of money put in reserve by a company to meet tax demands at the end of the financial year.

TB or **T.B.** or **t.b** Torpedo Boat; Trial Balance; Treasury Bill, q.v.; Atmospheric Pressure Correction altitude; Air Temperature; True Bearing nautical.

T&B Top and Bottom.

T.B.A. *or* **t.b.a.** To be agreed; To be arranged; To be announced; To be advised.

T.B.A.F. *or* **t.b.a.f.** To be accounted for.

TBC Tug Barge Combination.

T.B.D *or* **t.b.d** To be determined.

TBL Test Breaking Load.

T.B.L *or* **T.B/l** Through Bill of Lading.

TBM Tunnel Boring Machine.

TBN *or* **T.B.N.** To be named/notified; Total Base Number, *q.v.*

TBO *or* **T.B.O.** Time Between Overhauls *aviation*.

TBR *or* **T.B.R.** To be reported.

TBRN To be re-named.

TBS *or* **T.B.S.** Talk Between Ships *by radio*; Tubeshaft.

TBT Tributyl *ingredient in toxic paints*.

TC *or* **T.C.** Tariff Commission *USA*; Till Countermanded; Time Charter; Total Correction *nautical*; Transcontainer; Travellers Cheques; True Course *nautical*; Temperature Control.

TC Traders Combined *insurance*.

t.c. Temperature Control.

Tc Tierce, *q.v.*

T/C Till Countermanded; Time Charter, True Course *nautical*.

TCA Terminal Control Area *aviation*.

TCAS Traffic Alert and Collision Avoidance System *aviation*.

TCATLVO *or* **T.C.A.T.L.V.O.** Total or Constructive or Arranged Total Loss of Vessel Only.

TCB Testing and Certificate Branch, of the National Institute for Occupational Safety and Health at Morgantown, West Virginia, *USA*.

TCBM Transcontinental Ballistic Missile.

TCC Total Crew Concept *crew agreement on foreign ships* negotiated with the ITF.

TCC *or* **T.C.C.** Transport and Communications Commission *United Nations*.

T. & C.C. Technical & Clauses Committee.

TCD Tramp Chartering Documentary.

TCDC Technical Co-operation amongst Developing Countries.

Tce. Tierce, *q.v.*

TCE *or* **T.C.E.** *or* **t.c.e.** Ton of Coal Equivalent.

TCF Total Currency Flow.

TCFB Tanzania Central Freight Bureau.

TCH *or* **T.C.H.** Time Charter Hire.

TCI *or* **T.C.I.** Temporary Customs Import; Time Charterers' Interest;

TCL *or* **Trik** Trichloroethylene, for drycleaning and degreasing purposes.

TCO *or* **T.C.O.** Test Control Officer.

TCP Tubing-Conveyed Perforation *oil drilling*.

TCPA Target or Time of Closest Point of Approach *nautical*. *See also* **CPA**.

TCR *or* **T.C.R.** Timber Cargo Regulation.

TCV *or* **T.C.V.** Temperature Check Value; Temperature Control Value.

TCXO Temperature Compensated Crystal Oscillator.

TD *or* **T.D.** Tilbury Docks *UK*; Treasury Department; Traffic Director.

TDA Transport Distribution Analysis.

TDC Technical Development Corporation; Top Dead Centre; Total Distribution Cost *aviation*.

TDCC Transportation Data Co-ordinating Committee.

TDDL Time Diversion Data Link.

TDED Trade Data Elements Directory.

TDG *or* **T.D.G.** Twist Drill Gauge.

TDI Trade Date Interchange.

T.Dks Tween Decks.

TDM Time Division Multiplexing *radio communication*.

TDMA Time-Division Multiple Access System *electronics*.

TDR *or* **T.D.R.** Treasury Deposit Receipt.

t.d.r. Tous droits réservés *French*—All rights reserved.

TDRS Tracking and Data Relay Satellite.

TDS Tactical Data System; Total Dissolved Solids—This refers to the Reverse Osmosis Purification Plant System.

TDW *or* **T.D.W.** *or* **tdw** *or* **t.d.w.** Total Deadweight; Tonnage Deadweight, *similar to* **Total Deadweight Capacity.**

TDWAT *or* **Tdwat** *or* **T.D.W.A.T.** *or* **t.d.w.a.t.** Tonnage Deadweight All Told or Total Deadweight All Told, *similar to* **TDW**, *q.v.*, also *similar to* **Deadweight**, *q.v.*

TE *or* **T.E.** Telecommunications Engineer; Telecommunications Engineering; Trade Expenses, *q.v.*; Task Element *USA*.

TE Triple Expansion *engine*; Turbo-Electric; Air New Zealand *airline designator*.

t.e. *or* **t/e** Thermal Efficiency; Time Expired; Twin Engine; Turbine Engine.

T & E Telecommunications and Electronics; Test and Evaluation.

TEA Triethanolamine *cleaning solution for oil*.

TEC Transport Europeéns Combinés *French*—European Combined Transport.

TECE Trans-European Container Express, one of the container services available in Europe.

TECH Cargo *or* **T.E.C.H. Cargo** Toxic, Explosive, Corrosive or Hazardous Cargo.

TECHEVAL Technical Evaluation *US Navy*.

Technical and Clauses Committee A group of Marine Underwriters, *q.v.*, Lloyd's and company underwriters who meet to discuss, formulate and amend marine insurance clauses which are then recommended to the London market for general use.

TECO Technical Co-operation Committee.

TEDIS Trade Electronic Data Interchange System.

TEE *or* **T.E.E.** Trans-Europe Express, a train service connecting principal cities in Europe. *See* **TEEM**.

TEH Twin-engined helicopter.

TEEM Trans-Europe or European Express Merchandise. One of the categories of container services in Europe.

TEFRA Tax Equity and Fiscal Responsibility Act *USA*.

Teh. Teheran, Iran.

Te judice *Latin*—You being the judge; In your opinion.

T.E.U. Twenty-foot Equivalent Unit container.

Tel. Telegraph; Telegram; Telephone.

Tel add Telegraphic Address, *q.v.*

Telecon Telephone Conversation.

Telegraph Buoy A buoy at sea marking the position of a submarine telegraph cable.

Telegraphic Address *or* **Cable Address** An abbreviated or other registered address for use in telegrams.

Telematics Long distance transmission of computerised information.

Telemeter To record readings of an instrument at a distance, usually by means of radio devices.

Telemex *or* **TELEMEX** Telefonos de Mexico SA *Telephone Monopoly of Mexico*.

Telemotor Machinery which operates the steering system of a ship.

Telephone Answering Service A recording system connected to a telephone which answers the callers when the subscribers are away. It relays a message that no one is present and invites the caller to leave any messages for a reply when someone returns.

Teleprinting Over Radio A marine communication system linking ship at sea with any telex terminal anywhere in the world. *Abbrev.* TOR *or* T.O.R.

Telex A system of telegraphy in which printed signals or messages are exchanged by teleprinters connected to a public telecommunication network. *Abbrev.* Tx.

Teller Any of four persons appointed (two for each side) to count votes in the House of Commons; a person appointed to receive or pay out money in a bank, etc.

Tell–Tale (1) Device near the steering wheel of a ship, a semi-circular indicator showing the Rudder's *q.v.*, angle (2) A compass repeater in the Captain's Cabin to show the ship's heading. (3) Mechanical means of communicating between the bridge and the engine room.

TEM Transient Electromagnetic; Transmission Electron Microscope *electronics*.

tem. Temperature; Template, *q.v.*; Tempore *Latin*—In the time of; Temporary.

Template A pattern cut out of metal or other material to the required shape for a flat object; timber or a plate used to distribute weight in a wall or under a beam, etc.; a wedge for a building block under a ship's keel. *Abbrev.* tem.

Temporary Import Permit A permit obtained from the customs authorities to enable essential articles for personal use, e.g. tools, samples, vehicles, personal effects etc., to be imported on a temporary basis. These are allowed entry into the country subject to certain conditions laid down by the customs and to their being re-exported within a prescribed time. *Abbrev.* TIP. *Same as* **Temporary Import Licence**.

Tempori parendum *Latin*—One must yield to or move with the times.

Tempus omnia revelat *Latin*—Time reveals all.

TEMPSC Totally Enclosed Motor-Propelled Survival Craft.

Tenge Currency unit of Kazakhstan.

Tenn. Tennessee, *USA*.

Tenor *or* **Usance Bill** A Bill of Exchange, *q.v.*, payable on a date fixed by usage or habit of dealing.

Tensile Strength The capability of metals being subjected to stress and stretch. Generally referred to wire, synthetic and other ropes as well as steel sheets. *Abbrev.* T/S *or* TS *or* T.S.

Tensile Test Testing an allotted cross-section of metal which is subjected to a regular increase of tension up to the point of fracture.

Tension Leg Platform/s A type of platform having a system of anchoring by the vertical movement of the bottom of the platform itself. *Abbrev.* TLP.

Tension Motion Compensator A piece of equipment installed at each corner of an oil platform in the mooring compartments to heave up and down in compensated motions until they are safely locked. *Abbrev.* TMC.

Tenure The act or right to hold or possess property.

Terbapas *Indonesian*—Limited.

Terebene A disinfectant liquid obtained from turpentine and used as a paint thinner.

Teredo A ship-worm or mollusc that bores into wooden hulls.

Term Insurance time cover; an insurance policy period of 12 months in contrast to Short Term Policy, *q.v.*

Term Draft A Bill of Exchange, *q.v.*, payable on a credit term of a specified period after sight.

Terminal A station or terminus with facilities for passengers and freight, and with road and rail communications. The customs authorities are part of its organisation.

Terminal Benefits *Similar to* **Golden Handshakes** and **Redundancy Payment**.

Terminal Markets *Similar to* **Future Markets**, *q.v.*

Terminal Operator A port authority or port management in charge of a Container Terminal, *q.v.*

Terminal Storage Storage facility for containers, trailers or other cargoes.

Terminal Through-Put A measure of all movements through a Terminal, *q.v.*

Term Insurance Also called Level Temporary Insurance. As the name implies it is an Insurance, *q.v.* effected for a short period. It could be for a week or so to cover a trip or for a period of time while working at a risky job.

Terminus ad quem *Latin*—A legal expression meaning the finishing point. In the Voyage Policy, it refers to the termination of the risk at the place of destination as opposed to terminus a quo, *q.v.*

Terminus a quo *Latin*—The starting point. The beginning of an insurance risk in a voyage policy as opposed to terminus ad quem, *q.v.*

Term of a Bill The period of time at which a Bill of Exchange, *q.v.*, etc., is due.

Terms and Conditions A common terminology in the shipment of merchandise. These are given by the shippers to the receivers laying down their mutual general acceptance. The word Term represents the general procedure of the shipment, say CIF or FOB. Conditions concern the kind of payment, such as 'Cash Against Documents', Letter of Credit, or payment After 30 Days.

Terms of Credit Conditions of payment against credit.

Terms of Trade The relative position between the export and the import prices of a country.

Terns Trans-European Networks, *airline coordination.*

Terra incognita *Latin*—Unexplored region.

Terrestrial Pertaining to the earth, as opposed to celestial.

Territorial Waters That portion of the sea, up to a limited distance, which is immediately adjacent to the shores of any country, and over which the sovereignty and exclusive jurisdiction of that country extend. The generally recognised limit is three miles, which was the range of cannon in the 17th century, but of recent years certain countries have claimed greater distances. Territorial waters are considered as territory to the extent that fishing in such waters is reserved for the exclusive benefit of the subjects of the adjacent country. *See* **Hot Pursuit**, **Freedom of the Seas**, **Mare Clausum**.

Tes. Tierces. *See* **Tierce**.

Test. Testament; Testimony; Testator.

Tête-à-tête *French*—Confidential conversation; Private interview.

Tetralin A colourless liquid being a substitute for turpentine.

Tetsudo *Japanese*—Ship Transport.

TEU Twenty-foot Equivalent Unit, *q.v.*

Tex. Texas, USA.

Text. rec. Textus receptus *Latin*—The received text.

TF *or* **T.F.** Tropical Fresh Water, Loadline marking.

TFA Total Flow Area.

t.f.a. Total Fatty Acids.

tfd. Transferred.

TFL Through Flowline *oil drilling*.

Tfr. Transfer.

TG Task Group *USA*; Thai Airways International airline designator.

Tg. Ground Wave Reading *nautical*; Tail Gear.

t.g. Type Genus; Tail Gear.

T&G Tongued and Grooved, *q.v.*

TGB *or* **T.G.B.** Tongued, Grooved and Beaded *in timber trade. See* **Tongued and Grooved**.

TGED Tarif general europeén pour les expéditions de detail *French—Similar to* **EPLT**, *q.v.*

TGH *or* **T.G.H.** Tongued, Grooved and Headed, *in timber trade. See* **Tongued and Grooved**.

Tgs. Ground Wave Reading *nautical*; Sky Wave Reading *weather observation*.

TGWU *or* **T.G.W.U.** Transport and General Workers Union.

TH *or* **Th.** True Heading *nautical*; Thermal.

TH *or* **T.H.** Trinity House; True Heading *nautical*.

Th Thorium; Thursday.

Thai. Thailand.

Thames Division, Metropolitan Police The first uniformed police force in the world, formed in 1798 as the Marine Police to control the pilferage of cargo in the port of London, using oared galleys and a sailing cutter. After some years, it acquired three warship hulks as stations in the river. The force became part of the Metropolitan Police in 1839 and by the end of the 19th century acquired steam launches. It now has operational bases ashore and a fleet of patrol craft to handle every sort of crime or civil emergency on the tidal Thames.

Thames Measurement *or* **Thames Measurement Tonnage** A tonnage measurement for unregistered vessels which is needed for port dues purposes and the average cost calculation for building yachts.

THB *or* **THV** Trinity House Boat or Trinity House Vessel.

Th.B. Theologiae Baccalaureus *Latin*—Bachelor of Theology.

THC Terminal Handling Charges.

Th.D. Theologiae Doctor *Latin*—Doctor of Theology.

THE Technical Help to Exporters *division of BSI*.

The American Club *See* **American Steamship Owners Mutual Protection & Indemnity Association Inc. (The)**

The Bends *Similar to* **Staggers**, *q.v.*

The Line Seafarers' term for the Equator, as in 'Crossing the Line'.

The Malta Group *See* **Malta Group**.

The Near Clause *Similar to* **As Near As She Can Safely Get**, *q.v.*

Theodolite A surveying instrument for measuring horizontal and vertical angles.

Therm A statutory unit of calorific value in gas supply. A British thermal unit (B.Th.U.) is the amount of heat required to raise 1lb. of water at maximum density through 1 degree Fahrenheit.

Thermit (Thermite) A mixture of finely powdered aluminium and oxide of iron that produces a very high temperature on combustion. Used in welding and for incendiary bombs.

Thermoelectricity Electricity generated by friction of metals.

Thermograph A rotating recording instrument which automatically registers various temperatures on a graph.

Thermology The science of heat.

Thermometer A glass pencil-like instrument for measuring temperatures.

Thermomotor A motor which is made to function by means of heat or hot air.

Thermoscope An instrument to detect temperature variations without actual measurements.

Thermostat An instrument which automatically controls temperature in incubators, refrigerators, stoves, vehicles, hot water tanks, kettles, irons, etc.

thes Thesis.

THHM *or* **T.H.H.M.** Trinity House High Water Mark.

THI *or* **T.H.I.** Temperature Humidity Index; Time Handed In.

t.h.i. Time Handed in.

Thick Weather Weather conditions which cause extremely poor visibility due to fog, mist, rain, hail, etc.

Thieves A shipping word for robbery at sea with violence. In Marine Insurance, *q.v.*, theft from ship's crew members as well as from passengers is not covered.

Thieving A term used in connection with oil tankers. It is the quantity of water present underneath the oil in a cargo tank. *See* **Thieving Paste**.

Thieving Paste A chemical grease smeared on sounding tapes, rods, etc., to determine the presence of liquids or the levels of oil and water in oil cargo tanks. The thieving paste changes colour when it touches water. *See* **Thieving**.

Thin Market A term used when stock market trading is slack. This could give rise to sharp quotations.

Thinner A diluent for paint, such as turpentine, white spirit or acetone.

Third Carrier IATA term for the third aircraft which carries passengers or cargo. *See* **Second Carrier**.

Third Class Paper A Bill of Exchange, *q.v.*, graded as third class because of its lesser importance or reputation and popularity in the buying and selling market.

Third Flag US state-owned shipping lines are authorised to cut their Freight Rates, *q.v.*, to compete with the Third Flag controlled lines of all Communist and Socialist states.

Third-Liner Small company quoted on the Stock Exchange.

Third Party *or* **Third Person** A third party or person in a contract. An insurance term signifying another person as distinct from the insured party. A common term in motor vehicle insurance which has become compulsory in nearly all countries to cover any damage and injuries sustained by third parties, especially when the insured person concerned inflicted the damage and at the same time is at fault. *See* **Third Party Insurance**.

Third Party Insurance Insurance covering damage and expenses incurred through the fault of one party or person. *See* **Third Party Liability**.

Third Party Liability Legal liability to anyone other than another party to a contract, e.g. liability of one ship to another consequent upon a collision. *Abbrev.* TPL *or* T.P.Liab.

Thirds An insurance expression referring to claims where a deduction of one-third of the claim is retained to make good for the new spares or materials replacing the old ones.

Thirty Days Clause A Hull Insurance, *q.v.*, phrase giving an extra extension of 30 days insurance coverage on the vessel beyond the day of her arrival at the port of destination.

Thks. Thickness.

THLS Trinity House Lighthouse Service, *q.v.*

Thorite A black compact mineral found in Norway.

Thorium A radioactive metallic element.

THORP Thermal Oxide Reprocessing Plant.

Thou. 1/1000th of an inch.

THP *or* **T.H.P.** Torque Hub Propeller; Thrust Horse Power.

Three Island Ship A ship with a forecastle, bridge and poop.

Three Leader Agreement An Insurance Slip, *q.v.*, signed by three leading underwriters who guarantee conditions for the subsequent acceptance by the subscribing underwriters. Any amendments or additions are only to be authorised by the three leading under-writers. See **First Three**, *also* **Leading Underwriters**.

Three Leader Agreement (Cargo) A standard Institute clause, for cargo slip use only, which binds following Underwriters, *q.v.*, to acknowledge amendments agreed by the three leading underwriters on the Slip, *q.v.*, or if both Lloyd's and Institute Companies are on the slip, by the first two Lloyd's underwriters and the first two companies.

Thro' Through.

Thro. B/L Through Bill of Lading.

Through Fare Passenger fare covering the total passage from beginning to end, by land, sea or air.

Through Transport Club An insurance club for commercial transportation covering and protecting the vehicle and its driver during the passage on land or sea. It is more or less identical to the Protection and Indemnity Club, *q.v. Abbrev.* T.T. Club.

Thru. Through.

Thunderer Nickname for *The Times*, the oldest UK newspaper in circulation.

THV Trinity House Vessel.

Thw. Thwartship, *q.v.*

Thwartship Ship's transverse; Across the ship. *Abbrev.* Thw.

THWM *or* **T.H.W.M.** Trinity High Water Mark.

THY Tur Avrupa Hava Yollari, Turkish National Airlines.

Ti Symbol for Titanium.

T.I. Technical Institute; Transport Institute, *q.v.*; Technical Inspection.

TIA *or* **T.I.A.** Texas International Airlines; Tax Institute of America.

TIB *or* **T.I.B.** Tourist Information Bureau; Trimmed in bunkers *in the coal trade.*

Ticket A written or printed piece of card or paper entitling the holder to admission to a place of entertainment, etc., or to conveyance by train, ship, aircraft, etc.

Ticket Day A Stock Exchange, *q.v.*, expression meaning the day before settling day, when the names of actual purchasers are handed to Stockbrokers, *q.v.*

TID Technical Investigation Department.

Tidal Atlas A book of chartlets for an area illustrating the direction and rate of tidal streams for each hour before and after high water at a standard port.

Tidal Berths *or* **Mud Berths** Berths located in tidal harbours used by coasters and small tankers.

Tidal Dock A dock subject to the tidal movements of the water.

Tidal Fall The height difference between high and low water. Similar to Tidal Range.

Tidal Harbour A harbour affected by the tides.

Tidal Wave Large waves resulting suddenly from a big storm or an earthwake.

Tide(s) The periodical rise (flood-tide) and fall (ebb-tide) of the sea due to the attraction of the moon and sun. In some Charterparties, *q.v.*, the owners permit their vessels to lie on soft ground at ebb-tide. See **Always Afloat** *or* **Always Safely Afloat**.

Tide Bound When a ship is prevented from sailing or moving because of insufficient depth of water during low tide.

Tide Gauge A device used to measure the height of the tide.

Tide Sail When a ship is ready to sail by the next tide.

Tideway A channel or canal affected by the rise and fall of the tides.

Tie-in In sales practice an offer where the customer has to buy one or more other items besides that preferred, albeit at a price below the sum of the normal prices for all items, which could be to advantage.

Tier (1) One of the rows or levels in a stowage of containers, bales, barrels, casks, etc. (2) The place for stowing hemp cables in an old sailing ship. (3) The row or battery of guns on a gun deck. (4) A line of mooring buoys between which vessels, typically barges, lie alongside each other; craft so moored.

Tierce 42 gallons. *Abbrev.* Te *or* TCE.

TIF Transit International par fer *French*—International Transit by train.

TIF *or* **T.I.F.** *or* **t.i.f.** Telephone Interference Factor.

TIG *or* **T.I.G.** Tungsten Inert Gas.

Tight *See* **Making Water**.

Tight Fiscal Policy *See* **Fiscal Policy**.

Tight *or* **Dear money** *See* **Monetary Policy**.

Tight, Staunch and Strong in Every Way *or* **Tight, Staunch and Strong Every Way Fitted for the Voyage** A general expression seen in Contracts of Affreightment, *q.v.*, meaning that the ship is Seaworthy in all respects.

Tiller A lever fitted to the head of a rudder for steering *nautical*.

TIM Tubing Integrity Monitor *oil drilling*.

Time-and-a-Half Basic for overtime pay at $1\frac{1}{2}$ times the normal hourly wage rate.

Time and Motion Study Work study by calculations. The way the work is carried out by an employee against the time taken to facilitate procedure and increase efficiency.

Time Bill A Bill of Exchange, *q.v.*, payable at a given time.

Time Deposit Money deposited in a bank for a stated period. Withdrawal is accepted on the lapse of this period, but it may require the depositor to give the bank advance notice if interest is not to be lost.

Time Draft A Draft, *q.v.*, that is to be paid at a stated time after presentation.

Time Hire Freight Time Charter Hire. The money paid by a charterer to a shipowner for the hire of his ship on Time Charter.

Time Insurance Policy Similar to Time Policy, *q.v.*

Time Lost Waiting for Berth to Count A charterparty expression which denotes that the Laytime, *q.v.*, or Laydays, *q.v.*, are to start from the time the Letter of Readiness or Notice of Readiness, *q.v.*, is presented by the master to the charterer's agent, even though vessel is held up because no berth is available. *Similar to* **Time to Begin on Arrival**, *q.v.*

Timepiece Instrument for keeping and showing time, a chronometer, clock or watch.

Time Policy or Time Insurance Policy A marine insurance policy protecting a particular object for a period of time, as agreed with Underwriters, *q.v.* When a ship is insured on a Time Policy, and that time expires during a voyage at sea, the assured is allowed an extra month's grace or until she arrives at her destination. One of the main warranties to be adhered to during the whole period of insurance is that she is to be kept classed as per details submitted to underwriters when accepted to be insured. Also called Period Policy. *See* **Marine Insurance and Marine Policy**.

Time Rate Wages paid according to the time taken to do the work.

Time Reversible Overall calculations of Laydays, *q.v.*, in loading and discharging in a Voyage Charterparty. Time saved in loading can be made up for time lost, if any, in discharging or vice versa. Also expressed as **Reversible Laydays**, or **Total Days All Purposes**, *q.v.*, or **Total Days All Told**.

Time Signal The exact time signal relayed from radio stations regularly and daily for mariners to check their Chronometers, *q.v.*, and to determine their position in the course of navigation.

Time to Begin on Arrival A charterparty clause referring to the Laytime or Laydays in the strict sense that time is to count immediately on arrival at the port of loading or discharging irrespective whether a berth is available or not. A point of disadvantage for the Charterer, *q.v.*, as opposed to the term As Customary, *q.v.* Also expressed **Time Lost Waiting for Berth to Count**, *q.v.*

Time Work Work performed on a time wage basis, at so much per hour or per day.

Timon Code name for a Time Charter Party.

TiN Titanium Nitrate.

TIN Taxpayer Identification Number *USA*.

Tin Can US Naval slang for a destroyer.

Tin Fish US Naval slang for a torpedo.

TIP Taking Inward Pilot; Temporary Import Permit, *q.v.*

Tipstaff Tipstaves are officers, in the nature of constables, attached to the Supreme Court. They were originally appointed for the Court of King's Bench by the Marshal of the King's Bench Prison, and for the Courts of Chancery, Common Pleas and Exchequer by the Warden of the Fleet Prison (Queen's Prison Act, 1842); but by that Act the functions, in this respect, of those officials were transferred to the Lord Chief Justice (of the Queen's Bench as he then was) and to the Lord Chancellor, the Lord Chief Justice of the Common Pleas, and the Lord Chief Baron of the Exchequer. The appointment now lies with the Lord Chancellor and the Lord Chief Justice. Since the abolition of imprisonment on mesne process, the functions of the tipstaves have been confined to arresting persons guilty of contempt of court, except in the Queen's Bench Division, where the tipstaff also has in theory charge of any prisoner brought before the court, or committed by the court, in the exercise of its criminal jurisdiction. Writs of attachment are executed by the sheriff and the tipstaff does not usually act unless the offender is actually in court.

Tip Vortex Free A type of propeller estimated to save an appreciable amount of fuel consumption. *Abbrev.* TVF *or* T.V.F.

TIR *or* **T.I.R.** Transport International des Marchandises par la Route, Target Information Report.

TISI Thai Industrial Standards Institute.

TISTR Thailand Institute of Scientific and Technological Research.

Titanium A metallic element resembling iron applied to making a special high grade stainless steel which is abundantly used in oil drilling and in process industries such as chemicals, pulp and paper, especially where chlorimated fluids are present; in nuclear facilities to prevent corrosion; and in ship building. Symbol Ti.

Title of the Goods The Bill of Lading, *q.v.*, shows the Title of the Goods or the Owners of the Goods.

TIV Total Insured Value.

T. I/V Total Increased Value *insurance*.

TK *or* **T.K.** Telephone Kiosk.

Th Thanks; Thick.

TKO Technical Knock-Out *boxing*.

TKU Thank you *telex abbreviation*.

TKVM Thank You Very Much *telex abbreviation*.

TL Threshold Limits.

TL *or* **T.L.** Turk Loydu; Turkish Lloyd's *classification society*.

T.L. *or* **T/L** *or* **t.l.** Total Load; Total Loss, *q.v.*

TLACV *or* **T.L.A.C.V.** Track Laying Air Cushion Vessel.

TLB *or* **T.L.B.** *or* **t.l.b.** Temporary Lighted Buoy.

TLC Tender Loving Care *parents towards children*.

TLL *or* **T.L.L.** Timber Load Line.

TLO *or* **T.L.O.** *or* **t.l.o.** Total Loss Only, *q.v.*; Transport Law Association.

T.L.O. & Exs. Total Loss Only and Excess. *See* **Total Loss Only**.

T.L.O. R/I Total Loss Only *reinsurance clause*.

TLP Tension Leg Platform, *q.v.*

Tlr. Trailer.

TLR Times Law Reports; Top Level Requirement *USA*.

TLRV or **T.L.R.V.** Track Levitated Research Vessel *air cushion vessel*.

TLV Threshold Limit Value.

T.L.V.O. Total Loss of Vessel Only *insurance*.

TLV-STEL Threshold Limit Value—Short Term Exposure Limit.

TLV-TWA Threshold Limit Value—Time Weighted Average.

TLW or **T.L.W.** Time Lost Waiting *in voyage charterparties*.

TLWP Tension Leg Well Platform. *See* **Tension Leg Platform**.

TM or **T.M.** Thames Measurement, *q.v.*; Ton Measurement, *q.v.*, Ton Mile, *q.v.*; True Motion; Torpedoman's Mate *USA*.

T/M or **tm** Ton Measurement, *q.v.*

TMAP Textile Mills Association of the Philippines.

TMBS Texas Marine Bureau of Survey.

TMC Tension Motion Compensator, *q.v.*

TMG or **T.M.G.** Track Made Good *nautical*.

T.Mk. Tonnage Mark.

T.M.L. Three-Mile Limit, *see* **Territorial Waters**; Transport Manager's Licence.

TML Tetramethyl Lead; Transportable Moisture Limit.

TMO or **T.M.O.** Telegraph Money Order.

TMT or **T.M.T.** Thames Measurement Tonnage; Turbine Motor Train.

TMV or **T.M.V.** Tank Motor Vessel; True Mean Value.

TN or **Tn** Ton; Tin.

TN or **T.N.** True North *compass reading*.

Tnk. Tank.

TNR or **T.N.R.** or **t.n.r.** Ton Net Register. *Similar to* **Net Tonnage**, *q.v.*

TNS or **T.N.S.** Thames Navigation Service, Port Information Centre; Transcutaneous Nerve Stimulation.

TNT or **T.N.T.** Trinitrotoluene *high explosive*.

TO or **T.O.** Table of Organisation *USA*; Telegraphic Order; Telegraphic Office; Telephone Office; Tonnage Opening, *q.v.*; Transport Office; Transport Officer; Turn Over.

T/o Transfer Order *banking*.

To. Togo.

t.o. Take Off *aviation*; Turn Over.

To Arrive *See* **Delivery Terms**.

TOB Total Number on Board.

TOC or **T.O.C.** Technical Order Compliance; Terms of Credit.

TOD or **T.O.D.** Terms of Delivery; Time of Delivery.

Tod An old unit of weight for wool, about 28 lbs.

TODA Take-Off Distance Available *aviation*.

TOE Ton Oil Equivalent.

TOEFL Test of English as a Foreign Language *USA*.

TOFC or **T.O.F.C.** Trailer on Flat car. Also expressed as Piggyback system where conventional trailers are carried on wagons.

Tok Tokyo, Japan.

Token or **Token Money** Paper or coin money which has no actual face value but is legally considered to be worth the wording and figures on the statements inserted by the issuing country.

Token Payment Similar to Earnest Money, *q.v.*

Tokkyo *Japanese*—Express.

T.O.L. Tower of London.

Tola Indian unit of weight for precious metals, equal to 11.1g.

To Load as Customary always Afloat *See* **At All Times of the Tides and Always Afloat**.

Toll Tax or charge imposed for the use of highway roads, tunnels, bridges, etc.

Tolls Dock and canal charges which are borne by shippers and not by the carriers.

Tomming Off A stowage phrase. The act of inserting wedges in Broken Stowage, *q.v.*, of cargo to prevent it from moving.

Ton A measure of weight (2,240lb. avoirdupois). A short ton is 2,000lb. *Abbrev.* T *or* Tn *or* tn.

TONAC Total Navigation Control.

Ton by Ton Delivered Payment of freight to be paid currently with the discharge of the merchandise. This method is not always practicable and therefore normally a sum of money is deposited prior to commencement of discharge and the balance is finalised on completion before the vessel sails.

Tongued and Grooved Before wood boards are taken out of a factory they are cut and prepared for use as flooring. *Abbrev.* T & G *See also* **Prepared Timber**.

Ton Measurement *Similar to* **Measurement Ton**. *Abbrev.* TM *or* tm *or* T.M. *or* T/M.

Ton Mile Unit cargo ton/distance carried and covered. This is arrived at as follows: If two tons of merchandise are carried for 25 miles, then the Ton Mile will be 2 × 25 or 50 Ton Miles. Similar to Tonne Kilometre. *Abbrev.* TM *or* T.M. *or* T/M.

Tonnage Boards Strong boards/planks to close temporarily the Tonnage Opening(s), *q.v.* of a ship.

Tonnage Deck The uppermost continuous deck in ships having less than three decks, or the second continuous deck from below.

Tonnage Dues Port Dues, *q.v.*

Tonne Kilometre *Similar to* **Freight Tonne Kilometre**, *q.v. See* **Ton Mile**.

T.O.O. Time of Origin.

Toolpusher A rig superintendent or senior driller. *See* **Driller**.

Topgallant The mast fitted above a topmast in a sailing ship and its yards and sails.

Top Hamper Deck cargo or any other objects adding weight above the upper deck.

'Top Hat' Pensions Special pensions allotted to management personnel after their term of work has terminated.

Topographical Map Detailed description of the physical features in a map.

TOPS *or* **T.O.P.S.** Training Opportunities Schemes.

Topsides That part of the hull of a ship between the waterline and the deck. *Abbrev.* T/S or T.S.

TOR *or* **T.O.R.** *or* **t.o.r.** Teleprinter Over Radio, *q.v.*; Telex Over Radio; Time of Receipt; Time On Risk.

TORA Take-off Run Available *aviation*.

Tornado A violent whirlwind or rotary storm occurring during the summer months on the west coast of Africa and in the southern region of the USA.

Torps Ship's Torpedo Officer.

Torque The mechanical force that tends to produce torsion or rotation.

TORS Trade Opportunity Referral System *USA*.

Torsiometer An instrument registering the horsepower, through the power transmission of the moving propeller shaft to find the SHP, *q.v.*

Tort Accepted legal proceedings taken against a wrong action which is outside the scope of the contract; breach of contract which requires compensation.

Tortfeasor Person committing a Tort, *q.v.*; a wrongdoer.

Tortious Nature of a Tort, *q.v.*, relating to a Tort.

TOS *or* **T.O.S.** Terms of Service.

TOSCA Thames Oil Spill Clearance Association, formed by the Port of London Authority and the oil industry.

TOT *or* **T.O.T.** Terms of Trade.

Total Base Number This refers to the number allocated to the alkalinity level in lubricating oils and greases and its strength to counteract the acids produced by the combustion process. *Abbrev.* TBN *or* tbn *or* T.B.N.

Total Days All Purposes A charterparty term wherein the total number of days are given to be used for both loading and discharging. *Similar to* **Reversible Laydays**, *q.v.*, *or* **Time Reversible** *or* **Total Days All Told.**

Total Days All Told *Similar to* **Total Days All Purposes**, *q.v.*

Totalitarian State When all of a country's economical, political and financial activities are controlled by the state.

Total Loss When the subject-matter insured ceases to exist it becomes a total loss. It falls under two categories: (a) Constructive Loss and (b) Actual Total Loss, *q.v. Abbrev.* TL *or* T.L. *or* t.l. *See* **Arranged Total Loss.**

Total Loss Only Insurance Underwriters are only liable in case the object insured suffers a Total Loss, *q.v.*, or a Constructive Total Loss, *q.v. Abbrev.* TLO *or* T.L.O. *or* t.l.o.

Total Revenue The entire income derived from a sale. Also the total taxes collected from the general public in a country.

Toties quoties *Latin*—As often as the occasion arises.

To/To *or* **TO/TO** Trot On/Trot Off. A Livestock Carrier, *q.v.*

Touch and Stay A Marine Insurance, *q.v.*, term covering the owners for any damage which may happen during a vessel's call or stay at any customary ports and not named in the Policy, *q.v.*, but excluding deviation. *Abbrev.* T&S *or* TANDS *or* tands.

Touch Down Landing fees in aviation.

Touching Soft Ground *Similar to* **Sitting Aground**, *q.v. See* **Always Afloat.**

Tour de force *French*—A feat of strength or skill.

Tourist Floater An American Insurance Policy on inland luggage and personal effects.

Tour Operator Person or firm organising itineraries for groups of tourists at home and abroad including hotels, transport, excursions and entertainment.

Tout à fait *French*—Entirely.

Tout à l'heure *French*—Shortly; immediately.

Tout de suite *French*—Immediately.

Tout ensemble *French*—All together.

TOVALOP Tanker Owners' Voluntary Agreement Concerning Liability for Oil Pollution, *q.v.*

TOW *or* **T.O.W.** Take-Off Weight, *q.v.* Tube-Launched, optically tracked, wire guided *US Army missile.*

Tow (1) When one or more vessels are being towed. (2) When a tug is towing one or more floating objects. (3) To pull an object in the water by means of a rope.

Towage (1) Charges for the services of tugs assisting a ship or other vessels in ports or other locations. (2) The act of towing a ship or other object(s) from one place to another. In Admiralty law, a towage service is where one vessel is employed to expedite the voyage of another, when nothing more is required than to accelerate her progress, as opposed to a salvage service, which implies danger or loss. A towage service gives a right to remuneration; but, although in many cases where salvage has been claimed the court has decreed only towage remuneration, there are comparatively few cases in which suits have been instituted for mere towage. Claims for towage remuneration are generally enforced in courts having admiralty jurisdiction.

Towage Dues Remuneration due for services rendered by tugs.

Towage of the Insured Vessel An insurance clause making it a breach of Warranty, *q.v.*, for the assured vessel to be towed or assisted contrary to the insurance terms of the policy.

Tow and to be Towed A charterparty clause, authorising the master of a ship to tow or to be towed, to assist or to be assisted in case of distress and to deviate from the anticipated normal course to save life and property singly or jointly. Sometimes phrased as 'Towing Salving Clause'.

Towarzystwo *Polish*—Company.

Towboat *Similar to* **Tugboat**, *q.v.*

Tower's Liability Liability incurred by any ship or vessel when she is towing another ship, vessel or other object.

Towing Hawser *Similar to* **Towrope**, *q.v.*

Towing Light A white light placed as nearly as practicable at the stern showing an unbroken light over an arc of the horizon of 135 degrees and so fixed as to show the light 67.5 degrees from right aft on each side of the vessel.

Towing Salving Clause *See* **Tow and to be Towed**.

Towline An extra heavy flexible Hawser, *q.v.*, or wire rope generally used for towing purposes. Also called **Towing Hawser** or **Towrope**.

Towrope *Similar to* **Towline**, *q.v.*

Tow-Rope Horsepower *See* **Effective Horse Power**.

Toxic Of poison; poisonous.

Toxin A poison, especially one secreted by a microbe and causing some particular disease.

TP *or* **T.P.** Teleprinter; True Position *nautical*; Third Party, *q.v.*

Tp Trooper, *q.v.*

T&P Theft and Pilferage *insurance*.

TPA Tampa Port Authority.

TPC *or* **T.P.C.** Tonne Per Centimetre Immersion; Telecommunications Policy.

TPCFW *or* **T.P.C.F.W.** Ton Per Centimetre Immersion in Freshwater.

TPD *or* **T.P.D.** *or* **t.p.d.** Tons per Day.

TPFCJK Trans-Pacific Freight Conference of Japan and Korea.

TPFT *or* **T.P.F.T.** Third Party Fire and Theft *car insurance*.

TPH *or* **T.P.H.** *or* **t.p.h.** Tons per Hour.

TPI Third Party Insurance; Tropical Products Institute.

TPI *or* **T.P.I.** *or* **t.p.i.** Tons per Inch Immersion; Town Planning Institute.

TPL *or* **T.P.L.** Transferred Position Line; Third Party Liability, *q.v.*

TP Liab *or* **T.P. Liab.** Third Party Liability.

TPM *or* **T.P.M.** *or* **t.p.m.** Tour par minute, *French* —Revolutions per Minute; Tons per Minute.

TPND *or* **T.P.N.D.** Theft, Pilferage & Non Delivery *marine insurance*.

TPNSD *or* **T.P.N.S.D.** Theft, Pilferage & Non and/or Short Delivery *marine insurance*.

TPO *or* **T.P.O.** Travelling Post Office.

TPP Third Party Property *insurance*.

TPRDC & P & I *or* **T.P.R.D.C. & P. & I.** Third Party Running Down Clause and Protection & Indemnity *yacht insurance*.

TPS Towed Production System.

TPV Third Party Vehicle *insurance*.

TQ *or* **T.Q.** *or* **Tq** Tale Quale, *q.v.*

t/q Tel quel *French*—Just as it is; Tale quale, *q.v.*

TQC Total Quality Control.

TQM Total Quality Management.

TQS Total Quality Supervision.

TR *or* **T.R.** Tempora regis *Latin*—The Times of the King; Tons Registered. *See* **Registered Tonnage**; Toutes risques *French*—All Risks *insurance*; Track, *q.v.*; Tracking Radar; Transmit/Receive; Radio/Transmitting/Receiving; Trust Receipt.

Tr *or* **tr** Transit; Triple Screw; Trustee, *q.v.*; Track.

T/R Tons Registered. *See* **Registered Tonnage**; Transmitter/Receiver.

TRACALS Traffic Control and Landing System *aviation*.

Trace To investigate the whereabouts of cargo found to be missing when a ship has unloaded. This is the duty of the agent who will send Tracers, *q.v.*

Trade Barrier An obstacle in the way of the import and export trade of a country presented by prohibitive taxes or duty, or the withholding of import and export licences.

Trade Credit Credit allowed in business when selling merchandise on credit. The credit given may extend for one month or more.

Trade Discount Commission or discount allowed to buyers by manufacturers or exporters to importers to sell and resell on a commercial basis. Usually when trade discounts are shown on invoices, customs duties are exempted on such discounts, especially when the merchandise is imported by a bona fide commission agent, whose profits are specifically from commission given him by manufacturers, exporters or principals. Trade Discount is also allowable to retailers by wholesalers as an incentive profit. Sometimes it is called Concealed Discount. *See* **Trade Price**.

Trade Expenses Expenses involved in business transactions. *See* TE *or* T.E.

Trade Gap The difference between the value of imports and exports, both visible and invisible, resulting in either a Favourable or an Unfavourable Balance of Trade. *See* **Active Balance of Payments**.

Trade Marks At common law 'trade mark' denotes any means of showing that a certain trade or occupation is carried on by a particular person or firm, including, therefore, not only trade marks in the narrower sense of the word, but also trade names and marks which are not in themselves, or in their origin, distinctive, but which have become known by custom and reputation as showing that goods or implements of trade are made, sold, or employed by a particular person or firm.

In a narrower common-law sense of the word, a trade mark is a distinctive mark or device affixed to or accompanying an article intended for sale, for the purpose of indicating that it is manufactured, selected, or sold by a particular person or firm.

Trade Name *See* **Trade Marks**.

Trade Price Market price less a special discount or Trade Discount, *q.v.*, allowed for reselling.

Trade Reference Reference made out by bankers, special agents or popular organisations dealing with the financial and commercial stability of a company. The questioner is given the references in strict confidentiality before he commits himself to any business arrangements.

Trade Routes Implies regular sailings from various ports or places to others and vice versa. Also called Traffic Lanes. Also applicable to land and air routes.

Trade Ullage Loss of liquid cargo because of evaporation.

Trade Unions These were originally mere friendly societies consisting of artisans engaged in a particular trade, such as carpenters, bricklayers, etc.; but in course of time they acquired the character of associations for the protection of the interests of workmen against their employers. The principal objects of such societies are to increase the rate of wages, reduce the hours of labour, and bring about an equal division of work among a large number of workmen by establishing uniform minimum rates of wages. They attain these objects principally by negotiation but also by the legal means of strikes i.e., stoppage of work by all the members until their demands are complied with, and the process known as picketing or rattening. The members are supported during strikes by the funds of the society, which are obtained by weekly subscriptions during periods of work. Many trade unions are also friendly societies.

Formerly trade unions were not recognised by law; and, as a general rule, their regulations, being in restraint of trade, were illegal, and incapable of being enforced; but, by the Trade Union Act, 1871, this doctrine was abolished, and provisions (analogous to those applying to friendly societies) are made for the registration of trade unions, for the regulations to be contained in their rules, and for the appointment of trustees in whom the property of the union is to vest, etc.

Trade Week Period for the promotion of a country's or an industry's products which provides a trader with a special opportunity to display his wares to the public.

Trade Wind A wind blowing continually towards the thermal equator within parallels 30 degrees North and 30 degrees South in the Atlantic and Pacific Oceans and deflected westwardly by the rotation of the earth.

Trading Account Book-keeping account showing the gross profit.

Trading Estate Industrial Estate.

Trading with the Enemy Dealings with enemy aliens or persons resident in enemy or enemy-occupied territory. The Trading with the Enemy Act, 1914, provided that any person who traded with the enemy during the war of 1914–18 should be guilty of a misdemeanour. The Trading with the Enemy Act, 1939, made similar provision as regards the war of 1939–45. Provision was made for vesting enemy property in the Custodian of Enemy Property.

Traffic (1) Collective goods and persons moving regularly on roads, at sea or by air. (2) Exchange of uniform buying and selling in business (3) Movements of vehicles, ships, aircraft and people. (4) Flow of business in general. (5) Word in radio-transmitting referring to messages. (6) Frequent messages by any means to and fro.

Traffic Lane *Similar to* **Trade Routes**, *q.v.*

Traffic Separation Scheme A navigational arrangement, internationally agreed, whereby vessels navigating straits or through congested turning points are restricted in the tracks they may follow to avoid traffic in the opposite direction. The limits of these routes are shown on charts as well as any Separation Zone that divides them. Many such schemes are now in operation around the world since the first one established in the Strait of Dover in 1972. Radar Surveillance of vessels in transit is maintained by Coast Guards or other authorities to detect infringements of the rules.

Trailer A platform-shaped non-mechanical vehicle supported by four or six heavy wheels with a clamping device in front, towed by a tractor or by a similar strong vehicle. Trailers carry various loads and are transported by rail, road and sea. They are frequently to be seen within the port areas ready to be loaded and unloaded on Roll-on/Roll-off, *q.v.*, ships. *Abbrev.* Tlr.

Train Ferry Ship plying a short route between two ports and having rails on its deck for railway carriages and wagons which are easily moved on and off.

Training within Industries A government system introducing senior school leavers into private industries to work as apprentices. In some countries they are also taken into government departments. *Abbrev.* TWI.

TRAL Tape-Radio Auto Loop.

Trampship A merchant ship plying the oceans seeking cargoes wherever they are to be found.

Tran. Gir. Transverse Girders.

Trans. Transfer; Transfers; Transformer.

Trans. Alt. Transition Altitude.

Transcript Official copy of court proceedings.

Transfer To pass over the right of property from one person to another.

Transfer Manifest A customs transhipment form for air cargo purposes.

Transfer of Sale Entry into a promise of sale.

Transformer An electrical implement which converts voltage power from high to low and vice versa.

Transire A customs document used when a vessel is coasting, giving full cargo details. It serves as clearance from the port of issue.

Transit Clause A clause in the Institute Cargo Clauses, specifying the attachment and termination of cover. *See* **Jetty Clause**.

Transit Port A port where various cargoes are discharged and warehoused until such time as another ship arrives to pick them up for onward shipment to final ports of destination.

Transit Shed Customs sheds warehousing transit cargo pending re-exportation or Transhipment, to the final destination.

Trans. Lev. Transition Level.

Translocation An American form of insurance policy covering the entire production (up to the finished product of a factory).

Transonic Very near to the speed of sound.

Transparent Word used on Wall Street, New York, when a market maker will not confirm nor indicate the price for Bonds, *q.v.*

Transponder A small electronic instrument with many uses in the field of offshore oil drilling; one of which is tracking divers on the seabed.

Transport A state-owned or requisitioned ship also called a troopship; the carriage of cargo from one place to another by land, sea or air.

Transportaktiebolag *Swedish*—Transport Company Limited.

Transport Institute An American research organisation based in Washington with member companies interested in inland water and deep sea transportation. *Abbrev.* TI *or* T.I.

Transport International de Marchandise par la Route *French*—International Goods Road Transport. An association that worked hard for the easing of numerous lengthy procedures at the customs frontiers. There are in effect certain facilities to avoid delays, thus saving time and expense. *Abbrev.* TIR *or* T.I.R. It is a transportation system of containers carried on trailers shipped on ro/ro ships. Then they are either towed or carried to their destination after crossing various countries on land. This TIR or Container Vehicle Carrier is a common means of transportation and deals with all kinds of cargo including refrigerated stores.

Transport International Routier *Similar to* **Transport International de Marchandise par la Route**, *q.v. Abbrev.* TIR *or* T.I.R.

Transtainer A vehicle used for carrying cargo containers during loading or discharge operations or within port or terminal areas.

Transverse Across a ship at right angles to a line drawn from bow to stern.

Trating Refers to tainting by chemical products in transit.

TRAXIMPEX Foreign Trade Organisation for Import and Export. The ship supplying, bunkering and shiprepairing company of Bulgaria with head offices in Sofia.

Traveller (1) A person going from one place to another on foot, by road, by train, by sea, or by air. (2) A person employed by a commercial firm in a particular line of business. He may be paid on a salary or commission basis or both.

Travelling Bill A document produced by the UK customs authorities instead of a Jerque Note, *q.v.* A Travelling Bill refers to the discharge or part discharge of cargo.

Trawler A fishing vessel constructed of wood or steel which uses a trawl-net or nets. A trawl-net is a large bag-net with a wide mouth, held open by a beam or otherwise, and which is dragged along the seabed to catch large quantities of fish. Trawlers are used for both inshore and deep-sea fishing.

T.R.C Tax Reserve Certificates; Thames Rowing Club.

Tr. Co *or* **Tr. Coy.** Trust Company. *See* **Trusts**.

T.R. & D.A. Timber Research & Development Association.

Tread The length of the keel of a ship.

Treasure Trove Money or coin, gold, silver, plate or bullion found hidden in the earth or other private place, the owner thereof being unknown or unfound, in which case it belongs to the Crown, unless the owner appears to claim it. The Crown may grant the right to treasure trove found within a certain district to a private person, e.g. the lord of the manor. In the absence of such a grant anyone concealing treasure trove which he may find is guilty of a common law misdemeanour.

Treasury Bill A Bill of Exchange, *q.v.*, issued by the Treasury of a state to raise money for temporary needs and sold to the highest bidder. They are payable not more than 12 months after date. Such bills have superseded Exchequer bills.

Tret An allowance calculated on certain articles for wastage or depreciation while in the course of a voyage.

t.r.f. Tuned Radio Frequency.

TRI Toxic Release Inventory. A report on the total quantities of certain chemical emissions to the environment by an industrial process.

Trial Balance A book-keeping phrase shown in the ledger in a two–column list, one for credit and the other for debit balances. Checking for any errors is done from all the account books and the totals of both columns must agree.

413

TriCat Design of a vessel with three parallel submerged hulls, commonly used for fast ferries with speeds up to about 50 knots.

Tributyltin A very effective biocide in antifouling paints used as one of the ingredients of nearly all self-polishing copolymer paints. *Abbrev.* TBT.

Trick A seaman's duty at sea. A turn of duty at the wheel or as a look-out, which normally lasts for two hours.

Trimariner *See* **Landing Ship Dock** and **Capricorn**.

Trim a Boat, to (1) To distribute passengers, crew and loose articles so as to keep the boat level and on an even keel; (2) to level off cargo.

Trim Sails, to To adjust the setting of sails, using braces and sheets, so as to use the force of the wind to the best advantage.

Trinity House The Corporation of Trinity House had its origins in guild of mariners in the Middle Ages at Deptford. In 1514 it was incorporated by Royal Charter of King Henry VIII as a Guild or Fraternitie of the Trinity and St Clement under the management of a Master, Wardens and Assistants with responsibilities for pilotage. The first Master had been master of the *Mary Rose* and the *Henri Grace à Dieu*, and was also Comptroller of the Navy. In 1566, by Act of Parliament of Queen Elizabeth I, the Corporation took over beacons and buoyage. Its powers in maritime matters were extended by various Acts and Royal Charters of which that of James II in 1685 defined the constitution. Nowadays, this consists of the Master, Deputy Master and the Court of Elder Brethren, not exceeding 31 in number, who are experienced Master Mariners. The Corporation's affairs are in the hands of two Boards. The Corporate Board, consisting of the Deputy Master and a Board of Wardens and Assistants, all Elder Brethren, is concerned with the Corporation's business, Deep Sea Pilotage and charitable matters, which include funds to relieve aged and needy Merchant Navy officers, widows and dependants, maintaining an estate of retirement homes and cadet training scholarships. The Lighthouse Board, consisting of the Deputy Master, Elder Brethren, nominees of the Minister of Transport and Lighthouse Service staff manages the Trinity House Lighthouse Service, *q.v.* The Chief Administrative Officer is both Secretary of the Corporation and Clerk to the Court and sits on both Boards. Elder Brethren may be appointed, as Trinity Masters, to assist in the Royal Courts of Justice as Nautical Assessors, *q.v.*

Trinity House Lighthouse Service In England and Wales, the Channel Islands and Gibraltar, Trinity House is the General Lighthouse Authority with responsibility for lights and seamarks to aid navigation as well as the Decca Navigator System and radio beacon. It administers one lighthouse in the West Indies and one in the Falklands. It is also responsible for the marking and dispersal of wrecks outside port limits. Its powers are derived from the Merchant Shipping Acts 1894, 1979 and 1988. It maintains 68 lighthouses of which 16 (1994) are manned, 13 automated major floating aids (MFA) and 24 beacons. There are over 400 buoys of various types, excluding those laid by port authorities who act as Local Lighthouse Authorities under the Port Act 1991, but which nevertheless require Trinity House superintendence. 14 radio beacons or racons and 11 Decca Navigation Stations are established. The Service's resources comprise over 600 personnel, two tenders, five launches and a helicopter. It has the policy for navigational aids continually under review in view of new navigational and equipment technologies. An example is the development of solar power for light buoys. Trinity House is a member of the International Association of Lighthouse Authorities.

Trip Insurance Voyage Insurance *American term.*

Triptyque A customs permit for a motor car.

Trolley-bus Electric bus which receives power via a trolley-wheel from an overhead wire.

Trolley-car Carriage on rails laid on main roads, driven by electricity supplied from a trolley behind it.

Tromometer A device to detect and measure earth tremors.

Trooper A troopship. A vessel, generally a passenger liner, requisitioned by a government during wartime for the transportation of armed men and equipment.

Tropopause A narrow layer of atmosphere between the troposphere and the stratosphere.

Troposphere A layer of atmospheric air extending about seven miles upwards from the earth's surface, in which temperature falls with height.

Trot A line of buoys for mooring a number of small vessels head and stern.

Troy Weight A weight of 12 troy ounces to the pound, having its name from Troyes, a city in Aube, France. An ounce troy now consists of 480 of the 7,000 grains which go to the imperial pound (Weight and Measures Act 1878). The only customary subdivision of the ounce troy is the pennyweight, 20 of which go to the ounce.

Under the Weights and Measures Act 1878, precious metals (gold, silver, platinum and other precious metals), and gold and silver lace, and precious stones may be sold by the ounce troy of 480 grains; and all other articles must be sold by avoirdupois weight of 437·5 grains to the ounce avoirdupois, the metric equivalents for the respective ounces being 31·103496 grams (troy), and 28·34954 grams (avoirdupois). The pound troy does not seem to be referred to by statute. Drugs may be sold by apothecaries weight.

TRS Tropical Revolving Storm.

Trs. Truss, *q.v.*

Tr.s.m.s. Triple-screw motor ship.

TRSSSV Tubing-Retrievable Sub-Surface Controlled Seals and Valves *oil drilling*.

Truck A wooden disc at the top of a mast with holes for halyards *nautical*.

Truck to Keel From the extreme top of a vessel to the extreme bottom. Also termed Mast Top to Keel.

True Airspeed The actual speed of an aircraft in relation to the temperature and its height. *See* **Airspeed**.

True Bearing The actual true bearing as opposed to the magnetic bearing.

True & Fair In reference to the final report made out by the auditors in the Balance Sheet, *q.v.*, as being 'True and Fair' as far as the financial conditions stand at the end of the year.

Trunk A large box with a hinged lid, often covered with leather, for carrying clothes, etc., on a journey. This is only one of the many meanings of this word.

Trunk Roads Highways.

Truss Among the many meanings of this word there are (1) A bundle of old (56 lb.) or new (60 lb.) of hay or (36 lb.) of straw. (2) A supporting structure or framework of a roof, bridge, etc. (3) A heavy iron fitting securing the lower yards to a mast *nautical*.

Trust In the legal sense, this word means the confidence reposed in a person by making him or her the nominal owner of property to be used for another's benefit; the right of the latter to benefit by such property; property so held, the legal relationship between the holder and the property so held.

Trustee A person who holds property or money in trust for another.

Trusts In the commercial sense, these are organised associations of several companies for the purpose of defeating competition, etc., the shareholders in each transferring all or most of the stock to a central committee and losing their voting power while remaining entitled to profits.

TS Territorial Sea.

TS *or* **T.S** Tail Shaft, *q.v.*; Tailshaft Survey *Lloyd's Register notation*; Target Ship; Tensile Strength, *q.v.*; Time Sheet; Topsides, *q.v.*; Transhipment.

T/S *or* **t.s.** Tensile Strength, *q.v.*; Training Ship; Turbine Steamer; Twin Screw; Time Sheet, *q.v.*

t&s Toilet and Shower.

T&S Touch and Stay, *q.v.*; Transport and Supply.

TSA Techno Superliner *air cushion*; Test of Spoken English *USA*.

TSA *or* **T.S.A.** Training Services Agency *in insurance industry*.

TSAB Towing Safety Advisory Committee *USA*.

TS(CL) *Lloyd's Register notation* for survey of a tailshaft fitted with continuous liner.

T.S.F. Télégraphie Sans Fil *French*—Wireless Telegraphy; Téléphonie Sans Fil *French*—Wireless Set.

TSG Tanker Safety Guide, *prepared by the International Chamber of Commerce*.

TSI *or* **T.S.I.** *or* **t.s.i.** Tons per Square Inch; Total Sum Insured.

TSIC Tianjin Shipbuilding Industry Corporation.

T.S.M.S *or* **t.s.m.s** Twin-Screw Motor Vessel. *See* **Twin Screw**.

TSN Tailshaft Renewed *Lloyd's Register notation*.

TS(OG) Tailshaft with Oil Glands Survey *Lloyd's Register notation*.

TSPP Tanker Safety and Pollution Prevention.

TSR Tactical Strike Reconnaisance; Temporary Safe Refuge; Terminate, Stay Resident *computer*; Traders Single Rate *insurance*; Trans-Siberian Railway.

TSS Tailshaft Survey approved to be corrosive resistant *Lloyd's Register notation*; Traffic Separation Scheme *Maritime Safety Committee of IMO, q.v.*

T.S.S. Turbine Steamship; Twin-screw Steamer.

T.S.T.I. This side to the iron.

T.S.U. This side up.

Tsuhso *Japanese*—Commercial.

Tsunami A sea wave caused by disturbance of ocean floor or seismic movement (Japan).

TSVP *or* **T.S.V.P.** *or* **t.s.v.p.** Tournez s'il vous plaît *French*—Please Turn Over. *Abbrev.* PTO or P.T.O.

TT *or* **T.T.** Tank Top, *q.v.*; Teetotaller; Telegraphic Transfer; Telex Transfer; Tourist Trophy; Transit Time; Tubercular Tested.

TTA *or* **T.T.A.** Travel Trade Association.

TTC Titre de Transport Combiné *French*—Combined Transport Document, *q.v.*

T.T.Club Through Transport Club, *q.v.*

TTE Turning Telescope Elevator *automatic unloading/loading method*.

TTF *or* **T.T.F.** Timber Trade Federation.

TTL *or* **Ttl.** Total.

t.t.l. To take leave.

ttl. com. Total commission.

TTP Tripod Tower Platform *oil rig*.

TTT Navigational warning signal.

Tty. Treaty.

TU Tunis Air *airline designator*.

TU *or* **T.U.** Thermal Unit; Trade Union; Transmission Unit.

TUC *or* **T.U.C.** Trades Union Congress; Trades Union Council.

TUCC *or* **T.U.C.C.** Transport Users Consultative Committee.

TUCGC *or* **T.U.C.G.C.** Trades Union Congress General Council.

TUCP Trade Union Congress of the Philippines.

Tue *or* **Tues** Tuesday.

Tug *or* **Tugboat** A comparatively small vessel with powerful engines and constructed in such a way as to be able to manoeuvre easily for towage and/or to assist in salvage operations at sea. She can easily move about in restricted port areas to assist large vessels entering and leaving harbours, lakes, rivers and other navigational waterways. There are also tugs towing powerless barges, lighters and small craft. Also called Tow Boat.

Tug and Tow A sea-towage term since it was originally considered that the tug was the servant of the tow. However, it has been held in various court cases that the fault lies as follows: (1) With the tug if it solely collides. (2) With the tow if it solely collides. (3) If the collision occurs because of the tug then the tugowners are liable. (4) If the collision happens because of the tow then the tow owners are liable. (5) If fault lies between tow and tug both owners are jointly responsible. It is always difficult to prove who is the culprit in some cases and lengthy verification takes place before decisions are taken.

Tugrik Unit of currency used in Mongolia equal to 100 Mongos.

TUM or **T.U.M.** Trade Union Movement.

Tun. Tunis; Tunisia.

Tun A cask for wine, beer, etc., especially formerly as a measure of capacity (252 wine gallons).

Tundra Barren Arctic regions where the subsoil is frozen.

Tung-oil An oil used chiefly for varnishing woodwork, obtained from the Chinese tung tree.

Tungsten Wolfram, a steel-grey heavy metallic element with a very high melting point, used for the filaments of electric lamps and for alloying steel, etc.

Turbine A type of engine where kinetic energy is transformed from fuel or water to high pressure gas or steam into a rotary motion. Many of the fast ships, particularly passenger ships, have turbine engines installed.

Turbojet An engine having a turbine-driven compressor for supplying compressed air to the combustion chamber.

Turbometer An instrument often found in the engine room to indicate the revolutions per minute of the turbine-driven engines.

Turboprop Implies propellers driven by turbines.

Turbulence Irregular and vigorous weather.

Turk. Turkey; Turkish.

Turn-A-Round The time taken for a return or round trip by a ship, aircraft or road vehicle.

Turnkey Contract A contract to manage a project which includes responsibility for appointing the sub-contractors.

Turn Out Similar to Outturn, q.v.

Turnover The amount of money turned over in a business; the changing of labour in a factory, etc.

Turn Turtle To capsize nautical.

Turn Up To secure a rope or wire on a bollard.

Turp. Turpentine, q.v.

Turpentine A distilled liquid commonly used as a solvent or thinner in paints. Abbrev. Turps.

Turret Anchored Production System An oil production platform storage tanker with an off-loading buoy. Abbrev. TAPS.

TV or **T.V.** Television; Terminal Velocity; True View.

T.V.A. Taxe à la valeur ajoutée French—Value Added Tax; Tennessee Valley Authority USA.

TVD True Vertical Depth oil drilling.

TVF or **T.V.F.** Tip Vortex Free, q.v.

TVO or **T.V.O.** Tractor Vaporising Oil.

TVOR or **T.V.O.R.** Terminal VHF Omnirange.

TVP or **T.V.P.** True Vapour Pressure.

T.V.R. or **t.v.r.** Temperature.

TVWB or **T.V.W.B.** Thames Valley Water Board.

TWA or **T.W.A.** Time Waited Average.

TwD Tweendecker, q.v.

Tween Between.

Tweendecker A vessel designed to carry dry general goods—e.g. bagged sugar—having one or more decks between the main, weather deck and the tank top.

TWh Terawatt-hour.

TWI Training Within Industries.

TWIMC or **T.W.I.M.C.** To whom it may concern.

TWMC or **T.W.M.C.** Transport, Wages Maintenance and Cure insurance.

Twin Hatches Two hatches side by side serving a hold.

Twin Screw A ship with two propellers. Abbrev. Tw. Sc.

T.W.M.C. Transport, Wages, Maintenance and Cure. Insurance coverage included in the P. & I. Club.

Two-Class Vessel A passenger carrying vessel with two types of cabins such as 1st and 2nd class. 1st and tourist or 1st and Dormitory Class.

Two-Compartment Ship More or less similar to a one-compartment ship but instead the bulkhead compartments are again subdivided into other compartments so as to give more protection against sinking if the ship starts making water because of a collision or other damage. See **Bulkhead**.

TWR Torpedo Weapons Receiver(s) USA.

TWRA Transpacific Westbound Rate Agreement.

TWS or **T.W.S.** Tsunami, q.v., Warning System.

Tw.sc. Twin-Screw, q.v.

TWT Travelling Wave Tube transmitting amplifiers.

TYC or **T.Y.C.** Thames Yacht Club; Thousand Yards Course.

Tycoon A business magnate. A person who has accumulated large amounts of money during the course of business activity.

Typhoon A circular storm in the China Seas occuring especially from July to October.

TX Texas, USA; Telex.

Tx Radio Transmitter; Telex.

tx Tonneaux de jauge *French*—Ship's Tonnage; Telex.

T.Zn True zenith *nautical*.

T.Z.D. True Zenith Distance.

U

u Ugly and threatening weather *weather observation*; Upper *Lloyd's Register notation*; University; Unionist; Uncle.

U International code flag or morse signal meaning 'You are running into danger' *nautical*.

U *or* **u** Unmanned; Uranium; Burmese title (not an abbreviation)

U1 Classification grade for chain cables. Grades are numbered 1, 2 and 3, 1 being the highest.

UA Unauthorized Absentee *USA*; United Air Lines *airline designator*.

U/A *or* **U/a** *or* **u/a** Underwriting Account; Unit of Accounts, *q.v.*

u.a. Under Age; Usque ad *Latin*—As far as.

UAA *or* **U.A.A.** United Arab Airlines.

UAB *or* **U.A.B.** Union des Armateurs Belges, Belgian Ship Association.

UAE *or* **U.A.E.** United Arab Emirates. Formerly the Trucial States, Arabian Gulf.

UAM *or* **U.A.M.** Underwater to Air Missile; Afro-Malagasy Union.

u.a.m. Und anderes mehr *German*—And others.

UAR *or* **U.A.R.** United Arab Republic.

UARC *or* **U.A.R.C.** Underwriting Agents' Registration Committee.

UARS Upper Atmosphere Research Satellite.

u.a.s. Upper Airspace.

UAS *or* **U.A.S.** United Arab States, *q.v.*

UASC United Arab Shipping Company.

UATP Universal Air Travel Plan *credit cards*.

UAW United Automobile Workers.

UAWB Universal Air Waybill.

UBC Universal Bulk Carrier.

Uberrimae fidei *Latin*—Utmost good faith, *q.v.*

Ubi jus incertum ibi jus nullum *Latin*—Where the law is uncertain there is no law.

Ubique *Latin*—Everywhere.

Ubi remedium ibi jus *Latin*—Where there is a remedy there is a right.

U-Boat Unterseeboot *German*—Literally an under-sea boat (submarine).

UBS *or* **U.B.S.** Universal Bulk Ship.

U/C *or* **u/c** Under Construction.

U.C. Under Charge(d); Upper Canada; Useful Capacity, *q.v.*

u/c Upper Case.

UCATT Union of Construction, Allied Trades and Technicians.

U.C.B. *or* **u.c.b.** Unless Caused By, *q.v.*

UCC *or* **U.C.C.** Uniform Commercial Code; Universal Copyright Convention.

UCCC *or* **U.C.C.C.** Uniform Consumer Credit Code *USA*.

UCL *or* **u.c.l.** Upper Cylinder Lubricant.

UCMJ Uniform Code of Military Justice *USA*.

UCP Uniform Customs and Practice, *q.v.*

UCS Unified Connection Skid *oil drilling*.

U.C.T.A. Unfair Contract Terms Act 1977.

U/D *or* **u/d** Underdeck; Upperdeck.

UD *or* **U.D.** Upper Deck.

u.d. Ut dictum *Latin*—As directed; Underdeck, *q.v.*

UDATS Underwater Damage Assessment Television System.

UDC Upper Dead Centre.

UDE Under-Water Development Establishment; Union Douanière Equatoriale *French*—Equatorial Customs Union.

UDEAC Union Douanière et Economique de l'Afrique Centrale *French*—Central African Customs and Economic Union.

UDF Ulster Defence Force; Unducted Fan; Union Defence Force *South Africa*.

u.d.f. Und die folgende *German*—And the following.

UDI Unilateral Declaration of Independence.

U.Dk. Upperdeck.

UDT *or* **U.D.T.** Underwater Demolition Team; United Dominions Trust.

UDTC Underwater Dry Transfer Chamber.

UE University Engineer *USA*.

UEB *or* **U.E.B.** Union Economique Benelux *French*—Benelux Economic Union.

UEL *or* **U.E.L.** Upper Explosive Limit, *q.v.*

UEO Union de l'Europe Occidentale *French* —Western European Union.

UER Union Européenne de Radio-diffusion *French*—European Broadcasting Union.

UES Uniform Emission Standards.

UFL *or* **U.F.L.** Upper Flammable Limit. *See* **Upper Explosive Limit**.

UFO *or* **u.f.o.** Unidentified Flying or Floating Object.

u.f.p. Unemployed Full Pay.

UFR Union Frêt Route, *q.v.*

UFTAA Universal Federation of Travel Agents Associations.

UG Aero Uruguay *airline designator.*

UGC University Grants Committee.

UGS Union of Greek Shipowners, *q.v.*

UHF *or* **U.H.F.** Ultra High Frequency *radio frequency.*

UHT *or* **U.H.T.** Ultra High Temperature; Ultra Heat Tested (long-life milk).

UHV *or* **U.H.V.** Ultra High Voltage.

U/I Under Instructions.

U.i. *or* **Ut inf.** Ut infra *Latin*—As below.

UIA Union of International Associations; Union Internationale des Avocats *French*—International Association of Lawyers.

UIEO *or* **U.I.E.O.** Union of International Engineering Organisations.

UIFTPO University's Institute of Foreign Transportation and Port Operations.

UINF Union Internationale de la Navigation Fluviale *French*—International Union for Inland Navigation.

U.I.O.T. Unione Internazionale dei Organizzatori del Turismo *Italian*—Touring Organiser's International Association.

UIR Upper Flight Information Region *aviation.*

UIS *or* **U.I.S.** Union of Insurance Staffs.

UIT *or* **U.I.T.** Union International Télécommunications *French*—International Telecommunications Union.

U.I.T.P. Union Internationale des Transports Publicques, *French*—International Union of Public Transport.

U.J.D. Ultriusque Juris Doctor *Latin*—Doctor of Civil and Common Law.

UK *or* **U.K.** United Kingdom.

UK AF United Kingdom Amphibious Force.

UKAF United Kingdom Air Forces *NATO term for the Royal Air Force.*

UKAEA United Kingdom Atomic Energy Authority.

UKAP United Kingdom Air Pilot.

UKAPE *or* **U.K.A.P.E.** United Kingdom Association of Professional Engineers.

UKASTA United Kingdom Agricultural Supply Trade Association

UK/c *or* **U.K/cont.** United Kingdom and/or Continent—relating to a range of ports of call for loading and/or discharging in a charterparty.

UKCAA *or* **U.K.C.A.A.** United Kingdom Civil Aviation Authority, *q.v.*

UKCATR *or* **U.K.C.A.T.R.** United Kingdom Civil Aviation Telecommunications Representative.

UK/C (BH) *or* **U.K. Cont. BH** United Kingdom and/or Continent, Bordeaux/Hamburg Range—a charterparty term in connection with the range of loading/discharging ports.

UK/Cont/GH *or* **UK/Cont/G.H.** United Kingdom and/or Continent or Gibraltar/Hamburg Range—a charterparty term in connection with the range of loading/discharging ports.

UK/Cont/HH *or* **UK/Cont/H.H.** United Kingdom and/or Continent or Havre/Hamburg Range—a charterparty term in connection with the range of loading/discharging ports.

UKCOSA United Kingdom Council for Overseas Students' Affairs.

UKCS United Kingdom Continental Shelf *area of offshore oil and gas exploration.*

UKCTA *or* **U.K.C.T.A.** United Kingdom Commercial Travellers' Association.

UK fo *or* **U.K.f.o.** United Kingdom for orders.

UKG *or* **U.K.G.** United Kingdom Government.

UK/HAD *or* **U.K./H.A.D.** United Kingdom and/or Havre, Antwerp/Dunkirk Range—a charterparty term in connection with the range of loading/discharging ports.

UKHMA The Harbour Master's Association of the United Kingdom, the Channel Islands and the Isle of Man, formed in 1993.

UKIB *or* **U.K.I.B.** United Kingdom Insurance Brokers.

UKIBEC *or* **U.K.I.B.E.C.** United Kingdom Insurance Brokers European Committee.

UK/LLEC *or* **U.K./L.L.E.C.** Load Line Exemption Certificate for United Kingdom.

UKMP United Kingdom Major Ports Group, *q.v.*

UKNC *or* **U.K.N.C.** United Kingdom National Committee.

UKONA *or* **U.K.O.N.A.** United Kingdom Offshore Navigators' Association.

UKOOA *or* **U.K.O.O.A.** United Kingdom Offshore Operators' Association.

UKOP *or* **U.K.O.P.** United Kingdom Oil Pipelines.

UKOSRP *or* **U.K.O.S.R.P.** United Kingdom Offshore Steel Research Project.

UKPA *or* **U.K.P.A.** United Kingdom Pilots' Association.

UKPIA United Kingdom Petroleum Industry Association.

Ukr Ukraine.

UKREP United Kingdom Representative *NATO and EEC.*

UKSATA United Kingdom South Africa Trade Association.

UKT *or* **U.K.T.** United Kingdom Tariff.

UKTOTC *or* **U.K.T.O.T.C.** United Kingdom Tariff and Overseas Trade Association.

UKTTSMA *or* **U.K.T.T.S.M.A.** United Kingdom Timber Trade Shipowners' Mutual Association.

UL *or* **U.L.** Underwriters' Laboratories *USA*; Upper Limb *nautical.*

ul *or* **Unltd** Unlimited, *q.v.*

ULC Unitised Load Council.

ULCC or **U.L.C.C.** Ultra Large Crude Carrier; Ultra Large Cargo Carrier, *q.v.*

ULD Unit Load Device.

ulf or **u.l.f.** Ultra Low Frequency.

Ullage The empty space in a tank.

Ullage Tape Similar to Sounding Tape, *q.v.*

u.l.m. Ultrasonic Light Modulator.

ULMS Undersea Long-Range Missile System *USA*.

UL/S Uniform Law on the International Sale of Goods.

ULS or **U.L.S.** Unsecured Loan Stock.

Ult. Ultimate; Ultimo *Latin*—In or of the last month.

Ultimate Net Loss *Abbrev.* UNL or U.N.L. *insurance.*

Ultimo In, or of, the last month. *Abbrev.* Ult.

Ultra Large Crude Carrier 300,000/550,000 dwt—used for carrying crude oil on long-haul routes.

Ultrasonics Ultra high sound waves which are used in technical applications. They are applied to echo-sounding devices at sea, the drilling of special glass, etc.

Ultra vires *Latin*—Beyond one's power or authority.

u.l.v. Ultra Low Volume.

u/m or **u.m.** Undermentioned.

Umbrella Liabilities *Similar to* **Bumbershoot**, *q.v.*

UMC Underwater Manifold Centre *Lloyd's Register notation.*

UMDC Universal Modular Colour Display *computer.*

Umpire In the legal sense, an umpire is a third person called in to decide between arbitrators who disagree. *See* **Arbitrator**.

UMS or **U.M.S.** Unmanned Machinery Space, *q.v.*; Unattended Machinery Space. *Similar to* **Unmanned Machinery Space,** *q.v.*

UN or **U.N.** United Nations, *q.v.*

UNA or **U.N.A.** United Nations Association.

UNACC or **U.N.A.C.C.** United Nations Adninistrative Committee and Co-ordination.

Unauthorised Clerk Employee of a stock brokerage company who is not permitted to deal in Stocks and Shares, *q.v.*

Unberthed Passenger Ships Passenger ships trading in Far Eastern oceans carrying Indian natives and other pilgrims as deck passengers bound for Mecca. These passengers provide their own food for the entire voyage.

UNC or **U.N.C.** United Nations Conference; United Nations Command.

Uncalled Capital *See* **Call-up Capital** or **Called-up Capital**.

UNCFA or **U.N.C.F.A.** United Nations Conference on Food and Agriculture.

UNCITRAL or **U.N.C.I.T.R.A.L.** United Nations Commission on International Trade Law, *q.v.*

UNCL or **U.N.C.L.** United Nations Liaison Commission.

UNCLOS or **U.N.C.L.O.S.** United Nations Convention on the Law of the Sea.

Unconditional No conditions are attached to the contract in question.

Unconscionable Bargain A bargain for a loan or payment of money made on oppressive, extortionate or unconscionable terms, between a person having money and another person having little or no property immediately available but having property in reversion or expectancy. Relief against the unreasonable part of such a bargain is generally granted to the borrower. Formerly, the rule was, that mere inadequacy of price was a sufficient ground for rescinding a sale or other dealing with a reversion; but this rule has been abolished, and it is now a question in each case whether there has been fraud or unfair dealing.

Under the Money-lenders Acts in any proceedings by a money-lender to recover money lent or to enforce any agreement or security, the court may reopen the transaction and give relief if satisfied that the bargain made was harsh and unconscionable.

Uncontrollable Cost (1) Expenses arising from Force Majeure, for example, a storm which causes damage to a factory or building. (2) Unexpected change in the rate of exchange on foreign money. (3) Increase in freight and insurance charges following declaration of a war zone.

Und. Under; Underneath.

Under Bond Goods left in the custody of the customs against payment of store rent, from where they will in future be re-exported. Some of these goods may even be supplied to ships in harbour as ship's stores without payment of duty or tax. *Similar to* **In Bond**, *q.v.*

Undercarriage The landing gear under the Belly, *q.v.*, of an aeroplane. *Abbrev.* U/C.

Undercarrier *Opposite to* **Overcarrier**.

Under Deck Tonnage The cubic capacity of the vessel below the tonnage deck in measurement tons of 100 cubic feet per ton.

Under Insurance Insuring for less than the full value of the subject-matter insured.

Underlap When a chartered vessel is delivered to the charterers by the owners before the stipulated time of agreement.

Under Protest A phrase inserted when payment or any statement is produced or signed in disagreement or with the intention of disputing claims.

Undertaker Entrepreneur; Contractor; Funeral Director.

Under the Bow Sides A sign printed or painted on the extreme fore part of a ship on both sides just above the Light Load Line, *q.v.*, to indicate that she has a bulbous bow which is partly submerged and invisible and to serve as a warning or guide to ships approaching her.

Under the Wind Leeward. Direction towards which the wind is blowing *nautical*.

Undertow An undercurrent resulting from waves blowing against the coast or seashore.

Under Tow When a floating object, usually a vessel, is being towed.

Underway Said when a ship is moving and is not at anchor.

Underwriter One who agrees to compensate another person for loss from an insured peril in consideration of payment of a premium. *See* **Lloyd's**.

Underwriting Agent One who acts for an underwriter either in accepting business on his behalf (e.g. a company underwriting agent) or in taking care of his financial affairs in relation to underwriting (e.g. a Lloyd's underwriting agent). *See* **Lloyd's**.

UNDP *or* **U.N.D.P.** United Nations Development Programme.

Undsgd. Undersigned.

Unearned Income Income which is not earned from wages, salaries or profits; income not earned for services rendered, i.e., income from investment.

Unearned Premium Premium already paid to an underwriter which is in respect of a period when he was not at risk.

UNEC *or* **U.N.E.C.** United Nations Education Conference.

UNECE United Nations Economic Commission for Europe.

UNECLA *or* **U.N.E.C.L.A.** United Nations Economic Commission for Latin America.

UNEDA *or* **U.N.E.D.A.** United Nations Economic Development Administration.

UNEF *or* **U.N.E.F.** United Nations Emergency Forces.

UNEP *or* **U.N.E.P.** United Nations Environment Programme.

UNESCO *or* **U.N.E.S.C.O.** United Nations Educational Scientific and Cultural Organisation.

UNETAS *or* **U.N.E.T.A.S.** United Nations Emergency Technical Aid Service.

UNFAO *or* **U.N.F.A.O.** United Nations Food and Agricultural Organisation. Commonly known as FAO.

UNFC *or* **U.N.F.C.** United Nations Food Conference.

Unfounded Debt Short term government debt.

UNGA *or* **U.N.G.A.** United Nations General Assembly.

UNHQ *or* **U.N.H.Q.** United Nations Headquarters.

UNHRC *or* **U.N.H.R.C.** United Nations High Commissioner for Refugees Council.

UNI *or* **U.N.I.** Ente Nazionale Italiano di Unificazione—Italian Standards Association.

UNIC *or* **U.N.I.C.** United Nations Information Centre.

UNICE Union des Industries de la Communauté Européene *French*—European Common Industries Union.

UNICEF *or* **U.N.I.C.E.F.** United Nations International Children's Emergency Fund.

UNIDF United Nations Industrial Development Fund.

UNIDO *or* **U.N.I.D.O.** United Nations Industrial Development Organisation.

UNIDROIT Institut International pour L'Unification de Droit Privé *French*—International Institute for the Unification of Private Law.

Uniform International Radio Telephone phonetic alphabet for 'U'.

Uniform Customs and Practice Documentary credit forms enacted by the International Chamber of Commerce. *Abbrev.* UCP.

Union des Armateurs Belges Belgian Shipowners Association.

Union de Officiales de la Marina Mercanta Merchant Marine Officers' Association in Spain.

U.N.I.O. United Nations Information Organisation.

Union Frêt Route *French*—Freight Route Union. *Abbrev.* UFR.

Union of Greek Shipowners International association of Greek dry cargo shipowners in Piraeus, London, New York and Monte Carlo. *Abbrev.* U.G.S.

Union Professionelle Belge des Approvisionneurs de Navire *French*—Belgian Shipsuppliers Association. The French version for Belgische Beroepsvereniging der Scheepsbevooraders.

Union Purchase Two adjacent port and starboard side derricks linked together in such a manner as one derrick is used to plumb a selected point on the quayside and the other a selected point in the ship's hold. The derricks can handle cargo quickly but only light cargo weights can be handled in this way at slightly below half the capacity of the lowest rated derrick. Thus for two derricks of, say, 10 tons, the union purchase capacity is probably around 4.5 tons.

UNIS *or* **U.N.I.S.** United Nations International School.

UNITAR United Nations Institute for Training and Research.

Unit of Account It was originally the unit exchange rate practised in some countries who converted the German mark, Dutch florin, Belgian franc and Danish krone units. It was later found practicable to compute exchange rates between U/A and US dollar, sterling, French franc, etc. This unit of account is also referred to as Snake. *Abbrev.* U/A.

Unit of Force Equivalent to 1 dyne representing 1/981 gram.

Unit of Power Equivalent to 1 watt representing 10,000,000 ergs.

Unit of Velocity One centimetre in one second.

Unit of Work Equals 1 erg representing 1 Dyne —Centimetre per second.

Unit Retailer A businessman who is the sole proprietor and having no connection with any other person.

Unit Trust Company Pooled money subscribed by many people and companies, being registered under the provision of a trust deed. The fund invested is taken care of and managed by the formed company. Some of these companies or trusts extend their active operations abroad.

United Arab Emirates A Federation of seven Sheikdoms with a central government. The largest of the Emirates is Abu Dhabi, the others being Dubai, Sharjah, Ajman, Fujairah and Umm Al-Qaiwain *Abbrev.* UAE *or* U.A.E.

United Arab Republic Union of Egypt with Syria proclaimed on 1 February 1958 and ratified on 21 February 1958. After Egypt and Israel signed the Camp David Agreement in 1979, Syria stopped its recognition of the Union. *Abbrev.* UAR *or* U.A.R.

United Arab States Union of the Kingdom of Yemen with Egypt and Syria proclaimed in Damascus on 8 March 1958. The permanent seat of the Union is Hodeidah. *Abbrev.* UAS *or* U.A.S.

United Kingdom Civil Aviation Authority The civil aviation department which is the counterpart of the United States Civil Aeronautics Board. This department has the responsibility of issuing certificates to the state and commercial airlines operating within the airspace of the United Kingdom. It authorises foreign airlines to operate regular schedules to and from airports in the UK. *Abbrev.* UKCAA or U.K.C.A.A. It is more commonly referred to as the Civil Aviation Authority (CAA or C.A.A.)

United Kingdom Major Ports Group Consists of the following: Associated British Ports, Belfast Harbour Commissions, Port of London Authority, Medway Ports Ltd., The Mersey Docks and Harbour Company, Tees and Hartlepool Port Authority and Port of Tilbury London Ltd.

United Nations A group of nations forming an organisation on an international scale with the objective of maintaining peace and security wherever possible. Originally convened in San Francisco on 25 April 1945 it was officially launched on 26 June of the same year and has its headquarters in New York. Today most countries of the world are represented in the UN. Apart from its main branches (the General Assembly, the Economic and Social Council, the Security Council, the Secretariat, and the International Court of Justice) there are many agencies and commissions, some of which are: the International Children's Emergency Fund (UNICEF); the International Labour Organisation (ILO); the Food and Agriculture Organisation (FAO); the International Atomic Energy Agency (IAEA); the UN Educational, Scientific and Cultural Organisation (UNESCO); the UN Conference on Trade and Development (UNCTAD); the World Health Organization (WHO); the Organisation for Economic Cooperation and Development (OECD); the World Meteorological Organisation (WMO); the International Civil Aviation Organisation (ICAO); the International Telecommunications Union (ITU); the International Association of Airport and Seaport Police; the International Maritime Bureau (IMB); the International Hydrographic Bureau (IHB); the International Council for the Exploration of the Sea (ICES); the UN Conference on the Law of the Sea (UNCLOS); the Office of the UN High Commissioner for Refugees. In the United Kingdom, by the United Nations Act 1946, effect may be given by Order in Council to measures of the organisation not involving the use of armed force.

United Nations Commission on International Trade Law An international commission established in 1966 as a result of a UN General Assembly resolution. Its object is to promote the progressive harmonisation and unification of International Trade Law. *Abbrev.* UNCITRAL *or* U.N.C.I.T.R.A.L.

United Nations Conference on Trade and Development *Abbrev.* UNCTAD. A United Nations organisation which had its first meeting in 1964 when its Shipping Committee was formed to discuss a number of topics concerning the conduct of shipping and co-operation in commercial and economic aspects largely for the benefit of developing nations. In 1992 it was reorganised with one of its four standing committees dealing with the development of competitive service sectors in developing countries, including shipping. Its emphasis is on reviewing shipping policies and technical developments in sea transport as well as conditions which will facilitate trade between and within regions.

United States Civil Aeronautics Board The United States department in charge of issuing air licences to state and commercial aircraft. It has the jurisdiction of allowing foreign aviation companies to operate within US airspace and also regulates the air fares. Similar in a way to the **UK Civil Aviation Authority**, *q.v. Abbrev.* USCAB *or* CAB for Civil Aeronautics Board.

United States Coast Guard US Naval Department in charge of the following:—Inspection and issuing of licences to merchant vessels as well as officers; the installation and maintenance of marine aids to navigation such as beacons, lighthouses, buoys, etc; lists of lights in US waters and all American islands and enforcing regulations for pilots, safety of life and many other activities. *Abbrev.* USCG.

United States Government Life Insurance The assurance by the US Government to services personnel in war zones and to cover also the veterans of war. *Abbrev.* USGLI *or* U.S.G.L.I.

United States Gulf Ports Baton Rouge, Corpus Christi, Galveston, Houston, Mobile, New Orleans, Pascagoula and Port Arthur. *Abbrev.* USGP *or* U.S.G.P.

United States Marine Underwriters *See* **US Marine Underwriters**.

United States Merchants All US flag merchant ships of over 1,000 tons gross, engaged in foreign commerce and uncontrolled by the Military Sealift Command (MSC), are obliged to submit their whereabouts to the appropriate US Naval authorities who keep a location filing system about US merchant ships. This information is made available to people concerned in shipping and the US military authorities. *Abbrev.* USMER.

United States Pacific Coast Ports Astoria, Longview, Portland (Oregon), San Francisco, Stockton, Seattle, Tacoma, Vancouver, *Abbrev.* USPCP *or* U.S.P.C.P.

United States Ship Financing *See* **US Ship Financing**.

Universal Time *Similar* to **Greenwich Mean Time,** *q.v. Abbrev.* UT *or* U.T.

UNL *or* **U.N.L.** Ultimate Net Loss, *q.v.*

Unlay To untwist; to undo the strands of a rope.

UNLC *or* **U.N.L.C.** United Nations Liaison Committee.

Unless Caused By A cargo insurance policy clause covering the holder from loss or damage caused by the ship sinking, stranding, being in collision, etc. *Abbrev.* U.C.B *or* u.c.b.

Unlimited A word inserted in marine insurance policies meaning unlimited during the currency. *Abbrev.* ul *or* Unltd.

Unlimited A firm or company which does not add the initials Ltd (Limited) or PLC (Public Limited Company) after its name. This type of company is not common nowadays, especially as all the shareholders in the company become automatically liable for eventual debts incurred.

Unmanned Machinery Space A space where alarm bells are installed on the bridge of a ship to trace or rectify any machinery faults. The computerised devices will report any fault immediately it appears and the engineers on board can attend to the necessary rectifications. *Abbrev.* UMS *or* U.M.S.

Unmanned Refrigerated Store Installation The classification of cold storage in some countries by Lloyd's Register, mostly in Middle East countries. *Abbrev.* URSI.

UNMC *or* **U.N.M.C.** United Nations Mediterranean Commission.

Unmoor To unloose a ship from her moorings.

UNO *or* **U.N.O.** United Nations Organisation.

Uno animo *Latin*—Unanimously; with one mind.

Unofficial Strike A strike without the consent of the union.

UNPA *or* **U.N.P.A.** United Nations Postal Administration.

unpd. Unpaid.

UNPOC *or* **U.N.P.O.C.** United Nations Peace Observation Commission.

UNREF *or* **U.N.R.E.F.** United Nations Refugee Emergency Fund.

Unregistered Luggage Hand baggage.

UNREP Underway Replenishment, replenishment at sea.

Unrepaired Damage Depreciation value for unrepaired damage. Underwriters, *q.v.*, do not pay a claim for unrepaired damage in addition to a Total Loss, *q.v.*, on the same policy. On the other hand, it is possible that a vessel has unrepaired damage falling on one policy, in which case the owner could claim for the total loss on the running policy and also for the estimated cost of unrepaired damage on the old policy.

UNRISD *or* **U.N.R.I.S.D.** United Nations Research Institute for Social Development.

UNRRA *or* **U.N.R.R.A.** United Nations Relief and Rehabilitation Administration.

UNRWA *or* **U.N.R.W.A.** United Nations Relief Works Agency for Palestine Refugees.

UNSC *or* **U.N.S.C.** United Nations Social Commission; United Nations Security Council.

UNSCC *or* **U.N.S.C.C.** United Nations Standards Coordination Committee.

UNSCCUR *or* **U.N.S.C.C.U.R.** United Nations Scientific Conference on the Conservation and Utilisation of Resources.

UNSCER *or* **U.N.S.C.E.R.** United Nations Scientific Committee on the Effects of Radiation.

UNSCOB *or* **U.N.S.C.O.B.** United Nations Special Committee on the Balkans.

Unsecured Creditors Creditors of a company or any other organisation who are not covered for the debts owed to them which can therefore only be settled if and when funds are available.

UNSF *or* **U.N.S.F.** United Nations Special Fund for Economic Development.

UNSG *or* **U.N.S.G.** United Nations Secretary General.

Unshackle To unloosen a shackle from two lengths of cable.

UNSM United Nations Standard Messages *Electronic Data Interchange.*

UNSR *or* **U.N.S.R.** United Nations Space Registry.

Unstuffing Removing merchandise out of a container. *Opposite to* **Stuffing**, *q.v.*

UNTA *or* **U.N.T.A.** United Nations Technical Assistance.

UNTAA *or* **U.N.T.A.A.** United Nations Technical Assistance Administration.

UNTAB *or* **U.N.T.A.B.** United Nations Technical Assistance Board.

UNTAM *or* **U.N.T.A.M.** United Nations Technical Assistance Mission.

UNTC *or* **U.N.T.C.** United Nations Trusteeship Council.

UNTDID United Nations Trade Data Interchange Directory.

UNTSO *or* **U.N.T.S.O.** United Nations Truce Supervision Organisation in the Middle East.

UNTT *or* **U.N.T.T.** United Nations Trust Territory.

UNU United Nations University.

UO A flag code sign hoisted at the entrance of a port or navigable waterway prohibiting entry for some reason, say movements inside; Use and Occupancy, *q.v.*

u/o Use and Occupancy, *q.v.*; Underwriting Officer.

U&O *or* **u/o** Use and Occupancy, *q.v.*

U of S.A. Union of South Africa.

UOMM Union de Officiales de la Marina Mercanta, Spanish Merchant Marine Officers' Association.

Up Unsaturated Polyester.

UP *or* **U.P.** United Press; Underproof Spirits.

u.p. Underproof *spirits.*

Up Anchor An order to weigh the anchor so that the ship can proceed.

Up and Down Vertical position of lifting gear. It refers to the anchor chain cable when in a perpendicular position as the Anchor, *q.v.*, is being lifted from the bottom of the sea; it is also the perpendicular position of the cargo hoist having its hook with or without cargo.

U.P.B.A.N. Union Professionelle Belge des Approvisionneurs de Navires, Members of ISSA, *q.v.*

UPC *or* **U.P.C.** Underwriters Pay Claims.

upd. Unpaid.

UPF *or* **U.P.F.** Underwriters Pay Fee. *Abbrev.* U.P. Fee.

U.P. Fee *or* **UPF** *or* **U.P.F.** Underwriters Pay Fee meaning Underwriters, *q.v.*, settle claims that had been paid abroad by the assured. *Also abbrev.* C.P.A. for Claims Payable Abroad or C.S.A. for Claims Settled Abroad.

Upgrading Improving the state of cleanliness of oil tanks by removing sludge from the bottom and clinging oil residues from the sides to enable the tanks to receive clean or higher grade oil or other liquid cargo without contamination.

Uplift Weight The entire weight of an aircraft when airborne including the passengers, their luggage and all the cargo. *Abbrev.* U/W *or* U.W. *or* u.w.

Up-Line A railway line from the provinces leading to the main terminus.

Upper Berth The uppermost sleeping place in a train or cabin of a ship.

Upper Deck Topmost continuous deck above the maindeck of a ship.

Upper Explosive Limit A rich proportion of about 10% or over of a gas mixture which may cause an explosion. *Abbrev.* UEL *or* U.E.L. Also called **Upper Flammable Limit** and *abbrev.* UFL *or* U.F.L.

Upper Flammable Limit *Similar to* **Upper Explosive Limit**, *q.v.*

UPS United Polish Shipyards; Uninterrupted Power Supply.

U.P.S. Underwriters Pay Stamps *referring to stamp duties*; Uninterrupted Power Supply.

u.p.s. Uninterrupted Power Supply.

Upset Price The lowest and starting point of a bidding price in an Auction Sale. The alternative phrase is Reserve Price, *q.v.*

Upswing Recovery from a time of economic slump. Also called Upward Phase.

U.P. Tax Underwriters Pay Tax.

Uptown Warehouse In certain places, either because of the nature of the goods or because of insufficient allocation at the harbour, bonded stores are allowed to be based away from the quayside.

UPU *or* **U.P.U.** Universal Postal Union.

Upward Phase *Similar to* **Upswing**, *q.v.*

U/R *or* **U.R.** Under Repair; Ultraviolet Radiation; Unsatisfactory Report.

Uranium A radioactive white metallic element, the heaviest of the elements occurring in nature, used as a source of atomic energy and (in the isotope U 235) in atomic bombs.

Uranometry Astronomical measurement which may be represented on a chart or map depicting the positions and distances of celestial bodies.

urb. area. Urban Area.

U. Rep. Under Repair.

URG Underway Replenishment Group. Unit composed of auxiliary vessels supplying fuel, ammunition and stores to warships at sea.

urg. Urgent.

URN Unique Reference Number.

U/rs. Underwriters, *q.v.*

Ursa Major The Great Bear. A constellation forming an outline of seven clear stars, very familiar and helpful during night navigation. The seamen's slang word is Dipper.

Ursa Minor The Little Bear *constellation*.

URSI Unmanned Refrigerated Store Installation, *q.v.*

Urug. Uruguay.

U.S. *or* **US** Uncle Sam, referring to United States of America; Under Seal *customs seal*.

U/S Unsorted *timber*.

u.s. Ultimo scorso *Italian*—Ultimo.; ut supra *Latin*—As above; Ubi supra *Latin*—In the place above.

u/s Unserviceable; Unsorted *in timber trade*.

USA *or* **U.S.A.** United States of America.

USAEC United States Atomic Energy Commission.

USAF *or* **U.S.A.F.** United States Air Force.

Usage The usual procedure; common practice; customs.

USAID *or* **U.S.A.I.D.** United States Agency for International Development.

Usance A time limit fixed by law to pay foreign bills of exchange. *See* **Days of Grace**.

USASI *or* **U.S.A.S.I.** United States of America Standards Institute or ANSI for American Natural Standards Institute.

USBC *or* **U.S.B.C.** United States Bureau of Census.

USBL Ultra Short Base Line, *similar to* SSBL, *q.v.*

USC *or* **U.S.C.** United States Courts; United States Currency; United States Code.

USCAB *or* **U.S.C.A.B.** United States Civil Aeronautics Board, *q.v.*

USCD *or* **U.S.C.D.** United States Commerce Department.

USCFR *or* **U.S.C.F.R.** United States Code of Federal Regulations.

USCG *or* **U.S.C.G.** United States Coast Guard.

USCGA *or* **U.S.C.G.A.** United States Coast Guard Academy.

USCGAD *or* **U.S.C.G.A.D.** United States Coast Guard Air Attachment Department.

USCGC *or* **U.S.C.G.C.** United States Coast Guard Cutter.

USCGR *or* **U.S.C.G.R.** United States Coast Guard Reserve.

USCP *or* **U.S.C.P.** United States Commodity Prices.

USD *or* **U.S.D.** United States Dollar.

USDA *or* **U.S.D.A.** United States Department of Agriculture.

USDCNBS *or* **U.S.D.C.N.B.S.** United States Department of Commerce, National Bureau of Standards.

USDI *or* **U.S.D.I.** United States Department of the Interior.

USDV *or* **U.S.D.V.** Ultra Shallow Draft (Draught) Vessel.

USEA United States East Coast.

Use and Occupancy An American term in non-marine insurance covering the loss from normal income derived from buildings and other establishments due to unforeseen circumstances. *Abbrev.* U & O *or* u/o *or* U.O.

USEC *or* **U.S.E.C.** United States East Coast.

USED *or* **U.S.E.D.** United States Energy Department.

Useful Capacity *Similar to* **Deadweight** *or* **Deadweight Capacity**, *q.v. Abbrev.* u.c. *Also called* **Useful Deadweight**.

Useful Deadweight *Similar to* **Useful Capacity**, *q.v.*

User Pays System of light dues or charges for the provision of aids to navigation whereby the shipowner shares in the cost of the maritime service.

USES *or* **U.S.E.S.** United States Employment Service.

USF *or* **U.S.F.** United States Forces.

USFC *or* **U.S.F.C.** United States Federal Court.

USFEA *or* **U.S.F.E.A.** United States Federal Energy Authority.

USFG *or* **U.S.F.G.** United States Federal Government.

USFR *or* **U.S.F.R.** United States Federal Reserve.

USG *or* **U.S.G.** United States Gallon; United States Government; United States Standard Gauge.

USGAS *or* **U.S.G.A.S.** United States Coast Guard Air Station.

USGLI *or* **U.S.G.L.I.** United States Government Life Insurance, *q.v.*

USGP *or* **U.S.G.P.** United States Gulf Ports.

USGS United States Geological Survey.

USHR *or* **U.S.H.R.** United States House of Representatives.

USI *or* **U.S.I.** United States Industries.

USIA *or* **U.S.I.A.** United States Information Agency.

Usi giudici *Latin*—Legal custom.

USIS *or* **U.S.I.S.** United States Information Service.

USM *or* **U.S.M.** Ultrasonic Machinery; Underwater to Surface Missile; Unlisted Securities Market *Stock Exchange London*; United States Marines; United States Mail.

USMA United States Marine Safety Association.

USMA *or* **U.S.M.A.** United States Military Academy.

USMARF *or* **U.S.M.A.R.F.** United States Maritime Administration's Reserve Fleet.

US Marine Underwriters The earliest record that can be traced of marine insurance in America dates back to 1721. A notice in the weekly *Mercury Gazette* announced the opening of an insurance office to deal with vessels, goods and merchandise in High Street, Philadelphia. In 1958 nearly all insurance companies were absorbed in a consortium named as the American

Institute of Marine Underwriters—*abbrev.* AIMU *or* A.I.M.U.

USMC *or* **U.S.M.C.** United States Marine Corps; United States Maritime Commission.

USMER United States Merchant(s).

USMMA *or* **U.S.M.M.A.** United States Merchant Marine Academy, Kings Point.

USMS *or* **U.S.M.S.** United States Maritime Service.

USMU *or* **U.S.M.U.** United States Marine Underwriters, *q.v.*

USN *or* **U.S.N.** United States Navy.

USNA *or* **U.S.N.A.** United States Naval Academy.

USNC *or* **U.S.N.C.** United States National Committee.

USNG *or* **U.S.N.G.** United States National Guards.

USNH *or* **U.S.N.H.** United States North of Hatteras.

USNHO *or* **U.S.N.H.O.** United States Navy Hydrographic Office.

USNI *or* **U.S.N.I.** United States Naval Institute.

USNR *or* **U.S.N.R.** United States Naval Reserve; United States Navy Regulations.

USNS *or* **U.S.N.S.** United States Navy Ship.

USNSB *or* **U.S.N.S.B.** United States Naval Submarine Base.

USNUSL *or* **U.S.N.U.S.L.** United States Navy Underwater Sound Laboratory.

USO *or* **U.S.O.** United Service Organisation *USA.*

USP *or* **U.S.P.** United States Patent; United States Pharmacopoeia.

USPCP *or* **U.S.P.C.P.** United States Pacific Coast Ports.

USPG *or* **U.S.P.G.** United Society for the Propagation of the Gospel.

USPHS *or* **U.S.P.H.S.** United States Public Health Service.

USPO *or* **U.S.P.O.** United States Post Office.

USPS *or* **U.S.P.S.** United States Postal Service.

USR *or* **U.S.R.** United States Reserves.

USS *or* **U.S.S.** United States Ship; United States Standards; United States Seamen.

USSA *or* **U.S.S.A.** United States Salvage Association, New York.

USSAF *or* **U.S.S.A.F.** United States Strategic Air Force.

USSADC United States System Acoustic Display Console.

USSB *or* **U.S.S.B.** United States, Shipping Board.

USSC *or* **U.S.S.C.** United States Supreme Court.

U.S. Ship Financing Under the Merchant Shipping Acts of USA, this is a Federal Ship Financing Programme guaranteeing credits by the Government to US citizens under the Title XI of the Merchant Marine Acts of 1936–1970. It covers a wide scope of finance to US shipowners primarily in the new building of ships in local shipyards. The US Marine Administration offers advantageous financial subsidies to US shipowners subject to the strict conditions laid down.

USSL *or* **U.S.S.L.** United States Shipping Legislation.

USSO *or* **U.S.S.O.** United States Service Organisation.

USSS *or* **U.S.S.S.** United States Secret Service.

UST *or* **U.S.T.** United States Tanker.

USTS *or* **U.S.T.S.** United States Travel Service.

usu. Usual; Usually.

Usufruct The right of reaping the fruits (*fructus*) of things belonging to others, without destroying or wasting the subject over which such right extended.

Usury Originally usury had the same meaning as interest has at the present day, *viz.*, a periodical payment in consideration of a loan. Contracts could be avoided for usury above ten in the hundred.

Many attempts were made to evade the statutes against usury by making the interest payable in the form of rents, annuities, etc., and in such cases the question was whether the stipulated payment was bona fide a rent or annuity, etc., or whether it was usury and the contract a usurious one within the statute.

Hence 'usury' acquired the sense of interest above the rate allowed by those statutes, and the statutes making all such usurious contracts void became known as the Usury Laws. Eleven statutes from the statute 1545, 37 Hen. 8, c. 9, to the statute 1850, 13 & 14 Vict. c. 56, were repealed by the Usury Laws Repeal Act, 1854; but the Money-lenders Acts, 1900 to 1927, give the courts certain powers with regard to oppressive rates of interest and the interest which pawnbrokers may take is still restricted.

USV Underwater Shutdown Valve.

USV *or* **U.S.V.** United States Vessel; United States Volunteers.

USW Underwater Sea Warfare; Ultra Short Wave.

u.s.w. Ultra Short Wave.

USWB *or* **U.S.W.B.** United States Weather Bureau.

USWC *or* **U.S.W.C.** United States West Coast.

USWD Undersurface Warfare Division.

USWI *or* **U.S.W.I.** United States West Indies.

UT *or* **U.T.** Unlimited Transhipment; Universal Time, *q.v.*

Ut. Utah, *USA.*

UTA Ulster Transport Authority; Union des Transports Aériens *French airline*; Unit Trust Association; Upper Control Area *aviation.*

UTC United Technology Corporation.

Ut dict. Ut dictum *Latin*—As directed.

utg. Utgave *Norwegian*—Edition.

Ut infra *Latin*—As under; as stated below.

Utmost Good Faith (Uberrimae fidei) A contract is said to be uberrimae fidei when the promisee is bound to communicate to the promisor every fact and circumstance which may influence him in deciding whether to enter into the contract or not. Contracts which require uberrima fides are those entered into between persons in a particular relationship, as guardian and ward, solicitor and client, insurer and insured. *See also* **Material Fact**.

UTP Unshielded twisted pair *copper cabling linking computers*.

Ut res magis valeat quam pereat *Latin*—It is better for a thing to have effect than to be void.

UTS *or* **U.T.S.** Underwater Telephone System; Underwater Telephone Service; Ultimate Tensile Strength.

Ut sup. Ut supra *Latin*—As above.

Ut videtur *Latin*—As it seems.

Utter In criminal law, to utter a forged document, die or seal, etc., or counterfeit coin, is to pass or attempt to pass it off as genuine when it is known to be forged. Uttering a forged document, seal or die is in general the same offence as that of forging the same document.

UU Unless Used; Ulster Unionist.

UUV Unmanned undersea vehicle.

UV *or* **Uv** Ultraviolet.

UVA Ultraviolet Alpha—rays from the sun which cause tanning.

UVB Ultraviolet Beta—rays from the sun which cause burning.

U/W Underwriter, *q.v.*

UWIST *or* **U.W.I.S.T.** University of Wales Institute of Science and Technology.

Ux Uxor *Latin*—Wife

V

v Velocity.

v. Vice *Latin*—In place of; Vide *Latin*—See.

V Symbol for Vanadium, *q.v.*, Vertex; Vertical, *Lloyd's Register notation*; Vice; Volunteers; Volt(s); International code flag signal showing diagonal red cross on white background meaning 'I require assistance'; Versus *Latin*—Against; Roman numeral for figure 5.

Va. Virginia, *USA*.

VA *or* **V.A.** Value Analysis; Veterans Administration *USA*; Vice-Admiral; Vicar Apostolic.

V/A Voucher Attached.

v.a. Value Analysis; Vixit . . . annos *Latin*—lived (so many) years.

vac Volts Alternating Current.

Vacation The vacations are the periods of the year during which the courts and chambers of the Supreme Court of Judicature in the United Kingdom are closed for ordinary business. There are, however, certain kinds of business which may be transacted during vacation, e.g. applications for injunctions, for extensions of time, etc., and for this purpose two vacation judges and a staff of officials attend in court periodically during the vacations. The vacations are four, namely the Easter Vacation, the Spring Vacation, the Long Vacation and the Christmas Vacation, the dates of which vary slightly from year to year. *See* **Sittings**.

VACBI Video and Computer Based Instruction.

Vaccination Clause A standard clause for Time Charters, *q.v.*, approved by the Documentary Council of BIMCO in May 1981. Owners of the vessel are to ensure that all officers and crew have vaccination certificates in their possession at their expense. The charterers, *q.v.*, are to give sufficient time for the inoculations and/or injections to be taken.

Vacuity Broken Stowage, *q.v.*, left by the ovality of barrels when stowed; the space left in a cask or container when liquid is poured in.

Vacuum Pressure Impregnated A system of wrap-insulation of coils formed by applying a solventless varnish. *Abbrev.* VPI *or* V.P.I.

Vacuvators Small portable pneumatic machines used to suck cargo (e.g. grain) from a ship's hold and to discharge overside through tubes.

VADR Vessel's Arrival and Departure Report.

VADS Value Added Data Service. *Cost-saving method of passing information electronically.*

Val. Value; Valuation.

VALM *or* **V.A.L.M.** Vertical Angle Leg Mooring.

Valuation Certificate A Broker's, *q.v.*, assessment certificate of the market value of a vessel or other object.

Valuation Clause A Marine Insurance, *q.v.*, clause. In ascertaining whether a vessel is a Constructive Total Loss, *q.v.*, the insured value shall be taken as the repaired value, and nothing in respect of the break-up value of the vessel or wreck shall be taken into account. Before this clause came into operation a loophole existed through which one could claim a Constructive Total Loss by showing that the vessel would not be worth more than the cost of the repairs. With this clause, however, the estimated cost of repairs and other expenses must reach the insured value. *Abbrev.* V.C.

Valuation Report Surveyor's report and assessment of the value of property for the purpose of insurance claims, taxation, estate purchase and others.

Value The utility of some particular object; the power of purchasing other goods which the possession of that object conveys. The one may be called value in use, the other value in exchange. The things which have the greatest value in use have frequently little or no value in exchange; and, on the contrary, those which have the greatest value in exchange have frequently little or no value in use. Nothing is more useful than water, but it will purchase scarcely anything; scarcely anything can be had in exchange for it. A diamond, on the contrary, has scarcely any value in use, but a very great quantity of other goods may frequently be had in exchange for it.

Value is often used as an abbreviation for 'valuable consideration', expecially in the phrases 'purchaser for value', 'holder for value', etc. The question whether a person acting in good faith has given value for property is often of importance when the person from whom he acquired it had not a perfect title as against some other person.

Value Added Tax An indirect tax applied at each stage in the exchange of goods and services between production and final sales as a percentage of the sale value. The intermediate manufacturers and traders through whom goods and services pass add the tax to their selling prices while reclaiming the tax they have paid to their suppliers. Thus from them is collected the tax on the value they have added, while the end customer pays the full tax. In the UK, it was introduced in 1973 to replace Purchase Tax and is the basis of contributions to the European Community budget. *Abbrev.* VAT *or* V.A.T.

Value as Original *See* **Valued As Original**.

Valued as Original *or* **Value as Original in Policy** In the valuation of a claim in the reinsurance policy the value remains as that of the original amount insured. *Abbrev.* VOP *or* V.O.P. *or* v.o.p.

Valued Policy The agreed value of the object insured is stated in the insurance policy. Hull and Goods policies are invariably of this type.

Value in Bond *Similar to* **Bonded Value**, *q.v.*

Value in Use An article which is valued or appreciated in accordance to its particular requirements. A case in point is that equipment for certain work could be of immense value to a manufacturing company but could be absolutely useless for an administrative office.

Valuer A person whose business is to appraise, or set a value upon, property. Some valuers are officially recognised by the state. Also called Assessor, *q.v.*

Valuta *Latin/Italian*—Worth; exchange currency value.

VAMA *or* **Vama** Voyage as may appear *insurance*.

VAN Value Added Network.

Van To pack a container *American term*.

Vanadium A hard grey metallic element used in small quantities for strengthening some steels.

Vanc Vancouver, *Canada*.

Van Carrier American term for a Straddle Carrier.

Vap *or* **VAP** Very Advanced Propulsion; Voluntary Assistance Programme *USA*.

Vapeur *French*—Steam; Steamer; Steamship.

Vapor *Portuguese, Spanish*—Steam; Steamer; Steamship.

Vapore *Italian*—Steam; Steamer; Steamship.

Vapour Pressure Liquid evaporation taking place together with the pressure formed in an enclosed or restricted space in relation to the heating temperature. The more heat put in, the more the pressure increases. *See* **Saturation Vapour Pressure**.

Var. Variation *nautical*; Variant; Variety.

Varec Seaweed.

Variable Injection Timing Mechanism for adjusting the interval between the injection of fuel and the piston stroke of an engine. *Abbrev.* VIT.

Variable Pitch *See* **Controllable Pitch Propeller**.

Variables Shifting winds. A region between the north-east and sount-east trade winds. The area where these winds meet near the equator is referred to as the doldrums.

Variation The angle between True North and the direction of the magnetic pole, i.e. between the true and magnetic meridians. Variation must be applied to courses and bearings by the magnetic compass as a correction to obtain true readings.

Variform Having various forms.

VARIG-SA Empresade Viacào Aérea Rio Grandense, Brazilian State Airlines.

Variometer A device for varying the inductance in an electric circuit.

VAS Visual Augmented System *electronics*.

VASI Visual Approach Slope Indicator, *q.v.*

Vat A large tub, cistern or other vessel, especially for holding liquids or holding something in liquid in the process of manufacture.

VAT *or* **V.A.T.** Value Added Tax, *q.v.*

Vatu Currency unit of the Republic of Vanuatu.

VB Vertical Boiler.

V. Band A radio frequency.

VBS Vessel Berthing System *USA*.

v.c. Verbi causa *Latin*—For example; Vinyl chloride.

VC Vendor Control.

VC *or* **V.C.** Valuation Clause, *q.v.*; Victoria Cross; Vice-Chairman; Vice-Chancellor; Vice-Consul.

V/C Vehicle Container; Voyage Charter, *q.v.*; Voyage Clause.

VCB *or* **V.C.B.** Vertical Centre of Buoyancy.

VCG *or* **V.C.G.** Vertical Centre of Gravity; Vice-Consul General.

VCM Vinyl Chloride Monomer *flammable and toxic chemical gas which can be liquefied for transport*.

VCO Voltage Controlled Oscillator.

VCR Video Cassette Recorder.

Vd Void.

v.d. Vapour Density; Various Dates.

VDA *or* **V.D.A.** Vertical Danger Angle *aviation*.

VDC Volatile Organic Compound.

VDL *or* **V.D.L.** Ventilation Dead Light.

VDI Virtual Design Interface *computer*.

VDR Variable Diameter Rotor; Verband Deutscher Reeder, *q.v.*

VDT Visual Data Terminal; Visual Display Terminal.

VDU Visual Display Unit.

VEC Variable Exhaust Valve Closing *engines*.

VECS Vapour Emission Control Systems.

Vector Course of an aircraft; To direct an aircraft in flight to the correct destination.

Veer To change direction, especially of wind in a clockwise direction; to change the course of a ship.

Veer and Haul To slacken a rope or cable preparatory to giving a big heave.

Veerdienst *Dutch*—Ferry Service.

Vegetable Oils These are usually shipped in bulk or in barrels and consist mainly of bean, coconut, cottonseed, palm and tung oils.

Vel Velocity; Vellum.

Velocity The rate of motion, usually of inanimate things; speed in a given direction.

Veluti in speculum *Latin*—Just as in a mirror; a perfect reflection.

Ven. *or* **Venez.** Venezuela.

Vendee The buyer of lands or goods. *See* **Vendor**.

Venditioni exponas *Latin*—That you expose for sale. A judicial writ addressed to the sheriff, commanding him to expose to sale goods which he has already taken into his hands, to satisfy a judgment creditor.

After delivery of this writ the sheriff is bound to sell the goods, and have the money in court on the return day of the writ. When a writ of *fieri facias* has been issued and the sheriff returns that he has taken goods, but that they remain in his hands for want of buyers, a writ of *venditioni exponas* may be sued out to compel a sale of the goods for any price they will fetch.

When a writ of extent in chief or in aid had been returned and no one appeared to claim the goods, etc., mentioned in the inquisition, a *venditioni exponas* issued directing the sheriff to sell them.

Vendor One who sells anything, especially land. In the case of goods he is usually called a seller.

Vendor and Purchaser The law relating to vendor and purchaser includes such subjects as the particulars and conditions of sale of land, the contract of sale, the abstract of title, requisitions, searches, etc., and the preparation and completion of the conveyance.

Vendor's Lien Where a vendor conveys land and the purchase money or part of it has not been paid, a lien arises as between the vendor and the purchaser, and persons claiming as volunteers, for so much of the purchase money as remains unpaid. The mere giving of security will not prevent the lien arising, unless it appears that the security was to be substituted for the lien. Similarly a purchaser will have a lien for prematurely paid purchase money.

It must be registered or it will be void against a purchaser, even if he has actual notice.

Vendue An auction sale. *American term.*

Vendue Master An auctioneer. *American term.*

Vent Pipe *See* **Air Pipe**.

Ventilators Natural ventilators are usually metal tubes rising vertically from the weather deck and fitted with adjustable cowls or with mushroom-shaped covers preventing water ingress whilst assisting free natural circulation of air in the cargo spaces below. Certain vessels—e.g. reefers—are equipped with sophisticated electric ventilation systems.

Venture Capital Capital raised for the start-up of a new business venture accompanied by a high risk and a targeted high return.

Venue The place where a jury are to come for trial of causes.

Venture To undertake a risk for financial or other gain; business speculation.

VER Voyage Event Recorder, *q.v.*

Verbal Note A memorandum or note, in diplomacy, not signed, and sent when an affair has continued a long time without any reply, in order to avoid the appearance of an urgency which, perhaps, is not required, and, on the other hand, to guard against the supposition that it is forgotten, or that there is an intention of not prosecuting it any further.

Verba volant scripta manent *Latin*—Words fly away, but writings remain.

Verband Deutscher Reeder *German*—Shipowners' Association. *Abbrev.* VDR.

Verbatim et literatim *Latin*—Word for word and letter for letter.

Verborum obligatio *Latin*—A verbal obligation, contracted by means of a question and answer.

Verborum obligatio verbis dissolvitur *Latin*—An obligation created by words is dissolved by words; and so a verbal release of a verbal contract is effectual.

Verbum satis sapienti *or* **Verbum sapienti sat est** *Latin*—A word to the wise man suffices.

Verdict The opinion of a jury on a question of fact in a civil or criminal proceeding.

Verdigris Green rust on copper.

Vereenigde *Dutch*—Associated; United.

Vereeniging *Dutch*—Association; Union.

Vereinigung *German*—Association; Union.

Verein *German*—Association.

Veritas nihil veretur nisi abscondi *Latin*—Truth fears nothing but concealment.

Vermin In shipping this is expressly applied to the infestation of rats, mice and to a lesser degree cockroaches. *Abbrev.* Verm. *See* **Deratting Certificate**.

Verst Russian measure of length, 3,500 feet.

Vert. Vertical.

Vertex The highest point, top, apex; each angular point of a triangle, polygon, etc. *geometry.*

VERTOL Vertical Take-off and Landing *aviation.*

Verwaltung *German*—Management.

Very Large Crude Carriers 200,000/299,000 dwt. On similar long haul routes to Ultra Large Crude Carriers, *q.v.*, but with greater market flexibility due to smaller size.

Very Long Range This refers to an aircraft's fuel capacity and the distance to be covered. *Abbrev.* VLR *or* V.L.R.

Very Low Frequency *See* **VLF** *or* **V.L.F.**

VESDA Very Early Smoke Detection and Alarm.

Vessel Crossing A vessel making her way to the Fairway, *q.v.*

Vessel Inward A vessel entering a harbour or dock.

Vessel Leaving A vessel making preparations to leave a port.

Vessel Not Under Command The term 'vessel not under command' means a vessel which through some exceptional circumstance is unable to manoeuvre as required by the International Regulations for Preventing Collisions at Sea and is therefore unable to keep out of the way of another vessel. A vessel not under command must sound at intervals of not more than two minutes three blasts in succession, namely one prolonged followed by two short blasts. With regard to lights and shapes, she must exhibit two all-round red lights in a vertical line where they can best be seen, two balls or similar shapes in a vertical line where they can best be seen, and in addition, when making way through the water, sidelights and a sternlight.

Vessel Response Plan A plan submitted by ship's owners for dealing with oil spill emergencies which is demanded by the US Coast Guard to meet the requirements of OPA 90, *q.v. Abbrev.* V.R.P.

Vet. *or* **VS** Veterinary Surgeon.

Veto The constitutional right of a sovereign, president, governor, upper house of legislature, etc., to reject a legislative enactment. As applied to the charter of the United Nations the term means the power of the permanent members of the Security Council to prevent certain decisions by a refusal to concur in the vote when such concurrence is necessary.

VFR *or* **V.F.R.** Visual Flight Rules *aviation*.

Vfy Verify.

V.G. Very Good; Vicar General; Votre Grâce *French*—Your Grace.

V.g. *or* **v.g.** Verbi gratia *Latin*—For the sake of example; Very Good.

V.G.C. *or* **v.g.c.** Very Good Condition; Viscosity Gravity Constant.

VGR *or* **V.G.R.** Vertical Grab Rod.

V.H.C. *or* **v.h.c.** Very Highly Commended.

VHF *or* **V.H.F.** Very High Fidelity; Very High Frequency.

VHF/UHF *or* **V.H.F./U.H.F.** Very High and Ultra High Frequency.

VHLV Very Heavy Lift Vessel, with jumbo derricks or cranes.

VHO *or* **V.H.O.** Very High Output.

VI *or* **V.I.** Vancouver Island, *Canada*; Virgin Islands; Viscosity Index.

V.I. Vide infra *Latin*—See below.

Via Through; By way of.

Viability Capability to survive.

VIAS *or* **Vias** Voice Interference Analysis System.

VIASA Venezolana International de Aviacion S.A. Venezuelan International Airlines.

VIC Very Important Cargo.

Vice propre *French*—Inherent defect.

Victor International Radio Telephone phonetic alphabet for 'V'.

Victualler A shipsupplier or shipchandler; supplier of provisions and stores to ships.

Victualling Bill One of the customs forms needed to obtain permission from the authorities to supply stores from a Bonded Warehouse, *q.v.*, as ship's stores.

Victualling Officer A Catering Officer; Catering Superintendent. The Purser, *q.v.*, of a passenger ship has the victualling under his charge apart from other departments.

Victualling Scale The official maximum scale allowed by every maritime country for provisions to be supplied and served out daily during the engagement of their national seafaring men on their own registered ships.

VID Volunteers for International Development.

Vide *Latin*—See.

Videlicet *Latin*—Namely. *Abbrev.* Viz.

Videotex Generic term for services which allow computerised information to be transmitted and displayed in a particular format, either by broadcast or through a two-way network.

Vide supra *Latin*—See above.

V.I.I. Viscosity Index Improver.

V.imp. Verb impersonal.

Vincit omnia veritas *Latin*—Truth conquers all.

VIO Very Important Object; Veterinary Investigation Officer.

VIP *or* **V.I.P.** Value Improvement Project; Very Important Person; Vias Internacionales de Panama, Panama International Airlines.

VIPRE Visual Precision.

Virtute et labore *Latin*—By virtue and toil. Motto of Mississippi, USA.

Virtute non verbis *Latin*—By virtue not by words.

Virtute officii *Latin*—By virtue of (his) office.

Visa A document or endorsement on a passport. It can also be a separate document issued to a person or a group of persons for entry into a foreign country.

VISA Voluntary Intermodal Sealift Agreement.

Viscosity Resistance to flow of a fluid.

Vis et armis *Latin*—By force and arms, by main force.

Visibility The degree of clearness or transparency of the atmosphere in relation to the maximum distance that the naked eye can possibly perceive. Daily visibility reports are relayed by the authorities for air and sea navigation purposes, as well as for the general public.

Visible Balance Balance of Trade. That amount resulting from the money received for exports against the money paid for imports. *See* **Trade Gap.**

Visible Trade Import, export and re-export transactions.

Vis Major Such a force as it is practically impossible to resist, e.g. a storm, an earthquake, the acts of a large body of men, etc. The doctrine of vis major is that a person is not liable for damage if it was directly caused by vis major. It includes many things described as an act of God.

VISTA Viewing Instantly Security Transactions Automatically; Volunteers in Service to America.

Visual Approach Slope Indicator A system of red and white lights set at specific angles alongside the runways of airports to direct the pilots of approaching aircraft to determine the correct and safe landing course. *Abbrev.* VASI.

Visual Ray An imaginary line representing a ray of light from the eye to the point of the object.

Vis vitae *Latin*—Vital force.

VIT Variable Injection Timing, *q.v.*

VITA Volunteers for International Technical Assistance *USA*.

Vita brevis, ars longa *Latin*—life is short, art endures.

Vitesse *French*—Speed.

vitr Vitreum, *Latin*—glass.

Vivat *Latin*—Long may he or she live.

VLWT *or* **V.L.W.T.** Very Large Water Tanker.

Viz. Videlicet *Latin*—Namely.

VK Air Tungaru Corporation *airline designator*.

VK *or* **V.K.** Vertical Keel; Volume Kilos, *q.v.*

VKO *or* **V.K.O.** Vereeniging, van Kapiteins en Officieren Ter Koopvaardij; *Dutch* Mercantile Marine Captains' and Officers' Association.

v.l. Varia lectio *Latin*—A variant reading.

VLBC *or* **V.L.B.C.** Very Large Bulk Carrier.

VLCC *or* **V.L.C.C.** Very Large Crude Carrier, *q.v.*

VLF *or* **V.L.F.** Very Low Frequency *radio frequency.*

VLGC *or* **V.L.G.C.** Very Large Gas Carrier.

VLN *or* **V.L.N.** Very Low Nitrogen.

VLO *or* **V.L.O.** Very Low Oxygen.

VLPC *or* **V.L.P.C.** Very Large Product Carrier.

VLR *or* **V.L.R.** Very Long Range, *q.v.*

VLS *or* **V.L.S.** Volume, Loadability and Speed *aviation.*

VLSI Very Large Scale Integration, *computer.*

VLSS Very Large Submersible Ship.

Vltg Voltage.

VM *or* **V.M.** Velocity Modulation; Volatile Matter.

V/M Volts per Metre; Volts per Mile.

VMAA Vancouver Maritime Arbitrators' Association.

VMC Visual Meteorological Conditions *aviation.*

VMD *or* **V.M.D.** Veterinariae Medicinae Doctor *Latin*—Doctor of Veterinary Medicine.

V.M.H. Victoria Medal of Honour, *Royal Horticultural Society.*

VMIS Vessel Management Information System *computer.*

VMP *or* **V.M.P.** *or* **v.m.p.** Valued as Marine Policy.

V.M.T. *or* **v.m.t.** Very many thanks.

vo Verso *Latin*—left-hand page.

V/O Vneshnetorgovoe ob Edineniew *Russian*—Foreign Trade Organisation.

VOA Voice of America.

VOC Volatile Organic Compound; hydrocarbons that vaporise naturally. Used also to describe the vapours found in petroleum or product cargo tanks.

Vogue la galère! *French*—Come what may.

Void An agreement or other act is said to be void when it has no legal effect, or not the legal effect which it was intended to produce. But an agreement which is void may subject the parties to penal consequences. Thus, an agreement amounting to a conspiracy would be admissible as evidence in criminal proceedings.

An act may be void either *ab initio* or *ex post facto*. Thus, if a contract is made without the true consent of the parties or for an immoral consideration it is void *ab initio*. No person's rights can be affected by it, whether he is a party or a stranger. In the case of a contract which is void for illegality, immorality, or on a similar ground, if money has been paid as the consideration of its performance, the party who has paid it may repudiate the contract and recover it back at any time before performance. But when an illegal contract has been executed, money paid under it cannot usually be recovered back.

If a deed is properly entered into, and is afterwards altered in a material point by the fraud or laches of either party, it becomes void *ex post facto* as against him, so that he cannot enforce or take advantage of it. *See* **Voidable**.

Void Policy An Insurance Policy, *q.v.*, in which the assured has no insurable interest in the matter as in Policy Proof of Interest, *q.v.* This is legally void or invalid. *Abbrev.* Void.

Voidable An agreement or other act is said to be voidable when one of the parties is entitled to rescind it, while until that happens it has the legal effect which it was intended to have. It can, however, be disputed only by certain persons and under certain conditions, and the right of rescission may be abandoned by the party entitled to exercise it. If third persons acquire rights under a voidable contract or other transaction without notice and for value, they cannot afterwards be put in a worse position by its being set aside. Herein a voidable contract differs from a void contract, for in the latter case no third person can acquire rights under the contract unless the party against whom it is void elects to affirm it.

The principal examples of voidable transactions occur in the case of misrepresentation. *See* **Void**.

Voidable Policy A policy in respect of which the Underwriter, *q.v.*, is entitled to avoid liability.

Voilà tout *French*—That is all.

Voit. Voiture *French*—Wagon.

Voith-Schneider Abbrev. VSP, *q.v.*

Vol. Volcano; Volume.

Volatile Liquids Commodities which rapidly lose some weight or volume through evaporation. Such commodities include gasolene, turpentine, spirits, etc.

Volatility Said of 'Standard deviations' for gold and silver.

VOLCOA Code name for a BIMCO standard Volume Contract of Affreightment for the transportation of dry bulk cargoes.

Volenti non fit injuria *Latin*—To a willing person no injury is done. That to which a man consents cannot be considered an injury. This maxim is illustrated by the rule that no man can maintain an action of tort in respect of any act to the doing whereof he has consented. Consent or leave and licence is a defence in actions of tort or prosecutions unless the act amounts to the infliction of a serious physical injury or where the rights of the public as well as the individual sustaining harm have intervened. The public are interested in preventing one of their number from grievous bodily harm and from exhibitions which alarm the public conscience, such as prize-fights without gloves, duels, etc.

The maxim has also been invoked in cases where the person injured is alleged to have contracted to absolve the defendant from a known risk or from obligations, such as not to be negligent in the course of a contract. Knowledge of the risk is not conclusive; it is only evidence from which consent may be inferred.

The performance of a duty in the face of imminent risk is not consent disentitling a plaintiff from his remedy.

Volmet Derived from the French word 'vol' meaning Flight. Weather reports relayed to most airports by aircraft.

Volo, non valeo *Latin*—I am willing, but not able.

Volt A unit of electromotive force, the force that would carry one ampere of current against one ohm resistance.

Voltmeter An instrument for measuring electric currents.

Voltameter An instrument for measuring electric currents by their electrolytic effects.

Volume Kilos The chargeable air cargo freight. This is calculated as follows: Volume in inches divided by 427. If the result becomes bigger than the gross weight then it will be charged on the result known as Volume Kilo. If, on the other hand, it is calculated in cubic metres, then divide that volume by 7000 and again if the result is bigger than the gross weight that Volume Kilo is charged as airfreight. *Abbrev.* VK or V.K.

Volumometer An instrument for measuring the volume of a solid body by the quantity of liquid, etc., displaced.

Volumeter Kinds of instrument for measuring the volume of gases.

Voluntary Liquidation When a company is being wound up by resolution of its shareholders it is said to be in voluntary liquidation.

Voluntary Observing Ship A merchant ship which assists the Meteorological agencies by transmitting regular reports of surface weather conditions and oceanographical observations. Liaison with these ships is maintained by Port Meteorological Officers. This international programme involves ships of 49 countries and is under the auspices of the World Meteorological Organisation.

Voluntary Patient A person who voluntarily submits to treatment for mental illness.

Voluntary Sacrifice One of the actions the master has to be involved in to constitute a General Average Act, *q.v.* in the maritime adventure.

VOP *or* **V.O.P.** Value as in Original Policy.

VOR Very-High-Frequency Omnidirectional Range *aviation*.

VOR *or* **V.O.R.** Vehicle Off the Road.

VORTAG Very-High-Frequency Omnidirectional Range and Tactical Air Navigation *USA*.

Vortex A mass of whirling fluid, especially a whirlpool; the centre of a revolving storm in the tropics where the barometric pressure is lowest.

VOR VHF Omni Range, used in airways navigation.

VOS Voluntary Observing Ship, *q.v.*

Voy. Voyage.

Voyage Estimating The art of accurately estimating the likely loss or profit from a proposed charter.

Voyage Event Recorder Instrumentation contained in a fireproof buoyant box mounted above the upper deck continuously recording on tape a vessel's heading, speed, machinery readings, radar detections, audio transmissions and ambient noise. Similar to the Black Box, *q.v.*, carried by aircraft. *Abbrev.* VER.

VOYLAYRULES 93 Voyage Charterparty Laytime Interpretation Rules 1993, issued by BIMCO, *q.v.*

VOYWAR 1993 BIMCO Code name for Standard War Risks Clause for Voyage Charter.

VP *or* **V.P.** *or* **v.p.** Vapour Pressure; Variable Pitch, *q.v.*, Vice President; Vibrator Point *geophysical*.

VPI *or* **V.P.I.** Vacuum Pressure Impregnated, *q.v.*

VPP *or* **V.P.P.** Variable Pitch Propeller, *Similar to* **Controllable Pitch Propeller,** *q.v.*

VPS Vacuum Packing System.

v.p.s. Vibrations per second.

v.r. Various Readings; Voltage Regulator.

V.R. Valtionrautatiet, Finnish State Railways.

Vraic A seaweed found in the Channel Islands, at one time used for fuel and manure.

VRI *or* **V.R.I.** Visual Rule Instrument—for landing purposes by aircraft.

VRP Vessel Response Plan, *q.v.*

VS Veterinary Surgeon; Visual Signalling *flags or lamp*; Vital Statistics.

vs. Versus *Latin*—Against.

v.s. Vide Supra *Latin*—See above; Volti Subito *Italian*—Turn over quickly (music).

VSA *or* **V.S.A.** Vertical Sextant Angle *nautical*.

VSAT Very Small Aperture Terminals *telecommunications*.

VSC *or* **V.S.C.** Variation Setting Control *nautical*.

VSD *or* **V.S.D.** Vendor's Shipping Document.

VSE Vancouver Stock Exchange.

VSI Vertical Speed Indicator; Very Soft Iron.

Vsl Vessel.

VSO *or* **V.S.O.** Voluntary Service Overseas.

VSOP *or* **V.S.O.P.** Very Special Old Pale, cognac liqueur grade.

VSP Vertical Seismic Profiling *geophysical*; Voith–Schneider Propeller, also called Cyclodial Propeller.

VSQ *or* **V.S.Q.** Very Special Quality.

VSR *or* **V.S.R.** Very Short Range; Very Special Reserve.

V/Stol *or* **V/S.T.O.L.** Vertical and Short Take-Off and Landing *aviation*.

VSWR *or* **V.S.W.R.** Voltage Standing Wave Ratio *electricity*.

Vt *or* **V.T.** *or* **v.t.** Variable Time; Various Time.

VT Vermont, *USA*.

VTC *or* **V.T.C.** Verified With Test Certificate *Lloyd's Register notation*.

VTIS Vessel Traffic Information System.

VTMS Vessel Traffic Management Service.

VTO *or* **V.T.O.** Vertical Take-Off.

VTOL *or* **V.T.O.L.** Vertical Take-Off and Landing *aviation*.

VTR *or* **Vtr.** Video Tape Recorder.

VTS Vessel Traffic System/Service. The primary aim of VTS is to give a navigator advance warning of the movement of other vessels, traffic congestion, weather conditions and other hazards to navigation. However in certain conditions such as adverse weather conditions, or when the harbour is particularly congested, the VTS control centre can and will issue directions to control and supervise traffic by specifying times when vessels may enter, move within or through, or depart from locations within the VTS area.

VTSS Vessel Traffic Separation Scheme. *Similar to* **Traffic Separation Scheme**, *q.v.*

VTU Volunteer Reserve Training Unit *US Coast Guard*.

v.u. Volume Unit.

Vulgo *Latin*—Commonly.

v.v. Vice versa *Latin*—Reverse; Viva voce *Latin*—By word of mouth; Spoken.

v/v Volume per Volume; Volume/Volume.

VVSi Very, very small inclusion *diamond clarity*.

VWF Vibration-induced White Finger. Also known in the medical profession as Reynaud's Phenomenon.

VWG *or* **V.W.G.** Vibrating Wire Gauge.

V.Y. Victualling Yard.

W

w Weather symbol for Dew; wooden *ship*.

W Wales; Warden; Watch Time, *see* **Bell** *and* **Watch Duties**; Wave-height Correction *altitude*; Watt; Wednesday; West *compass point*; Winter mark for Load Line, *q.v.*; Worldscale; Tungsten *formerly wolfram*.

WA *or* **W.A.** West Africa; Western Africa; Western Australia; With Average, *q.v.*; Washington, *USA*.

W/A Wait for an Answer.

WAC World Aeronautical Chart.

WACCC *or* **W.A.C.C.C.** World Wide Air Cargo Community Classification.

WACD World Association for Christian Democrats.

WACL World Anti-Communist League.

W.A. Clauses Institute Cargo Clauses with Average.

WACO World Air Cargo Organisation.

WAD *or* **W.A.D.** With All Despatch or Dispatch. *See* **Custom of the Port**.

Wadi *or* **Wady** A channel or stream which gets flooded during rainy days and subsequently dries up.

WAE *or* **W.A.E.** When Actually Employed.

WAEA World Airline Entertainment Association.

WAEMA Western and English Manufacturers' Association.

W.A.F. *or* **w.a.f.** With all Faults.

Wage Drift The result of wages rising ahead of productivity, or the cost of living.

Wage Hourly Basis for daily or weekly payment of wages to an employee.

Wage Incentives Bonuses.

Wage Rates Wage rates in the category of piece-time, daily, weekly and monthly periods.

Wage Restraint During Inflation, *q.v.*, the state endeavours to stop the usual cycle of wage increases. One of the first moves will be to restrain high wage increases by reaching some form of agreement with the unions.

Wagering Policy A Policy, *q.v.*, effected on a life, ship, house, etc., in which the insurer has no Insurable Interest, *q.v.*

Wages A remuneration for services or work rendered by individuals or earnings derived from work done. Wages are paid according to agreements. *See* **Wage Rates**.

Wages, Stores, Provisions and Insurance A shipping term referring to the Running Expenses, *q.v.*, of the ship. *Abbrev.* WSPI or W.S.P.I.

Wagon-Lit Compartments in trains used as sleeping carriages.

Waist Part of the Upper Deck, *q.v.*, of a ship positioned between the forecastle and quarterdeck; Amidships.

Waiters Persons who act as attendants at a Stock Exchange, *q.v.* Also attendants in the Room at Lloyd's of London.

Waiting On Cement An oil drilling expression. Time allowed for waiting for cement to harden in a hole or when a cement plug is placed and hardened. *Abbrev.* WOC or W.O.C.

Waiting Time Time lost in waiting through no fault of the employee or labour force. In most cases workers are paid for Waiting Time.

Wait-Listed In the course of booking cargo or passengers the ship, aircraft or train may become fully booked and the Carrier, *q.v.*, will be committed to accept none further. Alternatively cargo or passengers may be accepted as 'wait-listed' so that, in the event that booked cargo or passengers do not turn up or their bookings are cancelled, the wait-listed cargo or passengers will fill the vacancies.

Waive To reject an advantage or benefit.

Waiver A person is said to waive a benefit when he renounces or disclaims it, and he is said to waive a tort or injury when he abandons the remedy which the law gives him for it. A waiver may be either express or implied.

Waiver Clause A Marine Insurance, *q.v.*, clause where both the Underwriters, *q.v.*, and the insured are covered within the legality of the policy in the process of negotiations over Abandonment, *q.v.*, claims. *See* **Subrogation**.

Wake A strip of smooth water left behind a moving ship.

Wake Speed The volume or strength of the Wake, *q.v.*, or stream left under the stern of the ship.

Wale A heavy timber along a jetty to serve as a fender; a thick steel plate rivetted or welded around the sides of a ship as a reinforcement for the ship itself and also to withstand heavy impact.

Walk Back To loosen or unwind a rope or cable by reversing the Capstan, *q.v.*, Windlass, *q.v.*, or Winch, *q.v.*

Walking Boss A ship's superintendent or ship's inspector *USA*.

Walking Off A Peril, *q.v.*, not usually covered by insurance. Deck cargo and livestock are cases in point.

Walk-Out A strike by labour.

Wall Sided A flat-bottomed ship.

Wall Street A well known street in the commercial area of New York where most of the important banks, insurance underwriters and other financial and business organisations are located including the New York Stock Exchange. *See* **Curb Market** *or* **Street Market**.

Waning Moon The gradual diminishing of the moon in size and illumination after full moon. *Opposite to* **Waxing Moon**, *q.v.*

WAR World Administrative Radio.

WARC World Administrative Radio Conference.

War Cancellation Clause *or* **War Risk Cancellation** A Marine Insurance, *q.v.*, clause in which the policy becomes null and void in the event of a declaration of war.

Wardroom The dining or mess room in a warship used by the officers. *Abbrev.* WR *or* W.R.

War etc. War, Strikes, Riots and Civil Commotions *insurance.*

Warehouse A building where goods are stored for future use; a large store where goods are sold on a wholesale basis; a store for goods to be deposited without payment of import duty. This latter type of store or warehouse may belong to the government or to a private person or company on special permission by the authorities. *See* **Bonded Warehouse** *or* **Bonded Stores**. *Abbrev.* Whse.

Warehouse Books Registers kept by warehousing officials to record the details of merchandise stored. *Abbrev.* WB *or* W.B. *or* W/B.

Warehouse Entry A customs house form of application to store merchandise in the government warehouse.

Warehouse Keeper A person who undertakes to store articles of any kind; a person who is in charge of a warehouse.

Warehouseman A person in charge of a warehouse.

Warehouse to Warehouse Clause A Marine Insurance, *q.v.*, clause covering any damage occurring in the period of transit from one warehouse to others until the merchandise has been withdrawn for its final destination. *Abbrev* W/W *or* whse-whse.

Warehouse Warrant An official document confirming the goods deposited in a government warehouse. These goods are transferable by endorsement. *Abbrev.* W/W.

Warehousing Entry. *See* **Entry for Warehousing**.

Warks. Warwickshire *UK*.

Warrant (1) An authority; a precept under hand and seal to some officer to arrest an offender, to be dealt with according to due course of law; also a writ conferring some right or authority, a citation or summons. (2) A voucher, written authorisation to receive money.

Warranted Surety, *q.v.*, has been given by the Underwriter, *q.v.*

Warranted Free from Average A marine insurance clause where partial loss is not covered. *Abbrev.* WFA *or* W.F.A. This is similar to Free of Particular Average, *q.v.*, and Warranted Free from Particular Average, *q.v.*

Warranted Free from Particular Average A Marine Insurance, *q.v.*, clause exempting indemnification by Underwriters, *q.v.*, of any Partial Losses, *q.v.*, except those that come under the category of General Average, *q.v. Similar to* Warranted Free from Average, *q.v.*; and Free of Particular Average. *Abbrev.* WFPA *or* W.F.P.A.

Warrantee A person to whom a Warranty, *q.v.*, is made.

Warranty A term of a contract collateral to the main purpose of the contract; any agreement either accompanying a transfer of property, or collateral to the contract for such transfer or to any other agreement or transaction. A warranty may be express or implied by law or statute.

Implied warranties have been said to underlie or to be the gist of actions for negligence or breach of duty arising out of the special relations between parties.

As a general rule an express warranty, or a known usage, excludes any implication of warranty in regard to the same subject-matter: *expressum facit cessare tacitum*. But upon a sale of goods if the express warranty has been superadded for the benefit of the buyer or is not inconsistent with the implied warranty, the latter is not excluded.

In the law relating to the sale of goods, a warranty is an agreement with reference to goods which are the subject of a contract of sale, but collateral to the main purpose of such contract, the breach of which gives rise to a claim for damages, but not to a right to reject the goods and treat the contract as repudiated. Whether a stipulation in a contract of sale is a condition, the breach of which may give rise to a right to treat the contract as repudiated, or a warranty, depends in each case on the construction of the contract; a stipulation may be a condition though called a warranty in the contract.

A warranty of goods may constitute a defence to a criminal charge.

In the law of marine insurance, what is called a warranty is in fact a condition. The Marine Insurance Act 1906 provides for the release of the insurer from liability, unless the policy contains a provision to the contrary, if the warranty is not exactly complied with. The most usual express warranties are that a vessel is safe on a given day, and that she will sail or depart on a given day. The most important implied warranty is that the vessel is seaworthy at the commencement of the risk. Warranty of documentation is a warranty that she has those documents which are required by international law or treaty to establish her national character. A continuing warranty is one which applies to the whole period during which the contract is in force: thus, an undertaking in a charterparty that a vessel shall continue to be of the same class as she was at the time the charterparty was made is a continuing warranty.

When a man holds himself out as the agent of another he is deemed to warrant his authority to act as agent, and if, and in so far as, he is without such authority as he purports to have, an action for breach of warranty of authority will lie against him for one damnified by such absence of authority. *Abbrev.* Wty.

Warranty of Attorney Power of Attorney, *q.v.* Authority given by a person to another empowering him to act on his behalf in legal or other situations.

Warranty of Legality A Charterparty, *q.v.*, implied warranty ascertaining that the adventure to be undertaken will be a lawful one throughout.

Warranty of Neutrality A Marine Insurance, *q.v.*, implied warranty on the part of the insured guaranteeing to underwriters that throughout the whole period of insurance both the ship and the cargo carried must be of a neutral character and does not involve the hostility of any nation.

Warranty of Workmanlike Service *or* **Warranty of Workmanlike Performance** Shipowner's liability in the USA covering the seaworthiness of a vessel *vis-à-vis* Longshoremen, *q.v. Abbrev.* WWLP.

War Risk Cancellation *Similar to* **War Cancellation Clause**, *q.v.*

Warsaw Convention 1929 An international convention for the Unification of Certain Rules Relating to International Carriage by Air.

War Zone An area on the High Seas, *q.v.*, so declared by nations that are at war. An area so defined by the competent authorities for insurance or for pay agreements.

WASA West African Shippers' Association.

Washed Overboard When cargo or other objects are washed overboard from a ship at sea. *Abbrev.* WOB *or* W.O.B.

Wash Primer A paint containing phosphoric acid mostly applied on bare metal and in a very thin film. It has an improved adhesive power and is applied prior to other paint coatings. Also called Etch Primer.

WASP Wind Assisted Ship Propulsion.

Waste Book A book used for daily entries in the book-keeping system. The contents are eventually entered into other complementary books.

Waste Cube Cubic space in a Container, *q.v.*, which is lost or unusable when awkward-shape cargo is stacked in it.

Wasting Assets Assets that become wasted in time or of no value because of the nature of the work or production.

WAT Weight Altitude Temperature *aviation.*

WATA *or* **W.A.T.A.** World Association of Travel Agents.

Watch Below A section of a ship's crew who are at rest or off duty while their colleagues keep watch.

Watch (1) The day of 24 hours starting from noon and divided into seven watches as follows: Noon or afternoon watch (start) 12 00 to 16 00 hours; first dog watch 16 00 to 18 00; second dog or last dog watch 18 00 to 20 00; first watch 20 00 to midnight; second watch or middle watch Midnight to 04 00 hours; morning watch 04 00 to 08 00; forenoon watch 08 00 to 12 00 (noon). Each of these watches is a period of duty for a seaman (when he is 'on watch') at the end of which he will be relieved by another. (2) In former times the seamen were organised in two Watches, 'port' and 'starboard', doing duty for alternative periods or 'watch and watch'. It is still the way of organising a warship's complement in two halves. (3) A buoy or other object is said to Watch when it is visible on the surface of the sea as opposed to being submerged.

Watch on Deck *Opposite to* **Watch Below**, *q.v.*

Water Below Leaks in the hold(s) of a ship.

Water Biscuits. *See* **Sea Biscuits**.

Waterborne (1) A floating vessel at sea. *Abbrev* WB *or* W.B. *or* W/B. (2) The carriage or transportation of goods by sea. (3) Cargo shipped on board a vessel. *Abbrev* WB *or* W.B. *or* W/B.

Waterborne Agreement An understanding in the Marine Insurance, *q.v.*, market whereby underwriters will cover goods against war risks only while they are on board an overseas vessel. Limited cover is allowed while goods are in craft en route between ship and shore and, also, during transhipment.

WATERCOM Waterway Communication System *USA.*

Water Fog Small droplets of water under high pressure injected through sprayers to extinguish fire in a ship's compartment or machinery.

Water-gate A gate to check or release the flow of water in a canal, river or dock.

Watergauge An instrument to measure the height of water in boilers, tanks, etc. *Abbrev* WG *or* W.G. *or* w.g. It may be a Gauge Glass.

Water Guard A customs officer engaged in anti-smuggling operations *USA. Abbrev.* WG.

Waterguard Customs officers in the UK charged with the prevention of smuggling, examining duty free stores and searching passengers' baggage.

Waterlogged Said of a ship so filled up with water that she will become unmanageable or impossible to operate.

Waterman A licensed boatman; a ferryman; a bargee.

Water Mark (1) A distinctive mark impressed in paper during its manufacture to identify the brand or the type, or for security as in bank notes. (2) A visible line on the shore of the observed or notional level reached by the tide.

Waterplane The surface area covered by a floating vessel gauged by taking the measurements from the whole waterline.

Water Quality Insurance Syndicate A US insurance covering American Shipowners and Ship Managers Syndicate on sea pollution. *Abbrev.* WQIS *or* W.Q.I.S.

Watertight Doors Doors of a ship which when closed will not allow water to enter compartments, cabins, engine rooms, etc. *Abbrev.* WTD *or* W.T.D.

Waterway (1) A navigable channel. (2) Gutter or gully along the side of a deck to carry water away to the *Scuppers, q.v.*

Watt The watt is the power of a current of one ampere under a pressure of one volt = 44.23 approximate foot lb. per minute. A kilowatt is one thousand watts, while 746 watts equals one horsepower.

W.Aus *or* **W.Aust.** Western Australia.

Wave Crest The highest part of a wave.

Wave Hollow The trough or valley-shaped form produced by two waves.

Wave Length The distance in the direction of advance between the same phase of consecutive waves.

Wave-Quelling Oil *Similar to* Storm Oil, *q.v.*

Waveson Flotsam, *q.v.*

Waxing and Waning of the moon. Between new moon and full moon its apparent face waxes, grows larger or increases, and thereafter it wanes or decreases in size.

WAY *or* **W.A.Y.** World Assembly of Youth.

Wayfarer A person travelling from one place to another; Traveller.

Wb Weber *SI Unit of magnetic flux.*

WB *or* **W.B.** *or* **W/B** Water Ballast *Lloyd's Register notation*; Waterboard; Waterborne, *q.v.*; Warehouse Book(s); World Bank, *q.v.*

WBAFC Weather Bureau Area Forecast Center *USA.*

WBAN Weather Bureau Airforce and Navy *USA.*

WB/EI *or* **W.B/E.I.** West Britain/East Ireland, ports of call range in a charterparty.

WBF World Boxing Federation.

WBGTI *or* **W.B.G.T.I.** Wet Bulb-Globe Temperature Index.

W.b.n. *or* **W. by N.** West by North.

w.b.p. Weather and Boil Proof.

WBS *or* **W.B.S.** Wide Body Short *aviation*; Without Benefit of Salvage, *q.v.*

W.b.s. *or* **W. by S.** West by South.

WC *or* **W.C.** *or* **w.c.** War Credits; Water Cock; Water Closet; Without Charge; Wing Commander; Work Council.

WCA *or* **W.C.A.** West Coast of Africa; Workman Compensation Acts.

WCB *or* **W.C.B.** *or* **w.c.b.** With Cargo on Board.

WCC World Council of Churches.

WCCHON *or* **W.C.C.H.O.N.** Whether Cleared at Customs House or Not.

WCCON *or* **W.C.C.O.N.** Whether Cleared Customs or Not, similar to WCCHON or W.C.C.H.O.N., *q.v.* Bill of Lading clause.

WCE *or* **W.C.E.** West Coast of England.

WCEC Western Canada and Europe Conference.

WCL World Confederation of Labour.

WCNA *or* **W.C.N.A.** West Coast of North America.

WCOTP World Conference of Organisations of the Teaching Profession.

WCP *or* **W.C.P.** West Canadian Ports.

WCS *or* **W.C.S.** Worshipful Company of Shipwrights.

WCSA *or* **W.C.S.A.** West Coast South America.

WCUK *or* **W.C.U.K.** West Coast United Kingdom.

WCUSA *or* **W.C.U.S.A.** West Coast of the United States of America.

WD Wood Covers.

WD *or* **W.D** *or* **W/D** Wind Direction; War Department; Warranted, *q.v.*; Working Day.

Wd. Warranted.

WDA *or* **W.D.A.** Welsh Development Agency; Working Day and Weather Permitting. *Similar to* **Weather Working Day**; Writing Down Allowance *tax.*

WDF *or* **W.D.F.** Wireless Direction Finder, *similar to* **Radar,** *q.v.*

w.d.f. Wood Door and Frame.

WDP Wire Drill Pipe.

wd.sc. Wood Screw.

WDV *or* **W.D.V.** Written Down Value.

W/E *or* **W.E.** Watch Error *nautical* referring to Chronometer error; Week Ending.

w.e. *or* **WE** Week Ending.

Wea. Weather.

WEA *or* **W.E.A.** Worker's Educational Association.

WEAA Western European Airports Association.

WEAAC Western European Airports Association Conference.

Weak Market *or* **Poor Market** When prices of commodities are down as opposed to a Strong Market.

Wear and Tear Normal deterioration of an article or a working part during the course of its use or age. Wear and Tear is one of the conditions laid down in Insurance, *q.v.*, and charterparties whereby Underwriters, *q.v.*, are exempted from claims and shipowners in a Charterparty, *q.v.*, have no right of claim against the normal wear and tear during the redelivery of the vessel. *Abbrev* W & T.

WEARCON Weather Observation and Forecasting Control System.

Weather The day-to-day state of the temperature, pressure of atmosphere, clouds, winds, humidity or moisture and other meteorological conditions.

Weather a Storm When a ship successfully withstands a storm.

Weatherboard Same as Weather Side, *q.v.*, or windward side. The side of a ship on which the wind blows.

Weather Bureau A section of the US Department of Commerce being the base of all meteorological reports.

Weather Deck The main, topdeck covering the hull of the vessel.

Weatherman Familiar name for a meteorological forecaster.

Weather Services for Shipping

The Meteorological Office The Meteorological Office was first set up by the British Government as a Department of the Board of Trade and under the superintendence of Admiral Robert Fitzroy in 1855. Six years later, the first direct weather forecasts to mariners were introduced. These took the form of the familiar visual gale warning signals hoisted at various points around the coasts of the British Isles. The development of radio led to the first transmission by wireless telegraphy in 1921 of a forecast for the western coastal waters of the British Isles from Poldhu in Cornwall. This was followed in 1925 by the first forecasts broadcast from the BBC by Radio Telephony. In 1948 the forecasts were extended into the Eastern North Atlantic and thus was born the North Atlantic Weather Bulletin. At the same time the coastal sea areas were given a new format very similar to that of today.

For the next 30 years, only minor changes were made to the North Atlantic Weather Bulletin. A major change to coastal shipping bulletins was in 1984, when all the northern European countries bordering the North Sea agreed to use common area boundaries, although different countries gave different names to the areas. The particular stations whose reports are included have to be altered from time to time to take into account the facilities provided by the station owners or managers. The Meteorological Office is constantly reviewing its meteorological services to the shipping and fishing industries and endeavours to ensure that these services are maintained to the satisfaction of the user.

Communications The biggest change in maritime communications since the introduction of radio, affecting weather information for mariners is the phasing in of the IMO Global Maritime Distress and Safety System (GMDSS) between February 1992 and February 1999. GMDSS aims to end the problems associated with short and medium range radio communications by Wireless Telegraphy and Radio Telephony, and to increase the reliability of distress alerting and rescue co-ordination. Marine weather information can now be transmitted via satellite or shore station, to be received automatically by NAVTEX or satellite communications equipment on board ship.

NAVTEX is a narrow-band telex direct-printing system by which all maritime safety messages are automatically received up to a range of about 100 nautical miles from the transmitting stations. Beyond that distance, the system called Digital Selective Calling enables the automation of most terrestrial radio communications, which is achieved via geostationary satellite for the greatest distances. These messages are received by a print-out or on a VDU. There is also a selective function to the system, allowing for specialist groups only to be targeted to receive certain messages, as desired. Weather bulletins for shipping can thus be received automatically by radio in any part of the world.

Reports from ships As the oceans occupy three-quarters of the world's surface, the value of regular weather observations from oceanic areas is obvious. Despite the current depressed state of the shipping industry, the number of British ships which are equipped with meteorological instruments and which regularly make and transmit weather messages to the various centres throughout the world has decreased only marginally. With the trend now being towards faster container ships, which spend more time at sea than their predecessors, the number of observations received has gradually increased over recent years. Although the value of satellite pictures to synoptic meteorology has increased, the observations received from the Voluntary Observing Ships are still of vital importance both for synoptic meteorology and climatology; without these reports, forecasts for sea areas would still be virtually impossible. Reports from 'sparse' sea areas, where few ships trade, are of particular value in filling the gaps from which too few weather reports are received. Thus, the radio weather messages provided voluntarily by merchant ships are of the utmost assistance to the Meteorological Office in its task of providing information for shipping, aviation, land transport, industry, agriculture and the whole community.

Ocean weather ships The original Ocean Weather Ship Service was founded as a result of an International Agreement signed in London in the summer of 1946, under the auspices of the International Civil Aviation Organisation (ICAO). The 13 stations at first established across the North Atlantic were gradually reduced for financial reasons. The stations were originally manned by the USA, the UK, Canada, France, the Netherlands and Norway. The UK contribution from the beginning in 1947 was in the form of four specially modified 'Flower'-class corvettes; these were replaced 12 years later by four converted 'Castle'-class frigates. By 1976 there remained only four stations, manned by the UK, France and the former USSR, with the fourth station manned jointly by the Netherlands and Norway. These were operated under the North Atlantic Ocean Stations (NAOS) agreement of the World Meteorological Organisation (WMO).

A UK presence was maintained by the two remaining 'Castle'-class frigates, refurbished for the purpose and renamed *Admiral Fitzroy* and *Admiral Beaufort*, in honour of the former captain of the *HMS Beagle* and first Superintendent of the Meteorological Office on its founding in 1854, and for the originator of the famous Beaufort Wind Scale in 1805, still in use today. In early 1982, these were replaced by one ship, the converted trawler *Starella*, operating out of Fleetwood, Lancashire, on monthly voyages to Ocean Station 'Lima' in position 57°00'N, 20°00'W, where the ship spent 27 days before returning to port for replenishment and crew change. Despite the disbanding of NAOS at the end of 1985, some countries decided to continue independent operation of their own Ocean Weather Ships, but with WMO still keeping a watching brief. At this time, the UK took over the former Netherlands 1963 purpose-built *OWS Cumulus*, on the understanding that the ship would be returned to the Netherlands when no longer required, for deployment at Ocean Station 'Lima'.

In 1994, only the United Kingdom and Norway continue to deploy one ship each. The *Cumulus* is deployed as convenient to the west of the United Kingdom on 33-day round voyages from her home port of Greenock, often southwards of the station to latitude 52°00'N, as found convenient for the forecasters who use the data transmitted from the ship. It is hoped to continue the cost-effective operation of *Cumulus* up to the turn of the century. Norway operate their own ship, *Polarfront*, on a similar basis, from the port of Bergen to the original Ocean Station 'Mike' off Norway's west coast in position 66°00'N, 02°00'E.

The duties of Ocean Weather Ships include the making of hourly surface observations and upper air radio sonde ascents up to a height of 24 km every six hours, with immediate telex-over-radio transmission to Bracknell. Oceanographic, biological and other scientific observations are also carried out on behalf of various technical institutions as agreed with them; and, in the case of *Cumulus*, dedicated training on board for up to 20 students, in meteorological and oceanographic subjects.

The existence of Ocean Weather Ships does not lessen the importance of frequent reports from merchant ships at sea. But they have certain advantages for the meteorologist in being established in fixed locations, like islands in the ocean from which they can transmit frequent and regular reports day and night, while providing reliable information on the upper air as well as surface conditions.

Ship Routeing Service Known as METROUTE, this is operated by the Meteorological Office at Bracknell, Berks, UK. The service offers ship routeing and sea ice information, with the object of enabling ships to reach their destinations in the shortest possible time, commensurate with the avoidance of damage to ship and cargo and with maximum fuel economy. The service is operated by experienced shipmasters on a worldwide basis.

Weather Side That part or side of a ship which is open to the weather.

Weather Stations Stations on land and at sea recording and reporting readings of atmospheric conditions prevailing about every six hours. The weather reports are transmitted by radio or by telex to places where weather forecasts are given and weather maps and charts are prepared.

Weather Vane A rotating device which indicates the direction of wind.

WEC *or* **W.E.C.** Wartime Extension Clause *insurance*.

WECCON *or* **W.E.C.C.O.N.** Whether Entered Customs Clearance or Not. *Similar to* Whether Cleared at Customs House or Not, *q.v.*

WECM Warranted Existing Class Maintained *insurance*.

WEDAR Weather Damage Reduction *insurance*.

Wedge Piece of wood thick at one end and thin at the other, a chock, with many applications in tightening or making things secure. Used to fasten the battens holding a tarpaulin over a hatch; to stop drums and rolling objects from shifting; to firm up shores; and elsewhere.

Wedge of Emersion The part of the ship's side which comes out of the water when she is listed. *Opposite to* **Wedge of Immersion**, *q.v.*

Wedge of Immersion The side of the ship which becomes immersed when she is listed. *Opposite to* **Wedge of Emersion**, *q.v.*

Weeping A shipping expression referring to slight leakage.

Weevil Free Certificate An official certificate confirming that no live insects injurious to grains, like weevils, exist in the hold of the ship.

W.E.F. *or* **w.e.f.** With effect from.

W.& E.Fl. Wages and Effects Form 1.

WEIA *or* **W.E.I.A.** Wife's Earned Income Allowance *income tax*.

Weigh The act of lifting the anchor *nautical*.

Weigh Anchor To lift the anchor from its anchorage. *Opposite to* **Cast Anchor**, *q.v.*, or **Drop Anchor**.

Weight, Contents and Value Unknown A clause often seen inserted on Bills of Lading, *q.v.*, meaning that the weights and other descriptions are only stated by the shipper and in no way confirmed by the master or the agents signing the bills of lading. In this way the owner may be exonerated from liability in case of any discrepancy in the weight, contents, etc., of the closed packages if disclosed after unloading.

Weight or Measure The method on which liner cargo may be charged—w/m.

Welfare Economics Public understanding and use of economics to advance the general welfare.

Welfare Services The following lists the Welfare Services provided in the United Kingdom for seafarers and the location of their offices:

Apostleship of the Sea	Tilbury
King George's Fund for Sailors, *q.v.*	London
Merchant Navy Welfare Board, *q.v.*	London
NUMAST Welfare Funds	Wallasey
Royal Alfred Seafarer's Society	Banstead
Royal Merchant Navy School Foundation	Wokingham
Royal Provident Fund for Sea Fishermen	Hull
Sailor's Families Society, *q.v.*	Hull
The Shipwrecked Mariner's Society, *q.v.*	Chichester
Ex-Service Fellowship Centres	London
British Sailor's Society, *q.v.*	Southampton
The Marine Society, *q.v.*	London
Missions to Seamen, *q.v.*	London
Queen Victoria Seamen's Rest	London
Royal Liverpool Seamen's Orphan Institution	Liverpool
Royal National Mission to Deep Sea Fishermen	London
Royal Seamen's Pension Fund	Ewell
Seamen's Hospital Society	Greenwich
Trinity House Hull	Hull
Ex-Services Mental Welfare Society	London
Buckie and District Fishermen's Benevolent Society	Buckie
Fleetwood Fishing Industry Benevolent Fund	Fleetwood
Glasgow Seamen's Friend Society	Ayr
Hull Fishermen's Trust Fund	Hull
Lowestoft Fishermen's Widows and Orphans Benevolent Society	Lowestoft
Royal British Legion	London
Sailor's Orphan Society of Scotland	Glasgow
Corporation of Trinity House, *q.v.*	London
Institute of Marine Engineers' Guild of Benevolence	London
National Union of Seamen Benevolent Fund	London
Cornwall Seamen's Benevolent Trust	Falmouth
Glasgow Aged Seamen's Relief Fund	Glasgow
Grimsby Fishermen's Dependants' Fund	Grimsby
Leith Aged Mariners' Fund	Edinburgh
Manx Marine Society	Isle of Man
Royal British Legion Scotland	Edinburgh
Tyne Mariner's Benevolent Institution	North Shields
Honourable Company of Master Mariners	London
Merchant Seamen's War Memorial Society	Guildford

There are a number of smaller societies providing education, residential care and financial assistance.

Welfare State General conditions and welfare of the people of a country. These include social insurance and benefits, old age pensions, national health schemes, family allowances, education, etc.

Well Deck The deck space between higher decks or between them and the superstructure.

WEMT West European Co-operation in Marine Technology *conference*.

WESA West Europe Airports Association.

West Cardinal point of the compass, directly opposite to East.

Westbound Describes a ship or aircraft proceeding in a Westerly direction towards its destination.

Westerlies Winds existing in the temperate zones. The northern hemisphere is an area where westerlies are common. *See* **Roaring Forties**.

Western Hemisphere Commonly describes the half of the globe which includes the North and South American continents and the West Indies.

Westpac *or* **WESTPAC** West Pacific.

Wet Bulk Cargo Any liquid cargo carried in bulk condition such as petroleum, vegetable oils and water.

Wet Cargo *Similar to* **Wet Goods**, *q.v.*

Wet Charter *or* **Wet Lease** *or* **Wet Hire** An aircraft charter or lease or hire consisting of the full complement of the crew but not necessarily with the stewards.

Wet Dock An artificial sea enclosure having basins with sliding gates at the entrance which become watertight when closed. The basins provide water at the required level to accommodate ships.

Wet Fish Uncured fresh fish, excluding shellfish.

Wet Gas Natural Gas, *q.v.*, containing condensate.

Wet Goods Liquids contained in barrels, bottles, tins, etc. Also called Wet Cargo.

Wet Hire *or* **Wet Lease** *Similar to* **Wet Charter**, *q.v.*

Wet Lease *or* **Wet Hire** *Similar to* **Wet Charter**, *q.v.*

Wet Oil *Similar to* **Cut Oil**, *q.v.*

Wet Paint Thickness The thickness of paint when applied and still not dry, calculated in Microns or Mills. *Abbrev* WPT *or* W.P.T.

Wetted Surface The underwater surface area of a ship.

WEU *or* **W.E.U.** Western European Union.

441

Wf. Wharf, *q.v.*

WF *or* **W.F.** Wheat Flour.

w.f. Wrong Fault.

WFA *or* **W.F.A.** Warranted Free From Average, *q.v.* With Following Alterations, *usually in chartering or insurance.*

WFC World Fisheries Conference.

WFF Wave Feed-Forward.

WFFS World Fleet Forecast Service.

WFFTU World Federation of Free Trade Unions.

WFP World Food Programme *FAO.*

WFPA *or* **W.F.P.A.** With Free of Particular Average. *Similar to* **Free of Particular Average**, *q.v.*, and Warranted Free from Average.

WFR *or* **W.F.R.** World Freight Revenue *in connection with civil aviation.*

WFTU *or* **W.F.T.U.** World Federation of Trade Unions.

WFUNA World Federation of United Nations Associations.

WG *or* **W.G** *or* **w.g.** Waterguard, *q.v.*; Water-gauge, *q.v.*; Weight Guaranteed; Wire Gauge.

W.Ger. West Germany.

WH Workable Hatch.

Wh. Watt Hour.

Whaler Vessel used for hunting whales, a whale-catcher, armed with a harpoon gun; A whale boat, a double-ended boat formerly used under oars for catching, now usually a warship's motor boat; A man employed in the whaling industry.

Wharfboat A floating office or warehouse *USA.*

Wharfinger Warrant *or* **Receipt** The receipt given by a Wharfinger for goods in his charge.

Wheeler Dealer A person buying and selling shrewdly and often ruthlessly in business.

Wheelhouse Deckhouse, *q.v.*, enclosing the steering wheel. Now usually part of the Bridge, *q.v.*, of a ship where the navigational aids and controls are found. During manoeuvring the Master is always there and at any time a navigating officer. There will also be an Able Seaman or Able-bodied Seaman, *q.v.*, on duty as Helmsman, *q.v.*, except under One Man Bridge Operation.

Whether in Roads or Not *Similar to* **Whether in Port or Not**.

Whf Wharf.

Whip A wire or rope rove through one block (single whip) or two (double whip) for hoisting.

Whipstock A cylindrical steel billet with a tapering face which, when inserted in a borehole, will deflect the direction of a drilling bit from the vertical.

Whirley Bird Nickname for a Helicopter, *q.v.*

Whiskey International Radio Telephone phonetic alphabet for 'W'.

Whitecaps *Similar to* **White Horses**, *q.v.*

White Collar Worker A person employed in clerical administration, a professional career, senior management, etc. *Opposite to* **Blue Collar Worker**, *q.v.*

White Gasoline White gas; Gasoline or petrol without tetraethyl lead.

White Gold White-coloured gold having an alloy of platinum and nickel.

White Goods Household appliances.

White Horses Small white broken wave crests on the surface of the sea, lakes, rivers, commonly seen when the water is agitated by winds. Also called whitecaps.

White Paper An official memorandum issued by the government in which a problem and various considerations bearing on it are set out, and the policy which the government advocates, or is disposed to advocate, is stated.

White Sale A sale of household linen, towels, sheets, bed-covers, pillow cases and suchlike wares.

Whitesmith A person, often a tinsmith, who finishes and polishes metals; particularly tinplate and galvanised iron.

White Spirit Liquid mixture of hydrocarbons obtained from petroleum used as a solvent for paint and as a substitute for turpentine.

WHO *or* **W.H.O.** White House Office *USA*; World Health Organisation.

Whole Gale Same as a Storm, Force 10 on the Beaufort Scale, *q.v.*, with winds of 48 to 55 knots.

Whole Life Policy An Insurance (assurance), *q.v.*, in which the sum assured is paid by the insurance company after the death of the assured, whether the full amount of premiums has been paid or not. *Abbrev.* WLP *or* W.L.P. *See* **Life Insurance** *or* **Assurance**.

Wholesale The act of buying or selling in bulk, or in large quantities, to be sold on a retail basis. *See* **Wholesaler**.

Wholesaler A person who negotiates the sale of items in bulk. He acts in between the manufacturer or the principal and the retailer.

WHP *or* **W.H.P.** *or* **w.h.p.** Water Horse Power.

Whr. Watt Hour *electricity*; Weather.

WHRA World Health Research Centre.

WHSA Whampoa Harbour Superintendency Administration.

Whse Warehouse, *q.v.*

Whs. Rec. Warehouse Receipt.

Whse/Whse *or* **W/W** Warehouse to Warehouse Clause, *q.v.*

WI Wisconsin *USA.*

W.I. West Indies; Windward Islands; When issued; Wrought Iron.

W&I Weighing and Inspection.

w.i. When Issued; Wrought Iron.

w.i.a. Wounded in Action.

WIAS West Indies Associated States.

WIBON *or* **W.I.B.O.N.** Whether in Berth or Not.

WICE World Industry Council for the Environment.

WIDER World Institute for Development in Economic Research.

Wider Terms Referring to more coverage against perils in insurance.

WIF *or* **W.I.F.** West Indies Federation.

WIFPON *or* **W.I.F.P.O.N.** Whether in Free Pratique or Not.

WIG *or* **W.I.G.** Wing-In Ground *air cushion vehicle.*

WILCO *or* **Wilco** Will Comply.

Wild-cat A speculative oil well drilled without a complete geological exploration.

Wildcat Strike An unofficial strike; sudden stoppage of work or unexpected strike.

Wild Well When the blow out of fuel or gas in an oil well gets out of control.

Wilful Misconduct An insurance term meaning the fraudulent action by the assured to the detriment of the underwriters.

Wilful Misrepresentation Deliberately giving false information.

Williwaw A strong squall often happening around the Strait of Magellan.

Wilts. Wiltshire *UK.*

WINAC *or* **Winac** West coast of Italy, Sicilian and Adriatic Ports, North Atlantic Range Conference.

Winch A geared machine, hand operated or powered, with one or more barrels or drums on a horizontal axis for lifting or hauling with wires and ropes.

Windfall Profit An unexpected profit or gain.

Winding Up As applied to a partnership or company, this is the operation of stopping the business, realising the assets and discharging the liabilities of the concern, settling any questions of account or contribution between the members and dividing the surplus assets (if any) among the members. The term 'winding up' is sometimes applied to the estates of deceased persons, but 'administration' is the usual expression.

The winding up of a partnership is either voluntary (i.e. by agreement between the partners) or by order of the court made in an action for the dissolution of the partnership.

Windjammer Familiar name for a large square-rigged sailing ship.

Windlass *Similar to* a Winch, *q.v.* Most usually found in the forepart of a ship for working anchors and cables. To handle chain cable it has a gipsy or gypsy, a drum with snugs that fit the links of the chain.

Windsail Canvas funnel rigged to direct air down through a hatch for ventilation.

Wind Scale *See* **Beaufort Scale**.

Wind Tunnel A tunnel-like apparatus for producing an air-stream of a known velocity past model aircraft, etc., to investigate the effect of wind pressure on the structure.

Windward Describes anything in the direction from which the wind blows.

Windward Flood A flood tide against the wind.

Wing and Wing Describes a square-rigged vessel running before the wind with studding sails set both sides, or, if she be fore-aft-rigged, with foresail and mainsail boomed out on opposite sides.

WIP *or* **W.I.P.** Work in Progress, *q.v.*

WIPO World Intellectual Property Organization.

WIPON *or* **W.I.P.O.N.** Whether in Port or Not.

WIRDS Weather Information Reporting and Display System.

Wireless Literally without wire(s); especially in wireless telegraphy; wireless (radio) waves produced by a transmitting station by electric current passing through the air over various distances.

Wire Netting Wire network made out of flexible wire rope for cargo lifting nets, gangway safety nets, etc.

Wire Rope *See* **Hawser**.

Wis *or* **Wisc.** Wisconsin, *USA.*

WITAG West Indies Trade Advisory Group.

With All Despatch *or* **With All Dispatch**. *Similar to* **Customary Despatch** *or* **Customary Dispatch**, *q.v. Abbrev* WAD *or* W.A.D.

With Average A marine insurance clause covering both Particular Average, *q.v.*, and General Average, *q.v. Abbrev.* WA *or* W.A.

With Customary Despatch *or* **With Customary Dispatch**. *Similar to* **Customary Despatch** *or* **Customary Dispatch**, *q.v.*

Withdrawal of Services In travel insurance this covers failures due to force majeure, such as a transport strike, water shortage or lack of hotel accommodation because staff are on strike.

Within Commercial Area Referring to the conditions and time of the laydays which are to be accountable as soon as the ship has arrived within the commercial area of the port.

Within Reach of Tackle The ship is only bound to receive or deliver the cargo within reach of Ship's Tackle, *q.v.*, or shore crane's tackle. *Similar to* **Alongside**, *q.v.*

Withholding Tax A form of income tax levied by a government on dividends or interest payable by a company to a foreign investor.

Without Benefit of Salvage A term in a Marine Insurance, *q.v.*, policy whereby the Underwriter, *q.v.*, forgoes his Subrogation, *q.v.*, rights.

Without Prejudice A phrase used in offers, in order to guard against any waiver of right; also for the purpose of negotiating a compromise.

Without Reserve An expression used in an auction sale meaning that an object can be sold without a fixed price being reached. Opposite to reserve price, meaning that price below which a bid will not be accepted.

With Profits Policy A policy of insurance under which bonuses from the insurance company's profits are allocated to the policy and increase its value, the premium being higher for such a policy than the premium for a 'without profits' policy.

Witness A person who has knowledge of an event. As the most direct mode of acquiring knowledge of an event is by seeing it, 'witness' has acquired the sense of a person who is present at and observes a transaction. When a deed or other instrument is executed in the presence of a person, and he or she records the fact by signing his or her name on it, that person is said to witness it, or to be an attesting witness.

Wkg. Working.

WL *or* **W.L.** Waiting List; Water List; Water Line; Wavelength.

Wl. coef. Waterline Coefficient.

WLL Working Load Limit.

W.Long. West Longitude.

WLP *or* **W.L.P.** Whole Life Policy, *q.v.*

Wly. Westernly.

WM *or* **W.M.** Watt Meter; Wave Meter; White Metal; Wire Mesh.

W.M. *or* **W/M** Weight/Measurement.

W & M War and Marine *insurance.*

WMARC *or* **W.M.A.R.C.** World Maritime Administrative Radio Conference.

W.M. fees Winter Mooring Fees.

Wmk. Watermark.

WMO *or* **W.M.O.** World Meteorological Organisation.

W.M.P. With Much Pleasure.

W.M.U. World Maritime University, *q.v.*

w.n.d.p. With no down payment.

WNL *or* **W.N.L.** Within Normal Limit.

WNW *or* **W.N.W.** West North West.

W/o *or* **w/o** Without; Write Off; Written Off.

W.O. *or* **W/O** War Office; Warrant Officer; Welfare Officer; Wireless Operator; Written Order; Write Off *or* Written Off.

WOA *or* **W.O.A.** Wharf Owners' Association.

WOB *or* **W.O.B.** Washed Overboard.

WOC *or* **W.O.C.** Waiting on Cement, *q.v.*; Without Compensation.

WOCOL World Coal Study.

WOG *or* **W.O.G.** Without Guarantee; With Other Goods; Water, Oil or Gas.

WOL *or* **W.O.L.** Wharf Owner's Liability.

WOM *or* **W.O.M.** Wireless Operator Mechanic.

Won A unit of currency used in North and South Korea.

W.O.O. World Oceangraphic Organization.

Woodbar *or* **Woolibar** A heavy hardwood found in Australia, perhaps the heaviest of its kind, having a specific gravity of 1.3.

Wood Tar Dark-coloured product extracted from wood and widely applied as a preservative to rope and timber.

WOP *or* **W.O.P.** With Other Property.

WOR *or* **W.O.R.** Without Our Responsibility.

Worcs. Worcestershire *UK.*

Work Force The total number of workers available to or employed in a business. Also the manpower of a country whether gainfully employed or otherwise.

Working Capital Funds needed in hand to keep the business running, meet daily demands and settle bills.

Working Hatch A hatch actually being used for loading or discharging cargo.

Working Party Group of people picked from various sections or interests and committed to a specific project, or to examine a problem and make recommendations or estimates.

Workman Clause A standard time charter clause approved by BIMCO. If workmen are carried on board for cargo operations, they are to be carried free of charge and on deck. All liability, including expenses for these workmen, are for charterers' account.

Work Measurement Assessment of the labour required to do a job by timing how long it takes an average person to perform it.

Works Committee A committee of workers regularly kept in close contact with the proprietor and administrators of a factory, company, etc.

Works Manager A person who supervises the works and the workmen in a factory, etc.

Work Study The study of working methods in connection with certain products or the administration of an office or company, etc., in order to improve efficiency.

Work to Rule *or* **Working to Rule** An alternative to striking in furtherance of an industrial dispute. Normal work and timing are disrupted by over-zealous and time-wasting interpretation of every minor regulation. *See also* **Go-Slow** and **Slow down**.

World Bank Alternatively known as the International Bank for Reconstruction and Development, the World Bank is one of the specialised agencies of the United Nations. Founded 1946.

World Court *See* **International Court of Justice**.

World Health Organization One of the specialised agencies of the United Nations inaugurated in 1948 for the purpose of minimising disease and improving health conditions all over the world. Prior to this establishment there was the International Office of Public Health formed in 1923 under the protection of the League of Nations with administrative offices in Paris. Apart from its successful campaign to eradicate smallpox, the World Health Organization is currently fighting diarrhoeal diseases, vitamin deficiency, acute respiratory infections, such as pneumonia, a host of tropical diseases, alcoholism, drug abuse, cancer and

Aids which, taken together, kill millions of people every year. *Abbrev.* WHO *or* W.H.O.

World Maritime University *Abbrev.* WMU. A major initiative of the International Maritime Organisation (IMO) was to reach agreement with the Swedish Government as to the founding of the University at Malmö in 1983 with considerable assistance from Sweden, the United Nations Development Programme (UNDP) and the City of Malmo itself. The Chancellor of the WMU is the Secretary-General of IMO. Its object is to provide advanced training to senior personnel involved in various maritime activities, particularly those from developing countries where resources are limited. Such personnel will already have gained high seagoing or academic qualifications and be engaged in, or destined for, important positions such as in maritime administration, shipping company management, instructing in training institutions, naval architecture and marine science, port management, marine accident investigation or as surveyors, examiners, inspectors and so on. The courses contribute to a deeper knowledge of the problems of marine safety, protection of the marine environment and the efficiency of international shipping and navigation, and help in developing internationally uniform standards of training, as well as promoting co-operation. In this way they support the purposes and objectives of the IMO. All the teaching is in the English language. Generally, the various courses are of two years' duration for some 100 students who are nominated by their governments. In addition, a number of seminars are conducted. Several marine academics and training institutions around the world have been accorded the status of branches of the WMU. The budget of the WMU is furnished by governments, national and international non-governmental agencies, associations and individual companies. A large part of the contributions is in the form of fellowships for the students, a source of income to which some developing countries are able to commit allocations from the UNDP. *See also* **International Maritime Organisation**.

World Meteorological Organisation, The *Abbrev.* WMO. The United Nations specialised agency, with headquarters in Geneva, through which the national meteorological and hydrological services of the world plan and co-ordinate and facilitate weather, climate and related environmental and geophysical services. It encourages co-operation in the making and exchange of observations, data analysis, interpretation and predictions and the application of information for the benefit of human existence and the environment. It traces its origin to the First International Meteorological Conference in Brussels in 1853, primarily concerned with services to mariners; and the First International Meteorological Conference in Vienna in 1873, when the International Meteorological Organisation (IMO) was established.

After the Second World War, with the formation of the United Nations, it became the WMO.

The supreme governing body is the World Meteorological Conference which meets every four years. The Executive Council of 36 national directors of service meets annually. Its eight technical commissions include the Commission for Marine Meteorology, first formed in 1907. There are six Regional Associations. Day-to-day affairs are managed by the Secretariat under the Secretary-General.

WMO's programmes in the fields of world weather, climate, atmospheric research, applications including marine, water resources, education and technology recognise that the atmosphere knows no boundaries while every nation needs global information; and that services can be provided to all countries at costs far below those that would pertain if each acted alone.

World to World Cargo insurance cover.

WORM Write Once Read Many times Memory *computers*.

WOW *or* **W.O.W.** Waiting on Weather.

WP *or* **W.P.** Water Plane; Weather Permitting; White Paper, *q.v.*, Without Prejudice.

WPA *or* **W.P.A.** Water Plane Area; With Particular Average, *See* **With Average**: Works Project Administration; World Psychiatric Association.

WP—AGW Weather Permitting—All Going Well.

w.p.b. Waste Paper Basket.

WPC *or* **W.P.C.** World Petroleum Congress; Weather Permitting Clause, *q.v.*; Woman Police Constable.

WPCA *or* **W.P.C.A.** Water Pollution Control Association *USA*.

WPE *or* **W.P.E.** White Porcelain Enamel.

WPF Widows' Pension Fund.

Wpg. Waterproofing.

WPI *or* **W.P.I.** Wholesale Price Index; World Port Index; World Press Institute.

WPL *or* **wpl** Warning Point Level.

WPM *or* **W.P.M.** *or* **w.p.m.** Words per Minute.

WPO World Packing Organisation, Hong Kong.

WPP *or* **W.P.P.** *or* **w.p.p.** Waterproof Paper Packing.

WPRL *or* **W.P.R.L.** Water Pollution Research Laboratory.

w.p.s. With Prior Service.

WPT *or* **W.P.T.** Windfall Profits Tax, Wet Paint Thickness.

WQIA *or* **W.Q.I.A.** Water Quality Improvement Act 1970 *USA*.

WQIS *or* **W.Q.I.S.** Water Quality Insurance Syndicate, *q.v.*

WR *or* **W.R.** Wardroom; War Risk *insurance*; Western Region; Warehouse Receipt.

w.r. War Risk *insurance*; Water Repellant.

W.R.A.C. *or* **WRAC** Women's Royal Army Corps.

W.R.A.F. or **WRAF** Women's Royal Air Force.

WRC Water Research Centre.

Wreck Such goods, including the ship or cargo or any part, as, after a shipwreck, are afloat or cast upon the land by the sea. Formerly they were not wreck so long as they remained at sea in the jurisdiction of the Admiralty. By the Merchant Shipping Act 1894 'wreck' includes Jetsam, *q.v.*, Flotsam, *q.v.*, and Derelict, *q.v.*, found in or on the shores of the sea or any tidal water. In the insurance sense, 'wreck' is whatever may remain of property which has been severely damaged in a maritime adventure. An underwriter who has paid a total loss on the property is entitled to take over the wreck, dispose of it as he thinks fit and retain the whole of the proceeds, if any, even if these exceed the claim paid.

Wreck Commissioners Persons appointed by the Lord Chancellor under the Merchant Shipping Act 1894 to hold investigations into shipping casualties. Similar investigations are in some cases made by stipendiary magistrates. Preliminary investigations are also held by various officers of the coastguard and customs services.

Wrecker Salvor, *q.v.*

Wrecking Tug US term for Salvage Tug, *q.v.*

W.Ref. or **w.ref.** With reference to.

W.Reg. or **w.reg.** With regard to.

WRHA or **W.R.H.A.** Winch Resistor House Aft.

WRHF or **W.R.H.F.** Winch Resistor House Forward.

WRHM or **W.R.H.M.** Winch Resistor House Middle.

WRIO War Risks Insurance Office.

Writ A document under the seal of the Crown, a court or an officer of the Crown, commanding the person to whom it is addressed to do or forbear from doing some act.

Writ of Assistance Court order to a receiver granting him authority to take possession of chattels, securities and documents.

Writ of Error A legal Writ, *q.v.*, formerly allowing a suspected error on a point of law in a previous action to be re-examined and corrected if necessary. Abolished in England by the Criminal Justice Act of 1907.

Writ of Prohibition A Writ, *q.v.*, issued by the superior courts to arrest the legal proceedings and prosecutions of the inferior courts. Abolished in England in 1938 when an Order of Prohibition was substituted.

Writ of Summons In actions in the High Court, the writ of summons is a process issued at the instance of the plaintiff for the purpose of giving the defendant notice of the claim made against him and of compelling him to appear and answer it if he does not admit it. It is the first step in the action.

Write-off A book-keeping expression referring to the transfer of debts unlikely to be settled to the debit side of the profit and loss account.

Writers to the Signet Designated by the letters W.S., Writers to the Signet are the senior society of solicitors or law agents in Scotland. Formerly many privileges were extended to this society, but most of these have disappeared.

Writing In any Act of Parliament, unless the contrary intention appears, this is construed as including printing, lithography, photography, and other modes of representing or reproducing words in a visible form. In the most general sense of the word 'writing' denotes a document, whether manuscript or printed, as opposed to mere spoken words. A contract of guarantee and a contract relating to land are unenforceable by action unless evidenced by writing. Writing is essential to the validity of certain contracts and other transactions, e.g. a bill of exchange or a will. In Scotland the word is used as meaning any writing possessing legal significance.

Writings Obligatory Bonds.

Written Line *See* **Full Signed Line**.

W.R.N.S. or **WRNS** Women's Royal Naval Service.

WRO or **W.R.O.** or **w.r.o.** War Risk Only *insurance*.

Wrong The privation of right, an injury; that which takes place when a right is violated or infringed. Wrongs are generally divided into private and public. A private wrong is one which confers a remedy or right of redress on an individual; such are breaches of contract, torts, etc. A public wrong is one which renders the wrongdoer responsible to the community; such are crimes and other offences and generally all infringements of the rules of public law. The maxim that 'no man can take advantage of his own wrong' means that a man cannot enforce against another a right arising from his own breach of contract or breach of duty.

Wrongful Dismissal An unjustifiable dismissal of a servant by the master from an engagement for services for a fixed time or, if upon notice, before expiration of the period of notice. The servant may elect to treat the contract as repudiated, in which case he can recover remuneration for what he has actually done on a *quantum meruit* or, if he treats the contract as continuing, he may sue for damages for loss of the employment and such wages as he has lost the opportunity of earning, taking into account the probability of finding another employment of the same kind and degree. By custom a domestic servant may be dismissed at any time on a month's notice or on payment of a month's wages. Punitive damages are not recoverable and the plaintiff must take diligent steps to find other suitable employment. The damages are the amount which the plaintiff has lost on the footing that the remuneration he would have received but for the dismissal would have been liable to tax.

Wrought Iron A pure form of iron which is forged or rolled, not cast, with the least amount of carbon content averaging 3.5% silicon and carbon. It can be welded.

WRRC War Risk Rating Committee *insurance*.

W.R.S. Rm. Wardroom Stateroom.

WRTD *or* **W.r.t.d.** Without reference to date *insurance*.

WRVS *or* **W.R.V.S.** Women's Royal Voluntary Service, formerly WVS.

WS *or* **W.S.** Wharf Superintendent; Writer to the Signet, *q.v.*; Worldscale, *q.v.*

WSA War Shipping Administration *USA*, in connection with war risk insurance policies; Wetted Surface Area, underwater parts of a ship.

WSB Wage Stabilising Board.

WSP Water Spray Protection; Water Supply Point.

WSPA World Society for the Protection of Animals *Switzerland*.

WSPI *or* **W.S.P.I.** Wages, Stores, Provisions and Insurance.

WSHTC *or* **W.S.H.T.C.** Working Sundays/Holidays to Count, Laydays.

WSW *or* **W.S.W.** West South West.

WT *or* **W.T.** War Time; Water Tight; Waterline; Water Tube *Lloyd's Register notation*.

wt. Warranted; Weight.

Wt *or* **wt.** Weight.

W/T Wireless Telegraphy.

W&T Wear and Tear, *q.v.*

WTA *or* **W.T.A.** World Transport Agency.

W.T. Aux. B. Water Tube Auxiliary Boiler(s) *Lloyd's Register notation*.

WTB Water Tube boiler *Lloyd's Register notation*.

WTBA *or* **W.t.b.a.** Wording to be agreed *insurance*.

WTCA World Trade Center's Association *New York*.

WTCR Worldwide Tanker Casualty Returns.

WTD *or* **Wtd.** Warranted; Watertight Door(s), *q.v.*

W.T.D. Watertight Door(s), *q.v.*

WTDB. Water Tube Domestic Boiler(s) *Lloyd's Register notation*.

Wtd. C.M. Warranted Class Maintained *insurance*.

WTF *or* **W.T.F.** Water Tight Flat.

W'ties. Warranties.

WTIS World Trade Information Service.

WTMH *or* **W.T.M.H.** Watertight Manhole.

WTO World Trade Organisation, established 1995.

WTT *or* **W.T.T.** Watertight Trunk.

WTUC *or* **W.T.U.C.** World Trade Union Congress.

Wty. Warranty, *q.v.*

WUA *or* **W.U.A.** Western Underwriters' Association *USA*.

Wud. Would, *telex abbreviation*.

WUF World Underwater Federation.

WUS World University Service.

WV *or* **W.V.** Water Value; West Virginia, *USA*.

W.VA *or* **W.Va** West Virginia, *USA*.

WVS Women's Voluntary Service, now WRVS.

W/W Warehouse Warrant, *q.v.*

W/W *or* **Whse/Whse** Warehouse to Warehouse referring to Warehouse to Warehouse Clause, *q.v.*

WW Worldwide.

WWD *or* **W.W.D.** Weather Working Days.

WWDFHEX *or* **W.W.D.F.H.E.X.** Weather Working Days Fridays and Holidays Excepted.

WWDSHEX *or* **W.W.D.S.H.E.X.** Weather Working Days Sundays and Holidays Excepted.

WWF Waterside Workers Federation *Australia*; World Wildlife Fund; World Wrestling Federation.

WWLP Warranty of Workmanlike Service, *q.v.*, or Warranty of Workmanlike Performance, *q.v.*

WWREADY *or* **WWR** When Where Ready.

WWSSN World Wide Standard Seismograph Network.

WX Outsize Women's Extra (clothing size).

WY Wyoming, *USA*.

Wy *or* **Wyo** Wyoming, *USA*.

X

x. Hoar frost *meteorology.*

X Greek letter chi, for Christos (Christ); Distance below deck; Mathematical sign for multiplication; the first unknown quantity (*maths.*); Parallactic Angle *nautical*; Roman numeral for figure 10; Crystal.

X-band Radio frequency from 5,200 up to 10,000 megaHertz per second.

X.C. *or* x.c. *or* x.cp. Ex (without) Coupon.

XC Roman numerals for 90.

XCIX Roman numerals for 99.

XCL *or* xcl Excess Current Liability.

XD Ex-Directory.

X.D. *or* x.d. *or* x/d *or* x. div. Ex (without) Dividend, *q.v.*; Examined.

X. dr. Ex (without) Drawings.

Xe Symbol for xenon (chemistry), a heavy inert gaseous element.

Xerography Electrostatic process for printing from or copying documentary material or film.

X.hvy Extra Heavy.

X.I. *or* X-in Ex (without) next Interest, *q.v.*

XL *or* X.L. Extra Large; Roman numerals for 40.

XLIX Roman numerals for 49.

XL & UL *or* xl & ul Exclusive of Loading and Unloading.

Xmas Christmas.

x.n. *or* x.new Ex New, without the right to new shares.

X Note US 10 dollar note.

Xnty Christianity.

XO Crystal Oscillator.

XOD *or* X.O.D. X-Class One Design *yachting.*

XP Christos *Greek*—Christ.

X.P. *or* X.p. Exprèss Payé Poste *French*—Express Paid *Post.*

X.pr *or* x.pr *or* x.pri(s) Without Privileges, Stock Exchange.

X.R. *or* xr Without Rights, *Stock Exchange.*

X-Ray International Radio Telephone phonetic alphabet for 'X'.

X-Rays Very short wavelengths which penetrate where normal light does not. When printed on sensitive photographic paper the picture of the object is reproduced.

X.S. *or* X.s. *or* x.s. Excess; Expense.

XSL *or* xsLoss Excess Loss *reinsurance.*

xs.pt. Excess Point.

X.Str. Ex Store.

Xt. Christ.

Xtra Extra.

Xty Christianity.

X Whf. Ex Wharf.

X Wks Ex Works, *q.v.*

XX Roman numerals for 20.

XX Note US 20 dollar note.

XXX Roman numerals for 30; Wireless Telegraphy emergency signal.

Xylography The printing of wood-block books.

Xylonite A celluloid.

Xylophone A percussion instrument made of bars vibrating when struck *music.*

Xylophonist The player of a xylophone.

Xystus A covered portico used by athletes for exercise in classical antiquity; a garden walk or terrace.

Y

y Yacht, *q.v.*, Yard, *q.v.*; the second unknown quantity (algebra).

Y Distance above deck; Yacht, *q.v.*; Yard; Year; Yen, the monetary unit of Japan.

Y.A. *or* **Y/A** *or* **Y.A.R.** York Antwerp Rules.

Yacht A light sailing vessel kept and usually specially built and rigged for racing; a vessel sometimes of considerable size which is propelled by sails, steam or electricity and used for private pleasure, excursions, cruising, travel, etc. *Abbrev.* Ycht *or* Yt.

Yankee International Radio Telephone phonetic alphabet for 'Y'.

YAR Yemen Arab Republic.

Yardage A charge made for the use of the livestock enclosure before boarding a cattle carrier or after unloading.

Yardarm Either end of a yard, a yard in this case being a spar slung or attached across a mast to carry a sail or halyards or aerials.

Yard Craft A collective word for vessels being used in a repair yard. Some of these are tugs, barges, lighters, pontoons, etc.

Yardmaster The overseer of a railway yard.

Yard Number *or* **Hull Number** A shipbuilder's identification number allotted to every vessel under construction in the shipyard. This number invariably appears on the plans and building contracts as well as in the Register of Shipping Records book.

Yaw *or* **Yawing** A ship or aircraft unable to keep a straight course because of bad steering or bad weather; an aircraft movement to the left or right.

Yawl A small rowing boat used as a ship's tender; Type of two-masted sailing vessel.

YB *or* **Y.B.** Year Book; an air fare symbol for Eurobudget.

YBDSA *or* **Y.B.D.S.A.** Yacht Brokers and Designers Association *UK*.

Y&BSC Yacht and Boat Safety Scheme *UK*.

YC *or* **Y.C.** Yacht Club; Yale College *USA*; Youth Club.

YCA *or* **Y.C.A.** Yacht Charter Association; Youth Camping Association.

Ycht. Yacht, *q.v.*

Yd/s Yard(s).

Y.D. Yugoslav dinar, the unit of currency used in Yugoslavia. One hundred para equals one dinar.

Yday. Yesterday.

YE A symbol for an excursion air fare available on many international air routes.

Year book An annual publication usually containing a directory, list of fixtures, and a review of immediate past and current topics of special interest to the organisation issuing it. For example *Lloyd's Nautical Year Book* covers a wide range of information of immediate interest to, and for reference by, mariners and shipping concerns.

Year of Grace An extension of a further year granted by a ship Classification Society, *q.v.*, before a special survey is due in order that the shipowner can complete any commitments he may have regarding a charter, etc. *Abbrev.* YoG.

Yellow Dog Contract A contract forced upon an employee by his employer which precludes the former from joining a union. This procedure is falling into disuse since it is illegal in many countries.

Yellow Duster Another word for the yellow 'Q Flag' hoisted on the ship's masthead on her arrival at a port signifying Quarantine, *q.v.*, before she is cleared by the authorities as being free from any contagious disease. *See* **Quarantine Restrictions**.

Yellow Fever *See* **Yellow Jack**.

Yellow Jack A colloquial expression for yellow fever, a contagious tropical fever with jaundice.

Yellow Metal Brass of 60 parts copper and 40 parts zinc.

Yest. Yesterday.

Yen The unit of currency used in Japan. One hundred sen equals one yen.

YGS Year of Grace Survey.

YH *or* **Y.H.** Youth Hostel.

YHA *or* **Y.H.A.** Youth Hostels Association.

Yield Point *or* **Yield Stress** Deformation resulting during constant stress for a Tensile Test, *q.v. Abbrev.* YP *or* Y.P. and YS *or* Y.S. for Yield Stress.

Yield Stress *Similar to* **Yield Point**, *q.v. Abbrev.* YS *or* Y.S.

YLA *or* **Y.L.A.** Yachtmen's Lifeboat Association.

YM *or* **Y.M.** Yacht Measurement. *See* **Thames Measurement**.

YMBA *or* **Y.M.B.A.** Yacht and Motor Boat Association.

YMCA *or* **Y.M.C.A.** Young Men's Christian Association.

YOG Year of Grace, *q.v.*

Yoke A cross-bar on which a bell swings; Cross-bar of a rudder to whose ends ropes are fastened.

YOM *or* **Yom** Yellow Oxide of Mercury.

Yorks Yorkshire, *UK.*

Young Ice Newly formed ice.

yr. Year.

YRA *or* **Y.R.A.** Yacht Racing Association.

YRBB Yacht Repair and Berthing Barges.

YRTEL *or* **Yrtel** Your telephone or telegram or telex.

YRTX/yrtx. Your telex.

Yrs. Yours.

Ys. Yugoslavia.

YSA Yugoslav Shipowners' Association.

YST *or* **Y.S.T.** Yukon Standard Time.

Yt Yacht, *q.v.*

YU *or* **Y.U.** Yale University *USA.*

Yuan The unit of currency used in China. Ten jiao or 100 fen equals one yuan.

Yugo. Yugoslavia.

Yuk Yukon.

Yukon Yukon Territory.

YUL Yale University Library.

YWCA *or* **Y.W.C.A.** Young Women's Christian Association.

Z

z Zenith, *q.v.*; Distance *nautical*; the third unknown quantity *algebra*.

Z Azimuth Angle; Coriolis Correction, being the vertical density of the earth's magnetic field; Symbol for impedance *electrical*; Zenith, *q.v.*

ZA *or* **Z.A.** Zero Absolute; Zuid Africa (South Africa).

Zaffre An impure oxide of cobalt used in making cobalt-blue and as a blue pigment in enamelling and porcelain-painting.

Zag. Zagreb, *Croatia*.

Zaire The unit of currency used in Zaire. One hundred makatu equals one zaire.

Zakat An Islamic direct tax on property and income, paid annually by citizens and companies in Saudi Arabia.

Zam. Zambia.

Zan *or* **Zanz.** Zanzibar.

Z.C.L. Zona di Commercio Libero, *Italian*—Free Trade Area.

z.d. Zenith Distance. *See* **Zenith**.

Z.D. Zone Description *nautical*.

Z-Drive Propulsion gearing for small vessels where the horizontal thrust from the engine is converted first to a vertical drive and then again to the horizontal to drive the propeller or rudder–propeller. The advantage is that the engine can be installed right aft almost above the propeller with a saving in shaft space and easier deck access for maintenance.

Zen. Zenith.

Zenith (1) The point in the heavens directly above the observer. (2) The highest point (opposite to nadir). *Abbrev.* Z *or* z.

Zentralverband Deutscher Schiffsmakler The National Shipbrokers Association of the Federal Republic of Germany.

Zeolite Any one of a number of minerals consisting mainly of hydrous silicates of lime, soda and alumina, commonly found in the cavities of igneous rocks.

Zernocon Code name for a charterparty form for the carriage of grains from Russia, Black Sea and Sea of Azoff ports to the United Kingdom and Continental ports.

Zero Figure 0, cipher; no quantity or number, nil; the starting point in scales from which positive and negative quantity is reckoned.

Zero *or* **Zero Coupon** An investment trust preference share which pays no dividend.

Zero Hour The planned time for the start of synchronised actions or manoeuvres. Also H-hour, *q.v.*, in military terms.

Zero Meridian The meridian of Greenwich can be described as zero meridian as it is at 0 degrees longitude.

Zero Population Growth American term indicating a nil rate of increase in the population.

Zero Rated Goods Goods or merchandise which is not subject to state tax, such as Value Added Tax (VAT), for certain reasons.

Zero-Zero Clear sky *weather condition*

Z.F. Zero Frequency.

Zig-Zag With abrupt alternate right and left turns, with alternating salient and re-entrant angles, with motion as of tacking vessel, e.g. a zig-zag line, or course; representation of flash of lightning.

Zinc A white metallic element much used in the arts, especially as a component of brass and German silver; as roofing material; as a coating for sheet iron.

Zincograph A zinc plate with a design etched in relief on it for printing from; a picture taken from it.

Zinc Oxide A white powder used as a paint pigment (Chinese white) and in antiseptic ointment. Once known as 'philosopher's wool'. Symbol ZnO.

Zinc Plates *or* **Zinc Rings** Corrosion prevention metals fastened to shell plating and other underwater parts of ships. They are placed near propellers, stern tubes, etc. Also known as zinc anodes.

Zinc Rich Primer A modern corrosion inhibiting paint primer widely used on bare metals.

Z.I.P. Zoning Improvement Plan.

ZIPEC Zone Industrielle Portuaire Euro-Calvados.

ZJR Zusen Juki Roren, Japanese Engineering Workers Union.

Zl. Zloty, *q.v.*

Zloty The unit of currency used in Poland. One hundred groszy equals one zloty.

Zn Symbol for zinc; Zenith.

ZnCl₂ Zinc Chloride.

ZnO Zinc Oxide, *q.v.*

Zodiac A belt of the heavens limited by lines about eight degrees from the ecliptic on each side, including all apparent positions of the sun and planets as known to the ancients, and divided into 12 equal parts called signs of the zodiac (Aries, Taurus, Gemini, Cancer, Leo, Virgo, Libra, Scorpio, Sagittarius, Capricorn(us), Aquarius, Pisces.

Zodiacal Of, in, the zodiac. Zodiacal light is a luminous tract of sky shaped like a tall triangle, occasionally seen in the east before sunrise or in the west after sunset, especially in the tropics.

Zoic Of animals. In geology it applies to rocks, etc., containing fossils, with traces of animal or plant life.

Zombie A corpse said to be revived by witch-craft.

Zone An encircling band or stripe distinguishable from the rest of the object encircled. In geology, any of five divisions of the earth bounded by circles parallel to the equator (frigid zones, north of the arctic and south of the antarctic circles; torrid zone, between the tropics; north, south and temperate zones, between the frigid and torrid zones). An area enclosed between two exact or approximate concentric circles; part of the surface of a sphere enclosed between two parallel planes, or of cone or cylinder between such planes cutting it perpendicularly to the axis; any well-defined tract of more or less beltlike form.

Zone Allowance The estimated weight of bunkers, domestic and boiler water, lubricating oils, provisions and stores to be consumed before a ship loaded with cargo reaches the restricted zone while on her way to her destined ports. For example, a ship loaded with cargo sails from Italy to the UK in November. She can be loaded to her summer marks for the Mediterranean and by the time she passes Gibraltar the consumption of fuel, water, etc., will allow her to rise up to the winter load line assigned for the Atlantic zone.

Zone Industrielle Portuaire Euro–Calvados A French syndicate formed by several organisations spreading over three departments and two regions for the development of harbour works in general.

Zone Time Local time for any longitude, as opposed to Greenwich time. *Abbrev.* Z.T.

Zooid Of incompletely animal nature, a cell.

Zoolatry Religious worship of animals.

Zoolite Fossil animal, fossilised animal substance.

Zoology The natural history of animals, the science of their structure, physiology, classification, habits and distribution.

Zoom To force an aircraft to mount at high speed and steep angle; an aircraft's steep climb.

Zoom Lens A lens which by variation of the focal length enables a quick transmission from long shot to close-up *photography.*

Zoophytes Kinds of plantlike animals, especially holothurians, starfishes, jelly-fishes, sea anemones and sponges.

Zoril A carnivorous quadruped of Africa and Asia Minor allied to the skunk and weasel, a zorro.

ZR Freezing rain *meteorological.*

ZRG Zero Rated Goods, *q.v.*

Z.T. Zone Time, *q.v.*

Zulu A member of and the language of a South African Bantu people; International Radio Telephone phonetic alphabet for 'Z'.

Zur. Zurich, *Switzerland.*

Zygal H-shaped, from the Greek zygon, a yoke.

Zymoscope An instrument for studying the fermentation of yeast.

Zymotic Of fermentation. Zymotic diseases are epidemic, endemic, contagious, or sporadic diseases regarded as caused by the multiplication of germs introduced from without.